# Biology Now
## with Physiology

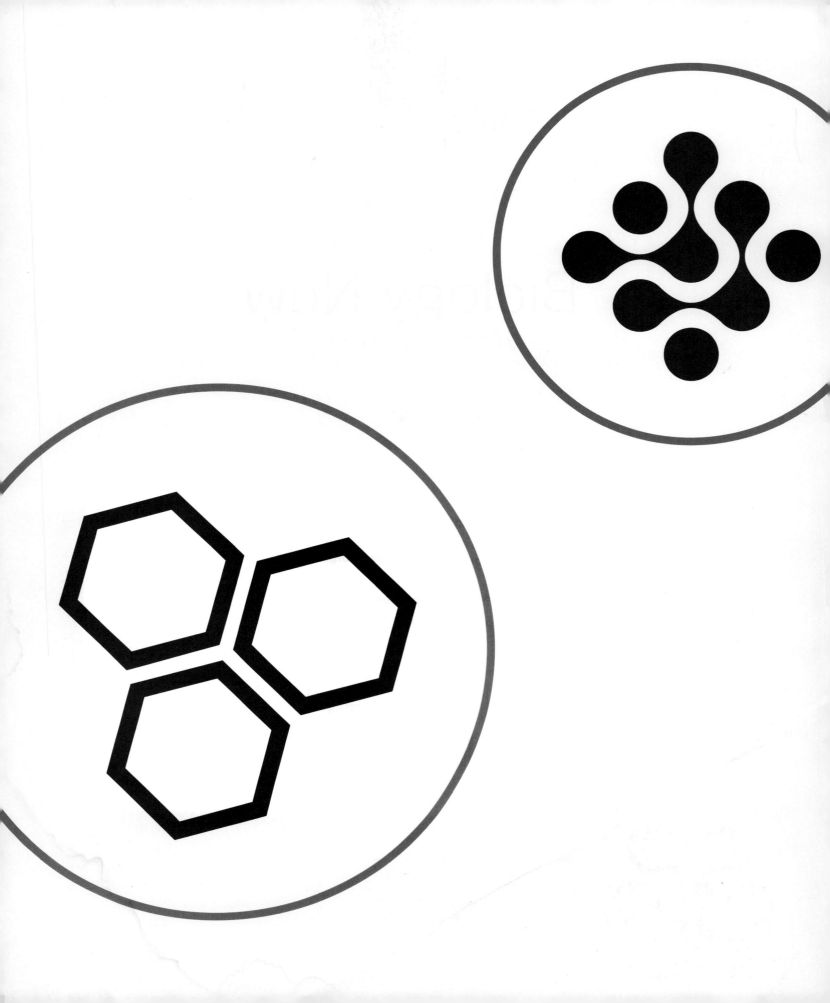

# Biology Now
## with Physiology

**ANNE HOUTMAN**
ROSE-HULMAN INSTITUTE OF TECHNOLOGY

**MEGAN SCUDELLARI**
SCIENCE JOURNALIST

**CINDY MALONE**
CALIFORNIA STATE UNIVERSITY, NORTHRIDGE

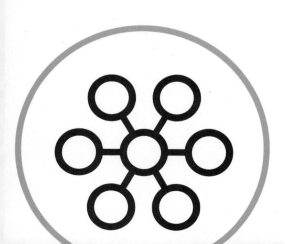

W. W. NORTON
NEW YORK • LONDON

W. W. Norton & Company has been independent since its founding in 1923, when William Warder Norton and Mary D. Herter Norton first published lectures delivered at the People's Institute, the adult education division of New York City's Cooper Union. The firm soon expanded its program beyond the Institute, publishing books by celebrated academics from America and abroad. By midcentury, the two major pillars of Norton's publishing program—trade books and college texts— were firmly established. In the 1950s, the Norton family transferred control of the company to its employees, and today— with a staff of four hundred and a comparable number of trade, college, and professional titles published each year— W. W. Norton & Company stands as the largest and oldest publishing house owned wholly by its employees.

Editor: Betsy Twitchell
Assistant Editor: Taylere Peterson
Developmental Editor: Andrew Sobel
Project Editor: Christine D'Antonio
Manuscript Editor: Stephanie Hiebert
Managing Editor, College: Marian Johnson
Managing Editor, College Digital Media: Kim Yi
Production Manager: Ashley Horna
Media Editor: Kate Brayton
Associate Media Editor: Cailin Barrett-Bressack
Media Project Editor: Jesse Newkirk
Media Editorial Assistant: Gina Forsythe
Marketing Manager, Biology: Todd Pearson
Design Director: Hope Miller Goodell
Photo Editor: Ted Szczepanski
Photo Researcher: Fay Torresyap
Permissions Manager: Megan Schindel
Permissions Clearer: Elizabeth Trammell
Composition by MPS North America LLC
Illustrations by Dragonfly Media Group
Manufacturing: Quad Graphics—Versailles, Kentucky

Permission to use copyrighted material is included in the backmatter of this book.

ISBN 978-0-393-62335-2

W. W. Norton & Company, Inc., 500 Fifth Avenue, New York, NY 10110-0017
wwnorton.com

W. W. Norton & Company Ltd., 15 Carlisle Street, London W1D 3BS

2 3 4 5 6 7 8 9 0

# Brief Contents

# Contents

 **INTRODUCTION**

# UNIT 1: CELLS

## CHAPTER 6: Cell Division 97

### *Toxic Plastic*

 **UNIT 2: GENETICS**

## CHAPTER 7: Patterns of Inheritance 117

### *Dog Days of Science*

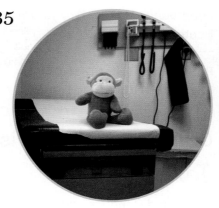

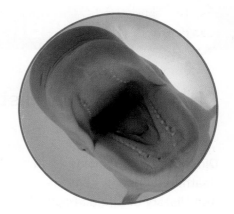

## CHAPTER 11: Evidence for Evolution  191

### *Whale Hunting*

## CHAPTER 12: Mechanisms of Evolution  211

### *Battling Resistance*

## CHAPTER 13: Adaptation and Species  231

### *Fast Lizards, Slow Corals*

# UNIT 4: BIODIVERSITY

## CHAPTER 17: Animals and Human Evolution   301
### *Neanderthal Sex*

 **UNIT 5: ECOLOGY**

## CHAPTER 18: General Principles of Ecology   323
### *Amazon on Fire*

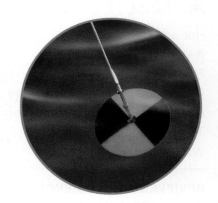

# UNIT 6: PHYSIOLOGY

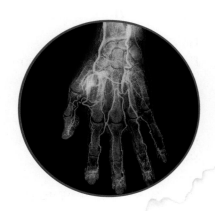

# About the Authors

**ANNE HOUTMAN** is Provost and Vice President of Academic Affairs at Rose-Hulman Institute of Technology, where she is also a full professor of biology. Anne has over 20 years of experience teaching nonmajors biology at a variety of private and public institutions, which gives her a broad perspective of the education landscape. She is strongly committed to evidence-based, experiential education and has been an active participant in the national dialogue on STEM (science, technology, engineering, and math) education for over 20 years. Anne's research interests are in the ecology and evolution of hummingbirds. She grew up in Hawaii, received her doctorate in zoology from the University of Oxford, and conducted postdoctoral research at the University of Toronto.

**MEGAN SCUDELLARI** is an award-winning freelance science writer and journalist based in Boston, Massachusetts, specializing in the life sciences. She has contributed to *Newsweek, Scientific American, Discover, Nature,* and *Technology Review,* among others, and she was a health columnist for the *Boston Globe.* For five years she worked as a correspondent and later as a contributing editor for *The Scientist* magazine. In 2013, she was awarded the prestigious Evert Clark/Seth Payne Award in recognition of outstanding reporting and writing in science. She has also received accolades for investigative reporting on traumatic brain injury and a feature story on prosthetics bestowed with a sense of touch. Megan received an MS from the Graduate Program in Science Writing at the Massachusetts Institute of Technology and worked as an educator at the Museum of Science, Boston.

**CINDY MALONE** began her scientific career wearing hip waders in a swamp behind her home in Illinois. She earned her BS in biology at Illinois State University and her PhD in microbiology and immunology at UCLA. She continued her postdoctoral work at UCLA in molecular genetics. She is currently a distinguished educator and a professor at California State University, Northridge, where she is the director of the CSUN-UCLA Stem Cell Scientist Training Program funded by the California Institute for Regenerative Medicine. Her research is aimed at training undergraduates and master's degree candidates to understand how genes are regulated through genetic and epigenetic mechanisms that alter gene expression. She has been teaching nonmajors biology for almost 20 years and has won teaching, mentorship, and curriculum enhancement awards at CSUN.

# Preface

A good biology class can improve the quality of students' lives. Biology is a part of so many decisions that students will need to make as individuals and as members of society. It helps parents to see the value of vaccinating a child, because they will understand what viruses are and how the immune system works. It helps homeowners in Texas, Florida, and Puerto Rico as they decide how to respond to the ongoing cleanup from 2017's Hurricanes Harvey, Irma, and Maria, because they understand how an ecosystem functions. It helps students make more informed decisions about their own nutrition because they understand the effects of fat, cholesterol, and vitamins, and minerals on our health. The examples are endless. Making informed decisions on these real-world issues requires students to be comfortable with scientific concepts and the process of scientific discovery.

How do we instill that capability in students? The last decade has seen an explosion of research on how students learn best. In a nutshell, they learn best when they see the relevance of a subject to their lives, when they are actively engaged in their learning, and when they are given opportunities to practice critical thinking.

In addition, most faculty who teach nonmajors biology would agree that our goal is to introduce students to both the key concepts of biology (for example, cells, DNA, evolution) and the tools to think critically about biological issues. Many would add that they want their students to leave the class with an appreciation for the value of science to society, and with an ability to distinguish between science and the nonscience or pseudoscience that bombards them on a daily basis.

How can a textbook help combine the ways students learn best with the goals of a nonmajors biology class? At the most basic level, if students don't read the textbook, they can't learn from it. When students read them, traditional textbooks are adept at teaching key concepts, and they have recently begun to emphasize the relevance of biology to students' lives. But students may be intimidated by the length of chapters and the amount of difficult text, and they often cannot see the connections between the story and the science. More important, textbooks have not been successful at helping students become active learners and critical thinkers, and none emphasize the process of science or how to assess scientific claims. It was our goal to make *Biology Now* relevant and interactive, and to be sure that it emphasized the process of science in short chapters that students *want* to read, while still covering the essential content found in other nonmajors biology textbooks.

Following the model of the first edition, each chapter in our book covers a current news story about people *doing* science, reported firsthand by Megan, an experienced journalist who specializes in reporting scientific findings in a compelling and accurate way, and fleshed out with a concise introduction to the science by Anne and Cindy. For this second edition we decided to direct our energies toward writing five current stories that will help instructors keep their courses grounded in real world events, and toward adding content requested by our first-edition adopters. Specifically, we've added a full unit—comprising two new chapters and two revised chapters—on the amazing diversity of life on planet Earth. Not only was more substantial coverage of this topic a common request in feedback about the first edition; it is also essential material for non-biology-major students, for it is partly through an appreciation of the diversity of life that students develop a personal relationship with the natural world.

Finally, we are thrilled for our book to be part of the online-assessment revolution! The second edition is accompanied by two excellent online homework platforms: a formative system called InQuizitive, and a summative system called Smartwork5. We no longer worry that our students aren't seeing the forest for the trees when they read the textbook. These systems are rich learning environments for students and automatically graded assignment platforms for instructors.

We sincerely hope you enjoy the fruits of our long labors.

*Anne Houtman*
*Megan Scudellari*
*Cindy Malone*

# What's New in the Second Edition?

- New chapter stories on current, fun, and unexpected topics like the Zika virus outbreak, the human microbiome, and the discovery of a CRISPR gene editing technology. New stories include:

**Chapter 5: How Cells Work—Rock Eaters**

Unusual electricity-"eating" microbes could someday provide a new way to store and produce energy as "bacterial batteries."

**Chapter 9: What Genes Are—Pigs to the Rescue**

CRISPR is perhaps one of the most exciting discoveries of the last century. Chapter 9 describes one application of the CRISPR genome editing technology: creating organs for transplant . . . in pigs.

**Chapter 15: Bacteria and Archaea—Navel Gazing**

A team at North Carolina State University leads a citizen science project to sequence the human belly button microbiome and gets some surprising results.

**Chapter 16: Plants, Fungi, and Protists—The Dirt on Black-Market Plants**

Poaching is illegal, and trafficking of tropical plants such as orchids threatens their survival. A group of scientists is tracking illegal plants from the United States to their source.

**Chapter 19: Growth of Populations—Zika-Busting Mosquitoes**

The spread of Zika throughout the Americas quickly became a health crisis. Genetically modifying mosquitoes is one of the ways that scientists are using to try to control Zika's spread.

- A new unit on biodiversity, which significantly expands coverage of the vast diversity of life on Earth, with two completely new chapters and two significantly revised chapters. Instructors who wish to continue teaching a brief introduction to biodiversity can do so with the "overview" chapter (Chapter 14). But for those wishing to spend time exploring life on Earth, Chapters 15, 16, and 17 provide thorough science coverage and lively stories.
- New, earlier placement of the chapter on applying science to making critical choices. The "capstone" final chapter in the second edition is now Chapter 2: Evaluating Scientific Claims. Introducing the concept of scientifically literate evaluation of scientific claims early in the book gives students the maximum amount of time to benefit from that skill.
- A new end-of-chapter question type—Challenge Yourself—which encourages students to think critically about the chapter's important biological concepts.
- New animation, interactive, and visually based questions in Smartwork5 and InQuizitive that promote critical thinking, interaction with data, and engagement with biology in the real world.
- New resources in the Ultimate Guide to Teaching with Biology Now, which will be accessible through the online Interactive Instructor's Guide platform, providing instructors with the ability to easily search and sort for active learning resources by topic, objective, and type of resource.

# The perfect balance of science and story

Every chapter is structured around a story about people doing science that motivates students to read and stimulates their curiosity about biological concepts.

**Dynamic chapter-opening spreads** inspired by each chapter's story draw students in to the material.

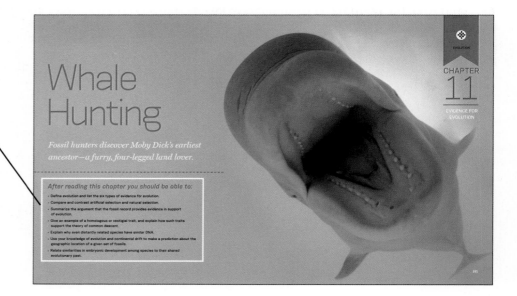

**"After reading this chapter you should be able to"** introduces learning outcomes that preview the concepts presented in each chapter.

## Cast-of-character bios
highlight the scientists, researchers, and professors at the center of each story.

### GORDON LARK

A geneticist at the University of Utah in Salt Lake City, Gordon Lark initiated the Georgie Project in 1996 to study the genetics of Portuguese water dogs. The national research project has led to valuable knowledge about the genetic basis of health and disease in humans and dogs.

### J. G. M. "HANS" THEWISSEN

Paleontologist and embryologist J. G. M. "Hans" Thewissen is a professor and whale expert at Northeast Ohio Medical University in the Department of Anatomy and Neurobiology. He and his lab study ancestral whale fossils and modern whale species.

### LISA COOPER

Lisa Cooper is an assistant professor at Northeast Ohio Medical University in the Department of Anatomy and Neurobiology. She earned her PhD in Thewissen's lab.

### MICHAEL HELLBERG AND CARLOS PRADA

Michael Hellberg (right) is an evolutionary biologist at Louisiana State University who studies how species evolve in marine environments. Carlos Prada (left) was a graduate student in Hellberg's lab, and is now a postdoctoral researcher at Penn State studying how organisms cope with changes in the environment.

### XU XING

Xu Xing is a paleontologist at the Chinese Academy of Sciences in Beijing. He has discovered more than 60 species of dinosaurs and specializes in feathered dinosaurs and the origins of flight.

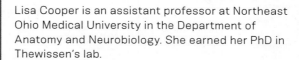

# An inquiry-based approach that builds science skills—asking questions, thinking visually, and interpreting data.

Most **figures** in the book are accompanied by three questions that promote understanding and encourage engagement with the visual content. Answers are provided at the back of the book, making the questions a useful self-study tool.

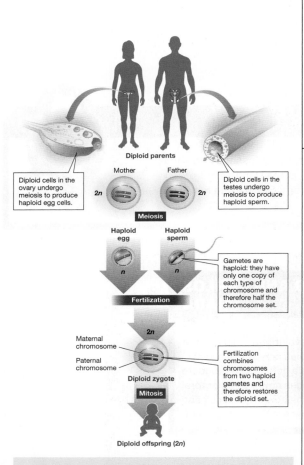

Diploid parents

Mother    Father

Diploid cells in the ovary undergo meiosis to produce haploid egg cells.

Diploid cells in the testes undergo meiosis to produce haploid sperm.

2n    2n

**Meiosis**

Haploid egg    Haploid sperm

n    n

Gametes are haploid: they have only one copy of each type of chromosome and therefore half the chromosome set.

**Fertilization**

2n

Maternal chromosome

Paternal chromosome

Diploid zygote

Fertilization combines chromosomes from two haploid gametes and therefore restores the diploid set.

**Mitosis**

Diploid offspring (2n)

**Q1:** Is a zygote haploid or diploid?

**Q2:** Which cellular process creates a baby from a zygote?

**Q3:** If a mother or father was exposed to BPA prior to conceiving a child, how might that explain potential birth defects in the fetus?

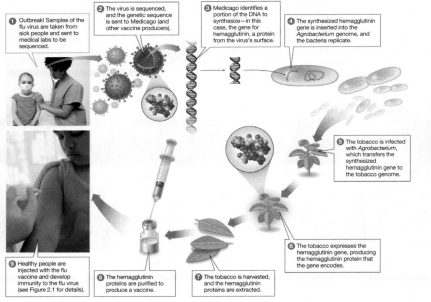

1 Outbreak! Samples of the flu virus are taken from sick people and sent to medical labs to be sequenced.

2 The virus is sequenced, and the genetic sequence is sent to Medicago (and other vaccine producers).

3 Medicago identifies a portion of the DNA to synthesize—in this case, the gene for hemagglutinin, a protein from the virus's surface.

4 The synthesized hemagglutinin gene is inserted into the *Agrobacterium* genome, and the bacteria replicate.

5 The tobacco is infected with *Agrobacterium*, which transfers the synthesized hemagglutinin gene to the tobacco genome.

6 The tobacco expresses the hemagglutinin gene, producing the hemagglutinin protein that the gene encodes.

7 The tobacco is harvested, and the hemagglutinin proteins are extracted.

8 The hemagglutinin proteins are purified to produce a vaccine.

9 Healthy people are injected with the flu vaccine and develop immunity to the flu virus (see Figure 2.1 for details).

**Q1:** In which of the step(s) illustrated here does DNA replication occur? In which step(s) does gene expression occur?

**Q2:** Why do vaccine producers not simply replicate the entire viral genome, instead isolating the gene for one protein and replicating only that gene?

**Q3:** What role do the bacteria play in this process? Why are they needed?

Engaging, data-driven **infographics** appear in every chapter. Topics range from global renewable energy consumption (Chapter 5) to genetic diseases affecting Americans (Chapter 8) and many more. The infographics expose students to scientific data in an engaging way.

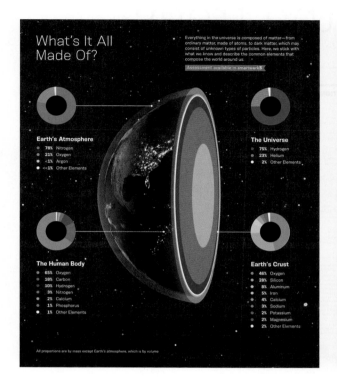

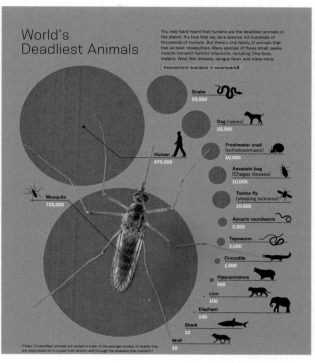

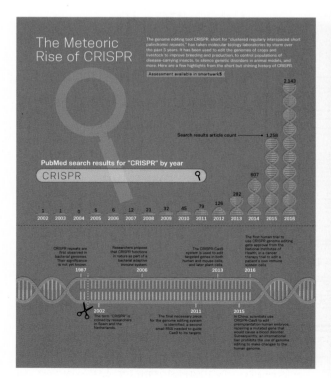

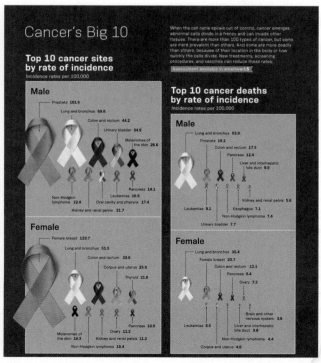

# Extensive end-of-chapter review ensures that students see the forest for the trees.

**Reviewing the Science** identifies each chapter's key science concepts, providing students with a guide for studying.

## End-of-chapter

**questions** follow Bloom's taxonomy, moving from review (The Basics), to synthesis (Try Something New), to critical thinking (Challenge Yourself), to application (Leveling Up).

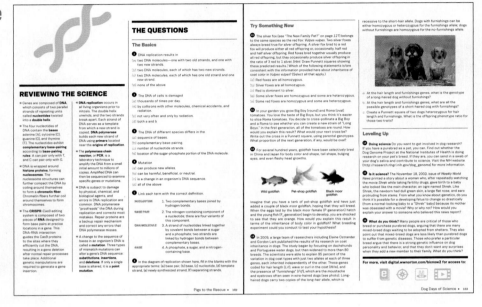

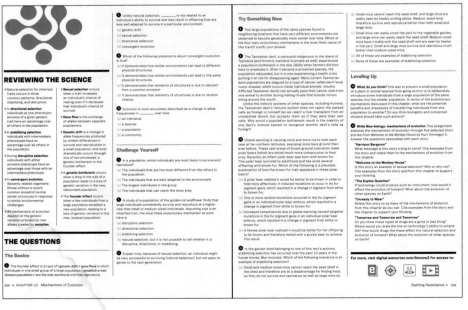

**Leveling Up questions**, based on questions the authors use in their classrooms, prompt students to relate biology concepts to their own lives. The questions focus on one of the following themes: "Doing science," "Is it science?," "Life choices," "Looking at data," "What do *you* think?," and "*Write Now* biology."

# Powerful resources for teaching and assessment

## Laura Zapanta, University of Pittsburgh
## Tiffany Randall, John Tyler Community College

The **Ultimate Guide** helps instructors bring *Biology Now*'s inquiry-based approach into the classroom through a wealth of resources, including activities useful in a variety of classroom sizes and setups, suggested online videos with discussion questions, clicker questions, sample syllabi, and suggested lecture outlines. The second-edition *Ultimate Guide* has been thoroughly reviewed and updated with new activities, Leveling Up rubrics, and descriptions of animations with discussion questions.

The **Interactive Instructor's Guide** is a searchable database of all the valuable teaching and active learning resources available in the *Ultimate Guide*. Instructors can easily filter by chapter, phrase, topic, or learning objective to find activities with downloadable handouts, streaming video with discussion questions, animations with discussion questions, lecture PowerPoints, and more.

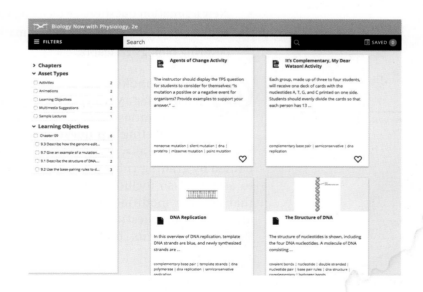

# Other presentation tools for instructors

**InQuizitive** InQuizitive is Norton's easy-to-use adaptive-learning and quizzing tool that improves student understanding of important learning objectives. Students receive personalized quiz questions on the topics they need the most help with. When instructors assign InQuizitive, students come better prepared to lectures and exams. The second-edition course includes new animation questions, story-based questions, and critical-thinking questions.

**Smartwork5** Smartwork5 delivers engaging, interactive online homework to students, helping instructors and students reach their teaching and learning goals. The second edition features:

- New infographic questions, which promote interaction with data and engagement with biology in the real world, while making this popular visual feature of the text an assignable activity.
- New story-based questions, which help students to learn and understand the science behind the stories in the text.
- New critical-thinking questions, which prompt students to think critically about important concepts in biology.
- New animation questions, which engage students with the book-specific animations covering biology concepts.

**Coursepacks** Norton's free coursepacks offer a variety of concept-based opportunities for assessment and review. The Leveling Up questions from the text are available as writing activities, accompanied by grading rubrics, making them easy to assign. Also included are reading quizzes that contain modified images from the text and animation questions, infographic quizzes that help students build skills in reading charts and graphs, and flashcards for student self-study of key terms.

**Ebook** Norton ebooks give students and instructors an enhanced reading experience at a fraction of the cost of a print textbook. Students are able to have an active reading experience and can take notes, bookmark, search, highlight, and even read offline. Instructors can even add their own notes for students to see as they read the text. Norton ebooks can be viewed on—and synced among— all computers and mobile devices.

**Animations** Key concepts and processes are explained clearly through high-quality, ADA-compliant animations developed from the meticulously designed art in the book. These animations are available for lecture presentation in the Interactive Instructor's Guide, PowerPoint outlines, and the coursepacks, as well as within our ebook, InQuizitive, and Smartwork5.

**Test Bank** The test bank is based on an evidence-centered design that was collaboratively developed by some of the brightest minds in educational testing. Each chapter's test bank now includes 75 or more questions structured around the learning objectives from the textbook and conforms to Bloom's taxonomy. Questions are further classified by text section and difficulty, and are provided in multiple-choice, fill-in-the-blank, and short-answer form. New infographic questions in every chapter help test student interpretation of charts and graphs.

**Art Files** All art and photos from the book are available, in presentation-ready resolution, as both JPEGs and PowerPoints for instructor use.

**Lecture Slides** Comprehensively revised by book author Cindy Malone, complete lecture PowerPoints thoroughly cover chapter concepts and include images and clicker questions to encourage student engagement.

# Acknowledgments

We could not have created this textbook without the enthusiasm and hard work of many people. First and foremost, we'd like to thank our indefatigable editor, Betsy Twitchell, for her keen eye to the market, terrific visual sense, and endless author-wrangling skills. Andrew Sobel has done far more than ought to be required of a developmental editor to ensure that our book is both accurate and readable (not to mention his tireless work on the eye-catching infographics you'll see in these pages), and for that he has our eternal gratitude.

Thank you to our supremely focused and talented project editor, Christine D'Antonio, for creating such a superior layout and for keeping our chapters moving. Thank you to our talented copy editor, Stephanie Hiebert, for being so meticulous with our manuscript, and so pleasant to work with.

We are grateful to photo researcher Fay Torresyap for her reliable and creative work, and to Ted Szczepanski for managing the photo process. Production manager Ashley Horna skillfully oversaw the translation of our raw material into the beautiful book you hold in your hands; she, too, has our thanks. Special thanks to book designer Hope Miller Goodell and cover designer Jennifer Heuer for creating such an extraordinary and truly gorgeous book.

Media editor Kate Brayton, associate editor Cailin Barrett-Bressack, and media assistant Gina Forsythe worked tirelessly to create the instructor and student resources accompanying our book. Their determination, creativity, and positive attitude resulted in supplements of the highest quality that will truly make an impact on student learning. Jesse Newkirk's commitment to quality as media project editor ensured that every element of the resource package meets Norton's high standards. Likewise,

assistant editor Taylere Peterson contributed in myriad ways, large and small, and for that she has our thanks.

We appreciate the tireless enthusiasm of marketing manager Todd Pearson and his colleagues, director of marketing Steve Dunn and marketing director Stacy Loyal. We thank director of sales Michael Wright and every single one of Norton's extraordinary salespeople for spreading the word about our book. Finally, we thank Marian Johnson, Julia Reidhead, Roby Harrington, Drake McFeely, and everyone at Norton for believing in our book.

Thank you to our accuracy reviewers Erin Baumgartner and Mark Manteuffel. We would be remiss not to thank also all of our colleagues in the field who gave their time and expertise in reviewing, class testing, and contributing to *Biology Now* and its many supplements and resources. Thank you all.

## Reviewers

### Second Edition

Anne Artz, Preuss School, UC San Diego
Allan Ayella, McPherson College
Erin Baumgartner, Western Oregon University
Joydeep Bhattacharjee, University of Louisiana, Monroe
Rebecca Brewer, Troy High School
Victoria Can, Columbia College Chicago
Lisa Carloye, Washington State University, Pullman
Michelle Cawthorn, Georgia Southern University
Craig Clifford, Northeastern State University

Beth Collins, Iowa Central Community College
Julie Constable, California State University, Fresno
Gregory A. Dahlem, Northern Kentucky University
Danielle M. DuCharme, Waubonsee Community College
Robert Ewy, SUNY Potsdam
Clayton Faivor, Ellsworth Community School
Michael Fleming, California State University, Stanislaus
Kathy Gallucci, Elon University
Kris Gates, Pikes Peak Community College
Heather Giebink, Pennsylvania State University
Candace Glendening, University of Redlands
Sherri D. Graves, Sacramento City College
Cathy Gunther, University of Missouri
Meshagae Hunte-Brown, Drexel University
Douglas P. Jensen, Converse College
Ragupathy Kannan, University of Arkansas–Fort Smith
Julia Khodor, Bridgewater State University
Jennifer Kloock, Garces Memorial High School
Karen L. Koster, University of South Dakota
Dana Robert Kurpius, Elgin Community College
Joanne Manaster, University of Illinois
Mark Manteuffel, St. Louis Community College
Jill Maroo, University of Northern Iowa
Tsitsi McPherson, SUNY Oneonta
Kiran Misra, Edinboro University of Pennsylvania
Jeanelle Morgan, University of North Georgia
Lori Nicholas, New York University
Fran Norflus, Clayton State University
Christopher J. Osovitz, University of South Florida
Christopher Parker, Texas Wesleyan University
Brian K. Paulson, California University of Pennsylvania
Carolina Perez-Heydrich, Meredith College
Thomas J. Peri, Notre Dame Preparatory School

Kelly Norton Pipes, Wilkes Early College High School
Gordon Plague, SUNY Potsdam
Benjamin Predmore, University of South Florida
Jodie Ramsey, Highland High School
Logan Randolph, Polk State College
Debra A. Rinne, Seminole State College of Florida
Michael L. Rutledge, Middle Tennessee State University
Celine Santiago Bass, Kaplan University
Steve Schwendemann, Iowa Central Community College
Sonja Stampfler, Kellogg Community College
Jennifer Sunderman Broo, Saint Ursula Academy
J. D. Swanson, Salve Regina University
Heidi Tarus, Colby Community College
Larchinee Turner, Central Carolina Technical College
Ron Vanderveer, Eastern Florida State College
Calli A. Versagli, Saint Mary's College
Mark E. Walvoord, University of Oklahoma
Lisa Weasel, Portland State University
Derek Weber, Raritan Valley Community College
Danielle Werts, Golden Valley High School
Elizabeth Wright, Athenian School
Steve Yuza, Neosho County Community College

**First Edition**

Joseph Ahlander, Northeastern State University
Stephen F. Baron, Bridgewater College
David Bass, University of Central Oklahoma
Erin Baumgartner, Western Oregon University
Cindy Bida, Henry Ford Community College
Charlotte Borgeson, University of Nevada, Reno
Bruno Borsari, Winona State University
Ben Brammell, Eastern Kentucky University
Christopher Butler, University of Central Oklahoma
Stella Capoccia, Montana Tech

Kelly Cartwright, College of Lake County
Emma Castro, Victor Valley College
Michelle Cawthorn, Georgia Southern University
Jeannie Chari, College of the Canyons
Jianguo Chen, Claflin University
Beth Collins, Iowa Central Community College
Angela Costanzo, Hawai'i Pacific University
James B. Courtright, Marquette University
Danielle DuCharme, Waubonsee Community College
Julie Ehresmann, Iowa Central Community College
Laurie L. Foster, Grand Rapids Community College
Teresa Golden, Southeastern Oklahoma State University
Sue Habeck, Tacoma Community College
Janet Harouse, New York University
Olivia Harriott, Fairfield University
Tonia Hermon, Norfolk State University
Glenda Hill, El Paso Community College
Vicki J. Huffman, Potomac State College, West Virginia University
Carl Johansson, Fresno City College
Victoria Johnson, San Jose State University
Anthony Jones, Tallahassee Community College
Hinrich Kaiser, Victor Valley College
Vedham Karpakakunjaram, Montgomery College
Dauna Koval, Bellevue College
Maria Kretzmann, Glendale Community College
MaryLynne LaMantia, Golden West College
Brenda Leady, University of Toledo
Lisa Maranto, Prince George's Community College
Roy B. Mason, Mt. San Jacinto College
Gabrielle L. McLemore, Morgan State University
Paige Mettler-Cherry, Lindenwood University
Rachel Mintell, Manchester Community College
Kiran Misra, Edinboro University of Pennsylvania

Lori Nicholas, New York University
Louise Mary Nolan, Middlesex Community College
Fran Norflus, Clayton State University
Brian Paulson, California University of Pennsylvania
Carolina Perez-Heydrich, Meredith College
Ashley Ramer, University of Akron
Nick Reeves, Mt. San Jacinto College
Tim Revell, Mt. San Antonio College
Eric Ribbens, Western Illinois University
Kathreen Ruckstuhl, University of Calgary
Michael L. Rutledge, Middle Tennessee State University
Brian Sato, UC Irvine
Malcolm D. Schug, University of North Carolina at Greensboro
Craig M. Scott, Clarion University of Pennsylvania
J. Michael Sellers, University of Southern Mississippi
Marieken Shaner, University of New Mexico
David Sheldon, St. Clair County Community College
Jack Shurley, Idaho State University
Daniel Sigmon, Alamance Community College
Molly E. Smith, South Georgia State College, Waycross
Lisa Spring, Central Piedmont Community College
Steven R. Strain, Slippery Rock University of Pennsylvania
Jeffrey L. Travis, SUNY Albany
Suzanne Wakim, Butte College
Mark E. Walvoord, University of Oklahoma
Sherman Ward, Virginia State University
Lisa Weasel, Portland State University
Jennifer Wiatrowski, Pasco-Hernando State College
Rachel Wiechman, West Liberty University
Bethany Williams, California State University, Fullerton
Satya M. Witt, University of New Mexico
Donald A. Yee, University of Southern Mississippi

## Focus Group Participants

Michelle Cawthorn, Georgia Southern University
Marc Dal Ponte, Lake Land College
Kathy Gallucci, Elon University
Tamar Goulet, University of Mississippi
Sharon Gusky, Northwestern Connecticut Community College
Krista Henderson, California State University, Fullerton
Tara Jo Holmberg, Northwestern Connecticut Community College
Brenda Hunzinger, Lake Land College
Jennifer Katcher, Pima Community College
Cynthia Kay-Nishiyama, California State University, Northridge
Kathleen Kresge, Northampton Community College
Sharon Lee-Bond, Northampton Community College
Suzanne Long, Monroe Community College
Boriana Marintcheva, Bridgewater State University
Roy B. Mason, Mt. San Jacinto College
Gwen Miller, Collin College
Kimo Morris, Santa Ana College
Fran Norflus, Clayton State University
Tiffany Randall, John Tyler Community College
Gail Rowe, La Roche College
J. Michael Sellers, University of Southern Mississippi
Uma Singh, Valencia College
Patti Smith, Valencia College
Bethany Stone, University of Missouri
Willetta Toole-Simms, Azusa Pacific University
Bethany Williams, California State University, Fullerton

## Class Test Participants

Bruno Borsari, Winona State University
Jessica Brzyski, Georgia Southern University
Beth Collins, Iowa Central Community College
Christopher Collumb, College of Southern Nevada
Jennifer Cooper, University of Akron
Julie Ehresmann, Iowa Central Community College
Michael Fleming, California State University, Stanislaus
Susan Holecheck, Arizona State University
Dauna Koval, Bellevue College
Kiran Misra, Edinboro University of Pennsylvania
Marcelo Pires, Saddleback College
Michael L. Rutledge, Middle Tennessee State University
Jack Shurley, Idaho State University
Uma Singh, Valencia College
Paul Verrell, Washington State University
Daniel Wetzel, Georgia Southern University
Rachel Wiechman, West Liberty University

## Instructor and Student Resource Contributors

Holly Ahern, SUNY Adirondack
Steven Christenson, Brigham Young University–Idaho
Beth Collins, Iowa Central Community College
Julie Ehresmann, Iowa Central Community College
Jenny Gernhart, Iowa Central Community College
Julie Harless, Lone Star College
Janet Harouse, New York University
Vedham Karpakakunjaram, Montgomery College
Dauna Koval, Bellevue College
Brenda Leady, University of Toledo
Boriana Marintcheva, Bridgewater State University
Paige Mettler-Cherry, Lindenwood University
Lori Nicholas, New York University
Christopher Osovitz, University of South Florida
Tiffany Randall, John Tyler Community College
Lori Rose, Sam Houston State University
Suzanne Wakim, Butte College
Bethany Williams, California State University, Fullerton

This book wouldn't have happened without Anne's husband, Will, who took care of every single other thing in her life so that she could write. His support, and that of her children, Abi and Ben, are what keep her going every day. With great patience, Megan's husband, Ryan, bore many dinner conversations about bats, algae, wolves, and more, and for that he has her thanks. To Megan's children: May you read this book and share your mother's joy about all things biology. Cindy thanks her husband, Mike; children, Ben and Lily; and their numerous pets for the chaotic lifestyle that inspired her to step up her game. Also, Cindy thanks her friends and students who laugh at her jokes and keep her grounded in reality.

Perhaps most of all, we are indebted to the many scientists and individuals who shared their time and stories for these chapters. To the men and women we interviewed for this book, we cannot thank you enough. Your stories will inspire the next generation of biologists.

# Biology Now

## with Physiology

# Caves of Death

*Scientists scramble to identify a mysterious scourge decimating bat populations.*

- - - - - - - - - - - - - - - - - - - - - - - - - - - - - - - - - - - - - - - - - - - - - - -

## After reading this chapter you should be able to:

- Caption a diagram of the scientific method, identifying each step in the process.
- Develop a hypothesis from a given observation and suggest one or more predictions based on that hypothesis.
- Design an experiment using appropriate variables, treatments, and controls.
- Give specific examples of a scientific fact and a scientific theory.
- Create a graphic showing the levels of biological organization.
- Determine whether something is living or nonliving based on the characteristics of living things.

CHAPTER
01

THE NATURE
OF SCIENCE

Every spring for 30 years, Alan Hicks laced up his hiking boots, packed his camera, and set out to count bats in caves in upstate New York. A biologist with the New York State Department of Environmental Conservation, Hicks leads one of the few efforts in the country to collect annual data on bat populations. Since 1980, he had never missed the annual cave trip—until March 17, 2007.

"That day, of all days in my entire career, I stayed at my desk," recalls Hicks, who had remained behind to write a report for his supervisor. A couple of hours after his crew left to inspect some local caves, 15 miles from the Albany office, Hicks's cell phone rang.

"Hey, Al. Something weird is going on here," said a nervous voice. "We've got dead bats. Everywhere."

The line went quiet. "What are we talking here?" asked Hicks. "Hundreds of dead bats?"

"No," said the voice. "Thousands."

At first, Hicks conjectured that the bats had died in a flood, which had happened in that particular cave before. But the next day, a young volunteer who had been out with the team told Hicks to check his e-mail. The volunteer had sent him a picture taken the day before of eight little brown bats (*Myotis lucifugus*) hanging from a cave outcropping. Each one had a fuzzy white nose. This was a surprise because little brown bats do not have white noses.

Hicks e-mailed the picture to every bat researcher he knew. The fuzzy white material looked like a fungus, but there was no previous record of a fungus killing bats. As scientist after scientist looked at the picture, they all replied the same way: "What is that?" Hicks resolved to find out what was killing the bats and whether the white fuzz was involved.

Why was Hicks so interested in saving the bats? And why should any of us care, apart from valuing the preservation of all of Earth's creatures? For one thing, bats help us by devouring insects that would otherwise destroy agricultural crops and forests (see "Bug Zappers" on page 15). And mosquitoes, which bats eat, are the world's most deadly animal to humans: through malaria transmissions, mosquitoes kill hundreds of thousands of people each year.

As a biologist, Hicks took a scientific view of the world—logical, striving for objectivity, and valuing evidence over other ways of discovering the truth. **Science** is a body of knowledge about the natural world, but it is much more than just a mountain of data. Science is an evidence-based process for acquiring that knowledge.

- Science deals with the natural world, which can be detected, observed, and measured.
- Science is based on evidence that can be demonstrated through observations and/or experiments.
- Science is subject to independent validation and peer review.
- Science is open to challenge by anyone at any time on the basis of evidence.
- Science is a self-correcting enterprise.

To gather knowledge, Hicks would apply the **scientific method** (**Figure 1.1**). The scientific method is not a set recipe that scientists follow in a rigid manner. Instead, the term is meant to capture the core logic of how science works. Some people prefer to speak of the **process of science** rather than the scientific method. Whatever we call it, the practices that produce scientific knowledge can be applied across a broad range of disciplines—including bat biology.

Keep in mind that, as powerful as the scientific method is, it is restricted to seeking natural causes to explain the workings of our world. There are other areas of inquiry that science cannot address. The scientific method cannot tell us what is morally right or wrong. For example, science can inform us about the differences between humans and other animals, but it cannot identify the morally correct way to act on that information. Science also cannot speak to the existence of God or any other supernatural being. Nor can it tell us what is beautiful or ugly, which poems are most lyrical, or which paintings are most inspiring. So, although science exists comfortably alongside different belief systems—religious, political, and personal—it cannot answer all questions.

**ALAN HICKS**

Alan Hicks is a retired bat specialist who began the investigation of a mysterious bat illness while working for the New York Department of Environmental Conservation.

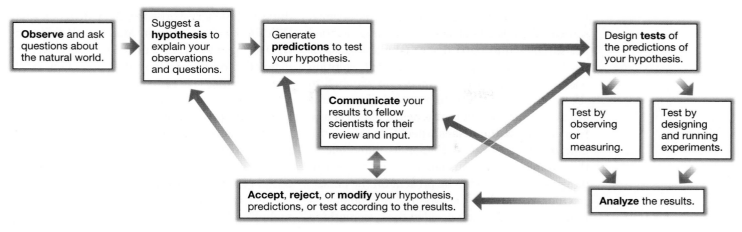

Figure 1.1

**The scientific method**

The scientific method is a logical process that helps us learn more about the natural world. ▶️

**Q1:** What were the original observation and question of the scientists studying the sick bats?

**Q2:** At what point in the scientific method would a scientist decide on the methods she should use to test her hypothesis?

**Q3:** How might you explain the scientific method to someone who complains that "scientists are always changing their minds; how can we trust what they say?"

But science is the best way to answer questions about the natural world. The first two steps of the scientific method are to *gather observations* and *form a hypothesis*. Hicks didn't waste a moment of time before applying the scientific method to the question of the white fuzz. Bats were dying. "Bats are part of the planet and vital members of the ecosystem," says Hicks. "They play an important role in the environment in which we live."

# Bat Crazy

On March 18, the day after the first dead bats were discovered, Hicks entered the cave to make observations—a key part of the scientific process. An **observation** is a description, measurement, or record of any object or phenomenon. Hicks's team observed that the sick bats had not only white noses, but also depleted fat reserves, meaning that the bats did not have enough stored energy to get through the winter. The bats also had white fuzz on their wings with scarred and dying wing tissue, and they were behaving

abnormally, waking up early from hibernation and leaving the cave when it was still too cold outside to hunt.

Hicks's team also observed that the illness cut across species—many different types of bats were getting sick—and the bats exhibited a high rate of death: in some cases, up to 97 percent of infected bats died. Hicks and others began to call the illness white-nose syndrome (WNS). They still didn't know what caused the syndrome, but its characteristics led them to the assumption that the cause was a living organism (see "The Characteristics of Living Organisms" on page 6).

"For the first few years, we were just sleuthing," says Paul Cryan, a research biologist with the U.S. Geological Survey (USGS), and one of the scientists who received the original e-mailed picture from Hicks. From that first picture, Cryan was involved in trying to pinpoint the cause. "We were trying to understand something that had never happened before in a group of animals that was poorly understood."

In the caves, Hicks began collecting dead bats and sending them to laboratories around the nation. In those labs, technicians scraped samples from the bats' noses and wings, rubbed the samples into petri dishes (shallow glass or plastic plates containing a nutrient solution used to grow microorganisms), and watched to see whether the white fuzz would grow. Time after time, many different types of bacteria and fungi grew on the dishes, speckling them with dots of different-colored colonies, but none of the samples were unusual. Nothing special or dangerous appeared to be present on the bats.

One researcher, a young microbiologist named David Blehert, decided to try something different. Blehert worked at the USGS National Wildlife Health Center in Madison, Wisconsin. In December 2007, Hicks called Blehert. Blehert listened carefully as Hicks described how WNS was spreading. "He said, 'We have a major problem on our hands,'" recalls Blehert. "It turns out he was 100 percent right."

Hicks described to Blehert the conditions under which the bats lived during hibernation—caves in upstate New York, where the temperature was often between 30°F and 50°F. Blehert realized that most of the laboratories, including his, were trying to grow the samples

# The Characteristics of Living Organisms

All living things share certain features that characterize life.

1. *They are composed of one or more cells.* The **cell** is the smallest and most basic unit of life; all organisms are made of one or more cells. Larger organisms are made up of many different kinds of specialized cells and are known as *multicellular organisms*.

2. *They reproduce autonomously using DNA.* All living organisms are able to **reproduce**, to make new individuals like themselves. **DNA** is the genetic material that transfers information from parents to offspring. A segment of DNA that codes for a distinct genetic characteristic is called a *gene*. Life, no matter how simple or how complex, uses this inherited genetic code to direct the structure, function, and behavior of every cell.

3. *They obtain energy from the environment to support metabolism.* All organisms need **energy** to survive. Organisms use a wide variety of methods to capture this energy from their environment. The capture, storage, and use of energy by living organisms is known as *metabolism*.

4. *They sense the environment and respond to it.* Living organisms **sense** many aspects of their external environment, from the direction of sunlight to the presence of food and mates. All organisms gather information about the environment by sensing it, and then respond appropriately.

5. *They maintain a constant internal environment.* Living organisms sense and respond to not only the external environment, but also their internal conditions. All organisms maintain constant internal conditions—a process known as **homeostasis**.

6. *They can evolve as groups.* **Evolution** is a change in the genetic characteristics of a group of organisms over generations. When a characteristic becomes more or less common across generations, evolution has occurred within the group.

| | Rock | Virus | Fungus | Plant | Animal |
|---|---|---|---|---|---|
| Composed of one or more cells | ✗ | ✗ | ✔ | ✔ | ✔ |
| Autonomously reproduce themselves | ✗ | ✗ | ✔ | ✔ | ✔ |
| Obtain energy from their environment | ✗ | ✗ | ✔ | ✔ | ✔ |
| Sense their environment and respond to it | ✗ | ✗ | ✔ | ✔ | ✔ |
| Maintain a constant internal environment (homeostasis) | ✗ | ✗ | ✔ | ✔ | ✔ |
| Can evolve as groups | ✗ | ✔ | ✔ | ✔ | ✔ |
| Living | ✗ | ? | ✔ | ✔ | ✔ |

from the bats at room temperature—a method conducive to the growth of many fungi. But in the caves, any living thing would have to grow at cold temperatures, so Blehert and his technicians took samples from dead bats, put them on petri dishes, and placed the dishes in the fridge.

At the same time, Melissa Behr, an animal disease specialist at the New York State Health Department, accompanied Hicks on a trip to a local cave (**Figure 1.2**). Behr swabbed a sample of the white fuzz directly from a bat in the cave, immediately spread it onto a glass slide, and looked at it under a microscope. A unique fungus was on the plate. The fungus was visible in little white fuzzy patches of cells, and up close, the individual spores of the fungus appeared crescent-shaped—different from all the other "normal" microbes growing on the bats' skin, and different from any fungus known to the researchers.

But Behr's single observation wasn't enough evidence to convince anyone that the strange-looking fungus was the cause of WNS. To be of use in science, an observation must be repeatable, preferably by multiple techniques. Independent observers should be able to see or detect the same object or phenomenon, at least some of the time.

In this case, Blehert was able to reproduce Behr's results by an independent technique. After letting his plates sit in the fridge for a few weeks, Blehert removed them and observed white patches of the same strange, crescent-shaped fungal spores. "OK, we now have in laboratory culture what Melissa captured when she collected white material in the caves," thought Blehert. "We've got it."

# Prove Me Wrong

In science, just as in everyday life, observations lead to questions, and questions lead to potential explanations. For example, if you flip on a light switch but the light does not turn on, you wonder why, and then you look for an explanation: Is the lamp plugged in? Has the lightbulb burned out? You then identify one of these explanations as the most likely hypothesis for why the light did not turn on.

Figure 1.2

**Preparing to enter the bat cave**
Scientists suit up to collect more observations on the infected bats and the environmental conditions in the bats' roosting cave.

Q1: Which step(s) in the scientific method does this photograph illustrate?

Q2: What types of environmental data might the researchers have collected?

Q3: Why do you think the researchers are wearing protective gear?

A scientific **hypothesis** (plural "hypotheses") is an informed, logical, and plausible explanation for observations of the natural world. From the start, Hicks hypothesized that a new, cold-loving fungus was the primary cause of death in the bats. After observing the unique crescent-shaped fungal spores, Behr and Blehert agreed with this hypothesis. "It was the simplest

**DAVID BLEHERT**

A microbiologist and branch chief of the Wildlife Disease Diagnostic Laboratories at the National Wildlife Health Center, David Blehert studies a variety of fungal and bacterial pathogens that are harmful to bats, humans, and other species.

solution," says Blehert. "We had bats with a white fungus that nobody had ever seen before growing on them, so that was the most likely thing that was doing it."

But other scientists disagreed. A fungus itself is rarely deadly to a mammal; more often, a fungus causes an annoying, but not lethal, skin infection or is a secondary response after an animal gets sick from a viral or bacterial infection. So scientists proposed other hypotheses for the cause of WNS. Some suggested the fungus was a secondary effect of an underlying condition, such as a viral infection. Others hypothesized that an environmental contaminant, such as a pesticide, was the cause of death. "There were so many different hypotheses," says Cryan. "But that's what is beautiful about the scientific process. You observe as much as you can, and from those observations you can form multiple hypotheses. Science doesn't proceed by just landing on the right hypothesis the first time."

One of the joys, and challenges, of the scientific method is that after scientists suggest competing hypotheses, they then test their own hypotheses against those of others. A scientific hypothesis must be constructed in such a way that it is potentially **falsifiable**, or refutable. In other words, it must make predictions that can be clearly determined to be true or false, right or wrong (**Figure 1.3**). A well-constructed hypothesis is precise enough to make predictions that can be expressed as "if . . . then" statements.

For example, *if* WNS is caused by a transmissible fungus, *then* healthy bats that hibernate in contact with affected bats should develop the condition. *If* the fungus is secondary to an underlying condition, *then* the infection will occur in bats only after the primary underlying condition is present. *If* an environmental contaminant is the cause, *then* bats with WNS symptoms will have elevated levels of that contaminant in their blood or on their skin.

In each "if . . . then" case, it is possible to design tests able to demonstrate that a prediction is right or wrong. Although predictions can be shown to be true or false, the same is not true of hypotheses. Hypotheses can be *supported*, but no amount of testing can *prove* a hypothesis is correct with complete certainty (**Figure 1.4**).

The reason a hypothesis cannot be proved is that there might be another factor, unmeasured or unobserved, that explains why the prediction is true. For example, consider the first prediction stated in the previous paragraph—that healthy bats hibernating in contact with affected bats will develop WNS. If this is true, the reason might be that the healthy bats were infected by a fungus from their neighbor, supporting the hypothesis that the disease is caused by a transmissible fungus. Alternatively, related bats may tend to hibernate together in the same cave, and the disease, or at least vulnerability to the disease, might be genetically based. The hypothesis that the disease is fungal is *supported* but not *proved* by the correctness of this prediction.

Blehert set out to test the hypothesis that he, Behr, and Hicks had put forward—that a unique, cold-loving fungus was the primary cause of death in the bats. One can test a hypothesis through observational studies or experimental studies. Blehert's first study was observational. Observational studies can be purely **descriptive**—reporting information (**data**) about what is found in nature. Observational studies can also be **analytical**—looking for (analyzing)

① **Observations and questions:** Bats are observed with white noses. What is causing the white fuzz? These bats are dying at higher rates than bats without white noses. Why?

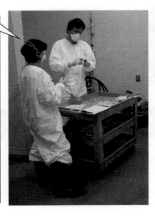

③ **Predictions:** *If* the white noses are caused by a transmissible fungus, *then* healthy bats that hibernate in contact with affected bats should develop the condition. *If* the white noses are caused by a deadly fungus, *then* healthy bats inoculated with the fungus should develop white noses and die at higher rates.

② **Hypothesis:** Bats with white noses are infected with a fungus, and this fungus is causing death.

Figure 1.3

**From observation to hypothesis to testable prediction**

Figure 1.4

**Hypotheses are supported or not supported, but never proved**
Although the claim of scientifically confirmed mildness in this vintage advertisement for cigarettes seems ridiculous, "science" is still used to sell products today. Most Americans see thousands of advertisements every day, and many of these make "scientific" claims that are exaggerated or inaccurate.

**Q1:** State the hypothesis that this advertisement is claiming was scientifically tested.

**Q2:** State a prediction that comes from this hypothesis. Is it testable? Why or why not?

**Q3:** Explain in your own words why the hypothesis cannot be "proved."

patterns in the data and addressing how or why those patterns came to exist. The tools of **statistics**—a branch of mathematics that can quantify the reliability of data—help scientists determine how well those patterns support a hypothesis. Observational studies usually rely on both descriptive and analytical methods to test predictions made by a hypothesis.

In 2009, Blehert, Behr, and Hicks published a scientific paper in which they described the results from inspecting 117 dead bats. They identified microscopic damage caused by a specific kind of fungus in 105 of the bats, and isolated and identified the fungus from a subset of 10 of them. It was a type of cold-loving fungus belonging to a group of fungi called *Geomyces*. They named this new species *Geomyces destructans*.

Their observational study revealed a correlation between white fungus on the noses of bats and bat illness and death. Observational studies suggest possible causes for a phenomenon, but they do not establish a cause-effect relationship. To demonstrate that the fungus was actually causing the illness—and not just correlated with it—Blehert designed and conducted an experiment. Testing scientific hypotheses often involves both observational and experimental approaches (**Figure 1.5**).

# Catching the Culprit

An **experiment** is a repeatable manipulation of one or more aspects of the natural world. Blehert's experiment was to take healthy bats into his laboratory and expose them to the fungus. Like analytical observational studies, experimental studies use statistics to determine whether the experimental results support or refute the hypothesis being tested.

In studying nature, whether through observations, experiments, or both, scientists focus on **variables**, characteristics of any object or individual organism that can change. In a scientific experiment, a researcher typically manipulates

**MELISSA BEHR**

Melissa Behr, formerly with the New York Department of Health, is now a doctor of veterinary medicine at South Dakota State University. She conducts research on the pathology and biology of bats and teaches at the UW veterinary school.

SCIENCE

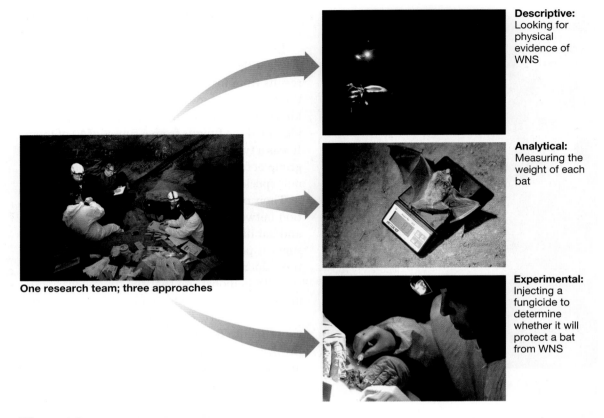

**Descriptive:** Looking for physical evidence of WNS

**Analytical:** Measuring the weight of each bat

**Experimental:** Injecting a fungicide to determine whether it will protect a bat from WNS

One research team; three approaches

Figure 1.5

**Testing hypotheses using multiple approaches**
Scientists set up an underground laboratory in Tennessee's New Mammoth Cave to test hypotheses about white-nose syndrome (WNS) using descriptive, analytical, and experimental approaches.

**Q1:** Give a possible hypothesis that could be tested by weighing the bats.

**Q2:** State the hypothesis being tested in the photo on the bottom right.

**Q3:** Explain in your own words why an experimental study is the only way to show a cause-effect relationship.

a single variable, known as the **independent variable**. In this case, Blehert's independent variable was fungal exposure. Some bats were exposed; others were not. A **dependent variable** is any variable that responds, or could potentially respond, to changes in the independent variable. Blehert's dependent variable was any sign of WNS on the healthy bats.

If we think of the independent variable as the cause, then the dependent variable is the effect. In the most basic experimental design, a researcher manipulates a single independent variable and tracks how that manipulation changes the value

of a dependent variable. Blehert manipulated his independent variable—exposing some bats to the fungus but not others—and then tracked his dependent variable, whether the bats showed symptoms of WNS.

Blehert made sure his experiment was a controlled experiment. A **controlled experiment** measures the value of the dependent variable for two groups of subjects that are comparable in all respects, except that one group is exposed to a change in the independent variable and the other group is not. In this case, healthy bats were either exposed to the fungus

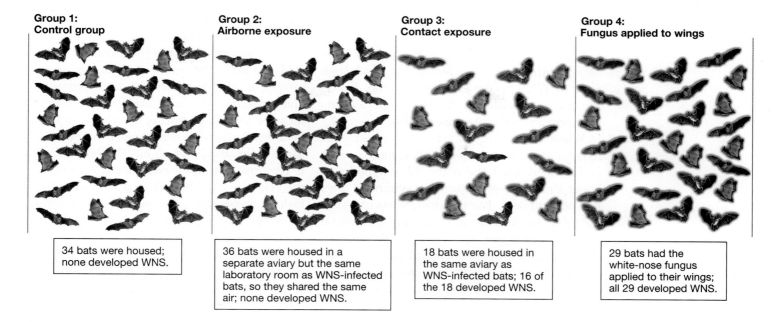

**Group 1:**
**Control group**

34 bats were housed; none developed WNS.

**Group 2:**
**Airborne exposure**

36 bats were housed in a separate aviary but the same laboratory room as WNS-infected bats, so they shared the same air; none developed WNS.

**Group 3:**
**Contact exposure**

18 bats were housed in the same aviary as WNS-infected bats; 16 of the 18 developed WNS.

**Group 4:**
**Fungus applied to wings**

29 bats had the white-nose fungus applied to their wings; all 29 developed WNS.

**Figure 1.6**

**Blehert's experimental design**

Blehert and his colleagues captured 117 healthy bats and brought them into the laboratory. They divided the bats into control and treatment groups and observed them for 102 days.

**Q1:** Which is the control group in this experiment, and what are the three treatment groups?

**Q2:** What is the hypothesis being tested in this experiment?

**Q3:** In one or two sentences, state the conclusions you can draw from the experiment. Was the hypothesis supported? Why or why not?

or not exposed. Typically, a researcher obtains a sufficiently large sample of study subjects and assigns them randomly to two groups. Randomization helps ensure that the two groups are comparable to start with.

One group, the **control group**, is maintained under a standard set of conditions with no change in the independent variable. Blehert had 34 healthy bats in his control group; he kept these bats in the laboratory, under the same conditions as the other bats, except for one difference: he did not expose them to *Geomyces destructans*.

The other group, known as the experimental or **treatment group**, is maintained under the same standard set of conditions as the control group, but the independent variable is manipulated. Blehert exposed 83 healthy bats to the fungus. Of these, 36 were exposed to the

fungus through the air, 18 were put into close contact with naturally infected bats, and 29 had the fungus applied directly to their wings (**Figure 1.6**).

As noted earlier, when scientists test a prediction of a hypothesis and find it upheld, the hypothesis is said to be supported. Scientists can be relatively confident in a supported hypothesis, but they cannot say that the hypothesis has been proved true. Even well-established scientific ideas can be overturned if new evidence against the prevailing view comes to light. Albert Einstein is famously reported to have said, "No amount of experimentation can ever prove me right; a single experiment can prove me wrong."

After watching both the control group and the treatment group, Blehert saw what he had hypothesized: physical exposure to the fungus caused white-nose syndrome, but exposure through the air did not. Healthy bats that had fungus applied directly to their wings or were caged with naturally infected bats had high rates of WNS by the experiment's end. It was the first direct evidence that the fungus was the primary cause of white-nose syndrome.

When a prediction is not upheld, the hypothesis is reexamined and changed, or it is discarded. Over the years, other hypotheses about the cause of white-nose syndrome have not been upheld. For example, scientists were not able to identify a single environmental contaminant at elevated levels in infected bats. A follow-up study by Blehert and others showed that the fungus not only leads to symptoms of WNS in bats, but is sufficient to cause death (**Figure 1.7**).

The story of the bats with white noses is just a single example of the scientific process. One of the greatest strengths of science is that scientific knowledge is tentative and therefore open to challenge at any time by anyone. An absolute requirement of the scientific method is that evidence be based on observations or experiments, or both. Furthermore, the observations and experiments that furnish the evidence must be subject to testing by others; independent researchers should be able to make the same observations, or obtain the same results, if they use the same conditions. In addition, the evidence must be collected in as objective a fashion as possible—that is, as free of personal or group bias as possible. Blehert's experiment fit all these conditions.

The main mechanism for policing personal or group bias and even outright fraud in science is **peer-reviewed publication**. Peer-reviewed publications are found in scientific journals that publish original research after it has passed the scrutiny of experts who have no direct involvement in the research under review. Before Blehert's research was published, it was reviewed by numerous scientists who had not participated in the experiment. If reviewers have concerns during the peer-review process, such as whether the evidence is strong enough to support a hypothesis, they can ask the paper's authors to address those concerns (for example, by gathering additional evidence) and to resubmit the paper. Blehert's paper passed the peer-review process and was published in the scientific journal *Nature* in 2011. At that point, says Blehert, the evidence that *G. destructans* causes WNS was strong enough that "I think we'd convinced most people." (In 2013, scientists renamed *G. destructans* as *Pseudogymnoascus destructans* when the fungus's genus was reclassified.)

But identifying the cause of WNS did not stop the disease from spreading. By March 2008, just a year after Hicks's team found the thousands of dead bats near Albany, more bats were found dead and dying in caves across Vermont, Massachusetts, and Connecticut. Within a year, the disease had spread as far as Tennessee and Missouri. The spread of WNS is a fact: bats around the United States are dying. In casual conversation, we typically use the term "fact" to mean a thing that is known to be true. A scientific **fact** is a direct and repeatable observation of any aspect of the natural world.

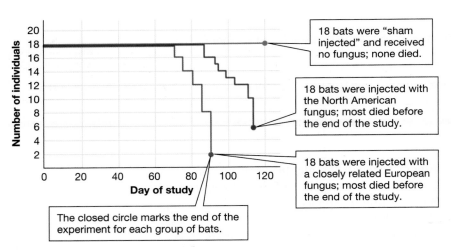

**Figure 1.7**

**Experiments support the hypothesis that the fungus causes WNS in bats**

The experiment whose results are plotted here was conducted by some of the same researchers interviewed for this story.

**Q1:** What is the control group in this experiment, and what are the two treatment groups?

**Q2:** At day 40, approximately how many individuals were alive in each treatment group? At day 80? At day 100?

**Q3:** In one or two sentences, state the conclusions you can draw from the experiment. Was the hypothesis supported? Why or why not?

Scientific fact should not be confused with scientific **theory**. Outside of science, people often use the word "theory" to mean an unproven explanation. In science, a theory is a hypothesis, or a group of related hypotheses, that has received substantial confirmation through diverse lines of investigation by independent researchers. Scientific theories have such a high level of certainty that we base our everyday actions on them. For example, the *germ theory of disease*, formally verified by Robert Koch in 1890, is the basis for treating infections and maintaining hygiene in the modern world (**Figure 1.8**).

# No End in Sight

According to the U.S. Fish and Wildlife Service, white-nose syndrome has killed more than 6 million bats across the United States since 2007 and shows no signs of slowing. As of this writing, the disease has spread to 29 U.S. states and 5 Canadian provinces. In March 2016, hikers found a bat in Washington State with WNS, and Rhode Island's first case was confirmed two months later. Almost all species of bats that hibernate in these regions have been affected, including little brown bats and

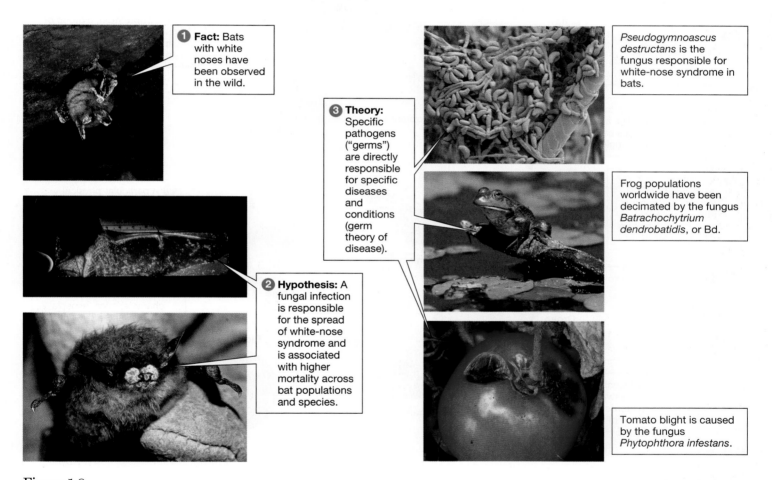

❶ **Fact:** Bats with white noses have been observed in the wild.

❸ **Theory:** Specific pathogens ("germs") are directly responsible for specific diseases and conditions (germ theory of disease).

❷ **Hypothesis:** A fungal infection is responsible for the spread of white-nose syndrome and is associated with higher mortality across bat populations and species.

*Pseudogymnoascus destructans* is the fungus responsible for white-nose syndrome in bats.

Frog populations worldwide have been decimated by the fungus *Batrachochytrium dendrobatidis*, or Bd.

Tomato blight is caused by the fungus *Phytophthora infestans*.

Figure 1.8

**Facts, hypotheses, and theories**
It is important to distinguish among facts, hypotheses, and theories when thinking and talking about science.

**Q1:** Give one *fact* about bats that you learned from this chapter.

**Q2:** What is another example of evidence for the *germ theory of disease*? (*Hint:* Think about human diseases.)

**Q3:** Explain in your own words the difference between a fact and a hypothesis, and between a hypothesis and a theory.

endangered Indiana bats, both of which have been particularly hard-hit.

The fungus appears to be related to a type of fungus common in caves in Europe, so a human traveler from Europe most likely carried it across the Atlantic and into the Albany cave, where it infected its first bat in the United States. Researchers continue to explore exactly how the fungus kills the bats. It appears to wake them from hibernation too many times during the winter, so the bats use up their fat reserves too soon and do not survive the months of cold weather. The fungus also eats through bats' delicate wings, which are important not only for flight, but also for maintaining healthy levels of water, oxygen, and carbon dioxide in the bats' bodies. The impact of this fungus on bats is a powerful example of how a microorganism can affect many levels of life, from individual tissues and organs up to whole populations, communities, and even the ecosystem itself (**Figure 1.9**).

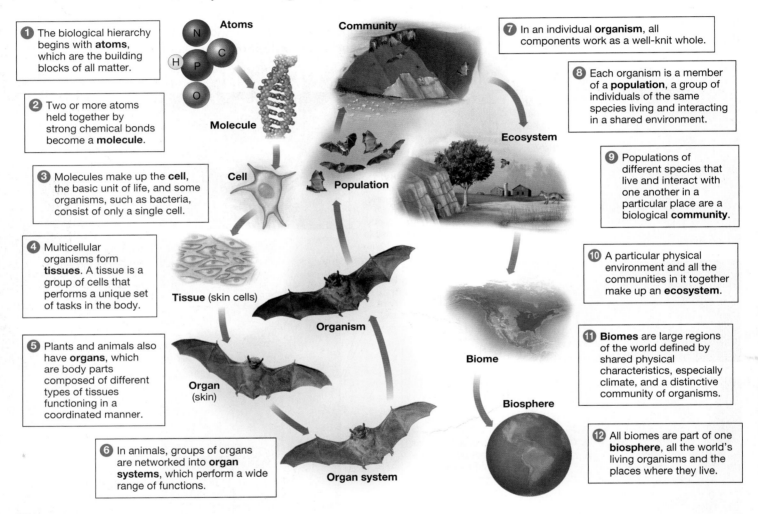

**1** The biological hierarchy begins with **atoms**, which are the building blocks of all matter.

**2** Two or more atoms held together by strong chemical bonds become a **molecule**.

**3** Molecules make up the **cell**, the basic unit of life, and some organisms, such as bacteria, consist of only a single cell.

**4** Multicellular organisms form **tissues**. A tissue is a group of cells that performs a unique set of tasks in the body.

**5** Plants and animals also have **organs**, which are body parts composed of different types of tissues functioning in a coordinated manner.

**6** In animals, groups of organs are networked into **organ systems**, which perform a wide range of functions.

**7** In an individual **organism**, all components work as a well-knit whole.

**8** Each organism is a member of a **population**, a group of individuals of the same species living and interacting in a shared environment.

**9** Populations of different species that live and interact with one another in a particular place are a biological **community**.

**10** A particular physical environment and all the communities in it together make up an **ecosystem**.

**11** **Biomes** are large regions of the world defined by shared physical characteristics, especially climate, and a distinctive community of organisms.

**12** All biomes are part of one **biosphere**, all the world's living organisms and the places where they live.

Atoms · N · H · P · C · O
Molecule
Cell
Tissue (skin cells)
Organ (skin)
Organism
Organ system
Community
Population
Ecosystem
Biome
Biosphere

Figure 1.9

**The biological hierarchy**

The **biological hierarchy** is a way to visualize the breadth and scope of life, from the smallest structures to the broadest interactions between living and nonliving systems.

**Q1:** Give examples of other kinds of organs that mammals such as bats have. (*Hint:* Think of the organs in your own body.)

**Q2:** Are bats in California part of the community of bats in upstate New York if they are of the same species? Why or why not?

**Q3:** Is the soil in a cave where bats live a part of the bats' population, community, or ecosystem? Explain your reasoning.

# Bug Zappers

Bats are skilled natural exterminators, consuming billions upon billions of insects each year, including crop pests and mosquitoes. For this reason, bats play a critical role in agriculture and potentially in human health with the onset of increased mosquito-borne viruses such as West Nile and Zika. The loss of bat populations to white-nose syndrome would have a significant impact on farms, forests, and people around the country.

Assessment available in smartwork5

## 1.3 MILLION

A single colony of **150 big brown bats** *(Eptesicus fuscus)* in Indiana is estimated to eat nearly **1.3 million pest insects each year.**

## 4–8 GRAMS

A single little brown bat *(Myotis lucifugus)* can consume **4–8 grams of insects each night** during the active season.

## 660–1,320 METRIC TONS

Extrapolating the diet of a single bat to 1 million bats estimated to have died from WNS, between **660 and 1,320 metric tons of insects** are no longer being consumed each year in WNS-affected areas.

## $22.9 BILLION

If bats disappeared entirely from the United States, it would cost the agricultural industry roughly **$22,900,000,000 per year** to save crops by dealing with the insects no longer being eaten by bats.

Today, WNS continues to spread, and scientists have not yet found a way to stop it, though many are trying. Recently, an international team of scientists identified a few populations of bats in Asia and the United States that survived the fungal infection. Evidence from those bats suggests they evolved resistance to the deadly disease. Researchers have also had some luck treating infected bats with a type of bacteria that kills the fungus, but they are still working on ways to deliver the treatment to large groups of bats.

For now, bats continue to die across America every winter. Hicks worries that students, hikers, and tourists will visit caves in the United States, not see any bats, and think that's normal. "In 2006, in one big cave in the Adirondacks, we counted over 185,000 bats. Anywhere you shined a light, there was a bat," says Hicks. "You go in now, and there's not a bat in sight."

# REVIEWING THE SCIENCE

- **Science** is both a body of knowledge about the natural world and an evidence-based process for generating that knowledge.

- The **scientific method** represents the core logic of the process by which scientific knowledge is generated. The scientific method requires that we (1) make **observations**, (2) devise a **hypothesis** to explain those observations, (3) generate predictions from that hypothesis, (4) test those predictions, and (5) share the results of the tests for **peer review** by fellow scientists.

- A hypothesis cannot be proved true; it can only be supported or not supported. If the predictions of a hypothesis are not supported, the hypothesis is rejected or modified. If the predictions are upheld, the hypothesis is supported.

- We can test hypotheses by making further observations or by performing **experiments** (controlled, repeated manipulations of nature) that will either uphold the predictions or show them to be incorrect.

- In a scientific experiment, the **independent variable** is manipulated by the investigator. Any variable that can potentially respond to the changes in the independent variable is called a **dependent variable**.

- A scientific **fact** is a direct and repeatable observation of any aspect of the natural world. A scientific **theory** is a major idea that has been supported by many observations and experiments.

- The term **biological hierarchy** refers to the many levels at which life can be studied: atom, molecule, cell, tissue, organ, organ system, organism, population, community, ecosystem, biome, biosphere.

- All living organisms have the following characteristics in common: (1) They are built of **cells**; some are single-celled and some are multicellular. (2) They **reproduce**, using **DNA** to pass genetic information from parent to offspring. (3) They take in **energy** from their environment. (4) They **sense** and respond to their environment. (5) They exhibit **homeostasis**, maintaining constant internal conditions. (6) They can **evolve** as groups.

# THE QUESTIONS

## The Basics

**1** Which of the following is a living organism? (Select all that apply.) If it is not living, which criterion, or criteria, does it not meet?

(a) an oak tree

(b) an influenza virus

(c) the fungus that causes white-nose syndrome in bats

(d) a diamond

(e) your teacher

**2** When scientists use the word "theory," they mean

(a) an educated guess.

(b) an overarching explanation of an interrelated set of observations.

(c) wild speculation.

(d) an experimental prediction.

(e) a fact proved by many experiments.

**3** Select the correct terms:
The process of science begins with a(n)

(**prediction / observation**) about the natural world. A scientist then proposes a (**hypothesis / prediction**), which is the basis of one or more testable (**observations / predictions**).

**4** Place the following steps of the scientific method in the correct order by numbering them from 1 to 7.

_____ a. Make observations about the natural world.

_____ b. Test the predictions by designing an experiment or collecting observational data.

_____ c. Run the experiment and analyze the results.

_____ d. Generate predictions to test the hypothesis.

_____ e. Share the results with fellow scientists so that they can review and evaluate them.

_____ f. Develop a hypothesis to explain the observations.

_____ g. Accept, reject, or modify your hypothesis depending on the results.

**5** Identify the level of biological organization for each of the following.

_____ a. the kidney of a bat

_____ b. an oak tree outside a cave in upstate New York

_____ c. bats in a cave in upstate New York

_____ d. the physical and biological components of a cave in upstate New York

_____ e. the respiratory system of a bat

_____ f. all the species living and interacting within a cave in upstate New York

## Challenge Yourself

**6** Describe one observation, one hypothesis, and one experiment from the white-nose syndrome research discussed in this chapter.

**7** Which of the following statements is a scientific hypothesis (that is, it makes testable predictions)? (Select only one.)

(a) Even though no one else can see him, the ghost of my dog lives in my backyard.

(b) The Atkins diet helps people lose more weight and keep it off than Weight Watchers does.

(c) People born under the sun sign Aquarius are kinder and cuter than those born under Scorpio.

(d) It is unethical to text while driving.

(e) none of the above

**8** Consider an experiment in which subjects are given a pill to test its effectiveness in reducing the duration of a cold. Which of the following is the best way to treat the control group?

(a) Give the control group two pills instead of one.

(b) Do nothing with the control group.

(c) Give the control group a pill that looks like the test pill but does nothing.

(d) Let the control group choose whether or not to take any pills.

(e) Expose the control group to the cold virus.

## Try Something New

**9** If a virus was discovered that had the ability to reproduce itself without the help of a cell and its machinery, several experiments could be performed in order to show that it should be categorized as "alive." For each experiment, specify which one of the six criteria for being classified as "alive" would be tested.

_____ a. Place a set number of viral particles in a nutrient broth and check for an increase in viral particles after several days.

_____ b. Place a set number of viral particles in a broth where all the nutrients are corralled to one side of the container and check for viral particle accumulation in the nutrient-rich area of the container.

_____ c. Place a set number of viral particles in a nutrient broth and check for waste products (broken-down nutrients) in the broth after several days.

_____ d. Place a set number of viral particles in a nutrient broth for several months, adding different nutrients daily. Randomly isolate viral particles and sequence their DNA to determine whether there have been changes in genetic characteristics of this population of viral particles compared to the DNA of the original population before culturing.

_____ e. Perform biochemical and microscopic analyses on a number of viral particles to determine whether they have the main components of a living cell, including a plasma membrane, ribosomes, and DNA.

**10** Mad cow disease, or bovine spongiform encephalopathy, appears to be caused by a novel infectious agent: a protein that replicates by causing related proteins to modify their structure from a harmless shape to a dangerous one. These prions (short for "proteinaceous infectious particles") also appear to be the cause of several other spongiform encephalopathy diseases, such as scrapie in sheep, and kuru and Creutzfeldt-Jakob disease in humans. Which of the following observations or experiments would *not* support the hypothesis that a prion causes spongiform encephalopathy?

(a) The brains of many sheep with scrapie contain prion proteins, but the brains of most sheep without scrapie do not.

(b) There is a high incidence of kuru in populations of people who consume brain tissue from prion-infected animals.

(c) Coyotes that feed on cows with mad cow disease do not subsequently develop spongiform encephalopathy.

(d) When introduced into sheep brain cells in culture, prions cause the normal proteins to change shape into dangerous prion proteins.

(e) When prions are fed to sheep, most of them subsequently develop scrapie, whereas sheep not fed prions do not develop scrapie.

11  Label each of the following statements as an observation, a hypothesis, an experiment, or a result.

_____ a. People who ingest a liquid contaminated with a rotavirus will subsequently experience acute diarrhea.

_____ b. Seventeen out of twenty students given rotavirus-contaminated chocolate bars subsequently experienced acute diarrhea.

_____ c. Ten students were given a liquid containing a rotavirus, and 10 other students were given uncontaminated liquid.

_____ d. People exposed to rotaviruses often experience acute diarrhea.

_____ e. The latest outbreak of acute diarrhea on campus was due to a rotavirus-contaminated elevator button in the dorms.

12  Review the map showing the confirmed and suspected outbreaks of WNS. Is each statement that follows a testable, properly formulated scientific hypothesis for the spread of WNS to bats in Washington State? Explain why or why not. (*Hint:* Think about the criteria for a scientific hypothesis.)

a. Did bats infected with WNS arrive in Washington State via human smuggling for research purposes?

b. The WNS fungus was transported from the Midwest to Washington State via a bat found in Flight 1701's cargo hold.

c. Bats in Washington State were infected by contaminated clothing and equipment from visiting researchers checking for WNS in these colonies.

d. How did bats end up with WNS in Washington State, so far away from all other sites of contamination?

e. A mysterious cloud of fungal particles must have floated over Washington State in early 2016.

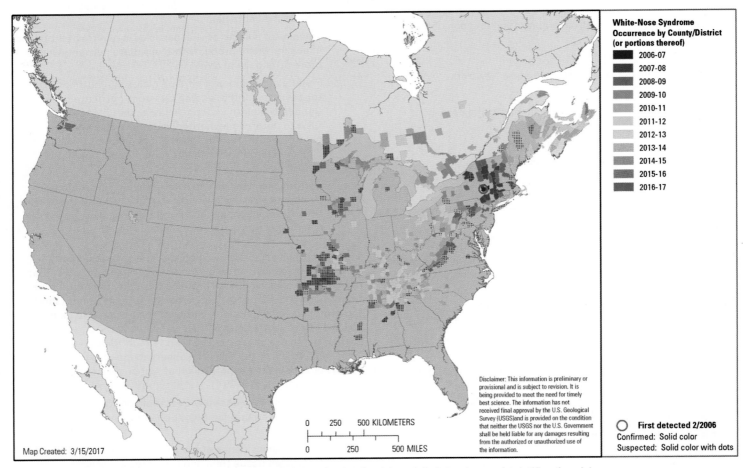

Map depicts the first time WNS is reported suspect or confirmed in a county or district (or portions thereof); each time period in the legend spans a winter bat hibernation period.

Citation: White-nose syndrome occurrence map - by year (2017). Data Last Updated: 3/15/2017. Available at: https://www.whitenosesyndrome.org/resources/map.

## Leveling Up

**13** **What do *you* think?** Insectivorous bats save the agriculture industry billions of dollars each year because they eat insects that damage crops. Other species of bats pollinate some crops. What would you predict about crop yields in areas that are experiencing heavy bat mortality from WNS? Does your prediction affect your level of concern about WNS? Why or why not?

**14** **Doing science** Although you are an expert on white-nose syndrome in bats, you have been asked to contribute your scientific expertise to understanding the fungal infection that is decimating frog populations worldwide: *Batrachochytrium dendrobatidis*, a.k.a. Bd.

a. Read the May 2013 *National Geographic* article on how Bd has been spread around the world: http://news.nationalgeographic.com/news/2013/13/130515-chytrid-fungus-origin-african-clawed-frog-science.

b. The graph provided here, based on data from a scientific paper published in 2013, shows that frog species with higher average body temperatures tend to have lower Bd infection rates. It also shows that, compared to males of the same species, females tend to have higher average body temperatures and lower Bd infection rates. Is this observational study descriptive or analytical? Explain your answer.

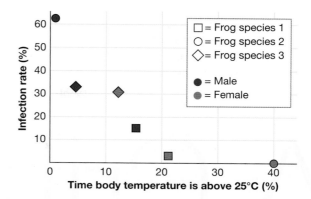

c. Propose a hypothesis to explain the results depicted in the graph, and then identify a testable prediction from that hypothesis. Design an experiment to test the prediction, identifying your independent and dependent variables, the control group, and the treatment conditions. Create a graph to show the results you expect to find (1) if your hypothesis is supported and (2) if your hypothesis is not supported.

**For more, visit digital.wwnorton.com/bionow2 for access to:**

# A Critical Choice

*A mother peels back the controversy to decide whether to vaccinate her children.*

- - - - - - - - - - - - - - - - - - - - - - - - - - - - - - - - - - - - - - - - - - - - - - - - - - - - -

## After reading this chapter you should be able to:

◆ Evaluate a scientific claim, using the process described in the chapter.

◆ Explain the importance of scientific literacy for making informed decisions.

◆ Distinguish between secondary and primary literature, and explain the role of peer review in the latter.

◆ Compare and contrast basic and applied research, and give an example of each.

◆ Distinguish between correlation and causation, and give a clear example of each.

◆ Determine whether a scientific claim is based on real science or pseudoscience.

CHAPTER

02

EVALUATING
SCIENTIFIC
CLAIMS

In 2009, Anna Eaton discovered she was pregnant. Several of her friends were also expecting or had recently had children, so the friends' afternoon chats were soon about all things baby: nursery colors, pediatricians, car seats. Every once in a while, however, one of the mothers broached a subject far more contentious than strollers and baby carriers. One of their hot topics of discussion became vaccination.

Vaccination is the injection of material—typically an inactivated and harmless infectious organism or parts of such an organism (e.g., a single protein)—that stimulates the immune system to protect against future exposure to that pathogen (**Figure 2.1**). When the body's immune system is exposed to a vaccine, it recognizes the inactivated organism (or its parts) as an invader and mounts an attack against it. Upon later contact with the infectious organism, a vaccinated individual's immune system remembers the inactivated organism from the vaccine and is already armed and bristling. The inactivated organism from the vaccine cannot cause disease because it was made harmless by the process of creating the vaccine. The vaccine itself cannot cause disease because it is not a functional infectious organism and therefore cannot replicate.

Over the past 200 years, scientists have developed vaccines to protect against dozens of pathogens, starting with the smallpox virus. In England in 1796, Dr. Edward Jenner first vaccinated his gardener's 8-year-old son against smallpox using the cowpox virus, because it was known that individuals who had previously suffered cowpox were immune to smallpox. Jenner's breakthrough ultimately led to a worldwide smallpox vaccination campaign, which was so successful that in 1980 the disease was officially declared eradicated by the World Health Organization. Vaccines have also been developed for the Tic Tac–shaped bacteria that cause the upper respiratory infection diphtheria, the round and highly contagious rotavirus that causes vomiting and diarrhea, the tiny airborne bacteria that cause pertussis (also known as whooping cough), and many more pathogens.

Before vaccines, children died in large numbers from smallpox, diphtheria, whooping cough, polio, and many other diseases (**Figure 2.2**). The infectious pathogens that cause those diseases still exist today in our environment, but because vaccines protect people, we almost never see those infections. Today, the Centers for Disease Control and Prevention (CDC), the public health branch of the U.S. government, recommends that children receive 10 vaccines, given over a total of 24 doses, between birth and 15 months of age. Additional vaccinations are recommended between the ages of 18 months and 18 years, including the annual influenza ("flu") vaccine, also recommended for all adults each year (see "Flu Shot" on page 24).

Eaton, pregnant with her first baby, was a microbiologist by training. She worked at the Cleveland Clinic in Ohio, an academic medical center, before moving on to teach high school biology and chemistry, as well as bioscience manufacturing at a local community college. Some of her friends were also scientists, and one pregnant friend adamantly commented, "Of course we're going to vaccinate. We're scientists."

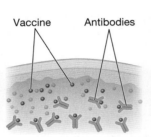

Vaccine

A vaccine with a harmless form of a virus (or other organism) is injected under the skin.

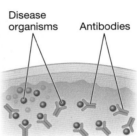

Vaccine    Antibodies

The vaccine stimulates the immune system to produce antibodies (in green) that recognize the virus.

Disease organisms    Antibodies

When the individual is exposed to the virus after vaccination, the new antibodies are primed to attack the invader.

Figure 2.1

**How vaccines work**
A vaccine trains the body's immune system to fight infection.

**Q1:** Describe in one sentence how a vaccine creates immunity to a virus.

**Q2:** Why is it impossible to become infected with a virus from a vaccine composed of viral proteins?

**Q3:** Natural immunity occurs without a vaccine, just by exposure to a particular stimulus, like the chicken pox virus. Explain why people don't get chicken pox twice.

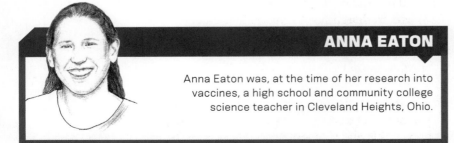

**ANNA EATON**

Anna Eaton was, at the time of her research into vaccines, a high school and community college science teacher in Cleveland Heights, Ohio.

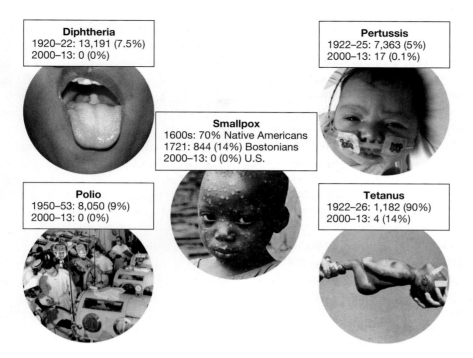

Figure 2.2

**Vaccines save lives**

The numbers in each box represent each disease's average annual number of deaths and percent mortality rate in the United States during prevaccination periods (for example, 1950–53 for polio) and the postvaccination twenty-first century.

**Diphtheria**
1920–22: 13,191 (7.5%)
2000–13: 0 (0%)

**Pertussis**
1922–25: 7,363 (5%)
2000–13: 17 (0.1%)

**Smallpox**
1600s: 70% Native Americans
1721: 844 (14%) Bostonians
2000–13: 0 (0%) U.S.

**Polio**
1950–53: 8,050 (9%)
2000–13: 0 (0%)

**Tetanus**
1922–26: 1,182 (90%)
2000–13: 4 (14%)

**Q1:** Before vaccinations, which diseases had the highest and lowest mortality rates? What are these mortality rates?

**Q2:** After vaccinations, which diseases have the highest and lowest mortality rates? What are these mortality rates?

**Q3:** If there were a sudden outbreak of pertussis at a university where pertussis vaccinations were not required and no one was protected, how many students would die? What is the probability or chance that you would die if infected?

But Eaton was not so confident. She had seen media stories suggesting that vaccines might cause autism spectrum disorders, and when she searched online, she read scary accounts from parents who had had a child vaccinated and then watched as their child developed autism. "I'm on message forums, encountering a hundred different moms with stories," recalls Eaton. "It was frightening."

Stories and anecdotes are powerful. They can change how we feel or think about a subject. But anecdotal evidence is not scientific evidence. Anecdotes are not representative of collected data or collected scientific observations, and therefore they cannot reliably give us the big picture of a subject or phenomenon. Scientific evidence can.

Eaton's scientist friend said there were data backing the safety of vaccines, but Eaton had not seen those data and wasn't about to inject her new baby girl with something she knew nothing about. "Feelings of distrust are in all of us, and possibly the strongest in a new mother," says Eaton. "Even though she was a friend of mine, it was still not enough for me to just go in the office and have my daughter vaccinated."

# True or False?

The statement that vaccines cause autism, or the opposite statement that vaccines are safe, is a **scientific claim**, a statement about how the world works that can be tested using the scientific method. We are exposed to scientific claims every day—dozens of them, in fact. They are not all true. In 2004, editors at the magazine *Popular Science* asked one of their journalists to write down every scientific claim he heard in a day and evaluate each one. He recorded a whopping 106 claims and dug into each of them, from Cheerios' claim that it reduces cholesterol (supported by scientific evidence) to the claim that a face cream infused with vitamin A would revitalize skin (not supported). Of the 106 claims, most were bogus.

The majority of scientific claims come from advertisers. Though companies are legally bound to tell the truth, not all of them do. For example, in 2005 an Italian footwear company named Vibram introduced the $100 FiveFingers running shoes, glove-like shoes that mimic the feel of barefoot running (**Figure 2.3**). In a marketing campaign, the company touted the health benefits of these unique shoes, advertising that they could reduce foot injuries and strengthen foot muscles. In 2013, however, two peer-reviewed studies of over a hundred joggers found that the shoes actually increased the likelihood of injury. Soon, Vibram was sued for false advertising,

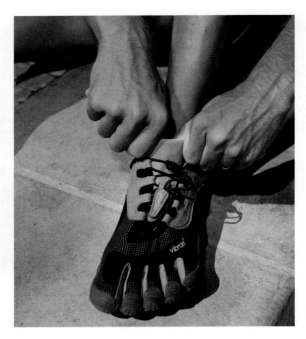

Figure 2.3

**Shoes designed to mimic bare feet**
FiveFingers running shoes did not live up to the manufacturer's scientific claims.

and in 2014 the company settled that lawsuit for $3.75 million, agreeing to refund FiveFingers shoe purchases and remove all health claims about the shoes from its advertisements.

Scientific claims also come from special-interest groups and organizations that exist to advance certain causes, often for political or religious reasons. These claims include statements about global warming, evolution, and medical care. Again, such claims are often untrue. Therefore, it is important to question the "truth" of a claim when you hear it.

We, the public, are not simply consumers of science and technology. We are participants. By voting on issues that have a scientific underpinning, we shape the course of science and influence which technologies are used, as well as where and how they're used. Although our personal values and political leanings are likely to influence how we vote, the underlying science should also be taken into consideration. **Scientific literacy**, an understanding of the basics of science and the scientific process, enables us to make informed decisions about the world around us and to communicate our knowledge to others. Our hope is that this book will help you become scientifically literate.

## Flu Shot

**W**hy do we need a flu shot every year? Why doesn't it protect us for life, like other vaccines do? In fact, each year's flu vaccine *does* protect you for life from a particular strain of the flu virus, but the virus mutates rapidly, and every year's flu strain is likely to be different from the previous year. To the immune system, the new strain looks like a completely new virus. But that's not all! That pesky flu virus can even combine itself with other flu viruses, from animals like birds and pigs, to create deadly flu pandemics like the H1N1 strain, or "swine flu," in 2009.

Each flu season, the CDC uses current flu cases to try to determine what the next season's flu viruses will be. It then creates a vaccine to protect us from the anticipated viruses. Sometimes the experts are extremely accurate and the flu vaccines are highly effective against that season's virus. Sometimes, however, they are not so accurate, and the vaccine is not as protective. It's a viral genetics puzzle and a guessing game, with a little attempt at clairvoyance thrown in. Still, even in years when the vaccine does not provide full coverage for that season, it is likely to protect you from future manifestations of the flu.

Here's the bottom line: The more yearly flu vaccines you receive, the more flu viruses your immune system will be ready to fight, and the less likely you will be to catch the seasonal flu. More important, you will be less likely to succumb to a deadly flu pandemic.

Eaton didn't know the source of the original claim that vaccines cause autism, but she knew that in order to be scientifically literate and make an informed decision for her daughter, she needed to find it. Scientific claims directly affect our lives because we make decisions based on them. Some of these are small decisions: Should I take a multivitamin every morning? How often should I exercise?

Others are larger decisions: Should I vote to support carbon taxes? Does a cell phone cause tumors from radiation? Should I vaccinate my children? The good news is that you can learn to evaluate scientific claims. You can be skeptical about claims and, using critical thinking, make scientifically literate decisions for yourself.

## Credentials, Please

Caroline was born in January 2010 (**Figure 2.4**). Eaton had Caroline vaccinated when she was two months old, but then Eaton found herself gripped with fear. "I worried I was doing more harm than good," she says. Because of her rising fears, she did not get Caroline's four-month vaccinations.

Eaton sought out a pediatrician who would discuss her fears about vaccination. Several pediatricians refused to talk with her about a delayed vaccination schedule; some even refused to treat Caroline if she was not vaccinated. Finally, Eaton found a pediatrician who was willing to listen and discuss. After Eaton shared her concerns about vaccine safety, the pediatrician handed Eaton a book, *Vaccines and Your Child*, by a pediatrician named Paul Offit. In the book, Offit laid out how vaccines work, how they are made, and which risks are real and which are false. He also provided documentation, with detailed references to scientific papers, showing that there is no scientific evidence that vaccines cause autism.

Eaton found the book to be well written and informative, but she wondered whether she could trust Offit. She checked his **credentials**—a first step toward assessing the strength of a person's scientific claim. Does the person making a scientific claim have a PhD or MD? Is the degree in the field in which they are making the claim? PhDs in physics do not have training in germ theory, for example, and medical doctors do not have training in atmospheric science.

Figure 2.4

**Anna and Caroline Eaton**
Anna Eaton had concerns about vaccinations for her first child, Caroline.

Offit had an MD from the University of Maryland and was a professor of vaccinology and pediatrics at the University of Pennsylvania. He was also chief of the Division of Infectious Diseases and the director of the Vaccine Education Center at the Children's Hospital of Philadelphia. In other words, Offit had an advanced degree in an appropriate field and held a job at a reputable university. While good credentials alone do not guarantee that a source is trustworthy, scientists practice for many years in their area of expertise, and their scientific claims tend to be based on that expertise and carefully stated.

In addition to evaluating someone's credentials, it is important to assess whether the person making a scientific claim has an agenda or **bias** (a prejudice or opinion for or against something). Does the person have an ideological, political, or religious belief that will be supported by the scientific claims being made? Does the person have a

**PAUL OFFIT**

Paul Offit is a pediatrician and chief of the Division of Infectious Diseases at the Children's Hospital of Philadelphia.

conflict of interest? Does he or she stand to make money in any way if others accept those claims?

Doing scientific research almost always requires money, so it is important to take into account where the money comes from. In North America, the vast majority of **basic research** in science is funded by the federal government—that is, by taxpayers. Basic research is intended to expand the fundamental knowledge base of science. In the United States, Capitol Hill appropriates more than $30 billion each year for basic and medical research in the life sciences, including biomedicine and agriculture. Researchers must compete vigorously for the limited funds, and this competition helps ensure that public money goes toward supporting high-quality science. Research funded by the government is normally not considered biased, since the funding comes from taxpayers.

But industries and businesses spend a great deal of money funding science as well, often in areas of **applied research**, in which scientific knowledge is applied to human issues and often commercial applications. In some cases, researchers funded by industry may have a bias in favor of whatever that industry is selling. Funding from industry does not necessarily mean that a scientific claim is incorrect, but the claim should be looked at closely to rule out possible bias.

Offit, Eaton read, had worked with other researchers to develop the RotaTeq vaccine against rotavirus, a virus that can cause death from severe diarrhea. The pharmaceutical company Merck had purchased the vaccine, and Offit had received an unspecified amount of money from that transaction. This financial compensation raised the possibility that Offit might have some bias toward the use of the rotavirus vaccine. Though she had enjoyed Offit's book, Eaton decided she did not want to take his word alone as an answer to her questions about vaccines. "In the end, I decided to just read about it, do a lot of research, and talk to a lot of people," she says.

## To the Books

Eaton dove into the secondary literature on vaccines. When investigating a scientific claim, your first stop should be the Internet or the library to get a basic overview of the topic from the **secondary literature**, which summarizes and synthesizes an area of research. Textbooks, review articles, and popular science magazines such as *National Geographic*, *Popular Science*, and *Scientific American* are good secondary sources.

Eaton went to her local library and came home with a stack of books, including a thick vaccine textbook so that she could learn the underlying science of how a vaccine works inside the body. She read about how vaccines stimulate cells of the immune system to protect a person from a virus or bacteria. Once, while she was reading, Eaton's husband walked into the room and looked at her with surprise. "What are you doing?" he asked. "I'm looking for answers!" she replied.

For secondary literature on the Internet, try to visit sites that are affiliated with the government, a university, or a respected institution like a major hospital or museum. Wikipedia often has overview articles that link to science blogs and review articles in science journals. Like Eaton did, it's important to check the credentials of the person or people behind a resource, especially on the Internet. Anonymous sources are not to be trusted.

When evaluating a scientific claim, you may need more detailed information than is available in the secondary literature, especially if you're dealing with a particularly important life decision or if the area of science involved is changing rapidly. In that case, you should next review the **primary literature**, where scientific research is first published (**Figure 2.5**). Primary sources include technical reports, conference proceedings, and dissertations, but the most important primary sources are peer-reviewed scientific journals such as *Science*, *Ecology*, and the *Journal of the American Medical Association* (*JAMA*).

Pulling from references she found in the secondary literature, Eaton compiled a stack of primary literature on vaccines. If her library didn't carry a particular journal, she acquired a copy through interlibrary loan or, in some cases, even e-mailed the journal directly and got articles for free. She found dozens and dozens of papers about vaccines.

## Correlation or Causation?

One of the first papers that Eaton came across in her investigation—one that had made huge waves in the media—was published in the *Lancet* in

**Figure 2.5**

**Scientific claims in the media and literature**
It is easy to find and read scientific claims in social media. However, this is not a good source of scientific information. For help in making important life choices, it is important to go to the secondary literature or even the primary scientific literature for accurate and reliable information.

Q1: Why are we less confident of scientific claims made over social media?

Q2: Where would you place a blog in this figure? Would it matter whether or not it was written by a practicing scientist? Explain your reasoning.

Q3: Give an example of when you would rely on secondary literature to evaluate a scientific claim and an example of when you would go to the primary literature. What is the basis of that decision?

1998. The *Lancet* is a well-known peer-reviewed medical journal and therefore a reputable source. The paper was unremarkably titled "Ileal-Lymphoid-Nodular Hyperplasia, Non-specific Colitis, and Pervasive Developmental Disorder in Children." It was a study of 12 children ranging in age from 3 to 10 years old who had experienced a loss of language skills—a symptom of autism spectrum disorders—as well as diarrhea and abdominal pain. Parents of 8 of the 12 children said the onset of symptoms occurred shortly after the child's immunization with the measles, mumps, and rubella (MMR) vaccine.

The authors of the paper, a team of 12 researchers, concluded that more research was needed to study a possible relation between the observed brain dysfunction, bowel problems, and the MMR vaccine. In a press conference when the paper was published, one of the authors, a British doctor named Andrew Wakefield, stated that he believed single vaccines, rather than the MMR triple vaccine, were likely to be "safer" for children. The study and press conference sparked widespread fear among parents that the MMR vaccine could cause autism.

The study was published around the same time that officials began documenting an increase in rates of autism. Since the early 1970s, when organizations started counting the number of people with autism for the first time, the incidence of autism has increased 20- to 30-fold in the United States and other countries. In 2002, the estimate was that one in 150 children aged eight years (the age of peak prevalence) had the disorder. In 2004, that number had risen to one in 125. In 2006, one in 110. By the year 2012, the most recent year with available data, an estimated one in every 68 eight-year-olds had an autism spectrum disorder. The CDC has said that the rise in rates is likely due to heightened disease awareness, more screening

**The real cause of increasing autism prevalence?**

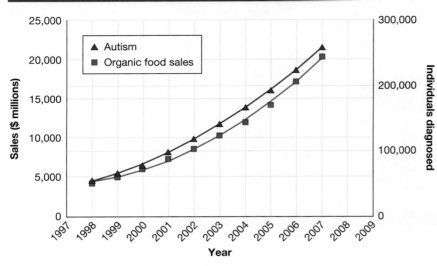

Figure 2.6

**Correlation is not causation: organic food and autism**

Reddit user Jasonp55 created a tongue-in-cheek demonstration of why it is important not to assume causation from correlation. Jason used real data on organic food sales and the prevalence of autism from 1998 to 2007. The two are highly correlated, but one does not cause the other.

**Q1:** How much did organic food sales grow during the period covered in the graph? How much did the incidence of autism grow?

**Q2:** Why might both organic food sales and autism prevalence have increased during this time period? A Reddit user in the original discussion thread suggested that both might be affected by increasing wealth in the United States. How might increased wealth affect these variables?

**Q3:** In what way has the vaccine-autism debate been confused by people misinterpreting correlation as causation?

within schools, and a willingness to label the condition. But after Wakefield suggested that the MMR vaccine might be causing autism, the press and other organizations began to report that the rise in autism rates was caused by the increased use of vaccines.

Linking rising rates of autism with increased use of vaccines is a correlation. **Correlation** means that two or more aspects of the natural world behave in an interrelated manner: if one shows a particular value for one aspect, we can predict a value for the other aspect. But correlation does not establish **causation**, in which a change in one aspect *causes* a change in another. Correlations may suggest possible causes for a

phenomenon, but they do not establish a cause-and-effect relationship. For example, there is also a correlation between organic food sales and the increase in autism. From 1998 to 2007, rates of autism increased hand in hand with organic food sales. They are correlated, but there is no scientific evidence that eating organic food causes autism (**Figure 2.6**).

Another correlation that spurred fears about vaccines is that the onset of autism symptoms occurs at about the same age that children receive vaccinations. Most children receive the MMR vaccine at about 15 months old, which is shortly before the first symptoms of autism are often noticed. Parents of children with autism saw their children begin to exhibit symptoms of the illness after their vaccination and therefore directly observed a correlation between the administration of the vaccine and the onset of an autism spectrum disorder.

Yet, again, correlation does not prove causation. Only scientific experiments can demonstrate causation. So, was Wakefield's claim based on good science? Did the MMR vaccine cause autism?

# Real or Pseudo?

Unfortunately, sometimes a claim that superficially looks like science is actually **pseudoscience**. Pseudoscience is characterized by scientific-sounding statements, beliefs, or practices that are not actually based on the scientific method. Asking a few simple questions at each step of the progression toward a claim purported to be "scientific" can help you distinguish science from pseudoscience (**Figure 2.7**).

Using these criteria, Eaton analyzed Wakefield's claim. It was quickly obvious that his study did not live up to the standards of good scientific research (**Figure 2.8**). First, the study was small—only 12 children participated—yet the conclusions were grand: Wakefield suggested that all children should stop receiving the MMR vaccine. Sample size is extremely important in observational studies, where small samples may skew data to one extreme. Large sample sizes are more representative of the population, are less likely to be affected by outliers, and provide the power to draw more accurate conclusions.

A second problem with the study was that it was not made up of a random sample of

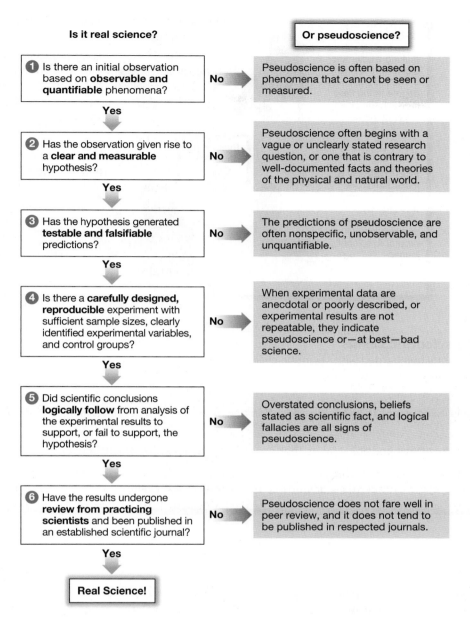

Figure 2.7

## Science or pseudoscience?

A series of simple questions based on the scientific method can help you determine whether a "scientific" study is real science or pseudoscience.

**Q1:** State the hypothesis of the people who believe that vaccines cause autism. What is an alternative hypothesis?

**Q2:** What part(s) of the figure show(s) where Wakefield's study failed to meet the standards of the scientific method?

**Q3:** Why does only one arrow point to "real science," whereas multiple arrows point to pseudoscience?

Figure 2.8

**Evaluating scientific claims**
We are constantly bombarded with scientific claims. A few simple questions will help you evaluate which of these claims are valid and which are not.

**Q1:** Why is it important to know the education and expertise of a person making a scientific claim?

**Q2:** List at least five possible biases that people making scientific claims might have.

**Q3:** Describe a situation in which you might not dismiss the scientific claim of a person who did not have appropriate credentials, or who had a bias toward the claim.

children. Instead, the subjects were picked specifically for their symptoms. Third, the study did not have a control group, such as children who had been vaccinated but did not show signs of autism, or children who had autism but had not been vaccinated. Finally, and most important, the finding could not be repeated. In fact, as of October 2016, at least 110 published papers, studying millions of children, have found no evidence of a link between vaccines and autism. Vaccination does not cause autism, and the two are not even directly correlated. In fact, studies first published in March 2014 provide evidence that autism begins in the womb, before birth.

In one of the most recent papers, researchers at the University of Sydney in Australia, who stated no conflicts of interest, performed a **meta-analysis** (work combining results from different studies) of five *cohort studies*—observational studies of a group of people over a certain period of time—involving 1,256,407 children. The researchers also looked at five *case-control studies*—studies comparing patients with a disease to those without—involving 9,920 children. The data showed no relationship between vaccination and autism, autism and MMR, or autism and thimerosal (a mercury-containing preservative in some vaccines). The work, published in the journal *Vaccine*, included data from the United States, United Kingdom, Japan, and Denmark on vaccines for measles, mumps, diphtheria, tetanus, and pertussis.

Many more peer-reviewed studies with good control groups and sufficient sample sizes, conducted by researchers without a conflict of interest, have shown that the number of vaccines given from birth to 15 months of age under the CDC's recommended vaccination schedule is safe. Researchers have also found that delaying the vaccination schedule increases a child's risk of contracting a disease.

Today it is clear that the paper that started it all, Wakefield's 1998 *Lancet* paper, proposed a claim that is incorrect. All told, it's OK for science to be wrong. There is no expectation that every single one of the thousands of studies published each year will be correct. Yet in the case of the *Lancet* paper, there was bias and wrongdoing. Years after publication of the paper, it came to light that Andrew Wakefield had received large amounts of money as a paid expert for lawyers who were suing vaccine manufacturers. Wakefield had also applied for a patent on a vaccine that would rival the most commonly used MMR vaccine.

Because of these conflicts of interest, in 2010 the General Medical Council, an organization that licenses all medical doctors in the United Kingdom, concluded that Wakefield's conduct had been "irresponsible and dishonest" and revoked his medical license. Eventually, 10 of the paper's 11 other authors retracted their support of the study and its conclusions. Ultimately, in February of 2010, the *Lancet* officially retracted the paper—a very rare action for publishers (**Figure 2.9**). A peer-reviewed paper is retracted (withdrawn as untrue or inaccurate) by a publisher or author when its findings are no longer considered trustworthy because of error, plagiarism, a violation of ethical guidelines, or other scientific misconduct.

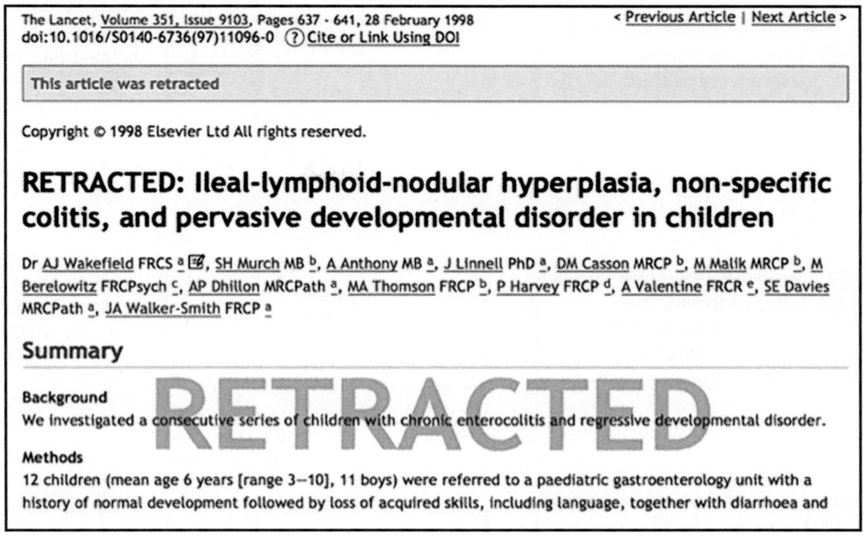

# Fears versus Facts

Wakefield's paper had not yet been retracted when Eaton was making her decision about whether to vaccinate Caroline. But Eaton followed the guidelines to distinguish science from pseudoscience, and she concluded that Wakefield and his colleagues had looked at too small a sample of children. Therefore, the paper "did not make a lasting impression," recalls Eaton. So she looked beyond that one paper to others. The multitude of papers she found, all

The Lancet, Volume 351, Issue 9103, Pages 637 - 641, 28 February 1998
doi:10.1016/S0140-6736(97)11096-0 (?) Cite or Link Using DOI

< Previous Article | Next Article >

This article was retracted

**RETRACTED: Ileal-lymphoid-nodular hyperplasia, non-specific colitis, and pervasive developmental disorder in children**

Dr AJ Wakefield FRCS [a], SH Murch MB [b], A Anthony MB [a], J Linnell PhD [a], DM Casson MRCP [b], M Malik MRCP [b], M Berelowitz FRCPsych [c], AP Dhillon MRCPath [a], MA Thomson FRCP [b], P Harvey FRCP [d], A Valentine FRCR [e], SE Davies MRCPath [a], JA Walker-Smith FRCP [a]

**Summary**

**Background**
We investigated a consecutive series of children with chronic enterocolitis and regressive developmental disorder.

**Methods**
12 children (mean age 6 years [range 3–10], 11 boys) were referred to a paediatric gastroenterology unit with a history of normal development followed by loss of acquired skills, including language, together with diarrhoea and

Figure 2.9

**The paper that precipitated the vaccine-autism scare is debunked and retracted**

with controls and larger sample sizes, assured her that vaccination does not cause autism. In the end, her research and scientific literacy led Eaton to fully vaccinate all of her children on the CDC's recommended schedule.

In making her decision, Eaton was aware that vaccines can have side effects. Most are minor, like a sore arm or a low-grade fever. In some very rare cases, however, vaccination can result in a serious reaction. An allergic reaction to the hepatitis B vaccine, for example, is estimated to occur once in 1.1 million doses. A serious allergic reaction to the MMR vaccine is seen less than once in a million doses. The CDC maintains a list of side effects and constantly monitors the safety of vaccines (**Figure 2.10**).

Weighing the benefits of avoiding disease against the small risk of side effects made the choice to vaccinate her children clear, says Eaton. "We [parents] are often so focused on possible effects of the vaccines that we don't think about our kids getting these diseases" that vaccines prevent, she says. The same was not true for your great-grandparents; in the early 1900s, one in five children worldwide died of a vaccine-preventable disease.

When making life choices based on scientific claims, you sometimes need to consider nonscientific aspects of your decision. Your

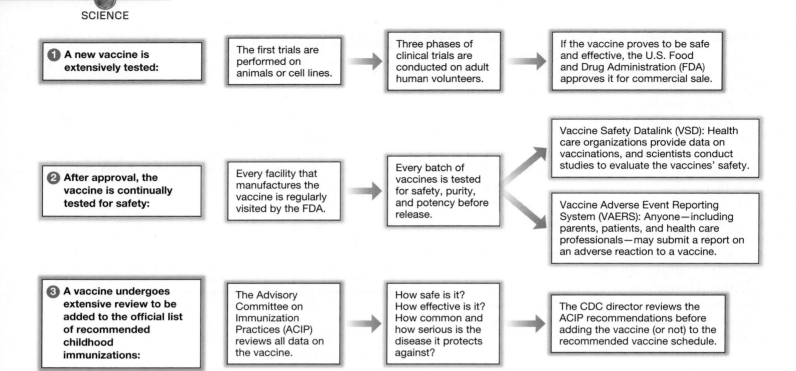

Figure 2.10

**Evaluating vaccines: an ongoing process**
Vaccines are continually tested and evaluated for effectiveness, safety, and side effects. The severity of the disease a vaccine works against, as well as how likely a child is to be infected with the disease, is also taken into consideration in determining whether to recommend the vaccine.

**Q1:** Why do vaccine manufacturers begin with tests on animals or cell lines before moving on to adult human subjects?

**Q2:** What ongoing testing and reporting are vaccines subjected to?

**Q3:** What do ACIP, FDA, and CDC stand for, and what is the role of each in evaluating vaccines?

values, ethical stances, and religious beliefs will make some choices more acceptable than others. For example, if you believe it is not ethical to eat meat, the scientific finding that leaner meats are healthier than fattier meats won't matter to you.

It is also important to consider how your choices will affect other people. Choosing not to vaccinate a child, for instance, affects the entire community. Rates of vaccination in the United States have been hovering around 75 percent because of misconceptions about the safety of vaccines, and this low level of immunization puts at risk not only the children who are not vaccinated, but also those who *cannot* be vaccinated, such as infants too young for a vaccine

or people who are genetically unable to respond to a vaccine. When a critical portion of a population is vaccinated, then the spread of disease is contained—a concept known as **herd immunity** (**Figure 2.11**). In other words, vaccinating a large number of people keeps germs out of circulation and protects the vulnerable members of the community.

When parents opt not to vaccinate their children, herd immunity disappears. And because of that loss of herd immunity, infectious diseases of the past, some thought to be eliminated, are roaring back. In August 2013, a megachurch in Texas whose founder had spoken out against vaccines made headlines after 21 members of its congregation contracted

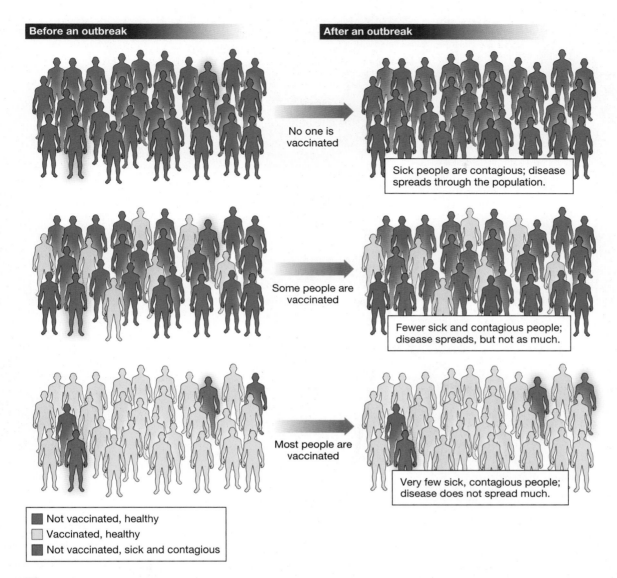

Before an outbreak

After an outbreak

No one is vaccinated

Sick people are contagious; disease spreads through the population.

Some people are vaccinated

Fewer sick and contagious people; disease spreads, but not as much.

Most people are vaccinated

Very few sick, contagious people; disease does not spread much.

■ Not vaccinated, healthy
□ Vaccinated, healthy
■ Not vaccinated, sick and contagious

## Figure 2.11

### Vaccine prevalence and herd immunity

When the majority of the population is not vaccinated, an outbreak of disease will spread further and may cause disease and death in members of the population who are too young to be vaccinated or have a compromised immune system. However, when most of the population is vaccinated against a contagious disease, the disease will spread little during an outbreak; this tighter containment of a disease is the result of what is known as herd immunity. Image modified from the National Institute of Allergy and Infectious Diseases (NIAID).

**Q1:** What happens to an immunized person when a disease spreads through a population? (*Hint:* In the graphic, follow an immunized individual before and after a disease spreads.)

**Q2:** Explain why a disease is less likely to spread to vulnerable members of a population if most people are immunized.

**Q3:** How does vaccination help an individual person? How does it help that person's community?

# Safety in Numbers

In 1999, the United Kingdom began a meningitis vaccination program for children up to the age of 18. Meningitis rates rapidly fell off throughout the population—children and adults alike—thanks to both the effectiveness of the vaccine and the protective effects of herd immunity. Models show that without herd immunity, the number of meningitis cases would have rebounded. The U.S. CDC recommends meningitis vaccination up to the age of 23, as outbreaks have been reported on college campuses in recent years.

Assessment available in smartw⬡rk5

## Meningitis cases in England and Wales

19 years old and younger

20 years old and older

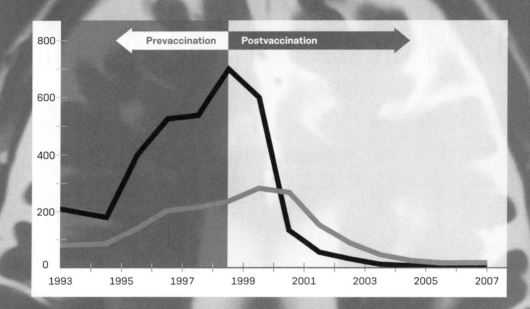

Observed cases

Predicted cases (with herd immunity)

Predicted cases (no herd immunity)

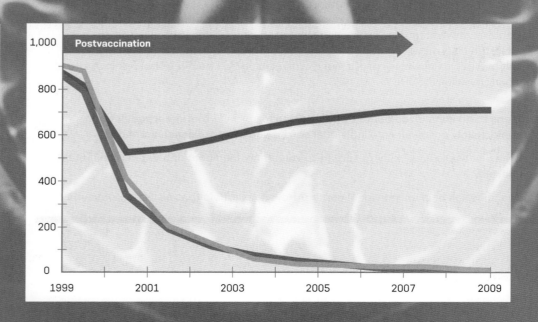

measles. Sixteen of them were unvaccinated. In 2014, measles cases in the United States hit a 20-year high, with 667 reported cases. And in the 2013–14 flu season, 90 percent of the 100 American children who died of complications from the flu had not received the flu vaccine. Decisions based on scientific claims have serious consequences.

Today, Eaton is thankful that all of her children are fully vaccinated (**Figure 2.12**). She encourages her friends, relatives, and colleagues to foster scientific literacy and use their critical-thinking skills to distinguish science from pseudoscience. Not everyone will want to read a vaccine textbook, Eaton admits with a laugh, but anyone can think critically and follow the process of evaluating scientific claims.

"Look to see what has credible backing and what doesn't. This issue transcends vaccines," says Eaton. "We should all have the tools to become scientifically literate citizens."

Figure 2.12

**Anna Eaton and her children**
Although Eaton initially had concerns about vaccinations for her first child, Caroline, her research gave her confidence that vaccines are a safe and responsible choice. Caroline and her younger brothers are now all up-to-date on their vaccinations.

# REVIEWING THE SCIENCE

- A **scientific claim** is a statement about how the world works that can be tested using the scientific method. To evaluate scientific claims, it is important to look at the **credentials** and **bias** of those making the claim.

- **Scientific literacy** requires a basic understanding of scientific facts and theories, and of the process of science. It is important to be scientifically literate in order to make well-informed life decisions.

- Scientific claims can be found in advertising, in social media, in the popular press, and in scientific publications. The best source for a review or overview of a scientific topic is the **secondary literature**, including the popular press, reputable websites, or review articles in scientific journals.

- The actual experimental or observational results related to a scientific claim are found in the **primary literature**, articles published in scientific journals that have undergone peer review.

- With **correlation**, two or more aspects of the natural world behave in an interrelated manner. With **causation**, a change in one aspect *causes* a change in another.

- **Basic research** explores questions about the natural world and expands the fundamental knowledge base of science. **Applied research** seeks to use the knowledge gained from basic research to address human issues and concerns. It may involve developing commercial applications.

- **Pseudoscience** is characterized by scientific-sounding statements, beliefs, or practices that do not meet the standards of the scientific method.

# THE QUESTIONS

## The Basics

**1** Which of the following should receive the *least* consideration when evaluating a scientific claim?

(a) the scientific credentials of the person making the claim

(b) your personal beliefs and values

(c) whether the study supporting the claim has been published in a peer-reviewed scientific journal

(d) whether the study supporting the claim meets the standards of the scientific method

(e) any possible biases of the person making the claim

**2** Scientific literacy means that you

(a) are able to easily read and understand a scientific journal article.

(b) have taken a university-level science course.

(c) understand the process of science and basic scientific facts and theories.

(d) enjoy reading current science news in newspapers and blogs.

(e) are a good critical thinker.

**3** "Correlation does not prove causation" means that

(a) if two variables are correlated, one is likely to have caused the other.

(b) if changes in one variable cause changes in another, the variables are not correlated.

(c) only experimental research can answer questions about the natural world.

(d) although two variables are interrelated, changes in one do not necessarily cause changes in the other.

(e) none of the above

**4** Link each term with the correct definition.

| SCIENTIFIC LITERACY | 1. Uses scientific knowledge to address human issues. |
| BASIC RESEARCH | 2. Helps in making informed life choices. |
| APPLIED RESEARCH | 3. Consists of peer-reviewed scientific journal articles. |
| SECONDARY LITERATURE | 4. Gives an overview of scientific findings on a subject. |
| PRIMARY LITERATURE | 5. Contributes to fundamental science knowledge. |

**5** Select the correct terms:

Evaluating a scientific claim begins with reviewing the (**credentials / fame**) of the person making the claim. It is also important to know whether those making a particular claim have a (**detachment / bias**), a vested interest in whether or not the claim is true. To gain an overview of scientific studies related to the claim, it is helpful to read the (**primary literature / secondary literature**).

**6** You are trying to determine whether a scientific claim is based on real science or pseudoscience. Place the following questions you will address in the correct order by numbering them from 1 to 6.

_____ a. Are the study's claims observable and quantifiable?

_____ b. Has the study been reviewed by practicing scientists and published in an established scientific journal?

_____ c. Are the predictions specific, testable, and falsifiable?

_____ d. Is the hypothesis clearly stated, measurable, and aligned with current scientific facts and theories?

_____ e. Are the experimental design and analysis well described, well designed, reproducible, and conducted with a large sample size?

_____ f. Are the study conclusions logical, based on evidence, and justified, given the study results?

## Challenge Yourself

**7** An example of basic research is

(a) a study of how hummingbirds learn song.

(b) an investigation of how the melting of polar ice caps affects agriculture.

(c) looking at possible genetic contributions to autism spectrum disorder.

(d) designing more effective vaccines for dangerous infectious diseases.

(e) exploring how agricultural waste can be turned into fuel.

**8** For each of the following, select the term that best describes the type of literature it represents: primary, secondary, or neither.

_____ a. A research study from Dr. Drake and colleagues on the blood sugar levels of diabetic rats that eat only kale, compared to control diets, is published in a peer-reviewed scientific journal.

_____ b. In the infomercial selling his own health care products, Dr. Horton states, "I believe my personal cure of daily meditation and yoga for diabetes is more effective than any drug I've ever prescribed in 20 years of practicing medicine."

_____ c. Dr. DeBellard insists in her blog that people will lose weight in a healthy manner if they acquire some of her personal parasitic tapeworm.

_____ d. In an article published in the *Annual Review of Nutrition*, a peer-reviewed journal, Dr. Pepper summarizes the last 10 years of basic research on diet and diabetes.

**9** Which of the following situations has the greatest potential for biased or inaccurate results in an experimental process?

(a) Ms. Ochoa-Bolton is an outside consultant who is conducting a health and wellness survey for a pharmaceutical company. She does not know the name of the company, nor does she know the name of the drug being tested during the survey.

(b) Ms. Adamian is a research technician surveying a study group on response to a new cold remedy. She knows only the e-mail address of each subject and asks them identical questions by computer.

(c) Dr. Wisidagama is evaluating cancer patients for their responses to a new therapeutic drug. She knows which patients are receiving the placebo and which are receiving the drug.

(d) Dr. Waters is analyzing biopsy samples from rats that have been given either a placebo or an experimental drug believed to reduce inflammation. Each sample is identified by a code number such that she cannot tell which treatment each rat received.

(e) Ms. Nuno is conducting a survey of weight loss regimens as part of her master's degree project. Her online survey is anonymous and asks each participant the same questions.

## Try Something New

**10** Determine whether each of the following statements is likely to represent real science or pseudoscience. In each case of likely pseudoscience, identify which scientific standard is not met.

_____ a. Dr. Oz says that green coffee beans will burn fat, so you can lose weight without dieting.

_____ b. The *Wall Street Journal* reports that climate scientists have conspired to exaggerate the effects of global climate change.

_____ c. A group of researchers reports at a national scientific conference that they have found a genetic link to autism.

_____ d. Many astrologers agree that people born under the sun sign Aquarius are more intelligent than those born under Scorpio.

_____ e. A study published in the scientific journal *Diabetes* finds that sleeping in a cooler room may increase metabolic rate and insulin sensitivity.

**11** The following graphs illustrate the incidence of pertussis (whooping cough) cases in the United States. The first graph organizes the data by year from 1922 to 2012, with the inset showing a zoomed-in view of 1990–2012. The second graph organizes the data by age group from 1990 to 2012. (The 2013 data are not complete in either graph.) DTP, Tdap, and DTaP are different formulations of the vaccine that covers tetanus, diphtheria, and pertussis.

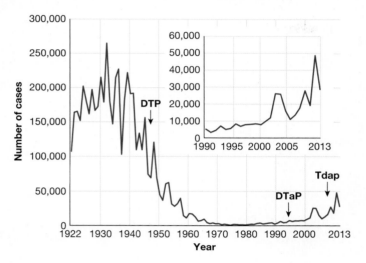

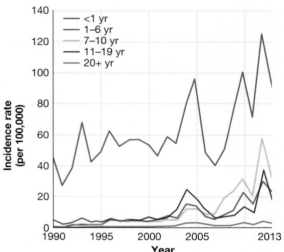

a. Describe what the first graph shows. What do each of the axes represent? What does any point on the line show? What general trend is seen, if all the data shown in the graph are considered?

b. Why do you think there was an increase in cases of pertussis in the first decade of the twenty-first century?

c. Describe what the second graph shows. What do each of the axes represent? What does any point on each of the different lines show? What general trend is seen, if all the lines are considered? What differs among the lines?

d. Compare the incidence of pertussis cases in children under 1 year old and people over the age of 20.

e. Summarize your reflections on reviewing these graphs and the relative risks of pertussis and the pertussis vaccine. What would you recommend to someone trying to decide whether to vaccinate a child?

**12** Human papillomavirus (HPV) is the most common sexually transmitted disease in the United States; almost all sexually active people are infected with HPV at some point. HPV can be contagious even when an infected person shows no symptoms, and symptoms may not appear until years after infection. Some but not all strains of HPV can cause genital warts and cancer. The HPV vaccine is very effective and is recommended for all young people from the ages of 11 to 26 years. What steps would you go through to decide whether to vaccinate yourself or your child with the HPV vaccine? (*Hint:* Use the method for evaluating scientific claims, including assessing real science versus pseudoscience.)

## Leveling Up

**13** **Life choices** Most American universities require students to have the following immunizations: MMR (measles, mumps, and rubella), varicella (chicken pox), and Tdap (tetanus, diphtheria, and pertussis/whooping cough). Most also recommend or require hepatitis A and B, meningococcal conjugate (meningitis), HPV (human papillomavirus), poliovirus, and the annual flu vaccine.

a. Which of these vaccines have you received?

b. If you have not received one or more of them, why not?

c. Some school districts allow parents to "opt out" of vaccines because of their belief system. Do you think this option should be allowed for university students? Why or why not? What are the possible consequences of allowing people to opt out of vaccinations?

**14** *Write Now* **biology: evaluating scientific claims** John Oliver, in a 2014 episode of *Last Week Tonight*, discussed a poll finding that one in four Americans is skeptical of global climate change. He dismissed the poll results and compared it to a poll asking, "Which number is bigger, 15 or 5?" or "Do owls exist?" Why does Oliver feel that what the American public believes about climate change is not relevant? (*Hint:* It could be argued that each of the people polled was making a scientific claim, either in support of or against the scientific consensus on climate change.)

**15** **Is it science?** Watch the 2011 movie *Contagion*, which depicts the spread of a fictional virus. While watching the movie, make a list of the scientific concepts that are presented, and note whether you think they represent real science or pseudoscience. Then use your textbook and Internet research to determine whether you were correct or incorrect in each case. If you evaluated any of the concepts incorrectly, reflect on how you came to your initial conclusions.

**16** **What do *you* think?** Read Peg Kehret's novel *Small Steps: The Year I Got Polio*, the author's true account of contracting polio as a seventh-grader. Keep a reflective journal while reading the book, documenting your emotional reactions to each chapter in this story. While reading, also note where you learned something new that you did not know previously about the poliovirus, the symptoms of polio, and the treatment options (or lack thereof) in the late 1940s. In your final entry, please reflect on how this novel did or did not change your feelings toward vaccination.

**For more, visit digital.wwnorton.com/bionow2 for access to:**

# Ingredients for Life

*A dusty box, rediscovered after 50 years, sparks a scientific treasure hunt for the origins of life on Earth.*

**After reading this chapter you should be able to:**

- Compare and contrast the types of bonds used to form molecules.
- Diagram the hydrogen bonds that form between water molecules and explain how those bonds produce the unique properties of water.
- Explain the difference between hydrophilic and hydrophobic molecules, and between acidic and basic molecules.
- Predict whether a solution has a high or low concentration of free hydrogen ions by using its pH number.
- Describe the chemical qualities of carbon that make it the basis of life on Earth.
- Create a graphic depicting how each of the four main classes of biomolecules and their subunits are related.

# CHAPTER
# 03

CHEMISTRY OF
LIFE

Jim Cleaves hauled yet another box to the dumpster. The young, dark-haired researcher at UC San Diego had the unenviable task of cleaning out the laboratory of his PhD adviser, the recently retired chemist Stanley Miller. It was 2003. Cleaves tossed box after box—20 years' worth of chemical samples from experiments done by Miller and his former students. As the lab emptied, a small cardboard box high on a shelf caught Cleaves's eye. It was labeled "Electric Discharge Samples."

"That box looked like something I'd really regret throwing out," says Cleaves. So instead of making a trip to the trash, Cleaves gave the box to another former student and friend of Miller's, marine chemist Jeffrey Bada at the Scripps Institution of Oceanography. Once again, the box was relegated to a shelf, where it sat, unopened, for years. Neither Bada nor Cleaves had any idea there was a scientific treasure trove inside, preserved for them by Miller.

In 1952, Miller was a thin, bespectacled graduate student at the University of Chicago, looking for a thesis idea. He approached Harold Urey, a Nobel Prize–winning chemist who, a year and a half earlier, had proposed a radical hypothesis about Earth's early atmosphere and the origins of life. Scientists knew that several key types of **matter** (anything that has mass and occupies a volume of space) existed on the early Earth, but they debated which types were necessary for the emergence of life 3–4 billion years ago. One type of matter is an **element**, a pure substance that has distinct physical and chemical properties, and that cannot be broken down into other substances by ordinary chemical methods. There are 98 natural elements known to us, and another 20 have been created in laboratories.

An **atom** is the smallest unit of an element that retains the element's distinctive properties. Atoms make up all common materials, including this book, the air, and you. Every atom has a dense core called a **nucleus** (plural "nuclei") made up of positively charged **protons** and electrically neutral **neutrons**. A cloud of negatively charged **electrons** surrounds the nucleus (**Figure 3.1**). Electrons have significantly less mass than protons and neutrons have: if an electron weighed as much as a 1-liter bottle of water, a proton or neutron would be as heavy as a car.

The number of protons in an atom's nucleus is called its **atomic number** and is unique to that element. **Isotopes** of an element have the same number of protons but different numbers of neutrons. The sum of the number of protons and the number of neutrons is the **atomic mass number** of an isotope. The atomic mass number is how much mass is in an element or, in other words, how much it weighs. For example, the most common isotope of carbon has 6 protons and 6 neutrons, giving it an atomic mass number of 12, and we call it carbon-12 ($^{12}$C). The isotope of carbon with 6 protons and 8 neutrons is carbon-14 ($^{14}$C).

Atoms interact with other atoms via electrons; they can donate electrons, accept electrons, and even share electrons. When two atoms share electrons, they form a **covalent bond**. Atoms linked by covalent bonds form **molecules**. Molecules that include at least one carbon atom are referred to as **organic molecules**. Urey and others suspected that gas molecules in early Earth's atmosphere combined to form the earliest organic molecules, which later assembled into the first living organism.

Urey proposed that Earth's early atmosphere resembled that of other planets in our solar system, including Jupiter, Saturn, and Uranus. He suggested that, like the atmospheres of those planets, our atmosphere was once rich with important **chemical compounds**, molecules that contain atoms from two or more elements. Earth's early atmosphere, he hypothesized, was made up of mainly four compounds: the gases methane ($CH_4$), ammonia ($NH_3$), hydrogen ($H_2$), and water vapor ($H_2O$).

The young Miller was intrigued by Urey's hypothesis, and he asked Urey if he could perform an experiment to test it, to see whether such a combination of gases could be used to create other simple compounds. Urey dismissed the idea, saying that it was too difficult for a student's thesis and that, if it didn't work, Miller would have nothing to show in order to graduate. But Miller pestered him until at last Urey

## STANLEY MILLER

Stanley Miller was an American chemist who designed the first experiment to mimic Earth's early atmosphere and pioneered the study of the origins of life. He died in 2007 at the age of 77.

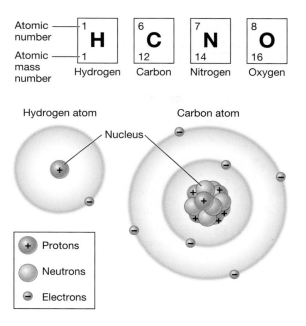

Figure 3.1

**Atomic structure**

The electrons, protons, and neutrons of these hydrogen and carbon atoms are shown greatly enlarged in relation to the size of the whole atom.

**Q1:** How many protons, electrons, and neutrons does the hydrogen atom shown here have? What are the atomic number and the atomic mass number of the hydrogen atom?

**Q2:** What are the atomic number and the atomic mass number of the carbon isotope shown?

**Q3:** Nitrogen-11 is an isotope of nitrogen that has 7 protons and 4 neutrons. What are the atomic number and atomic mass number of nitrogen-11?

agreed to let Miller work on the project for six months, a year at the most. As it turned out, all Miller needed was a few weeks.

Miller mixed the four gases together in a large glass apparatus. He then zapped the swirling cloud of compounds with a continuous electrical spark to mimic lightning strikes on early Earth, which he surmised would break apart the compounds (**Figure 3.2**). The process of breaking existing chemical bonds and creating new ones is known as a **chemical reaction**. The **reactants** (the gases, in this case) undergo a chemical change and form new molecules, called **products**. Some chemical reactions, like the one Miller devised, require energy in order to proceed. Others release energy.

After the first day of the experiment, a pink liquid pooled in the bottom of the apparatus. By the end of a week, it had turned "deep red and turbid," Miller reported. The color changes, he later discovered, reflected the formation of new products through chemical reactions.

To Miller and Urey's pleased surprise, the resulting red broth contained several biologically significant compounds, including five **amino acids**, small molecules important to life.

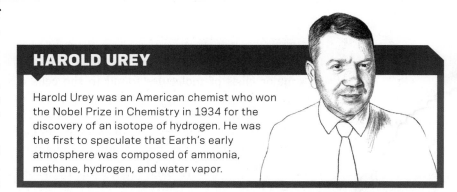

**HAROLD UREY**

Harold Urey was an American chemist who won the Nobel Prize in Chemistry in 1934 for the discovery of an isotope of hydrogen. He was the first to speculate that Earth's early atmosphere was composed of ammonia, methane, hydrogen, and water vapor.

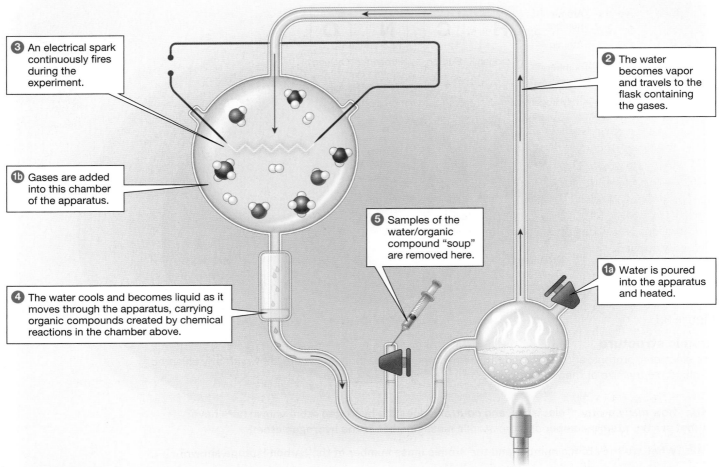

**③** An electrical spark continuously fires during the experiment.

**②** The water becomes vapor and travels to the flask containing the gases.

**①b** Gases are added into this chamber of the apparatus.

**⑤** Samples of the water/organic compound "soup" are removed here.

**①a** Water is poured into the apparatus and heated.

**④** The water cools and becomes liquid as it moves through the apparatus, carrying organic compounds created by chemical reactions in the chamber above.

## Figure 3.2

### Miller's original spark discharge experiment

When Stanley Miller zapped a mixture of methane ($CH_4$), ammonia ($NH_3$), hydrogen ($H_2$), and water vapor ($H_2O$) with electrical sparks, he produced compounds called amino acids.

**Q1:** Before the experiment was run, the apparatus was sterilized and then carefully sealed. Why was this an important thing to do?

**Q2:** Why is inclusion of methane in the gas flasks an essential part of the hypothesis that complex organic molecules were formed in the early atmosphere of Earth? (*Hint:* What makes a molecule organic?)

**Q3:** Answer this question after reading about Miller's "steam injection" experiments: Where was the steam injected in the experimental apparatus?

There are hundreds of known amino acids, but only 20 are the building blocks for proteins, a major class of molecules found in the cells of living things. Of the five amino acids that Miller isolated, three were part of that group of 20. Thus, from a simple mixture of gases Miller had indeed created some, but not all, of the amino acids used in life on Earth.

At a crowded seminar, Miller presented his results. Urey sat in the front row. Colleagues at the meeting recalled that after the presentation, the famed physicist Enrico Fermi turned to Urey and said, "I understand that you and Miller have demonstrated that this is one path by which life might have originated. Harold, do you think it was the way?" Urey looked at Fermi and replied, "Let me put it this way, Enrico. If God didn't do it this way, he overlooked a good bet!"

Miller and Urey's experiment, which became an instant classic in the scientific community, demonstrated that the basic chemicals for life could arise under natural conditions. Miller continued to perform spark discharge experiments, tweaking the experimental conditions and types of gases in the hope of producing more amino acids. But for an unknown reason, perhaps lack of time, Miller never published or followed up on many of his results—that is, until the mysterious box sitting on Jeffrey Bada's shelf was finally opened.

# One Picture, a Thousand Experiments

In 2007, Bada was visiting the University of Texas to give a lecture. His talk was scheduled immediately after a talk by another close friend of Miller's, Antonio Lazcano, a biologist at the School of Sciences at the National Autonomous University of Mexico in Mexico City. The two men agreed to review each other's lecture slides to make sure their talks didn't overlap.

One of Lazcano's slides caught Bada's eye. It was a picture of a small glass vial, labeled as containing a residue from Miller's early experiments. Bada asked Lazcano about it. Lazcano explained that during a visit to his friend, Miller had pulled a cardboard box off a shelf, lifted out the vial, and let Lazcano take a picture. At the time, Miller told Lazcano that it was one of the leftover samples of his spark discharge experiments. Miller had saved them all.

"I was flabbergasted," says Bada. "I'd known Stanley since 1965, and he never once mentioned it." It dawned on Bada that the box might still exist. He called his lab and asked if anyone had seen the box that Cleaves had given him four years earlier. As soon as Bada returned to the lab, he found and opened the box. It was like a scientist's Christmas; the box was full of carefully labeled plastic boxes containing thin glass

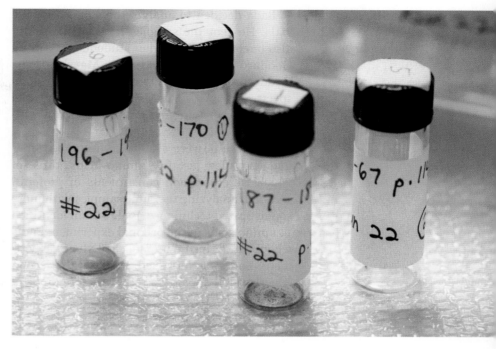

Figure 3.3

**Electric discharge samples from Miller's experiments**

vials, many with films of dried brown, tar-like gunk in the bottom. "It was on the order of 200–300 vials," says Bada (**Figure 3.3**). "It was extracts from experiments throughout the course of his life."

Luckily, Miller kept notebooks detailing the specific contents of each vial. He had performed two other experiments shortly after his original spark discharge work, using variations on the original apparatus. In one, a different method generated the spark. In another, which caught Bada's attention, hot steam was injected directly into the spark chamber.

## HENDERSON (JIM) CLEAVES

Jim Cleaves is an organic geochemist at the Carnegie Institution for Science in Washington, DC. As a graduate student of Stanley Miller in 2003, he discovered old vials from Miller's 1950s experiments. Today, Cleaves continues to study how life arose on Earth.

Bada decided to analyze the contents of the vials. He sent the samples to Jason Dworkin, also one of Miller's former students, at NASA's Goddard Space Flight Center in Maryland. The Goddard Center is home to an advanced mass spectrometer, an instrument that measures the weights of tiny amounts of matter as a way to identify them.

The results were stunning. In 1953, Miller had identified just five amino acids, three of which were common constituents of proteins. In 2008, analyzing the same samples with more sophisticated techniques, Bada's team identified 14 amino acids in the original experiment and a whopping 22 in the unpublished steam experiment. Bada was intrigued. Why did the second experimental apparatus result in more kinds of amino acids?

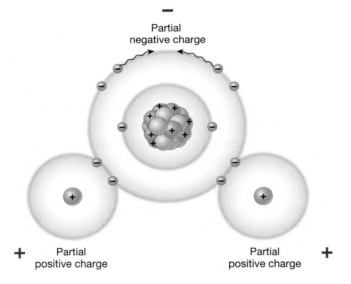

**Figure 3.4**

### Water molecules are polar

The polarity of water molecules is the cause of water's unique properties.

**Q1: Where are the covalent bonds in this figure?**

**Q2: This figure shows a water molecule ($H_2O$). A hydrogen molecule ($H_2$) consists of two hydrogen nuclei that share two electrons. Draw a simple diagram of a hydrogen molecule indicating the positions of the two electrons.**

**Q3: When table salt (sodium chloride, NaCl) dissolves in water, it separates into a sodium ion ($Na^+$) and a chloride ion ($Cl^-$). Which portion of a water molecule would attract the sodium ion, and which portion would attract the chloride ion?**

# The World of Water

The major difference between the two experiments was steam—hot water vapor shooting through the gas chamber. Water is essential for life because of its unique chemical properties, which enable molecules to dissolve and interact in special ways through chemical reactions.

A water molecule is made up of two hydrogen atoms and one oxygen atom held together by shared electrons—that is, by covalent bonds. Electrons moving around an atom's nucleus have different energy levels, so we can think of the electrons as segregated into rings or shells, each of which can contain up to a fixed number of electrons. The innermost shell, closest to the nucleus, can hold up to two electrons. Moving outward, the next two shells can hold up to eight electrons each.

A shell needs to be full to be stable, so in an effort to fill their shells, atoms may form covalent bonds. In a covalent bond, atoms share electrons in their outermost shell of electrons, also called the valence shell. In a water molecule, the oxygen atom uses an electron from each of two hydrogen atoms to fill its outer shell, increasing its count of electrons from six to eight. The hydrogen atoms also benefit from the bond: they fill their outer (and only) shell with the two electrons needed to be stable.

However, the electrons are not shared equally; they spend more time near the oxygen atom than near the hydrogen atoms. Because electrons are negatively charged particles, the oxygen end of a water molecule therefore takes on a slightly negative charge, and the hydrogen ends become slightly positively charged. This lopsided electron sharing means that water is a **polar molecule** (**Figure 3.4**).

In addition to covalent bonds, two other types of **chemical bonds** attach atoms to one another: ionic bonds and hydrogen bonds. Atoms that have lost or gained electrons are called **ions**. Since electrons are negatively charged, an atom that has gained an electron is a negative ion, and an atom that has lost an electron is a positive ion. When a negatively charged ion and a positively charged ion are in the same vicinity, they will chemically attract each other and form an **ionic bond**. Common table salt, sodium chloride (NaCl), is composed of sodium and chlorine held together by ionic bonds. Unlike a covalent bond, no electrons are shared in an ionic bond.

The third means of attaching atoms to one another is called a **hydrogen bond**. Hydrogen bonds are weak electrical attractions between a hydrogen atom with a partial positive charge and a neighboring atom with a partial negative charge. Molecules of water bind to each other through hydrogen bonds because the negatively charged oxygen end of one water molecule weakly attracts one of the positively charged hydrogen ends of another water molecule. A single hydrogen bond is about 20 times weaker than a covalent bond, but water makes up for that lack of strength with sheer quantity. The collective cross-linking of many, many water molecules through hydrogen bonds amounts to a potent force.

The polarity of water molecules and hydrogen bonding explain nearly all of the special properties of water, which were critical in Miller's experiments. The foremost of these was that water was able to break apart the compounds in the flask. As you may have noticed the last time you soaked a dirty dish, water has an incredible ability to dissolve other materials. This is because water molecules form hydrogen bonds with other polar molecules, like sugars or, in Miller's experiment, ammonia. The formation of hydrogen bonds with polar molecules causes those compounds to dissolve in water. Such compounds are said to be **soluble**; that is, they mix completely with the water.

A **solution** is any combination of a **solute** (a dissolved substance, such as sugar) and a **solvent** (the fluid, such as water, into which the solute has dissolved). Water is called the "universal solvent" because it successfully dissolves so many substances. The polar nature of water molecules, however, means that they will *not* interact with uncharged or nonpolar substances, such as fat or oil. Molecules that are soluble in water (such as salt) are called **hydrophilic** ("water-loving"); molecules that don't dissolve well in water (such as oil) are called **hydrophobic** ("water-fearing"). **Figure 3.5** shows these processes in action.

When Bada and his colleagues published their results from reanalyzing Miller's vials, some scientists proposed that Miller's second experiment, in which he shot a jet of steam

Oil molecules are hydrophobic. They are excluded from water and tend to clump together.

Olive oil

Vinegar molecules are hydrophilic. They are held in solution by water molecules.

Vinegar

Figure 3.5

**Hydrophilic substances dissolve in water, but hydrophobic substances do not**

**Q1:** Describe what will happen to the molecules of olive oil if you shake the bottle and then leave it alone for an hour. What about the molecules of vinegar?

**Q2:** What would happen if you added another fat to the bottle, such as bacon grease, and shook it?

**Q3:** Given how sugar behaves when it is mixed into coffee or tea, would you predict that it is hydrophobic or hydrophilic?

into the spark, had resulted in more amino acids because the hot water enabled a wider variety of chemical reactions. Whether or not that was the case, water was central to Miller's success.

**JEFFREY BADA**

Jeffrey Bada is a chemist at the Scripps Institution of Oceanography at UC San Diego. With Jim Cleaves, he closely analyzed and then duplicated the Miller-Urey experiments. He is also a leading scientist studying organic compounds beyond Earth, including in meteorites and on Mars.

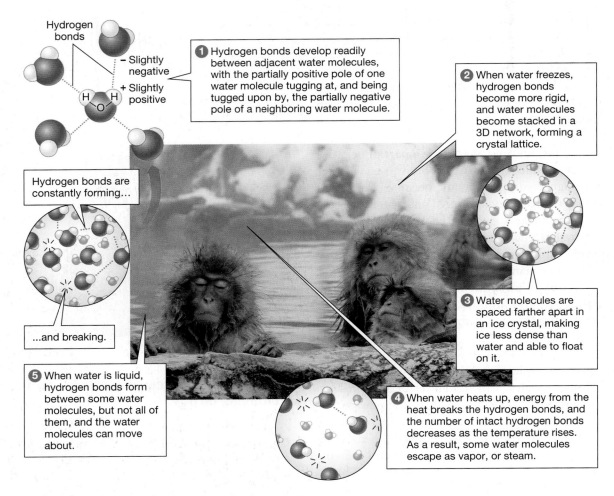

Hydrogen bonds

− Slightly negative

+ Slightly positive

**1** Hydrogen bonds develop readily between adjacent water molecules, with the partially positive pole of one water molecule tugging at, and being tugged upon by, the partially negative pole of a neighboring water molecule.

**2** When water freezes, hydrogen bonds become more rigid, and water molecules become stacked in a 3D network, forming a crystal lattice.

Hydrogen bonds are constantly forming...

...and breaking.

**5** When water is liquid, hydrogen bonds form between some water molecules, but not all of them, and the water molecules can move about.

**3** Water molecules are spaced farther apart in an ice crystal, making ice less dense than water and able to float on it.

**4** When water heats up, energy from the heat breaks the hydrogen bonds, and the number of intact hydrogen bonds decreases as the temperature rises. As a result, some water molecules escape as vapor, or steam.

Figure 3.6

**Water molecules change state as hydrogen bonds increase or decrease**

Japanese snow macaques escape the cold with a daily dip in natural hot springs. Water can be seen here in its liquid, solid, and gas states.

**Q1:** Identify where in the picture water can be seen in its liquid, solid, and gas states.

**Q2:** In the gas state, water molecules move too rapidly and are too far apart to form hydrogen bonds. Compare the volumes occupied by an equal number of water molecules in the liquid, solid, and gas states.

**Q3:** Explain in your own words how ice floats on water.

Water can exist in all three states of matter: liquid, solid, and gas. Hydrogen bonds explain the physical properties of water in these three states. Though water is composed of two elements that are gases at room temperature (hydrogen and oxygen), it forms a liquid at room temperature because hydrogen bonds stick water molecules together, keeping them close. Those hydrogen bonds are constantly forming and breaking in water, creating a nonstop jostling that gives water its liquid form (**Figure 3.6**, left).

When water chills, its molecules cannot move about as vigorously (**Figure 3.6**, right). As water turns into ice at 0°C (32°F), a stable network of hydrogen bonds emerges. The molecules become spaced farther apart, locked into an orderly pattern known as a crystal lattice. That spacing is the reason ice occupies more space than liquid water. Normal ice is 9 percent less dense than liquid water (density is mass divided by volume), which explains why ice floats on water. This property of water is quite unusual—most substances are *more* dense in the solid state than in the liquid state—but it has helped shape life as we know it. If ice did not float on water, then each time a lake or river froze in the winter, the frozen top layer would sink to the bottom and the new, liquid top layer would freeze. This process would repeat until the entire lake or river was frozen solid, no matter the depth. All the aquatic creatures and plants would be killed in the process, and terrestrial plants and animals would no longer have access to the freshwater.

To boil water, as Miller did to inject steam into his experiments, requires the addition of a significant amount of energy to snap the water molecules' network of hydrogen bonds before the molecules can move fast enough to escape as steam (**Figure 3.6**, bottom). That transition from the liquid to the gas state is a phenomenon known as **evaporation**. By the opposite reaction, as water vapor cools, molecules slowly re-form hydrogen bonds and return to the liquid state—a process known as **condensation**. Without evaporation and condensation, both of which are essential parts of the water cycle, life on Earth could not exist (see Figure 18.12 for more details).

Figure 3.7

**Miller's experiments were like volcanic eruptions**
An erupting volcano releases all of the gases included in Miller's mixtures, and the volcanic ash contains iron and other metals. Steam is produced when the magma (superheated rock) comes in contact with groundwater, and lightning bolts lance through the cloud of gas and ash.

**Q1:** Suppose you were going to repeat Miller's experiments. How would you decide how much of each gas to include in the chamber?

**Q2:** Why did the addition of steam to the gases in Miller's second set of experiments increase the yield of amino acids?

**Q3:** Miller used electrical energy in his experiment. What other forms of energy were present in the early atmosphere of Earth that could have led to the formation of complex molecules?

# The Smell of Success

Water was key to the success of Miller's spark discharge experiments, but another compound would turn out to be almost as important. In 2011, three years after reanalyzing the first of Miller's samples, Bada, Cleaves, and a group of scientists teamed up again to analyze another set of vials in the cardboard box.

In 1958, Miller had performed spark discharge experiments that included a new gas: hydrogen sulfide ($H_2S$). You may know hydrogen sulfide

gas as the cause of the awful smell released by rotten eggs, but it is also released by volcanoes and so may have been present in Earth's early atmosphere (**Figure 3.7**). Miller's lab notes show that he performed the experiment and isolated the amino acids, but then never analyzed them. "Why he never analyzed it, I don't know," says Bada. Bada's team finished the job, identifying 22 amino acids in Miller's test tube, including six that contained sulfur. One of those sulfur-containing amino acids, methionine, initiates the construction of all proteins in cells.

Miller's experiments support the hypothesis that volcanoes—a major source of hydrogen

CELLS

sulfide today—coinciding with lightning, which is often focused around volcanoes, may have played a role in making large and varied quantities of biologically crucial molecules, setting the stage for the evolution of life on Earth. But this is not the only hypothesis about the origins of life. There are other ideas, including one that has gained traction in recent years: that the building blocks for life arrived from space.

Although proteins use only 20 amino acids, scientists have identified 90 amino acids in meteorites, suggesting that the first organic materials could have come to Earth from outer space. At NASA's Ames Research Center in California, astrophysicist Scott Sandford and colleagues have been able to make amino acids from gases using conditions like those that exist in interstellar space.

In several experiments, Sandford's team froze different combinations of gases to simulate ices found on comets and in interstellar clouds (where new stars and planets form), and then bombarded those ices with radiation. And what do their icy-hot recipes create? "We make amino acids all the time," says Sandford. In many cases, they identified amino acids similar to those that Miller's experiment created. "The universe seems to be an organic chemist," says Sandford. "It is hardwired to turn simple molecules into more complicated ones. The consequence of that is that it makes a whole host of products, some of which are biologically interesting."

On October 7, 2008, an asteroid entered Earth's atmosphere over Africa, and a NASA satellite photographed its impact (**Figure 3.8**). Initially 6–15 feet in diameter, the asteroid exploded over eastern Africa, and its fragments landed in Sudan. NASA scientists were able to find pieces of the asteroid on the ground because the dark rocks stood out against the light-colored desert sand. These fragments are called the "Almahata Sitta" or "Station Six" meteorites, named after a train station near the location where pieces were recovered.

When the fragments were analyzed, scientists found 19 different amino acids. The extraterrestrial origin of these amino acids is confirmed by a particular characteristic of amino acid molecules: They have two forms, left-handed and right-handed, that are mirror images of each other. All of the amino acids produced by living organisms on Earth are of the left-handed form; right-handed amino acids are produced only in laboratories. The amino acids in the Almahata Sitta meteorites are a mixture of left-handed and right-handed forms. The presence of right-handed amino acids shows that the amino acids in the fragments really came from space, not from terrestrial organic contamination after the fragments landed.

A NASA satellite photographed the impact of the asteroid on October 7, 2008. The yellow arrow traces the path the asteroid followed, and the reddish-orange blob shows the point at which it exploded after entering Earth's atmosphere.

The asteroid broke into fist-sized pieces that fell into the Nubian Desert in Sudan. The dark color of the fragments made them conspicuous on the desert sand, and many of the fragments were recovered by NASA scientists in February 2009.

Figure 3.8

**Asteroid 2008 TC3 contains amino acids that formed in outer space**

**Q1:** How did the NASA scientists find the fragments of the meteorite that exploded over eastern Africa?

**Q2:** What piece of evidence suggests that amino acids found in the meteorite fragments originated in outer space?

**Q3:** Speculate on the significance of finding extraterrestrial amino acids.

# Getting the Right Mix

Miller never stopped trying to make complex molecules from simple ones. Though his 1953 experiment was originally met with fanfare, scientists later began to dispute the usefulness of his experiments. Methane and ammonia, they argued, didn't exist in large amounts on the early Earth. Instead, new evidence suggested that the atmosphere contained nitrogen gas ($N_2$) and carbon dioxide ($CO_2$). "Most people agree now [that Miller and Urey] didn't have the right composition," says Sandford. So, in 1983, 30 years after his original experiment, Miller repeated it using nitrogen and carbon dioxide. But instead of a deep red/brown broth, the liquid produced was clear and seemingly barren. The experiment looked like a failure.

In 2007, Bada and Cleaves decided to revisit that experiment to see what had gone wrong. Instead of just reanalyzing samples, they redid the experiment and discovered that the reactions between the new gases were producing chemicals called nitrites, which destroy amino acids. The solution also became acidic because of the presence of nitrous acid ($HNO_2$). An **acid** is a hydrophilic compound that dissolves in water and loses one or more hydrogen ions ($H^+$). By donating $H^+$ ions to water, acids increase the concentration of free $H^+$ ions in an aqueous solution. $H^+$ ions are extremely reactive and can disrupt or alter other chemical reactions. The acidic solution, Bada and Cleaves realized, was preventing amino acids from forming.

To counteract that acidity, Bada added a **base**. Acids and bases are chemical opposites. Unlike acids, bases *accept* hydrogen ions from aqueous surroundings. Because a base removes $H^+$ ions from solution, it has the overall effect of *reducing* the concentration of free $H^+$ ions in an aqueous solution (strong bases, like strong acids, can be dangerous because they disrupt chemical reactions important to life). Acids react with bases to have an overall neutralizing effect, reducing the concentration of reactive $H^+$ ions.

Hydrogen ion concentration is commonly expressed on a scale from 0 to 14, where 0 represents an extremely high concentration of free $H^+$ ions and 14 represents the lowest concentration. This scale, called the **pH scale**, is logarithmic: each pH unit represents a 10-fold increase or decrease in the concentration of hydrogen ions (**Figure 3.9**). Pure water is said to be neutral at pH 7, in the middle of the pH scale. The addition of acids to pure water raises the concentration of free hydrogen ions, making the solution more acidic and pushing the pH below

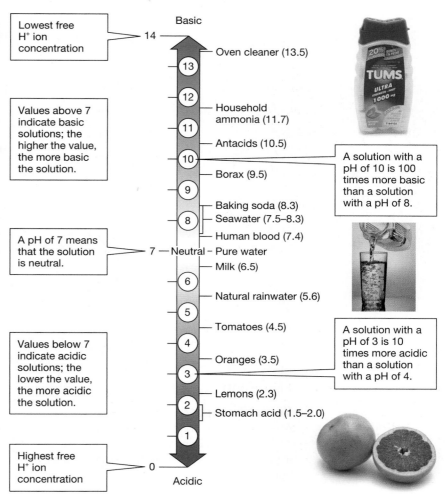

Figure 3.9

**The pH scale indicates hydrogen ion concentration**

**Q1:** Which has a higher concentration of free hydrogen ions: vinegar, pH 2.8; or milk, pH 6.5?

**Q2:** What happens to the concentration of free hydrogen ions in your stomach when you drink a glass of milk?

**Q3:** Black coffee has a pH of 5. Does adding coffee to water (pH 7) increase or decrease the concentration of free hydrogen ions in the liquid?

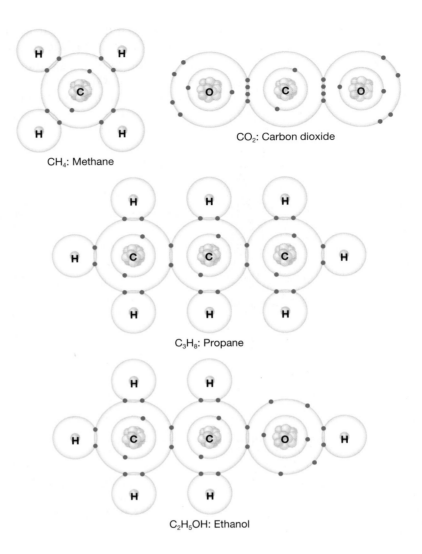

CH₄: Methane

CO₂: Carbon dioxide

C₃H₈: Propane

C₂H₅OH: Ethanol

## Figure 3.10

### Versatile carbon

All organic compounds are built from carbon atoms interacting with atoms of other elements through strong covalent bonds. A carbon atom can bond with up to four other atoms because it has four valence electrons in its outer shell. Recall that atoms must fill their outer shells to be stable and that covalent bonds require the sharing of valence electrons. Count the electrons circling each of the atoms in the molecules shown; all the outer shells are full.

**Q1:** In methane gas, how many electrons is each hydrogen atom sharing? How many is the carbon atom sharing?

**Q2:** In carbon dioxide, how many electrons is each oxygen atom sharing? How many is the carbon atom sharing?

**Q3:** Draw a molecule of formaldehyde ($CH_2O$). How many electrons is the oxygen atom sharing with the carbon atom? How many is the carbon atom sharing with the oxygen atom and with each hydrogen atom?

the neutral value of 7. Adding a base lowers the concentration of free hydrogen ions in the solution, making the resulting solution more basic and raising the pH above 7.

Bada added a simple base, calcium carbonate, to the experiment in order to raise the pH of his solution. This time, the resulting brew was bursting with amino acids, suggesting that the building blocks of life could indeed have originated on Earth and not been solely delivered to Earth by meteorites or comets. Today, many scientists agree that it is likely that both processes—amino acids arriving from space and originating on Earth—contributed to life as we know it. But no matter where amino acids came from, the logical follow-up question is one that continues to stump scientists to this day: What happened next?

# Life's First Steps

"No one really knows how life got started," says Sandford. But researchers do agree on what life requires. If all the water in any living organism were removed, four major classes of large organic molecules, or **biomolecules** (sometimes called "macromolecules"), would remain, all of them critical for living cells: proteins, carbohydrates, nucleic acids, and lipids.

Each of these biologically important molecules is built on a framework of covalently bonded carbon atoms. Carbon is the predominant element in living systems, partly because it can form large molecules that contain thousands of atoms. A single carbon atom can form strong covalent bonds with up to four other atoms (**Figure 3.10**). Carbon atoms can also bond to other carbon atoms, forming long chains, branched molecules, and even rings. No other element is as versatile as carbon in the sheer diversity of complex molecules that can be assembled from it. In fact, while there are only about 4,500 known naturally occurring inorganic molecules (molecules that do not contain a carbon atom) on Earth, the number of known organic molecules is in the range of millions, and the number of those that are not yet uncharacterized is likely several orders of magnitude larger.

Proteins, carbohydrates, and nucleic acids are **polymers**, long strands of repeating units of small molecules called **monomers**. Amino acids are the monomers making up proteins, simple sugars are the monomers in carbohydrates, and nucleotides are the basis of nucleic acid polymers.

All three of these polymers have essential functions for every life-form. **Proteins** are known to be the most numerous and versatile of the biomolecules. Different combinations of the twenty amino acid monomers allow for countless proteins that vary in size and shape, and therefore function (**Figure 3.11**, top left). For example, enzymatic proteins, like polymerases, enable us to copy our DNA (see Chapter 9 for more on DNA replication). Structural proteins give our cells shape. Hormone and receptor proteins, like insulin and its receptor, allow our cells to take up sugars for use as energy. Other equally important protein categories include membrane transport proteins that help move substances into and out of cells, antibody molecules that protect us from disease, storage proteins like LDL and HDL (low-density and high-density lipoproteins, respectively) that carry cholesterol, and venoms and toxins such as the tetanus toxoid.

**Carbohydrates** are the next-most-versatile biomolecules. They range in size from simple sugar monomers (monosaccharides) and two-monomer sugars (disaccharides) to complex carbohydrates that may contain thousands of monomers (**Figure 3.11**, top right). Simple sugars are the cell's direct fuel to make ATP (adenosine triphosphate), the molecular energy source essential for all cellular work (see Chapter 4 for more on ATP). Other carbohydrates, such as glycogen in animals and starch in plants, are used for energy storage. Three additional complex carbohydrates provide structural support to cells: cellulose, also known as fiber, helps plants to grow tall; chitin forms a hard outer covering to protect organisms without an internal skeleton, such as insects, spiders, and crustaceans; and peptidoglycan is a major component of bacterial cell walls.

The third and most crucial category of polymer, the **nucleic acids**—DNA (deoxyribonucleic acid) and RNA (ribonucleic acid)—form the basis of life itself. Nucleic acids are polymers of nucleotide monomers: DNA is composed of deoxyribonucleotides, and RNA is composed of ribonucleotides (**Figure 3.11**, bottom left). DNA provides living organisms with long-term, stable genetic information storage in a form that is easily copied and passed on to future generations (see Chapter 9). Our genes are DNA. That sounds important, but what about RNA? Without RNA,

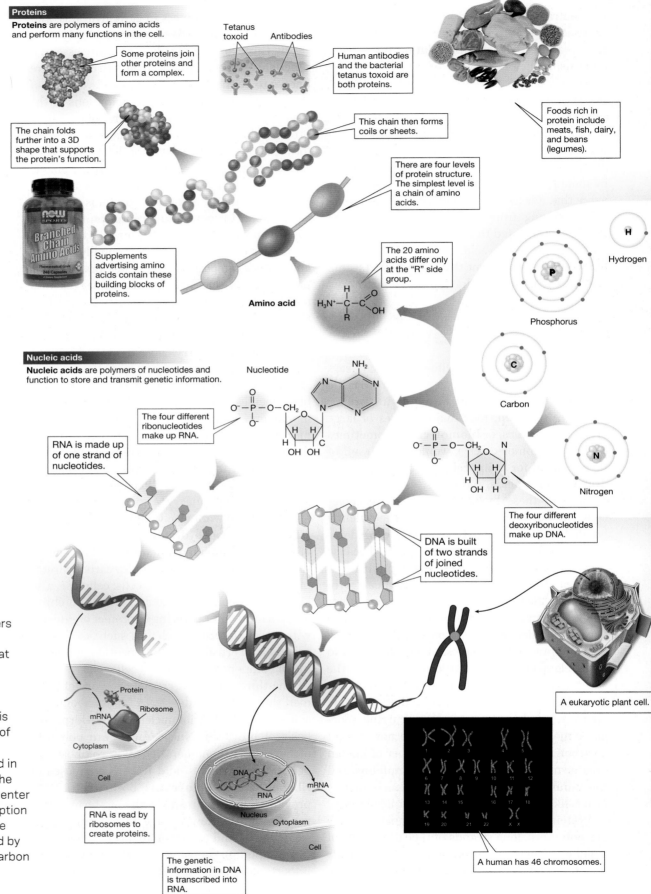

**Proteins**

**Proteins** are polymers of amino acids and perform many functions in the cell.

Some proteins join other proteins and form a complex.

Tetanus toxoid    Antibodies

Human antibodies and the bacterial tetanus toxoid are both proteins.

Foods rich in protein include meats, fish, dairy, and beans (legumes).

The chain folds further into a 3D shape that supports the protein's function.

This chain then forms coils or sheets.

There are four levels of protein structure. The simplest level is a chain of amino acids.

Supplements advertising amino acids contain these building blocks of proteins.

The 20 amino acids differ only at the "R" side group.

**Amino acid**

$H_3N^+ - C - C$
with H on top, R below, and OH on the carboxyl

H · Hydrogen

P · Phosphorus

C · Carbon

N · Nitrogen

**Nucleic acids**

**Nucleic acids** are polymers of nucleotides and function to store and transmit genetic information.

Nucleotide

The four different ribonucleotides make up RNA.

RNA is made up of one strand of nucleotides.

The four different deoxyribonucleotides make up DNA.

DNA is built of two strands of joined nucleotides.

A eukaryotic plant cell.

Protein
mRNA    Ribosome
Cytoplasm
Cell

RNA is read by ribosomes to create proteins.

DNA
RNA
mRNA
Nucleus
Cytoplasm
Cell

The genetic information in DNA is transcribed into RNA.

A human has 46 chromosomes.

**Figure 3.11**

**Biomolecules are critical for life**

A handful of monomers (*mono*, "one") can be assembled into a great variety of polymers (*poly*, "many"). Each of these biologically important molecules is built on a framework of repeating monomers made up of some, and in several cases all, of the atoms shown in the center circle—with the exception of the lipids, which are instead characterized by repeating atoms of carbon and hydrogen.

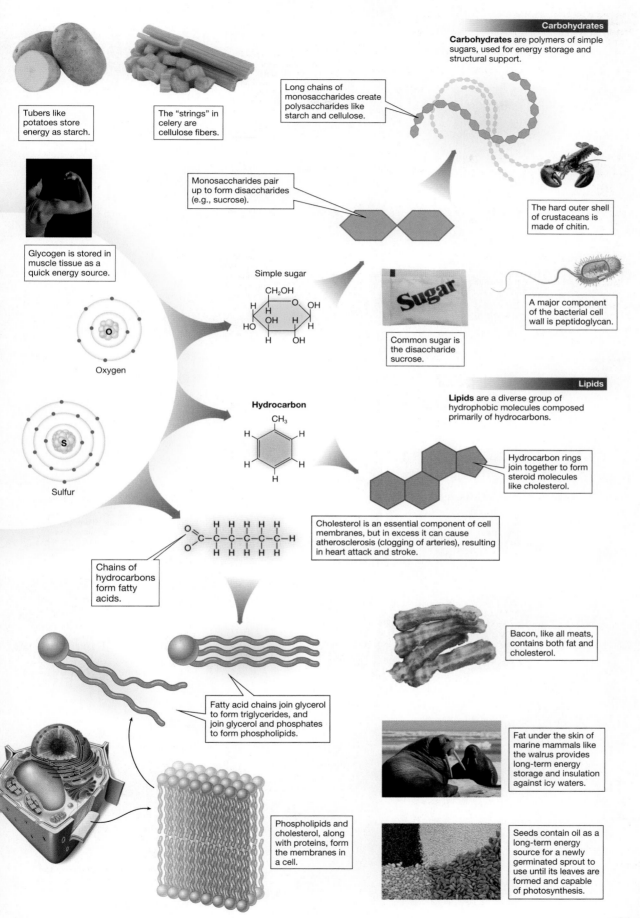

**Carbohydrates**

**Carbohydrates** are polymers of simple sugars, used for energy storage and structural support.

Tubers like potatoes store energy as starch.

The "strings" in celery are cellulose fibers.

Long chains of monosaccharides create polysaccharides like starch and cellulose.

Monosaccharides pair up to form disaccharides (e.g., sucrose).

The hard outer shell of crustaceans is made of chitin.

Glycogen is stored in muscle tissue as a quick energy source.

Simple sugar

CH₂OH

Oxygen

A major component of the bacterial cell wall is peptidoglycan.

Common sugar is the disaccharide sucrose.

**Lipids**

**Lipids** are a diverse group of hydrophobic molecules composed primarily of hydrocarbons.

**Hydrocarbon**

Sulfur

Hydrocarbon rings join together to form steroid molecules like cholesterol.

Cholesterol is an essential component of cell membranes, but in excess it can cause atherosclerosis (clogging of arteries), resulting in heart attack and stroke.

Chains of hydrocarbons form fatty acids.

Bacon, like all meats, contains both fat and cholesterol.

Fatty acid chains join glycerol to form triglycerides, and join glycerol and phosphates to form phospholipids.

Fat under the skin of marine mammals like the walrus provides long-term energy storage and insulation against icy waters.

Phospholipids and cholesterol, along with proteins, form the membranes in a cell.

Seeds contain oil as a long-term energy source for a newly germinated sprout to use until its leaves are formed and capable of photosynthesis.

# What's It All Made Of?

Everything in the universe is composed of matter—from ordinary matter, made of atoms, to dark matter, which may consist of unknown types of particles. Here, we stick with what we know and describe the common elements that compose the world around us.

Assessment available in smartwork**5**

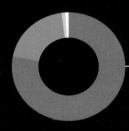

## Earth's Atmosphere

- **78%** Nitrogen
- **21%** Oxygen
- **<1%** Argon
- **<<1%** Other Elements

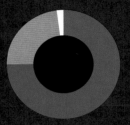

## The Universe

- **75%** Hydrogen
- **23%** Helium
- **2%** Other Elements

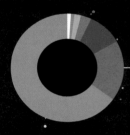

## The Human Body

- **65%** Oxygen
- **18%** Carbon
- **10%** Hydrogen
- **3%** Nitrogen
- **2%** Calcium
- **1%** Phosphorus
- **1%** Other Elements

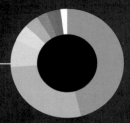

## Earth's Crust

- **46%** Oxygen
- **28%** Silicon
- **8%** Aluminum
- **5%** Iron
- **4%** Calcium
- **3%** Sodium
- **2%** Potassium
- **2%** Magnesium
- **2%** Other Elements

All proportions are by mass except Earth's atmosphere, which is by volume

the information stored in our genes would be stuck there, like a blueprint in a foreign language that no one can decipher. RNA comes in many forms and plays many roles, but its most important job is providing a readable genetic language that enables genes to be expressed as proteins.

A final valuable type of biomolecule is the **lipids**, which are better known as fats, oils, and steroids. Lipids are not polymers, because their structure is not composed of a chain of monomers (**Figure 3.11**, bottom right). Nevertheless, the lipids form a diverse group of biomolecules made up of combinations of hydrocarbons (carbon- and hydrogen-only molecules), fatty acids, or glycerol molecules. For example, triglycerides— composed of three fatty acid molecules linked to a glycerol molecule—provide long-term energy storage in both plants (as oils) and animals (as fats), and serve as insulation against the cold in animals. And phospholipids, which have a hydrophilic half and a hydrophobic half, are the main component of all cellular membranes (see Chapter 4 for more on phospholipids).

Miller, Bada, and other scientists have not yet been able to identify any biomolecules in their prebiotic soups. "It's still very much of an unknown how we go from simple to complex molecules," says Bada. Lipids and carbohydrate-like compounds, however, have been found in meteorites, and Sandford's team at NASA is now attempting to create carbohydrates using its space ice simulations. Other scientists are trying to re-create lipids and nucleic acids in the lab (see Chapter 4).

# Fifty More Years

Sadly, Miller died in 2007, never having seen the results of Bada's reanalysis. Today, researchers continue to perform spark discharge experiments with equipment that is essentially the same as in Miller's original design. "We're not done with it," says Jim Cleaves. "There are just so many variations to look at—different combinations of gases, changing the pH, adding metals, et cetera."

"This experiment lives on," agrees Bada. "Not many people can say that an experiment done over 50 years ago is still being investigated today, and finding a wealth of new information." And with better and better instruments to detect the results, who knows what scientists might discover over the next 50 years?

# REVIEWING THE SCIENCE

- The physical world is composed of **matter**, which is anything that has mass and occupies space. There are 98 distinct **elements** found naturally on Earth. An **atom** is the smallest unit of an element that maintains its unique properties; it contains positively charged **protons**, uncharged **neutrons**, and negatively charged **electrons**.

- The chemical interactions that cause atoms to associate with each other are known as **chemical bonds**. When an atom loses or gains electrons, it becomes, respectively, a positively or negatively charged **ion**. Ions of opposite charge are held together by **ionic bonds**.

- **Covalent bonds** are formed by the sharing of electrons between atoms. A **molecule** contains at least two atoms that are held together by covalent bonds.

- **Hydrogen bonds** are weak associations between two molecules such that a partially positive hydrogen atom within one molecule is attracted to a partially negative region of the other molecule.

- Partial electrical charges result from the unequal sharing of electrons between atoms, giving rise to **polar molecules**. The polarity of individual water molecules and the hydrogen bonding across water molecules explain nearly all of the special properties of water.

- A **solution** is any combination of a dissolved substance, known as the **solute**, and a fluid into which the solute has dissolved, known as the **solvent**.

- Ions and polar molecules are **hydrophilic**; they readily dissolve in water. Nonpolar molecules cannot associate with water and are therefore **hydrophobic**.

- In **chemical reactions**, bonds between atoms are formed or broken. The participants in a chemical reaction (**reactants**) are modified to give rise to new ions or molecules (**products**).

- The concentration of free hydrogen ions in water is expressed by the **pH scale** and reflects whether a solution is **acidic**, **basic**, or neutral.

- **Chemical compounds** are molecules that contain atoms from at least two different elements. Carbon atoms can link with each other and with other atoms to generate a great diversity of chemical compounds called **organic molecules**. The four main types of large organic molecules, or **biomolecules**, are **proteins**, **carbohydrates**, **nucleic acids**, and **lipids**.

# THE QUESTIONS

## The Basics

**1** The atomic number of an atom is determined by the number of _____ in the atom.

(a) protons

(b) neutrons

(c) electrons

(d) electrons plus neutrons plus protons

**2** The atomic mass number of an atom is determined by the sum of the number of _____ in the atom.

(a) protons plus electrons plus neutrons

(b) protons plus electrons

(c) neutrons plus electrons

(d) neutrons plus protons

**3** Link each term with the correct definition.

| | |
|---|---|
| ION | 1. The smallest subunit of an element. |
| MATTER | 2. A molecule made up of repeating monomers. |
| SOLUTION | 3. An atom that has gained or lost an electron. |
| ELEMENT | 4. Consists of two or more atoms chemically bonded together. |
| CHEMICAL COMPOUND | 5. Consists entirely of atoms with the same atomic number. |
| MOLECULE | 6. Anything that has mass and occupies space. |
| ISOTOPE | 7. Has the same atomic number but a different atomic mass than its original element. |
| POLYMER | 8. A molecule that contains atoms from two or more different atoms. |
| ATOM | 9. A combination of a solvent and a solute. |

**4** The partial negative charge at one end of a water molecule is attracted to the partial positive charge of another water molecule. What is this attraction called?

(a) a hydrogen bond

(b) a van der Waals interaction

(c) an ionic bond

(d) a covalent bond

(e) a hydrophilic bond

**5** Select the correct terms:

Proteins are (**polymers / monomers**) of amino acids. A common carbohydrate is (**sugar / fat**). Nucleic acids are composed of (**nucleotides / DNA**). Lipids (**are / are not**) polymers. All of these organic molecules contain (**carbon / nitrogen**).

## Challenge Yourself

**6** You are asked to determine the classification of an unknown biomolecule. You are told that it can be broken down into only carbon, oxygen, and hydrogen atoms, and nothing else. Using your knowledge of biomolecules and their components, identify which of the following statements about your unknown sample are true. (Select all that apply.)

(a) It is an organic compound.

(b) It contains amino acids.

(c) It contains sugar.

(d) It is a nucleic acid.

(e) It is a carbohydrate.

**7** Explain why life on Earth is carbon-based rather than, for example, hydrogen- or oxygen-based.

**8** Why did Miller's addition of hydrogen sulfide ($H_2S$) to the mixture of molecules in the spark discharge experiments increase the number of amino acids he found?

## Try Something New

**9** In the accompanying figure, a carbon atom resides at each unlabeled corner of the hexagons and wherever a "C" is shown. Each "O" is an oxygen atom, and each "H" is a hydrogen atom. What is this structure?

(a) a fatty acid monomer

(b) two simple sugar molecules linked together

(c) a nucleotide polymer

(d) a hydrocarbon ring polymer

(e) an amino acid

**10** Lipoproteins are relatively large, combined clumps of both protein and lipid molecules that circulate in the blood of mammals. They come in two forms, called HDL and LDL, and they act like suitcases to move cholesterol, fatty acid remnants, triglycerides, and phospholipids from one place to another through the bloodstream. (LDL recirculates lipids throughout the body, while HDL takes lipids to the liver to excrete them in feces.) Given that lipids are hydrophobic and proteins can be hydrophilic, which of the following statements is correct?

(a) The lipid portion of LDL does not dissolve in the bloodstream, while the lipid portion of HDL does.

(b) The protein portions of both LDL and HDL can dissolve or interact with the water molecules in the bloodstream.

(c) Neither the protein nor the lipid portions of LDL molecules can interact with water molecules in the bloodstream.

(d) Both the protein and the lipid portions of HDL molecules can interact with water molecules in the bloodstream.

(e) none of the above

**11** You have found an unknown cleaning solution in your roommate's under-sink cabinet. You have several spring cleaning projects to take care of and would like to use this product if it has the correct pH for the job. You know you need an extremely basic solution to clean your oven, but an acidic solution like vinegar to clean your coffee maker. You borrow a few pH indicator strips from your biology lab TA and test the solution. For each of the following scenarios, select the best answer.

a. The solution tested at a pH of 11; you will be able to clean your (**oven / coffee maker / both / neither**).

b. The solution tested at a pH of 2; you will be able to clean your (**oven / coffee maker / both / neither**).

c. The solution tested at pH 7; you will be able to clean your (**oven / coffee maker / both / neither**).

d. The solution tested at pH 9; you will be able to clean your (**oven / coffee maker / both / neither**).

e. The solution tested at pH 4; you will be able to clean your (**oven / coffee maker / both / neither**).

**12** Laundry detergent molecules have a short polar end and a long nonpolar end, enabling them to bind to water (on the polar end) and oils (on the nonpolar end). Suppose you spill some salad dressing on your shirt. Explain how washing your shirt with detergent will help remove the dressing.

## Leveling Up

**13** **What do *you* think?** Read the articles "Two Miles Underground, Strange Bacteria Are Found Thriving" in *News at Princeton* (http://www.princeton.edu/main/news/archive/S16/13/72E53/index.xml?section=newsreleases) and "Martian Underground Could Hold Clues to Life's Origins" (http://www.studentnews.eu/s/3229/68257-Articles/4014087-Martian-underground-could-hold-clues-to-lifes-origins.htm). Then answer the following questions.

a. How are bacteria able to stay alive below Earth's surface or (potentially) below the surface of Mars?

b. How does the discovery of bacteria living 2 miles beneath Earth's surface relate to the importance of the McLaughlin Crater on Mars?

c. What research do you think should next occur to answer the questions brought up by these studies?

d. From what you've read, what do you think is the probability of life on other planets? What would you expect that life to look like?

**14** **Is it science?** Life on Earth evolved as carbon-based. Silicon shares many of the chemical properties of carbon, yet it is not a building block of life on Earth. Watch the episode of the TV series *The X-Files* titled "Firewalker" (http://www.hulu.com/watch/158588), or read through its story line (https://en.wikipedia.org/wiki/Firewalker_%28The_X-Files%29). While doing so, make a list of the scientific concepts presented, and note whether you think they represent real science or pseudoscience. Then use your textbook and Internet research to determine whether you were correct or incorrect in each case. If you evaluated any of the concepts incorrectly, reflect on how you came to your initial conclusions.

**For more, visit digital.wwnorton.com/bionow2 for access to:**

# Engineering Life

*In 2003, scientists began trying to build an artificial cell from scratch. Today, they're closer than you might think.*

After reading this chapter you should be able to:

- Explain cell theory and why it is central to the study of life.
- Describe the differences in the structures of viruses, prokaryotes, and eukaryotes.
- Diagram a plasma membrane, showing how the structure allows some substances in and keeps others out.
- Compare and contrast passive and active transport of materials into and out of cells.
- Differentiate between exocytosis and the three types of endocytosis.
- Describe the role of any given organelle in a eukaryotic cell.
- Identify the main differences between a plant cell and an animal cell.

CHAPTER
04

LIFE IS
CELLULAR

The sky was still dark when Daniel Gibson hurried into the J. Craig Venter Institute (JCVI) in La Jolla, California. At 5:00 a.m., his footsteps echoed through the empty halls of the building. He reached a laboratory door and slipped inside. There, Gibson peered into a warm incubator, his eyes scanning rows of palm-sized petri dishes. His stomach was in knots. For 3 months the experiment had failed. Would this day—Monday, March 29, 2010—be any different?

Gibson is part of a team at the JCVI with a single, audacious goal: to create life. For more than a decade, this team of scientists and engineers has attempted to build a synthetic, or human-made, **cell**. Cells are the smallest and most basic unit of life—microscopic, self-contained units enclosed by a protective membrane (**Figure 4.1**). The human body is composed of approximately 100 trillion ($10^{14}$) cells. On that day in 2010, however, the JCVI was trying to synthesize just one cell—a single-celled bacterium.

Gibson's team had sequenced a bacterium's complete genetic information, its **genome**; built a synthetic version of that genome using basic laboratory chemicals; and, finally, replaced the natural DNA of another species of bacterium with the synthetic DNA. **DNA** (deoxyribonucleic acid) is a large and complex molecule that acts as a set of instructions for building an organism, like a blueprint. Almost every cell of every living organism contains DNA. DNA transfers information from parents to offspring, which is why it is essential for reproduction. Life, no matter how simple or how complex, uses this inherited genetic code to direct the structure, function, and behavior of every cell. DNA is made up of many nucleotides held together in a structure called the double helix, a ladderlike assembly twisted along its length into a spiral (see Figure 3.11).

Gibson's boss, the famous geneticist J. Craig Venter, worked for more than 15 years and spent millions of dollars to construct a synthetic DNA helix from chemicals in the laboratory, but Gibson and the team had been unable to get that synthetic DNA to work inside a cell. Every Friday for 3 months, they transplanted the synthetic DNA into a bacterial cell whose own DNA had been removed. The synthetic DNA included a gene to make the cells turn bright blue, so every Monday, Gibson hurried to the incubator and checked the petri dishes for a colony of blue cells. But Monday after Monday, the dishes were barren. "We did the genome transplantation again and again," he recalls, "but nothing was working."

Then, in mid-March, Gibson identified an error in a single gene in the synthetic DNA. A **gene** is a segment of DNA that codes for a distinct genetic characteristic, such as having O-type blood or a dimpled chin. In Gibson's bacterium, the gene with the error was responsible for DNA replication. When it wasn't working, the bacterium couldn't replicate its DNA, and it died. So in late March, Gibson fixed the DNA error, transplanted the genome yet again, and waited.

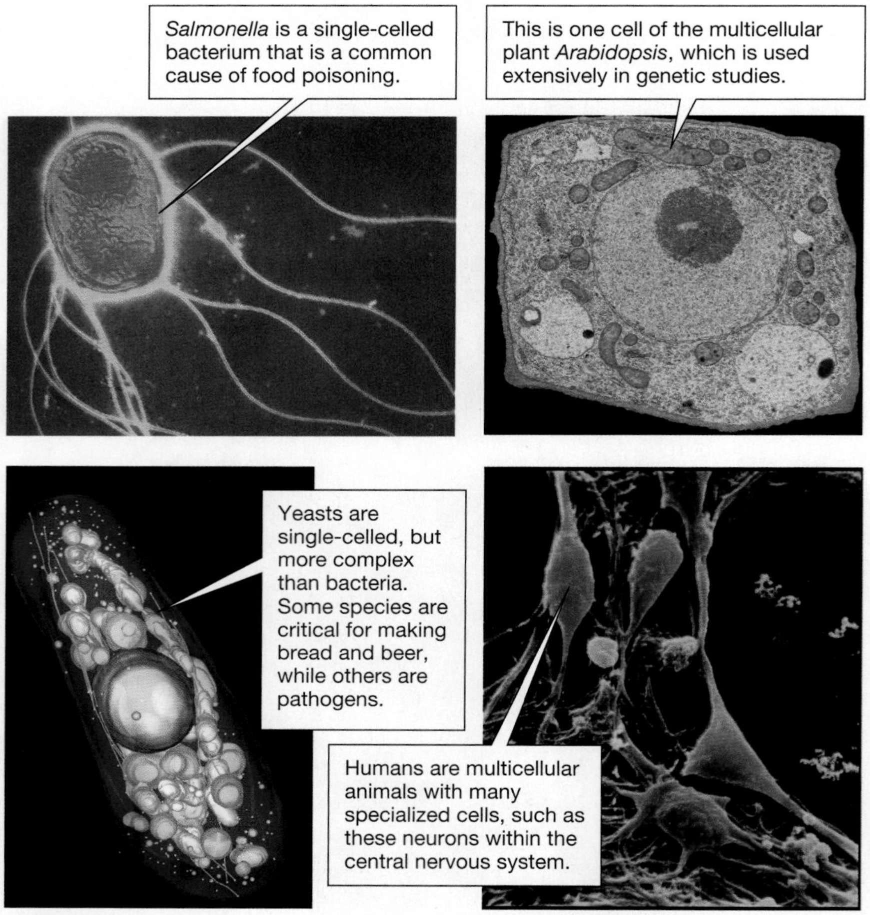

*Salmonella* is a single-celled bacterium that is a common cause of food poisoning.

This is one cell of the multicellular plant *Arabidopsis*, which is used extensively in genetic studies.

Yeasts are single-celled, but more complex than bacteria. Some species are critical for making bread and beer, while others are pathogens.

Humans are multicellular animals with many specialized cells, such as these neurons within the central nervous system.

Figure 4.1

**An individual organism may consist of a single cell or many cells**

All of these photos of cells were taken using electron microscope technology. Color has been added to the images to differentiate structures within the cells.

# Life, Rewritten

The JCVI is just one institution among a large group of universities and companies pursuing synthetic biology, a field that aims to design and construct new biological entities with novel and useful functions. Scientists are working to create algae that digest trash and produce energy, microbes that use light and water to create hydrogen gas, and bacteria that produce new kinds of antibiotics to treat infections. With synthetic biology "we can harness what nature has made, but repurpose it," says James Collins, a synthetic biologist at Boston University. "We can reprogram organisms and endow them with novel functions." In addition to useful tools, the pursuit of artificial life sheds light on the very origins of life. "It's going to be a big challenge to create a totally synthetic cell," says Collins, "but it's fundamentally intriguing to explore how life may have arisen on the planet."

To make their own cell, scientists are pushing the boundaries of **cell theory**, one of the unifying principles of biology. Cell theory has two main parts: every living organism is composed of one or more cells, and all cells living today came from a preexisting cell. By trying to engineer a cell in the laboratory, Venter, Gibson, and others are challenging the second part of the definition.

# Starting Small

The JCVI's first step toward a synthetic cell was a small one. In 2003, Venter's team flexed its scientific muscles by synthesizing the 11-gene, 5,386-base-pair genome of phiX174, a virus that infects bacteria. A **virus** is a small, infectious agent that can replicate only inside a living cell. Most viruses are little more than stripped-down genetic material wrapped in proteins, yet these pathogens attack and devastate organisms in every kingdom of life, from bacteria to plants and animals. Though the JCVI team successfully created a virus with a synthetic genome, it was not considered the first synthetic life, because scientists debate whether viruses are alive. (For more on this debate, see "Viruses— Living or Not?," page 66.)

Next, Venter and his colleagues moved up to a bacterium, a living organism that consists of a single cell. In 2010, they sequenced and built the genome of a bacterium called *Mycoplasma mycoides* (*M. mycoides*), an organism that can cause the mammary glands of goats to swell. They constructed the genome—a 1.1-million-base-pair DNA sequence—using four necessary ingredients: the nucleotides **adenine** (**A**), **thymine** (**T**), **guanine** (**G**), and **cytosine** (**C**), the building blocks of DNA. A, T, G, and C, organized in different combinations, carry all the instructions for everything a cell does.

The team used a machine to read the nucleotide sequence of the *M. mycoides* genome and then "print out" little bits of that code, creating strands of DNA about 50–80 bases long. They then strung these pieces together using living cells as factories, inserting the short segments into yeasts and *Escherichia coli*—small, single-celled organisms. These organisms interpreted the strands as broken pieces of DNA and stitched them together, creating longer and longer sequences. It was like building the Eiffel Tower from a massive box of Legos, constructing a single support beam at a time. The effort—with many mistakes along the way—took years. "It was very complex," said Venter. "It was a long, involved process."

# Congratulations, It's a . . . Cell

Once the DNA sequence of *Mycoplasma mycoides* was complete and intact, it was up to the JCVI team to transfer it into another species and make it work. This was the experiment that almost drove Gibson crazy. The researchers removed all the DNA from a cell of a closely related bacterium, *Mycoplasma capricolum*, and replaced it with the *M. mycoides* synthetic DNA.

After months of trying, on that Monday morning at 5:00 a.m., Gibson cautiously scanned the

**J. CRAIG VENTER**

J. Craig Venter is an American biologist and founder and CEO of the J. Craig Venter Institute. He led a team to fully sequence and publish the human genome in 2001, and initiated the effort to create the first cell constructed with synthetic DNA.

petri dishes. There, on a single dish, was a group of bright-blue cells—proof that the *M. capricolum* cell had "booted up" the *M. mycoides* DNA and transformed itself into an *M. mycoides* cell (**Figure 4.2**). Gibson was ecstatic. Moments later, he sent a text message to Venter, waking him up. Within the hour, Venter was in the lab with a camera, taking pictures of the tiny blue dollop in the dish. "How does it feel to create life?" Venter asked Gibson. They opened champagne and toasted their success.

Over the following weeks, Gibson repeated the experiment hundreds of times to make sure the blue cells were not an accident or a fluke, confirming that they contained only

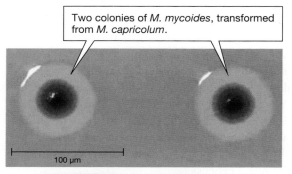

Two colonies of *M. mycoides*, transformed from *M. capricolum*.

100 µm

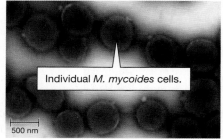

Individual *M. mycoides* cells.

500 nm

Figure 4.2

### The first synthetic organism

A colony of bacteria was transformed from *M. capricolum* by the insertion of synthetic DNA of a closely related species of bacterium, *M. mycoides*. The investigators inserted a gene into the synthetic DNA that codes for blue pigment.

**Q1:** What was the purpose of inserting the gene that codes for blue pigment into the synthetic DNA?

**Q2:** What part of the transformed bacterium is synthetic?

**Q3:** Did this experiment create life?

*M. mycoides* DNA. Every time, the cells with a synthetic genome survived. The team had done it—created the first synthetic cell. "They are living cells," Venter told *The Scientist* magazine when the research was published 2 months later. "The only difference is that they have no natural history. Their parents were the computer."

Once the research was published, the reaction from the academic community, captured in *Nature* magazine, was swift and divided. Some called it a significant advance: "We now have an unprecedented opportunity to learn about life," said Mark Bedau, a professor of philosophy at Reed College in Oregon. Arthur Caplan, a bioethicist at the University of Pennsylvania, said, "Venter's achievement would seem to extinguish the argument that life requires a special force or power to exist. In my view, this makes it one of the most important scientific achievements in the history of mankind."

Others were more hesitant. "Has [Venter] created 'new life'?" asked George Church, a prominent geneticist at Harvard Medical School. "Not really. . . . Printing out a copy of an ancient text isn't the same as understanding the language." Gibson and Venter agree that they did not create life from scratch, but they argue that they did create new life from existing life.

## A Different Approach

While Venter and Gibson were making headlines, other scientists were quietly pursuing a different approach to building a cell, working from the bottom up rather than from the top down. One young scientist in California decided to start by building one of the simplest, yet most vital, components of a cell: the layer of molecules that surrounds it.

In the chemistry department at UC San Diego, assistant professor Neal Devaraj was fascinated by the idea of building life and approached synthetic biology from the perspective of chemistry. "Most molecular biologists study what exists," says Devaraj, "but when you're a chemist and constantly make new compounds, you want to engineer something from scratch." So instead of taking a cell apart and determining how

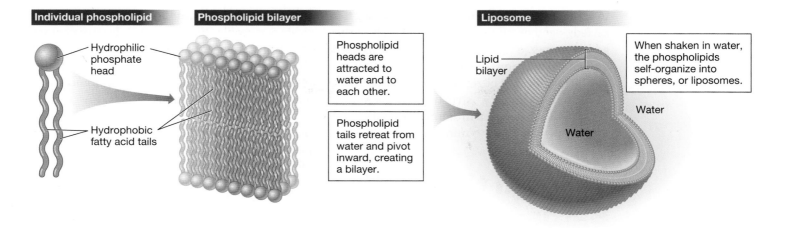

**Individual phospholipid**

Hydrophilic phosphate head

Hydrophobic fatty acid tails

**Phospholipid bilayer**

Phospholipid heads are attracted to water and to each other.

Phospholipid tails retreat from water and pivot inward, creating a bilayer.

**Liposome**

Lipid bilayer

Water

Water

When shaken in water, the phospholipids self-organize into spheres, or liposomes.

Figure 4.3

**Liposomes form when phospholipids and water are shaken together**

When you shake a mixture of phospholipids and water, the phospholipid bilayers bend and link together to form spheres called liposomes. This simple structure is remarkably similar to the basic structure of a cell.

**Q1:** Why is it important that the phosphate head of a phospholipid is hydrophilic?

**Q2:** What essential component of a cell do liposomes lack, and why is that omission important?

**Q3:** Could the tendency of phospholipid bilayers to spontaneously form spheres have played a role in the origin of life? (*Hint:* Refer to "The Characteristics of Living Organisms" on page 6 of Chapter 1.)

it works, Devaraj decided to try to build a cell artificially, using materials not typically found in nature. "If you want to really understand the principles by which life operates and evolves, the best way to do so is to build a cell from the ground up," he says.

Scientists suspect that one of the first events at the beginning of life on Earth was the formation of a **plasma membrane**, a barrier separating a cell from its external environment. A plasma membrane is made of two layers of **phospholipids**, organic molecules with a water-loving, or *hydrophilic*, head, and a water-fearing, or *hydrophobic*, tail. In water, these molecules form a double layer with heads out and tails in, a barrier that separates the contents of the cell from what lies outside the cell. Thus, the membrane is a **phospholipid bilayer**, a mostly impermeable barrier. When a phospholipid bilayer forms a sphere, or **liposome**, the fluid inside the liposome can have a different composition from the fluid outside (**Figure 4.3**). The ability to maintain an internal environment separate from the external environment is one

of the most critical functions of the plasma membrane of a cell.

Given a container of phospholipids, anyone can make a membrane, says Devaraj. "Making membranes is almost a trivial thing," he says. "You take natural or synthetic phospholipids, add water, and they form membranes." Yet researchers had been unable to form a membrane from scratch—without using preexisting phospholipids. In nature, new phospholipids are created by enzymes embedded in the cell.

Instead of trying to engineer new phospholipids, Devaraj wanted to start with something simpler. He worked with graduate student Itay Budin, who was then at Harvard University and is now a postdoctoral researcher at UC Berkeley,

**NE AL DEVARAJ**

JNeal Devaraj is a biochemist at UC San Diego who is working to make an artificial cell from the bottom up, starting with the membrane and then building other organelles.

to create a self-assembling membrane. They first mixed together oil and a detergent. Then they added copper, a metal ion, as a catalyst to spark a chemical reaction. With the addition of copper, sturdy membranes begin to bud off the oil; these were self-assembling structures. "There's no equivalent whatsoever in nature," says Devaraj. "Our goal was simply to mimic biology."

# Through the Barrier

Devaraj admits that his artificial membrane is far simpler than a real cell's plasma membrane, which is dotted with numerous proteins, including transport proteins. **Transport proteins** are gates, channels, and pumps that allow molecules to move into and out of the cell, making it selectively permeable (**Figure 4.4**). **Selective permeability** means that some substances can cross the membrane, others are excluded, and still others can pass through the membrane when aided by transport proteins.

All movement of substances through the plasma membrane occurs by either active or passive transport. Some transport proteins facilitate **active transport**, the movement of a substance that requires an input of energy (**Figure 4.4**, left). Molecules move across the plasma membrane by active transport when they need to move from a region of *lower* concentration to a region of *higher* concentration. In contrast, **passive transport** is the movement of a substance without the addition of energy (**Figure 4.4**, middle and right). Movement via passive transport is spontaneous.

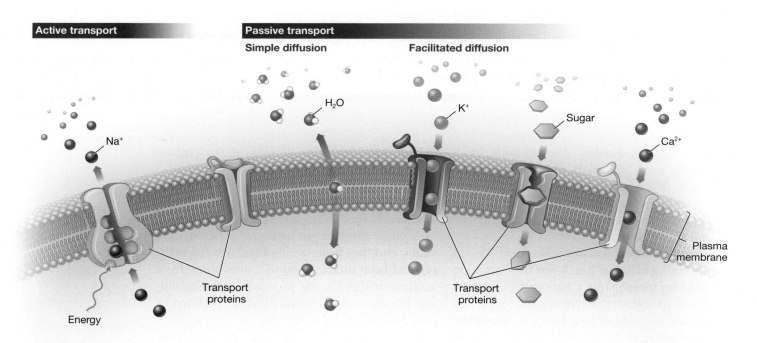

Figure 4.4

**The plasma membrane is a barrier and a gatekeeper**
The plasma membrane moves substances in a highly selective fashion, determined in large part by the types of membrane proteins embedded in the phospholipid bilayer. ▶️

**Q1:** In what ways is the plasma membrane a barrier, and in what ways is it a gatekeeper?

**Q2:** Why can't ions (such as Na⁺) cross the plasma membrane without the help of a transport protein?

**Q3:** If no energy were available to the cell, what forms of transport would not be able to occur? What forms of transport could occur? (*Hint:* Look ahead at Figures 4.5 and 4.6.)

A primary type of passive transport is **diffusion**, the movement of a substance from a region of *higher* concentration to a region of *lower* concentration (**Figure 4.5**). Water, oxygen, and carbon dioxide usually enter and leave cells by **simple diffusion**: These small, uncharged molecules slip between the large molecules in the phospholipid bilayer without much hindrance.

Water moves in and out of cells (and compartments inside cells) by **osmosis**. Osmosis is a form of simple diffusion because the water molecules are moving from areas of higher concentration to areas of lower concentration (**Figure 4.6**). Osmosis is critical for cellular processes because most cells are at least 70 percent water, and nearly all cellular processes take place in a water-rich environment.

Cells must maintain a stable internal water concentration to function properly, but the concentration of water in most cells changes from moment to moment. For example, when salt molecules move into a cell, the concentration of water in the cell decreases because additional molecules have been added. In response, water

At first, the molecules of food coloring are concentrated in one region.

Net movement of the food coloring is from regions of high concentration to regions of low concentration.

Diffusion ceases when food-coloring molecules are evenly distributed. At equilibrium, just as many molecules move into any given region as leave that region, so there is no net change in concentration.

Figure 4.5

**Food coloring in water illustrates diffusion**

Sugar has just been added to a beaker of water. This beaker is divided by a semipermeable membrane—that is, a membrane with pores large enough to allow water molecules to pass through, but too small for sugar molecules to pass.

Sugar molecule  Water molecule  Semipermeable membrane  Water molecule

Semipermeable membrane

After a period of time, water molecules have moved by osmosis from the right side of the membrane (which had a higher concentration of water) to the left side of the membrane. The concentration of water on the two sides of the membrane is now the same, and the movement of water molecules is now the same in both directions.

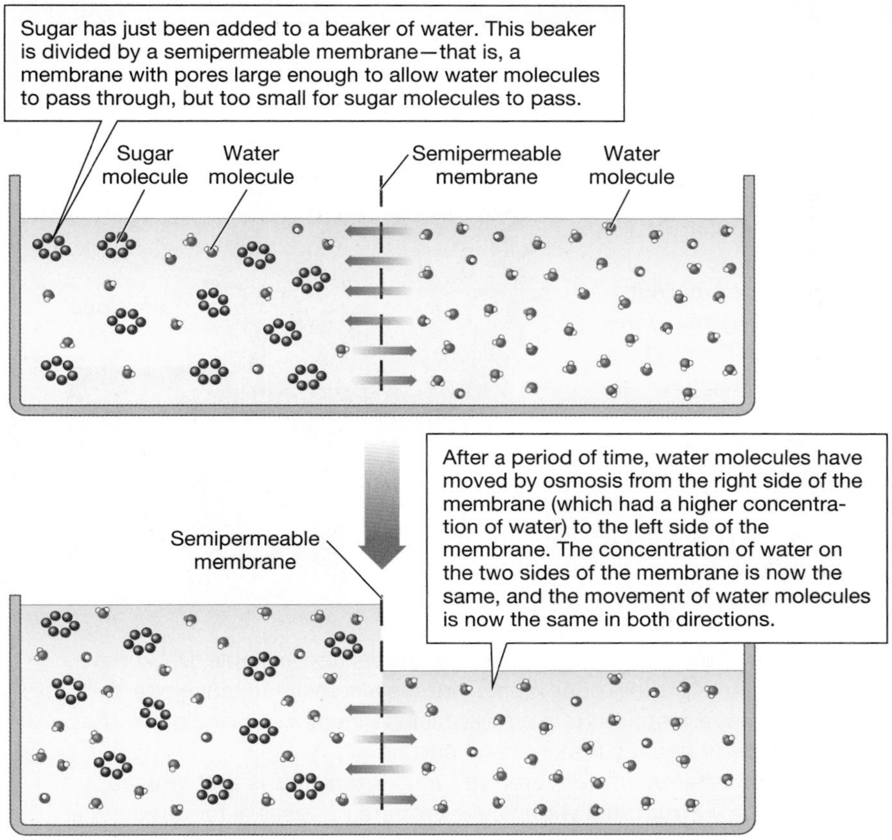

Figure 4.6

**In osmosis, water diffuses across a semipermeable membrane**

Osmotic movement of water between a cell and the external environment is critical for maintaining a constant water concentration in the cell, which it needs to function properly.

Q1: Is the dye at equilibrium in any of these glasses? Describe how the first glass will look when the dye is at equilibrium with the water.

Q2: Will diffusion mix the molecules of dye evenly through the water, or is it necessary to shake the container to get a uniform mixture?

Q3: Will diffusion mix the dye faster in hot water than in cold water? Why or why not? (*Hint:* Review the discussion of the behavior of water molecules at different temperatures in Chapter 3.)

Q1: What would the second diagram look like if the pores in the semipermeable membrane were too small to allow water molecules to pass through?

Q2: What would the second diagram look like if the pores were large enough to let both water molecules and sugar molecules through?

Q3: The fluid in an IV bag is isotonic to blood. What change would you see in the red blood cells of a patient if a bag of a hypertonic solution was used in error?

molecules immediately move by osmosis—that is, they diffuse—across the plasma membrane into the cell until the concentration of water inside the cell is the same as the concentration on the outside. On the other hand, when salt molecules move out of a cell, the water molecules become more concentrated, and osmotic movement of water out of the cell then restores the concentration of water in the cell to its correct level.

You can imagine, then, that the concentration of solutes within a cell in relation to the concentration *outside* the cell is of critical importance. When cells are surrounded by fluid with the same solute concentration as the cell interior, the extracellular and intracellular environments are said to be **isotonic** to each other (*iso*, "equal"). If the extracellular environment has a higher solute concentration, it is **hypertonic** to the cell interior (*hyper*, "more"). If it has a lower solute concentration than the cell's interior, it is said to be **hypotonic** to it (*hypo*, "less").

The careful balance of concentrations is particularly vital in human blood. Red blood cells are typically in an isotonic solution in our bloodstream, where the solute concentrations inside and outside the cells are the same. Doctors and nurses are taught to administer an IV drip of saline, a solution of salt in water, to dehydrated patients. If a dehydrated patient was administered water instead of a saline solution, the water would dilute the patient's blood, making it hypotonic. Osmosis would then occur, causing water to rapidly diffuse into red blood cells to the point where the cells could burst and die.

Most hydrophobic molecules, even fairly large ones, can pass through the plasma membrane via simple diffusion because they mix readily with the hydrophobic tails that form the core of the phospholipid bilayer. But hydrophilic substances such as sodium ions ($Na^+$), hydrogen ions ($H^+$), and larger molecules, including sugars and amino acids, cannot cross the plasma membrane without assistance. These substances move across the plasma membrane by **facilitated diffusion**, a type of passive transport that requires transport proteins (**Figure 4.4**, right).

Devaraj's artificial membrane does not currently contain any transport proteins, he says, so it is impermeable to large hydrophilic molecules. But small molecules such as water can pass through his membrane via simple diffusion.

The plasma membrane also contains **receptor proteins**, which are sites where molecules released by other cells can bind. The binding of a molecule to a receptor protein starts a chain of

# Viruses—Living or Not?

You've heard their names: Ebola virus, Zika virus, H5N1, dengue virus. Viruses—microscopic, noncellular infectious particles—are perhaps the smallest biological agents with the greatest impact on human health. Like living organisms, viruses reproduce and evolve, yet they lack some of the key characteristics of life—which is why most scientists today regard viruses as nonliving. For one thing, viruses are not made up of cells. A virus is much simpler than a cell, usually consisting of a small piece of genetic material (for example, DNA) that is wrapped in a protein coat. Some viruses also have an envelope, a lipid layer usually stolen from a cell's plasma membrane, enclosing the central core of genetic material and protein.

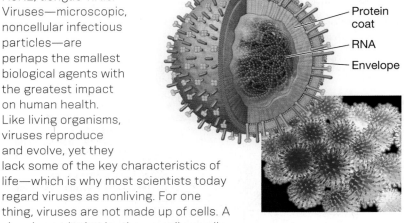

Protein coat

RNA

Envelope

Another difference, compared to living organisms, is that viruses lack the many structures within cells that are necessary for critical cellular functions such as homeostasis, autonomous reproduction, and metabolism. To gain these functions, they become "body snatchers": Viruses use their genetic material to make the cells of the organisms they infect do their work for them. They accomplish this feat by invading cells, releasing their genetic material into the cell interior, and "hijacking" the host cell's machinery. Viruses multiply to huge numbers, and viral offspring escape from a host cell either by causing it to burst open or by budding off from the cell, wrapped in a layer of the host cell's plasma membrane.

Uniquely, unlike the case with living organisms, the genetic material that viruses pass from one generation to the next is not always DNA; sometimes it is RNA. Viruses are generally classified by the type of genetic material they possess (type of DNA or RNA molecule), their shape and structure, the type of organism (host) they infect, and the disease they produce. The variant forms of a particular type of virus are called **viral strains** or serotypes. Viruses evolve new strains within a host so quickly that sometimes an antiviral drug or vaccine developed to fight an older strain becomes useless against a new strain.

events inside the cell that causes the cell to do something. For example, the receptors in nerve cells receive molecular signals from other nerve cells that cause the cells to fire. Receptor proteins are key components of a cell's communication system, enabling it to respond appropriately to changes in the organism.

# Another Way Through

In addition to transport proteins, there is another way that molecules move into and out of a cell. Sections of the plasma membrane can bulge inward or outward to form packages called **vesicles**. Vesicles move molecules from place to place inside a cell but also transport substances into and out of the cell (**Figure 4.7**).

Cells expel materials in vesicles via **exocytosis** (**Figure 4.7**, top left). The substance to be exported from the cell is packaged into a vesicle, and as the vesicle approaches the plasma membrane, a portion of the vesicle's membrane fuses with the plasma membrane. The inside of the vesicle then opens to the exterior of the cell, discharging its contents.

**Endocytosis** (**Figure 4.7**, right and lower left) is the opposite of exocytosis. In this process, a section of plasma membrane bulges inward to form a pocket around extracellular fluid,

**Exocytosis**

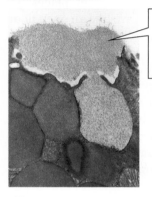

Exocytosis is used to eject substances from the cell. Here, a cell ejects waste material into the outside environment.

**Receptor-mediated endocytosis**

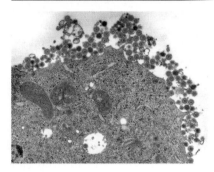

Receptor-mediated endocytosis is a selective process in which only certain molecules bind to receptor proteins. Here, low-density lipoprotein (LDL) particles bind to LDL receptors and are transported to the cell interior.

**Endocytosis**

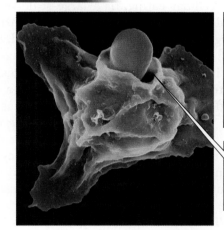

Endocytosis is the reverse of exocytosis, bringing material from the outside of the cell to the inside, enclosed in vesicles. Endocytosis may be nonselective, as shown here, drawing in any and all molecules near the opening of the vesicle.

Here, a white blood cell engulfs a yeast cell through phagocytosis, a form of nonspecific endocytosis.

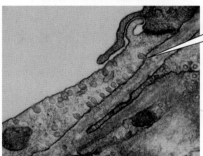

Cells lining blood vessels have created fluid-filled vesicles through pinocytosis, nonspecific endocytosis of fluid.

Figure 4.7

**Substances move into cells by endocytosis and out of cells by exocytosis**

**Q1:** If endocytosis itself is nonspecific, how does receptor-mediated endocytosis bring only certain molecules into a cell?

**Q2:** What sorts of molecules could be moved by endocytosis or exocytosis, but not by diffusion?

**Q3:** How does the fluid that enters a cell via pinocytosis differ from the fluid that enters by osmosis?

molecules, or particles. The pocket deepens until the opening in the membrane pinches off and the membrane breaks free as a closed vesicle, now wholly contained within the cell. Endocytosis can be nonspecific or specific. In nonspecific endocytosis, all of the material in the immediate area is surrounded and enclosed; in specific endocytosis, one particular type of molecule is enveloped and imported.

There are three types of endocytosis. **Receptor-mediated endocytosis** (**Figure 4.7**, bottom left) is a form of endocytosis in which receptor proteins embedded in the membrane recognize specific surface characteristics of substances to be incorporated into the cell. For example, our cells use receptor-mediated endocytosis to take up cholesterol-containing packages called low-density lipoprotein (LDL) particles. **Phagocytosis** (**Figure 4.7**, top right), or "cellular eating," is a large-scale version of endocytosis in which particles considerably larger than biomolecules are ingested. Specific cells in the immune system use phagocytosis to ingest an entire bacterium or virus. **Pinocytosis** (**Figure 4.7**, bottom right) is a form of endocytosis that is often described as "cellular drinking" because it involves the capture of fluids. However, the cell does not attempt to collect particular solutions. Pinocytosis is nonspecific: the vesicle budding into the cell contains whatever happened to be dissolved in the fluid when the cell "drank."

There is a long way to go before an artificial plasma membrane will perform processes like endocytosis and exocytosis, says Neal Devaraj. "But just because a research problem is difficult doesn't mean we shouldn't tackle it," he adds.

# Prokaryotes versus Eukaryotes

Devaraj's team continues to pursue that ideal. Prior synthetic membranes, their own included, did not have the ability to grow by adding new phospholipids, but in 2015 the team succeeded in designing and synthesizing an artificial membrane that sustained continual growth, just like a living cell. Membranes are important not only because they form the structure of a cell, but because they compartmentalize processes within eukaryotic cells. Depending on the fundamental structure of their cells, all living organisms can be sorted into one of two groups: **prokaryotes** or **eukaryotes** (**Figure 4.8**). *Mycoplasma mycoides* and all other bacteria are prokaryotes, but virtually all the organisms you see every day, including all plants and animals, are eukaryotes.

Eukaryotic cells are larger and more complex than prokaryotic cells: they are roughly ten times wider, with a cell volume about a thousand times greater. Unlike prokaryotic cells, eukaryotic cells have a membrane-enclosed **nucleus** (plural "nuclei") that contains the organism's DNA, and they have a variety of membrane-enclosed subcellular compartments called **organelles**. Through specialization and division of labor, these organelles act like cubicles in a large office, allowing the cell to localize different processes in different places. In contrast, prokaryotic cells are like an open floor plan: they lack a cell nucleus or any membrane-encased organelles.

Though the first fully artificial cell will most likely be a simple prokaryotic cell, synthetic biologists aspire to build a eukaryotic cell. "We're not quite there yet, but it's interesting to think about making complex structures like organelles," says Devaraj. Having achieved a self-assembling and growing artificial membrane, his team is working to create membranes inside preexisting vesicles, mimicking organelles like mitochondria that are made up of membranes within membranes. It is the first step toward creating the variety of organelles inside eukaryotic cells.

# What's in a Cell?

The nucleus is the control center of the cell. It contains most of the cell's DNA and may occupy up to 10 percent of the space inside the cell (**Figure 4.9**, top left). Inside the nucleus, long strands of DNA are packaged with proteins into a remarkably small space. The boundary of the nucleus, called the **nuclear envelope**, is made up of two concentric phospholipid bilayers. The nuclear envelope is speckled with thousands of small openings called **nuclear pores**. These pores allow chemical messages to enter and exit the nucleus.

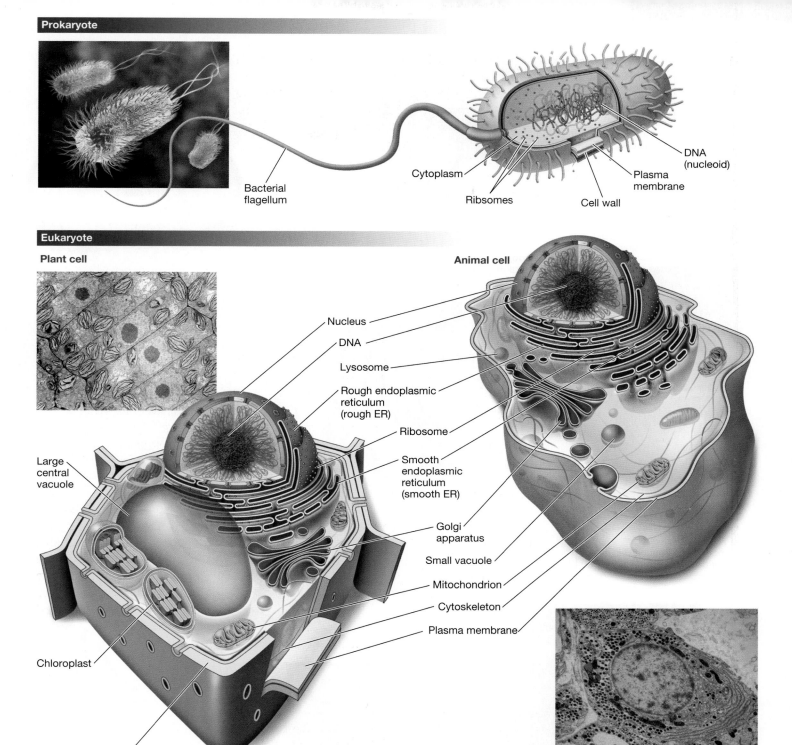

**Prokaryote**

Bacterial flagellum

Cytoplasm

Ribsomes

Cell wall

Plasma membrane

DNA (nucleoid)

**Eukaryote**

**Plant cell**

**Animal cell**

Nucleus

DNA

Lysosome

Rough endoplasmic reticulum (rough ER)

Ribosome

Smooth endoplasmic reticulum (smooth ER)

Golgi apparatus

Small vacuole

Mitochondrion

Cytoskeleton

Plasma membrane

Large central vacuole

Chloroplast

Cell wall

## Figure 4.8

### Prokaryotic and eukaryotic cells

A prokaryotic cell, like all cells, contains DNA, cytoplasm, ribosomes, and a plasma membrane. Many prokaryotes also have a cell wall that serves as a kind of exoskeleton. To enable movement, some bacteria possess a flagellum, or several flagella. The components of eukaryotic cells are described in the text and in Figure 4.9.

**Q1:** What structures do prokaryotic and eukaryotic cells have in common?

**Q2:** What cellular processes occur in both prokaryotic and eukaryotic cells?

**Q3:** Both plants and animals are eukaryotes, but there are differences in their cellular structure. What are those differences?

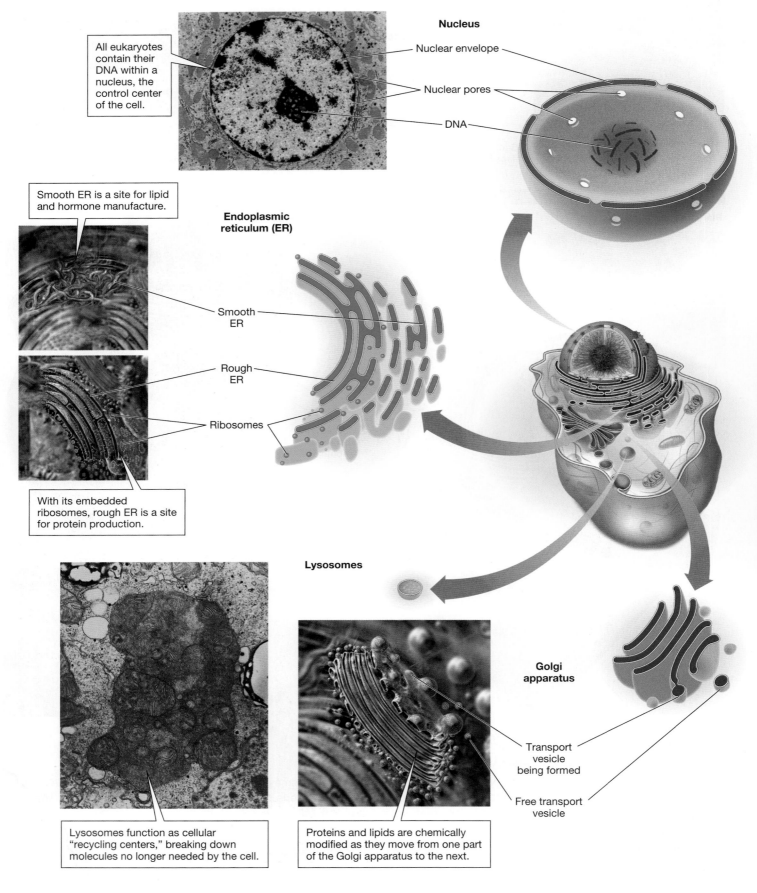

**Nucleus**

Nuclear envelope

Nuclear pores

DNA

All eukaryotes contain their DNA within a nucleus, the control center of the cell.

Smooth ER is a site for lipid and hormone manufacture.

**Endoplasmic reticulum (ER)**

Smooth ER

Rough ER

Ribosomes

With its embedded ribosomes, rough ER is a site for protein production.

**Lysosomes**

**Golgi apparatus**

Transport vesicle being formed

Free transport vesicle

Lysosomes function as cellular "recycling centers," breaking down molecules no longer needed by the cell.

Proteins and lipids are chemically modified as they move from one part of the Golgi apparatus to the next.

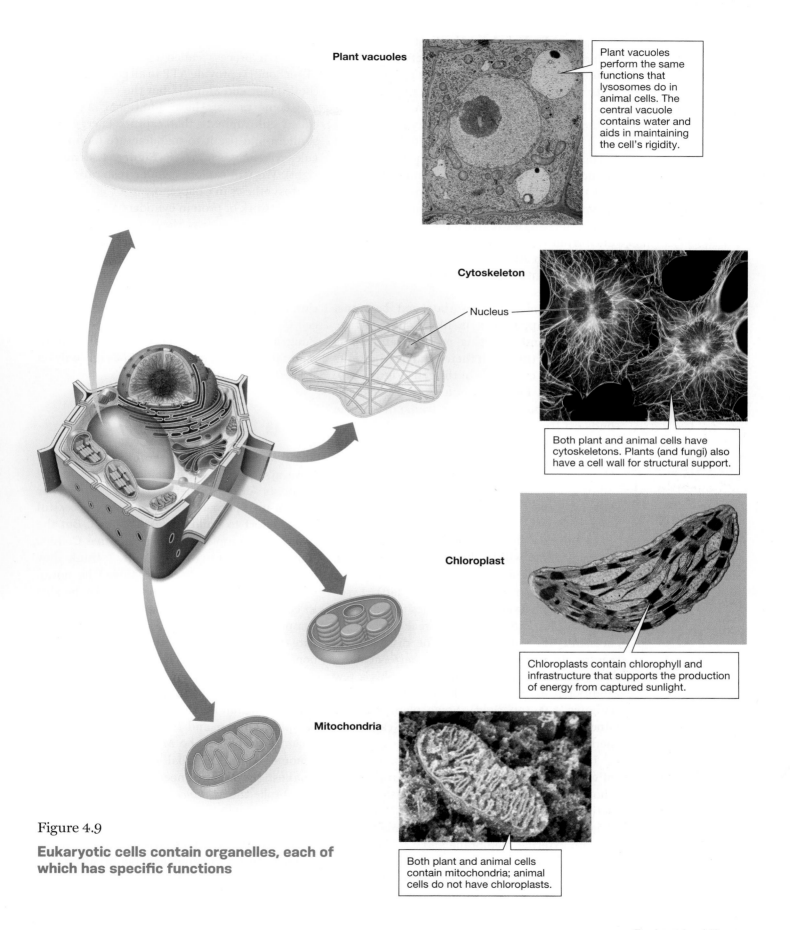

**Plant vacuoles**

Plant vacuoles perform the same functions that lysosomes do in animal cells. The central vacuole contains water and aids in maintaining the cell's rigidity.

**Cytoskeleton**

Nucleus

Both plant and animal cells have cytoskeletons. Plants (and fungi) also have a cell wall for structural support.

**Chloroplast**

Chloroplasts contain chlorophyll and infrastructure that supports the production of energy from captured sunlight.

**Mitochondria**

Both plant and animal cells contain mitochondria; animal cells do not have chloroplasts.

Figure 4.9

**Eukaryotic cells contain organelles, each of which has specific functions**

The **endoplasmic reticulum** (**ER**) is an extensive and interconnected network of sacs made of a single membrane that is continuous with the outer membrane of the nuclear envelope (**Figure 4.9**, middle left). The membranes of the ER are classified into two types based on their appearance: smooth and rough. Enzymes associated with the surface of the **smooth ER** manufacture lipids and hormones. In some cell types, smooth-ER membranes also break down toxic compounds. **Ribosomes** embedded in the **rough ER** give it the knobby appearance from which it gets its name. Ribosomes on the rough ER assemble proteins that will be inserted into the cell's plasma membrane or its organelles.

Resembling a pile of flattened balloons, the **Golgi apparatus** is like a post office, packaging and directing proteins and lipids produced by the ER to their final destinations either inside or outside of the cell (**Figure 4.9**, bottom left). Each molecule is first packaged into a transport vesicle. The transport vesicle buds off from the ER membrane and delivers the cargo to its destination by fusing with the membrane of the target compartment. The Golgi apparatus targets these destinations by adding a specific chemical tag to each molecule it receives, like attaching a shipping label to a package.

In animal cells, transport vesicles bring large molecules that will be discarded to **lysosomes**, organelles that act as garbage and recycling centers (**Figure 4.9**, bottom far left). Lysosomes contain a variety of enzymes that degrade biomolecules and release the breakdown products into the cell interior to be discarded or reused. In plant cells, **vacuoles** perform functions similar to those of lysosomes, plus a few additional functions, such as water storage (**Figure 4.9**, top right). Some plant vacuoles stockpile noxious compounds that can deter herbivores from eating plants.

In most eukaryotic cells, the main source of energy is the **mitochondrion** (plural "mitochondria"), a tiny power plant that fuels cellular activities (**Figure 4.9**, bottom right). Mitochondria are made up of double membranes—a smooth external membrane and a folded internal membrane—that form a mazelike interior. Mitochondria use chemical reactions to transform the energy of food molecules into ATP (adenosine triphosphate), the universal cellular fuel, in a process called *cellular respiration*. Mitochondria provide ATP to all eukaryotic cells (both plant and animal), but the cells of plants and some protists have additional organelles called **chloroplasts** that capture energy from sunlight and use it to manufacture food molecules via *photosynthesis* (**Figure 4.9**, middle right). We will explore cellular respiration and photosynthesis in Chapter 5.

A network of protein cylinders and filaments collectively known as the **cytoskeleton** forms the framework of a cell (**Figure 4.9**, middle right). The cytoskeleton organizes the interior of a eukaryotic cell, supports the intracellular movement of organelles such as transport vesicles, and enables whole-cell movement in some cell types. It also gives shape to wall-less cells. Fungi and plants separately evolved cell walls to maintain cell structure, but animals rely only on the cell's cytoskeleton.

# Life Goes On

From Venter's complex synthetic genome to Devaraj's simple self-assembling membrane, scientists debate whether the top-down or bottom-up approach will be more successful in the effort to build an artificial cell. But the two can be complementary, says James Collins. "I'm not sure one will win out. I think they bring different things to the table," he notes. Gibson agrees: "I would like to one day be able to combine all of a cell's parts from nonliving components, including the genome, and incubate them, and see if we can get life out of those nonliving components," he says. "It would help us better understand how cells work."

Today, the work on both ends continues. While Devaraj makes his membranes more complex, Gibson and his team (**Figure 4.10**) recently simplified the *Mycoplasma mycoides* genome. With the goal of determining the smallest set of genes needed to maintain life, they broke the *M. mycoides* genome into eight DNA segments and mixed and matched them to see which combinations would produce viable cells. Eventually, they narrowed the genome down to just 473 genes capable of sustaining life. Amazingly, the team could not identify the function of 149 of the 473 genes. "We don't

# Sizing Up Life

Cells dramatically range in size, and viruses are even smaller. The amoeba, a single-celled eukaryote that eats smaller unicellular organisms, is visible under a light microscope. But to observe a virus, one would have to use an electron microscope, which relies on a high-voltage beam of electrons to magnify an image.

Assessment available in smartw⊕rk**5**

←——— 500 μm ———→

**Amoeba proteus**

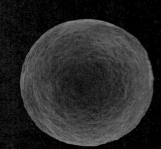

←——— 130 μm ———→

**Human egg cell**
*(shown next to an Amoeba proteus)*

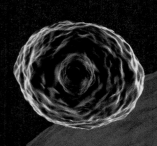

←——— 30 μm ———→

**Skin cell**
*(shown next to a human egg cell)*

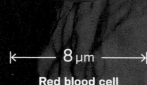

←——— 8 μm ———→

**Red blood cell**
*(shown next to a skin cell)*

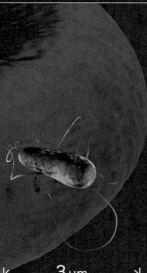

←——— 3 μm ———→

*E. coli* **bacterium**
*(shown next to a red blood cell)*

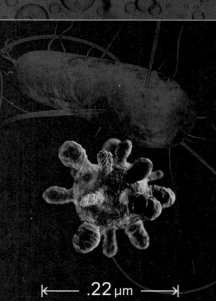

←——— .22 μm ———→

**Measle virus**
*(shown next to an E. coli bacterium)*

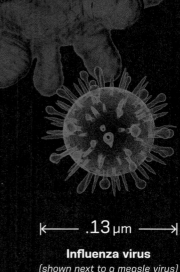

←——— .13 μm ———→

**Influenza virus**
*(shown next to a measle virus)*

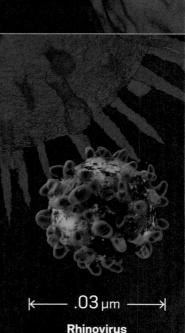

←——— .03 μm ———→

**Rhinovirus**
*(shown next to an influenza virus)*

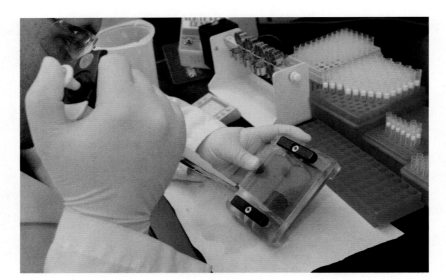

Figure 4.10

**Scientist in Action**
Laboratory technician Javier Quinones is setting up the beginning of the sequencing procedure used to sequence the *M. mycoides* genome in the sequencing laboratory at the J. Craig Venter Institute in Rockville, Maryland.

know about a third of essential life, and we're trying to sort that out now," Venter told *Nature* magazine in 2016.

Knowing both the identities and function of a core set of genes necessary for life will make it easier to build other synthetic cells with specific functions, says Gibson. "There's not a single organism in the world where we understand what every gene does," he notes. If they can reach such understanding with those 473 genes in the *M. mycoides* genome, "this would be the first example of that."

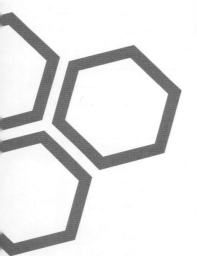

# REVIEWING THE SCIENCE

- **Cells** are the basic units of all living organisms. **Cell theory** states that all living things are composed of one or more cells, and that all cells living today came from a preexisting cell.

- Every cell is surrounded by a **plasma membrane** that separates the chemical reactions inside the cell from the surrounding environment. The plasma membrane is **selectively permeable** and formed by a **phospholipid bilayer** embedded with proteins that perform a variety of functions.

- In **passive transport**, substances move across the plasma membrane without the direct expenditure of energy. **Active transport** by cells requires energy. **Diffusion** is the passive transport of a substance from a region where it is at a higher concentration to a region where it is at a lower concentration.

- **Osmosis** is the diffusion of water across a selectively permeable membrane. In a **hypotonic** solution, a cell gains water. In a **hypertonic** solution, a cell loses water. In an **isotonic** solution, there is no net movement of water. Cells can actively balance their water content through osmosis.

- Cells export materials by **exocytosis** and import materials by **endocytosis**.

- In **receptor-mediated endocytosis**, **receptor proteins** in the plasma membrane recognize and bind the substance to be brought into the cell.

- **Prokaryotes** are single-celled organisms lacking a nucleus and complex internal compartments. **Eukaryotes** may be single-celled or multicellular, and their cells typically possess many membrane-enclosed compartments.

- By volume, eukaryotic cells can be a thousand times larger than prokaryotic cells. They require internal compartments, or **organelles**, that concentrate and organize cellular chemical reactions.

- The **nucleus** of a eukaryotic cell contains DNA. It is bounded by the **nuclear envelope**, which has **nuclear pores** that allow communication between the nucleus and the cell interior.

- Lipids are made in the **smooth endoplasmic reticulum**. Some proteins are manufactured in the **rough endoplasmic reticulum**.

- The **Golgi apparatus** receives proteins and lipids, sorts them, and directs them to their final destinations.

- In animals, **lysosomes** break down biomolecules such as proteins into simpler compounds that can be used by the cell. Plant **vacuoles** are similar to lysosomes but also store ions and molecules and lend physical support to plant cells.

- **Mitochondria** produce chemical energy for eukaryotic cells in the form of ATP.

- **Chloroplasts** harness the energy of sunlight to make sugars through photosynthesis.

- Eukaryotic cells depend on the **cytoskeleton** for structural support and for the ability to move and change shape. Plants and fungi also have a cell wall that provides structural support.

# THE QUESTIONS

## The Basics

**1** To be able to recognize a colony of bacteria that had grown from cells of *Mycoplasma capricolum* in which the DNA had been replaced by a synthetic DNA of *Mycoplasma mycoides*, Daniel Gibson added a _____ that coded for a blue pigment.

(a) cell

(b) chromosome

(c) gene

(d) bacterium

**2** A phospholipid is a biomolecule composed of a phosphate group that is bonded to two lipid chains. Which of the following correctly describes the nature of those two components?

(a) The phosphate group is hydrophobic, and the lipid chains are hydrophilic.

(b) Both the phosphate group and the lipid chains are hydrophilic.

(c) Both the phosphate group and the lipid chains are hydrophobic.

(d) The phosphate group is hydrophilic, and the lipid chains are hydrophobic.

**3** Link each process with the correct definition.

RECEPTOR-MEDIATED ENDOCYTOSIS    1. A cell ingests a large particle, such as a bacterial cell.

PHAGOCYTOSIS    2. Receptor proteins embedded in the membrane recognize specific surface characteristics of substances.

PINOCYTOSIS    3. A transport vesicle inside the cell approaches the plasma membrane of the cell, fuses with it, and releases its contents to the outside of the cell.

EXOCYTOSIS    4. A vesicle containing whatever molecules are in solution outside the cell bulges inward, pinches off, and enters the cell.

**4** Link each structure with the correct function.

CHLOROPLAST    1. Location of the cell's DNA.

GOLGI APPARATUS    2. Site of protein synthesis.

LYSOSOME    3. Site of lipid synthesis.

MITOCHONDRION    4. Adds chemical tags to newly synthesized proteins to direct them to their correct location.

NUCLEUS    5. Breaks down biomolecules by enzymatic action.

ROUGH ENDOPLASMIC RETICULUM    6. Site of cellular respiration.

SMOOTH ENDOPLASMIC RETICULUM    7. Site of photosynthesis.

**5** In the table below, indicate whether the specified cellular component is found in each type of cell by placing an "X" in the relevant columns.

| Component | Prokaryotes | Eukaryotes | |
|---|---|---|---|
| | | Animals | Plants |
| Plasma membrane | | | |
| Cellulose cell wall | | | |
| Nucleus | | | |
| Endoplasmic reticulum | | | |
| Golgi apparatus | | | |
| Ribosomes | | | |
| Cytoskeleton | | | |
| Mitochondria | | | |
| Chloroplasts | | | |

## Challenge Yourself

**6** How does the phospholipid bilayer of a liposome differ from the phospholipid bilayer of the plasma membrane of a cell?

(a) The phospholipid bilayer of a liposome contains only phospholipids, without the proteins that are embedded in the plasma membrane of a cell.

(b) The phospholipid bilayer of a liposome contains two bilayers of phospholipid molecules, whereas the plasma membrane of a cell contains only one.

(c) The phospholipid bilayer of a liposome completely envelops the liposome, whereas the plasma membrane of a cell does not completely envelop the cell.

(d) The phospholipid molecules in the phospholipid bilayer of a liposome are oriented with the lipid ends on the outside of the bilayer and the phosphate groups on the inside.

**7** Examine the graph presented here. Using your knowledge of how the processes work, label each curve as representative of either simple diffusion or facilitated diffusion. Explain how you determined which curve was which.

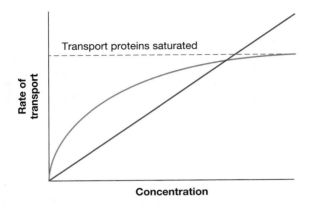

Transport proteins saturated

Rate of transport

Concentration

**8** Which of the following statements about transport through the plasma membrane of a cell is correct?

(a) Both passive and active transport require the input of energy.

(b) Passive transport does not require the input of energy, and active transport does.

(c) Passive transport does require the input of energy, and active transport does not.

(d) Neither active nor passive transport requires the input of energy.

## Try Something New

**9** Picture a beaker divided by a semipermeable membrane like the one in Figure 4.6. Imagine that you put 5 grams of sugar in the left side of the beaker and 10 grams of sugar in the right side, and then add water to bring both sides to the same depth. Select the correct terms in the following description of what happens next: The water will move by osmosis to the (**left side / right side**) of the beaker because there are (**more / fewer**) molecules of solute and (**more / fewer**) molecules of water on that side.

**10** When the chemical oligomycin is added to cells, the production of ATP is blocked. Which of the following processes would be most affected by oligomycin?

(a) simple diffusion

(b) facilitated diffusion

(c) active transport

(d) All of these processes would be equally affected by oligomycin.

(e) None of these processes would be affected by oligomycin.

**11** Use the accompanying image to select the correct terms in the statements below each of the cells shown.

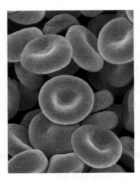

(a) This red blood cell is in a(n) (**hypertonic / hypotonic / isotonic**) solution.

(b) In this environment, normal red blood cells (**gain / lose / neither gain nor lose**) water.

(c) The total solute concentration outside the cell is (**lower than / higher than / equal to**) the total solute concentration inside it.

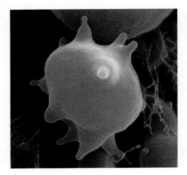

(d) This red blood cell is in a(n) (**hypertonic / hypotonic / isotonic**) solution.

(e) In this environment, normal red blood cells (**gain / lose / neither gain nor lose**) water.

(f) The total solute concentration outside the cell is (**lower than / higher than / equal to**) the total solute concentration inside it.

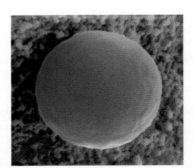

(g) This red blood cell is in a(n) (**hypertonic / hypotonic / isotonic**) solution.

(h) In this environment, normal red blood cells (**gain / lose / neither gain nor lose**) water.

(i) The total solute concentration outside the cell is (**lower than / higher than / equal to**) the total solute concentration inside it.

---

## Leveling Up

**12** **What do *you* think?** Viruses display many of the characteristics of living organisms. In particular, they reproduce, creating new virus particles. During reproduction, viruses make copies of their genetic material, and some of the copies contain mutations that are beneficial to the virus. For example, HIV (human immunodeficiency virus), the virus that causes the disease AIDS (acquired immunodeficiency syndrome) mutates so often that its surface proteins change faster than we can develop antiviral drugs. New drug-resistant strains of HIV are appearing constantly. However, viruses can reproduce only after entering a living cell of an organism, because viruses hijack the cell's machinery and use it to produce new viruses. Where does that combination of characteristics place viruses on the scale of nonlife to life? Are viruses living organisms? Nonliving? If neither of those categories fits the properties of viruses, how should they be classified?

**13** **What do *you* think?** Did Venter create "new life" by inserting synthetic DNA copied from the genome of *Mycoplasma mycoides* into a cell of *Mycoplasma capricolum* from which the DNA had been removed? Be prepared to support your opinion in a class discussion.

**14** **Doing science** Go to the National Science Foundation website (http://www.nsf.gov), and search for and read "Biologists Replicate Key Evolutionary Step in Life on Earth" (Press Release 12-009, January 16, 2012). State the hypothesis that the biologists were testing, describe their experimental design, and explain their results. Was their hypothesis supported? What experiment do you think they should try next? State a hypothesis that you would test using their multicellular yeast samples.

---

**For more, visit digital.wwnorton.com/bionow2 for access to:**

# Rock Eaters

*Unusual "electric" microbes could provide the energy storage solution we've been looking for.*

**After reading this chapter you should be able to:**

- Explain how photosynthesis and cellular respiration support life on Earth.
- Differentiate between anabolic and catabolic metabolic processes.
- Detail the role of ATP in a cell.
- Describe the importance of enzymes in metabolic pathways.
- Label a diagram of the light reactions and the Calvin cycle of photosynthesis.
- Define the three stages of cellular respiration and explain the function of each.
- Compare and contrast photosynthesis and cellular respiration.

CELLS

CHAPTER
05

HOW CELLS
WORK

Annie Rowe stands knee-deep in cold, clear salt water. Sweating, she pushes a shovel into the ground and lifts a mound of heavy mud. It's a sunny day on Santa Catalina Island, a rocky retreat off the coast of California, where Rowe has come to collect samples. Once the bucket is full, she sits on a rock and retrieves a mesh filter to sieve out any small invertebrates, such as snails and worms, from the sand. They are not part of the experiment.

A postdoctoral research fellow at the University of Southern California, Rowe pours the filtered sediment into one of her five 10-gallon aquariums. When each aquarium is about two-thirds full, she and a graduate student transport the tanks to a local laboratory on the island. There, they hook each aquarium up to a steady stream of seawater from the harbor.

To the untrained eye, the aquariums appear sterile, empty. In reality, both the water and sand in the tanks are teeming with microscopic organisms, or *microbes*, and Rowe is aiming to find a rare, strange one. Deep in the sand of each aquarium, she buries a metal electrode buzzing with electrical current to attract her quarry: a microbe that eats electricity (**Figure 5.1**).

"I went into microbiology because there's so much diversity in things that microbes can do," says Rowe. As a PhD student, Rowe began studying unusual microbes with the hope of someday putting them to use toward one of the greatest human challenges of our time: developing sources of alternative energy.

Humans rely heavily on fossil fuels such as oil and petroleum to power our cars, trains, planes, and more. Hundreds of millions of years ago, single-celled organisms used carbon dioxide and sunlight to make cell materials, which then became buried and concentrated into oil in Earth's crust. Burning that oil releases that ancient carbon dioxide back into the

Figure 5.1

**Annie Rowe burying electrodes in her sediment aquariums**

atmosphere, and a consequence of this increase in carbon dioxide is that the surface of our planet has begun to warm. To prevent further climate change and lessen our dependence on fossil fuels—many of which we pay a premium to import from other countries—scientists and engineers have developed clean, renewable ways to capture energy, such as solar, wind, and water power.

Yet to switch from a fossil fuel–based economy to renewable energy technologies, we need efficient and inexpensive ways to store renewable energy. With the right chemical stabilizers, oil can be left in a barrel for decades, but a direct current of electricity from a wind farm or solar panels needs to be used immediately or fed into the electrical grid. Today's batteries aren't a good storage option; they are bulky, expensive, and not terribly efficient.

Now imagine the ability to siphon electricity from a wind farm or solar panel and transform it into a liquid fuel. That's where electricity-eating microbes come in.

# Energy for Life

All living cells require energy. Organisms use energy for growth, reproduction, and defense, and to manufacture the many chemical compounds that make up living cells. They must obtain energy from the living or nonliving components of their environment, and at the very core of making and storing energy is the transfer of *electrons*—subatomic particles with a negative charge. Electrons play an essential role in electricity, magnetism, and many other physical phenomena that shape the world we know. The success of computers, solar cells, cell phones, and other devices is due to our ability to shape and control the flow of electrons.

But long before electrons flowed through computers, they moved through cells. The first law of thermodynamics says that energy cannot be created or destroyed, but it can be changed from one form to another. In other words, cells cannot create energy from nothing, so they must utilize one form of energy and change it to another form. However, organisms can't simply perform the biological equivalent of plugging into an electrical socket and sucking up electricity. That's because the cell membrane is an electrically neutral zone that prevents charged particles from sneaking through, including electrons. So, to move electrons into and out of a cell, living organisms attach electrons to molecules. Plants, for example, smuggle electrons into the cell via water molecules. Humans and other organisms obtain electrons from food (sugars, proteins, fats, etc.). As the Nobel Prize–winning physiologist Albert Szent-Györgyi reportedly said, "Life is nothing but an electron looking for a place to rest."

Cells use and store energy by transferring electrons among molecules via chemical reactions. Thousands of different types of chemical reactions are required to sustain life in even the simplest cell. The term **metabolism** describes all the chemical reactions that occur inside living cells, including those that store or release energy. Most chemical reactions in a cell occur in chains of linked events known as **metabolic pathways**. Metabolic pathways produce key biological molecules in a cell, including important chemical building blocks like amino acids and nucleotides.

Two metabolic pathways drive most of the life around us. The sun is the ultimate source of energy for most living organisms, and in the first process, known as **photosynthesis**, organisms capture energy from the sun and use it to create sugars from carbon dioxide and water (**Figure 5.2**). In this way, photosynthetic organisms such as plants transform light energy into chemical energy stored in the covalent bonds of sugar molecules. These sugar molecules, for example, glucose, fuel the cell's activities, and some are converted to fatty acids to help build cell membranes and to store energy for future needs.

The second important process is **cellular respiration**, a process reciprocal to photosynthesis. During cellular respiration, the cell breaks down sugars into usable energy

## ANNETTE ROWE

Annette ("Annie") Rowe is a postdoctoral research associate at the University of Southern California, where she studies microbes that take up electrons from inorganic surfaces.

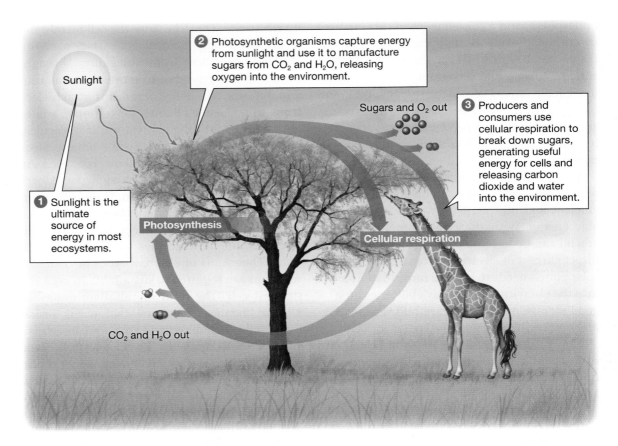

**2** Photosynthetic organisms capture energy from sunlight and use it to manufacture sugars from $CO_2$ and $H_2O$, releasing oxygen into the environment.

Sunlight

Sugars and $O_2$ out

**3** Producers and consumers use cellular respiration to break down sugars, generating useful energy for cells and releasing carbon dioxide and water into the environment.

**1** Sunlight is the ultimate source of energy in most ecosystems.

Photosynthesis

Cellular respiration

$CO_2$ and $H_2O$ out

Figure 5.2

**From sunlight to usable energy**

Photosynthesis transforms sunlight into sugar molecules within the cell, releasing oxygen as a by-product. Without photosynthesis, we would not have access to the sun's energy, and we wouldn't have oxygen in our atmosphere to support life. In a complementary process, cellular respiration breaks down these sugar molecules, allowing organisms to access the energy stored in them.

**Q1:** Why is photosynthesis called "primary production"?

**Q2:** How does animal life depend on photosynthesis?

**Q3:** Explain how photosynthesis and cellular respiration are "complementary" processes.

(**Figure 5.2**). Plants break down the glucose made during photosynthesis, whereas nonphotosynthetic organisms, including animals, eat plants or other animals and break down sugars in that food to release energy. Electrons are transferred back and forth during these two metabolic processes. Metabolism, at its core, is a feat of juggling electrons.

Rowe spent a week on Catalina setting up her tanks and then let them sit for 3 months, untouched. During that time, her instruments began to detect evidence of life: the negative charge flowing through the electrodes in the sand began to steadily increase. Something was stealing the electrons, consuming more and more electricity.

# An Unusual Pathway

Scientists in the 1980s first discovered two types of bacteria that expel—breathe, actually—electrons onto metal. Because it was assumed

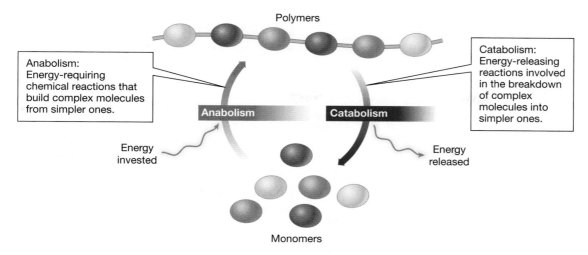

**Polymers**

Anabolism:
Energy-requiring
chemical reactions that
build complex molecules
from simpler ones.

Catabolism:
Energy-releasing
reactions involved
in the breakdown
of complex
molecules into
simpler ones.

**Anabolism**

**Catabolism**

Energy
invested

Energy
released

**Monomers**

Figure 5.3

## Anabolism builds biomolecules; catabolism breaks them down

Metabolic reactions either build or break down molecules. Molecule-building reactions (anabolism) cost energy, and molecule-breakdown reactions (catabolism) release energy.

**Q1:** What source of energy would plants use for anabolic reactions? Would an animal use the same kind of energy?

**Q2:** What source of energy would plants release in catabolic reactions? Would an animal release the same kind of energy?

**Q3:** Create a mnemonic or jingle that helps you remember the difference between anabolism and catabolism.

that electrons couldn't cross the cell membrane, the discovery was a big surprise. For decades, however, the way the electron transfer worked remained a mystery. Then, in 2006, a scientific team identified a trio of proteins that form a bridge in the bacterial cell membrane, allowing the transfer of electrons from the inside to the outside of the cell.

Inspired by these "rock breathers" that expelled electrons onto a mineral surface, scientists began looking for "rock eaters." If nature produced microbes that spit out electrons, why not microbes that directly ingest electrons from minerals?

All living cells have two main types of metabolism: anabolism and catabolism (**Figure 5.3**). **Anabolism** refers to metabolic pathways that create complex molecules from simpler compounds—a process used by all cells to make the basic building blocks of the cell. These processes often require an energy input. Photosynthesis is an example of anabolism where the energy is provided by sunlight: as plants sit in

the noonday sun, the cells in their green leaves capture energy to make glucose from carbon dioxide; a complex molecule is assembled from simpler ones. **Catabolism**, unsurprisingly, refers to the opposite: metabolic pathways that release chemical energy in the process of breaking down complex molecules. Plant cells rely on cellular respiration, a catabolic process, to break down the glucose made by photosynthesis.

All of this metabolic activity requires energy, so cells need **energy carriers** to deliver usable "on-demand" energy. Every living cell uses **ATP** (adenosine triphosphate), a small, energy-rich organic molecule, to carry energy from one part of the cell to another. ATP powers almost all activities in the cell, such as moving molecules and ions in and out of the cell, sending nerve impulses, triggering muscle contractions, and moving organelles around inside the cell. In addition to powering a cell's activities, ATP fuels metabolic reactions and most other enzymatic reactions. If a cell exhausts its ATP supply, it will die.

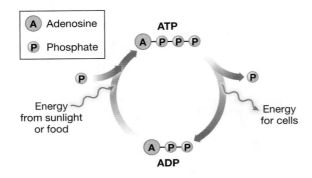

Figure 5.4

**Cells store and deliver energy using ATP**
Loading a high-energy phosphate (P) on ADP transforms the otherwise sedate molecule into the live wire that is ATP. But turning ADP and phosphate groups into ATP—the universal energy currency of cells—takes metabolic energy.

---

**Q1:** Define ATP in your own words.

**Q2:** How is ATP involved in anabolism and catabolism? (*Hint:* Review Figure 5.3.)

**Q3:** Arsenic disrupts ATP production. Why would this characteristic cause it to be a potent poison?

---

Much of the usable energy in ATP is held in its energy-rich phosphate bonds (**Figure 5.4**). Energy is released when a molecule of ATP loses its terminal phosphate group, breaking into a molecule of **ADP** (<u>a</u>denosine <u>di</u>phosphate) and a free phosphate. Converting ADP and phosphate back into ATP takes metabolic energy.

ATP is not the only energy carrier that cells rely on; NADPH, NADH, and $FADH_2$ are also energy carriers. Each one is a specialist in terms of the amount of energy it carries and the types of chemical reactions to which it supplies energy and from which it receives energy. ATP, because of its phosphate bonds, carries the most energy.

**ALFRED SPORMANN**

Alfred Spormann is a professor of civil and environmental engineering, and of chemical engineering, at Stanford University. He investigates novel microbial metabolism, including how microbes transport electrons between themselves and surfaces.

# Into the Light, Part 1

In 2009, environmental engineers at Pennsylvania State University identified the first elusive rock eaters by exposing a mix of microbes to a negatively charged electrode (**Figure 5.5**). One of the species, *Methanobacterium palustre*, survived on the electricity, taking in electrons and using them in a metabolic pathway to convert carbon dioxide to methane. One news article summed up the experiment with the headline BUG EATS ELECTRICITY, FARTS BIOGAS. The Penn State engineers proposed, but were unable to prove, that the bacteria ("bugs") were ingesting naked electrons straight from the surface of the electrode.

A few years later, Alfred Spormann launched his own investigation into bacteria that seemed to be "eating" electrons. Spormann, a microbiologist at Stanford University, has long been interested in outliers, microbes with unusual metabolisms. Tall and thin with a throaty German accent, Spormann has spent more than 25 years studying such microbes, from those in the human gut that contribute to irritable bowel syndrome to bacteria that could be used to help decontaminate groundwater. "I'm really interested in exceptions to the rules, and in understanding the plasticity of microbial metabolism," says Spormann. Typically, the microbes he studies are **anaerobic**, growing in places with little or no oxygen. These cells don't need oxygen

Figure 5.5

**A negatively charged electrode used to grow "rock eaters"**

for metabolism; in fact, some are even poisoned by it.

Oxygen plays an important role in photosynthesis, considered by many to be the most important life process on Earth; it is the way our planet stores energy from the sun and produces oxygen for animals to breathe. In the cells of algae and all plants, photosynthesis takes place inside chloroplasts, organelles that look like green, oval gumballs when viewed under a light microscope (**Figure 5.6**). Chloroplasts contain an extensive network of structures called thylakoids, piled up like stacks of pancakes, that contain enzymes needed for photosynthesis. Also embedded in those membranes is a green pigment, called **chlorophyll**, that is specialized for absorbing light energy.

Photosynthesis takes place in two principal stages: the *light reactions* and the *light-independent reactions*, or the *Calvin cycle* (**Figure 5.7**). During the **light reactions**, chlorophyll molecules absorb energy from sunlight and use that energy to split water (**Figure 5.7**, left). The splitting of water produces oxygen gas ($O_2$)—the oxygen that we breathe—as a by-product that is released into the atmosphere. More important for photosynthesis, electrons and protons ($H^+$) from the light reactions are handed over to other molecules via the **electron transport chain**, an elaborate chain of chemical events that ultimately generates ATP and NADPH. The light reactions depend on protein complexes embedded in the chloroplast membrane, including photosystems I and II, and ATP synthase.

In the next stage, the **light-independent reactions**, or **Calvin cycle**—a series of chemical reactions—convert carbon dioxide ($CO_2$) into sugar, using energy delivered by ATP, and electrons and hydrogen ions donated by NADPH (**Figure 5.7**, right). Enzymes catalyze these reactions at each step; the enzyme needed in the first step—and the most abundant enzyme on the planet—is **rubisco**. This process is also known as **carbon fixation**. By capturing inorganic carbon atoms from $CO_2$ gas and converting them into glucose, the Calvin cycle reactions make carbon from the nonliving world available to the photosynthetic organisms and eventually to other living organisms, including us.

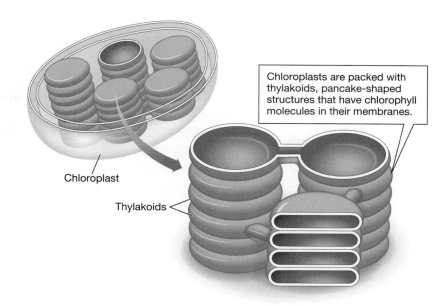

Chloroplasts are packed with thylakoids, pancake-shaped structures that have chlorophyll molecules in their membranes.

Chloroplast

Thylakoids

Figure 5.6

## Chloroplasts are the site of photosynthesis in eukaryotes

Most plants and algae have multiple chloroplasts to contain their chlorophyll. Photosynthetic bacteria embed chlorophyll directly into the plasma membrane.

**Q1:** Is chlorophyll found only within chloroplasts?

**Q2:** What could be an advantage of concentrating chlorophyll molecules in the membranes of chloroplasts?

**Q3:** What is the advantage of having multiple chloroplasts per cell?

# Catalyzing Reactions

In about 2014, Spormann's team began studying an "electric" microbe that lives in water, *Methanococcus maripaludis*, to see whether it was capable of directly ingesting electrons. It wasn't. The researchers found that the microbe actually excretes an enzyme onto the surface of an electrode to do its dirty work. Nearly all metabolic reactions are facilitated by enzymes. **Enzymes** are biological catalysts—molecules that speed up chemical reactions. Without the action of enzymes—most of which are proteins—metabolism would be extremely slow, and life as we know it could not exist. Enzymes work by positioning **substrates**—molecules that will react to form new products—in

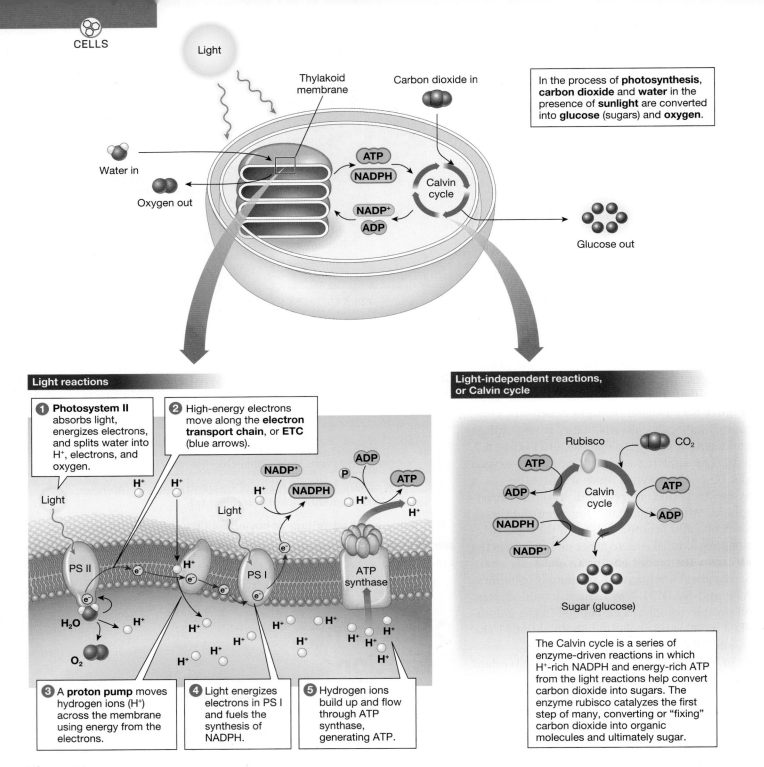

Light

Thylakoid membrane

Carbon dioxide in

In the process of **photosynthesis**, **carbon dioxide** and **water** in the presence of **sunlight** are converted into **glucose** (sugars) and **oxygen**.

Water in

Oxygen out

ATP
NADPH
NADP⁺
ADP

Calvin cycle

Glucose out

**Light reactions**

**Light-independent reactions, or Calvin cycle**

❶ **Photosystem II** absorbs light, energizes electrons, and splits water into H⁺, electrons, and oxygen.

❷ High-energy electrons move along the **electron transport chain**, or **ETC** (blue arrows).

Light

Light

ADP
P
ATP
NADP⁺
NADPH
H⁺
H⁺
H⁺
H⁺
H⁺

PS II
PS I
ATP synthase

e⁻

H₂O

O₂

H⁺ H⁺ H⁺ H⁺ H⁺ H⁺ H⁺ H⁺ H⁺ H⁺

❸ A **proton pump** moves hydrogen ions (H⁺) across the membrane using energy from the electrons.

❹ Light energizes electrons in PS I and fuels the synthesis of NADPH.

❺ Hydrogen ions build up and flow through ATP synthase, generating ATP.

Rubisco
CO₂

ATP
ADP
Calvin cycle
ATP
ADP
NADPH
NADP⁺

Sugar (glucose)

The Calvin cycle is a series of enzyme-driven reactions in which H⁺-rich NADPH and energy-rich ATP from the light reactions help convert carbon dioxide into sugars. The enzyme rubisco catalyzes the first step of many, converting or "fixing" carbon dioxide into organic molecules and ultimately sugar.

Figure 5.7

**Photosynthesis occurs in two stages**

The light reactions generate energy carriers; the light-independent Calvin cycle reactions create sugars. Rubisco contributes to fixing carbon during the Calvin cycle, but enzymes are critical throughout both stages of photosynthesis. ▶️

**Q1:** What is the source of the carbon dioxide used for photosynthesis?

**Q2:** Which products of the light reactions of photosynthesis does the Calvin cycle use?

**Q3:** What are the two major products of photosynthesis?

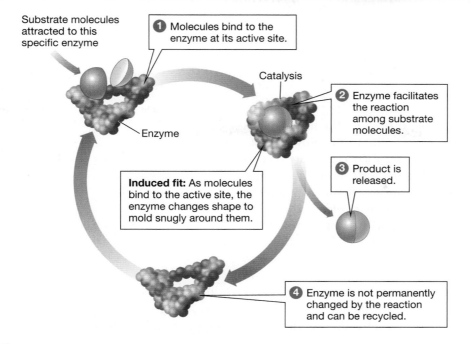

Substrate molecules attracted to this specific enzyme

**1** Molecules bind to the enzyme at its active site.

Catalysis

Enzyme

**2** Enzyme facilitates the reaction among substrate molecules.

**Induced fit:** As molecules bind to the active site, the enzyme changes shape to mold snugly around them.

**3** Product is released.

**4** Enzyme is not permanently changed by the reaction and can be recycled.

**Figure 5.8**

**Enzymes are molecular matchmakers**

Enzymes dramatically increase the rate of chemical reactions by positioning molecules so that they more easily form or break chemical bonds.

**Q1:** Why is it important that enzymes are not permanently altered when they bind with substrate molecules?

**Q2:** How would a higher temperature or higher salt concentration make it more difficult for an enzyme to function effectively?

**Q3:** If a cell was unable to produce a particular enzyme necessary for a metabolic pathway, describe how the absence of that enzyme would affect the cell.

an orientation that favors the making or breaking of chemical bonds (**Figure 5.8**).

Each enzyme binds only to a specific substrate or substrates and catalyzes a specific chemical reaction, such as rubisco catalyzing the first step of the Calvin cycle. An enzyme's function is based on its chemical characteristics and the three-dimensional shape of its **active site**—the location within the enzyme where substrates bind. When molecules bind to the active site, the enzyme changes shape—a process called **induced fit**. The enzyme's shape, and therefore its activity, can be affected by temperature, pH, and salt concentration. Because the enzyme's active site is not permanently changed as reactions are catalyzed, an enzyme, like rubisco, is used over and over.

The enzyme that Spormann identified was unusual because it was excreted to the exterior of the cell. Most enzymes work on the inside of cells. A multistep metabolic pathway can proceed rapidly and efficiently because the required enzymes are physically close together and the products of one enzyme-catalyzed chemical reaction serve as the basis for the next reaction in the series, as is especially true for metabolic pathway reactions.

In the case of *M. maripaludis*, the excreted enzyme's role was to grab an electron from the metal, pair it with a proton from water, and create a hydrogen atom—a familiar food for microbes, and easily passed through the cell membrane. "They [*M. maripaludis* bacteria] found a way to produce a compound that is easily metabolized by the cells," says Spormann. And although this particular microbe wasn't eating naked electrons, Spormann has since isolated a microbe that does directly take up electrons. His team has yet to publish the details.

# Into the Light, Part 2

As Spormann was unraveling the mechanisms of how electron-eating life works, Rowe's prospecting trip to Catalina uncovered more species of rock eaters (**Figure 5.9**). From her tanks, Rowe identified 30 new varieties of microbes sucking electrons from the electrodes. She was even able to grow a few of them on plates in the laboratory, using minerals like sulfur and iron as electron sources for the microbes to munch. Not all the species relied exclusively on electrons as a food source; that is, they didn't just live off mineral sediments. Like eukaryotic cells, most microbial cells use the traditional route of cellular respiration for obtaining energy, converting sugars into energy through catabolic reactions.

During cellular respiration, the carbon-carbon bonds in glucose molecules are broken,

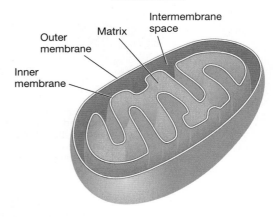

**Mitochondrion**

Figure 5.10

**Mitochondria are the site of cellular respiration in eukaryotes**

and each carbon atom is released into the environment in a molecule of $CO_2$, with water ($H_2O$) as a by-product. Thus, cellular respiration is the reciprocal process of photosynthesis, as illustrated earlier, in **Figure 5.2**. In most eukaryotes, cellular respiration is a multistep process that occurs in the cell's mitochondria (**Figure 5.10**). Cellular respiration is the main way that animal cells, which do not perform photosynthesis, obtain energy. We digest glucose from food sources; break it down into carbon dioxide, hydrogen, and electrons; and then process those captured electrons for energy.

Simply put, cellular respiration consists of three stages: glycolysis, the Krebs cycle, and oxidative phosphorylation (**Figure 5.11**). The first stage, **glycolysis**, takes place in the cytoplasm of the cell. During glycolysis, sugars (mainly glucose) are split to make a three-carbon compound called pyruvate. This process results in two useful molecules of ATP and two molecules of NADH for each glucose molecule that is split (**Figure 5.11**, top left). In other words, glycolysis converts some of the chemical energy of glucose into the chemical energy of NADH and ATP.

From an evolutionary standpoint, glycolysis was probably the earliest means of producing ATP from food molecules, and it is still the primary means of energy production in many prokaryotes, including some of Rowe's microbes. However, because glucose is only partially broken down through this process, the

Figure 5.9

**Sediment containing rock eaters, collected off the coast of Santa Catalina Island**

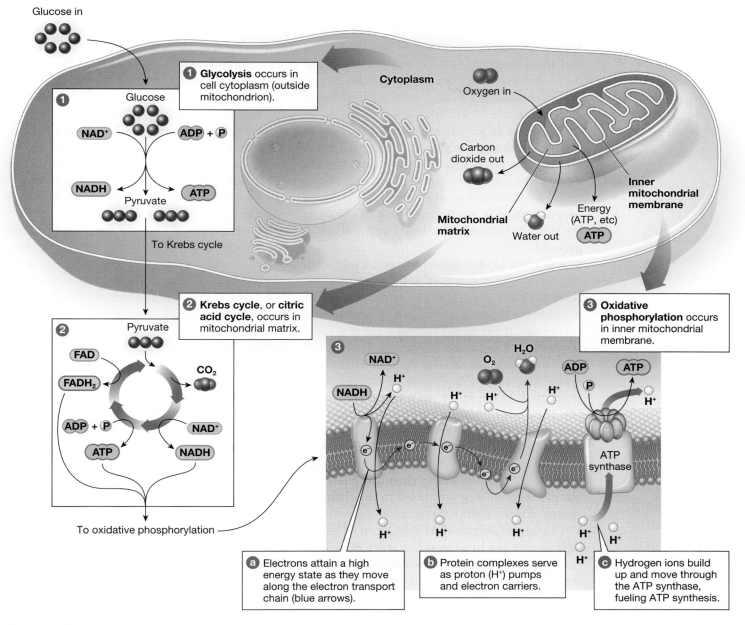

## Figure 5.11

### Cellular respiration is highly efficient in eukaryotes

In glycolysis, each six-carbon glucose molecule is converted into two molecules of pyruvate. The next two stages of cellular respiration require oxygen. The Krebs cycle (bottom left) releases carbon dioxide and generates high-energy molecules. Oxidative phosphorylation (bottom right), the last stage in cellular respiration, produces more ATP than any other metabolic pathway. ▣

**Q1:** What are the products of cellular respiration?

**Q2:** Considering the inputs and products of each process, why is cellular respiration considered the reciprocal process to photosynthesis?

**Q3:** Which of the three stages of cellular respiration—glycolysis, the Krebs cycle, or oxidative phosphorylation—could organisms have used 4 billion years ago, before photosynthesis by cyanobacteria released oxygen into the atmosphere?

energy yield from glycolysis is small. For most eukaryotes, glycolysis is just the first step in extracting energy from sugars, and the second and third stages of cellular respiration, which occur in the mitochondria, help to generate much more ATP than is possible through glycolysis alone.

Glycolysis is an anaerobic process; that is, it does not require oxygen. Cells experiencing low oxygen levels—such as our muscle cells during intense exercise or cancer cells in the interior of a tumor mass—use glycolysis alone because there is not enough oxygen available for the later stages of cellular respiration to proceed.

But when oxygen is present, eukaryotic cells rely on all three stages of cellular respiration. After pyruvate is made in the cytoplasm during glycolysis, it enters the mitochondria, where it is broken down during the second stage of cellular respiration: a sequence of enzyme-driven reactions known as the **Krebs cycle** or **citric acid cycle**. The carbon backbone of the pyruvate molecule is taken apart, releasing carbon dioxide (**Figure 5.11**, bottom left). The breakdown of carbon backbones by the Krebs cycle produces a large bounty of energy carriers, including ATP, NADH, and $FADH_2$. Essentially, during the Krebs cycle the remaining chemical energy of glucose is completely converted to the chemical energy of these energy carriers.

The largest output of ATP, however, is generated during the third and last stage of cellular respiration: **oxidative phosphorylation**. The electrons and hydrogen atoms are removed from NADH and $FADH_2$ and handed over to molecular $O_2$ through an electron transport chain, creating water ($H_2O$). In the process, a large amount of ATP is generated (**Figure 5.11**, bottom right). During oxidative phosphorylation, the chemical energy of NADH and $FADH_2$ is converted into the chemical energy of ATP. In fact, oxidative phosphorylation can generate 15 times as much ATP as can glycolysis alone.

ATP production in mitochondria is crucially dependent on oxygen; that is, the Krebs cycle and oxidative phosphorylation are strictly **aerobic** processes. Highly aerobic tissues, like your muscles, have high concentrations of mitochondria and a rich blood supply to deliver the large amounts of $O_2$ needed to support their activity. Even so, intense exercise exhausts this oxygen supply, at which point your cells switch to glycolysis, as mentioned already. But how do the cells extract sufficient energy via glycolysis alone?

Under anaerobic conditions, these muscle cells use a fermentation pathway. **Fermentation** begins with glycolysis, followed by a special set of reactions whose only role is to help perpetuate glycolysis. This process enables organisms to generate ATP through glycolysis alone, when aerobic ATP production is constrained by low oxygen levels (**Figure 5.12**). If glycolysis cannot be perpetuated, the cells will run out of ATP and die.

Many species of bacteria that live in places like oxygen-deficient swamps, sewage, or deep layers of soil never use oxygen or are actually poisoned by oxygen. Similar to rock-breathing microbes, many microbes can utilize other electron acceptors, such as iron or sulfur compounds. When these run out, anaerobic organisms use the fermentation pathway to regenerate molecules needed to perpetuate glycolysis, and exclusively generate ATP through glycolysis without ever using oxygen.

Most eukaryotic cells can perform both aerobic and anaerobic ATP production, depending on the oxygen levels in their environment. Because the Krebs cycle and oxidative phosphorylation produce magnitudes more ATP than glycolysis and the fermentation pathway yield, these cells always use oxygen and aerobic respiration when levels permit.

Together, photosynthesis and cellular respiration enable cells to store and utilize energy from the sun. They are two sides of the same coin; the products of one reaction are the ingredients for the other. Photosynthesis, an anabolic process, requires energy (sunlight) and $CO_2$, and releases $O_2$ and glucose. Cellular respiration, a catabolic process, requires $O_2$ and glucose, and releases $CO_2$ and energy.

Unlike the situation with photosynthesis and cellular respiration—metabolic processes that have been well studied for decades—researchers don't yet know the details of how rock eaters store and use energy. "There isn't really any pathway worked out yet," says Rowe. "Truthfully, it's amazing that it works." She's eager to find out how. Of the 30 species Rowe has isolated, she is currently focused on 8, working to develop tools to identify which enzymes are involved in their metabolism.

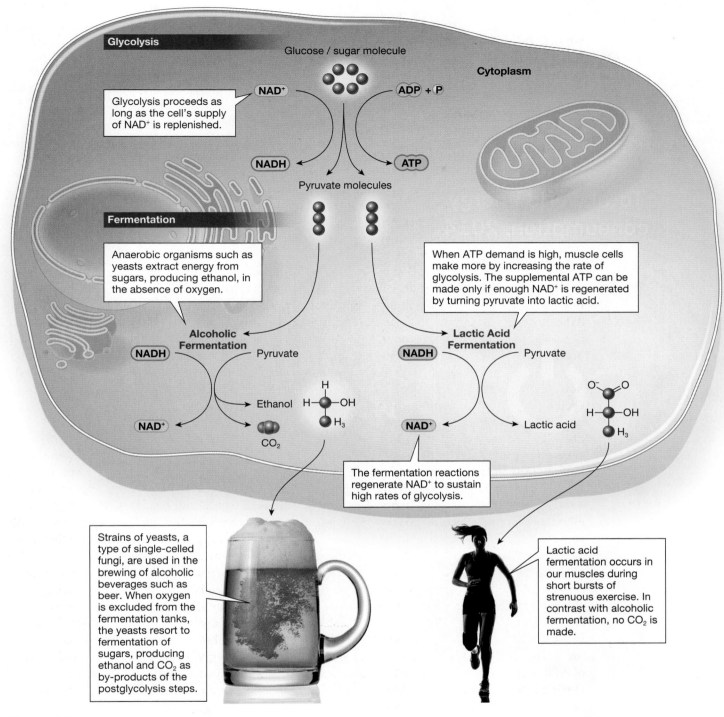

**Glycolysis**

Glucose / sugar molecule

Cytoplasm

NAD⁺

ADP + P

Glycolysis proceeds as long as the cell's supply of NAD⁺ is replenished.

NADH

ATP

Pyruvate molecules

**Fermentation**

Anaerobic organisms such as yeasts extract energy from sugars, producing ethanol, in the absence of oxygen.

When ATP demand is high, muscle cells make more by increasing the rate of glycolysis. The supplemental ATP can be made only if enough NAD⁺ is regenerated by turning pyruvate into lactic acid.

**Alcoholic Fermentation**

NADH

Pyruvate

Ethanol

NAD⁺

$CO_2$

H—C—OH
H₃

**Lactic Acid Fermentation**

NADH

Pyruvate

NAD⁺

Lactic acid

The fermentation reactions regenerate NAD⁺ to sustain high rates of glycolysis.

Strains of yeasts, a type of single-celled fungi, are used in the brewing of alcoholic beverages such as beer. When oxygen is excluded from the fermentation tanks, the yeasts resort to fermentation of sugars, producing ethanol and $CO_2$ as by-products of the postglycolysis steps.

Lactic acid fermentation occurs in our muscles during short bursts of strenuous exercise. In contrast with alcoholic fermentation, no $CO_2$ is made.

**Figure 5.12**

**Fermentation produces energy in the absence of oxygen**

When the oxygen supply is inadequate to support ATP production through cellular respiration, fermentation enables ATP production through glycolysis alone.

**Q1:** Which product released by fermentation accounts for the bubbles in beer?

**Q2:** Bakers of yeast breads also rely on fermentation, allowing bread to "rise" before baking. Describe what is occurring with the yeast as the bread rises.

**Q3:** Explain in your own words why lactic acid builds up in your muscles during strenuous physical activity.

# Making Way for Renewables

Microbes that "eat" electricity may someday provide an efficient way to store renewable energy. That's good, because we need such storage technologies as soon as possible. Technologies that produce renewable energy are no longer "alternative"—they're mainstream and compete financially with fossil fuel producers. From solar power to biofuels, the fuel economy is rapidly changing toward more sustainable solutions.

Assessment available in smartwork5

## Total world energy consumption, 2014

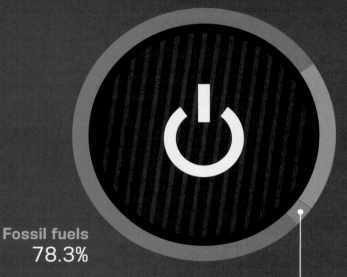

Total renewables
19.2%

Fossil fuels
78.3%

Nuclear power
2.5%

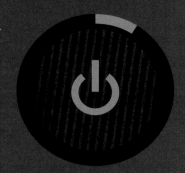

Traditional biomass
(wood and agricultural fuels)
8.9%

Modern renewables
10.3%

Renewable heat energy from biomass, geothermal, and solar technologies
4.2%

Hydropower
3.9%

Renewable power from wind, biomass, geothermal, and solar technologies
1.4%

Biofuels
0.8%

# Bacterial Batteries

Because cell membranes are insulated, researchers long believed that cells could not transport electrons directly across their membranes, but now it appears they can. In 2010, in fact, Rowe's post-doctoral adviser, Mohamed El-Naggar, a pioneer in the rock breather field, found that the tiny hairlike structures called pili (singular "pilus") on some bacteria are conductive (**Figure 5.13**)—just as conductive, in fact, as silicon, the basis of most electronics. One single nano-sized pilus, which researchers refer to as a bacterial "nanowire," can transport about 106 electrons per second—enough to sustain the respiration of a whole cell.

Today, Rowe, Spormann, and others are focusing on how to use these unique bacteria in "bacterial batteries." Energy absorbed during the day from a solar panel currently has to be either instantly used or siphoned off into the electrical grid. But what if it could be passed into a tank of bacteria instead, where the microbes would gobble up the electricity and store it in chemical bonds in the form of methane or another natural gas? In that case the gas could be stored until needed.

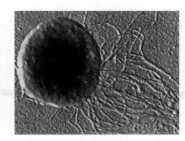

Figure 5.13

**The rock eater *Methanococcus maripaludis*, with electrically conductive hairlike pili**

"If you could convert electrical energy, through microbial processes, to a liquid fuel, that could be a really valuable way of storing energy," says Rowe. This electricity-to-fuel scenario could also be run in reverse, using microbes to digest a fuel source such as raw sewage and convert it into electricity.

There's also the intriguing scenario that someday our electronic devices will be powered by microorganisms that manipulate the flow of electrons. In this rapidly evolving field, the possibilities are endless. And so the hunt for rock eaters continues. "We don't know how many organisms do or don't do it," adds Spormann. "It's entirely possible that rock eating is more widespread than previously thought. We just don't know because it's so new."

# REVIEWING THE SCIENCE

- The sun is the source of energy fueling most living organisms. Plants, algae, and some bacteria gain energy from their environment through **photosynthesis**. Most organisms use **cellular respiration** to extract usable energy from sugar molecules. In chemical terms, photosynthesis is the opposite of cellular respiration.

- All the many chemical reactions involved in the capture, storage, and use of energy by living organisms are collectively known as **metabolism**. Energy-releasing breakdown reactions like cellular respiration are **catabolism**; energy-requiring synthesis reactions like photosynthesis are **anabolism**.

- A **metabolic pathway** is a multistep sequence of chemical reactions, with each step catalyzed by a different enzyme.

- **Energy carriers** store energy and deliver it for cellular activities. **ATP** is found in all cells and is the most commonly used energy carrier.

- Photosynthesis takes place in chloroplasts and occurs in two stages. In the **light reactions**, energy is absorbed using pigment molecules that include

**chlorophyll** as electrons flow along the **electron transport chain**. The light reactions create the energy carriers ATP and NADPH, splitting water molecules and releasing oxygen gas.

- The energy carriers are then used to convert carbon dioxide into sugar molecules during the **light-independent reactions**, or **Calvin cycle**. In the first of the reactions, the enzyme **rubisco** catalyzes the fixation of $CO_2$.

- **Enzymes** are biological catalysts, usually small proteins, that speed up chemical reactions. An enzyme's function is based on its chemical characteristics and the three-dimensional shape of its **active site**.

- Cellular respiration occurs in three stages: (1) **glycolysis**, which yields small amounts of ATP and NADH; (2) the **Krebs cycle**, which releases carbon dioxide and produces NADH, $FADH_2$, and ATP; and (3) **oxidative phosphorylation**, which generates many molecules of ATP.

- In the absence of oxygen, **fermentation** breaks down the products of glycolysis into alcohol or lactic acid.

# THE QUESTIONS

## The Basics

**1** Metabolic pathways

(a) always break down large molecules into smaller units.

(b) only link smaller molecules together to create polymers.

(c) are often organized as a multistep sequence of reactions.

(d) occur only in mitochondria.

**2** Enzymes

(a) speed up reactions that would otherwise occur much more slowly.

(b) spur reactions that would otherwise never occur.

(c) provide energy for anabolic but not catabolic pathways.

(d) are consumed during the reactions they catalyze.

**3** The most common energy-carrying molecule in all organisms is

(a) carbon dioxide.

(b) water.

(c) ATP.

(d) rubisco.

**4** The major product of photosynthesis is

(a) lipids.

(b) sugar.

(c) amino acids.

(d) nucleotides.

**5** Which of the following statements is *not* true?

(a) Glycolysis is the first stage of cellular respiration.

(b) Glycolysis can proceed under low oxygen levels with the assistance of fermentation.

(c) Glycolysis produces less ATP than does either the Krebs cycle or oxidative phosphorylation.

(d) Glycolysis produces most of the ATP required by aerobic organisms like us.

**6** Select the correct terms:

Photosynthesis and cellular respiration are chemically (**identical / opposite**) processes. Cellular respiration is an example of (**anabolism / catabolism**), which (**produces / expends**) energy.

**7** In the diagram of photosynthesis shown here, fill in each blank with the appropriate term: (a) Sunlight; (b) $CO_2$; (c) Oxygen; (d) $H_2O$; (e) ATP and NADPH; (f) ADP and $NADP^+$; (g) Sugar; (h) Light reactions; (i) Calvin cycle.

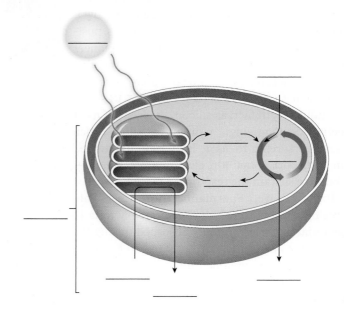

**8** Place the following steps of cellular respiration in the correct order by numbering them from 1 to 4.

_____ a. The Krebs cycle produces the energy carriers NADH, $FADH_2$, and ATP.

_____ b. If oxygen levels are adequate, pyruvate is transported into the mitochondrion. If oxygen levels are very low, fermentation proceeds.

_____ c. Glucose is broken down to produce ATP and NADH.

_____ d. An electron transport chain produces ATP from ADP.

## Challenge Yourself

**9** The Calvin cycle reactions are sometimes called the "light-independent reactions" or "dark reactions" to contrast them with the light (or light-dependent) reactions. Can the Calvin cycle be sustained in algae that are kept in total darkness for several days? Why or why not?

**10** After strenuous exercise, you may notice that your muscles burn and feel sore the next day. Which statement best explains this phenomenon?

(a) Proteins in muscle cells are being digested to provide energy.

(b) Carbon dioxide is building up in muscle cells and changing their pH.

(c) Spontaneous combustion occurs during strenuous exercise, so avoid it at all costs!

(d) ATP is accumulating in muscle cells, causing a burning sensation.

(e) Without adequate oxygen, muscle cells are fermenting pyruvate into lactic acid.

**11** The graph presented here depicts the activation energy, or the amount of energy needed for a reaction to proceed, with and without an enzyme.

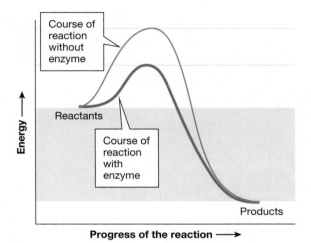

a. Which reaction requires more energy to proceed—the one with or without an enzyme? How do you know?

b. Is this reaction anabolic or catabolic? How do you know?

## Try Something New

**12** In 2012, an Illinois man was killed by cyanide poisoning after he won a million dollars in the lottery. Cyanide is a lethal poison because it interferes with the electron transport chain in mitochondria. What effect would cyanide have on cellular respiration?

(a) Glycolysis, the Krebs cycle, and oxidative phosphorylation would all be inhibited.

(b) The Krebs cycle would be inhibited, but oxidative phosphorylation would not.

(c) Oxidative phosphorylation would be inhibited.

(d) Glycolysis, the Krebs cycle, and oxidative phosphorylation would all be stimulated.

**13** Plants in the genus *Ephedra* have been harvested for their active substance ephedrine for centuries. Ephedrine is used to reduce the symptoms of bronchitis and asthma, as a stimulant and study aid, and as an appetite suppressant. It is also the main ingredient in the illegal production of methamphetamine. Since 2006, the sale of ephedrine and related substances has been limited and monitored in the United States. One effect of ingesting ephedrine is greatly increased metabolism, which has been known to kill users of ephedrine. How might an increased metabolic rate cause death?

**14** You have been transported into the future, where a nanosensor can be inserted into a living cell and subsequently travel into any organelle. The sensor relays information stating that it has lodged itself in a compartment of the mitochondrion, where there is a concentration gradient of hydrogen ions ($H^+$) and levels of ATP production are high. From what you know about cellular respiration in mitochondria, where is the sensor lodged?

(a) inner mitochondrial membrane

(b) mitochondrial matrix

(c) chloroplast stroma

(d) cell cytoplasm

(e) plasma membrane

## Leveling Up

**15** **What do *you* think?** What would happen if a virus destroyed all photosynthetic organisms on Earth?

**16** ***Write Now* biology: calorie-burning fat** Your friend has e-mailed you a link (see below) to a *New York Times* article on "brown" fat. He has been trying to lose weight and wants to know whether you think it would be a good idea for him to spend more time in the cold, rather than continuing to exercise regularly. He is also interested that the article mentions ephedrine's ability to stimulate brown fat, and he asks if you think he should begin to take ephedrine supplements. Compose an reply to your friend addressing the following points (using one to two paragraphs for each one). [*Note:* You may need to do further reading to answer (b) and (c).]

a. Explain in detail how brown fat burns calories when someone is chilled.

b. Explain how ephedrine affects metabolism and what its possible side effects are.

c. Contrast (a) and (b) with the effect of exercise on metabolism, both in the short term and in the longer term by increasing muscle mass.

d. In your final paragraph, advise your friend as to whether he should begin spending time in the cold to increase weight loss or take ephedrine supplements, and whether he should continue to exercise regularly. Justify your opinion with data and logic.

To research your answer, begin with the following *New York Times* article, published on April 8, 2009: "Calorie-Burning Fat? Studies Say You Have It" (http://www.nytimes.com/2009/04/09/health/research/09fat.html). Consult and reference at least two additional sources in your answer.

**17** **Is it science?** Wouldn't it be amazing if humans could manufacture their own food through photosynthesis like plants do? Use Internet research to determine whether this idea is real science or science fiction. Is there evidence that animals can use the sun's energy directly and/or perform photosynthesis in nature? In the lab? If so, how would that capability affect life as we know it?

**For more, visit digital.wwnorton.com/bionow2 for access to:**

# Toxic Plastic

*Two ruined experiments expose the health risks of chemicals in everyday products.*

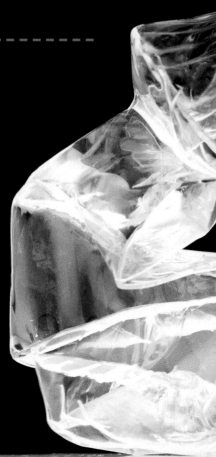

## After reading this chapter you should be able to:

- Label a figure of the major stages of the cell cycle, and explain the processes that occur during each of these stages.

- Compare and contrast cell division by binary fission, mitosis, and meiosis.

- Distinguish between sister chromatids and homologous chromosomes.

- Diagram, using the appropriate terms, the steps in mitosis and in meiosis.

- Explain the importance of the checkpoints in the cell cycle and the consequences of bypassing those checkpoints.

- Identify the ways in which meiosis and fertilization together produce genetically diverse offspring.

CHAPTER

06

CELL DIVISION

I t began as a run-of-the-mill experiment. In 1989, biologists Ana Soto and Carlos Sonnenschein at Tufts University in Massachusetts were studying how the hormone estrogen regulates the growth of cells in the female reproductive system. For their research, the duo developed an experimental setup consisting of human breast tumor cells growing in plastic bottles called cell culture flasks (**Figure 6.1**, top). The flasks were filled with a liquid containing a factor that prevented the cells from multiplying. But when estrogen was added to the flasks, the cells grew. When estrogen was absent, they didn't.

One day, suddenly and surprisingly, cells in the flasks began growing even when estrogen hadn't been added. "What had worked for years didn't work anymore," says Sonnenschein. The two scientists immediately stopped their experiments and began searching for the cause. "It smacked of contamination," recalls Soto, as if estrogen had somehow gotten into the flasks. But after weeks of searching, Soto and Sonnenschein still couldn't identify a source of contamination. They became so paranoid that they suspected someone was entering the lab at night and secretly dripping estrogen into their flasks.

Almost 10 years later, in August 1998, geneticist Patricia Hunt at Case Western Reserve University in Ohio stared dumbfounded at another experimental anomaly. Hunt was studying why older women are at increased risk of having children with chromosomal abnormalities, like Down syndrome, in which an individual has 47 **chromosomes**—the tiny, stringlike structures in cells that contain genes—instead of the usual 46 (see Chapter 8 for more on chromosomes). She hypothesized that hormone levels have an impact on that increased risk. To test her hypothesis, Hunt

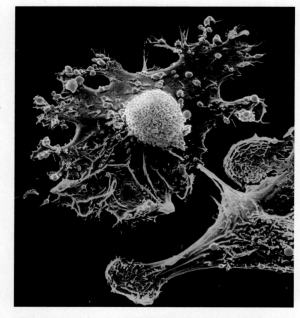

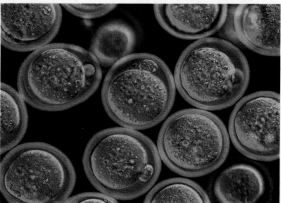

Figure 6.1

**Two research models, two unusual results**

Breast tumor cells (top) and mouse oocytes (bottom) grew in unexpected ways, leading scientists to study them more deeply.

raised groups of mice with varying levels of hormones and checked their egg cells for abnormal numbers of chromosomes (**Figure 6.1**, bottom).

The experiment was almost complete when Hunt went in to check on the control mice one last time. A control population is a necessary baseline for comparison against an experimental population; in this case, the control was a group of healthy mice whose hormone levels had not been altered. Using a light microscope, Hunt examined mouse oocytes—precursors to egg cells—at the moment just before the cells

**ANA SOTO AND CARLOS SONNENSCHEIN**

Ana Soto is a biologist at Tufts University who studies how cell division is affected by sex steroids. Carlos Sonnenschein, also a professor at Tufts University, studies chemicals that disrupt hormone systems in mammals.

undergo a specialized type of cell division that produces the eggs. She was shocked. The cells were a mess, the chromosomes scrambled. A whopping 40 percent of the resulting eggs had chromosomal defects. "The controls were completely bonkers," says Hunt. "One week they were fine, the next week they weren't. That's when we knew something was going on."

Like Soto and Sonnenschein, Hunt scrutinized every method and every piece of lab equipment used in the experiment, looking for the culprit. But as weeks passed, she couldn't figure out what had ruined her experiment. Soto, Sonnenschein, and Hunt didn't know it at the time, but their botched experiments would change the course of their scientific careers forever. The three would spend the next decade identifying, tracking, and investigating a toxic chemical that pervades our environment.

# Divide and Conquer

In Soto and Sonnenschein's experiment, the breast cells were multiplying under the wrong circumstances. In Hunt's experiment, the mouse oocytes were not producing egg cells correctly. In both cases, something was interfering with the ability of the cells to divide—disrupting the cell cycle. The **cell cycle** is a sequence of events that make up the life of a typical eukaryotic cell, from the moment of its origin to the time it divides to produce two daughter cells. The time it takes to complete a cell cycle depends on the organism, the type of cell, and the life stage of the organism. Human cells, for example, typically have a 24-hour cell cycle, while Hunt's mouse oocytes can take days to complete a cycle. Some fly embryos, on the other hand, have cell cycles that are only 8 minutes long.

There are two main stages in the cell cycle of eukaryotes: *interphase* and *cell division*—each marked by distinctive cell activities (**Figure 6.2**). **Interphase** is the longest stage of the cell cycle; most cells spend 90 percent or more of their life span in interphase. During this phase, the cell takes in nutrients, manufactures proteins and other substances, expands in size, and conducts special functions depending on the cell type. Neurons in the brain, for example,

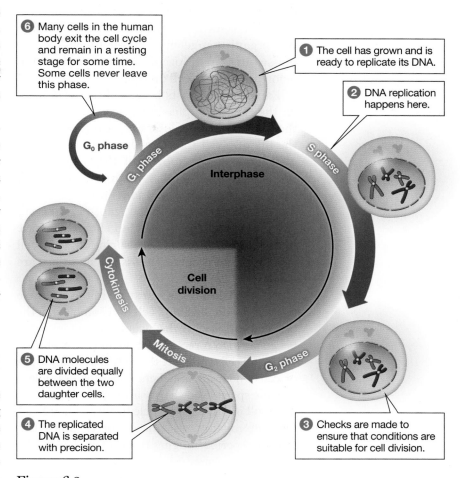

6 Many cells in the human body exit the cell cycle and remain in a resting stage for some time. Some cells never leave this phase.

$G_0$ phase

1 The cell has grown and is ready to replicate its DNA.

2 DNA replication happens here.

$G_1$ phase

S phase

Interphase

Cell division

Cytokinesis

Mitosis

$G_2$ phase

5 DNA molecules are divided equally between the two daughter cells.

4 The replicated DNA is separated with precision.

3 Checks are made to ensure that conditions are suitable for cell division.

Figure 6.2

## The cell cycle

The eukaryotic cell cycle consists of two major stages: interphase and cell division. ▶▦

Q1:   When is DNA replicated during the cell cycle?

Q2:   When in the cell cycle does DNA separate into the two genetically identical daughter cells?

Q3:   If a cell is not destined to separate into daughter cells, what phase does it enter? Is this part of the cell cycle?

transmit electrical impulses, while beta cells in the pancreas release insulin.

Interphase can be divided into three main intervals: $G_1$, S, and $G_2$. The **$G_1$ phase** (for "gap 1") is the first phase in the life of a newborn cell. In cells that are destined to divide, preparations for cell division begin during the **S phase**

("S" stands for "synthesis"). A critical event during the S phase is the copying, or *replication*, of all the cell's DNA molecules, which contain the organism's genetic information. The $G_2$ **phase** (for "gap 2") occurs after the S phase but before the start of cell division.

Early cell biologists bestowed the term "gap" on the $G_1$ and $G_2$ phases because they believed those phases to be less significant periods in the life of a cell than are the S phase and cell division. We now know that the "gap" phases are often periods of growth during which both the size of the cell and its protein content increase. Furthermore, each "gap" phase serves as a checkpoint to prepare the cell for the phase immediately following it, ensuring that the cell cycle does not progress unless all conditions are suitable.

**Cell division** is the last stage in the cell cycle of an individual cell. As cell division begins, the cell contains twice the usual amount of DNA because of DNA replication during the S phase.

Not all cells complete the cell cycle. Many types of cells—neurons and beta cells, for example—become specialized shortly after entering $G_1$, and they pull out of the cell cycle to enter a nondividing state called the $G_0$ phase. The $G_0$ **phase** can last for a period ranging from a few days to the lifetime of the organism.

Cells begin cell division for two basic reasons: (1) to reproduce and (2) to grow and repair a multicellular organism. Most single-celled organisms use cell division to produce offspring through **asexual reproduction**. Asexual reproduction generates *clones*, offspring that are genetically identical to the parent.

Cell division occurs in all living organisms (eukaryotes and prokaryotes) and involves copying the DNA in the parent cell, then delivering one copy of it to each of two daughter cells. Most prokaryotes carry their genetic material in just one loop of DNA. These cells reproduce through **binary fission**, a type of cell division in which a

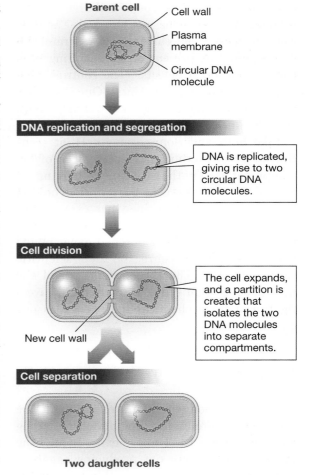

**Parent cell** — Cell wall
— Plasma membrane
— Circular DNA molecule

**DNA replication and segregation**

DNA is replicated, giving rise to two circular DNA molecules.

**Cell division**

New cell wall

The cell expands, and a partition is created that isolates the two DNA molecules into separate compartments.

**Cell separation**

**Two daughter cells**

Figure 6.3

### Cell division in a prokaryote

Many prokaryotes reproduce asexually in a type of cell division known as binary fission.

**Q1:** Name one similarity between cell division in prokaryotes and eukaryotes.

**Q2:** Why is binary fission referred to as "asexual reproduction"?

**Q3:** Name one difference between cell division in prokaryotes and eukaryotes.

### PATRICIA HUNT

Patricia Hunt is a reproductive biologist at Washington State University in Pullman, Washington, who studies meiosis in mammalian cells.

cell simply copies the circular chromosome and each daughter cell receives one copy of the DNA loop—resulting in two cells that are genetically identical (**Figure 6.3**).

Soto, Sonnenschein, and Hunt, however, were studying eukaryotic cells, where cell division is more complicated than binary fission. Eukaryotic DNA forms multiple, distinct linear chromosomes wrapped around proteins and coiled into fibers that have to be unwound, replicated, and equally distributed between the two daughter cells. In addition to the multiple-chromosome problem, eukaryotic DNA lies inside a nucleus, enclosed by a double layer of membranes that make up the nuclear envelope. In most eukaryotes, the nuclear envelope must be disassembled in the dividing cell and then reassembled in each of the daughter cells toward the end of cell division. To make things even more complicated, eukaryotic cells undergo two types of division, depending on cell type: asexual reproduction through *mitosis*, and sexual reproduction—the production of sperm and eggs—through *meiosis*.

# Trade Secret

Back in Massachusetts, Soto and Sonnenschein spent 4 months trying to figure out why their experiment had stopped working—how unknown estrogen was getting into their cell culture flasks and causing the cells to divide. By trial and error, they determined that a compound seemed to be shedding from the walls of the plastic tubes in which they stored the liquid being added to the cell culture flasks.

They called the tube manufacturer, who confirmed that an ingredient was added to make the tubes more impact resistant. But the company refused to reveal the identity of the "trade secret" ingredient. Soto and Sonnenschein spent a year purifying the secret ingredient and finally identified a compound called nonylphenol, a chemical used to make detergents and hard plastics. The reason for all their problems became clear: nonylphenol mimics the action of estrogen.

Like estrogen, nonylphenol activates **mitotic division**, a type of cell division that generates two genetically identical daughter cells from a single parent cell in eukaryotes. Mitotic division consists of two steps that occur after interphase: *mitosis* and *cytokinesis*. The first step, **mitosis**, refers to the division of the copied chromosomes in the nucleus. Mitosis is divided into four main

## Cancer: Uncontrolled Cell Division

Cancer accounts for almost 600,000 deaths in the United States each year—nearly one in every four deaths. Only heart disease kills more people. Over the course of a lifetime, an American male has a nearly one in two chance of being diagnosed with cancer; American women fare slightly better, with a one in three chance of developing cancer. There are more than 200 different types of cancer, but the big four—lung, prostate, breast, and colon cancers—combine to account for about half of all cancers. More than 15 million Americans alive today have been diagnosed with cancer and are either in remission or undergoing treatment. The National Cancer Institute estimates that the collective price tag for the various forms of cancer is more than $100 billion per year.

Every cancer begins with a single rogue cell that starts dividing without the checkpoints of a normal cell. This runaway cell division rapidly creates a cell mass known as a **tumor**. Tumors that remain confined to one site are **benign**. Because benign tumors can usually be surgically removed, they are generally not a threat to the patient's survival. However, an actively growing benign tumor is like a cancer-in-training. Because these tumor cells are not subject to the monitoring that occurs at checkpoints during the cell cycle of a normal cell, their descendants can become increasingly abnormal—changing shape, increasing in size, and ultimately ceasing normal cell functions. As tumor cells progress toward a cancerous state, they begin secreting substances that cause **angiogenesis**, the formation of new blood vessels. The resulting increase in blood supply to the tumor is important for delivering nutrients to it and whisking waste away from it, allowing the tumor to grow larger.

Most cells in the adult animal body are firmly anchored in one place and will stop dividing if they are detached from their surroundings—a phenomenon known as **anchorage dependence**. But some tumor cells may acquire anchorage *in*dependence, the ability to divide even when released from their attachment sites. When tumor cells gain anchorage independence and start invading other tissues, they are transformed into **cancer cells**, also known as **malignant cells**. Cancer cells may break loose from their attachment sites and enter blood or lymph vessels to emerge in distant locations throughout the body, where they form new tumors. The spread of a disease from one organ to another is known as **metastasis**. Metastasis typically occurs at later stages in cancer development. Once a cancer has metastasized to form tumors in multiple organs, it may be very difficult to fight.

Cancer cells multiply rapidly wherever they establish themselves, overrunning neighboring cells, monopolizing oxygen and nutrients, and starving normal cells in the vicinity. Without restraints on their growth and migration, cancer cells steadily destroy tissues, organs, and organ systems. The normal function of these organs is then seriously impaired, and cancer deaths are ultimately caused by the failure of vital organs.

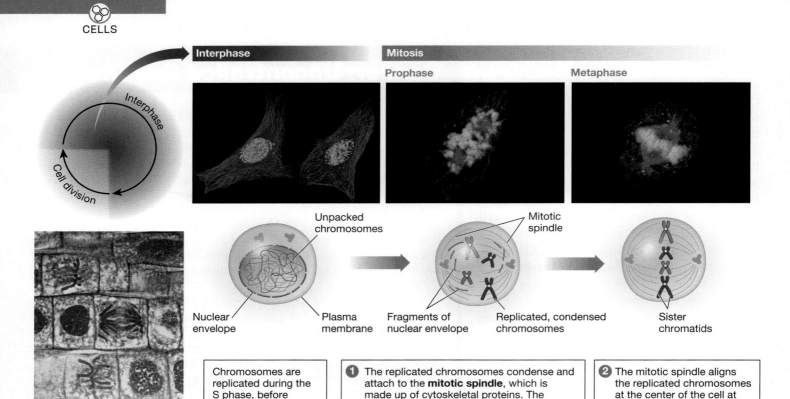

**Figure 6.4**

## Cell division in a eukaryote

Mitotic cell division is composed of two main stages: mitosis (with four substages) and cytokinesis. ▶

**Q1:** Do all cells in an organism enter each stage of mitosis at the same time? (*Hint:* See image of onion root tip at far left in the figure.)

**Q2:** What happens between the end of interphase and early prophase that changes the appearance of the chromosomes?

**Q3:** Explain in your own words the role of the mitotic spindle in mitosis.

phases: *prophase, metaphase, anaphase,* and *telophase.* Each phase is defined by easily identifiable events (**Figure 6.4**).

A parent cell sets up for an upcoming mitotic division by replicating its DNA during the S phase of interphase, well before mitosis gets under way. Note that DNA in the nucleus is not tightly packaged during the gap and synthesis phases. This is because the DNA must be accessible for replication and for conducting the business of the cell. Then, as cell division begins, each long, double-stranded DNA molecule is attached to proteins that help pack it for cell division into a chromosome. This packing is

necessary because every DNA molecule is enormously long, even in the simplest eukaryotic cells. When a chromosome is replicated, two identical DNA molecules called **sister chromatids** are produced. These sister chromatids are firmly attached at a central region of the chromosome called the **centromere** and do not separate until metaphase (**Figure 6.5**).

One of the main objectives of mitosis is to separate those sister chromatids, pulling them apart at the centromere and delivering one of each to the opposite ends of the parent cell. Eukaryotic cells have evolved an elaborate choreography to minimize the risk of mistakes during

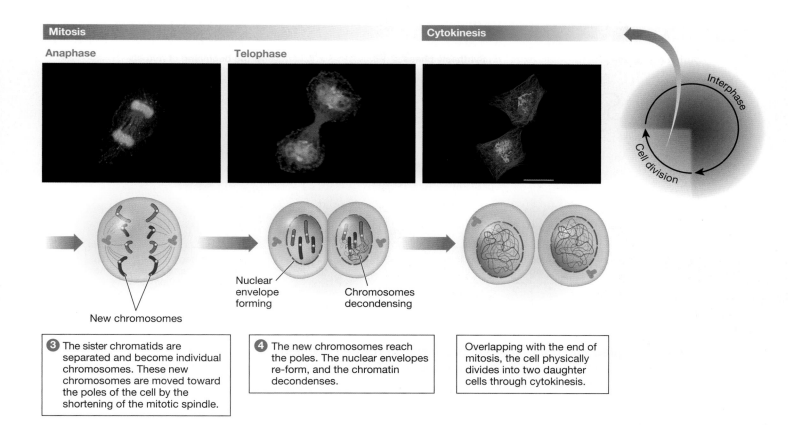

**Mitosis**

Anaphase

Telophase

**Cytokinesis**

Interphase

Cell division

Nuclear envelope forming

Chromosomes decondensing

New chromosomes

**3** The sister chromatids are separated and become individual chromosomes. These new chromosomes are moved toward the poles of the cell by the shortening of the mitotic spindle.

**4** The new chromosomes reach the poles. The nuclear envelopes re-form, and the chromatin decondenses.

Overlapping with the end of mitosis, the cell physically divides into two daughter cells through cytokinesis.

the equal and symmetrical partitioning of the replicated genetic material. Normally, no daughter cell winds up short a chromosome, nor does it acquire duplicates. Unless an error occurs, each daughter cell inherits the same genetic information that the parent cell possessed in the $G_1$ phase of its life.

After the replicated DNA has been divided in two, half to each end of the parent cell, the cytoplasm is divided by a process called **cytokinesis** ("cell movement"), like pulling apart a ball of Silly Putty into two halves. Cytokinesis gives rise to two self-contained daughter cells that are clones of each other.

Mitotic division can serve both the eukaryotic organism's need to replace itself (to reproduce) and its need to add new cells to its body. Many multicellular eukaryotes use mitotic division to reproduce asexually, including seaweeds, fungi, and plants, and some animals, such as sponges and flatworms. All multicellular organisms also rely on mitotic division for the growth of tissues and organs and the body as a whole, and for repairing injured tissue and replacing worn-out cells. Mitosis is why children grow taller and why skin closes over a cut.

# Good Cells Gone Bad

Cell division is not always a good thing. Runaway cell division can create a tumor (see "Cancer: Uncontrolled Cell Division" on page 101). In a developing organism, it can also cause an organ such as the heart or liver to form incorrectly and not function properly. It is little wonder, then, that the cell cycle is carefully controlled in healthy individuals. The decision to divide a cell is made during the $G_1$ phase of the cell cycle in response to internal and external signals. In humans, external signals that influence the commitment to divide include hormones and proteins called growth factors. Some hormones and growth factors act like the gas pedal in a car and push a cell toward cell division; others act like a brake and prevent cell division.

After a cell enters the cell cycle, special *cell cycle regulatory proteins* are activated. These proteins "throw the switch" that enables the cell to pass through critical checkpoints and proceed from one phase of the cell cycle to the

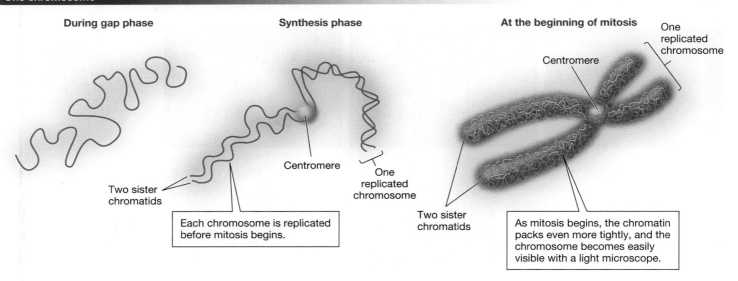

Figure 6.5

**Chromosomes are copied and condensed in preparation for cell division**
Chromosomes spend the majority of the cell cycle unpackaged (left). They are copied (replicated) during synthesis, then tightly packaged, or condensed, during early mitosis.

**Q1:** Why is it important for a chromosome to be copied before mitosis?

**Q2:** Are sister chromatids attached at the centromere considered to be one or two chromosomes?

**Q3:** Why is the chromosome's DNA tightly packed for mitosis and cytokinesis? (*Hint:* Think about what would happen if it were unpackaged, as during interphase.)

next (**Figure 6.6**). For example, upon receiving the appropriate signals, cell cycle regulatory proteins advance a cell from the $G_1$ phase to the S phase by triggering chromosome replication and other processes associated with it.

Cell cycle regulatory proteins also respond to negative internal or external control signals. Internal signals will pause a cell in the $G_1$ phase under any of the following conditions: the cell is too small, the nutrient supply is inadequate, or the cell's DNA is damaged. $G_2$ pauses in the same circumstances, as well as when chromosome duplication in the S phase is incomplete for any reason.

Nonylphenol interferes with the $G_0$ and $G_1$ checkpoints. In essence, it gives the cell a green light to enter the cell cycle at a time when the cell would not normally divide. So, when Soto and Sonnenschein realized that nonylphenol enables human breast cells and rat uterine cells to divide

inappropriately, they became concerned. If nonylphenol was being used in everyday plastics, it was possible that healthy human cells were exposed to it on a regular basis.

In a 1991 paper detailing their discovery, Soto and Sonnenschein wrote that nonylphenol might be interfering with science experiments like theirs and that, even more important, it could be harmful to humans. "From the very beginning, we realized that this could be a health problem," says Sonnenschein.

# Unequal Division

In Ohio, Patricia Hunt was having no success determining why her control mouse eggs divided abnormally, with either too many or too few chromosomes. After searching in vain for months, one day she noticed that something was wrong with

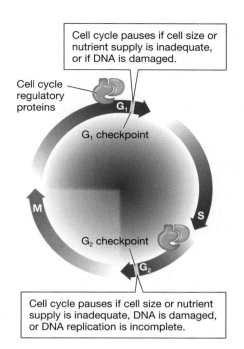

Cell cycle pauses if cell size or nutrient supply is inadequate, or if DNA is damaged.

Cell cycle regulatory proteins

$G_1$

$G_1$ checkpoint

M

S

$G_2$ checkpoint

$G_2$

Cell cycle pauses if cell size or nutrient supply is inadequate, DNA is damaged, or DNA replication is incomplete.

**Figure 6.6**

### The cell cycle must pass through checkpoints to proceed

Just two of the known cell cycle checkpoints are depicted in this diagram. Checkpoints also operate in both the S phase and the M phase (mitosis).

**Q1:** What could happen if the cell's checkpoints are disabled?

**Q2:** What is the advantage of stopping the cell cycle if the cell's DNA is damaged?

**Q3:** Which part of the cell cycle may have been influenced in Soto and Sonnenschein's breast tumor cell experiments?

the plastic mouse cages: the water bottles were leaking and the plastic cage walls were hazy.

She asked around and found out that, months earlier, a substitute janitor had used detergent with a high pH instead of the normal low-pH detergent to clean the cages and bottles. With some lab work, she identified the chemical oozing from the corroded plastic. "It took essentially one washing with the wrong detergent," says Hunt, "and once it was damaged, [the plastic] started to leach bisphenol A."

Bisphenol A, or BPA, was first synthesized in 1891. A synthetic hormone that, like nonylphenol, mimics estrogen, BPA was tested in the

1930s by clinicians seeking a hormone replacement therapy for women who needed estrogen, but it wasn't as effective as other substitutes. In the 1940s and '50s, the chemical industry found another use for BPA, as a chemical component of a clear, strong plastic. Manufacturers began to incorporate it into eyeglass lenses, water and baby bottles, the linings of food and beverage cans, and more. Unfortunately, as Hunt found out, BPA doesn't necessarily stay in those products. Not all the BPA used to make plastics gets locked into chemical bonds, and what doesn't get locked into bonds can work free—especially if the plastic is heated, such as when a baby bottle is warmed, or if the plastic is exposed to a harsh chemical, as Hunt's mouse cages were. Because of its prevalence in products and its ability to leach out of them, BPA is one of the most common chemicals we are exposed to in everyday life.

To confirm the hypothesis that BPA caused the mouse egg abnormalities, Hunt's team re-created the original event. They intentionally damaged a set of new cages and put healthy female mice in them. They also had the mice drink from damaged water bottles. Later, when they examined the eggs of these mice, they saw the same toxic effects as before: 40 percent of the eggs had abnormal chromosomes. The eggs showed errors in meiosis.

**Meiosis** is a specialized type of cell division that kicks off **sexual reproduction**, the process by which genetic information from two individuals is combined to produce offspring. Sexual reproduction has two steps: cell division through meiosis, and fertilization. BPA was affecting the first of these steps—meiosis.

We learned earlier that mitosis produces daughter cells with the same number of chromosomes as the parent cell. These non–sex cells are called **somatic cells**. In contrast, meiosis produces **gametes**, daughter cells that have half the chromosome count of the parent cell. The differences between mitosis and meiosis are the reason why the somatic cells of plants and animals have twice as much genetic information as their gametes have. The double set of genetic information possessed by somatic cells is known as the **diploid** set (represented by $2n$), and the single set possessed by gametes is called the **haploid** set (represented by $n$).

**Fertilization**, the fusion of two gametes, results in a single cell called the **zygote**. The

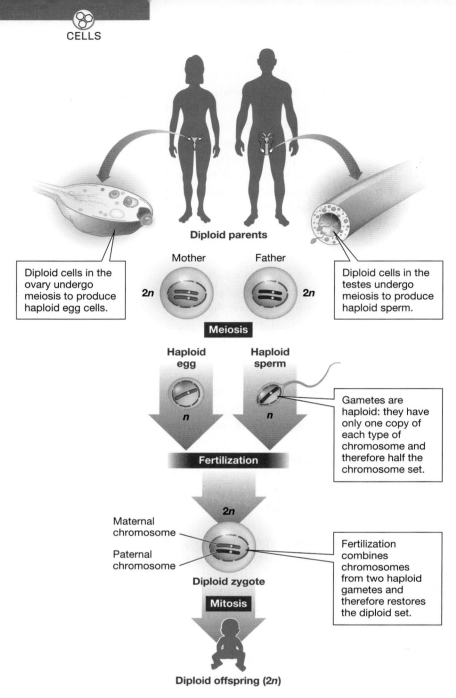

## Figure 6.7

### Fertilization creates a zygote from the fusion of two gametes

In species with two sexes, female gametes are *eggs* and male gametes are *sperm*. This figure shows only one of the 23 homologous pairs found in human cells.

**Q1:** Is a zygote haploid or diploid?

**Q2:** Which cellular process creates a baby from a zygote?

**Q3:** If a mother or father was exposed to BPA prior to conceiving a child, how might that explain potential birth defects in the fetus?

zygote inherits a haploid ($n$) set of chromosomes from each of the gametes, restoring the complete diploid ($2n$) set of genetic information to the offspring. Each **homologous pair** of chromosomes in the zygote consists of one chromosome received from the father and one from the mother. The zygote then divides by mitosis to create a mass of cells that will eventually develop into a mature organism (**Figure 6.7**). All cells in a mature organism are diploid, containing homologous pairs of chromosomes, except for gametes, which are haploid and contain only one of each homologous pair.

Meiosis occurs in two stages after interphase—*meiosis I* and *meiosis II*—each involving one round of nuclear division followed by cytokinesis, for a total of two cell divisions (**Figure 6.8**). **Meiosis I** reduces the chromosome number to haploid by *separating one of each homologous pair* into two different daughter cells. Each homologous chromosome lines up with its partner and then separates to the two ends of the cells, and cytokinesis occurs. **Meiosis II** *separates each sister chromatid* into two different daughter cells. This time, the phases of the division cycle are almost exactly like those of mitosis: sister chromatids separate to the two ends of the cells, and cytokinesis occurs, leading to an equal segregation of chromatids into two new daughter cells. In summary, meiosis I produces two haploid cells ($n$) with one of each pair of duplicated chromosomes, and in meiosis II these two haploid cells separate the sister chromatids and each divide, giving rise to a total of four haploid cells ($n$) with unduplicated chromosomes. Each haploid gamete now has half of the chromosome set found in the original diploid cell ($2n$) that underwent meiosis.

BPA is toxic because it disrupts the process of meiosis, hindering the ability of the chromosomes to separate into four haploid cells. Hunt realized that if BPA was disrupting meiosis in mice, it could be doing the same in humans. And if a human gamete (either the sperm or the egg) does not contain the correct number of chromosomes, fertilization typically results in a miscarriage.

Hunt was nervous about publishing the results of her experiment. "We knew we were stepping into a landmine," she says. "We knew the paper would get some press, because essentially we were publishing that this chemical—used in

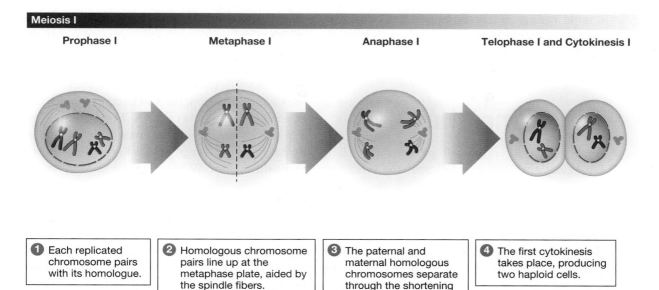

**Meiosis I**

**Prophase I**     **Metaphase I**     **Anaphase I**     **Telophase I and Cytokinesis I**

**1** Each replicated chromosome pairs with its homologue.

**2** Homologous chromosome pairs line up at the metaphase plate, aided by the spindle fibers.

**3** The paternal and maternal homologous chromosomes separate through the shortening of the spindle fibers.

**4** The first cytokinesis takes place, producing two haploid cells.

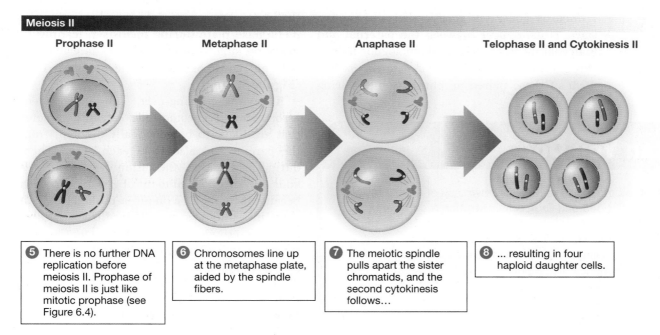

**Meiosis II**

**Prophase II**     **Metaphase II**     **Anaphase II**     **Telophase II and Cytokinesis II**

**5** There is no further DNA replication before meiosis II. Prophase of meiosis II is just like mitotic prophase (see Figure 6.4).

**6** Chromosomes line up at the metaphase plate, aided by the spindle fibers.

**7** The meiotic spindle pulls apart the sister chromatids, and the second cytokinesis follows…

**8** … resulting in four haploid daughter cells.

**Figure 6.8**

**Meiosis with cytokinesis creates haploid daughter cells**

The homologous chromosomes of a diploid cell after interphase occurs are paired and then separated into two haploid cells in meiosis I. Sister chromatids of these haploid cells are separated during meiosis II, each into two haploid cells. Cytokinesis occurs at the end of both meiosis I and meiosis II. ▶

**Q1:** Is a daughter cell haploid or diploid after the first meiotic division? How about after the second meiotic division?

**Q2:** What is the difference between homologous chromosomes and sister chromatids?

**Q3:** If the skin cells of house cats contain 38 homologous pairs of chromosomes, how many chromosomes are present in the egg cells they produce?

a wide variety of consumer products and that we are probably all exposed to—can cause an increased risk of miscarriage and babies with birth defects."

# Shuffling the DNA

As a glance at a pair of parents and their biological children will tell you, the offspring resulting from sexual reproduction are similar to their parents, but—unlike the clones resulting from asexual reproduction—they are *not* identical. Because each half of a sexually reproducing organism's DNA comes from a different parent, meiosis and fertilization maintain the constant chromosome number of a species while allowing for genetic diversity within the population.

Meiosis generates genetic diversity in two ways: *crossing-over* between the paternal and maternal members of each homologous pair of chromosomes, as well as *independent assortment* of these paired-up chromosomes during meiosis I. **Crossing-over** is the physical exchange of identical chromosomal segments during meiosis I between the nonsister chromatids in each duplicated homologous pair. These nonsister chromatids make physical contact at random sites along their length, and each exchange the exact same segments of DNA (**Figure 6.9**). The chromatids are said to be *recombined*, and the exchange of DNA segments is known as **genetic recombination**. Without crossing-over, every chromosome inherited by a gamete would be just the way it was in the parent cell.

The **independent assortment** of chromosomes—the random distribution of the homologous chromosomes into daughter cells during meiosis I—also contributes to the genetic variety of the gametes produced. Each homologous chromosome pair in a given meiotic cell orients itself independently when it lines up at the imaginary equatorial plane known as the metaphase plate during meiosis I, leading to many possible combinations of maternal and paternal chromosomes in the daughter cells (**Figure 6.10**).

As with crossing-over, the independent assortment of chromosomes creates gametes that are likely to be genetically different from the parent, and also from each other. Then, during fertilization, the fusion of two gametes adds a tremendous amount of genetic variation because it combines a one-in-a-million egg with a one-in-a-million sperm. These three processes together—crossing-over, independent assortment, and fertilization—give each of us our genetic uniqueness.

# Ten Years Later

Hunt and her team published their results in 2003. "We got a firestorm," she recalls with a grimace. The press reported the findings extensively, and many people and companies were upset over the allegations that BPA was toxic. Members of the plastics industry who did not agree with the paper's conclusions criticized Hunt's work. But there was more supporting research to come. Soto and Sonnenschein had also turned their attention to BPA because in everyday products it is a far more common chemical than nonylphenol and is therefore of greater concern.

## What Can You Do?

There are things you can do to reduce your own risk of exposure to BPA. Today, the U.S. Food and Drug Administration recommends that individuals not put hot or boiling liquid intended for consumption in plastic containers made with BPA. (Some, but not all, plastics that are marked with the recycling code 3 or 7 may be made with BPA.) The organization also recommends discarding all bottles with scratches, which may harbor bacteria and, if the plastic contains BPA, may lead to greater release of the chemical.

"Get educated about your world," says Heather Patisaul, a BPA researcher at North Carolina State University. "You can either become completely reliant on the information you get from media and government, or you can educate yourself, which is vastly more useful." If you are concerned about BPA exposure, it is possible to cut it down by making lifestyle changes like not eating canned food, not drinking bottled water, and not putting plastic in the microwave. "You can be empowered," says Patisaul. "Those types of things can effect great change."

Still, BPA and similar chemicals are ubiquitous in modern life, says Ana Soto. Ultimately, the best way to avoid them will be for government regulators to take a stand and outlaw the use of these chemicals in consumer products, she notes. She encourages individuals to contact their representatives and ask them to push legislation limiting the use of BPA in manufacturing.

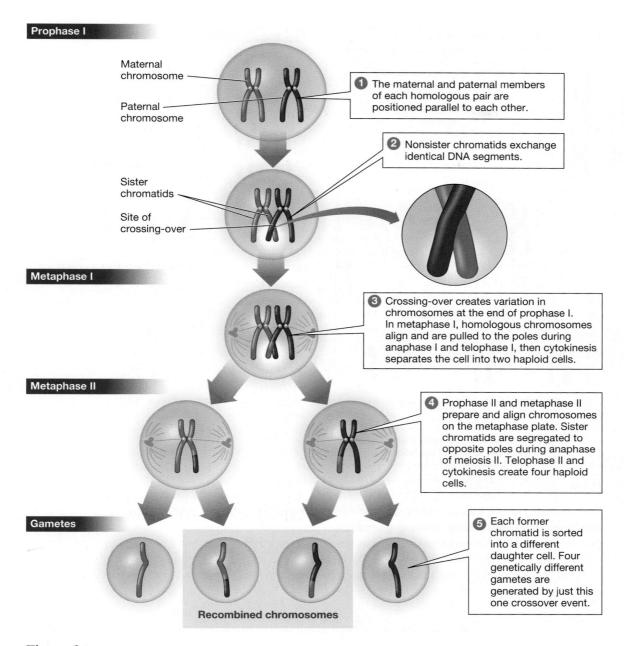

**Prophase I**

Maternal chromosome

Paternal chromosome

**1** The maternal and paternal members of each homologous pair are positioned parallel to each other.

**2** Nonsister chromatids exchange identical DNA segments.

Sister chromatids

Site of crossing-over

**Metaphase I**

**3** Crossing-over creates variation in chromosomes at the end of prophase I. In metaphase I, homologous chromosomes align and are pulled to the poles during anaphase I and telophase I, then cytokinesis separates the cell into two haploid cells.

**Metaphase II**

**4** Prophase II and metaphase II prepare and align chromosomes on the metaphase plate. Sister chromatids are segregated to opposite poles during anaphase of meiosis II. Telophase II and cytokinesis create four haploid cells.

**Gametes**

**5** Each former chromatid is sorted into a different daughter cell. Four genetically different gametes are generated by just this one crossover event.

Recombined chromosomes

## Figure 6.9

**Crossing-over produces chromosomes with new combinations of DNA**

Only one maternal and one paternal chromosome are depicted here, rather than the 23 pairs of homologous chromosomes found in humans. Only the resulting cells of prophase I, metaphase I, metaphase II, and the resulting gametes of the process of meiosis are depicted here.

**Q1:** Why is the term "crossing-over" appropriate for the exchange of DNA segments between homologous chromosomes?

**Q2:** At what stage of meiosis (I or II) does crossing-over occur?

**Q3:** What would be the effect of crossing-over between two sister chromatids?

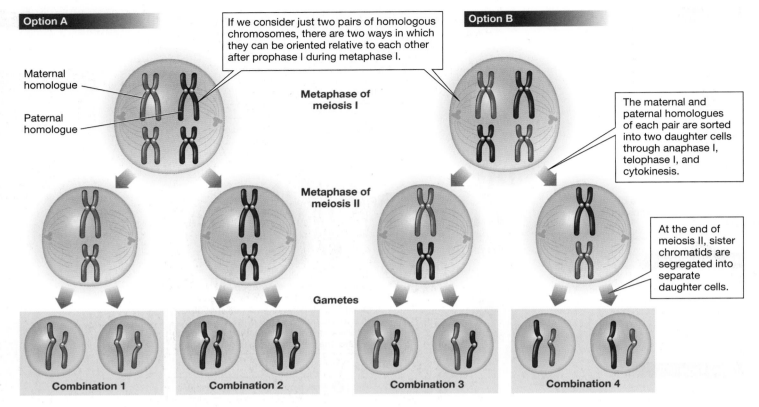

**Option A**

**Option B**

Maternal homologue

Paternal homologue

If we consider just two pairs of homologous chromosomes, there are two ways in which they can be oriented relative to each other after prophase I during metaphase I.

Metaphase of meiosis I

The maternal and paternal homologues of each pair are sorted into two daughter cells through anaphase I, telophase I, and cytokinesis.

Metaphase of meiosis II

At the end of meiosis II, sister chromatids are segregated into separate daughter cells.

Gametes

Combination 1

Combination 2

Combination 3

Combination 4

Four possible combinations of chromosomes in gametes generated by meiosis II.

## Figure 6.10

### The independent assortment of homologous chromosomes generates chromosomal diversity among gametes

Only two pairs of homologous chromosomes are shown here, rather than the 23 homologous pairs in human cells. Each gamete will receive either a maternal or a paternal homologue of each chromosome. Only the resulting cells of metaphase I, metaphase II, and the resulting gametes of the process of meiosis are depicted here.

**Q1:** During meiosis, does random assortment occur before or after crossing-over?

**Q2:** What would be the effect on genetic diversity if homologous chromosomes did not randomly separate into the daughter cells during meiosis?

**Q3:** With two pairs of homologous chromosomes, four kinds of gametes can be produced. How many kinds of gametes can be produced with three pairs of homologous chromosomes? What does this suggest for the 23 homologous pairs of chromosomes in human cells?

At the same time that Soto, Sonnenschein, and Hunt were doing their work, Frederick vom Saal at the University of Missouri found that male mice that had been exposed to BPA in utero—even at very low doses—had dramatically enlarged prostates in adulthood that were hypersensitive to hormones. This study suggests that men are also at risk of health effects from BPA.

In 2007, Hunt followed up her original work with a study that she says made the first paper look like "child's play." Her team exposed pregnant mice to BPA just as their female fetuses were producing a supply of eggs in their ovaries. When that second generation of females became adults, their eggs were also damaged, Hunt found, demonstrating that BPA exposure affects not just adult females, but two generations of their offspring.

Exposure aside, not everyone agrees that BPA is toxic. Numerous companies that manufacture plastics have conducted studies whose results do not match Hunt's and vom Saal's results. To reach a scientific consensus, on November 28, 2006, Soto, Sonnenschein, Hunt, vom Saal, and 34 other researchers from across the United States gathered at the University of North Carolina in Chapel Hill to summarize the research on BPA. The result of their two-day meeting was the "Chapel Hill Bisphenol A Consensus Statement," summarizing hundreds of studies done in vitro and in vivo over the previous 10 years. Completing this analysis, the group concluded firmly that BPA exposure at current levels in our environment presents a risk to human health (see "What Can You Do?" on page 108). "It was quite clear that there is a serious problem," says Soto.

Over time, many baby-bottle manufacturers took BPA out of their bottles, even as government regulators were slower to respond. "As scientists, our role is to call attention to what is wrong, but it is the role of the politicians to act on it and try to straighten it out," says Sonnenschein. Then, in July 2012, the FDA banned BPA from baby bottles and children's drinking cups, though the prohibition does not apply to the use of BPA in other types of containers (**Figure 6.11**). There is still concern, however, from many scientists about the chemicals that have replaced BPA, some of which are also estrogen mimics.

In the last several years, scientists all over the world have shown that BPA disrupts meiosis and mitosis and causes a plethora of health problems in mice and rats, including breast and prostate cancer, miscarriage and birth defects, diabetes and obesity, and even behavioral problems such as attention deficit hyperactivity disorder. Whether BPA is causing similar diseases in humans remains unknown; such hypotheses are hard to test because most people already have BPA in their bodies, making experimental control groups difficult to set up. BPA has been found in human blood, urine, breast milk, and amniotic fluid. In 2016, Hunt and her colleagues found "near-universal exposure" to BPA in a group of pregnant women in the United States. A major source of that exposure, they found, was cash register receipts, which can be hard to avoid touching.

Hunt, Soto, and Sonnenschein continue to explore the effects of BPA; they are studying how exposure to low doses of BPA affects monkeys, a model animal that more accurately represents the human system. "We're slowly raising awareness," says Hunt, "and slowly changing things."

Figure 6.11

**BPA-free bottles and cans are now widely available**
If you are concerned about being exposed to BPA, check labels before you buy.

# Cancer's Big 10

When the cell cycle spirals out of control, cancer emerges: abnormal cells divide in a frenzy and can invade other tissues. There are more than 100 types of cancer, but some are more prevalent than others. And some are more deadly than others, because of their location in the body or how quickly the cells divide. New treatments, screening procedures, and vaccines can reduce these rates.

Assessment available in smartwork5

## Top 10 cancer sites by rate of incidence
Incidence rates per 100,000

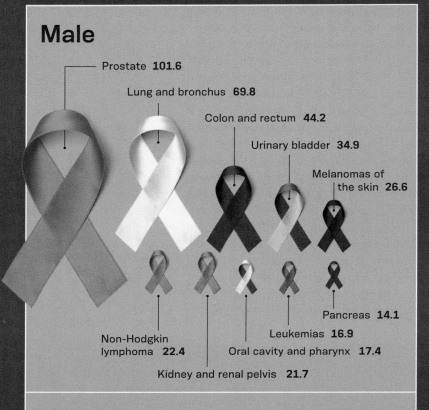

### Male

- Prostate **101.6**
- Lung and bronchus **69.8**
- Colon and rectum **44.2**
- Urinary bladder **34.9**
- Melanomas of the skin **26.6**
- Pancreas **14.1**
- Leukemias **16.9**
- Oral cavity and pharynx **17.4**
- Kidney and renal pelvis **21.7**
- Non-Hodgkin lymphoma **22.4**

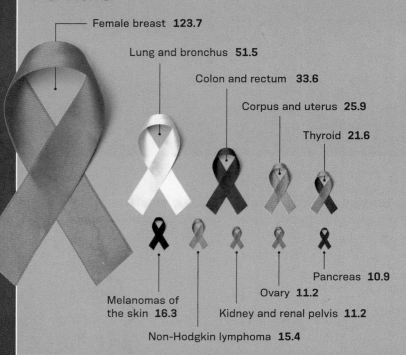

### Female

- Female breast **123.7**
- Lung and bronchus **51.5**
- Colon and rectum **33.6**
- Corpus and uterus **25.9**
- Thyroid **21.6**
- Pancreas **10.9**
- Ovary **11.2**
- Kidney and renal pelvis **11.2**
- Non-Hodgkin lymphoma **15.4**
- Melanomas of the skin **16.3**

## Top 10 cancer deaths by rate of incidence
Incidence rates per 100,000

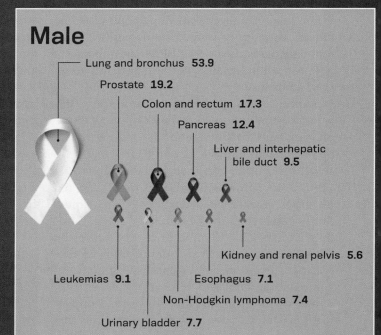

### Male

- Lung and bronchus **53.9**
- Prostate **19.2**
- Colon and rectum **17.3**
- Pancreas **12.4**
- Liver and interhepatic bile duct **9.5**
- Kidney and renal pelvis **5.6**
- Esophagus **7.1**
- Non-Hodgkin lymphoma **7.4**
- Urinary bladder **7.7**
- Leukemias **9.1**

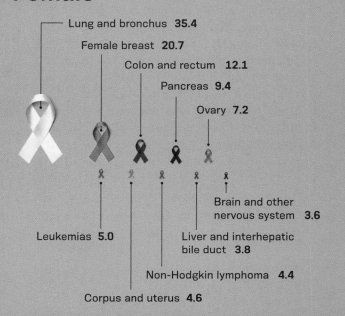

### Female

- Lung and bronchus **35.4**
- Female breast **20.7**
- Colon and rectum **12.1**
- Pancreas **9.4**
- Ovary **7.2**
- Brain and other nervous system **3.6**
- Liver and interhepatic bile duct **3.8**
- Non-Hodgkin lymphoma **4.4**
- Corpus and uterus **4.6**
- Leukemias **5.0**

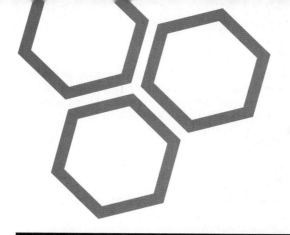

# REVIEWING THE SCIENCE

- The **cell cycle** is the set sequence of events over the life span of a eukaryotic cell that will divide. **Interphase** and **cell division** are the two main stages of the cell cycle. Interphase is longer, and consists of the **G₁**, **S**, and **G₂** **phases**. DNA is replicated in the S phase. Cells that will not divide exit the cell cycle and enter a **G₀ phase**.

- Cell division is necessary for growth and repair in multicellular organisms, and for **asexual** and **sexual reproduction** in all types of organisms. Many prokaryotes divide through **binary fission**, a form of asexual reproduction.

- Each **chromosome** in a cell contains a single DNA molecule compacted by packaging proteins. The **somatic cells** of eukaryotes have two of each type of chromosome matched together in **homologous pairs**. One chromosome in each pair is inherited from the mother; the other, from the father. Chromosomal replication produces two identical **sister chromatids** that are held together firmly at the **centromere**.

- Eukaryotes perform cell division through **mitosis** followed by **cytokinesis**, producing daughter cells that are genetically identical to each other and to the parent cell. The four main phases of mitosis are prophase, metaphase, anaphase, and telophase. Through these phases, the chromosomes of a parent cell are condensed and positioned appropriately, and the sister chromatids are separated to opposite ends of the cell. During cytokinesis, the cytoplasm of the parent cell is physically divided to create two daughter cells.

- The cell cycle is carefully regulated. Checkpoints ensure that the cycle does not proceed if conditions are not right. Unregulated cell growth and cancer occur when checkpoints fail.

- **Meiosis** is critical for sexual reproduction. In animals, the products of meiosis are sex cells, called **gametes**, that fuse during **fertilization** to give rise to a **zygote**. Meiosis—consisting of two rounds of nuclear and cytoplasmic divisions—produces **haploid** gametes containing only one chromosome from each homologous pair. When two gametes fuse during fertilization, a **diploid** zygote is formed, having one of each homologous chromosome from both parents.

- During **meiosis I**, the maternal and paternal members of each homologous pair are sorted into two daughter cells. **Meiosis II** is similar to mitosis in that sister chromatids are segregated into separate daughter cells at the end of cytokinesis.

- Meiosis produces genetically diverse gametes through **crossing-over** of homologous chromosomes, leading to **genetic recombination** and then the **independent assortment** of homologous chromosomes. Meiosis and fertilization together introduce genetic variation into populations.

# THE QUESTIONS

## The Basics

**1** Homologous chromosomes are

(a) the same thing as sister chromatids.

(b) a pair of chromosomes of the same kind.

(c) identical copies of the same chromosome.

(d) always haploid.

**2** Which of the following is *not* a contributor to genetic variation?

(a) binary fission

(b) crossing-over of homologous chromosomes

(c) random assortment of homologous chromosomes

(d) fertilization

**3** Link each cell phase to the event that occurs within it.

| CYTOKINESIS | 1. Each of the chromosomes in a human cell contains two sister chromatids by the end of this phase. |
| S PHASE | 2. Most cell growth occurs during this phase. |
| G₁ PHASE | 3. Cells that will never replicate leave the cell cycle and enter this phase. |
| G₀ PHASE | 4. Two separate daughter cells are produced at the end of this phase. |

**4** Select the correct terms in the following sentences: (**Mitosis / Meiosis**) produces daughter cells with half the number of chromosomes that the parent cell has. Cell division in prokaryotes is called (**mitosis / binary fission**). Meiosis I separates (**sister chromatids / homologous chromosomes**); meiosis II separates (**sister chromatids / homologous chromosomes**) into separate daughter cells.

**5** Place the following events of sexual reproduction in the correct order by numbering them from 1 to 5.

_____a. Separation of homologous chromosomes

_____b. Separation of sister chromatids

_____c. Mitosis within the zygote, leading to a multicellular organism

_____d. Cytokinesis, leading to four haploid daughter cells

_____e. Fusion of two gametes

**6** Which of the following is not a contributor to genetic variation?

(a) replication of sister chromatids

(b) crossing-over of homologous chromosomes

(c) random assortment of homologous chromosomes

(d) fertilization

**7** Loss of cell cycle control may lead to

(a) pregnancy.

(b) cancer.

(c) fertilization.

(d) crossing-over of homologous chromosomes.

## Challenge Yourself

**8** Suppose a scientist has recently identified a new protein that prevents a cell from entering mitosis if there are any signs of DNA damage. This protein would be classified as a type of _____ protein.

(a) chromatid

(b) cell cycle checkpoint

(c) benign

(d) malignant

(e) angiogenesis

**9** You've been reduced in size by a misfire from a "shrink-inator" gun! You realize that you are inside a cell during prophase I of meiosis. You see two linear molecules of DNA compacted and attached to one another by a centromere. What exactly are you looking at?

(a) a homologous pair of chromosomes

(b) sister chromatids

(c) the metaphase plate

(d) cell cycle checkpoints

(e) gametes

**10** Scientists are able to isolate cells in various phases of the cell cycle. During an experiment in which Dr. Patrick McGroyn is causing cultured tumor cells to go through mitotic cell division, he isolates a group of cells that contain one and a half times more DNA than cells isolated in the $G_1$ phase have. What phase must these cells be in? How do you know?

## Try Something New

**11** Domesticated sheep cells have a total of 54 chromosomes (versus 46 chromosomes in human cells).

(a) How many separate DNA molecules are present in a sheep liver cell at the end of the $G_1$ phase?

(b) How many separate DNA molecules are present in the daughter cells after meiosis I in the ovary of the sheep?

(c) How many separate DNA molecules are present in the daughter cells after meiosis II in the ovary of the sheep?

**12** Biopsies from aggressive cancers often have cells that contain several nuclei per cell when viewed through a microscope. Which scenario could explain how such a multinucleated cell might have come to be?

(a) The cell underwent repeated mitosis with simultaneous cytokinesis.

(b) The cell had multiple S phases before it entered mitosis.

(c) The cell underwent repeated mitosis, but cytokinesis did not occur.

(d) The cell underwent repeated cytokinesis but no mitosis.

(e) The cell actually went through meiosis and not mitosis.

**13** Describe the likely consequences of bypassing the $G_1$ and $G_2$ checkpoints in the cell cycle. Why do compounds like nonylphenol lead to the multiplication of abnormal cells?

## Leveling Up

**14** **What do *you* think?** Cancer begins with a single cell that breaks loose of normal restraints on cell division and starts dividing rapidly to establish a colony of rogue cells. As cancer cells spread through the body, they disrupt the normal functions of tissues and organs; unchecked, cancer can cause death through failure of multiple organ systems. Many cancers could be prevented by not smoking or chewing tobacco, eating less meat and processed foods, eating more fruits and vegetables, drinking alcohol only moderately if at all, exercising regularly, and maintaining a healthy weight. Only 5–10 percent of cancers are directly attributable to genetic causes.

A "sin tax" is a tax on a product or activity that has negative effects on others, as a way to offset some of those effects. Common targets of sin taxes are tobacco and alcohol because of their public health costs. Proponents of a sin tax on tobacco argue that such a tax would decrease the amount that people smoke (because of the increased cost) and could also partially fund the costs of medical care necessitated by increased rates of cancer and other diseases caused by smoking. Critics point out that sin taxes have historically triggered smuggling and black markets, and have a disproportionate effect on poor people because the wealthy can more easily afford to pay the higher prices.

What do you think? Should we institute taxes on tobacco? Would fewer people smoke, or would they smoke less, if tobacco were more expensive? What about higher health care premiums for smokers, based on their higher risk for cancer? Would increasing the premiums cause more people to stop smoking? Would you support such a policy?

**15** *Write Now* **biology: BPA effects** The studies of BPA described in this chapter used an inbred strain of mice that is known to be especially susceptible to estrogen and estrogen-like chemicals, such as BPA. The plastics industry maintains that the susceptibility of this strain of mice to estrogen renders these studies invalid as a basis for estimating the effects of BPA on humans. BPA researchers respond that the current situation, exposing millions of people to unknown levels of BPA, constitutes a massive uncontrolled experiment. They maintain that even a small risk of harm is too great to be allowed when so many people are exposed. Bills banning the use of BPA in food and beverage containers were introduced in Congress in 2009, 2011, 2013, 2014, and 2015. All failed to pass. Another bill was introduced to the House of Representatives on September 28, 2016.

a. Go to the Embryo Project Encyclopedia (http://embryo.asu.edu), an NSF-funded online repository of information about embryo research. Search for the term "BPA" to find the most up-to-date information on BPA research.

b. Go to the website of the American Chemistry Council (http://www.americanchemistry.com), a trade organization that advocates for the chemical industry. Search for the term "BPA" to find its most up-to-date policy statement on the safety of BPA.

c. Draft a letter urging your congressional representative to support or oppose the most recent BPA bill, explaining why you believe the weight of scientific evidence makes your position a prudent response to the situation.

**For more, visit digital.wwnorton.com/bionow2 for access to:**

# Dog Days of Science

*Two canine-loving researchers unravel the genetic secrets of man's best friend.*

## After reading this chapter you should be able to:

- Distinguish between the genotype and phenotype of a given genetic trait.
- Describe the importance of Gregor Mendel's experiments to our understanding of inheritance.
- Illustrate Mendel's laws of segregation and independent assortment.
- Create a Punnett square to predict the phenotype of offspring from parents with a known genotype—both for single genes and for two independent genes.
- Give examples of Mendelian traits and of traits with complex inheritance.
- Explain how an individual's phenotype may be determined by multiple genes that interact with one another and with the environment.

CHAPTER
07

PATTERNS OF
INHERITANCE

ordon Lark's best friend was dying. Soft and shaggy, with tousled black hair, Georgie hadn't left Lark's side in 10 years, since his daughter had first purchased Georgie as a puppy from two kids by the side of the road. But as she aged, Georgie had become ill with Addison's disease, a disorder in which her body's immune system began to attack and destroy her own tissues. Georgie passed away in 1996.

Lark, a scientist at the University of Utah in Salt Lake City, was heartbroken. To help heal the wound, he decided to adopt another dog of the same breed—a Portuguese water dog (PWD), named for its history of helping Portuguese fishermen with their work (**Figure 7.1**). He contacted Karen Miller, a PWD breeder on a farm in rural New York. As part of the owner screening process, Miller asked Lark about his profession. "I said I was a soybean geneticist," Lark recalls, "but all she heard was 'blah, blah, genetics, blah, blah.' And she got really excited."

As a breeder, Miller was keenly interested in how dogs inherit characteristics from their parents, so Miller and Lark began talking by phone each week about genetics. When it came time to pick up his new puppy, Mopsa, Lark requested the bill. But Miller didn't want Lark's money. She had something else in mind: she gave him Mopsa free of charge, in the hope that Lark would start researching dog genetics. "That's silly," Lark told her. He wasn't a dog researcher. Lark had spent a career studying the genetic traits of bacteria and soybeans.

A **genetic trait** is any inherited characteristic of an organism that can be observed or detected in some manner. Some genetic traits are **invariant**, meaning they are the same in all individuals of the species. All soybeans, for example, have pods that contain seeds. Other genetic traits are **variable**; for example, soybean seeds occur in various sizes and colors, including black, brown, and green.

Apart from his love for canines, Lark wasn't intimately familiar with dogs' physical and biochemical traits. **Physical traits**, such as the shape of a dog's face, are easy to observe. **Biochemical traits**, on the other hand, such as a dog's susceptibility to Addison's disease, are often more difficult to observe. It is easy to collect physical and biochemical information from a field of soybeans but far more difficult to collect it from domesticated animals in homes all over the country. And to study dogs, Lark would also need data on **behavioral traits**, such as shyness and extroversion—factors he didn't have to take into account with soybeans. All of these traits—physical, biochemical, and behavioral—are influenced by genes.

Figure 7.1

**A Portuguese Water Dog**

# Getting to the Genes

A **gene** is the basic unit of information affecting a genetic trait. At the molecular level, a gene consists of a stretch of DNA on a **chromosome**—a threadlike molecule, made of DNA and proteins, that is found in the nucleus of a eukaryotic cell (**Figure 7.2**). A gene contains the information, or "code," for a specific protein; the protein then causes or contributes to a particular genetic trait. To study PWD traits, Lark would need dog DNA, which can be obtained from blood or saliva. Once he had that DNA, he could begin to search for **alleles**—different versions of a given gene—and link them to genetic traits. Alleles of a gene arise by **mutation**, which is any change in the DNA that makes up a gene (see Chapter 9 for more on mutation). Genetic diversity in a species, whether it is soybeans or dogs or humans, comes about because the species as a whole contains many different alleles of its genes.

Dogs are the champions of genetic variation. All dogs are the same species, yet a Pekingese weighs only a couple of pounds, while a Saint Bernard can weigh over 180 pounds. Dogs, in fact, are reported to have more variation in the size and shape of their species than any other living land mammal on Earth, with the possible exception of humans.

On a whim, Lark agreed to dabble in dog genetics, but to do so, he would need to compare the **genotypes** of individual dogs (their genetic makeup, controlled by their combinations of alleles) with their **phenotypes** (the physical expression of their genetic makeup). The genotype of a given trait is the pair of alleles that codes for its phenotype. To identify genes responsible for dog traits, Lark would need both types of information for each of many dogs.

Miller was already on the case. Three months after Lark got Mopsa, Miller sent him 5,000 PWD pedigrees—detailed health and breeding records for individual dogs. Lark was astonished. It was the first of many times that the enthusiasm and generosity of dog owners would contribute to his research.

"That was literally how this started," says Lark. Today, Lark and Miller's unlikely partnership has blossomed into a national research project producing valuable knowledge about the genetic basis of health and disease in both man and man's best friend. What's more, their effort has demonstrated how tiny genetic changes can create huge variation in a single species.

# Pet Project

The Georgie Project, as Lark fondly named it, officially began in 1996. Lark's first task was to collect genotypes and phenotypes from PWDs. To his pleasant surprise, PWD owners were enthusiastic and began flooding him with pedigrees, blood samples, and X-rays taken by their veterinarians. In short order, Lark had DNA from more than 1,000 dogs and detailed body measurements for over 500. Then the hard work began.

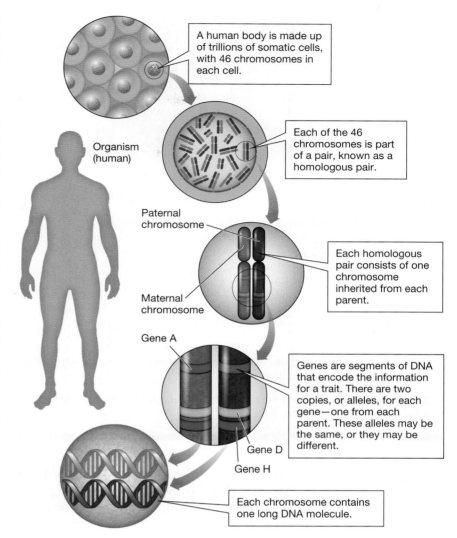

Figure 7.2

**Genes are segments of DNA that confer a genetic trait**
Somatic cells (cells of the body) have two copies of most genes.

**Q1:** What is the physical structure of a gene?

**Q2:** How many copies of each gene are found in the diploid cells in a woman's body?

**Q3:** With 46 chromosomes in a human diploid cell, how many chromosomes come from the person's mother and how many from the father?

Using the dogs' genotypes and phenotypes, Lark set out to pinpoint the alleles for particular traits. Some genes have alleles that are **dominant** when paired with another allele; that is, one allele prevents a second allele from affecting the phenotype when the two alleles are paired together. The

Phenotype:

Genotype:         *bb*                        *BB* or *Bb*

Figure 7.3

**Poodles illustrate variation in the coat color gene**

These poodles, close cousins of the Portuguese water dog, may have a black coat (dominant allele *B*) or a brown coat (recessive allele *b*). Other coat colors, with different inheritance patterns, are found in poodles and other dog breeds.

**Q1:** Which might you observe directly: the genotype or the phenotype?

**Q2:** Which poodle could be heterozygous: the one with the black coat or the one with the brown coat?

**Q3:** Can you identify with certainty the genotype of a black poodle? A brown poodle?

black-fur allele (*B*), for example, is dominant in dogs. An allele that has no effect on the phenotype when paired with a dominant allele is said to be **recessive**. In dogs, the brown-fur allele (*b*) is recessive. (When a gene has one dominant and one recessive allele, we generally use an uppercase letter for the dominant allele and a lowercase letter for the recessive allele.)

An individual that carries two copies of the same allele (such as *BB* or *bb*) is **homozygous** for that gene. An individual whose genotype consists of two different alleles for a given phenotype (*Bb*) is **heterozygous** for that gene. Having one dominant allele and one recessive allele, a heterozygous individual will show the dominant phenotype;

**GORDON LARK**

A geneticist at the University of Utah in Salt Lake City, Gordon Lark initiated the Georgie Project in 1996 to study the genetics of Portuguese water dogs. The national research project has led to valuable knowledge about the genetic basis of health and disease in humans and dogs.

a dog that is heterozygous for fur color (*Bb*), for example, will be black (**Figure 7.3**).

The first dog trait that Lark decided to investigate was size. What makes a Great Dane large and a Chihuahua small? To find out, Lark asked for help from the "mother of all dog projects," as Lark calls her—a researcher named Elaine Ostrander, whose entry into dog research was almost as strange as Lark's.

# Crisscrossing Plants

In 1990, Ostrander was a young, enthusiastic researcher who had just completed her postdoctoral studies in molecular biology at Harvard University and was ready to start her own laboratory in California. But first she had to decide which organism to study. Typical choices included fruit flies, worms, or plants—organisms that are easy to grow and manipulate. Ostrander picked plants, just as Gregor Johann Mendel, an Austrian monk who later became known as the "father of modern genetics," had done in the mid-1800s.

Mendel famously bred pea plants in a garden at his monastery. Through his work with pea plants, Mendel discovered patterns of inheritance that today form the foundation of genetics for scientists like Ostrander. "Mendel's laws," as they are now called, describe how genes are passed from parents to offspring. These laws allow us to use parental genotypes to predict offspring genotypes and phenotypes.

Each time Mendel bred two pea plants together, he was performing a **genetic cross**, or just "cross" for short. A genetic cross is a controlled mating experiment performed to examine how a particular trait is inherited. In a series of genetic crosses, the organisms involved in the first cross are called the **P generation** ("P" for "parental").

For example, Mendel investigated the inheritance of flower color by crossing pea plants that had different flower colors (**Figure 7.4**). He had noticed that some plants always "bred true" for flower color; that is, the offspring always produced flowers that had the same color as the parents and were therefore homozygous. He performed a genetic cross with a P generation

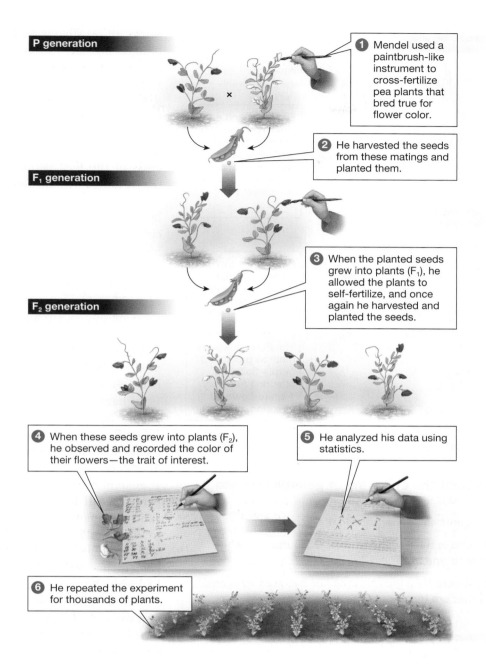

**P generation**

**①** Mendel used a paintbrush-like instrument to cross-fertilize pea plants that bred true for flower color.

**②** He harvested the seeds from these matings and planted them.

**F₁ generation**

**③** When the planted seeds grew into plants ($F_1$), he allowed the plants to self-fertilize, and once again he harvested and planted the seeds.

**F₂ generation**

**④** When these seeds grew into plants ($F_2$), he observed and recorded the color of their flowers—the trait of interest.

**⑤** He analyzed his data using statistics.

**⑥** He repeated the experiment for thousands of plants.

Figure 7.4

## Mendel's careful experiments

Mendel was meticulous in conducting his research and making observations, following a very careful protocol. Controlled breeding is possible in flowering plants because they have both male and female reproductive structures. Mendel manipulated his P-generation plants by removing all the female structures from one plant and all the male structures from the other, thus preventing self-fertilization. He could then perform the initial cross by transferring pollen from the "male" plant to the flower of the "female" plant. ▶

**Q1:** What would you predict about the color of the $F_1$ plants' flowers?

**Q2:** Why was it important that Mendel begin with pea plants that he knew bred true for flower color? Why couldn't he simply cross a purple-flowered plant and a white-flowered plant?

**Q3:** Over the years, Mendel experimented with more than 30,000 pea plants. Why did Mendel collect data on so many plants? Why didn't he study just one cross? *Hint:* Read "What Are the Odds?" on page 124 before answering.

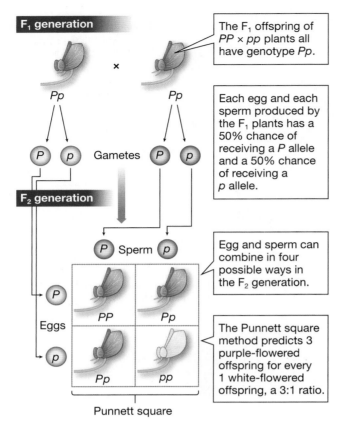

**F₁ generation**

The F₁ offspring of *PP* × *pp* plants all have genotype *Pp*.

*Pp*    ×    *Pp*

Each egg and each sperm produced by the F₁ plants has a 50% chance of receiving a *P* allele and a 50% chance of receiving a *p* allele.

P  p  Gametes  P  p

**F₂ generation**

P  Sperm  p

Egg and sperm can combine in four possible ways in the F₂ generation.

Eggs

P

p

*PP*    *Pp*

*Pp*    *pp*

The Punnett square method predicts 3 purple-flowered offspring for every 1 white-flowered offspring, a 3:1 ratio.

Punnett square

Figure 7.5

**Punnett squares predict the offspring of genetic crosses**

**Q1:** Why did Mendel's entire F₁ generation look the same?

**Q2:** The phenotype ratio in the F₂ generation is 3:1 purple-to-white flowers. What is the genotype ratio?

**Q3:** Draw a Punnett square for a genetic cross of two heterozygous, black-coated dogs. What are the phenotype and genotype ratios of their offspring?

shows all possible ways that two alleles can be brought together through fertilization. To create a Punnett square showing how a trait is inherited, list the alleles of the male genotype across the top of the grid, writing each unique allele just once. List the alleles of the female genotype along the left edge of the grid, again writing each unique allele only once. In the case of Mendel's cross of the F₁ generation, the male genotype (*Pp*) is mated with the female genotype (also *Pp*).

Next, fill in each box (or "cell") in the grid by combining the male allele at the top of each column with the female allele listed at the beginning of each row. The Punnett square shows all four ways in which the two alleles in the sperm can combine with the two alleles found in the eggs. The four genotypes shown within the Punnett square are all equally likely outcomes of this cross.

Using the Punnett square method, we can predict that ¼ of the F₂ generation is likely to have genotype *PP*, ½ to have genotype *Pp*, and ¼ to have genotype *pp*. Because the allele for purple flowers (*P*) is dominant, plants with *PP* or *Pp* genotypes have purple flowers, while plants with *pp* genotypes have white flowers. Therefore, we predict that ¾ (75 percent) of the F₂ generation will have purple flowers and ¼ (25 percent) will have white flowers—a 3:1 (¾:¼) ratio of phenotypes. This prediction is very close to the actual results that Mendel obtained. Of a total 929 F₂ plants that Mendel raised, 705 (76 percent) had purple flowers and 224 (24 percent) had white flowers.

Results like these supported Mendel's first law, the **law of segregation**, which, in modern terms (Mendel did not know about DNA), states that the two alleles of a gene are separated during meiosis, the specialized type of cell division during sexual reproduction that was discussed in Chapter 6, and end up in different gametes—egg or sperm cells. One of the two alleles is found on one of the chromosomes in a homologous pair, and the other allele is found on the other chromosome in the pair (see **Figure 7.2**). Remember, homologous chromosome pairs are partitioned into separate daughter cells during meiosis I. Mendel's law of segregation can be used to predict how a single trait will be inherited.

You can try this on your own by making a Punnett square to predict the ratio of black

in which one parent bred true for purple flowers (*PP*) and the other bred true for white flowers (*pp*). The first generation of offspring of a genetic cross is called the **F₁ generation** ("F" is for "filial," a word that refers to a son or daughter). When the individuals of the F₁ generation are crossed with each other, the resulting offspring are said to belong to the **F₂ generation**. Mendel allowed the F₁-generation pea plants to self-fertilize to produce the F₂ generation.

We can predict the results of an experimental cross by using a grid-like diagram called a **Punnett square** (**Figure 7.5**). A Punnett square

and brown offspring that would result if two heterozygous (*Bb*) black-coated dogs were mated. It is important to understand that the predicted ratios simply give the *probability* that a particular offspring will have a certain phenotype or genotype; the actual ratio will vary (see "What Are the Odds?" on page 124).

# Peas in a Pod

Mendel's research on pea seeds led to his second law, the **law of independent assortment**. This law states that when gametes form, the two alleles of any given gene segregate during meiosis independently of any two alleles of other genes. For example, pea seeds can have a round or wrinkled shape, and they can be yellow or green. Two different genes control the two different traits—the *R* gene, with alleles *R* (round) and *r* (wrinkled), controls seed shape, while the *Y* gene, with alleles *Y* (yellow) and *y* (green), controls the color of the seed—but neither gene affects the inheritance of the other.

Mendel tested the idea of independent assortment in the set of experiments illustrated in **Figure 7.6.** He tracked seed shape, the trait controlled by the *R* and *r* alleles, and seed color, controlled by the *Y* and *y* alleles. The test of his hypothesis came when he examined the phenotypes of the offspring produced by crossing the heterozygous F₁ plants (*RrYy*). As predicted by the hypothesis, two new phenotypic combinations were found among the F₂ offspring: plants with round, green seeds (*RRyy* or *Rryy*) and plants with wrinkled, yellow seeds (*rrYY* or *rrYy*). Figure 7.6 summarizes the ratios of the two parental phenotypes and the two novel, nonparental phenotypes.

Traits controlled by a single gene and unaffected by environmental conditions are called **Mendelian traits**. But when Mendel described his laws of inheritance, he had no idea what genes were made of, where they were located within a cell, or how they segregated and independently assorted. Today we know that genes are located on chromosomes and that these chromosomes are the basis for all inheritance. We call these assertions the **chromosome theory of inheritance**, which explains the mechanism underlying

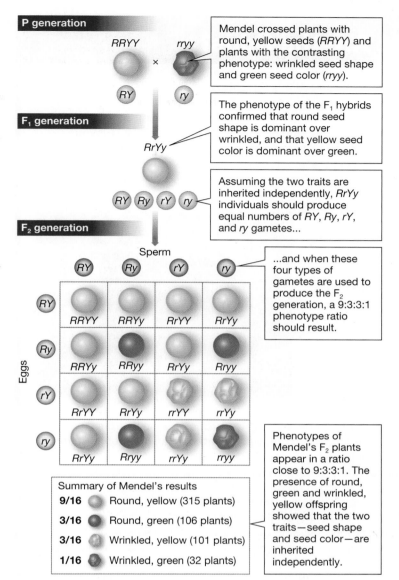

**Figure 7.6**

## Independent assortment of pea color and shape

Mendel used two-trait breeding experiments, called **dihybrid crosses**, to test the hypothesis that the alleles of *two different genes* are inherited independently from each other. ▶▤

**Q1:** List all the possible offspring genotypes and phenotypes.

**Q2:** What is the offspring phenotype ratio?

**Q3:** Complete a Punnett square for a genetic cross of two true-breeding Portuguese water dogs—one with a black, wavy coat (homozygous dominant, *BBWW*) and one with a brown, curly coat (homozygous recessive, *bbww*). What is the phenotype ratio of their offspring (F₁)? Now fill out another Punnett square, crossing two of the offspring. What is the phenotype ratio of the F₂ generation?

# What Are the Odds?

The *probability* of an event is the chance that the event will occur. For example, there is a probability of 0.5 that a coin will turn up "heads" when it is tossed. A probability of 0.5 is the same thing as a 50 percent chance, or ½ odds, or a ratio of one heads to one tails (1:1). If you toss the coin only a few times, the observed percentage of heads may differ greatly from 50 percent. But if you toss it many, many times, that observed percentage will be very close to 50 percent. Each toss of a coin is an independent event, in the sense that the outcome of one toss does not affect the outcome of the next toss. The probability of getting two heads in a row is a product of the separate probabilities of each individual toss: 0.5 × 0.5, which is 0.25. In our cross, the probability of getting a brown puppy is ¼, or 0.25. To go back to a Punnett square to predict the ratio of puppies from a genetic cross of two heterozygous (*Bb*) black-coated dogs, the probability of getting a black puppy is ³⁄₄, or 0.75.

We cannot know with certainty what the actual phenotype or genotype of a particular offspring is going to be, except when true-breeding individuals are crossed. For example, two brown dogs, both of whom have a *bb* genotype, will have only *bb*-genotype, brown-phenotype offspring. Moreover, the probability that a particular offspring will display a specific phenotype is completely unaffected by how many offspring there are. The likelihood that we will see the 3:1 black-to-brown outcome, however, increases when we analyze a larger number of offspring, just as Mendel analyzed thousands of pea plants.

Mendel's laws by identifying chromosomes as the paired factors (where each homologous chromosome in a pair has one allele for a gene) that are shuffled and recombined, and then separated randomly into sperm and egg cells during meiosis (see Figures 6.7 and 6.8 in Chapter 6). Then, during fertilization, a one-in-a-million sperm fuses with a one-in-a-million egg to create a unique individual. That is how offspring can have genotypes and phenotypes that were not present in either parent, such as a brown puppy born to two black dogs.

# Going to the Dogs

Like Mendel, Elaine Ostrander planned to study plants to unravel the secrets of genetics and inheritance. But when she arrived at UC Berkeley to open her lab, the space was not yet available. So she wandered down the hall and into the office of Jasper Rine, a geneticist who normally studied yeast but was looking for someone to start a mammalian genome research project. Ostrander volunteered.

But which mammal to study? "I was allergic to cats, and I didn't know enough about cows or pigs or horses," she recalls, so she picked dogs. Not only was Ostrander a dog lover, but the American Kennel Club had just begun offering funding to researchers trying to identify genes associated with dog diseases.

In 1993, Ostrander began identifying all the genes unique to dogs—that is, making a map of the dog **genome**. Some colleagues said she was nuts, that no one would give her money to support the research. But Ostrander is nothing if not persistent, and she knew the potential value of the research: dogs have more than 350 inherited diseases, and up to 300 of those are similar to conditions in people, including cancer, epilepsy, heart disease, and Addison's disease, the illness that killed Lark's dog Georgie. The genetics of bladder cancer is difficult to study in humans, for example, but the disease is quite common in Scottish terriers and would be easier to study in a dog species. By cracking the genetic code of dogs, Ostrander hoped to uncover causes and potential treatments for human diseases.

In 2005, she published the first full dog genome sequence, for a female boxer named Tasha. The achievement gained her scientific fame and raised awareness among scientists of the importance of dog genetics to human health. "Of the more than 5,500 mammals living today, dogs are arguably the most remarkable," Ostrander's coauthor, Eric Lander, a professor of biology at the Massachusetts Institute of Technology, said when the first dog genome sequence was published. "The incredible physical and behavioral diversity of dogs—from Chihuahuas to Great Danes—is encoded in their genomes. It can uniquely help us understand embryonic development, neurobiology, human disease, and the basis of evolution" (**Figure 7.7**).

Shadow, a standard poodle, was the first dog to have its genome partially (about 80%) sequenced.

Tasha, a boxer, was the first dog to have its complete genome sequenced. Boxers are vulnerable to hip, thyroid, and heart problems. Scientists identified a gene for cardiomyopathy in boxers, a heart disorder also found in humans.

Pembroke Welsh corgis may develop a fatal neurodegenerative condition similar to amyotrophic lateral sclerosis (ALS) in humans. The human gene mutation associated with ALS was also found in corgis with the condition.

Psychiatric disorders often have a genetic component. Doberman pinschers are susceptible to canine compulsive disorder, similar to obsessive-compulsive disorder in humans. The responsible gene in Dobermans has been linked to autism disorders in humans.

Golden retrievers are prone to cancers of the bone marrow. Ostrander's research group is analyzing the genomes of hundreds of goldens with and without cancer, hoping to identify the genes responsible.

## Figure 7.7

### Man's best friend

The Dog Genome Project has identified the genetic basis of several diseases and conditions in dogs, and in some cases it has been able to link the gene to a similar gene in humans.

**Q1:** Boxers are far more inbred than poodles. Why does that inbreeding make the former a better target for genetic studies of disease than the latter?

**Q2:** Explain why a geneticist interested in finding a gene linked to cancer would want to look at the DNA of senior golden retrievers with *and* without cancer?

**Q3:** Obsessive-compulsive disorder (OCD) in humans is characterized by obsessive thoughts and compulsive behavior, such as pacing. Canine compulsive disorder (CCD) is characterized by compulsive behavior such as "flank sucking," sometimes seen in Doberman pinschers. Would you predict that the medications given to humans with OCD would decrease compulsive behaviors in CCD dogs? Why or why not?

But years before Ostrander completed the dog genome, she had begun a different pet project. In 2001, Ostrander received a call from a scientist in Utah who wanted to talk about dogs. It was Gordon Lark, who told her he was collecting genetic trait information about PWDs. "The day I met Gordon was the best day of my life," says Ostrander. "I knew it was golden."

In 2002, the duo published a paper pinpointing genes that control dog body shape, from the tall, lanky look of a greyhound to the short, stocky frame of a pit bull. In the acknowledgments of the paper, they thanked Karen Miller and all the PWD owners who had contributed pedigree information.

In the spring of 2006, Lark and Ostrander began their second collaboration, this time to identify the genetic basis of dog size. Lark collected skeletal measurements of 92 PWDs and DNA samples from each dog. Ostrander used that genotype and phenotype information to identify a key gene for body size—*IGF1*, which controls the activity of a growth factor and is known to influence body size in mice and humans. This gene's two alleles are called *I* and *B*. Lark and Ostrander discovered that PWDs homozygous for allele *I* (*II*) were usually large dogs, and those homozygous for allele *B* (*BB*) were always small dogs. That single gene accounted for whether a PWD was large or small.

Interestingly, neither *IGF1* allele is dominant or recessive. Instead, heterozygous dogs, with an *IB* genotype, are medium-sized dogs. This is an example of a trait inherited by **incomplete dominance**—in which neither allele is able to exert its full effect, so a heterozygote displays an intermediate phenotype. Dogs with an *IB* genotype aren't large or small, but rather medium-sized (**Figure 7.8**).

Early in the twentieth century, geneticists identified yet another type of interaction among alleles—codominance—that Mendel had not observed among his pea plants. A pair of alleles shows **codominance** when the effect of the two alleles is equally visible in the phenotype of the heterozygote. In dogs, gum color is codominant. A dog's gums can be pink, black, or pink with black spots; in the latter case, both alleles are fully on display, and neither is diminished or diluted by the presence of the other allele (as in incomplete dominance) or suppressed by a dominant allele (as in the case of dominant and recessive alleles). In humans, the blood type AB is a codominant trait.

**Genotype:** *II*
**Phenotype:** Large

**Genotype:** *BB*
**Phenotype:** Small

**Genotype:** *IB*
**Phenotype:** Medium

Figure 7.8

**Incomplete dominance of body size alleles**
Great Danes and Chihuahuas illustrate the extreme size variation found in domestic dogs (see the chapter-opening photo). Unlike the case with Mendelian traits, dogs heterozygous for the main body size gene show an intermediate size like the Cocker Spaniel pictured here on the right.

**Q1:** What are the genotypes of a large and a small dog?

**Q2:** Is it possible to have a heterozygous large dog? Explain why or why not.

**Q3:** Crossing a Great Dane and a Chihuahua is likely to be unsuccessful, even though they are members of the same species (and thus have compatible sperm and egg). Why is that? What are some potential risks of such a cross?

# It's Complicated

Many of the traits people tend to be curious about—body weight, intelligence, athleticism, and musical talent, to name a few—are yet more complicated. A **complex trait** is a genetic trait whose pattern of inheritance cannot be predicted by Mendel's laws of inheritance. Complex traits do not fit the straightforward single-gene, single-phenotype pattern discussed so far.

Sometimes a *single gene* influences a number of *different traits*. Such cases are examples of

# Most Chronic Diseases Are Complex Traits

A disease is a condition that impairs health. It may be caused by external factors, such as infection by viruses, bacteria, or parasites, or injury from harmful chemicals or high-energy radiation. Nutrient deficiency can also lead to disease. Inadequate vitamin C consumption, for example, produces scurvy, once common among sailors and pirates. Disease may also be caused by the malfunction of one or more genes. Diseases caused exclusively by gene malfunction are described as genetic disorders, distinguishing them from infections and other types of diseases.

But many of the diseases that are most common in industrialized countries—heart disease, cancer, stroke, diabetes, asthma, and arthritis, for example—are caused by multiple genes interacting in complex ways with each other and with external factors. They are complex traits: malfunctions in key genes make a person susceptible to developing these diseases, but environmental factors affect whether the disease will actually appear and how severe the symptoms will be. A large percentage of the estimated risk of developing chronic diseases is preventable by lifestyle choices such as maintaining good nutrition, exercising regularly (see graph), and avoiding tobacco. (The word "chronic" means "unceasing"—a reference to the fact a person who develops one of these diseases will have it for the rest of their lives.)

A major goal of modern genetics is to identify genes that contribute to human disease. Researchers have identified alleles associated with increased risk of a number of common ailments, including high blood pressure, heart disease, diabetes, Alzheimer disease, several types of cancer, and schizophrenia. The hope is that one day soon, genetic tests will tell us whether we are predisposed to a disease before we become ill with it. Then, a person carrying a risky allele might take preventive measures to reduce the chance of developing the condition, and treatment could be customized to fit the particular allele involved. This tailored approach to treatment, called "personalized medicine," is already being used to treat breast cancer and other chronic diseases.

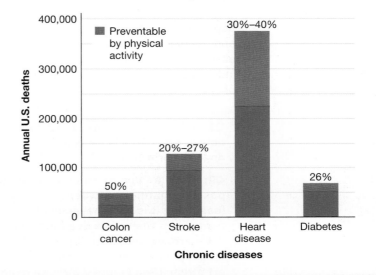

pleiotropy (*pleio*, "many"; *tropy*, "change"). In PWDs, Lark found that single genes can control multiple related skeletal traits. The shape of a dog's head and the shape of its limb bones are controlled by a single gene. That connection makes sense, says Lark, since a small head and long legs are advantageous for a fast dog, while a strong dog, like a pit bull, uses both its massive jaw and its short, thick legs for power.

Another good example of pleiotropy comes from a long-term breeding experiment to tame Russian silver foxes. Researchers found that as foxes became tamer and tamer, they also developed floppy ears instead of straight ones and had shorter legs and curlier tails than ordinary foxes have (see "The New Family Pet?," page 129).

Patterns of inheritance can get even more complicated. *Single traits* governed by the action of *more than one gene* are called **polygenic traits**. In humans, polygenic traits include eye and skin color, running speed, blood pressure, body size, and more. Of the thousands of human genetic traits, governed by an estimated 24,000 genes, fewer than 4,000 are known or suspected to be controlled by a single gene with a dominant and a recessive allele. The rest are polygenic traits.

Another twist on inheritance is **epistasis**, which occurs when the phenotypic effect of a gene's alleles depends on the presence of alleles for another, independently inherited gene. Labrador coat color, for example, is affected by epistasis (**Figure 7.9**). Dog fur, as mentioned earlier, has a dominant allele (*B*) that leads to black fur and a recessive allele (*b*) that produces brown fur. But the effects of these alleles (*B* and *b*) can be eliminated completely,

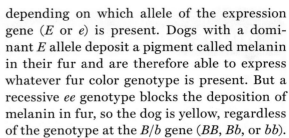

| Genotype: | B-, E- | bb, E- | --, ee |
|---|---|---|---|
| Phenotype: | Black | Brown | Yellow |

## Figure 7.9

**Epistasis in coat color**

These Labrador retrievers show complex inheritance of coat color. The yellow dog carries two alleles that interfere with the deposition of melanin in hair. Both the brown and the black dogs must carry at least one allele that allows melanin deposition. A dash indicates that the allele is unknown, based on phenotype.

**Q1:** What are the possible genotypes (at both genes) of the black dog? The yellow dog? The brown dog?

**Q2:** Draw a Punnett square showing possible matings between the black dog (assuming it is heterozygous at both genes) and the yellow dog (assuming it is heterozygous at the *B* gene). List all the possible phenotypes of their offspring. (See Figure 7.6 for an example of a Punnett square made with two traits.)

**Q3:** If you wanted the most variable litter possible, what colors of Labrador retrievers would you cross?

### ELAINE OSTRANDER

The "mother of all dog projects," Elaine Ostrander is chief and NIH distinguished investigator of the Cancer Genetics and Comparative Genomics Branch, as well as head of the Comparative Genetics Section at the National Institutes of Health. She studies genes important to growth, size variation, and cancer in dogs.

depending on which allele of the expression gene (*E* or *e*) is present. Dogs with a dominant *E* allele deposit a pigment called melanin in their fur and are therefore able to express whatever fur color genotype is present. But a recessive *ee* genotype blocks the deposition of melanin in fur, so the dog is yellow, regardless of the genotype at the *B/b* gene (*BB*, *Bb*, or *bb*).

If the environment affects the phenotype, it becomes nearly impossible to predict the phenotype when given only the genotype of an individual or its parents. The effects of many genes depend on internal and external environmental conditions, such as body temperature, carbon dioxide levels in the blood, external temperature, and amount of sunlight.

For example, cats have a gene that codes for an enzyme called tyrosinase, which is involved in melanin production. Siamese cats have a special $C^t$ allele of the gene. The $C^t$ allele codes for a tyrosinase that works well at colder temperatures ($\leq 35°C$) but does not function at warmer temperatures ($\geq 37°C$), so the production of melanin depends on the temperature of the surroundings (**Figure 7.10**). Because a cat's extremities tend to be colder than the rest of its body, melanin is produced there, and hence the paws, nose, ears, and tail of a Siamese cat tend to be dark. If a patch of light hair is shaved from the body of a Siamese cat and the skin is covered with an ice pack, the hair that grows back will be dark. Similarly, if dark hair is shaved from the tail and allowed to grow back under warm conditions, it will be light-colored.

# Man's Best Friend

After describing the inheritance of size in Portuguese water dogs, Lark and Ostrander looked at the *IGF1* gene in over 350 dogs representing 14 small breeds and 9 giant breeds. The genotype *BB* was common in small dogs and virtually nonexistent in large dogs. "All small dog breeds had them. It didn't matter when they were bred or how; they all had the exact same pattern," says Ostrander.

"It was amazing," adds Lark. Breeders have, over time, been selecting for these alleles to create smaller and smaller dogs. "What mankind can do, without any genetic tools but just knowledge of heritability, is just extraordinary," he says.

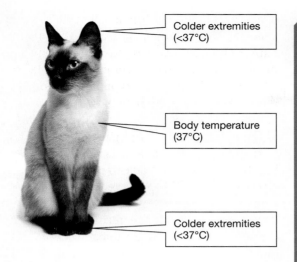

Colder extremities
(<37°C)

Body temperature
(37°C)

Colder extremities
(<37°C)

Figure 7.10

**The environment can alter the effects of genes**

Coat color in Siamese cats is controlled by a temperature-sensitive allele.

**Q1:** The gene that brings about the pale Siamese body fur is also responsible in part for the typical blue eyes of the species. What is the term for this type of inheritance?

**Q2:** Siamese kittens that weigh more tend to have darker fur on their bodies. Why might this be?

**Q3:** The Siamese cat pictured is called a "seal point" because it has seal-colored (dark brown) extremities. Some Siamese cats show the same color pattern, but the dark areas are of a lighter color or even a different shade—for example, lilac point, red point, blue point. What results would you predict if the experiments described in the text (shaving the cat and then increasing or decreasing temperature) were conducted on cats with these color patterns?

After their success with dog size, Lark and Ostrander identified genes responsible for other traits—fur color, leg length, skull shape, and more. They also identified genes related to cancer and other complex traits that might tell us something about human disease (see "Most Chronic Diseases Are Complex Traits," page 127). In border collies, Ostrander's team identified a gene

# The New Family Pet?

The silver fox is the same species as the more familiar red fox. Because of its soft, silver coat, it has been bred in captivity for over 100 years to provide fur coats, stoles, and hats for the wealthy.

In 1959, a Russian geneticist, Dmitry Belyaev, began to conduct breeding experiments on silver foxes he had purchased from a fur breeder, pairing only the tamest individuals of each generation. He determined how tame a fox was by observing its response when approached and offered food. As the foxes became tamer in each generation, they did not show a "fear response" until they were older—9 weeks instead of 6 weeks. (Domestic dogs develop a fear response at about 8–12 weeks.) In addition, in the tame foxes, the hormones associated with a fear response did not increase until later. These traits were all clearly influenced by the same gene or genes— an example of pleiotropy.

Another surprising result was that the foxes' appearance began to change along with the changes in behavior. They began to develop shorter tails, wider faces, and floppier ears. All of these made the adult

Only 50 years and 35 generations separate this tame silver fox from its wild relative

foxes look more puppy-like, and they are similar to the differences that can be observed when comparing the domestic dog to its ancestor, the wolf. Scientists conjecture that in both cases, tameness and associated changes in development, physiology, and anatomy were brought about by breeding for juvenile features.

**GEORGIE**

Georgie was Gordon Lark's first Portuguese water dog, a fiercely loyal and playful friend.

# Does Bigger Mean Better?

Genome size is the total amount of DNA in one copy of an organism's genome, typically measured in millions of base pairs (Mb). There is a huge range in genome sizes across plants and animals, from living organisms of 150,000 Mb to the *E. coli* genome of 4.6 Mb. Dogs and humans have smaller genomes than many species, making them easier to study genetically than, say, a lungfish.

Assessment available in smartwork5

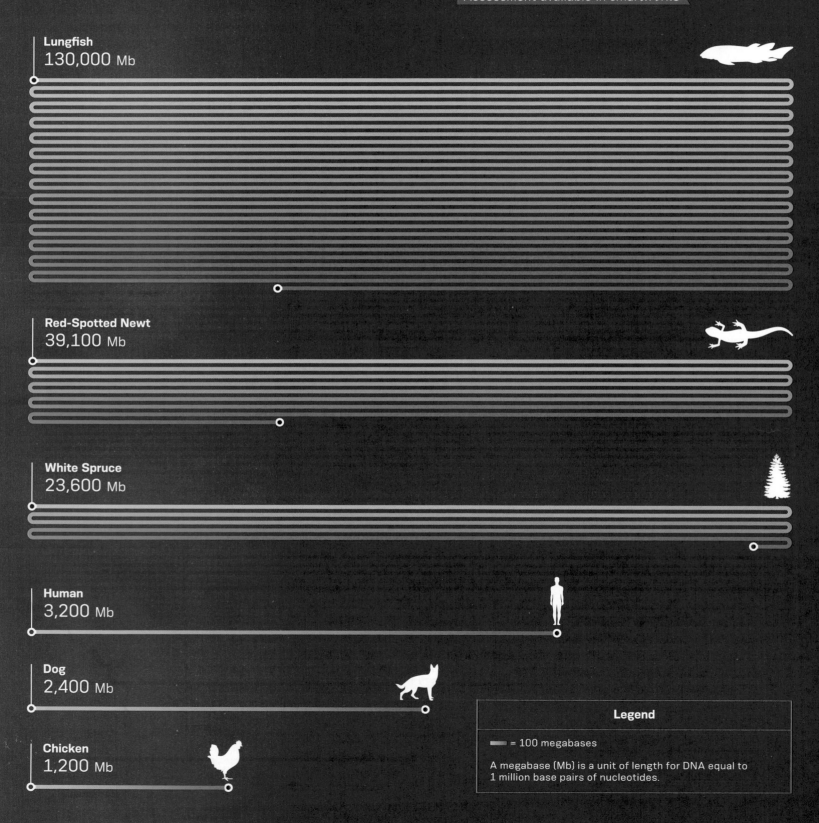

**Lungfish**
130,000 Mb

**Red-Spotted Newt**
39,100 Mb

**White Spruce**
23,600 Mb

**Human**
3,200 Mb

**Dog**
2,400 Mb

**Chicken**
1,200 Mb

### Legend

━━ = 100 megabases

A megabase (Mb) is a unit of length for DNA equal to 1 million base pairs of nucleotides.

involved in an eye disease that causes blindness in both humans and dogs. Her lab also identified a gene involved in kidney cancer in dogs that causes a similar syndrome in humans.

Today, Lark and Ostrander both continue their canine research, and the Georgie Project lives on. In one investigation, the team is collecting data from tissues of deceased PWDs to determine the dogs' state of health at time of death. From data gleaned, they hope to learn how the health of aging dogs (and maybe people) is associated with their genetic makeup. "I'm really happy with the growth of the field," says Ostrander, "and there's a lot of research coming down the pike."

Mopsa, the puppy that Karen Miller gave Lark in 1996 in return for studying PWDs, died in April 2012, just a week short of her sixteenth birthday (**Figure 7.11**). But Lark has a new best friend, a PWD puppy he named Chou (pronounced "shoo"), for the French *petit chou*, meaning "little cabbage."

"We often make a mistake and call Chou 'Mopsa,' because Chou looks so similar," says Lark. After all, he adds, PWDs share similar genotypes and thus similar phenotypes. And these genetic traits make them the cuddly, devoted pets that they are.

Figure 7.11

**Gordon Lark and Mopsa**

# REVIEWING THE SCIENCE

- A **genetic trait** is any characteristic that is inherited and may be **physical**, **biochemical**, or **behavioral**. All of these types of traits are either **invariant** (the same in all members of a species) or **variable** (different in different members of a species).

- A **gene** is a stretch of DNA that affects one or more genetic traits. Genes are formed on **chromosomes**, threadlike molecules made of DNA and proteins.

- The **genotype** is an individual's genetic makeup or, more specifically, the pair of different versions of a given gene (**alleles**) that determine a given trait. The **phenotype** is the physical expression of an individual's genetic makeup or, more specifically, the expression of a version of the given trait.

- An allele is **dominant** when it prevents a second allele from affecting the phenotype. This second allele is said to be **recessive** because it has no effect on the phenotype when paired with a dominant allele.

- When a genotype consists of two copies of the same allele, it is **homozygous** for that gene. A genotype that consists of two different alleles is **heterozygous** for that gene.

- A grid-like diagram called a **Punnett square** helps predict the probability of genotypes and phenotypes resulting from a **genetic cross**.

- Mendel's experiments enabled him to deduce two laws of inheritance: The **law of segregation** states that two alleles of a gene are separated during meiosis and end up in different gametes. The **law of independent assortment** states that during meiosis, the two alleles of any given gene segregate independently of any two alleles of any other gene.

- **Mendelian traits** are genetic traits controlled by a single gene and unaffected by environmental conditions.

- The **chromosome theory of inheritance** explains how Mendel's laws arise: genes occupy specific locations on chromosomes, and those chromosomes are randomly shuffled and recombined during meiosis.

- When a heterozygote displays an intermediate phenotype, neither allele is dominant—a condition known as **incomplete dominance**, in which neither allele is able to exert its full effect. When the effect of the two alleles is equally visible in the phenotype of the heterozygote, the pair of alleles shows **codominance**.

- **Complex traits** have patterns of inheritance that Mendel's laws cannot predict. In **pleiotropy**, for example, a single gene influences a number of different traits. **Polygenic traits**, by contrast, are governed by the action of more than one gene. And in **epistasis**, the phenotypic effect of the alleles of one gene depends on the presence of alleles for another, independently inherited gene.

# THE QUESTIONS

## The Basics

**1** Link each term with the correct definition.

GENOTYPE

PHENOTYPE

HETEROZYGOTE

HOMOZYGOTE

DOMINANT

RECESSIVE

1. An individual that carries one copy each of two different alleles (for example, an *Aa* individual or an *IB* individual).

2. An individual that carries two copies of the same allele (for example, an *AA*, *aa*, or *II* individual).

3. An allele that does not affect the phenotype when paired with a dominant allele in a heterozygote.

4. The genetic makeup of an individual; more specifically, the two alleles of a given gene that affect a specific genetic trait in a given individual.

5. The specific version of a genetic trait that is displayed by a given individual.

6. The allele that controls the phenotype when paired with a different allele in a heterozygote individual.

**2** Select the correct terms:
The (**gene / allele**) for coat color has two (**genes / alleles**)—one for brown coloring and one for black.

**3** Select the correct terms:
Cells undergo (**mitosis / meiosis**) to become gametes. This process sorts the alleles of a gene into separate gametes, which is the basis for Mendel's law of (**segregation / independent assortment**). Genes on different chromosomes also sort into separate gametes during this process, which is the basis for Mendel's law of (**segregation / independent assortment**).

**4** For each of the following cases, identify whether the described trait is an example of Mendelian inheritance (M) or a more complex form of inheritance (C).

_____ a. brown versus black coat color in dogs

_____ b. body size in dogs

_____ c. coat color in Siamese cats

_____ d. skin color in humans

_____ e. flower color in pea plants

**5** A single phenotype that results from a combination of two different genes in which one gene interferes with the expression of another gene is known as

(a) pleiotropy.

(b) complete dominance.

(c) incomplete dominance.

(d) codominance.

(e) epistasis.

**6** Before Mendel conducted his experiments with pea plants, people believed that offspring were a "blend" of their parents and would show intermediate levels of their parents' traits. What would Mendel's $F_1$ pea flowers have looked like if this were true?

(a) white

(b) purple

(c) red

(d) yellow

(e) light purple

## Challenge Yourself

**7** A riddle: In my type of inheritance, the $F_1$ offspring of a true-breeding black parent and a true-breeding white parent are all gray. What type of inheritance am I?

**8** Some time ago, you noticed that the sunflowers in your garden were either tall or short—nothing in between. You bred the tall sunflowers for many generations, until you felt confident that they "bred true," and you did the same for the short sunflowers. You then set up a parental cross (P), and all of the resulting $F_1$ offspring were short sunflowers. From this experiment, you conclude that the short phenotype is

(a) pleiotropic.

(b) recessive.

(c) true-breeding in the $F_1$ generation.

(d) dominant.

(e) incompletely dominant.

**9** In chickens, a mutant gene called *frizzle* causes not only feathers that curl outward like a Labradoodle's fur, but also an abnormal body temperature, an increased metabolism, and fewer eggs laid than by a normal chicken. From this information, you can conclude that the *frizzle* gene is _____.

## Try Something New

**10** The silver fox (see "The New Family Pet?" on page 127) belongs to the same species as the red fox: *Vulpes vulpes*. Two silver foxes always breed true for silver offspring. A silver fox bred to a red fox will produce either all red offspring or, occasionally, half red and half silver offspring. Red foxes bred together usually produce all red offspring, but they occasionally produce silver offspring in the ratio of 3 red to 1 silver. (*Hint:* Draw Punnett squares showing these predicted results.) Which of the following statements is/are consistent with the information provided here about inheritance of coat color in *Vulpes vulpes*? (Select all that apply.)

(a) Red foxes are all homozygous.

(b) Silver foxes are all homozygous.

(c) Red is dominant to silver.

(d) Some silver foxes are homozygous and some are heterozygous.

(e) Some red foxes are homozygous and some are heterozygous.

**11** In your garden you grow Big Boy (round) and Roma (oval) tomatoes. You love the taste of Big Boys, but you think it's easier to slice Roma tomatoes. You decide to cross-pollinate a Big Boy and a Roma to see whether you can create a new strain of "Long Boys." In the first generation, all of the tomatoes are round. How would you explain this result? What would your next cross be? Write out the cross in a Punnett square, using parental genotypes. What proportion of the next generation, if any, would be oval?

**12** For several hundred years, goldfish have been selectively bred in China and Japan for body color and shape, tail shape, bulging eyes, and even fleshy head growths.

Wild goldfish     Pet-shop goldfish     Black moor goldfish

Imagine that you have a tank of pet-shop goldfish and have just added a couple of black moor goldfish, hoping that they will breed. When the eggs laid by the black moor female (P generation) hatch and the young fish ($F_1$ generation) begin to develop, you are shocked to see that they are orange. How would you explain this result in terms of the inheritance of body color in goldfish? What breeding experiment could you conduct to test your hypothesis?

**13** In 2009, a large team of researchers including Elaine Ostrander and Gordon Lark published the results of its research on coat inheritance in dogs. The study began by focusing on dachshunds and Portuguese water dogs, but then widened to more than 80 breeds. The scientists were able to explain 95 percent of the variation in dog coat types with just two alleles at each of three genes, each inherited independently of the other. These genes coded for hair length (*L/l*), wave or curl in the coat (*W/w*), and the presence of "furnishings" (*F/f*), which are the moustache and eyebrows often seen in wire-haired dogs (see photo). Long-haired dogs carry two copies of the long-hair allele, which is

recessive to the short-hair allele. Dogs with furnishings can be either homozygous or heterozygous for the furnishings allele; dogs without furnishings are homozygous for the no-furnishings allele.

a. At the hair length and furnishings genes, what is the genotype of a long-haired dog without furnishings?

b. At the hair length and furnishings genes, what are all the possible genotypes of a short-haired dog with furnishings?

c. Create a Punnett square of two dogs heterozygous for hair length and furnishings. What is the offspring phenotype ratio for those two traits?

## Leveling Up

**14** **Doing science** Do you want to get involved in dog research? If you have a purebred as a pet, you can. Find out whether the Dog Genome Project at the National Institutes of Health is doing research on your pet's breed. If they are, you can send in a swab of your dog's saliva and contribute to science. Visit the NIH website (http://research.nhgri.nih.gov/dog_genome) for more information.

**15** **Is it science?** The November 18, 2003, issue of *Weekly World News* printed a story about a woman who, after repeatedly watching the movie *Shrek* while taking fertility drugs, gave birth to a baby who looked like the main character, an ogre named Shrek. Like Shrek, the newborn had dull green skin, a large flat nose, and ears protruding from stems. From what you know about genetics, do you think it's possible for a developing fetus to change so drastically (from a normal-looking baby to a "Shrek" baby) because its mother was obsessed with a movie? Why or why not? How would you explain your answer to someone who believed this news report?

**16** **What do *you* think?** Many people are critical of those who breed or purchase purebred dogs, arguing that there are many mixed-breed dogs waiting to be adopted from shelters. They also point out that mixed-breed dogs are less likely than purebred dogs to suffer from genetic diseases. Those who prefer a particular breed argue that there is a strong genetic influence on dog personality and behavior, and that they don't want any surprises when they add a new member to their family. What do you think?

**For more, visit digital.wwnorton.com/bionow2 for access to:**

# A Deadly Inheritance

*How researchers identified a mysterious genetic disorder, and their risky effort to develop a cure.*

## After reading this chapter you should be able to:

- Interpret a human pedigree to determine whether a given condition is recessive, dominant, or sex-linked.
- Review a human karyotype to identify the sex chromosomes and any abnormalities in chromosome number.
- Diagram a chromosome, identifying genes, alleles, and loci.
- Explain how sex is genetically determined in humans, and how sex determination relates to the inheritance of sex-linked traits.
- Describe the genotype and phenotype of a "genetic carrier" and use this term appropriately for sex-linked traits only.
- Compare and contrast the inheritance of recessive, dominant, and sex-linked genetic disorders.
- Calculate the probability of inheriting a particular genetic disorder by using a Punnett square.

# CHAPTER
## 08

CHROMOSOMES
AND HUMAN
GENETICS

Felix clutches his mother's side. Her arm is wrapped tightly around him. Lying in a rumpled white hospital bed, Felix looks away from the two red tubes protruding from his body. Blood pumps out one side of his body into a humming machine next to the bed and is pumped back into the other side. Dr. Christoph Klein steps into the room, clothed head to toe in blue scrubs, and offers a smile and a reassuring word (**Figure 8.1**).

Shortly after his birth in 2005, Felix began to bleed. His parents rushed him to an intensive care unit. The bleeding eventually stopped, but the hospital visits did not. Three years later, Felix was diagnosed with a rare and deadly disease: Wiskott-Aldrich syndrome (WAS). "Wiskott-Aldrich was the very last diagnosis I would ever want to receive," Felix's mother later said. "Every day I prayed, 'Please, Lord, let this chalice pass from us.'"

Felix's pediatrician feared the worst. WAS patients suffer from recurring infections, pneumonia, bleeding, and rashes; they often develop leukemia or lymphoma and die of complications due to infections. Some patients can be treated with a bone marrow transplant. However, if the donor is not a matched sibling or a close unrelated match, survival rates are low.

So, in 2009, Felix's family appeared in Klein's office at the Hannover Medical School in Germany, holding tightly to the hope that he might be able to save Felix's life. Klein, a pediatrician who has made a career of studying rare diseases, was running a clinical trial testing a new treatment for WAS. The treatment was risky and the results unknown, but it was Felix's only hope to be cured of his deadly disease—one that had perplexed scientists for decades.

# A Mysterious Malady

More than 70 years earlier, in 1937, three young brothers had come to see Dr. Alfred Wiskott, also a German pediatrician. At first, Wiskott had no idea what was causing their symptoms. The boys bled abnormally: their blood was unable to clot, and they had bloody diarrhea. They also had recurring ear infections and blistering, weeping rashes on their skin. Wiskott recorded their symptoms, but he could not help the boys. All three died at a young age.

What had killed the brothers? Their parents also had four daughters, all of whom were healthy, so it was unlikely that an infection, a toxin, or an environmental factor had caused the illness. Instead, Wiskott suspected, the boys might have inherited a disease from their parents.

Thanks to the chromosome theory of inheritance, Wiskott knew that the boys had inherited hereditary material, in the form of chromosomes, from their parents. Recall from Chapter 7 that offspring inherit one chromosome from the mother and one from the father. Wiskott suspected that a gene on one of the inherited chromosomes was causing the mysterious illness.

A disease caused by an inherited mutation, passed down from a parent to a child, is a **genetic disorder**. Wiskott recognized the importance of researching genetic disorders, since such studies could lead to the prevention of or cure for a disease. But daunting problems have long plagued the study of human genetic disorders. From a biological point of view, humans have a long generation time, select their own mates, and decide whether and when to have children. In addition, human families tend to be much smaller than would be ideal for a scientific study. From an ethical point of view, geneticists and physicians cannot intervene and perform experiments directly on humans to determine how genetic disorders are inherited.

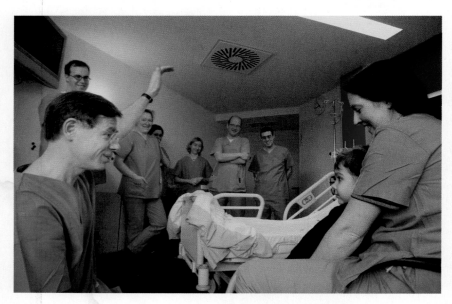

Figure 8.1

**Felix and Dr. Klein**

Dr. Christoph Klein examines Felix, who suffers from a rare genetic disease.

# Painful Pedigree

Twenty years after Wiskott described the first cases of the mysterious disease and determined that it was inherited, an American pediatrician, Dr. Robert Anderson Aldrich, solved the next piece of the puzzle. Aldrich met a six-month-old boy with anemia, bloody diarrhea, and general weakness. After several emergency room visits, the baby died.

Aldrich sat down with the boy's mother to review possible causes. After an hour of asking questions, he still had no idea what might have brought on the illness. Finally, he asked about other relatives who might have had a similar illness. The boy's grandmother, who had tagged along to the meeting, exclaimed sadly, "Just like all the rest of them." Other male infants in the family, it turned out, had died under similar circumstances.

Given that information, Aldrich worked with the mother and grandmother to trace the family's history back six generations by drawing a **pedigree**, a chart similar to a family tree that shows genetic relationships among family

**CHRISTOPH KLEIN**

A medical doctor and cancer researcher, Christoph Klein is now chair of pediatrics at the University of Munich. In 2010, Klein began testing a new gene therapy to treat young children with Wiskott-Aldrich syndrome, a rare and life-threatening disease. The therapy, though still experimental, has been very successful.

members over two or more generations of a family's medical history (**Figure 8.2**). Pedigrees provide scientists with a way to analyze information in order to learn about the inheritance of a particular genetic trait or disorder. Aldrich found that 16 male infants in the family, but no females, had died of the syndrome (**Figure 8.3**).

Because of the "remarkable family history," as he described it, Aldrich concluded, like Wiskott, that the illness was a genetic disorder, caused by a mutation passed down from parent to child. Genetic disorders can be caused by mutations in

**Generation**

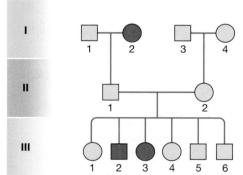

All geneticists who study human inheritance use a standard set of symbols in pedigrees:

• Roman numerals identify different generations

• Circles represent females; squares represent males

• Open symbols (blue here) represent unaffected individuals; filled symbols (red here) represent affected individuals

• Numbers listed below the symbols identify individuals of a given generation

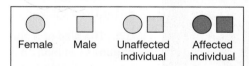

Female    Male    Unaffected individual    Affected individual

Figure 8.2

## Patterns of inheritance can be analyzed in family pedigrees

This cystic fibrosis pedigree shows six children (generation III), two of whom are affected with the disease.

**Q1:** Which two children in this pedigree have cystic fibrosis? How do you know?

**Q2:** Does either parent of these two children have cystic fibrosis? If so, which one(s)? How do you know?

**Q3:** Do any of the grandparents of these two children have cystic fibrosis? If so, which one(s)? How do you know?

**Generation**

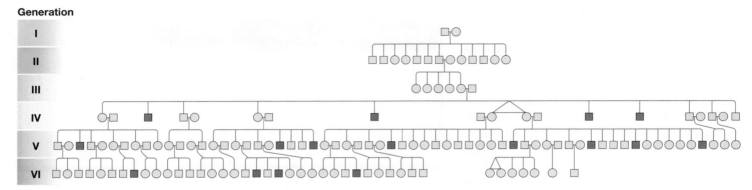

## Figure 8.3

### Pedigree of a family with a history of Wiskott-Aldrich syndrome

As in Figure 8.2, circles represent females and squares represent males in this pedigree. Individuals affected by WAS are shaded red.

**Q1:** How many male and how many female descendants (individuals that did not join the family by marriage) does generation IV of Aldrich's pedigree contain?

**Q2:** What proportions of the male and female descendants in generation IV were affected by the disorder?

**Q3:** Why did Aldrich hypothesize that the disease was X-linked? (You will need to read ahead to answer this question.)

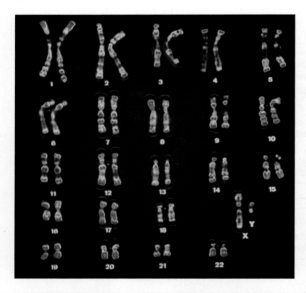

## Figure 8.4

### A human karyotype

To research chromosomal abnormalities, a scientist will take a photograph of a cell's chromosomes during mitosis and then pair up each set of homologous chromosomes to create a **karyotype**. In humans, the autosomes are numbered 1 through 22, and the sex chromosomes are designated X or Y.

**Q1:** Is this the karyotype of a male or a female?

**Q2:** How would the karyotype of a person with Down syndrome differ from this karyotype?

**Q3:** The size of a chromosome correlates roughly with the number of genes residing on it. Why are an extra copy of chromosome 21 and a missing Y chromosome two of the least damaging chromosomal abnormalities?

individual genes or by abnormalities in chromosome number or structure.

Every species has a characteristic number of chromosomes: humans have 23 pairs of homologous chromosomes, for a total of 46, while mosquitoes, for example, have only 3 pairs, or 6 in total. One of the 23 pairs of human chromosomes consists of the **sex chromosomes**; they determine whether a person is male or female. All other chromosomes are called **autosomes**.

Autosomes are homologous chromosomes exactly alike in terms of length, shape, and the genes they carry (**Figure 8.4**). Human autosomes are labeled with the numbers 1 through 22 (for example, chromosome 4). Sex chromosomes are assigned letter names; in humans, males have one X chromosome and one Y

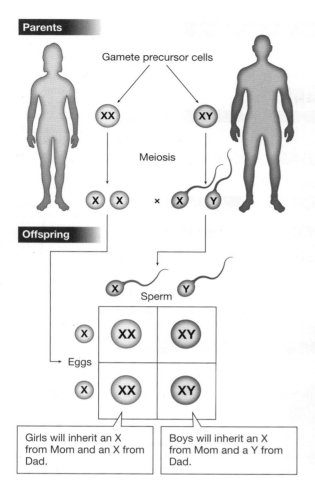

**Parents**

Gamete precursor cells

XX          XY

Meiosis

X  X  ×  X  Y

**Offspring**

X          Y

Sperm

| | X | Y |
|---|---|---|
| X | XX | XY |
| X | XX | XY |

Eggs

Girls will inherit an X from Mom and an X from Dad.

Boys will inherit an X from Mom and a Y from Dad.

Figure 8.5

## Dad's chromosomes determine baby's sex

**Q1:** What are the odds that a given egg cell will contain an X chromosome? A Y chromosome? What are those odds for a sperm cell?

**Q2:** If a couple has two daughters, does that mean their next two children are more likely to be sons? Explain your reasoning. (*Hint:* Refer back to "What Are the Odds?" on page 124.)

**Q3:** Sisters share the same X chromosome inherited from their father, but they may inherit different X chromosomes from their mother. What is the probability that brothers share the same Y chromosome? What is the probability that brothers share the same X chromosome?

### ALFRED WISKOTT

A German pediatrician who lived from 1898 to 1978, Alfred Wiskott described the cases of three young brothers with a serious bleeding disease—what would later be called Wiskott-Aldrich syndrome. Wiskott deduced that the boys had inherited a gene from their parents that was causing the illness.

chromosome, whereas females have two X chromosomes. The Y chromosome in humans is much smaller than the X chromosome.

Because human females have two copies of the X chromosome, all the gametes (eggs) they produce contain one X chromosome, passed on to their offspring. Males, however, have one X chromosome and one Y chromosome, so half of their gametes (sperm) will contain an X chromosome and half will contain a Y chromosome (**Figure 8.5**).

Each chromosome has a particular structure, with genes arranged on it in a precise sequence. Any change in the chromosome number or structure, compared to what is typical for a species, is considered a **chromosomal abnormality**. The two most common types of chromosomal abnormalities in humans are changes in the overall number of chromosomes and changes in chromosome structure, such as a change in the length of an individual chromosome (**Figure 8.6**).

Changes in the number of chromosomes in humans are usually lethal, although Turner syndrome, in which females receive only one X chromosome, is an exception to this rule. Turner syndrome individuals tend to live long, healthy lives with mild to moderate reproductive issues. Similarly, Down syndrome individuals receive three copies of chromosome 21 and can live relatively long lives, but they still suffer from mild to moderate intellectual and developmental disabilities. Changes in chromosome structure can also have dramatic effects. For example, cri du chat syndrome, caused by a deletion on chromosome 5, results in slowed growth, a small head, and developmental delays.

## Looking for Loci

In his 1954 paper describing the disease, Aldrich suggested that since primarily males inherit the syndrome, it is caused by a mutation on a sex chromosome. Aldrich recognized that the family's disease was tightly linked to gender and concluded that the disease was caused by

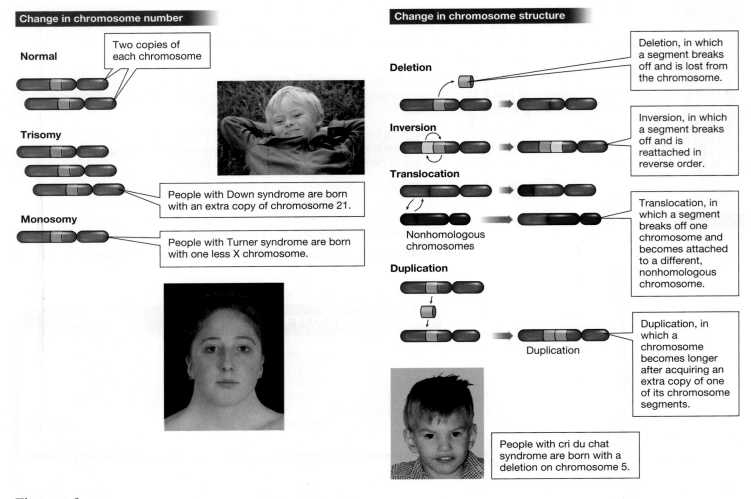

**Change in chromosome number**

**Normal**

Two copies of
each chromosome

**Trisomy**

People with Down syndrome are born
with an extra copy of chromosome 21.

**Monosomy**

People with Turner syndrome are born
with one less X chromosome.

**Change in chromosome structure**

**Deletion**

Deletion, in which
a segment breaks
off and is lost from
the chromosome.

**Inversion**

Inversion, in which
a segment breaks
off and is
reattached in
reverse order.

**Translocation**

Nonhomologous
chromosomes

Translocation, in
which a segment
breaks off one
chromosome and
becomes attached
to a different,
nonhomologous
chromosome.

**Duplication**

Duplication

Duplication, in
which a
chromosome
becomes longer
after acquiring an
extra copy of one
of its chromosome
segments.

People with cri du chat
syndrome are born with a
deletion on chromosome 5.

## Figure 8.6

### Chromosomal abnormalities can cause serious genetic disorders

Any increase or decrease in the number of chromosomes almost invariably results in spontaneous abortion of the fetus, which is estimated to occur in up to 20 percent of all pregnancies. Down syndrome and a missing or additional sex chromosome are exceptions. Changes in chromosome structure may have relatively minor or more severe effects, depending on the size and location of the change.

**Q1:** Why are changes in chromosome *number* almost always more severe than changes in chromosome structure?

**Q2:** In which part of meiosis would you predict that chromosomal abnormalities are produced? (Refer back to Chapter 6 if necessary.)

**Q3:** Create a mnemonic to help remember the four kinds of structural changes (for example, Doctors Improve Treatment Daily).

sex-linked inheritance—an inherited mutation on the X or Y chromosome. But where exactly was the mutation located?

The physical location of a gene on a chromosome is called its **locus** (plural "loci"). Because a gene can occur in different versions, or alleles, a diploid cell can have two different alleles at a given locus on a pair of homologous chromosomes. If the two alleles at a locus are different, the cell is heterozygous for the gene. If the two alleles at a locus are identical, the cell is homozygous for the gene (**Figure 8.7**).

The sex chromosomes, however, are different. Roughly 1,240 of the estimated 20,000 human genes are found on the X and Y chromosomes. Approximately 1,180 of those 1,240 genes are located on the X chromosome, while only about 60 are located on the much smaller Y chromosome. These 1,240 genes are said to be **sex-linked**. Sex-linked genes on the X chromosome are **X-linked**. Sex-linked genes on the Y chromosome are **Y-linked**. One of these Y-linked genes is the *SRY* gene (short for "<u>s</u>ex-determining <u>r</u>egion of <u>Y</u>"). *SRY* functions as the "master sex switch," committing the sex of the developing embryo to male. In the absence of this gene, a human embryo develops as a female.

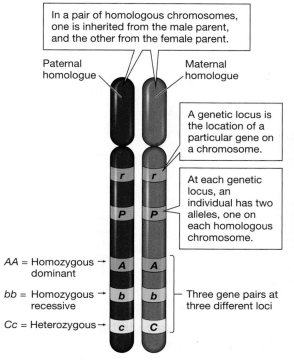

Figure 8.7

**Genetic loci on homologous chromosomes**

The genes shown here take up a larger portion of the chromosome than they would if they were drawn to scale. The average human chromosome has more than a thousand different genes interspersed with large stretches of noncoding DNA.

Q1: How do we know whether two chromosomes are homologous?

Q2: In one sentence, explain how the terms "gene," "locus," and "chromosome" are related.

Q3: If hair color were determined by a single gene, what would be an example of the gene's alleles?

The *SRY* gene does not act alone; in both males and females, other genes on the autosomes and sex chromosomes directly influence the development of the sexual characteristics that distinguish men and women. Still, the *SRY* gene plays a crucial role because when present, it causes other genes to produce male sexual characteristics, but when absent, those genes produce female sexual characteristics. Individuals that have XY chromosomal makeup but a nonfunctional SRY gene are considered intersex; they are not reproductively functional males or females, although they appear phenotypically female (**Figure 8.8**).

Interestingly, there are no well-documented cases of disease-causing Y-linked genes. X chromosomes, however, contain genes known to be involved in many human genetic traits and

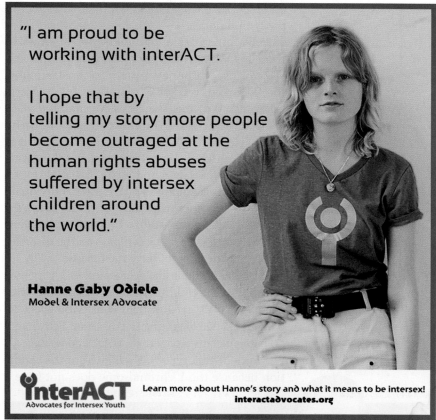

Figure 8.8

**Advocates for intersex youth: interACT**

Founded by attorney Anne Tamar-Mattis in 2006 under the name "Advocates for Informed Choice," interACT was created with a focused mission of ending harmful medical interventions on intersex children. Its mission statement reads "interACT uses innovative legal and other strategies, to advocate for the human rights of children born with intersex traits."

disorders. Aldrich correctly concluded that the gene leading to the family's bleeding disease was located on the X chromosome, so it was X-linked.

# X Marks the Spot

Thanks to advances in molecular biology tools, in 1994, researchers determined that the gene causing WAS is on the X chromosome and called it, unsurprisingly, *WAS* (gene names are typically italicized, while disorder names, in this case "WAS," are not). *WAS*, they discovered, is the genetic code for a protein crucial for the formation and function of blood cells and immune system cells. Without a healthy copy of this gene, individuals acquire blood and immune system disorders, are susceptible to infections, and have increased risk of lymphoma, cancer of the lymph nodes.

Felix inherited a mutated version of *WAS* from his mother and no allele from his father. We can use a Punnett square to illustrate how the X-linked recessive mutation for WAS is inherited. We label the recessive mutated *WAS* allele *a*, and in the Punnett square we write this allele as $X^a$ to emphasize the fact that it is on the X chromosome. Then we label the dominant, healthy allele *A* and write this allele as $X^A$ in the Punnett square (**Figure 8.9**). Individuals like Felix's mother, who have only one copy of a recessive allele, are said to be **genetic carriers** of the disorder. Carriers can pass on the disorder allele, but they do not have the disease.

If a carrier female like Felix's mother, with genotype $X^A X^a$, has children with a normal male (with genotype $X^A Y$), each of their sons will have a 50 percent chance of getting the disorder. Felix had a 50 percent chance of getting WAS, and he did.

Males of genotype $X^a Y$ suffer from the condition because the Y chromosome does not have a

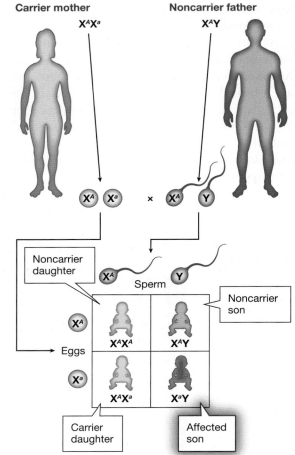

Figure 8.9

## X-linked recessive conditions are more common in males

The recessive disorder allele (*a*) is located on the X chromosome and is denoted by $X^a$. The dominant normal allele (*A*) on the X chromosome is denoted by $X^A$. ▶️

**Q1:** Which of the children specified in this Punnett square represents Felix? What is his genotype?

**Q2:** Explain why Felix is neither homozygous nor heterozygous for the *WAS* gene.

**Q3:** Create a Punnett square to illustrate the offspring that could result if Felix had children with a noncarrier woman. What is the probability that a son would have WAS? What is the probability that a daughter would be a carrier of WAS?

## ROBERT ALDRICH

Robert Aldrich was an American pediatrician who lived from 1917 to 1998. By creating a family pedigree, Aldrich demonstrated that the mysterious bleeding syndrome afflicting infant boys was a sex-linked, recessive disorder.

# Genetic Diseases Affecting Americans

Wiskott-Aldrich syndrome is only one of many genetic conditions seen in humans. Here are some of the most common genetic diseases in the United States, most of which can be identified in newborns through genetic testing.

## U.S. births per year: 4,000,000

• = 1 birth

**6,037 births**
Down Syndrome

**1,140 births**
Cystic Fibrosis

**800 births**
Marfan Syndrome

## Male U.S. births per year: 2,050,000

• = 1 male birth

**590 births**
Duchenne Muscular Dystrophy

**400 births**
Hemophilia

**400 births**
Fragile X Syndrome

**Gaucher disease**
Chronic enzyme deficiency; fairly common among Ashkenazi Jews

**Familial colon cancer**
One in 200 people has this allele; of those who have it, 65% are likely to develop the disease

**Retinitis pigmentosa**
Progressive degeneration of the retina

**ALD (adrenoleukodystrophy)**
Nerve disease portrayed in movie *Lorenzo's Oil*

**Huntington disease**
Neurodegenerative disorder tending to strike people in their forties and fifties

**Neurofibromatosis, type 2**
Tumors of the auditory nerve and tissues surrounding the brain

**Familial polyposis of the colon**
Abnormal tissue growths frequently leading to cancer

**Amyotrophic lateral sclerosis (ALS) (Lou Gehrig disease)**
Fatal degenerative nerve ailment

**Spinocerebellar ataxia**
Destroys nerves in the brain and spinal cord, resulting in loss of muscle control

**Adenosine deaminase (ADA) immune deficiency**
Metabolic disorder that damages the immune system

**Cystic fibrosis**
Mucus fills up the lungs, interfering with breathing; one of the most prevalent genetic diseases in the U.S.

**Familial hypercholesterolemia**
Extremely high cholesterol

**Multiple exostoses**
A disorder of cartilage and bone

**Amyloidosis**
Accumulation in the tissues of an insoluble fibrillar protein

**Malignant melanoma**
Tumors originating in the skin

**Breast cancer**
Roughly 5% of cases are caused by this allele

**Multiple endocrine neoplasia, type 2**
Tumors in endocrine glands and other tissues

**Polycystic kidney disease**
Cysts resulting in enlarged kidneys and renal failure

**Sickle-cell disease**
Chronic inherited anemia in which red blood cells sickle (form crescents), plugging small blood vessels

**Tay-Sachs disease**
Fatal hereditary disorder involving lipid metabolism; most common in Ashkenazi Jews and French Canadians

**Burkitt lymphoma**
A translocation between chromosomes 14 and 18 results in cancer of the white blood cells; most common in children and young adults

**Retinoblastoma**
Relatively common eye tumor, accounting for 2% of childhood malignancies

**PKU (phenylketonuria)**
An inborn error of metabolism that results in mental retardation if untreated

Figure 8.10

### Single-gene disorders

Mutations of single genes that lead to genetic disorders are found on the X chromosome and on each of the 22 autosomes in humans. In each of these mutations, the healthy allele at that locus codes for an important function; for example, the sickle-cell allele is a mutation in the gene that codes for the hemoglobin protein, critical for carrying oxygen in the blood. For clarity, only one such genetic disorder per chromosome is shown.

**Q1:** Which chromosome contains the gene for cystic fibrosis? For Tay-Sachs disease? For sickle-cell disease?

**Q2:** No known genetic disorders are encoded on the Y chromosome. Why do you think this is?

**Q3:** In your own words, explain why most single-gene disorders are recessive rather than dominant.

copy of that gene. In other words, because males cannot be heterozygous for any X-linked genes, the effects of an *a* allele cannot be masked. In general, males are more likely than females to get recessive X-linked disorders, because they need to inherit only a single copy of the disorder allele to exhibit the disorder. Females, on the other hand, must inherit two copies to be affected. X-linked recessive inheritance explains why boys are more likely to get WAS than girls are.

Other X-linked genetic disorders in humans include red-green color blindness, hemophilia, and Duchenne muscular dystrophy—a lethal disorder that causes muscles to waste away, often leading to death at a young age. All of these X-linked disorders are caused by recessive alleles.

# More Common, but No Less Deadly: Zoe's Story

X-linked disorders like WAS are rare compared to autosomal disorders. Both sexes are equally likely to be affected by **recessive genetic disorders** on autosomes, since both males and females have two copies of autosomal chromosomes and therefore identical odds of being homozygous or heterozygous for a disorder allele. Several thousand human genetic disorders are inherited as recessive traits on autosomes. These include sickle-cell disease, Tay-Sachs disease, and the most common fatal genetic disease in the United States, cystic fibrosis (**Figure 8.10**).

Scott and Jada first began to suspect something was wrong when their newborn daughter, Zoe, didn't put on any weight. Every time Zoe ate, her belly became hard and bloated. She screamed in pain. "I was beside myself," recalls Jada. "I knew something wasn't right."

Like Felix, Zoe spent the first year of her life in and out of the pediatrician's office, until the eve of her first birthday, April 6, 2005. On that day, Scott and Jada sat with Zoe in a children's hospital in Florida, waiting for a second opinion. The doctor came in and asked them to sit down. The diagnosis was cystic fibrosis (**Figure 8.11**).

Cystic fibrosis (CF) is a lethal recessive genetic disorder caused by one or more mutations in the

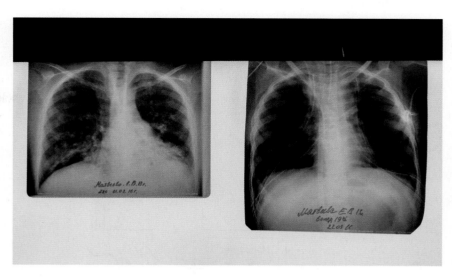

Figure 8.11

**Cystic fibrosis damages lung tissue**
The panel on the left shows a chest x-ray from a cystic fibrosis patient and the panel on the right shows a normal chest x-ray. Notice the damaged fibrotic lung tissue in the cystic fibrosis x-ray.

cystic fibrosis transmembrane regulator gene (*CFTR*). A *CFTR* mutation causes the body to produce abnormally thick, sticky mucus, which clogs the airways and leads to lung infections. The thick mucus also obstructs the pancreas, preventing enzymes from reaching the intestines, where they are needed to break down and digest food. The average life span of people with CF who live to adulthood is approximately 35 years. There is no cure.

Recessive genetic disorders vary in severity; some, like cystic fibrosis, are lethal, whereas others have relatively mild effects. Adult-onset lactose intolerance, for example, is caused by a single recessive allele that leads to a shutdown in the production of lactase, the enzyme that digests milk sugar.

The only individuals who get a disorder caused by an autosomal recessive allele (*a*) are those who have two copies of that allele (*aa*). Usually, when a child inherits a recessive genetic disorder, both parents are heterozygous; that is, they both have the genotype *Aa* (**Figure 8.12**). It is also possible for one or both parents to have the genotype *aa* and thus the disease. Because the *A* allele is dominant and does not cause the disorder, heterozygous individuals (*Aa*) like Scott and Jada, Zoe's

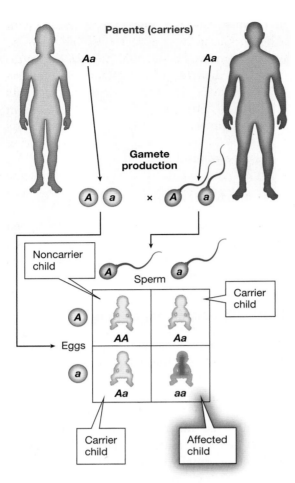

**Figure 8.12**

**Inheritance of cystic fibrosis, an autosomal recessive disorder**

The patterns of inheritance for a human autosomal recessive genetic disorder are the same as for any recessive trait (compare this figure with the pattern shown by Mendel's pea plants in Figure 7.5). Recessive disorder alleles are denoted *a*. Dominant, normal alleles are denoted *A*. Here, the parents are a carrier female (genotype *Aa*) and a carrier male (genotype *Aa*). ▶

**Q1:** Which of the children in this Punnett square represents Zoe? What is her genotype?

**Q2:** If Zoe's parents had another child, what is the probability that the child would have cystic fibrosis? That the child would be a CF carrier?

**Q3:** If Zoe is able to have a child of her own someday, and the other parent is not a carrier of cystic fibrosis (he would likely be tested before they chose to have children), what is the probability that the child would have cystic fibrosis? That the child would be a carrier?

parents, are genetic carriers of the disorder; they carry the disorder allele (*a*) but do not have the disease.

If two carriers of a recessive genetic disorder have children, the patterns of inheritance are the same as for any other recessive trait. Each child, male or female, has a 25 percent chance of not carrying the disorder allele (genotype *AA*), a 50 percent chance of being a carrier (genotype *Aa*), and a 25 percent chance of inheriting the disorder (genotype *aa*). Zoe did not beat the odds.

These percentages reveal one way in which lethal recessive disorders such as cystic fibrosis can persist in the human population. Although homozygous recessive individuals (with genotype *aa*) often die before they are old enough to have children, carriers (with genotype *Aa*) are not harmed by the disorder. In a sense, the *a* alleles can hide in heterozygous carriers, and those carriers are likely to pass the disorder allele to half of their children. An estimated one in 29 European Americans has a mutated *CFTR* gene. Recessive genetic disorders can also arise in the human population because new mutations can produce new copies of the recessive alleles.

# Deadly with One Allele

Cystic fibrosis is an example of a recessive genetic disorder, in which a child, like Zoe, inherits two recessive copies of a disorder allele. A more rare type of inherited disease is a **dominant genetic disorder**, caused by an autosomal dominant allele (*A*). In this case, the allele that causes a disorder cannot "hide" in the same way that a recessive allele can: *AA* and *Aa* individuals get the disorder; only *aa* individuals are symptom-free (**Figure 8.13**). Dominant genetic disorders are more rare than recessive disorders because a dominant disorder often produces serious negative effects immediately upon birth, and individuals with the *A* allele may not live long enough to reproduce. Hence, few people with a dominant genetic disorder pass the allele on to their children.

For this reason, most cases of a dominant genetic disorder are produced by a new mutation in a generation. For example, achondroplasia, a form of dwarfism, is caused by a mutation

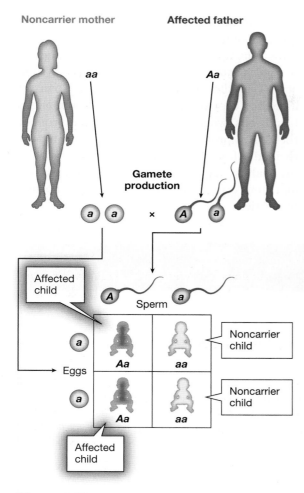

**Figure 8.13**

### Inheritance of an autosomal dominant disorder

The pattern of inheritance for a human autosomal dominant genetic disorder is the same as for any other dominant trait. This Punnett square shows the possible children of a normal female (genotype *aa*) and an affected male (genotype *Aa*).

**Q1:** What is the probability that a child with one parent who has an autosomal dominant disorder will inherit the disease?

**Q2:** Why are there no carriers with a dominant genetic disorder?

**Q3:** Because dominant genetic disorders are rare, it is extremely rare for both parents to have the condition (genotype *Aa*). Draw a Punnett square with two *Aa* parents. What proportion of the offspring would have the disorder? What proportion would be normal?

in a gene involved in bone growth. People with achondroplasia have a decreased life span, so few of them live long enough to pass the mutation on to their offspring. Instead, infants with achondroplasia are born to unaffected parents at a rate of between one in 10,000 and one in 100,000. Almost all of these births are to fathers older than 35 years, who produce this mutation during sperm production.

Huntington disease, a dominant genetic disorder, is an exception to that rule because symptoms of the disease—uncontrolled movements and loss of intellectual faculties caused by dying brain cells—arise later in life, in one's forties, after the person carrying that allele has had the opportunity to reproduce. In this way the allele is readily passed from one generation to the next. This is why some couples whose families have a history of Huntington's may choose to screen their developing fetus for the gene that causes the disorder.

## Replacing Deadly Genes: A Work in Progress

Most inherited genetic disorders, including cystic fibrosis and Huntington disease, have no cure. Patients and their families do everything they can simply to manage the disease. At the age of 7, Zoe was taking 25 pills each day just to digest food. At the age of 10, she underwent four extended hospitalizations, and almost her entire summer was spent on drip IVs. Zoe continues to take antibiotics and mucus thinners, uses nasal sprays, and spends 2 hours a day doing breathing therapy. "We don't talk about the future," says Jada, Zoe's mother. Her voice cracks. "But she is aware that she's not like everybody else."

Scientists have not given up the race to find effective treatments, even cures, for genetic disorders. Thanks to Christoph Klein, WAS is one of the few such disorders for which an effective treatment has been identified.

It was in 2003 that Klein, then at Boston Children's Hospital, saw the first glimmer of hope that there might be a therapy for boys afflicted with WAS. He and colleagues believed that they might be able to treat WAS using **gene therapy**, a technique for correcting defective genes responsible for disease development. Gene therapy is a type of **genetic engineering**, the permanent introduction of one or more genes into a cell, tissue, or organism. Klein hoped that by correcting the defective *WAS* gene in young affected boys, he could offer a short-term treatment, and potentially even a permanent cure, for the disorder.

First, Klein and his team tested their plan in mice. They bred mice that lacked the entire *WAS* gene and thus could not produce the important WAS protein. These mice had some of the same symptoms that Felix had, including a reduction in the number of blood and immune system cells. Then the researchers used a virus to insert a healthy copy of the *WAS* gene into the mice's blood cells, where they hoped it would produce the WAS protein. Months later, Klein was thrilled to find that the mouse cells were expressing healthy WAS protein, and that mature blood and stem cells were being produced in normal quantities.

# Prenatal Genetic Screening

How is the baby? This is one of the first questions we ask after a child is born. Usually everything is fine, but sometimes, as with Felix and Zoe, the answer can be devastating. Today, some parents choose to have prenatal genetic screening tests performed to check their baby's health before birth.

In **amniocentesis**, a needle is inserted through the abdomen into the uterus to extract a small amount of amniotic fluid from the pregnancy sac that surrounds the fetus. This fluid contains fetal cells (often sloughed-off skin cells) that can be tested for genetic disorders. Another method is **chorionic** (kohr-ee-AH-nik) **villus sampling** (**CVS**), in which a physician uses ultrasound to guide a narrow, flexible tube through a woman's vagina and into her uterus, where the tip of the tube is placed next to the villi, a cluster of cells that attaches the pregnancy sac to the wall of the uterus. Cells are removed from the villi by gentle suction and then tested for genetic disorders.

Risks associated with amniocentesis and CVS—including vaginal cramping, miscarriage, and premature birth—have declined dramatically in recent years because of advances in technology and more extensive training. Recent studies suggest that the risk of miscarriage after CVS and amniocentesis is essentially the same: about 0.06 percent. The tests are widely used by parents who know they face an increased chance of giving birth to a baby with a genetic disorder. Older parents, for example, might want to test for Down syndrome, since the risk of that condition increases with the age of the mother and perhaps the father too. A couple in which one parent carries an allele for a dominant genetic disorder (such as Huntington disease), or both parents are carriers for a recessive genetic disorder (such as cystic fibrosis), might also choose prenatal genetic screening.

Couples who elect to have such tests performed have only two choices if their fears are confirmed: they can abort the fetus, or they can give birth to a child with a genetic disorder. Prior to conception, however, couples at risk of having a child with a genetic disorder have options to minimize that risk.

- If they are willing and can afford the procedure, a couple can choose to have a child by **in vitro fertilization** (**IVF**), in which an egg is fertilized by a sperm in a petri dish, after which one or more embryos are implanted into the mother's uterus.

- During **preimplantation genetic diagnosis** (**PGD**), one or two cells are removed from the developing embryo in the dish, usually 3 days after fertilization. The cell or cells removed from the embryo are tested for genetic disorders. Finally, one or more embryos that are free of disorders are implanted into the mother's uterus, and the rest of the embryos, including those with genetic disorders, are frozen. Typically, parents who opt for PGD either have a serious genetic disorder or carry alleles for one.

Like all other genetic screening methods, the use of PGD raises ethical issues. People who support PGD feel that amniocentesis and CVS provide parents with a bleak set of moral choices: if the fetus has a serious genetic disorder, the parents can either abort the fetus or allow the birth of a child who will live a life that may be short and full of suffering. In their view, discarding an embryo at the 4- to 12-cell stage is morally preferable to aborting a well-developed fetus, or to giving birth to a child that will suffer the devastating effects of a serious genetic disorder. Those opposed to PGD agree that the moral choices are bleak, but they argue that once fertilization has occurred, a new life has formed and it is immoral to end that life, even at the 4- to 12-cell stage. What do you think?

Bolstered by these results, the scientists tested the technique in human cells and finally, in 2005, began a human clinical trial in Germany. The first two boys, both 3 years old, were admitted in 2006. Between 2006 and 2009, 10 boys were admitted to the clinical trial, with Felix coming on board in 2009.

During the first phase of the trial, Felix had to lie still for 9 hours as a machine pumped blood out of his body in order to extract cells. These cells were then taken to a laboratory where scientists used a genetically modified virus to insert a healthy copy of *WAS* into them (**Figure 8.14**). The doctors then pumped the genetically engineered cells back into Felix's body. And the waiting began.

Klein published the early results in 2010. After gene therapy, nine of the boys improved with regard to bleeding and infections. They were able to play soccer like their healthy peers, they responded to vaccinations, and they did not develop severe infectious diseases. Klein and his colleagues showed that gene therapy for the fatal WAS diagnosis is feasible, and that WAS can be corrected.

A new therapy does not come without risks, though. A few years after receiving the gene therapy, seven of the patients developed leukemia

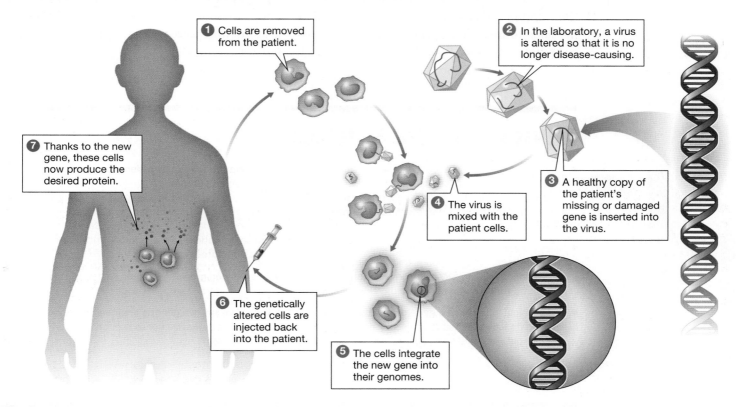

Figure 8.14

## Gene therapy

In gene therapy, genetic information is transferred into cells to achieve a desired effect. Gene therapy may be used to compensate for a genetic mutation in a cell that causes the cell to malfunction. ▶️

**Q1:** Which gene was missing or damaged in Felix's case? From what chromosome would a healthy copy be taken?

**Q2:** Why did Dr. Klein's group first conduct gene therapy on mice rather than on humans? What are the advantages and limitations of this approach?

**Q3:** If Felix has children of his own someday, will they run the risk of inheriting his disorder, or has gene therapy removed that possibility? Explain your reasoning.

from the type of virus used to insert the healthy gene into their cells. Three of the boys died, either from leukemia or from complications after bone marrow transplants. For this reason the trial was halted. Together with other physicians and researchers, Klein is investigating the causes of leukemia after gene therapy, and he is looking for safer ways to insert the healthy gene into cells.

# A Happy Ending for Felix

Numerous gene therapy trials continue around the world for other genetic diseases, using other methods to insert new genes as a way to avoid risks like leukemia. In 2012, European health officials green-lit gene therapy for a recessive genetic disorder called lipoprotein lipase deficiency (LPLD), the first commercially approved gene therapy treatment in either Europe or the United States. Many others nip at its heels, including treatments for hemophilia and ADA-SCID, also known as the "bubble boy disease."

As for Felix, a year after the gene therapy, he returned to the hospital for tests. The treatment had worked. Today, his body is producing healthy, functional blood cells, and most of his symptoms are completely resolved. Felix now enjoys a normal life. His inherited genetic disorder was cured by gene therapy. "The most beautiful moment is when my son smiles, hugs me, and says, 'Mum, I love you so much,'" said Felix's mother after the treatment. "We are grateful that there was a treatment that helped our son."

## REVIEWING THE SCIENCE

- A **genetic disorder** is a disease caused by an inherited mutation in a gene, passed down from a parent to a child. Genetic disorders can be caused by mutations in individual genes or by **chromosomal abnormalities** (changes in chromosome number or structure).

- Every person has two **sex chromosomes**: males have one X and one Y chromosome, and females have two X chromosomes. The *SRY* gene on the Y chromosome is required for human embryos to develop as males.

- The physical location of a gene on a chromosome is called its **locus**. Genes found solely on the X or Y chromosome are said to be **sex-linked**.

- A **dominant genetic disorder** is caused by a dominant allele on an **autosome** (non–sex chromosome). Dominant genetic disorders are more rare than **recessive genetic disorders**, which are caused by two autosomal recessive alleles. Several thousand human genetic disorders are inherited as recessive traits on autosomes.

- A family **pedigree** can be used to determine whether a given condition is recessive, dominant, or sex-linked.

- **Gene therapy** is a **genetic engineering** technique for correcting defective genes responsible for disease development.

## THE QUESTIONS

### The Basics

**1** Use these terms correctly in the following sentence: **alleles, chromosomes, genes, loci.**
Two homologous _____ contain the same _____, found at the same _____, but may have the same or different copies of _____.

**2** A particular person is said to be a carrier of a genetic trait. What does this tell you about their phenotype?

(a) They physically show the trait.

(b) They physically show the trait more than a noncarrier would show the trait.

(c) They are almost normal but show an intermediate phenotype for the trait.

(d) They are completely normal and do not physically show the trait.

**3** Match each of the following terms with the correct definition.

| | |
|---|---|
| **GENE THERAPY** | 1. A procedure in which cells are gently suctioned from a pregnant woman's uterus to test for genetic disorders in the fetus. |
| **IN VITRO FERTILIZATION** | 2. A procedure in which a small amount of fluid (and fetal cells within it) is carefully extracted from a pregnant woman's uterus to test for genetic disorders in the fetus. |
| **PREIMPLANTATION GENETIC DIAGNOSIS (PGD)** | 3. A treatment approach that seeks to correct a genetic disorder by inserting healthy copies of the mutated genes responsible for the disorder. |
| **CHORIONIC VILLUS SAMPLING (CVS)** | 4. A procedure in which one or two cells are removed from a developing embryo and tested for genetic disorders; embryos that are free of genetic disorders may then be implanted into a woman's uterus. |
| **AMNIOCENTESIS** | 5. A procedure in which an egg is fertilized in a petri dish, after which one or more embryos are implanted into a woman's uterus. |

**4** In the karyotype shown here, identify the sex chromosomes. Is this individual a male or a female?

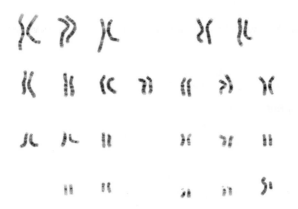

## Challenge Yourself

**5** Sickle-cell disease is inherited as a recessive genetic disorder in humans; the normal hemoglobin allele (*H*) is dominant to the sickle-cell allele (*h*). For two parents of genotype *Hh* (carriers), construct a Punnett square to show the possible genotypes of their children.

**6** Which of the following correctly predicts the possible genotypes of the children from question 5?

(a) ¼ *HH*, ¼ *Hh*, ½ *hh*

(b) ½ *HH*, ½ *Hh*

(c) ½ *HH*, ½ *hh*

(d) ¼ *HH*, ½ *Hh*, ¼ *hh*

(e) ¼ *HH*, ³/₄ *Hh*

**7** Which of the following correctly predicts the possible phenotypes of the children from question 5?

(a) ¼ normal, ³/₄ sickle-cell disease

(b) all normal

(c) all sickle-cell disease

(d) ½ normal, ½ sickle-cell disease

(e) ³/₄ normal, ¼ sickle-cell disease

**8** Given that the normal hemoglobin allele (*H*) is dominant to the sickle-cell allele (*h*), each time two *Hh* individuals have a child together, what is the chance that the child will have sickle-cell disease?

(a) 0%

(b) 75%

(c) 25%

(d) 50%

(e) 100%

**9** Which of the following karyotypic sex chromosome abnormalities result(s) in a male phenotype? (Select all that apply.)

(a) XO

(b) XXY

(c) XXX

(d) XYY

(e) XXXY

(f) XY with a complete deletion of the *SRY* gene

## Try Something New

**10** On the Indonesian island of Bali, about ³/₄ of the feral (stray) cats have a stumpy tail, while only ¼ of the cats have a long tail. Experiments in which a bunch of random stumpy-tailed cats were mated with each other yielded some stumpy-tailed kittens and some long-tailed kittens. The same experiments with long-tailed cats produced only long-tailed kittens. What do these results show about the genotype of the long-tailed cats? (*Hint:* Draw Punnett squares using the letter "t" for the tail trait to help you determine which allele is dominant and which is recessive.)

(a) They must all be *tt*.

(b) They are both *Tt* and *tt*.

(c) They are both *Tt* and *TT*.

(d) They must all be *Tt*.

**11** Recall that human females have two X chromosomes, while human males have one X chromosome and one Y chromosome.

a. Do males inherit their X chromosome from their mother or from their father?

b. If a female has one copy of an X-linked recessive allele for a genetic disorder, does she have the disorder?

c. If a male has one copy of an X-linked recessive allele for a genetic disorder, does he have the disorder?

d. Assume that a female is a carrier of an X-linked recessive disorder. With respect to the disorder allele, how many types of gametes can she produce?

e. Assume that a male with an X-linked recessive genetic disorder has children with a female who does not carry the disorder allele. Could any of their sons have the genetic disorder? How about their daughters? Could any of their children be carriers for the disorder? If so, which sex(es) could they be?

**12** For the pedigree shown here, the disorder is caused by a recessive (*g*) allele on the X chromosome. Label each of the following individuals with the correct genotype ($X^GY$, $X^gY$, $X^GX^G$, $X^GX^g$, $X^gX^g$).

**Generation**

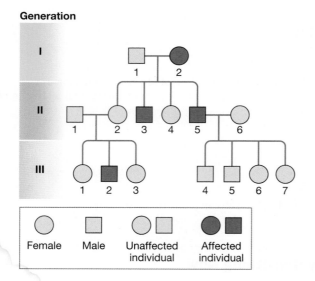

| Female | Male | Unaffected individual | Affected individual |

a. Generation I, number 2

b. Generation II, number 2

c. Generation II, number 5

d. Generation III, number 2

e. Generation III, number 6

**13** For the pedigree shown here, the disorder is caused by a recessive (*g*) autosomal allele. Label each of the following individuals with the correct genotype (*GG*, *Gg*, *gg*).

**Generation**

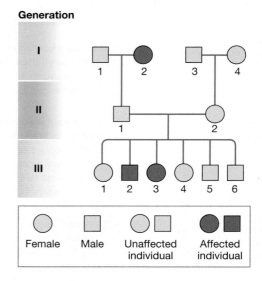

| Female | Male | Unaffected individual | Affected individual |

a. Generation I, number 2

b. Generation II, number 1

c. Generation II, number 2

d. Generation III, number 3

## Leveling Up

**14** *Write Now* **biology: debating preimplantation genetic diagnosis (PGD)** This assignment is designed to give you the experience of applying your knowledge of biology to a current controversy or topic of interest. You will apply the same sort of reasoning that you should be able to use as an informed citizen and consumer when making decisions that involve biology.

The scenario: The U.S. Senate Committee on Commerce, Science and Transportation (CCST) is considering proposing legislation on preimplantation genetic diagnosis (PGD; see "Prenatal Genetic Screening" on page 148). The chair of the committee has invited special-interest groups to present testimony on the pros and cons of PGD. You will contribute a position paper defining PGD, describing your group's position on the technology, and making a recommendation for legislation.

Choose one of the following special-interest groups to represent (or your instructor will assign one to you):

(a) **Reproductive Specialist Group (medical doctors).** You will describe how PGD and in vitro fertilization (IVF) can be used to screen for genetic disorders or for sex selection and "family balancing," and to increase fertility in women of advanced maternal age. You will argue that legislation is not necessary, since medical association guidelines are already in place, and that it is not the place of government to judge individuals' reasons for undergoing PGD.

(b) **Genetics and Public Policy Group.** You will present data on the beliefs of the American public in this matter. You will also present the status of current legislation in the United States and how it compares to PGD legislation internationally. You will propose limited legislation based on these findings.

(c) **Parents of Down Syndrome Children.** You will argue that people with disabilities already suffer from prejudice, and that widespread use of such testing will cause even more prejudice. You will propose legislation prohibiting PGD for almost all conditions, with the exception of deadly, infant-onset disease.

(d) **Americans with Cystic Fibrosis.** You will present the case of a couple, both of whom are carriers for CF, and discuss their options. You will propose limited legislation that disallows PGD for sex selection and nondisease conditions.

**15** **What do *you* think?** *Gattaca*, a science fiction movie from 1997, is based in a future in which preimplantation genetic diagnosis (PGD) is used to create genetically superior children, and social status is determined mainly by one's DNA. Watch the movie and take careful notes as you observe how PGD has been misused by society. Do you think these outcomes are likely, if we continue to develop PGD? Are the benefits worth this risk?

**16** **Life choices** While you cannot control the genes that you inherited from your parents, you can control your exposure to external factors that increase your chances of developing a disease that is only partially genetically determined. For this assignment, you will estimate your life expectancy, as determined mainly from your lifestyle choices. Go to the "Living to 100" website (http://www.livingto100.com) and complete the Life Expectancy Calculator. Answer all questions honestly (no one will see your answers), and go to the "feedback" page at the end of the questionnaire. Take notes so that you can answer the following questions.

a. What is your calculated life expectancy, given the answers you gave?

b. List the recommendations to increase your life expectancy, and how much they could increase your life span.

c. What change would increase your life span the most?

d. Identify one way that you could feasibly increase your life span. Will you do that? Why or why not?

**For more, visit digital.wwnorton.com/bionow2 for access to:**

# Pigs to the Rescue

*Using CRISPR, a hot new genome-editing tool, scientists hope to create a steady stream of transplant organs—from pigs.*

- - - - - - - - - - - - - - - - - - - - - - - - - - - - - - - - - - - - - - - - -

## After reading this chapter you should be able to:

- Describe the structure of DNA, using appropriate terminology.
- Use the base-pairing rules to determine a complementary strand of DNA based on a given template strand.
- Describe how the genome editing tool CRISPR-Cas9 works.
- Label a diagram of DNA replication, identifying the location of each step in the process.
- Identify when PCR and gene sequencing technology should be used in an experiment.
- Explain the cause of DNA replication errors, and describe how they are repaired.
- Give an example of a mutation and its potential effects on an organism.

CHAPTER
09

WHAT GENES
ARE

"Pigs," thought Marc Güell. "What if we could modify pigs?"

Güell, a young biochemist at Harvard Medical School in Massachusetts, had been working with colleagues on techniques to manipulate **DNA (deoxyribonucleic acid)**, the genetic code of life. In 2012, Güell and his colleague Luhan Yang, along with their boss, the geneticist George Church, had begun adding and deleting genes from organisms, a process known as genome editing, using a new technological breakthrough called CRISPR-Cas9—CRISPR for short.

Before the discovery of CRISPR (pronounced "crisper"), editing DNA was difficult and expensive, and it was typically done only in model organisms such as mice and fruit flies—species for which biologists had developed a solid tool kit for genetic manipulation. CRISPR-Cas9, a simple but creative combination of a specific protein and two single-stranded **RNA (ribonucleic acid)** molecules, made genome editing inexpensive and available to all, not just to well-funded genetics labs. It also enabled researchers to quickly and efficiently change the DNA of nearly any organism—like a "molecular scalpel," as one of the CRISPR inventors called it—making it possible to edit genes in fungi, plants, humans, you name it. "It was the biggest change in my career," says Güell. "With CRISPR, everything is easier, faster, and cheaper."

CRISPR also provided the new opportunity to edit more than one gene at a time, and the team picked a unique initial specimen: *Sus scrofa domesticus*, the domestic pig. "We started to think about what would be a good application to use this technology," says Güell. "One of the things that seemed like a very interesting, very difficult problem was the lack of organs for transplantation. We thought, what if we could modify pigs to make them compatible enough to be an unlimited source of organs?" (**Figure 9.1**).

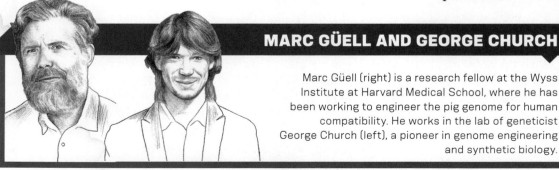

Figure 9.1

**Experimental pig-to-baboon lung transplant**

A lung from a pig engineered by CRISPR to prevent rejection is tested for safety and efficacy in primates by being transplanted into a baboon.

**MARC GÜELL AND GEORGE CHURCH**

Marc Güell (right) is a research fellow at the Wyss Institute at Harvard Medical School, where he has been working to engineer the pig genome for human compatibility. He works in the lab of geneticist George Church (left), a pioneer in genome engineering and synthetic biology.

## Deep in the DNA

Each day in the United States, an average of 22 people die waiting for an organ transplant, and 119,000 men, women, and children are currently on the national transplant waiting list, each hoping their name is called before it's too late.

Researchers have explored many ways to grow and store organs for transplantation—from freezing them to building them from scratch—but one of the most promising, if you can look past the mud and flies, is pigs. Our porcine friends have long been considered an excellent potential source of organs because their organs are relatively close in size to human organs—including the heart, liver, and kidneys—and because pigs and humans have similar organ anatomies (**Figure 9.2**). In addition, pigs are an easier sell to the public: in general, people tend to prefer the idea of transplants from pigs over transplants from mammals more closely related to us, such as baboons. If we were able to transplant organs from animals into humans, a process called *xenotransplantation*, healthy organs could be available on demand and in essentially limitless supply.

Yet there has been a barrier to harvesting pig organs for humans: the pig genome is dotted with DNA from a family of viruses called p̲orcine e̲ndogenous r̲etrov̲iruses, or PERVs. DNA is built from two parallel strands of repeating units called **nucleotides**. Each nucleotide is composed of the sugar deoxyribose, a phosphate group, and one of four **bases**: adenine, cytosine, guanine, or thymine. We identify nucleotides by their bases, using "adenine nucleotide" as shorthand for "nucleotide with an adenine base."

The nucleotides of a single strand are connected by covalent bonds between the phosphate group of one nucleotide and the sugar of the next nucleotide. The two DNA strands are connected by hydrogen bonds linking the bases on one strand to the bases on the other, like the rungs that connect the two sides of a ladder (**Figure 9.3**). The term **base pair** (or nucleotide pair) refers to two nucleotides held together by one of these bonds between their bases; that is, a base pair corresponds to one rung of the DNA ladder. The ladder twists into a spiral called a **double helix** (**Figure 9.4**). Within the long, winding double helix of the pig genome, short sections of DNA from PERVs are scattered about—sections made up of the same four nucleotides but encoding information for viral proteins rather than pig proteins.

Nucleotides do not form base pairs willy-nilly. The adenine (A) nucleotide on one strand can pair only with thymine (T) on the other strand (see **Figure 9.3**); similarly, cytosine (C) on one strand can pair only with guanine (G) on the other strand. These **base-pairing rules**, which provide **complementary base-pairing** between two nucleic acid strands, have an important consequence: when the sequence of nucleotides on one strand of the DNA molecule is known, the sequence of nucleotides on the other, complementary strand of the molecule is automatically known as well. The fact that A can pair only with T and that C can pair only with G allows the original strands

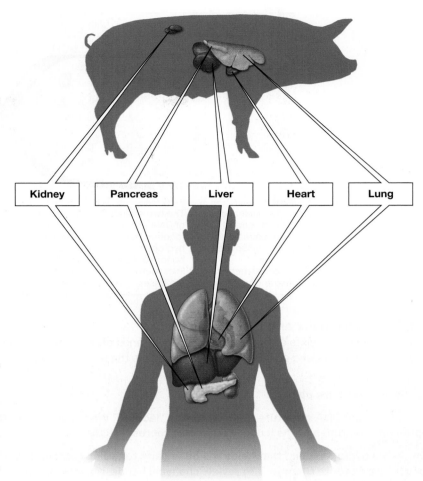

Figure 9.2

**Pig and human organs are remarkably similar in size**

**Q1:** Name one reason why, for a potential transplant, matching the size of organs shown would be important.

**Q2:** Name a tissue transplant for which size matching would not be as important.

**Q3:** Name a tissue transplant for which size matching would not be important at all.

to serve as "template strands" on which new strands can be built through complementary base-pairing. (We will see more in Chapter 10 about building new DNA strands, including how RNA can pair with DNA, which CRISPR takes advantage of.)

Still, the four nucleotides can be arranged in any order along a single strand of DNA, and each DNA strand is composed of millions of these nucleotides, so a tremendous amount of

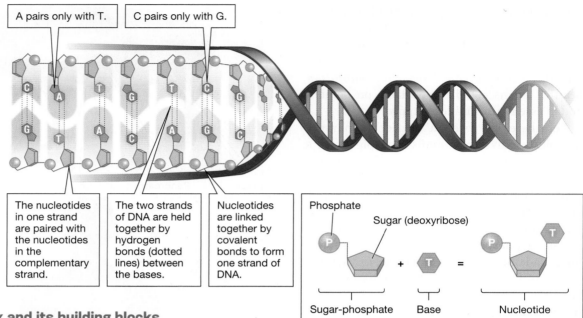

A pairs only with T.

C pairs only with G.

The nucleotides in one strand are paired with the nucleotides in the complementary strand.

The two strands of DNA are held together by hydrogen bonds (dotted lines) between the bases.

Nucleotides are linked together by covalent bonds to form one strand of DNA.

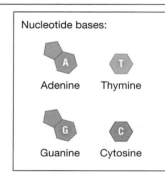

Phosphate

Sugar (deoxyribose)

Sugar-phosphate    Base    Nucleotide

Nucleotide bases:

Adenine    Thymine

Guanine    Cytosine

### Figure 9.3

## The DNA double helix and its building blocks

A molecule of DNA consists of two complementary strands of nucleotides that are twisted into a spiral around an imaginary axis, rather like the winding of a spiral staircase. ▶

**Q1:** Name two base pairs.

**Q2:** Why is the DNA structure referred to as a "ladder"? What part of the DNA represents the rungs of the ladder? What part represents the sides?

**Q3:** Is the hydrogen bond that holds the base pairs together a strong or weak chemical bond? (*Hint:* Refer to Chapter 3 to review chemical bonds, if necessary.)

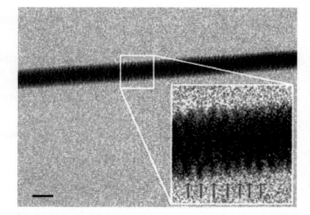

### Figure 9.4

## What DNA actually looks like

In November 2012, Italian researchers used an electron microscope to directly visualize DNA for the first time. This is the single thread of double-stranded DNA that they saw.

information can be stored in a DNA sequence and in a genome. The genome of the domestic pig, for example, has about 3 billion base pairs, the human genome has about 3.2 billion base pairs, a tomato has only about 900 *million* base pairs, and the bacterium *Escherichia coli* has a measly 4.6 million. The sequence of nucleotides in DNA differs among species and among individuals within a species, and these differences in genotype can result in different phenotypes (**Figure 9.5**).

In the mid-1990s, scientists became very excited about the idea of using pig organs in humans, but testing stalled because of the fear that humans would become infected with PERVs. Just breeding pigs in sterile conditions can't get rid of the virus; it's integrated right there in the double helix. The Harvard team believed CRISPR might be able to solve that

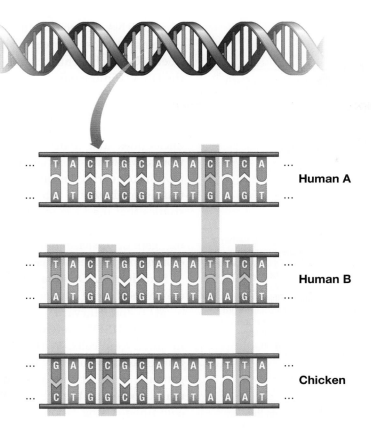

Figure 9.5

**The sequence of bases in DNA differs among species and among individuals within a species**

The sequence of bases in a hypothetical gene is compared for two humans (A and B) and a chicken. Base pairs highlighted in blue are variant; that is, they differ between the genes of persons A and B, and between the same genes in humans and chickens.

---

**Q1:** If all genes are composed of just four nucleotides, how can different genes carry different types of information?

**Q2:** Would you expect to see more variation in the sequence of DNA bases between two members of the same species (such as humans) or between two individuals of different species (for example, humans and chickens)? Explain your reasoning.

**Q3:** Do different alleles of a gene have the same DNA sequence or different DNA sequences?

---

problem by destroying the PERV DNA in pig cells once and for all.

## Precise Cuts

The DNA sequences called **CRISPR**, a blessedly easy acronym for *clustered regularly interspaced short palindromic repeats*, are actually part of a defense system used by bacteria. Bacteria are constantly under bombardment from viruses that try to sneak into and take over their genomes, so bacteria evolved a set of defensive measures, including a tool to recognize and cut foreign, interloping DNA. Pioneered by microbiologists Jennifer Doudna of UC Berkeley and Emmanuelle Charpentier at Umeå University in Sweden, the CRISPR-Cas9 system is composed

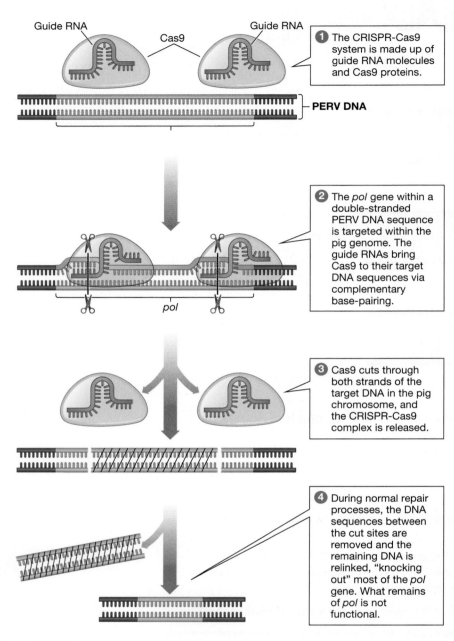

Guide RNA    Cas9    Guide RNA

**1** The CRISPR-Cas9 system is made up of guide RNA molecules and Cas9 proteins.

PERV DNA

**2** The *pol* gene within a double-stranded PERV DNA sequence is targeted within the pig genome. The guide RNAs bring Cas9 to their target DNA sequences via complementary base-pairing.

*pol*

**3** Cas9 cuts through both strands of the target DNA in the pig chromosome, and the CRISPR-Cas9 complex is released.

**4** During normal repair processes, the DNA sequences between the cut sites are removed and the remaining DNA is relinked, "knocking out" most of the *pol* gene. What remains of *pol* is not functional.

Figure 9.6

**Genome editing with CRISPR-Cas9, an efficient and cost-effective tool**

**Q1:** What common mechanism is employed by the guide RNA to find its target DNA sequence?

**Q2:** How many strands of DNA must Cas9 cut to be effective?

**Q3:** Does Cas9 also cause the deletion of DNA from the genome?

of two RNA molecules that guide two Cas9 proteins to chosen, precise sites in a genome. There, Cas9 efficiently cleaves both strands of the DNA (**Figure 9.6**).

The CRISPR system is like a "Delete" key or, in certain experiments, the cut-and-paste tool of a word processing program—except with nucleotides and genes instead of letters and words. While most new laboratory tools can take months or years to become widely used, CRISPR was immediately a lab favorite, with genetic studies that used the tool being published right and left: editing the zebra fish genome, correcting genetic disease mutations in adult mice, enhancing pest resistance in wheat, and more.

Shortly after Doudna and Charpentier's discovery, both Feng Zhang (at the Broad Institute of MIT and Harvard) and Church's lab engineered CRISPR systems for genome editing in eukaryotic cells. With that innovation, the Church team was ready to start editing the pig genome, but first they needed to identify how many copies of the PERV DNA the genome contained. Using laboratory tools to sequence the pig DNA, they identified 62 copies of PERVs scattered throughout the genome.

Sixty-two may sound like a lot, but it's a drop in the bucket compared to the number of genes in a mammalian genome. Pigs have over 1,600 genes for their sense of smell alone. An estimated 19,000 protein-coding genes are packed into the human genome, surrounded by even more DNA that does not code for proteins. Cells are adept at stuffing an enormous amount of DNA into a small space; they use a variety of packaging proteins to wind, fold, and compress the DNA double helix, going through several levels of packing to create the DNA-protein complex that we call a chromosome (**Figure 9.7**).

Short lengths of double-stranded DNA are wound around "spools" of proteins, known as **histone proteins**, to create a beads-on-a-string structure consisting of many histone beads, called **nucleosomes**, connected by strings of DNA. This beads-on-a-string structure is compressed and coiled into a more compact form, known as the **chromatin fiber**, by yet other types of packaging proteins. The chromatin fiber is then looped back and forth to further condense and coil again to form the resulting chromosome in the interior of the nucleus.

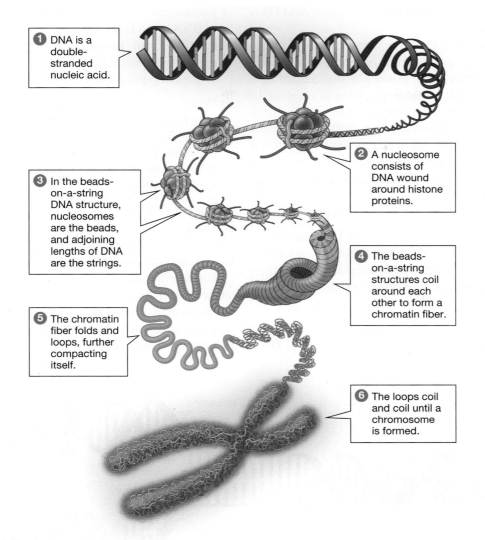

**Figure 9.7**

① DNA is a double-stranded nucleic acid.

② A nucleosome consists of DNA wound around histone proteins.

③ In the beads-on-a-string DNA structure, nucleosomes are the beads, and adjoining lengths of DNA are the strings.

④ The beads-on-a-string structures coil around each other to form a chromatin fiber.

⑤ The chromatin fiber folds and loops, further compacting itself.

⑥ The loops coil and coil until a chromosome is formed.

**Chromosomes are meticulously organized DNA-protein complexes**
The DNA double helix is continuously coiled and packaged around proteins until it is compacted into a chromosome.

**Q1:** What are the structures that result from the first level of coiling around proteins called?

**Q2:** What makes up a "bead" and what makes up the "string" in the beads-on-a-string structure of DNA?

**Q3:** What is the name for the structure that is more compact than the beads-on-a-string structure but less compact than an actual chromosome?

# Double or Nothing

DNA from viruses is actually present in all mammalian genomes that we know of, including the human genome, yet not all these viral instructions behave the same. In humans, viral DNA was integrated into the genome a long, long time ago, and although it remains in the human genome, it is no longer active. In pigs, viruses entered and wove their DNA into the pig genome much more recently, and that viral DNA is still active.

All this viral DNA remains in these genomes because it is passed from generation to generation via **DNA replication**, the duplication of a DNA molecule. DNA replication is ongoing in our bodies: It occurs right before a cell enters mitosis, so that there is a copy of the DNA to pass along to the new cell. It occurs feverishly when a new embryo is being formed. And it occurs when viruses hijack our cell machinery and copy their own DNA.

Cells replicate DNA in three steps:

1. The DNA molecule unwinds through special proteins that bind the DNA at sequences known as **origins of replication**, which then break the hydrogen bonds connecting the two strands of DNA.

2. Each strand is then used as a template for the construction of a new strand of DNA. **DNA polymerase**—a key enzyme in the replication of DNA—builds the two new strands of DNA, starting from **primers** that pair complementary nucleotides at specific sites near the origins of replication.

3. When construction is completed, there are two identical copies of the original DNA molecule. Each copy is composed of a template strand of DNA (from the original DNA molecule) and a newly synthesized strand of DNA.

This mode of replication is known as **semiconservative replication** because one "old" strand (the template strand) is retained, or "conserved," in each new double helix (**Figure 9.8**).

The mechanics of copying DNA are far from simple. More than a dozen enzymes and proteins are needed to unwind the DNA, to stabilize the separated strands, to start the replication process, to attach nucleotides that are complementary

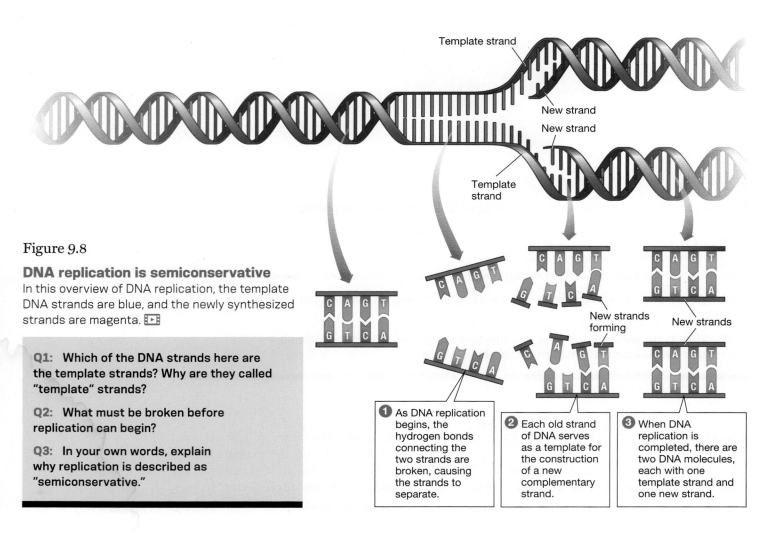

## Figure 9.8

**DNA replication is semiconservative**
In this overview of DNA replication, the template DNA strands are blue, and the newly synthesized strands are magenta.

**Q1:** Which of the DNA strands here are the template strands? Why are they called "template" strands?

**Q2:** What must be broken before replication can begin?

**Q3:** In your own words, explain why replication is described as "semiconservative."

**1** As DNA replication begins, the hydrogen bonds connecting the two strands are broken, causing the strands to separate.

**2** Each old strand of DNA serves as a template for the construction of a new complementary strand.

**3** When DNA replication is completed, there are two DNA molecules, each with one template strand and one new strand.

to the template strand, to "proofread" the results, and to join partly replicated fragments of DNA to one another, ultimately forming a completely replicated chromosome.

Despite the complexity of this task, cells can copy DNA molecules containing billions of nucleotides in a matter of hours—about 8 hours in humans (over 100,000 nucleotides per second). This speed is achieved in part by starting the replication of the DNA molecule at thousands of different origins of replication at once.

Before CRISPR was developed, the greatest genetic technology was a tool for replicating DNA quickly in a test tube. In the 1980s, scientists developed a way to mimic natural DNA replication in a laboratory—fast. The **polymerase chain reaction** (**PCR**) is a technique that makes it possible to produce millions of copies of a targeted DNA sequence—to "amplify" the DNA—in just a few hours, even with an extremely small initial amount of DNA (**Figure 9.9**).

PCR relies on heat, not special proteins, to cause the DNA to unwind and its strands to separate. PCR also uses targeted primers to start the replication process at points determined by the scientist, rather than at DNA's natural origins of replication. The researcher

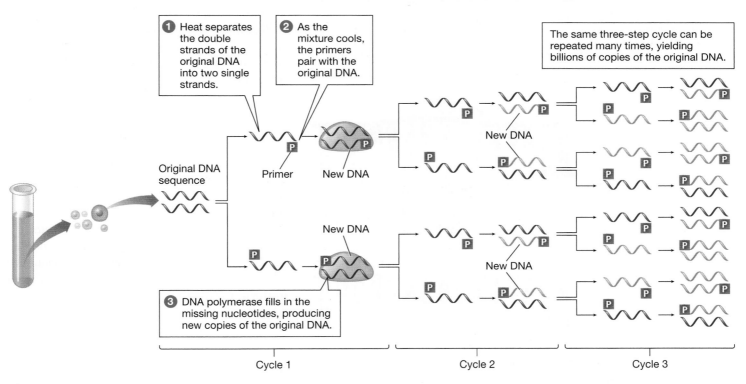

**Figure 9.9**

**PCR can amplify small amounts of DNA more than a millionfold**

Short primers consisting of synthetic DNA segments are mixed in a test tube with a sample of the target DNA, the enzyme DNA polymerase, and all four nucleotides (A, C, G, and T). The primers form base pairs with the two ends of a gene of interest. A machine then processes the mixture and doubles the number of double-stranded versions of the template sequence. The doubling process can be repeated many times (only three cycles are shown here).

**Q1:** PCR replicates DNA many times to increase the amount available for analysis. Why is this process called "amplification"?

**Q2:** During the PCR cycle, what causes the DNA strands to separate?

**Q3:** Identify a difference between how PCR and DNA replication are accomplished.

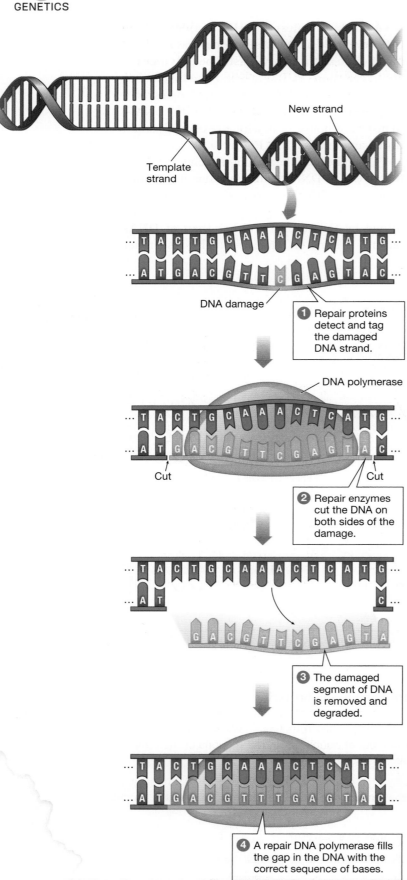

New strand

Template strand

... T A C T G C A A A A C T C A T G ...
... A T G A C G T T C G A G T A C ...

DNA damage

**1** Repair proteins detect and tag the damaged DNA strand.

DNA polymerase

... T A C T G C A A A A C T C A T G ...
... A T G A C G T T C G A G T A C ...

Cut                    Cut

**2** Repair enzymes cut the DNA on both sides of the damage.

... T A C T G C A A A A C T C A T G ...
... A T                              C ...

G A C G T T C G A G T A

**3** The damaged segment of DNA is removed and degraded.

... T A C T G C A A A A C T C A T G ...
... A T G A C G T T T G A G T A C ...

**4** A repair DNA polymerase fills the gap in the DNA with the correct sequence of bases.

creates a solution containing DNA polymerase, primers, and loose nucleotides, and the solution goes into a PCR machine, a device developed for this specific process. The machine cycles between heating, which separates the strands, and cooling, which enables the added enzymes to build two new strands of DNA through complementary base-pairing. These heating and cooling steps are repeated 25–40 times to replicate millions of exact copies of the targeted region of DNA.

PCR remains a central tool in most biology labs. The Harvard team, for example, used PCR to replicate pig DNA for their sequencing reactions, which require many copies of the same sequence in order to succeed. DNA sequencing machines, which produce a data file listing the complete sequence of nucleotides in a given strand of DNA, were used to identify all the places in the pig genome where PERV had been integrated.

# Making Mutations

To inactivate all 62 copies of PERV in the pig genome at once, the Church team decided to try to prevent DNA polymerase from successfully replicating the PERV DNA. To that end, the team planned to use CRISPR to *mutate* a small region of the PERV DNA: a gene named *pol* (see **Figure 9.6**). A change to the sequence of nucleotides in an organism's DNA is called a **mutation**. The extent of a mutation can range from the change of one nucleotide in a single base

Figure 9.10

**Repair proteins fix DNA damage**
Large complexes of DNA repair proteins work together to fix damaged DNA.

**Q1:** Summarize how DNA repair works and why the repair mechanisms are essential for the normal function of cells and whole organisms.

**Q2:** Is DNA repair 100 percent effective?

**Q3:** What would happen to an organism if its DNA repair became less effective?

pair (known as a **point mutation**) to the addition or deletion of one or more whole chromosomes (a *chromosomal abnormality*, described in Chapter 8).

Güell was developing techniques to artificially mutate genes, yet point mutations occur naturally and randomly often, especially during DNA replication. When DNA is copied right before mitosis occurs in a cell, there are many opportunities for mistakes to be made. The enzymes that copy DNA sometimes insert an incorrect nucleotide in the newly synthesized strand. In addition, DNA in cells is constantly being damaged by chemical, physical, and biological agents, including energy from radiation or heat, collisions with other molecules in the cell, attacks by viruses (like PERVs), and random chemical accidents (some of which are caused by environmental pollutants, but most of which result from normal metabolic processes).

Replication errors and damage to DNA—especially to essential genes—disrupt normal cell functions. If not repaired, DNA damage leads to malfunctioning proteins, such as Felix's WAS protein in Chapter 8. DNA damage can also cause the death of cells and, ultimately, the death of an organism. Thankfully, cells have a way to recover: DNA polymerase immediately corrects almost all mistakes during DNA replication, "proofreading" complementary base pairs as they form.

DNA polymerase is not infallible. When an incorrect nucleotide is added but escapes proofreading by DNA polymerase, a mismatch error has occurred. This happens about once in every 10 million nucleotides. But cells have another backup safety program: repair proteins that correct 99 percent of mismatch errors, reducing the overall chance of an error to one mistake in every *billion* nucleotides (**Figure 9.10**).

On the rare occasions when a mismatch error is not corrected by repair proteins, the DNA sequence is changed, and the new sequence is reproduced the next time the DNA is replicated. If the mutation occurs within a gene, it will result in the formation of a new allele. Most new alleles are either neutral or harmful, but occasionally a mutation may be beneficial.

Three types of point mutations can alter a gene's DNA sequence: substitutions, insertions, and deletions. In a **substitution** point mutation, one nucleotide is substituted for another in the DNA sequence of the gene. An **insertion** or **deletion** point mutation occurs when a nucleotide is, respectively, inserted into or deleted from a DNA sequence. Sickle-cell disease, a human genetic blood disorder, is caused by a substitution point mutation (**Figure 9.11**). Sometimes,

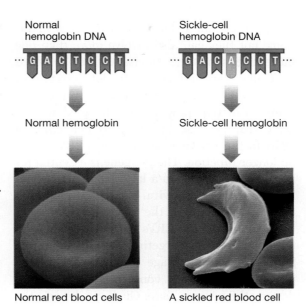

Normal hemoglobin DNA  Sickle-cell hemoglobin DNA

Normal hemoglobin  Sickle-cell hemoglobin

Normal red blood cells  A sickled red blood cell

Figure 9.11

### A point mutation in the hemoglobin gene leads to sickle-cell disease

In people with the genetic disorder sickle-cell disease, a single base in the gene that makes hemoglobin, an important protein involved in oxygen transport in red blood cells, is altered. The red blood cells of people with sickle-cell disease become curved and distorted under low-oxygen conditions and can clog blood vessels, leading to serious effects, including heart and kidney failure.

**Q1:** What are the three types of point mutations?

**Q2:** Sickle-cell disease is an autosomal recessive genetic disorder. How many mutated hemoglobin alleles do people with sickle-cell disease have?

**Q3:** Because of improved treatments, individuals with sickle-cell disease are now living into their forties, fifties, or longer. How might this extension of life span affect the prevalence of sickle-cell disease in the population?

however, changing a few nucleotides in a gene's DNA sequence has little or no effect. In such cases, a mutation is said to be "silent" because it produces no change in the function of the protein, and therefore no change in the phenotype of the organism. (We further explore mutations and their effect on the resulting proteins in Chapter 10.)

Insertions and deletions can be point mutations, but they can also involve more than one nucleotide; sometimes thousands may be added or deleted. Large insertions and deletions almost always result in the synthesis of a protein that cannot function properly. Güell's goal was to use CRISPR in the pig genome to delete bits of DNA in the *pol* gene so that the PERVs could no longer function. The *pol* gene is essential for PERVs to replicate, so a large mutation should halt the formation of viral particles. In addition to being essential to the virus, *pol* is present in all 62 copies of PERVs but not elsewhere in the pig genome, so targeting CRISPR to mutate *pol* would spare damage to other parts of the pig DNA. "My biggest concern is always not to destroy the genome," says Güell. So, he and the team designed two CRISPR RNA guides to target the complex right to *pol*, and to mutate the heck out of it.

After much trial and error, the team inserted its CRISPR-Cas9 construct into pig embryos, and the system began its work. The CRISPR system made 455 different cuts in the PERV *pol* genes throughout the genome, resulting in deletions ranging from 1 to 148 base pairs. About 80 percent of the engineered mutations were small deletions of fewer than nine base pairs—tiny, but enough to disrupt the gene.

# Pigs Are People Too?

In October 2015, Güell, Church, and the team presented the results of their initial work: Using CRISPR, they had inactivated 62 PERVs in pig embryos. Recall that in previous non-CRISPR-edited experiments, PERVs from the pig cells infected human cells when put in close proximity. This time, when the team placed the CRISPR-edited pig cells next to human cells, there was zero transmission of the virus. "The gene editing completely disrupted the ability of the virus DNA to replicate and form viral particles," says Güell.

Although the team has successfully destroyed the PERV genes in the pig genome, there are still a few more steps before pig organs will be ready and safe for human transplantation. Other genes need to be manipulated, says Güell, such as immune system proteins so that the human immune system doesn't reject the organs as foreign. However, the next immediate step, says Güell, is to edit an embryo, implant it back into a female pig, and raise a live, genetically engineered animal. "We're trying hard," says Güell. "It could be in the next year."

Other researchers are pursuing a different path: growing actual human organs in pigs. Again, CRISPR is the workhorse for the experiment. In 2016, a team at UC Davis used CRISPR to remove the genes that encode for the pig pancreas in a pig embryo. The procedure created a void in the embryo, which they then filled with human stem cells. Stem cells have the potential to develop into almost all human cells and organs—including, in this case, a fully functional human pancreas.

Some research groups are trying to use this method to grow other organs. **Figure 9.12** shows the process by which scientists could use CRISPR and human stem cells to grow a pig with human kidneys, which would then be used to donate a kidney to a patient in need of a transplant. No researchers have yet to complete this process, and the work remains controversial, especially because of concerns that the human cells might migrate to the developing pig's brain and make the pig, even in some small way, more human.

Overall, CRISPR has revitalized the idea of a safe, clean, limitless source of healthy organs. "It opens up the possibility of not just transplantation from pigs to humans but the whole idea that a pig organ is perfectible," Church told the BBC in 2016. "Gene editing could ensure the organs are very clean, available on demand, and healthy, so they could be superior to human donor organs."

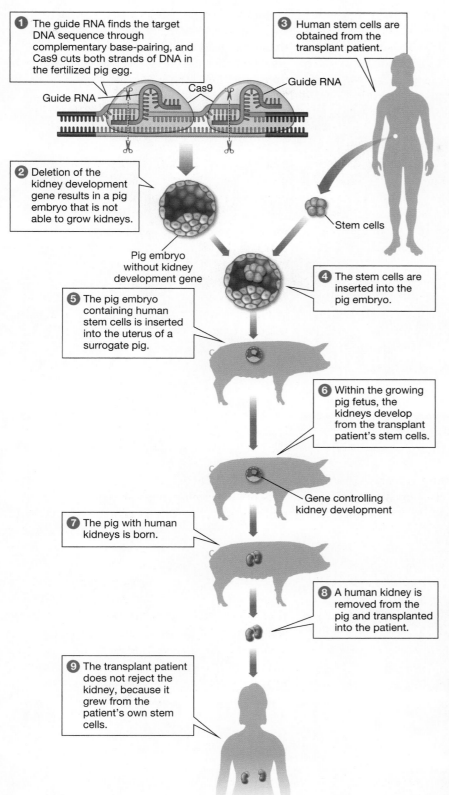

**1** The guide RNA finds the target DNA sequence through complementary base-pairing, and Cas9 cuts both strands of DNA in the fertilized pig egg.

Guide RNA    Cas9    Guide RNA

**2** Deletion of the kidney development gene results in a pig embryo that is not able to grow kidneys.

Pig embryo without kidney development gene

**3** Human stem cells are obtained from the transplant patient.

Stem cells

**4** The stem cells are inserted into the pig embryo.

**5** The pig embryo containing human stem cells is inserted into the uterus of a surrogate pig.

**6** Within the growing pig fetus, the kidneys develop from the transplant patient's stem cells.

Gene controlling kidney development

**7** The pig with human kidneys is born.

**8** A human kidney is removed from the pig and transplanted into the patient.

**9** The transplant patient does not reject the kidney, because it grew from the patient's own stem cells.

### Figure 9.12

**Human organs could grow up in pigs**
Growing a human organ (here, a kidney) in pigs modified by CRISPR to lack that organ could help meet transplant needs.

**Q1:** Name a step in this process that is similar to the original CRISPR method that removed the PERVs from the pig genome. (*Hint:* Review Figure 9.6.)

**Q2:** Name a step in this process that is not included in the original CRISPR method that removed the PERVs from the pig genome.

**Q3:** What parts of this process would scientists need to change in order to develop several different human organs in a single pig?

# The Meteoric Rise of CRISPR

The genome editing tool CRISPR, short for "clustered regularly interspaced short palindromic repeats," has taken molecular biology laboratories by storm over the past 5 years. It has been used to edit the genomes of crops and livestock to improve breeding and production, to control populations of disease-carrying insects, to silence genetic disorders in animal models, and more. Here are a few highlights from the short but shining history of CRISPR.

Assessment available in smartwork5

## PubMed search results for "CRISPR" by year

CRISPR 🔍

Search results article count ————●

| Year | Count |
|------|-------|
| 2002 | 1 |
| 2003 | 1 |
| 2004 | 0 |
| 2005 | 5 |
| 2006 | 6 |
| 2007 | 12 |
| 2008 | 21 |
| 2009 | 32 |
| 2010 | 45 |
| 2011 | 79 |
| 2012 | 126 |
| 2013 | 282 |
| 2014 | 607 |
| 2015 | 1,258 |
| 2016 | 2,143 |

**1987**
CRISPR repeats are first observed in bacterial genomes. Their significance is not yet known.

**2002**
The term "CRISPR" is coined by researchers in Spain and the Netherlands.

**2006**
Researchers propose that CRISPR functions in nature as part of a bacterial adaptive immune system.

**2011**
The final necessary piece for the genome editing system is identified: a second small RNA needed to guide Cas9 to its targets.

**2013**
The CRISPR-Cas9 system is used to edit targeted genes in both human and mouse cells, and later plant cells.

**2015**
In China, scientists use CRISPR-Cas9 to edit preimplantation human embryos, repairing a mutated gene that would cause a blood disorder. Subsequently, an international ban prohibits the use of genome editing to make changes to the human genome.

**2016**
The first human trial to use CRISPR genome editing gets approval from the National Institutes of Health, in a cancer therapy trial to edit a patient's own immune system cells.

# REVIEWING THE SCIENCE

- Genes are composed of **DNA**, which consists of two parallel strands of repeating units called **nucleotides** twisted into a **double helix**.

- The four nucleotides of DNA contain the **bases** adenine (A), cytosine (C), guanine (G), and thymine (T). The nucleotides exhibit **complementary base-pairing** according to **base-pairing rules**: A can pair only with T, and C can pair only with G.

- DNA is wrapped around **histone proteins**, forming **nucleosomes**. The nucleosome structures can further compact the DNA by coiling around themselves to form a **chromatin fiber**. Chromatin fibers further coil around themselves to form chromosomes.

- The **CRISPR**-Cas9 editing system is composed of two pieces of **RNA** designed to form base pairs at precise locations in a gene. This DNA-RNA interaction guides the Cas9 proteins to the sites where they efficiently cut the DNA, resulting in a gene deletion after normal repair processes take place. Additional genetic manipulations are required to generate a gene insertion.

- **DNA replication** occurs in all living organisms prior to mitosis. The double helix unwinds, and the two strands break apart. Each strand of DNA serves as a template from which a new strand is copied. **DNA polymerase** builds each new strand of DNA using **primers** located near the **origins of replication**.

- The **polymerase chain reaction**, or **PCR**, is a laboratory technique to amplify the DNA from a small initial amount to millions of copies. Amplified DNA can then be sequenced to examine specific genes or mutations.

- DNA is subject to damage by physical, chemical, and biological agents, and errors in DNA replication are common. DNA polymerase "proofreads" the DNA during replication and corrects most mistakes. Repair proteins are a backup repair mechanism and correct any errors that DNA polymerase misses.

- A change to the sequence of bases in an organism's DNA is called a **mutation**. Three types of mismatch mutations can alter a gene's DNA sequence: **substitutions**, **insertions**, and **deletions**. If only a single base is altered, it is a **point mutation**.

# THE QUESTIONS

## The Basics

**1** DNA replication results in

(a) two DNA molecules—one with two old strands, and one with two new strands.

(b) two DNA molecules, each of which has two new strands.

(c) two DNA molecules, each of which has one old strand and one new strand.

(d) none of the above

**2** The DNA of cells is damaged

(a) thousands of times per day.

(b) by collisions with other molecules, chemical accidents, and radiation.

(c) not very often and only by radiation.

(d) both a and b

**3** The DNA of different species differs in the

(a) sequence of bases.

(b) complementary base-pairing.

(c) number of nucleotide strands.

(d) location of the sugar-phosphate portion of the DNA molecule.

**4** Mutation

(a) can produce new alleles.

(b) can be harmful, beneficial, or neutral.

(c) is a change in an organism's DNA sequence.

(d) all of the above

**5** Link each term with the correct definition.

| | |
|---|---|
| NUCLEOTIDE | 1. Two complementary bases joined by hydrogen bonds. |
| BASE PAIR | 2. The nitrogen-containing component of a nucleotide; there are four variants of this component. |
| DNA MOLECULE | 3. A strand of nucleotides linked together by covalent bonds between a sugar and a phosphate; two strands are linked by hydrogen bonds between complementary bases. |
| BASE | 4. A phosphate, a sugar, and a nitrogen-containing base. |

**6** In the diagram of replication shown here, fill in the blanks with the appropriate terms: (a) base pair, (b) base, (c) nucleotide, (d) template strand, (e) newly synthesized strand, (f) separating strands.

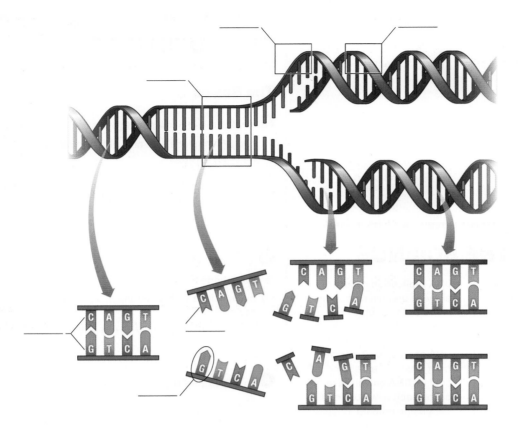

**7** Select the correct terms:

To work on removing PERV DNA from the pig genome, researchers first replicate the DNA many times, using (**PCR / CRISPR**). This process increases the amount of DNA so that they are able to run experiments to edit the genome, using (**PCR / CRISPR**).

---

## Challenge Yourself

**8** If a strand of DNA has the sequence CGGTATATC, then the complementary strand of DNA has the sequence

(a) ATTCGCGCA.

(b) GCCCGCGCT.

(c) GCCATATAG.

(d) TAACGCGCT.

**9** Place the following steps of DNA replication and repair in the correct order by numbering them from 1 to 5.

_____ a. A template strand begins to be replicated.

_____ b. If the incorrect base is not identified and replaced, it remains as a point mutation in the DNA.

_____ c. DNA polymerase identifies and replaces most incorrect bases with the correct base, complementary to the base on the template strand.

_____ d. An incorrect base is added to the growing strand of DNA.

_____ e. Proteins identify and replace any incorrect bases missed by DNA polymerase.

**10** Given the template DNA sequence GCAGCATGTT, identify each of the following mutations to the complementary strand as an insertion (I), a deletion (D), or a substitution (S).

_____ a. CGTCGTACA

_____ b. CGTGGTACAA

_____ c. CGTCGTACTAA

**11** If the target sequence for PCR on one strand of DNA is ATGCAAATCCTGG, what is the sequence of each of the two strands of the double-stranded DNA produced?

---

## Try Something New

**12** Using base-pairing rules to guide you, for a DNA double helix that contains 20 percent adenine (A), specify the percentage of (a) thymine, T; (b) guanine, G; and (c) cytosine, C.

**13** Why do scientists need to use CRISPR to remove PERVs from pig embryos, rather than just editing the genes in the organs of adult pigs?

**14** Scientists estimate that genes encoding proteins make up less than 1.5 percent of the human genome. Other genes in our cells encode different types of nonprotein RNA molecules. The rest of our genome consists of various types of **noncoding DNA**, defined as DNA that does not code for any kind of functional RNA. Some of the remaining DNA has regulatory functions—for example, controlling gene expression. Some of it has architectural functions, such as giving structure to chromosomes or positioning them at precise locations within the nucleus. Given that these sections of DNA are noncoding, does it matter if they contain replication errors? Explain your reasoning.

## Leveling Up

**15** **What do *you* think?** In another line of research directed at solving the shortage of organs available for transplantation, scientists have been working feverishly on generating human tissues and organs using human induced pluripotent stem cells (iPSCs). These cells, which have the ability to transform into many different cell types, start as a patient's own skin cells and are treated to attain an undifferentiated, or embryonic-like, state. Scientists then direct them with growth factors to become kidney cells, liver cells, or any other cell type needed. For more information, see https://www.cirm.ca.gov/patients/creating-new-types-stem-cells.

Read the article "Eight Tiny Organs Grown by Scientists," about the potential of iPSCs to fill the transplantable organ void (http://www.insight.mrc.ac.uk/2015/07/20/eight-tiny-organs-grown-by-scientists). Using the information in this article, make a list of pros and cons for both pig organs and iPSC organs. Reflect on which option you think is the better course of action.

**16** ***Write Now* biology: next-generation GMO** Read the article "CRISPR Brings an Early Harvest" (http://www.genengnews.com/gen-news-highlights/crispr-brings-an-early-harvest/81253507), which describes a new way to produce GMO tomatoes without using standard gene transfer techniques. Write a letter to your state or federal legislator expressing your opinion either for or against this new technology for creating GMO foods. Cover the following points in your letter.

a. Explain how CRISPR-Cas9 technology works, comparing and contrasting it to traditional GMO technology.

b. Describe how the CRISPR system was applied to the plants in the article.

c. In your final paragraph, try to persuade your government representative to either endorse or condemn this technology for improving crop yield and quality, using specific examples.

**For more, visit digital.wwnorton.com/bionow2 for access to:**

# Tobacco's New Leaf

*One company's quest to produce tomorrow's drugs in today's plants.*

## After reading this chapter you should be able to:

- Explain how an organism uses gene expression to create a phenotype from a genotype.
- Caption a diagram of transcription, identifying each step in the process and the relevant molecules.
- Caption a diagram of translation, identifying each step in the process and the role of each type of RNA.
- Determine the correct amino acid from a given codon.
- Justify how the genetic code fits the definitions of "redundant" and "universal."
- Describe how a cell can increase or decrease its expression of particular genes, and why this ability is important for an organism.

# CHAPTER

# 10

## HOW GENES WORK

The greenhouse is vast, the size of half a football field. A loud, steady thrum reverberates in the room as massive metal fans push air through the hot, humid space. Mike Wanner, a tall, serious man with gray hair and piercing brown eyes, walks through rows of leafy plants, each a foot and a half tall. He stoops to rub a leaf between his fingers, then raises his hand to his nose and sniffs. "That's what tobacco smells like," Wanner shouts, straining to be heard over the noise of the fans.

Wanner is the North Carolina site manager for operations at Medicago, a Canadian biotechnology company growing tobacco at a facility outside Durham, once home to Lucky Strike cigarettes (**Figure 10.1**). But these plants are not being grown to smoke, chew, or dip. Instead, they serve a dramatically different purpose: this tobacco will be used to make flu vaccines.

Influenza vaccines are normally grown in chicken eggs—a process that takes months. As an alternative, companies and researchers have begun experimenting with "biopharming"— manufacturing vaccine proteins in plants. Since

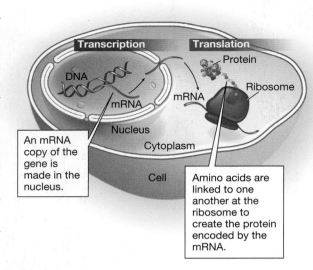

Figure 10.2

## An overview of gene expression

Genetic information flows from DNA to RNA to protein during gene expression, which occurs in two steps: first *transcription* and then *translation*. The transcription of a protein-coding gene produces an mRNA molecule, which is then transported to the cytoplasm, where translation occurs and the protein is made with the help of ribosomes.

most genes contain instructions for building proteins, scientists insert a gene that codes for the protein that interests them. This production of a protein from a gene occurs via **gene expression**—the process by which genes are transcribed into RNA and then translated to make proteins (**Figure 10.2**).

Proteins participate in virtually every process inside and between cells, so gene expression is the fundamental way by which genes influence the structure and function of a cell or organism. It is the process through which an organism's genotype gives rise to its phenotype (see Chapter 7). All prokaryotes, eukaryotes, and viruses utilize gene expression.

In addition to being involved in every process in a cell, proteins are the key components of some vaccines. Injected pieces of viral proteins activate the human immune system to defend against that virus in the future (see Figure 2.1). To make these vaccines, scientists must produce large quantities of viral proteins. The traditional method is to inject chicken eggs with a virus, let the virus multiply in the chicken cells, and then

Figure 10.1

## Growing tobacco to treat the flu

Tobacco plants grow in a Medicago greenhouse in Quebec City, Canada.

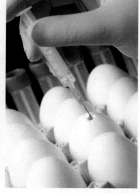

### Figure 10.3

**Tobacco or egg?**
Using tobacco plants to produce proteins for influenza vaccines has several advantages over the traditional approach using chicken eggs.

|  | Tobacco | Egg |
|---|---|---|
| Speed from outbreak to vaccine production: | 1 month | 6 months |
| Cost to produce 50 million flu vaccines: | $36 million | $400 million |
| Allergy risks: | Minimal | Individuals with egg allergies cannot receive the vaccine |
| Availability: | Not yet approved by the Food and Drug Administration | Approved and currently the most widely used technology to produce flu vaccines |

**Q1:** Which is the faster way to produce vaccines: biopharming with plants or creating vaccines in eggs? Why is this important?

**Q2:** How much cheaper is biopharming with plants than creating vaccines in eggs? Why is this important?

**Q3:** Why must tobacco-derived vaccines, or any new medications for that matter, be approved by the FDA?

extract the virus, remove its genetic material, and prepare a vaccine from the leftover viral proteins. It's a long, cumbersome process.

Unlike chicken eggs, however, plants can be grown in vast quantities, and they grow rapidly—often in just days or weeks. "The big advantage of plant systems is that they can produce massive amounts of proteins very inexpensively," says James Roth, director of the Center for Food Security and Public Health at Iowa State University, who studies biopharming in plants and animals (and is not associated with Medicago). In the event of a pandemic flu outbreak, Medicago could produce vaccines 6 times faster and 12 times cheaper than traditional egg manufacturing could, the company claims (**Figure 10.3**).

In April 2012, Medicago put its tobacco plants to the test, running the North Carolina manufacturing facility at full tilt for 30 days. The U.S. Department of Defense had given the company millions of dollars to test whether it could quickly produce enough pandemic flu vaccine from the tobacco to stem an outbreak. The pressure is on, said Wanner, standing inside the facility. He looked out over his crop. "We'll see what happens."

# Fighting the Flu with Tobacco

In 1997, Louis-Philippe Vézina, then a research scientist at Agriculture and Agri-Food Canada and an associate professor at Université Laval in Quebec City, Canada, decided to start a company. A plant biotechnologist, Vézina wanted to explore the possibility of manufacturing proteins in alfalfa plants, and thus he called his company Medicago, after the genus to which alfalfa belongs. Later, as the company grew, Vézina and his team discovered that tobacco produces higher yields of proteins in a shorter time frame than alfalfa does, so the company switched plants.

### MICHAEL WANNER

Michael Wanner is chief financial officer and U.S. site manager for Medicago, an innovative biotechnology company that develops vaccines in the leaves of tobacco plants.

Figure 10.4

## Swine flu

In 2009, people around the world donned masks as a precaution against influenza virus H1N1, nicknamed "swine flu" because it contained DNA from bird, swine, and human flu viruses. The virus caused a pandemic, killing an estimated 284,500 people.

By 2005, Medicago had begun to receive calls asking about its product and when it would begin clinical trials, the first step toward getting a drug approved by the U.S. Food and Drug Administration (FDA). "I was surprised. Even the FDA was asking, 'When are we going to see your vaccine?'" recalls Vézina. "They were keen on technologies like this because they know how much vaccines cost, and they're interested in anything that can decrease that cost."

One of the first vaccines that Medicago produced was a vaccine for influenza virus H1N1, or swine flu, the most common cause of the flu in 2009 (**Figure 10.4**). In the previous H1N1 pandemic, it had taken months for vaccines grown in chicken eggs to reach the market. By

### LOUIS-PHILIPPE VÉZINA

Plant researcher Louis-Philippe Vézina cofounded Medicago in 1997 and served as the company's chief scientific officer until 2014, overseeing all scientific research and development.

contrast, Medicago produced its vaccine, ready for testing, in just 19 days.

To make a flu vaccine in plants, Vézina and his team use genetic engineering in much the same way that it was used to create a gene therapy for Wiskott-Aldrich syndrome, as described in Chapter 8. They identify and synthesize a single viral gene that codes for hemagglutinin, a protein found on the surface of flu viruses. To make large quantities of that viral protein, the scientists at Medicago first insert the hemagglutinin gene into small, rod-shaped bacteria called "agrobacteria," which infect plants. Loaded with the hemagglutinin gene, these agrobacteria are then exposed to the tobacco plants: a robotic arm lifts a tray containing 5-week-old tobacco plants secured to the surface, flips the tray upside down, and dips the plants into a liquid solution swimming with the bacteria. Once the plants are immersed in the liquid, the technicians turn on a vacuum, sucking air out of the leaves and pulling bacteria into them—like dipping a sponge into water, squeezing it, and then releasing it to soak up the liquid.

When the plants are returned to the greenhouse, their leaves are floppy and almost translucent, like wet tissue paper. Now things are cooking. Once the bacteria are inside the leaves, the bacterial cells release the viral hemagglutinin gene into the plant cells, where it is transported into plant cell nuclei to begin the process of gene expression: making a protein from DNA via the two steps of transcription and translation.

"When you get the *Agrobacterium* solution into the leaf, it will invade the cells and there will be a burst of viral gene expression in the plant cell," said Vézina. "The plant takes over and uses its machinery to produce the protein" (**Figure 10.5**).

# Two-Step Dance, Transcription: DNA to RNA

The first step in gene expression—in this case, for a flu gene in a plant cell—is **transcription**, the synthesis of RNA based on a DNA template. In the nucleus, an enzyme called

**RNA polymerase** binds to a segment of DNA near the beginning of the gene that is called a **promoter**. The promoter contains a specific sequence of DNA bases that the RNA polymerase recognizes and binds. At Medicago, scientists attach specific promoters to the hemagglutinin gene so that the plant cell's RNA polymerase can identify the gene and actively transcribe it, maximizing the rate of transcription, says Vézina.

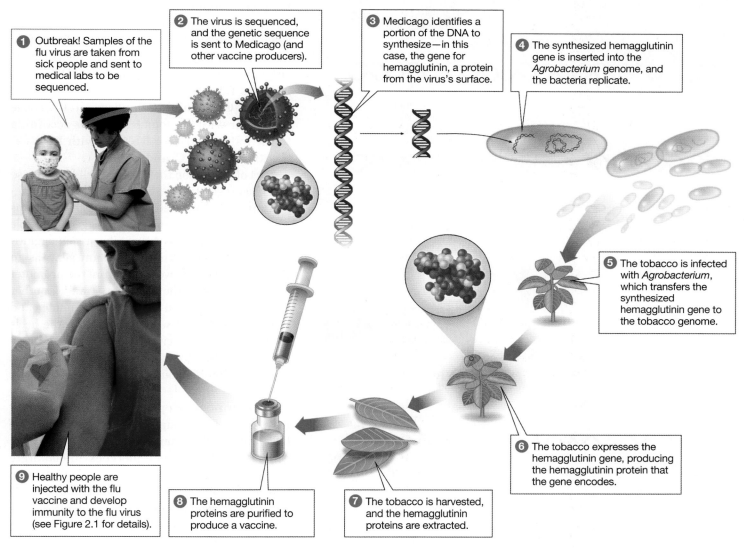

**Figure 10.5**

**From outbreak to vaccine**

During flu season or any other viral outbreak, a large network of medical professionals and scientists is activated to create a vaccine quickly and effectively. ▶▦

**Q1:** In which of the step(s) illustrated here does DNA replication occur? In which step(s) does gene expression occur?

**Q2:** Why do vaccine producers not simply replicate the entire viral genome, instead isolating the gene for one protein and replicating only that gene?

**Q3:** What role do the bacteria play in this process? Why are they needed?

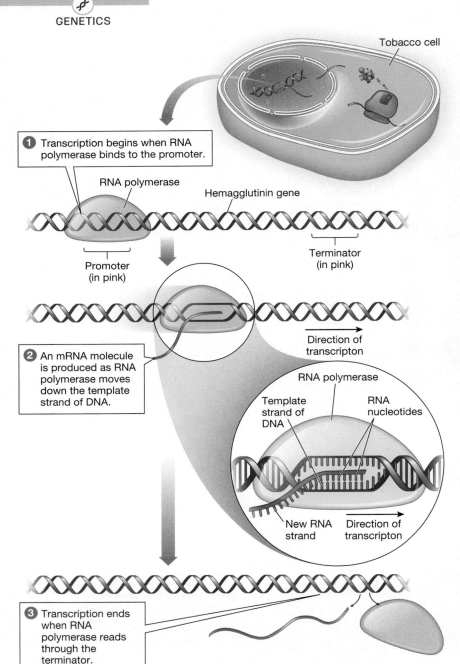

**1** Transcription begins when RNA polymerase binds to the promoter.

RNA polymerase

Hemagglutinin gene

Terminator (in pink)

Promoter (in pink)

Direction of transcripton

**2** An mRNA molecule is produced as RNA polymerase moves down the template strand of DNA.

RNA polymerase

Template strand of DNA

RNA nucleotides

New RNA strand

Direction of transcripton

**3** Transcription ends when RNA polymerase reads through the terminator.

Tobacco cell

Figure 10.6

**Plants making proteins, I: Transcription**

RNA polymerase transcribes the hemagglutinin gene into a molecule of RNA.

**Q1:** Why is only one strand of DNA used as a template?

**Q2:** If a mutation occurred within the promoter or terminator region, do you think it would affect the mRNA transcribed? Why or why not?

**Q3:** The template strand of part of a gene has the base sequence TGAGAAGACCAGGGTTGT. What is the sequence of RNA transcribed from this DNA, assuming that RNA polymerase travels from left to right on this strand?

Once bound to the promoter, the RNA polymerase unzips the DNA double helix at the beginning of the gene, separating a short portion of the two strands. Only one of the two DNA strands is used as a template, and thus it is called the **template strand**. The RNA polymerase begins to move down the DNA template strand, constructing a **messenger RNA (mRNA)** molecule, a strand of nucleotides complementary to the DNA, from free nucleotides floating around in the nucleus (**Figure 10.6**). RNA does not have all the same bases as DNA: its four bases are adenine (A), cytosine (C), guanine (G), and uracil (U). Those bases pair with the four DNA bases according to the following rules: RNA's A with DNA's T, C with G, G with C, and U with A.

Part of the reason that tobacco cells produce so much hemagglutinin so quickly is that the hemagglutinin gene inserted into the cells contains a special DNA sequence that triggers multiple RNA polymerases to transcribe a hemagglutinin gene at a single time. As for any gene, as an RNA polymerase moves away from the promoter and travels down the template strand, another RNA polymerase can bind at the promoter and start synthesizing a second mRNA on the heels of the first. At any given time, therefore, many RNA polymerases can be traveling down a DNA template simultaneously, each synthesizing an mRNA.

Transcription stops when the RNA polymerase reads through a special sequence of bases called a **terminator**. In eukaryotic cells, the mRNA then undergoes an elaborate sequence of modifications that prepare it to leave the nucleus. These steps include chemical modification of both ends of the mRNA, as well as a process called RNA splicing.

Most eukaryotic genes (and many viral genes) are embedded with stretches of sequences that don't code for anything, called **introns**. The stretches of DNA in a gene that carry instructions for building the protein are called **exons**. Because of this patchwork construction, with genes made of introns and exons interspersed, newly transcribed mRNA (pre-mRNA) is also a patchwork of coding sequences intermixed within noncoding sequences. During **RNA splicing**, the introns are snipped out of a pre-mRNA and the remaining pieces of mRNA—the exons—are joined to generate the

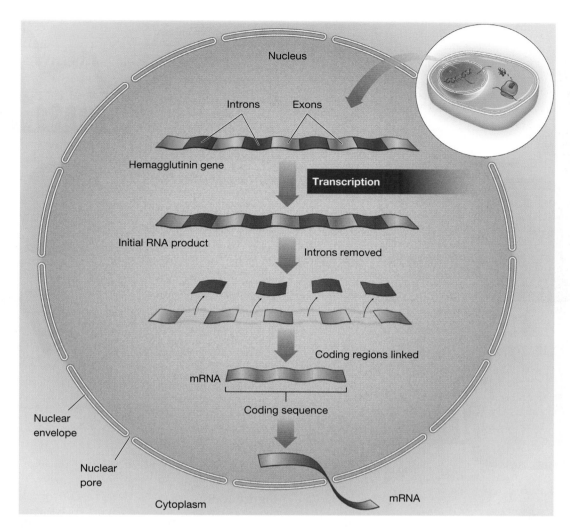

Figure 10.7

**Processing mRNA for export to the cytoplasm**
In eukaryotes, introns must be removed before an mRNA leaves the nucleus.

**Q1:** In your own words, define RNA splicing. When during gene expression does it occur?

**Q2:** What do you predict would happen if the introns were not removed from RNA before translation? Why would it be a problem if the introns were not removed?

**Q3:** Where is the mRNA destined to go once it has been transported out of the nucleus?

mature mRNA (**Figure 10.7**). This mRNA is then ready to leave the nucleus.

To review, transcription occurs when RNA polymerase binds to a promoter, unzips the DNA helix, and constructs a strand of mRNA based on the DNA template strand. Transcription ends at the terminator sequence, and the mRNA is then processed, at which time noncoding introns are spliced out of the sequence. Voilà! mRNA is created from a gene.

# Two-Step Dance, Translation: RNA to Protein

The microscopic molecular dance inside the tobacco cells continues with translation. Once the hemagglutinin gene has been transcribed into mRNA in the nucleus, it is time to make

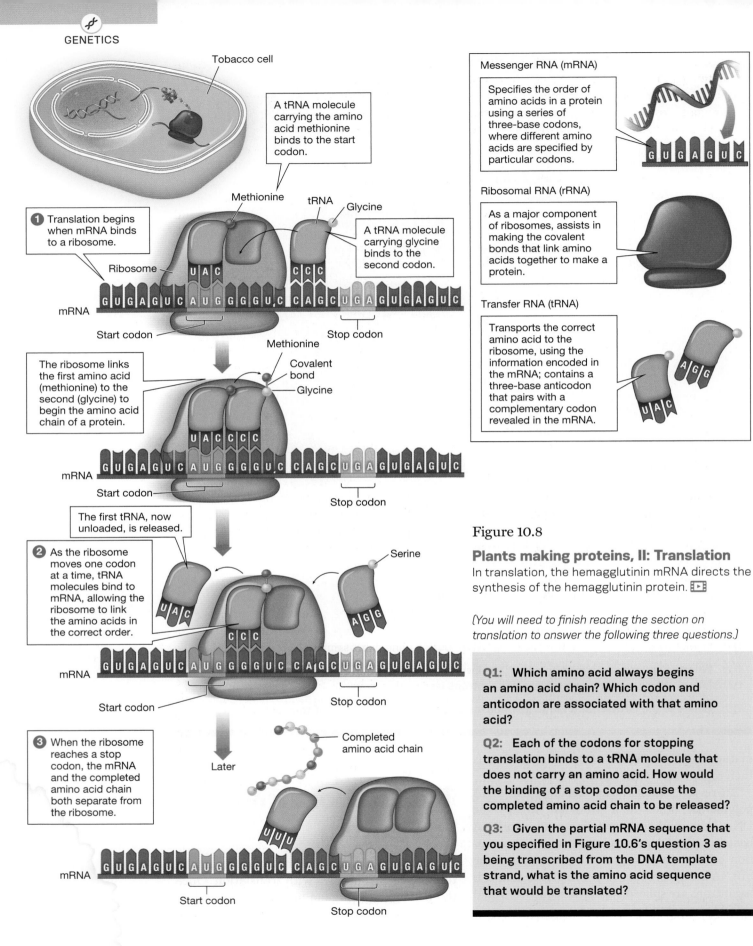

Tobacco cell

A tRNA molecule carrying the amino acid methionine binds to the start codon.

Methionine tRNA Glycine

A tRNA molecule carrying glycine binds to the second codon.

**1** Translation begins when mRNA binds to a ribosome.

Ribosome

U A C     C C C

mRNA    G U G A G U C A U G G G U C C A G C U G A G U G A G U C

Start codon                                    Stop codon

The ribosome links the first amino acid (methionine) to the second (glycine) to begin the amino acid chain of a protein.

Methionine

Covalent bond

Glycine

U A C C C C

mRNA    G U G A G U C A U G G G U C C A G C U G A G U G A G U C

Start codon                                    Stop codon

The first tRNA, now unloaded, is released.

**2** As the ribosome moves one codon at a time, tRNA molecules bind to mRNA, allowing the ribosome to link the amino acids in the correct order.

Serine

U A C          A G G

C C C

mRNA    G U G A G U C A U G G G U C C A G C U G A G U G A G U C

Start codon                                    Stop codon

**3** When the ribosome reaches a stop codon, the mRNA and the completed amino acid chain both separate from the ribosome.

Completed amino acid chain

Later

U U U

mRNA    G U G A G U C A U G G G U C C A G C U G A G U G A G U C

Start codon                                    Stop codon

**Messenger RNA (mRNA)**

Specifies the order of amino acids in a protein using a series of three-base codons, where different amino acids are specified by particular codons.

G U G A G U C

**Ribosomal RNA (rRNA)**

As a major component of ribosomes, assists in making the covalent bonds that link amino acids together to make a protein.

**Transfer RNA (tRNA)**

Transports the correct amino acid to the ribosome, using the information encoded in the mRNA; contains a three-base anticodon that pairs with a complementary codon revealed in the mRNA.

A G G

U A C

## Figure 10.8

### Plants making proteins, II: Translation

In translation, the hemagglutinin mRNA directs the synthesis of the hemagglutinin protein. ▶️

*(You will need to finish reading the section on translation to answer the following three questions.)*

**Q1:** Which amino acid always begins an amino acid chain? Which codon and anticodon are associated with that amino acid?

**Q2:** Each of the codons for stopping translation binds to a tRNA molecule that does not carry an amino acid. How would the binding of a stop codon cause the completed amino acid chain to be released?

**Q3:** Given the partial mRNA sequence that you specified in Figure 10.6's question 3 as being transcribed from the DNA template strand, what is the amino acid sequence that would be translated?

the protein, the actual product that will be extracted from the tobacco leaves. First the mRNA is transported from the nucleus, where it was made, to the sites of protein synthesis: the cell structures called **ribosomes** in the cytoplasm. To escape the nucleus, the long strand of mRNA passes through a nuclear pore, like a noodle slipping through the hole of a colander. Once the mRNA molecule arrives in the cytoplasm, the information it contains must be translated, with the help of ribosomes, from the language of mRNA (nitrogenous bases) to the language of proteins (amino acids). **Translation** is the process by which ribosomes convert the information in mRNA into proteins.

During translation, ribosomes "read" the mRNA code like a grocery list (bread, milk, etc.), and collect the corresponding amino acids, linking them in the precise sequence dictated by mRNA (**Figure 10.8**). Ribosomes read the mRNA information in sets of three bases at a time, and each unique sequence of three mRNA bases is called a **codon**. The hemagglutinin gene has about 1,770 bases, of which 1,695 code for the protein. That makes 565 codons (1,695 divided by 3), and therefore the hemagglutinin protein is composed of 565 amino acids.

The four bases of mRNA (A, C, G, U) can be arranged to create a three-base sequence in 64 different ways (because $4^3 = 64$). Therefore, there are 64 possible codons (**Figure 10.9**). Most of the 64 codons specify a particular amino acid. A couple of amino acids are specified by only one codon, while other amino acids are specified by anywhere from two to six different codons. Some codons do not code for any amino acid and instead act as signposts that communicate to the ribosomes where they should start or stop reading mRNA. The **start codon** (AUG) is the ribosome's starting point on the mRNA strand, and there are three possible **stop codons** (UAA, UAG, and UGA). By beginning and ending at fixed points, the cell ensures that the mRNA message is read in precisely the same way every time.

The information specified by all 64 possible codons is the **genetic code** (see **Figure 10.9**). The genetic code has several significant characteristics. First, it is *unambiguous*: each codon specifies only one amino acid. It is also *redundant*: since there are a total of 64 codons but only 20 amino acids, several different codons call for the same amino acid, as already mentioned. Finally, the genetic code is virtually *universal*: nearly every organism on Earth uses the same code, from agrobacteria to tobacco cells to human cells—a feature that illustrates the common descent of all organisms.

Making a protein from an mRNA strand requires two additional types of RNA: The first is **ribosomal RNA (rRNA)**, an important component of ribosomes. The second is **transfer RNA (tRNA)**, which is the caddy for the process, delivering specific amino acids to the ribosomes as the codons are read off the mRNA "list."

Each tRNA specializes in binding to a specific amino acid and recognizes and pairs with a specific codon in the mRNA, like a puzzle piece that fits one amino acid on one end and one codon on the other. At one end of a tRNA molecule, a special sequence of three nitrogenous bases, called an **anticodon**, binds the correct codon on the mRNA. At the other end, the specific amino acid attaches (see **Figure 10.8**).

Let's recap. For translation to occur, an mRNA molecule must first bind to a ribosome. The ribosomal machinery then "scans" through the mRNA until it finds a start codon (AUG). Next the ribosome recruits the appropriate tRNAs one by one, as determined by the codons read in the mRNA sequence. A special site on the ribosome facilitates the linking of one amino acid to another, like beads on a string. Finally, the ribosome reaches a stop codon. The amino acid chain cannot be extended further, because none of the tRNAs will recognize and pair with any of the three stop codons. At this point the mRNA molecule and the completed amino acid chain separate from the ribosome. The new protein then folds into its compact, specific three-dimensional shape and is ready to go to work in the cell.

But this process does not always go as planned. In Chapter 9 we learned that a mutation is a change in the sequence of DNA bases. Mutations affect an organism by disrupting or preventing the healthy formation of a protein. For example, a mutation can cause a DNA sequence not to be translated or transcribed, prompt the amino acid chain to end prematurely, or make the final protein fold incorrectly,

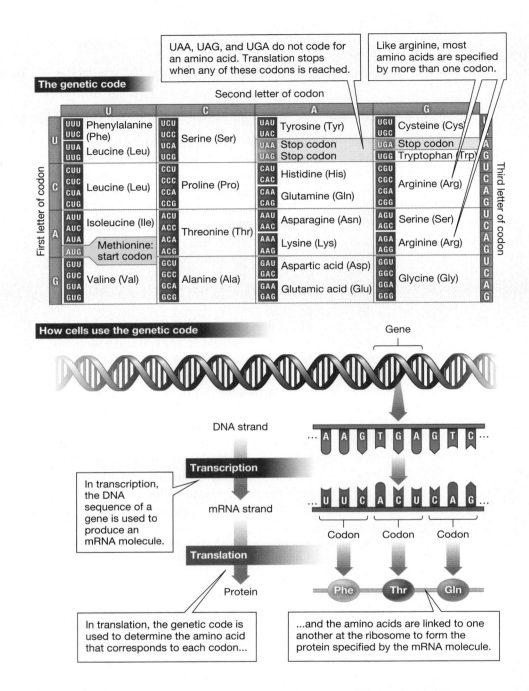

**Figure 10.9**

## The genetic code

(Top) The genetic code is composed of the 64 possible codons found in the mRNA. Each codon specifies an amino acid or is a signal that starts or stops translation. (Bottom) The genetic code is used during the translation of mRNA to protein.

**Q1:** How many codons code for isoleucine? For tryptophan? For leucine?

**Q2:** What codons are associated with asparagine? With serine?

**Q3:** From the partial mRNA sequence that you specified in Figure 10.6's question 3 as being transcribed from the DNA template strand, remove only the first A. What amino acid sequence would be translated as a result of this change? How does that sequence compare to the amino acid sequence you translated from the original mRNA sequence? *Bonus:* What kind of mutation is this? (*Hint:* See Chapter 9.)

among other possibilities. While single-base substitutions are not always a problem, single-base insertions and deletions cause a genetic "frameshift," shifting all subsequent codons "downstream" by one base (**Figure 10.10**). This shift scrambles the entire downstream DNA message, in turn scrambling the entire RNA message and causing the ribosomes to assemble a very different sequence of amino acids from the mutation point onward—as if every letter in this phrase was shifted to the left one space while the word length and spaces between words were retained (a si fever ylette ri nthi sphras ewa sshifte dt oth erigh ton espac ewhil eth ewor dlengt han dspace sbetwee nword swer eretaine d).

# Tweaking Gene Expression

Inside a tobacco cell, as in most living cells, the expression of many genes can be turned on or off, slowed down (**down-regulated**), or sped up (**up-regulated**). This **gene regulation** enables organisms to change which genes they express in response to internal signals (from inside the body) or external cues in the environment. In this way, by producing different proteins as needed, organisms can adapt to their surroundings.

All cells in a multicellular individual have essentially the same DNA, yet different cells express different sets of genes, and within a given cell the pattern of gene expression can change over time. Single-celled organisms, such as bacteria, face a more difficult challenge: they are directly exposed to their environment, and they have no specialized cells to help them deal with changes in that environment. One way they meet this challenge is to express different genes at different times.

The expression of most genes in prokaryotes and eukaryotes is regulated by both internal and external signals. Many genes are also developmentally regulated, meaning that their expression can change, sometimes dramatically, as an organism grows and develops. Gene expression is regulated at many different points in the cell, including DNA packing (the way DNA is compressed or unwound in the genome), transcription, mRNA processing, and several

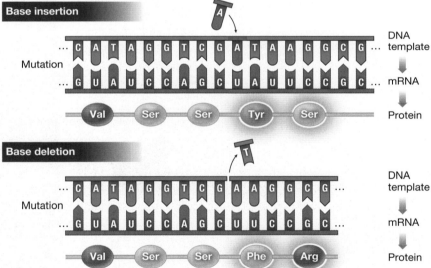

Figure 10.10

**Effects of point mutations**

**Q1:** Why is an insertion or a deletion in a gene more likely to alter the protein product than a substitution, such as A for C, would?

**Q2:** Which would you expect to have more impact on an organism: a point mutation as shown here, or the insertion or deletion of a whole chromosome (discussed in Chapter 8)?

**Q3:** Which mechanisms in a cell prevent mutations? (*Hint:* Refer back to Chapter 6 if needed.)

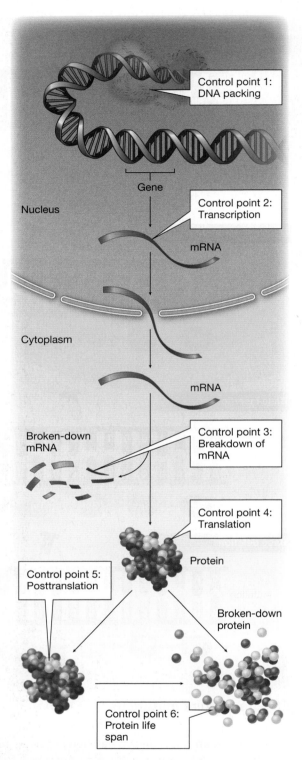

points during translation (**Figure 10.11**). But for all living cells, a few genes are always expressed at a low level; their transcription is not regulated, because these genes are needed at all times.

At Medicago, the company takes advantage of gene regulation in tobacco cells to produce as much hemagglutinin protein as possible. After the agrobacteria are vacuum-sucked into the leaves and transcription and translation begin to actively occur in the tobacco cells, the plants are sent to grow in an incubation room, where technicians can alter the humidity, temperature, and level of light to maximize the amount of protein expressed by the plants.

"We've been able to tweak environmental conditions of the plants to boost gene expression," says Vézina. "You name it, we've tried it." It's an important step in the process, adds Wanner. "We determined what the best conditions for protein expression are."

# To the Market

It's the end of April, and the final steps are being taken to isolate and purify the hemagglutinin protein from the tobacco plants, to see how much vaccine can be made in a month. After the plants have incubated for several days, the leaves are stripped and diced into green confetti, then digested with enzymes to

## Figure 10.11

### How gene expression is regulated

In eukaryotes, gene expression can be controlled at several points along the pathway from gene to protein to phenotype: before transcription, at transcription, during RNA processing, or at translation. Gene expression can also be regulated after translation, by control of the activity or life span of the protein.

**Q1:** As illustrated here, at what control point is transcription regulated?

**Q2:** What is a possible advantage of regulating gene expression before transcription, versus after?

**Q3:** If you wanted to up-regulate production of the hemagglutinin protein in a tobacco plant carrying the hemagglutinin gene, at which control point(s) would that be possible? Justify your reasoning.

# The Deadly Price of a Pandemic

The Spanish flu of 1918 devastated populations across Europe, killing not only the young and old, but also healthy adults. Since then, new strains of flu have reared their ugly heads, and doctors and vaccine manufacturers struggle to anticipate the next outbreak.

Assessment available in smartwork5

## 1918
### H1N1

**50**M
dead worldwide

**650**K
dead in U.S.

## 1957
### H2N2

**1.5**M
dead worldwide

**70**K
dead in U.S.

## 1968
### H3N2

**750**K
dead worldwide

**34**K
dead in U.S.

## 2009
### H1N1

**285**K
dead worldwide

**12**K
dead in U.S.

break up the leaf material so that the desired proteins are released into solution. The resulting solution, which resembles green-pea soup, is filtered several times to isolate clusters of hemagglutinin, which will then be processed into a vaccine product that is safe to inject into people.

Medicago is not the first company to produce a human drug using a plant. The first human-like enzyme was produced from tobacco back in 1992 at Virginia Polytechnic Institute. Numerous other plant biopharming companies have sprung up since then, experimenting with various plant species, including corn, soybean, duckweed, and more. But the field is not without risks and controversy. "The main concern and risk is spread—that the gene will get out into nature and spread," says Roth. "But there are techniques to make sure that doesn't happen, and it is closely regulated."

One of those techniques is to grow the plants in contained environments, where there is no risk of contaminating food crops. Israel-based biotech company Protalix Biotherapeutics, for example, grows carrot cells inside in large hanging bags of fluid and cells. In May 2012, the FDA approved the first biopharmed drug for humans (produced by Protalix)— a therapy for Gaucher disease, a rare genetic disorder—grown in the company's carrot cells. "This approval demonstrates a proof of concept for the power of this technology," the CEO of Protalix told the journal *Nature*.

Today, Medicago has completed safety trials for its pandemic flu vaccine and has had positive results in the first of two large clinical trials in humans that tested its seasonal flu vaccine's efficacy. When the April manufacturing test was completed, the company had produced an astounding 10 million doses of flu vaccine in a single month. It would have taken 5–6 months to produce the same amount using the traditional method of growing vaccines in chicken eggs. With that success under its belt, in 2015 the company began building a second production complex. Based in Quebec City, the manufacturing facility will have the capacity to deliver up to 50 million doses of seasonal flu vaccine.

Medicago is also developing novel vaccines against rotavirus and rabies virus, and the company recently received a U.S. government contract to manufacture antibodies to treat Ebola virus infection. "It might have taken a bit longer than we thought for biopharming to be accepted, but it deserves the visibility and attraction it has now," says Vézina. "Biopharming is here to stay. I'm convinced of this."

# REVIEWING THE SCIENCE

- Most genes contain instructions for building proteins. **Gene expression** is the process by which genes are transcribed into RNA and then translated into a protein.

- During **transcription**, which occurs in the nucleus, RNA polymerase binds the **promoter** of a gene and produces a **messenger RNA (mRNA)** version of the gene sequence from free nucleotides.

- Next, during **RNA splicing**, **introns** are snipped out of the pre-mRNA sequence, and the remaining **exons** are joined. The mRNA is transported out of the nucleus.

- During **translation**, which occurs in the cytoplasm, ribosomes convert the sequences of bases in an mRNA molecule to the sequence of amino acids in a protein, with the help of **ribosomes** composed of **ribosomal RNA (rRNA)** and proteins, and associated with **transfer RNA (tRNA)**.

- Ribosomes read the mRNA information in sets of three bases at a time, called **codons**. There are 64 possible codons, including a **start codon** (AUG) and three possible **stop codons** (UAA, UAG, and UGA).

- The **genetic code** is redundant because there are 64 codons and only 20 amino acids, and universal because nearly all organisms on Earth use the same code.

- Gene expression is regulated at many different points in the pathway from gene to protein. Organisms rely on **gene regulation** to respond to signals inside the body and to external cues in the environment.

# THE QUESTIONS

## The Basics

**1** Link each term with the correct definition.

| GENE EXPRESSION | 1. The production of RNA, using the information in the DNA sequence of a gene. |
| GENE REGULATION | 2. The flow of information from gene to protein. |
| TRANSCRIPTION | 3. The control of gene expression in response to environmental or developmental needs. |
| TRANSLATION | 4. The linking of amino acids in the precise sequence dictated by an mRNA base sequence. |

**2** For each of the following, identify the type of RNA involved (mRNA, rRNA, or tRNA).

_____ a. Transports the correct amino acid to the ribosome, using the information encoded in the mRNA.

_____ b. Is a major component of ribosomes.

_____ c. Specifies the order of amino acids in a protein, using a series of three-base codons, where different amino acids are specified by particular codons.

_____ d. Contains a three-base anticodon that pairs with a complementary codon revealed in the mRNA.

_____ e. Assists in making the bonds that link amino acids together to make a protein.

**3** Select the correct terms:
The genetic code demonstrates (**ambiguity / redundancy**) because some amino acids are coded by more than one codon. The lack of (**ambiguity / redundancy**) in the genetic code is evidenced by the fact that each codon codes for one, and only one, amino acid.

**4** In the diagram of transcription shown here, fill in the blanks with the appropriate terms: (a) gene; (b) promoter; (c) terminator; (d) RNA polymerase; (e) mRNA.

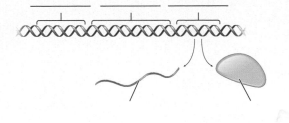

**5** Place the following steps of translation in the correct order by numbering them from 1 to 9.

_____a. A tRNA molecule carrying the amino acid methionine binds at its anticodon site to the mRNA start codon.

_____b. The ribosome links the first amino acid to the second amino acid to begin the amino acid chain.

_____c. The ribosome continues to link each amino acid to the growing amino acid chain.

_____d. The ribosome reaches a stop codon.

_____e. An mRNA binds to a ribosome.

_____f. The mRNA and the completed amino acid chain separate from the ribosome.

_____g. The first tRNA, separated from its amino acid, releases from the mRNA.

_____h. A tRNA molecule carrying the second amino acid binds to the second mRNA codon.

_____i. Each tRNA releases from the mRNA after it is separated from its amino acid.

**6** Which of the following are possible reasons that a cell would regulate its expression of a gene? (Select all that apply.)

(a) an increased need for a particular enzyme

(b) a decreased need for a particular enzyme

(c) increasing temperature in the external environment

(d) changing needs as an organism ages

(e) death

## Challenge Yourself

**7** Using the genetic code shown in Figure 10.9, find the amino acid coded by each of the following codons.

(a) AAU

(b) UAA

(c) AUA

(d) GGG

(e) CCC

**8** Using the genetic code shown in Figure 10.9, find a codon that codes for each of the following amino acids.

(a) arginine

(b) alanine

(c) methionine

(d) glycine

**9** During transcription, what RNA molecule will be made from the DNA template CGTTACG?

(a) CGTTAGC

(b) GCAAUGC

(c) GCATTGC

(d) CGUUAGC

**10** Which amino acid sequence will be generated during translation from the following small mRNA: ...CCC-AUG-UCU-UCG-UUA-UGA-UUG...? (*Hint:* Remember where translation starts and stops.)

(a) Met-Glu-Arg-Arg-Glu-Leu

(b) Met-Ser-Ser-Leu-Leu

(c) Pro-Met-Ser-Ser-Leu-Leu

(d) Pro-Met-Ser-Ser-Leu

(e) Met-Ser-Ser-Leu

## Try Something New

**11** Some diseases (for example, Huntington's and Parkinson's) appear to be related to increasing protein levels in brain cells, which lead eventually to cell death. At which of the control points shown in Figure 10.11 might a gene regulation error be occurring with these diseases? Identify one control point at which the error would result in up-regulation of gene expression, and one control point at which the error would result in down-regulation of gene expression.

**12** The following nucleic acid is an entire primary transcript (pre-mRNA not yet processed): ACGCAUGCGaugaugccccucag GUCUguuuccgugaUGCCGUUGACCUGA. The nucleotides in capitals are exons; and the nucleotides in lowercase type are introns. Appropriately splice this primary transcript.

(a) ACGCAUGCGGUCUUGCCGUUGACCUGA

(b) augccuuucagguuuccguga

(c) augccuuucagGUCUguuuccguga

(d) ACGCAUGCGaugGUCUUGCCGUUGACCUGA

(e) ACGCAUGCGaugccuagGUCUguuuccgugaUGCCGUUGACC UGA

**13** How is gene expression similar to DNA replication, and how is it different? Give at least one similarity and one difference.

## Leveling Up

**14** *Write Now* biology: genotype to phenotype Your roommate, who is also taking a biology class, has become a little confused. He informs you that the genetic code is known to be ambiguous because a given genotype may give rise to a variety of phenotypes during gene expression (for example, his twin brother is an inch taller and more tan than he is). You like your roommate and would like him to pass his next biology exam, so you decide to help him. Write him a brief note explaining (a) why the genetic code is not, in fact, ambiguous and (b) how gene expression derives a phenotype from a genotype.

**15** What do *you* think? Most people carry two copies of a normal gene that codes for an enzyme, glucosylceramidase, that is involved in breaking down lipids no longer needed in cells. (Enzymes are proteins that cause specific chemical changes; they are biological catalysts.) One in 100 people in the United

States carries a recessive mutation that codes for a defective glucosylceramidase enzyme. And about one in 40,000 people carries two copies of the mutation and displays the symptoms of Gaucher disease. These symptoms, caused by the accumulation of lipids in cells, include anemia, enlarged organs, swollen glands and joints, and, in severe cases, neurological problems and early death.

Enzyme replacement therapy is effective but very expensive—about $200,000 annually—and must be continued, every 2 weeks, for life. The Israeli biotech company Protalix Biotherapeutics, working with the U.S.-based Pfizer Pharmaceuticals, has developed a process to genetically modify carrots to produce a replacement enzyme. The biopharmed enzyme will cost about 25 percent less than the standard enzyme therapies, which are grown in mammalian cell lines. Protalix is now working on treatments for other enzyme deficiency diseases.

The FDA's May 2012 approval of the drug developed by Protalix alarmed some environmental activists and health advocates, who fear that the company's genetically modified carrot is just the beginning of a wedge that will lead to an underregulated and potentially dangerous industry. There is some legitimacy to their concerns: the U.S. Department of Agriculture (USDA) does not require an environmental impact assessment for biopharmed crops; nor does it require biotech companies to share the location of their test fields or the identity of the biopharmed molecules being produced. Furthermore, the USDA is not sufficiently staffed to effectively monitor companies involved in biopharming.

What do you think? Should biopharming be allowed in the United States? If so, under what conditions and with what limits? For example, should it be allowed to produce drugs for only life-threatening illnesses, or only under highly controlled conditions? Be prepared to discuss your observations and reflections in class.

**16** **Life choices** Go to the Centers for Disease Control and Prevention (CDC) influenza website (http://www.cdc.gov/flu) and read the pages "Key Facts About Influenza (Flu)" (under "Flu Basics") and "Key Facts About Seasonal Flu Vaccine" (under "Prevention—Flu Vaccine"). You can also go to the Mayo Clinic's influenza website (http://www.mayoclinic.org/diseases -conditions/flu/home/ovc-20248057). Then answer the following questions.

a. What is the flu? How is it passed on?

b. What are the possible symptoms and complications of the flu?

c. How can you decrease your chance of getting the flu, and what treatments are available if you become infected?

d. What are the benefits and risks of the flu vaccine?

e. Why is there a new flu vaccine every year?

f. Why is the flu vaccine more effective in some years than in others?

g. Who would you recommend should get the flu vaccine? Explain your reasoning.

h. Do you get a flu vaccine every year? Why or why not?

**For more, visit digital.wwnorton.com/bionow2 for access to:**

# Whale Hunting

*Fossil hunters discover Moby Dick's earliest ancestor—a furry, four-legged land lover.*

## After reading this chapter you should be able to:

- Define evolution and list the six types of evidence for evolution.

- Compare and contrast artificial selection and natural selection.

- Summarize the argument that the fossil record provides evidence in support of evolution.

- Give an example of a homologous or vestigial trait, and explain how such traits support the theory of common descent.

- Explain why even distantly related species have similar DNA.

- Use your knowledge of evolution and continental drift to make a prediction about the geographic location of a given set of fossils.

- Relate similarities in embryonic development among species to their shared evolutionary past.

CHAPTER

11

EVIDENCE FOR
EVOLUTION

Fossils break all the time. This time, the 50-million-year-old ear bone of a small, deerlike mammal called *Indohyus* snapped clean off the skull. Sheepishly, the young laboratory technician cleaning the fossil handed the broken piece to his boss, paleontologist and embryologist J. G. M. "Hans" Thewissen at Northeast Ohio Medical University. Thewissen tenderly turned the preserved animal remains over in his hand. Then, as the tech reached for the fossil to glue it back onto the animal's skull, Thewissen went rigid.

"Wow, that is weird," said Thewissen. The *Indohyus* ear bone, which should have looked like the ear bone of every other land-living mammal— like half a hollow walnut shell, but smaller—was instead razor thin on one side and very thick on the other (**Figure 11.1**). "Wow," repeated Thewissen. This wasn't the ear of a deer, or any other land mammal. Thewissen squinted closer. "It looks just like a whale," he said.

Although they live in the ocean like fish, whales are mammals like us. So are dolphins and porpoises (**Figure 11.2**). Like all mammals, whales are warm-blooded, have backbones, breathe air, and nurse their young from mammary glands. Numerous fossils have been found documenting whales' unique transition from land-living mammals to the mammoths of the sea, during which whale populations developed longer tails and shorter and shorter legs. But one crucial link in the fossil record was missing: the closest land-living relatives of whales. What did the ancestors of whales look like before they entered the water? Staring at the strange fossil in his hand, Thewissen realized he could be holding the ear of that missing link.

Whales are but one of the many organisms that share our planet. Every species is exquisitely fit for life in its particular environment: whales in the open ocean, hawks streaking through the sky, tree frogs camouflaged in the green leaves of a rainforest. There is a great diversity of life on Earth—animals, plants, fungi, and more—with each species well matched to its surroundings. This diversity of life is due to evolution.

"Evolution," in everyday language, means "change over time." In science, biological **evolution** is a change in the inherited characteristics of a

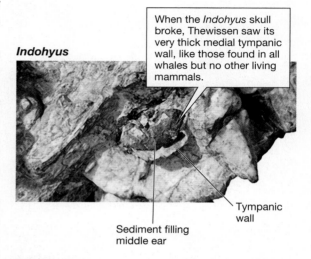

*Indohyus*

When the *Indohyus* skull broke, Thewissen saw its very thick medial tympanic wall, like those found in all whales but no other living mammals.

Sediment filling middle ear

Tympanic wall

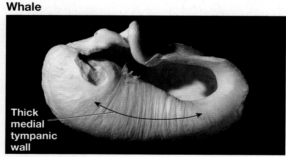

**Whale**

Thick medial tympanic wall

**White-tailed deer**

Thin medial tympanic wall

**Figure 11.1**

**The mysterious ear bone**
The *Indohyus* fossil ear bone (top) looks more like the ear bone of whales (middle) than that of any modern land mammal (bottom). (Source: *Indohyus* and whale photos courtesy of Thewissen Lab, NEOMED.)

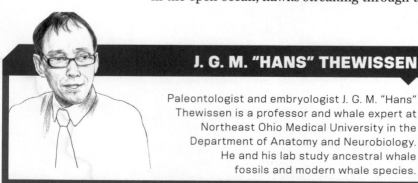

## J. G. M. "HANS" THEWISSEN

Paleontologist and embryologist J. G. M. "Hans" Thewissen is a professor and whale expert at Northeast Ohio Medical University in the Department of Anatomy and Neurobiology. He and his lab study ancestral whale fossils and modern whale species.

Figure 11.2

**A montage of mammals**
Earth boasts an amazing diversity of life, as evidenced here. These are only a few of the almost 5,500 species of mammals currently living on the planet.

group of organisms over generations. Whales, for example, evolved from four-legged land-living animals into sleek emperors of the ocean, slowly becoming suited for the water over tens of millions of years. Whales changed as a population. Populations evolve; individuals do not.

You may wonder how we can be sure of evolutionary change, especially for a transformation as extreme as that of a furry, four-legged beast into a whale. There is strong evidence for evolution not only from fossils like those that Thewissen studies, but also from features of existing organisms, common patterns of how embryos develop, DNA evidence, geographic evidence, and even direct observation of organisms evolving today—including man's best friend.

# Artificial to Natural

As we saw in Chapter 7, all dogs are a single species. That species is *Canis lupus familiaris*, a subspecies of the gray wolf. Scientists estimate that domestication of the gray wolf began about 16,000 years ago as humans habituated the

animals to civilization. From there, people bred the wolf for desired qualities, such as decreased agression and the ability to follow commands. Millennia later, dogs were selectively bred for specific traits like the long legs of a greyhound or the short snout of a bulldog. This selective process has led to incredible variation in the size and shape of dog breeds, from 6-pound Chihuahuas to 200-pound Great Danes (**Figure 11.3**). Dogs are an example of evolution that we can directly observe, and they evolved via artificial selection. **Artificial selection** is brought about by **selective breeding**, in which humans allow only individuals with certain inherited characteristics to mate.

Through selective breeding, humans have crafted enormous evolutionary changes within not only dogs, but many other domesticated organisms, including ornamental flowers, pet birds, and food crops. We can observe evolution happening through artificial selection via selective breeding.

Artificial selection happens when humans choose which individuals of a particular species are allowed to breed. Without the intervention of humans, can the environment itself

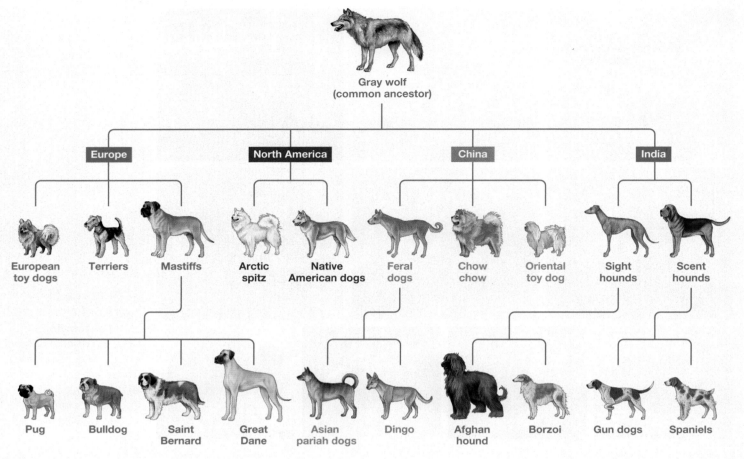

## Figure 11.3

### Selective breeding of dogs produces myriad traits

Dogs were domesticated only a few times, and always from gray wolves. Thus, the remarkable diversity of dogs represents the effects of selective breeding on a small number of lineages of domesticated wolves.

**Q1:** What is selective breeding, and how does it work?

**Q2:** Explain how selective breeding leads to artificial selection.

**Q3:** Name as many organisms as you can whose current characteristics are due to artificial selection.

"choose" who survives and breeds? It does. In nature, evolution occurs mainly via *natural* selection, as well as through other mechanisms we will discuss later. **Natural selection** is the process by which individuals with advantageous inherited characteristics for a particular environment survive and reproduce at a higher rate than do individuals with other, less useful characteristics. In other words, whoever has the most kids wins!

After the environment "chooses" the winners—those that successfully breed the most—the characteristics of those individuals become more common in successive generations because they have produced more offspring. For example, in 1977 a terrible drought struck the Galápagos Islands off the coast of Ecuador. One species of small ground finches—petite birds with sharp, pointy beaks—starved as the small, tender seeds they ate became scarce. But some heat-loving, drought-resistant plants still produced large, hard seeds. The finches with larger beaks could eat those seeds. They survived and reproduced, and by 1978, in just one generation, the average beak size in the population had increased (**Figure 11.4**).

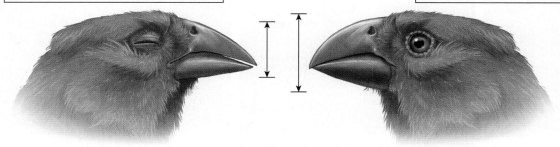

Birds with smaller beaks were less likely to find seeds to eat and died at higher rates, thus failing to have sex, produce offspring, and pass on their genetic traits.

Birds with large beaks were more likely to survive, and passed on their large beak size to their offspring.

Figure 11.4

**Natural selection results in larger beak size in finches**

After a drought, only birds with larger beaks were able to eat the available food: large, hard seeds. In the span of just one generation, the average size of the species' beak was visibly larger.

**Q1:** What is natural selection?

**Q2:** If heavy rains caused an abundance of small, tender seeds and fewer large seeds, what do you predict would happen to the average beak size of the finches?

**Q3:** Compare and contrast artificial selection and natural selection. Name two ways in which they are similar. How are they different?

This is one of many examples of how a population can evolve via natural selection so that more and more individuals have beneficial traits, and fewer and fewer have disadvantageous traits. This is called **adaptation**—an evolutionary process by which a population becomes better matched to its environment over time. The finch population quickly adapted to its new, drier environment. Over time, the small-beaked finches died off and the large-beaked birds survived and reproduced; the finch population adapted to its environment in just a few years. Other adaptations take millions of years, such as whale ancestors adapting to aquatic life.

It is important to realize that biological evolution includes human evolution. Surveys taken since 2013 reveal that about 40 percent of adults in the United States do not believe that humans evolved from earlier species of animals. This statistic is startling because evolution has been a settled issue in science for nearly 150 years. Scientists like Thewissen go to work every day and see evolution in action. In fact, the vast majority of scientists of all nations agree that the evidence for evolution is overwhelming.

Six lines of evidence provide compelling support for biological evolution:

1. Direct observation of evolution through artificial selection

2. Fossil evidence

3. Shared characteristics among living organisms

4. Similarities and differences in DNA

5. Biogeographic evidence

6. Common patterns of embryo development

Nowhere is all this evidence more present and intriguing than in one of the most dramatic transitions to occur on Earth: the evolution of small, land-living mammals into dolphins, porpoises, and mighty whales.

# Fossil Secrets

**Fossils** are the mineralized remains of formerly living organisms or the impressions of formerly living organisms (**Figure 11.5**). For Thewissen, getting his hands on the

Soft-bodied animals such as this one dominated life on Earth 600 million years ago (mya).

A fossil of a trilobite that lived between 410 and 355 mya.

Fossilized leaf of a 300-million-year-old seed fern.

This 20-million-year-old termite is preserved in amber, the fossilized resin of a tree.

A fossil of a *Velociraptor* entangled with a *Protoceratops*, which bit down on the predator's claw, locking both in a death grip.

Once solid wood has fossilized into solid rock, it is known as petrified wood.

## Figure 11.5

### Fossils through the ages

Myriad fossils exist, ranging from imprints of organisms, to preserved organisms, to completely mineralized bone and wood. Each fossil can be dated, and when the results are compiled, they can tell the life history of Earth.

*Indohyus* fossils was no easy task. Beginning in 2003, he made an annual pilgrimage to Dehradun, India, a city nestled in the foothills of the Himalayas. There he visited the widow of A. Ranga Rao, an Indian geologist who had hoarded piles of fossils excavated from Kashmir, a disputed border area between India and Pakistan. Most early whale fossils have come from the India-Pakistan region, where whales first evolved. But because of political tensions, in recent decades it has been too dangerous to travel to Kashmir, much less dig for fossils. Thewissen was frustrated by his inability to collect fossils in Kashmir.

The fossil record enables biologists to reconstruct the history of life on Earth, and it provides some of the strongest evidence that species have evolved over time. The relative depth or distance from the surface of Earth at which a fossil is found indicates its *order* in the fossil record. The ages of fossils correspond to their order: older fossils are found in deeper, older rock layers.

The fossil record contains excellent examples of how major new groups of organisms arose from previously existing organisms. The record includes numerous **transitional fossils**, evidence of species with some similarities to the ancestral group (land-living mammals) and some similarities to the descendant species (whales). Thewissen spent decades studying these intermediates—from the first known whale, the wolflike *Pakicetus* that waded in shallow freshwater; to the larger crocodile-like *Ambulocetus* that stalked its prey underwater; to the fully aquatic *Dorudon*, with its blowhole, flippers, and tail (**Figure 11.6**). Yet Thewissen and others had long been searching for the

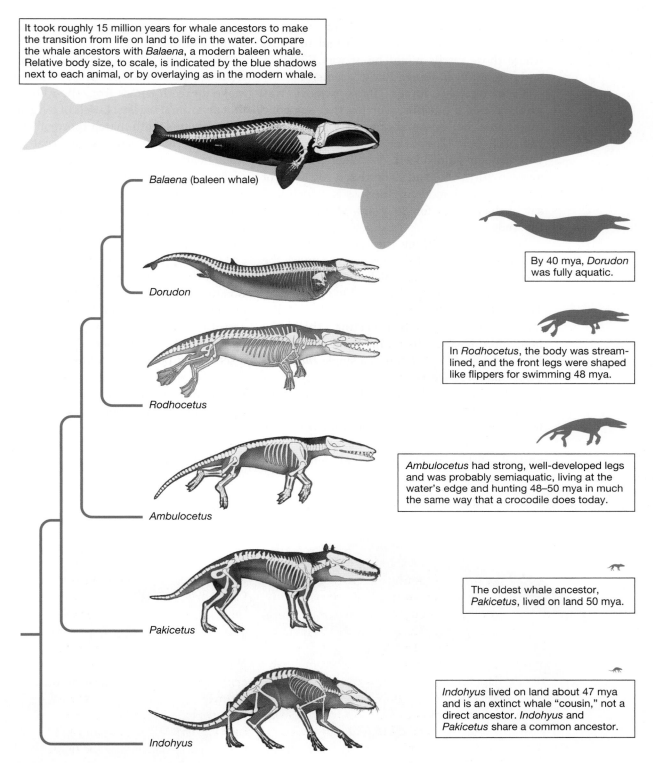

It took roughly 15 million years for whale ancestors to make the transition from life on land to life in the water. Compare the whale ancestors with *Balaena*, a modern baleen whale. Relative body size, to scale, is indicated by the blue shadows next to each animal, or by overlaying as in the modern whale.

*Balaena* (baleen whale)

*Dorudon*

By 40 mya, *Dorudon* was fully aquatic.

*Rodhocetus*

In *Rodhocetus*, the body was streamlined, and the front legs were shaped like flippers for swimming 48 mya.

*Ambulocetus*

*Ambulocetus* had strong, well-developed legs and was probably semiaquatic, living at the water's edge and hunting 48–50 mya in much the same way that a crocodile does today.

*Pakicetus*

The oldest whale ancestor, *Pakicetus*, lived on land 50 mya.

*Indohyus*

*Indohyus* lived on land about 47 mya and is an extinct whale "cousin," not a direct ancestor. *Indohyus* and *Pakicetus* share a common ancestor.

## Figure 11.6

### Skeletons and body sizes of modern whales and fossil ancestors

These reconstructed skeletons from modern whales (top) and various ancestors are in chronological order.

**Q1:** What is the general definition of a fossil?

**Q2:** How are the ancestors of modern whales different from their present form?

**Q3:** What is meant by the term "intermediate fossil" when referring to the fossil record?

animal that preceded them all—the ancestor that lived on land. If Thewissen had to guess where those fossils might be, it was Kashmir, and potentially in Ranga Rao's basement.

Unfortunately, Ranga Rao's widow was protective of the fossils, worried that someone might steal her husband's property and legacy. But each year, Thewissen visited her, chatted with her, and gained her trust. When she passed away in 2007, she made Thewissen cotrustee of her estate, and suddenly the fossils, which had sat in dusty piles for 30 years, were available for study.

"I focused on taking the rocks back to the U.S., and having my fossil preparers remove the fossils from the rocks, which is very difficult," says Thewissen. From Ranga Rao's collection, Thewissen identified more than 400 bones that belonged to *Indohyus*. By collecting a thighbone here and a jawbone there, his team compiled a Frankenstein-like skeleton of a single *Indohyus* individual (**Figure 11.7**). After the discovery of the whalelike ear, the researchers looked even more carefully at the features of the fossils and found additional evidence that *Indohyus* was a relative of whales. This unassuming little animal, with a pointy snout and slender legs

tipped with hooves, lived close to and had an affinity for the water.

The first clues about *Indohyus*'s lifestyle came from its teeth. Oxygen in the molecules that make up teeth comes from the water and food that an animal ingests. Levels of oxygen isotopes in *Indohyus*'s teeth match those of water-going mammals today, suggesting that *Indohyus* lived near and potentially spent a significant amount of time in the water. It also had large, crushing molars with levels of carbon isotopes that suggest it grazed on plants, as do hippopotamuses and muskrats that graze near and in water (**Figure 11.8**).

Lisa Cooper, a graduate student in Thewissen's lab at the time, identified another adaptation to the water: *Indohyus*'s leg bones. From the outside, the limbs of *Indohyus* look like those of any other mammal walking around on land. But on the inside, it's another story. Cooper cut out a section of bone from a limb, ground it down until she could see light through it, and then looked at the bone under a microscope. She saw that a thick layer of bone was wrapped around the bone marrow.

"Hans already had lots of bones of the earliest whales, and they all had extraordinarily

Figure 11.7

**Fossilized skeleton of *Indohyus*, oldest cousin of the whales**
A reconstructed fossilized skeleton of *Indohyus* was compiled from multiple sources and locations. The illustration is an artist's depiction of the living animal about 47 million years ago. (Source: Photo courtesy of Thewissen Lab, NEOMED.)

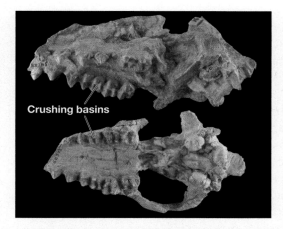
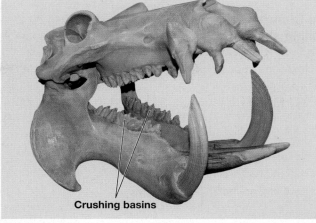
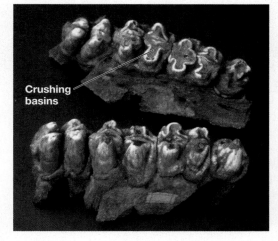
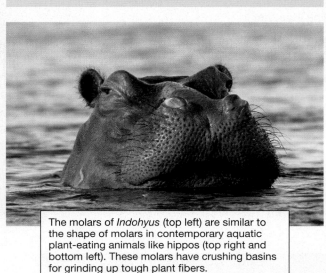

The molars of *Indohyus* (top left) are similar to the shape of molars in contemporary aquatic plant-eating animals like hippos (top right and bottom left). These molars have crushing basins for grinding up tough plant fibers.

## Figure 11.8

**Comparing the skulls and jaws of fossilized *Indohyus* and a modern hippopotamus**
These organisms' teeth indicate their ability to eat plant material.

thick bones," says Cooper, now an assistant professor at Northeast Ohio Medical University. Modern animals that live in shallow water, such as manatees and hippos, also have thick bones, which help prevent them from floating and enable them to dive quickly (**Figure 11.9**). "It isn't just isolated to whales," says Cooper. "Bones have thickened again and again as different groups of vertebrates entered the water. When you trace back through the fossil record, there is a pretty good correlation between thickness of bone and whether something was living in the water."

*Indohyus*'s thick bones are an example of an **adaptive trait**, a feature that gives an individual

improved function in a competitive environment. By being able to easily wade and dive in water, *Indohyus* had an advantage over other organisms in escaping predators and accessing plants to eat on the river floor. Adaptive traits take many forms, from an anatomical feature

### LISA COOPER

Lisa Cooper is an assistant professor at Northeast Ohio Medical University in the Department of Anatomy and Neurobiology. She earned her PhD in Thewissen's lab.

Figure 11.9

## Cross sections of the femurs of *Indohyus*, water-dwelling animals, and land-dwelling animals

Aquatic organisms have thick, dense bone around a narrow marrow space, while terrestrial organisms have thin bone and a large marrow space. *Indohyus* bone structure is a trait that is shared with other water-dwelling animals.

> **Q1:** Why do water-dwelling animals have thicker bones than land-dwelling animals?
>
> **Q2:** Why does this thick-bone adaptation suggest a water-dwelling lifestyle?
>
> **Q3:** How did this adaptation likely increase survival or reproduction in *Indohyus*?

*Pakicetus*, an extinct water-dwelling whale ancestor

Manatee, a water-dwelling mammal

*Indohyus*, an extinct water-dwelling whale cousin

Hippopotamus, a land- and water-dwelling animal

Polar bear, a land-dwelling mammal

Rat, a land-dwelling mammal

like *Indohyus*'s bones, to behaviors, to the functions of individual proteins. Echolocation in bats, for example, is an adaptation for catching insects in the dark. Stick insects physically and behaviorally mimic the plants they live on—an adaptation that helps them avoid detection by predators (**Figure 11.10**).

Figure 11.10

## Stick insects avoid detection by predators

Stick insects are well adapted to their environment; they move slowly and look just like the branches they live on.

# The Ultimate Family Tree

Thick bones are not restricted to just the water-loving ancestors of whales, as we noted, but can be seen in many other animals, such as hippos. This similarity across organisms is another type of evidence for evolution—shared characteristics among species. Many shared characteristics—such as thick bones for animals that take to shallow water, or sexual reproduction via egg and sperm, or eukaryotic cells—result from organisms sharing traits that evolved from a **common ancestor**, an organism from which many species have evolved. A group of organisms have **common descent** if they share a common ancestor.

When one species splits into two, the two resulting species share similar features, or **homologous traits**, because they have common descent—though these features may begin to look different from one another over time

(**Figure 11.11**). For example, whales are so different from humans that it can be difficult to identify similarities, but we evolved from a common mammalian ancestor and do share homologous traits: humans and whales both nurse their young and have a single lower jawbone because our common ancestor had those traits.

**Vestigial traits** are another type of trait that many organisms have because of a common ancestor. These features are a piece of the evolutionary past, inherited from a common ancestor but no longer used. Vestigial traits may appear as reduced or degenerated parts whose function is hard to discern (**Figure 11.12**). For example, many modern whales have vestiges of thighbones, also called femurs, embedded in the skin next to the pelvis. In land mammals, birds, and other tetrapod vertebrates, these bones are critical for walking, running, and jumping. Aquatic whales have no need of this bone, yet its traces remain.

Whales also have small muscles devoted to nonexistent external ears, likely from a time when they were able to move their ears as land animals such as dogs do for directional hearing. Vestigial traits are not adaptations. In fact, they can be detrimental. Most humans no longer need wisdom teeth to replace lost teeth during adolescence, yet most people still have them. They tend to erupt around the twentieth year of life, often causing severe pain and displacing other teeth, and they usually require removal.

# Clues in the Code

Within the cells of every organism is one of the strongest pieces of evidence for evolution: DNA. Living things universally use DNA as hereditary, or genetic, material (see Chapter 8 for review). The fact that all organisms on Earth—even those as different as bacteria, redwood trees, and humans—use the same genetic code is further evidence that the great diversity of living things evolved from a common ancestor.

Researchers have analyzed the DNA sequences of whales and other animals and shown that, of all animals, whales are most

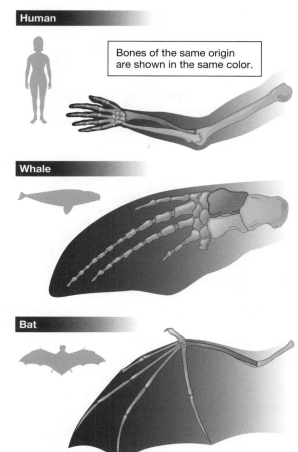

Figure 11.11

**Homologous traits are shared characteristics inherited from a common ancestor**

The human arm, the whale flipper, and the bat wing are homologous structures due to common descent. All three structures have a matching set of five digits and a matching set of arm bones that have been altered by evolution for different functions.

**Q1:** What is meant by the term "common ancestor"? Give an example.

**Q2:** Why are homologous structures among organisms evidence for evolution?

**Q3:** Aside from skeletal structural similarities, what other commonalities among organisms might be considered homologous?

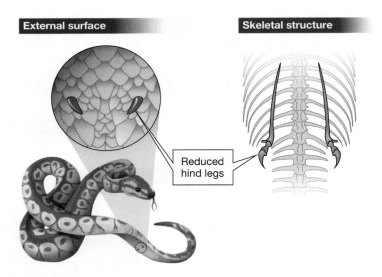

External surface

Skeletal structure

Reduced
hind legs

Figure 11.12

**Vestigial traits are reduced or degenerated remnants with no apparent function**
Snakes are limbless reptiles with no apparent use for the degenerated remnants of hind legs that
they still have. The python shown here has extremely reduced hind legs that are only barely visible
externally.

**Q1:** Why are vestigial structures among organisms evidence for evolution? Give an example of
another vestigial structure.

**Q2:** Are vestigial structures also homologous structures? Explain.

**Q3:** Why do vestigial structures still exist if they are no longer useful?

closely related to even-toed ungulates—a group
of hoofed mammals that includes modern deer,
giraffes, camels, pigs, and hippos. *Indohyus*
is an example of an extinct even-toed ungu-
late. Molecular studies therefore confirm the
prediction that whales and *Indohyus* share a
common ancestry.

According to DNA sequence similarity,
hippos are whales' closest living relatives.
Whale DNA is more similar to hippo DNA
than to the DNA of other marine mammals,
such as seals and sea lions. **DNA sequence
similarity** is a measure of how closely related
two DNA molecules are to each other. For
example, in the DNA sequences of the same
insulin gene in humans and mice, 83 percent
of the nucleotides are identical at corresponding positions (**Figure 11.13**). We share a more
recent common ancestor with mice than we do

with chickens, for whom the DNA sequence of
the same gene is only 72 percent identical to
ours. In fact, the insulin gene of our closest
living relative on Earth, the chimpanzee, has
98 percent similarity to the human insulin
gene. That comparability implies that humans
and chimpanzees share a very recent common
ancestor.

The fact that these separate lines of
evidence—anatomical features and DNA—
yield the same result over and over again for
diverse groups of organisms is strong evidence
for evolution.

Whale evolution is "one of the best case studies documenting how a vertebrate can go from a
terrestrial to an aquatic environment," says Lisa
Cooper. Another type of evidence supporting
whale evolution comes from the locations where
whale fossils have been found.

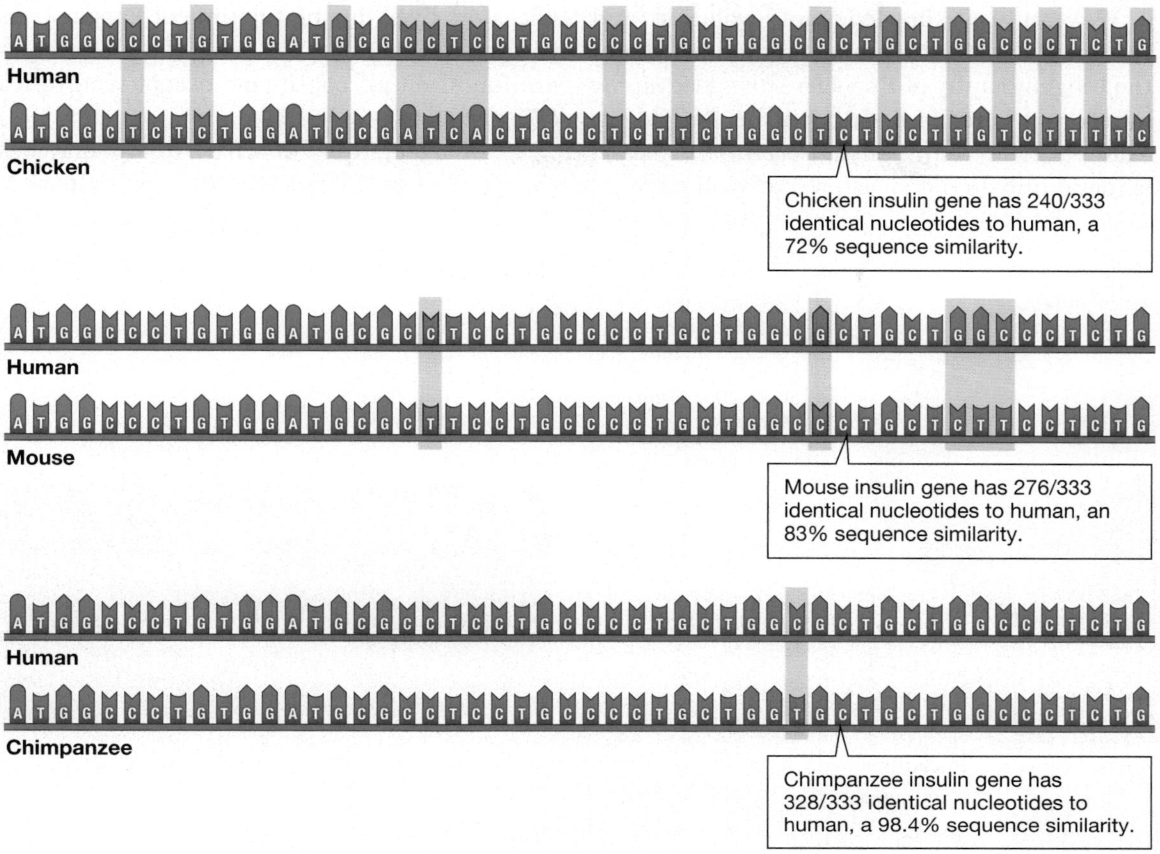

**Figure 11.13**

**DNA sequence similarities of the insulin gene**

The complete coding sequence of the human insulin gene contains 333 nucleotides. Only the first 50 nucleotides are shown here. Unshaded paired sequences are identical nucleotides at that position; those shaded in yellow are different.

**Q1:** If a sequence from another species showed a 96 percent sequence similarity to humans, would that species be more or less closely related to humans than chimpanzees are?

**Q2:** Should similarities in the DNA sequences of genes be considered evolutionary homology? Explain.

**Q3:** How is the increased similarity in the DNA sequences of genes between more closely related organisms—and the decreased similarity between less closely related organisms—evidence for evolution? Use the example in this figure to support your answer.

# Birthplace of Whales

Earth's continents are on massive tectonic plates, which slowly move over time in a process called continental drift or plate tectonics. About 250 million years ago, South America, Africa, and all of the other landmasses of Earth had drifted together to form one giant continent called Pangaea. About 200 million years ago, Pangaea slowly began to split up, separating to ultimately form the continents as we know them. That separation continues today; for example, each year South America and Africa drift farther apart by about an inch.

We can use knowledge of evolution and continental drift to make predictions about the **biogeography** of a species—the geographic locations where its fossils will be found. For example, today the lungfish *Neoceratodus fosteri* is found only in northeastern Australia, but its ancestors lived during the time of Pangaea. As predicted, therefore, fossils of those ancestors are found on all continents except Antarctica (**Figure 11.14**).

The biogeography of whale fossils matches the pattern predicted by evolution: all early

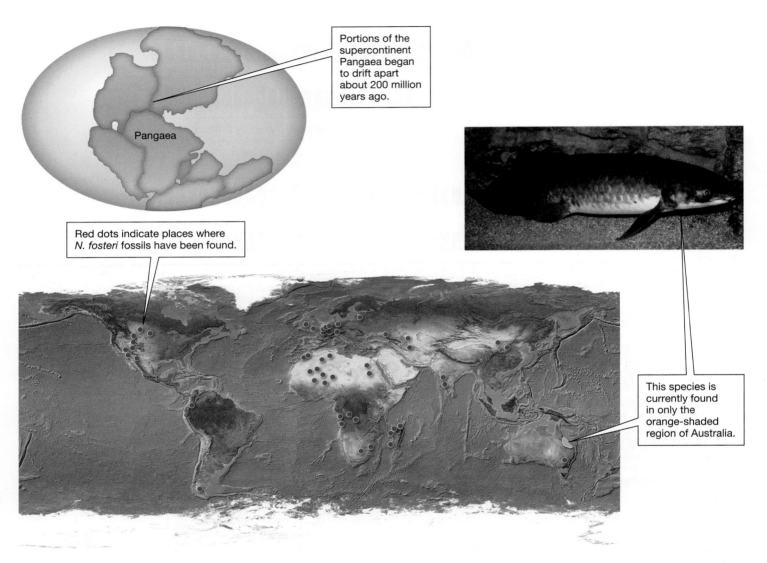

Portions of the supercontinent Pangaea began to drift apart about 200 million years ago.

Pangaea

Red dots indicate places where *N. fosteri* fossils have been found.

This species is currently found in only the orange-shaded region of Australia.

## Figure 11.14

### Biogeography can reveal a species' evolutionary past

Ancestors of the freshwater lungfish *Neoceratodus fosteri* lived during the time of Pangaea. *N. fosteri* fossils have been found on all continents except Antarctica.

Q1: Why should we expect to find *N. fosteri* fossils all over the world, given that it first evolved in Pangaea?

Q2: Can we use biogeographic evidence to support evolution without using fossil evidence? Give examples.

Q3: How might we use DNA sequence similarities together with biogeography as evidence for evolution?

species of whales, such as the crocodile-like *Pakicetus*, that lived in rivers and lakes but did not swim in the ocean are found near India and Pakistan. "It makes sense," says Thewissen. "You don't have crocodiles crossing the Atlantic." But fossils of fully aquatic protocetids, which emerged about 40 million years ago, are geographically much more widespread; they have been found as far away from Pakistan as Canada. "Protocetids are good swimmers, so we find their fossils all around the world," says Thewissen.

## Growing Together

Though Thewissen has built a career on finding and describing whale fossils, he has recently become enamored with another vein of evolutionary evidence: embryology. A major prediction of evolution is that organisms should carry within themselves evidence of their evolutionary past, and they do. Evidence of evolution can be observed in shared patterns of **embryonic development**.

Once again, these common patterns are caused by descent from a common ancestor. Rather than evolving new organs "from scratch," new species inherit structures that may have been modified in form and sometimes even in function.

Upon fusion of sperm and egg, an animal embryo begins to grow and develop. The manner in which an embryo develops, especially at the early stages, may mirror early developmental stages of ancestral forms. For example, anteaters and some baleen whales do not have teeth as adults, but as fetuses they do. And the embryos of fishes, amphibians, reptiles, birds, and mammals (including humans) all develop pharyngeal pouches or gill slits (**Figure 11.15**). In fish, the pouches develop into gills that adult fish use to absorb oxygen underwater. In human embryos, these same features become parts of the ear and throat.

"I was interested to get embryos to look at some of these processes that we see happen in evolution, to see if they happen in development," says Thewissen. The first trait he examined was

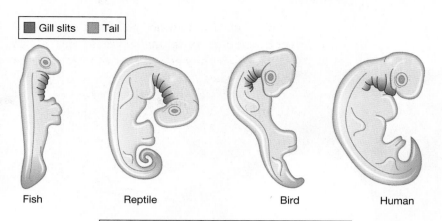

| Gill slits | Tail |

Fish    Reptile    Bird    Human

Embryos all share gill slits and tails because fishes, reptiles, birds, and humans all evolved from a common ancestor that had these features.

Figure 11.15

**Evolutionary history can be extrapolated from similarities in embryo development**

Complex structures in descendant species are generally elaborations of structures that existed in their common ancestor.

**Q1:** Why are the similarities among organisms during early development evidence for evolution? Give an example.

**Q2:** Are the similar structures among vertebrate species during embryogenesis homologous structures? Explain.

**Q3:** Why do embryonic structures still exist at points during embryogenesis if they are not used after birth?

hind limbs. "We know, from fossil evidence, that early whales lose their hind limbs," says Thewissen. So he wondered if hind limbs exist in whale or dolphin embryos (dolphins are also mammals and are closely related to whales). And if so, what makes them subsequently disappear before the animal is born.

Examining spotted dolphin embryos, Thewissen saw that when the embryos are the size of a pea, they do develop hind limb buds, but by the time they grow into the size of a bean, the limb buds are gone. In 2006, he and researchers at several other universities studied the genes that are active in whale and dolphin embryos and concluded that whales' hind limbs regressed over millions of years through small changes in

# Watching Evolution Happen

In the laboratory, scientists can manipulate populations of organisms to watch evolution in real time via artificial selection. In 2012, researchers at UC Irvine manipulated the growing environment of *Escherichia coli* bacteria, exposing them to far hotter temperatures than normal, to see if they would adapt. Most of the bacteria adapted via one of just two primary pathways.

Assessment available in smartwork**5**

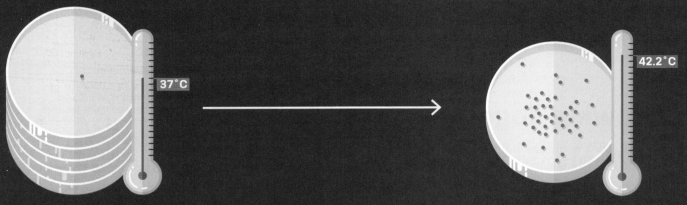

**1** Scientists grow **115** separate, genetically identical populations of *E. coli* at a comfortable temperature of **37°**C, or **98.6°**F.

**2** The populations reproduce through **2,000** generations while being subjected to an increased heat of **42.2°**C, or **107.96°**F.

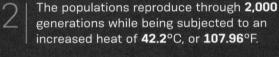

**3** One sample is taken from each of the **115** populations, and the genome is sequenced.

**4** **1,258** molecular changes, averaging **11** genetic mutations per clone, are detected. To help survive the heat, the *E. coli* tend to mutate along two different trajectories:

- Mutations in the RNA polymerase complex, an enzyme that transcribes RNA.

- Mutations in the *rho* gene, which encodes a protein that stops RNA transcription.

**5** The research continues. The next step is to figure out how the mutations in the RNA polymerase complex and in *rho* helped the *E. coli* survive the heat.

a number of genes relatively late in embryonic development (**Figure 11.16**). The loss of hind limbs in the embryo corresponds to the disappearance of hind limbs in the fossil record. "It was an awesome way to combine embryology with fossils," says Thewissen. In 2015, researchers at the Smithsonian Institution likewise found that the development of fetal ear bones in the womb paralleled changes observed in the whale fossil record.

With their discovery of *Indohyus*, Thewissen, Cooper, and their team bridged a 10-million-year gap in the fossil record, identifying an important transition species to whales. It is another rock in the mountain of evidence surrounding whale evolution, which was further fortified in April 2015 with the discovery of fossils from a new species of extinct pygmy sperm whale found in Panama.

Evolution is supported by mutually reinforcing, independent lines of evidence: direct observation, fossils, shared characteristics among living organisms, similarities and divergences in DNA, biogeographic evidence, and common patterns of embryo development. Just as the theory of gravity forms the foundation of physics, so evolution is the central tenet of biology.

There is no question that evolution happens. The intriguing, fascinating question is *how* does it happen? Through scientific research, we know that whales descended from a group

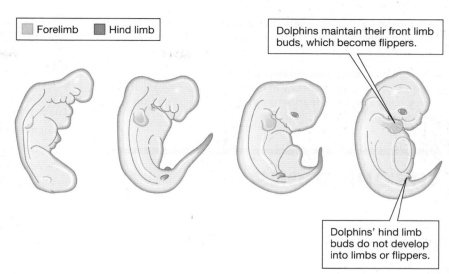

Figure 11.16

**Dolphin embryonic development**
The embryos of dolphins from weeks 4–9 of development show the formation and then subsequent loss of the hind limb buds. (Source: Based on a photo by Thewissen Lab, NEOMED, permission granted.)

of land-living mammals including the petite *Indohyus*, munching on freshwater plants and splashing through the shallows. But how did a population of small, furry animals become the mammoths of the sea? Next we investigate the mechanisms of evolution—how and why it works.

# REVIEWING THE SCIENCE

- Biological **evolution** is the change in characteristics of a population of organisms over inherited generations.

- **Artificial selection** results in biological evolution. Humans choose which organisms survive and reproduce—a process known as **selective breeding**.

- **Natural selection** is the process by which individuals with advantageous genetic characteristics for a particular environment survive and reproduce at a higher rate than competing individuals with other, less useful characteristics.

- **Adaptation** is an evolutionary process by which a population becomes better matched to its environment over time. An adaptation is also a trait—an **adaptive trait**—that has evolved as a result of this process.

- **Fossils** are the preserved remains (or their impressions) of formerly living organisms. The fossil record enables biologists to reconstruct the history of life on Earth, and it provides some of the strongest evidence that species have evolved over time.

- Many similarities among organisms are due to the fact that the organisms evolved via **common descent** from a **common ancestor**. When one species splits into two, the two resulting species share similar features, called **homologous traits**. If a homologous trait is no longer useful, it is called a **vestigial trait**.

- The fact that organisms as different as bacteria, redwood trees, and humans show **DNA sequence similarity** is evidence that the great diversity of living things evolved from a common ancestor.

- **Biogeography** uses knowledge about evolution and plate tectonics to make predictions about the geographic locations where fossils will be found.

- Similarities in **embryonic development** of different organisms show that the characteristics of modern organisms arose through evolutionary modifications of traits inherited from common ancestors.

# THE QUESTIONS

## The Basics

**1** Intermediate fossils

(a) share some similarities with their ancestral group.

(b) share some similarities with their descendant groups.

(c) both of the above

(d) none of the above

**2** If two different organisms are closely related evolutionarily, then they will

(a) be similar in size.

(b) share a recent common ancestor.

(c) have very different DNA sequences in their genes.

(d) be randomly located throughout the world.

**3** When two organisms are very distantly related in an evolutionary sense,

(a) they should have extremely similar embryonic development.

(b) they must share a very recent common ancestor.

(c) the sequences of DNA in their genes should be less similar (more different) than those of two more closely related organisms.

(d) they should share more homologous traits than two more closely related organisms share.

**4** Link each term with the correct definition.

| | |
|---|---|
| BIOGEOGRAPHY | 1. A reconstruction of the history of life on Earth. |
| FOSSIL RECORD | 2. The similarities in the nucleotide sequences among related organisms. |
| DNA SEQUENCE SIMILARITY | 3. The similarities among organisms that are due to the fact that the organisms evolved from a common ancestor. |
| EMBRYONIC SIMILARITY | 4. The geographic locations where related organisms and fossils are found. |
| HOMOLOGOUS TRAITS | 5. Specifically during development, complex structures in descendant species that are generally elaborations of structures that existed in the common ancestor. |

## Challenge Yourself

**5** All mammals have tailbones and muscles for moving a tail. Even humans have a reduced tailbone and remnant tail-twitching muscles, though these features have no apparent usefulness. These traits in humans would best be described as

(a) convergent structures.

(b) fossil evidence.

(c) evidence from biogeography.

(d) vestigial traits.

**6** Reduced tailbones and the associated remnant muscles in humans are an example of what type of evidence for common descent?

(a) artificial selection

(b) homologous traits

(c) biogeography

(d) fossil evidence

**7** When a population evolves to be better fitted to its environment, this is an example of _____, which is brought about by the process of _____.

**8** In one sentence each, identify the similarities and the differences between artificial selection and natural selection.

## Try Something New

**9** Cat DNA is much more similar to dog DNA than to tortoise DNA. Why?

(a) Cats and dogs are both carnivores and take in similar nutrients.

(b) Cats and dogs have lived together with humans for a long period of time, so they have grown more similar.

(c) Cats and dogs have more offspring during their lifetime than tortoises have, so their DNA changes less rapidly.

(d) Cats and dogs have a common ancestor that is more recent than the common ancestor of cats and tortoises.

**10** DNA sequences were analyzed from humans and three other mammals: species X, Y, and Z. Which of these mammals is most closely related to humans? (*Note*: Regions identical to human DNA are shown in bold type.)

**Human:**
**AATGCTTTGGGGGATCGCGAGCGCAGCGC**

**Species X:**
GGGTT**TTT**ATCGCTATATATATATA

**Species Y:**
**AATGCTTTGGGGGATCGCGAGCGCA**TATA

**Species Z:**
**AATGC**GGGTTTTT**ATC**TATATATATATA

**11** Which of the following is *not* an example of artificial selection?

(a) Your younger sibling got a hamster as a birthday present, and it turned out to be pregnant. Several of the offspring had long, silky hair, and your sibling put them together in an enclosure to try to produce more baby hamsters with long, silky hair. Your sibling continued to breed long- and silky-haired hamsters to each other and now, several years later, your sibling's bedroom is full of cages containing long- and silky-haired hamsters.

(b) Your mother has been saving the best seeds from her lima bean plants every summer and replanting them the next year. She likes seeds that are plump and bright green and saves only these each year. Within 10 years, almost all of her lima beans have become plump and bright green.

(c) Farmer Brown has a duck that can type. He sets up an online dating profile for his duck to find a female duck that can also type. Seven years after the arranged wedding of the two typing ducks, Farmer Brown has an entire flock of ducks, most of which can type.

(d) Female fish in a natural pond have variable skill at depositing their eggs near the murky shore, where the eggs are better hidden from predators. Eggs that are not deposited near the murky shore are quickly devoured by fish. Female hatchlings from the eggs deposited near the murky shore grow up to be good at depositing eggs near the shore and therefore have a survival advantage.

**12** The fossil record shows that the first mammals evolved 220 million years ago. The supercontinent Pangaea began to break apart 200 million years ago. On which continents would you predict that fossils of the first mammals will be found?

## Leveling Up

**13** **What do *you* think?** The prerequisites for medical school application always include courses in cell biology, genetics, and biochemistry but rarely include a formal course in evolution. Do you think medical schools should require a formal course in evolution as a prerequisite for admission? Why or why not? Research the issue and support your case using information you find from credible sources.

**14** ***Write Now* biology: evidence for evolution** This assignment is designed to expand your knowledge of the evidence for evolution. View the following videos, found at http://www.pbs.org/wgbh/evolution/educators/teachstuds/svideos.html, and answer the questions accompanying each one here.

### Video 1: "Isn't Evolution Just a Theory?"
Why is evolution not just a theory? Use specifics from the video to defend your answer.

### Video 2: "Who Was Charles Darwin?"
Why do you think Charles Darwin's ideas and book *On the Origin of Species* were so groundbreaking and "revolutionary"? Use specifics from the video to defend your answer.

### Video 3: "How Do We Know Evolution Happens?"
Describe how the video portrays whale evolution. Include specific examples of transitional fossils described in the video, and explain why the scientists at the time the video was made considered the fossils to be whale ancestors.

### Video 5: "Did Humans Evolve?"
DNA sequences of different species can be used to provide evidence of common descent. Using examples from the video, explain why DNA sequence similarity is the best evidence of evolution on Earth.

### Video 6: "Why Does Evolution Matter Now?"
Describe why the theory of evolution matters to the field of medicine and to individual doctors. Use the example of tuberculosis from the video to support your answer. How does this video affect your answer to question 13?

**For more, visit digital.wwnorton.com/bionow2 for access to:**

# Battling Resistance

*An antibiotic-resistant superbug is evolving to overcome our drugs of last defense. How do we beat a champion pathogen?*

**After reading this chapter you should be able to:**

- Clarify why evolution occurs only in populations, not in individuals.
- Distinguish among directional, stabilizing, and disruptive selection.
- Explain how natural selection brings about increased reproductive success of a population in its environment.
- Compare and contrast convergent evolution and evolution by common descent, and give an example of each.
- Describe how DNA mutations can create new alleles at random.
- Illustrate how gene flow works and how it may inhibit evolution.
- Relate the process of genetic drift to genetic bottlenecks and the founder effect, using examples.

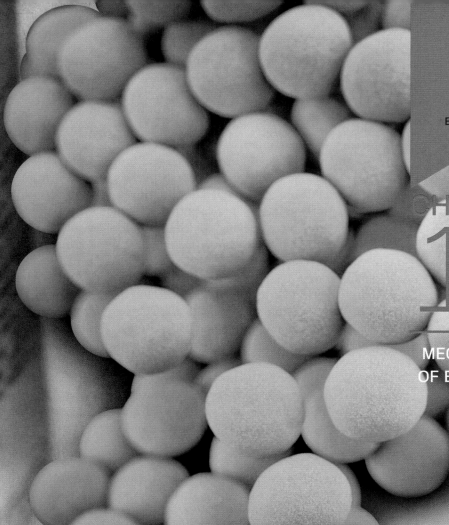

CHAPTER

12

MECHANISMS
OF EVOLUTION

Dawn Sievert vividly remembers June 14, 2002—the day disaster struck. Sievert is an epidemiologist, or "disease detective"—someone who studies the patterns and causes of human disease. At the time, she was working at the Michigan Department of Community Health, monitoring reported cases of antibiotic-resistant bacteria, a major health care concern. Sievert spent a significant amount of her time investigating outbreaks of an increasingly common and worrisome microbe called MRSA, short for "methicillin-resistant *Staphylococcus aureus*."

*S. aureus*, commonly known as "staph," is a small, round bacterium that usually lives benignly in our nostrils and on our skin. But on rare occasions, staph slips beneath the surface of a burn or cut and causes an infection, which can be especially dangerous for individuals with suppressed immune systems, such as the elderly and patients on chemotherapy.

Staph is one of the most common causes of hospital infections today, and it is treated with antibiotics, drugs that kill bacteria but not human cells. Penicillin (**Figure 12.1**) was the first antibiotic used against staph, but the wily microbe evolved resistance to penicillin even before the drug became commercially available to the public in the 1940s.

When penicillin stopped working against staph, doctors switched to an antibiotic called methicillin. Methicillin worked for about 20 years, until populations of staph evolved widespread resistance to that antibiotic as well. To the chagrin of doctors and patients everywhere, bacteria are able to adapt rapidly to new antibiotics, thanks to their short generation time and ability to share resistance genes among themselves. (See Chapter 15 for more on bacteria.) These tiny microbes are the Navy SEALs of evolution—the best of the best at evolving.

As we saw in Chapter 11, biological *evolution* is a change in the frequencies of inherited traits in a population over generations. Staph adapted to the presence of methicillin: Only bacteria with genetic traits protecting the microbe from the antibiotic survived. These traits were passed among populations and down from one generation to the next until methicillin resistance was frequent across staph populations (**Figure 12.2**).

Today, MRSA is rampant in hospitals, so doctors have been forced to turn to one of medicine's last lines of defense against the superbug. Vancomycin, a strong, blunt antibiotic that was first isolated from the mud of the Borneo jungle, is considered one of the "drugs of last resort" for fighting these serious infections—which brings us back to June 14, 2002, and Dawn Sievert.

On that day, a lab technician at a dialysis center in Detroit, Michigan, took two swabs of an infected foot ulcer belonging to a 40-year-old diabetic woman (**Figure 12.3**). The patient had previously suffered from numerous foot infections, including MRSA, which had been treated with vancomycin for 6½ weeks. The swabs of this latest infection were sent to a local laboratory, where technicians grew the bacteria in a dish to test its susceptibility to various antibiotics.

When the results of the first test came in, the laboratory staff immediately picked up the phone and called Sievert's office. The bacteria, they told Sievert and the health department, appeared to be resistant to vancomycin. "First, we needed laboratory confirmation and had to

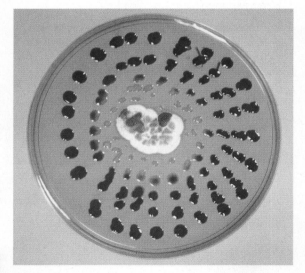

Figure 12.1

**Penicillin, produced by mold, kills bacteria**
The mold *Penicillium*, the fuzzy white growth with blue spores growing on this petri dish, secretes the antibiotic penicillin into the agar medium, killing the red bacteria surrounding it.

## DAWN SIEVERT

Dawn Sievert is an infectious disease epidemiologist who worked for the Centers for Disease Control and Prevention until 2016. While at the Michigan Department of Community Health previously, Sievert investigated the first-ever VRSA infection.

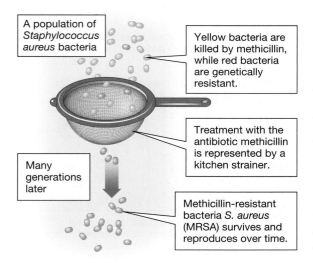

**Figure 12.2**

**Natural selection results in resistance to antibiotics**

The use of antibiotics allows any bacteria (in this case, *S. aureus*) that are resistant to the antibiotic to survive and reproduce. Over time, the frequency of resistant bacteria increases in the surviving populations.

A population of *Staphylococcus aureus* bacteria

Yellow bacteria are killed by methicillin, while red bacteria are genetically resistant.

Many generations later

Treatment with the antibiotic methicillin is represented by a kitchen strainer.

Methicillin-resistant bacteria *S. aureus* (MRSA) survives and reproduces over time.

**Q1:** What is natural selection selecting for here?

**Q2:** Why do bacteria that are not genetically resistant to antibiotics die out when exposed to antibiotics?

**Q3:** Why is the antibiotic represented by a kitchen strainer in this figure?

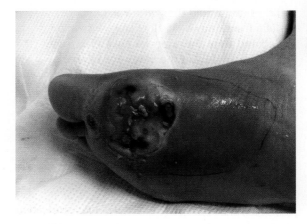

**Figure 12.3**

**MRSA infection in the foot of a patient with diabetes**

control any potential for panic," says Sievert. Her team asked the local lab to run its own test again and to send the health department a sample to independently test. Both teams waited.

The tests came back from each lab. Both were positive; the woman's foot was infected with the first reported vancomycin-resistant strain of *S. aureus*. VRSA had arrived.

# Birth of a Superbug

"At that point, it was the first-ever VRSA in the world," says Sievert. Unfortunately, it was not the last.

Through evolution, staph first survived penicillin, then methicillin, then vancomycin (**Figure 12.4**). The evolution of *Staphylococcus*

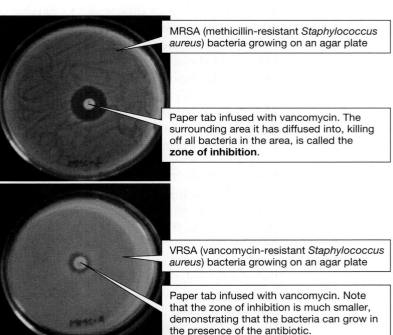

MRSA (methicillin-resistant *Staphylococcus aureus*) bacteria growing on an agar plate

Paper tab infused with vancomycin. The surrounding area it has diffused into, killing off all bacteria in the area, is called the **zone of inhibition**.

VRSA (vancomycin-resistant *Staphylococcus aureus*) bacteria growing on an agar plate

Paper tab infused with vancomycin. Note that the zone of inhibition is much smaller, demonstrating that the bacteria can grow in the presence of the antibiotic.

**Figure 12.4**

**MRSA versus VRSA in a vancomycin-resistance test**

In the presence of vancomycin, MRSA (top) does not grow, while VRSA (bottom) grows relatively well.

**Q1:** What is the difference between MRSA and VRSA?

**Q2:** Why is there a clear zone (the "zone of inhibition") around the paper disk in the top dish but not in the bottom dish?

**Q3:** Why is the lack of a clear zone around the paper disk in the bottom dish so alarming?

*aureus* is a profound example of a species changing over time.

In the mid-1800s, two English biologists, Charles Darwin and Alfred Russel Wallace, studied the diversity of life and concluded that species were not, as was generally thought at that time, the unchanging result of separate acts of creation. Instead, both men came to the bold new conclusion that species "descend with modification" from ancestor species; that is, new species arise from previous species.

Descent with modification, we know today, occurs not only in populations of large organisms like whales and finches, but also in the tiniest single-celled bacteria and viruses. We generally think of evolution happening over millions of years, but some evolutionary changes, such as adaptations for antibiotic resistance, take place over very short time spans as particular alleles spread rapidly through a population. Recall from Chapter 7 that *alleles* are different versions of the same gene (sequences of DNA) produced by random mutation, and therefore *allele frequencies* are the proportions of specific alleles in a population (**Figure 12.5**). Evolution corresponds to changes in the relative proportions or frequencies of alleles in a population over time.

When allele frequencies in a population change, becoming more or less common, the attributes or phenotypes of the population change as well; that is, the population evolves. As more and more staph containing the allele for methicillin resistance survived and reproduced, the whole population of staph evolved, becoming new, more powerful bacteria. But how exactly does this happen? Where do new alleles come from, and how do the frequencies of alleles in a population change?

Sievert and a host of other researchers and doctors experienced the emergence of a new allele firsthand. When the results of the foot ulcer test came in positive for VRSA, Sievert's team immediately called the Centers for Disease Control and Prevention (CDC), a government agency that investigates disease outbreaks and makes public health recommendations. The Michigan Health Department and the CDC converged on the dialysis center where VRSA had been found. They pored over the medical history of the patient, examined her wound, took swabs from the nostrils and wounds of anyone

who had come in contact with her, and then waited anxiously to see whether the dangerous microbe had spread.

Thankfully, they found that the vancomycin-resistant bug had not yet spread. "We're lucky.

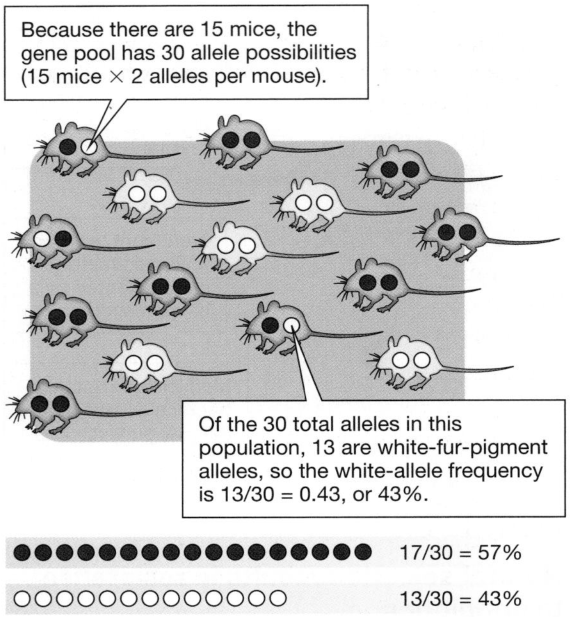

Because there are 15 mice, the gene pool has 30 allele possibilities (15 mice × 2 alleles per mouse).

Of the 30 total alleles in this population, 13 are white-fur-pigment alleles, so the white-allele frequency is 13/30 = 0.43, or 43%.

17/30 = 57%

13/30 = 43%

**Figure 12.5**

**Allele frequencies are calculated as percentages in a population**

These mice have two white-fur-pigment alleles and appear white, have two black-fur-pigment alleles and appear gray, or have one black and one white allele and appear gray. To calculate the white-fur-pigment allele frequency in the population, the number of white alleles is counted and divided by the total number of alleles.

**Q1:** What would the white-fur-pigment allele frequency be if three of the homozygous black allele mice (having two black alleles) were heterozygous (having one white and one black allele) instead?

**Q2:** What would the white-fur-pigment allele frequency be if all of the white mice died and were therefore removed from the population? Would the black-fur-pigment allele frequency be affected? If so, how?

**Q3:** What would the white-fur-pigment allele frequency be if all of the gray mice died and were therefore removed from the population?

If that highly resistant organism had the ability to spread rapidly, we'd have some very sick people at risk in hospitals and other health care settings," says Sievert.

Yet it was not the end of VRSA. Since 2002, there have been 13 additional cases of vancomycin-resistant staph infections in the United States: in the urine of a woman in New York with multiple sclerosis, in the toe wound of a diabetic man in Michigan, in the triceps wound of a woman in Michigan, and more. So far, each infection has been isolated; the microbe has never been transmitted from person to person, as MRSA has. VRSA, while dangerous, doesn't appear to spread through the human population.

But that's not to say it won't evolve to do so. When MRSA first emerged, it seemed restricted to hospital settings. Today, however, clinicians have been horrified to see cases of MRSA pop up from simple scrapes on a playground, suggesting that the microbe is out in the community, where it can do widespread damage. In theory, the same is possible for VRSA, raising the bone-chilling specter of the superbug evolving into an "apocalyptic bug," as one reporter called it.

How did staph acquire vancomycin resistance 14 separate times? How likely is vancomycin resistance to become more widespread? We can find answers to these questions by understanding four mechanisms of evolution:

1. Natural selection

2. Mutation

3. Gene flow

4. Genetic drift

Bacteria are ideal organisms for examining these evolutionary mechanisms, for the same reason that they are so dangerous: because they evolve incredibly fast.

# Rising Resistance

Harvard Medical School microbiologist Michael Gilmore has long tracked the ways that bacteria evolve antibiotic resistance. After staph evolved widespread resistance to methicillin in the 1980s and vancomycin began to be used in hospitals, "we waited and waited and waited" for vancomycin resistance

to emerge, says Gilmore. He knew that once vancomycin was widely used to kill staph, the microbe population would evolve resistance to the poison. The process by which a population gains one or more alleles that enable it to survive better than other populations is called *natural selection*, as we saw in Chapter 11. Darwin and Wallace were the first to propose natural selection, and today we know that it is the central driver of evolution.

During natural selection, individuals with particular inherited characteristics survive and reproduce at a higher rate than other individuals in a population. Natural selection acts by favoring some phenotypes over others (**Figure 12.6**). For example, in an environment where bacteria are exposed to vancomycin, the bacteria that can resist the antibiotic will continue to live and reproduce, while those that cannot will perish. Although natural selection acts directly on the phenotype, not on the genotype, of a population, the alleles that code for a phenotype favored by natural selection tend to become increasingly common in future generations. Bacteria that survive an antibiotic attack, for example, pass on alleles that confer that resistance to their offspring.

Natural selection acts on adaptive traits, and therefore a population can become better suited to survive and reproduce in its environment. And unfortunately for us, natural selection is the mechanism by which staph is adapting to our use of vancomycin.

Because staph's vancomycin-resistance adaptation is very recent, scientists are studying the patterns of how natural selection gives rise to VRSA. In nature we observe three common patterns of natural selection: *directional selection*, *stabilizing selection*, and *disruptive selection*. All types of natural selection operate by the same principle: individuals with certain forms of an inherited trait have better survival rates

**MICHAEL GILMORE**

Michael Gilmore is a microbiologist at Harvard Medical School. He and his laboratory uncovered the genetic basis for the recent emergence of VRSA.

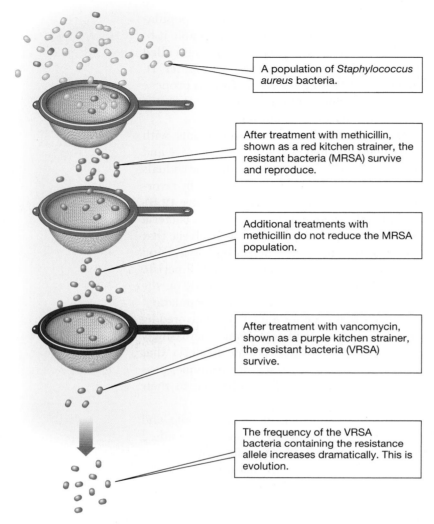

A population of *Staphylococcus aureus* bacteria.

After treatment with methicillin, shown as a red kitchen strainer, the resistant bacteria (MRSA) survive and reproduce.

Additional treatments with methicillin do not reduce the MRSA population.

After treatment with vancomycin, shown as a purple kitchen strainer, the resistant bacteria (VRSA) survive.

The frequency of the VRSA bacteria containing the resistance allele increases dramatically. This is evolution.

## Figure 12.6

### Evolution happens

Imagine there is a population of *Staphylococcus aureus* bacteria living on your skin. Most of them are susceptible to the antibiotic methicillin (red strainer). A few, however, are randomly resistant to methicillin (red), like the ones in Figure 12.2. A few of these bacteria, in turn, could also be resistant to vancomycin (purple).

**Q1:** Why does the population of *S. aureus* bacteria *not* pose a life-or-death health threat outright?

**Q2:** Why do vancomycin-resistant bacteria have a higher frequency in the population after treatment with vancomycin?

**Q3:** If this figure used the mouse example of allele frequency from Figure 12.5, and the white mice increased in numbers like the vancomycin-resistant bacteria here did, what would happen to the frequencies of the white-fur-pigment and black-fur-pigment alleles?

and produce more offspring than do individuals with other forms of that trait.

**Directional selection** is the most common pattern of natural selection, in which individuals at *one extreme* of an inherited phenotypic trait have an advantage over other individuals in the population. The peppered moth provides a vivid example: Before 1959, the number of dark-colored moths in both England and the United States increased after industrial pollution blackened the bark of trees, causing dark-colored moths to be harder for bird predators to find than light-colored moths. A reduction in air pollution following clean-air legislation, enacted in 1956 in England and in 1963 in the United States, caused the bark of trees to become lighter, and suddenly the reverse occurred: light-colored moths became harder for predators to find than dark-colored moths. As a result, the proportion of dark-colored moths plummeted because they were easily seen and eaten by predators (**Figure 12.7**). Similarly, when methicillin became widely used to fight staph, MRSA evolved via directional selection: the bacteria in hospital settings that were resistant to the antibiotic survived, while those that were not perished.

In cases of **stabilizing selection**, individuals with *intermediate values* of an inherited phenotypic trait have an advantage over other individuals in the population. Birth weight in humans provides a classic example of this pattern of natural selection (**Figure 12.8**). Historically, light or heavy babies did not survive as well as babies of average weight, and as a result there was stabilizing selection for intermediate birth weights. Today, however, this stabilizing trend is not as strong, because advances in the care of low-birthweight babies and an increase in the use of cesarean deliveries for large babies have allowed babies of all weights to thrive.

Finally, **disruptive selection** occurs when individuals with *either extreme of an inherited trait* have an advantage over individuals with an intermediate phenotype. This type of selection is the least commonly observed in nature, but one example of a trait affected by disruptive selection is the beak size within a population of birds called African seed crackers (**Figure 12.9**). During one dry season, birds with large beaks survived on hard seeds and

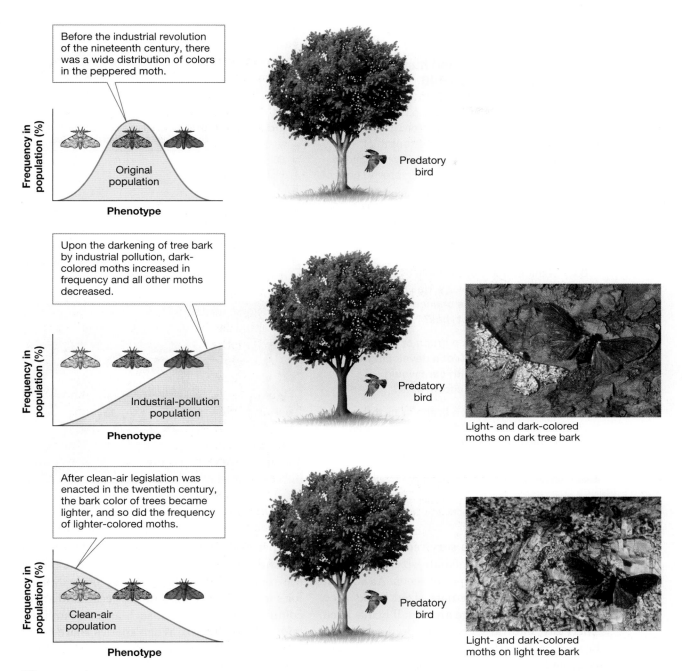

Before the industrial revolution of the nineteenth century, there was a wide distribution of colors in the peppered moth.

**Frequency in population (%)**

Original population

**Phenotype**

Predatory bird

Upon the darkening of tree bark by industrial pollution, dark-colored moths increased in frequency and all other moths decreased.

**Frequency in population (%)**

Industrial-pollution population

**Phenotype**

Predatory bird

Light- and dark-colored moths on dark tree bark

After clean-air legislation was enacted in the twentieth century, the bark color of trees became lighter, and so did the frequency of lighter-colored moths.

**Frequency in population (%)**

Clean-air population

**Phenotype**

Predatory bird

Light- and dark-colored moths on light tree bark

Figure 12.7

**The peppered moth has undergone directional selection two different times in the last 200 years**
The industrial revolution of the nineteenth century caused extreme air pollution in both England and the United States, due mostly to soot from the mass burning of coal.

**Q1:** If one extreme phenotype makes up most of a population after directional selection, what happened to the individuals with the other phenotypes?

**Q2:** What do you think would happen to the phenotypes of the peppered moth if the tree bark was significantly darkened again by disease or pollution?

**Q3:** What do you think would happen to the phenotypes of the peppered moth if the tree bark became a medium color, neither light nor dark? (You will need to read the next paragraph to answer this question.)

Figure 12.8

## Stabilizing selection and human birth weight from 1935 to 1946

This graph is based on data for 13,700 babies born between 1935 and 1946 in a hospital in London. In countries that can afford intensive medical care for newborns, the strength of stabilizing selection has been greatly reduced in recent years. However, even with improved care of low-birthweight babies and cesarean deliveries of large babies, babies at the extremes of newborn weight still survive at a lower rate than those closer to the median weight.

**Q1:** Think of another example of stabilizing selection in human biology. Has modern technology or medicine changed its impact on the resulting phenotypes?

**Q2:** How do you think a graph of birth weight versus survival for a developing country with little health care would compare to the graph shown here?

**Q3:** How do you think a graph of birth weight versus survival for an affluent area of the United States today would compare to the graph shown here?

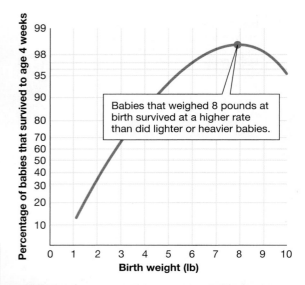

Babies that weighed 8 pounds at birth survived at a higher rate than did lighter or heavier babies.

birds with small beaks survived on soft seeds, but birds with intermediate-sized beaks fed inefficiently on both types of seeds. As a result, the birds with intermediate-sized beaks died, and the birds with large and small beaks lived. Therefore, natural selection favored both large-beaked and small-beaked birds over birds with intermediate beak sizes.

Any of these patterns of natural selection can cause distantly related organisms to evolve similar structures because they survive and reproduce under similar environmental pressures. This type of evolution, called **convergent evolution**, results in organisms that appear very much alike despite vastly dissimilar genetics. Cacti found in North American deserts and distantly related plants (euphorbias) found in African and Asian deserts offer an excellent example of convergent evolution. These two types of desert plants have very different genetics, but they look similar and function in a similar manner. In

another example, sharks and dolphins are only distantly related (sharks are fish, and dolphins are marine mammals), yet they both evolved for success as predators in the ocean and share common characteristics, such as a streamlined body (**Figure 12.10**). When species share characteristics due to convergent evolution and not because of modification by descent from a recent common ancestor, those characteristics are called **analogous traits** (instead of homologous traits).

# Enter *Enterococcus*

Intent on determining how VRSA was evolving after the first Michigan infection, Gilmore began to track the appearance of the bug. He and others wanted to know where the allele for vancomycin resistance came from, and how staph had acquired it 14 different times.

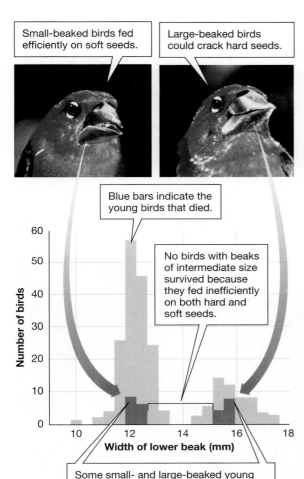

Small-beaked birds fed efficiently on soft seeds.

Large-beaked birds could crack hard seeds.

Blue bars indicate the young birds that died.

No birds with beaks of intermediate size survived because they fed inefficiently on both hard and soft seeds.

Some small- and large-beaked young birds (red bars) survived the dry season.

## Figure 12.9

**Disruptive selection for beak size**

Among a group of young African seed crackers hatched in one year, only those with small or large beaks survived the dry season, when seeds were scarce. Although many of the small- and large-beaked birds did not survive, none of the intermediate-beaked birds survived.

**Q1:** Almost all birds starved during the dry season depicted here. What type of selection would have been present if only the intermediate-beaked birds had survived (instead of the small- and large-beaked birds)?

**Q2:** Describe a scenario in which African seed crackers would experience directional selection for either smaller- or larger-beaked birds. What kind of environmental conditions might bring about such a situation?

**Q3:** Of the three patterns of natural selection presented in this discussion, which one always results in two different phenotypes in the following generations?

As Gilmore searched for patterns among the cases of VRSA infection, he noticed some peculiar things. First, most of the infections were turning up in people with diabetes, typically in bad foot wounds. Second, in most cases, when scientists looked closely at the samples, they saw not only staph but also a small spherical bacterium called *Enterococcus*, which had evolved resistance to vancomycin years earlier (**Figure 12.11**). Upon close observation, Gilmore and others noted that the enterococci, growing cozily side by side with staph, contained a vancomycin-resistance gene identical to the one in the staph. The presence of this gene suggested that staph had acquired vancomycin resistance directly from *Enterococcus*, rather than via random mutations in its own genome.

New alleles in a species emerge via mutation. A **mutation** is a change in the sequence of any segment of DNA in an organism, and it is the only means by which new alleles are generated. DNA mutations create new alleles at random, thereby providing the raw material for evolution. In this sense, all evolutionary change depends ultimately on mutation. Mutations can stimulate the rapid evolution of populations by providing new genetic variation—differences in genotypes between individuals within a population. Then, natural selection and other mechanisms of evolution act on the resulting phenotypes.

In sexually reproducing species, genetic mutations that occur in an organism's germ line cells—the cell lineage that produces gametes such as eggs and sperm—can contribute to evolution. Mutations in other cells of the body, such as skin or blood cells, can affect the individual by causing cancer or other problems, but those mutations are not passed to that individual's offspring. If mutations are not passed to offspring, they cannot contribute to evolution.

These three plants evolved from very distantly related plants via natural selection to adapt to life in the desert.

Natural selection has caused two distantly related animals—sharks (a type of fish, left) and dolphins (a type of mammal, right)—to look very similar.

**Figure 12.10**

**Natural selection can result in convergent evolution**

Q1: How is convergent evolution different from evolution by common descent?

Q2: What is the main difference between a homologous trait (see Figure 11.11) and an analogous trait?

Q3: Why are convergent traits considered evidence for evolution (see Chapter 11)?

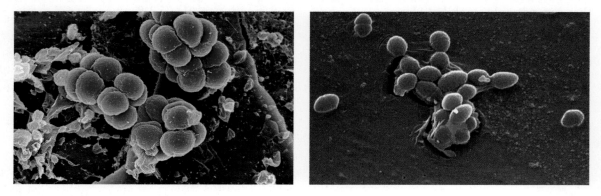

**Figure 12.11**

***Staphylococcus aureus*** (left) and ***Enterococcus*** (right)

The same does not hold true for bacteria, which are single cells and reproduce via asexual reproduction. All genetic mutations in a bacterial cell are passed on to the offspring. (See Chapter 6 for a review of cellular replication.) If a mutation passed to an offspring increases the individual's ability to reproduce, that mutation is favored by natural selection. These favored mutated genes are passed from parent to offspring, spreading through future generations in a way that alters the population as a whole.

But bacteria don't always wait for the right random mutations to pop up in their genomes. Sometimes they simply borrow new alleles from other organisms. That appeared to be the case with VRSA, researchers found. "In most cases, VRSA has developed in a perfect storm," says Sievert. It emerges within "a very bad wound that's not healing and is a soup mix of organisms coming together and sharing genes."

**Horizontal gene transfer** is the process by which bacteria pass genes to one another (**Figure 12.12**). Bacteria store these genes on small, circular pieces of DNA called *plasmids*. Some bacteria send these plasmids to each other through small tunnels, like dropping a package down a trash chute.

Horizontal gene transfer is one example of another mechanism by which evolution occurs: gene flow. **Gene flow** is the exchange of alleles between populations. Gene flow can occur between two different species—in this case, between a population of staph and a population of *Enterococcus*—or between two populations of the same species, such as strains of staph passing methicillin resistance among themselves in the community. An individual that migrates between two otherwise isolated populations of a species may facilitate gene flow as well (**Figure 12.13**). Gene flow can also occur when only gametes move from one population to another, as happens when wind or pollinators transport pollen from one population of plants to another.

The introduction of new alleles via gene flow can have dramatic effects. Two-way gene flow consists of an exchange of alleles between one population and another, so it tends to make the genetic composition of different populations more similar. If one strain of staph shares the methicillin-resistance allele with another strain of staph, for example, the two populations

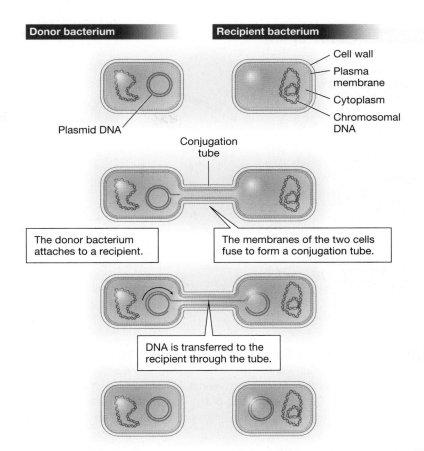

The donor bacterium attaches to a recipient.

The membranes of the two cells fuse to form a conjugation tube.

DNA is transferred to the recipient through the tube.

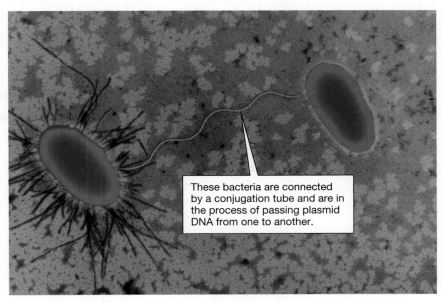

These bacteria are connected by a conjugation tube and are in the process of passing plasmid DNA from one to another.

## Figure 12.12

### Horizontal gene transfer accelerates the rate of evolution in prokaryotes

The diagram depicts the horizontal gene transfer of plasmid DNA through **conjugation**, the physical process of transferring genetic material through direct contact. In the case of VRSA, *Staphylococcus aureus* acquired the vancomycin-resistance gene through horizontal gene transfer from *Enterococcus*. ▶▮

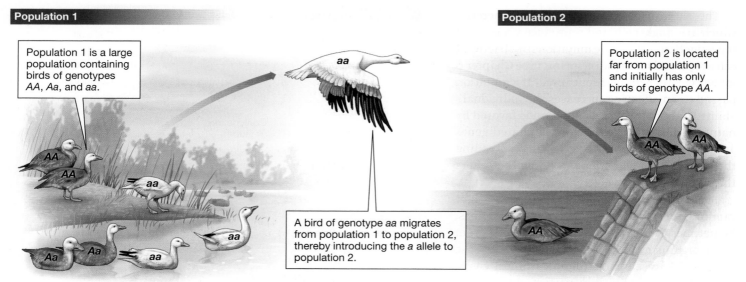

**Population 1**

Population 1 is a large population containing birds of genotypes *AA*, *Aa*, and *aa*.

**Population 2**

Population 2 is located far from population 1 and initially has only birds of genotype *AA*.

A bird of genotype *aa* migrates from population 1 to population 2, thereby introducing the *a* allele to population 2.

Figure 12.13

### Migrants can move alleles from one population to another

As a result of gene flow between these two genetically disparate populations of geese, the populations will eventually become more similar to each other. Mating between a migrant *aa* genotype goose and a population 2 goose of *AA* genotype will result in *Aa* genotype offspring. Continued mating of the *Aa* genotypes will result in all three possible genotypes (*AA*, *Aa*, and *aa*). The resulting population 2 gene pool will look much more similar to the gene pool of population 1.

**Q1:** If a goose with genotype *AA* had migrated instead of the goose with genotype *aa*, would the scenario described here still be considered gene flow? Why or why not?

**Q2:** If a goose with genotype *Aa* had migrated instead of the goose with genotype *aa*, would the scenario still be considered gene flow? Why or why not?

**Q3:** If the goose with genotype *aa* had migrated to population 2 as shown but had failed to mate with any of the *AA* individuals, would the scenario still be considered gene flow? Why or why not?

become more alike. A mutual exchange of alleles through gene flow can *counteract* the effects of the other mechanisms, such as mutation, that tend to make populations more different from one another.

Gene flow appeared responsible for the emergence of VRSA; staph picked up an allele for vancomycin resistance from *Enterococcus*. But Gilmore made another observation that suggested why this horizontal gene transfer was happening now, and not back in the 1980s when vancomycin had first begun to be widely used.

Gilmore noted that all the VRSA samples taken from patients were clonal cluster 5 (CC5) strains of *S. aureus*. "There were implications that there was something special about clonal

cluster 5 strains that was leading to this vancomycin resistance bubbling up in these strains," says Gilmore. Compared to other strains of staph, CC5 strains appeared to have evolved the ability to readily take up and use vancomycin-resistance alleles from *Enterococcus*.

## Primed for Pickup

In 2012, Gilmore and his team at Harvard analyzed the DNA of 11 of the 12 known cases of VRSA at the time (samples from one of the 12 weren't available, and the thirteenth and fourteenth cases had yet to occur). In the genome of every sample of VRSA, they found three traits that demonstrate how the CC5 strains of staph evolved to effectively pick up the allele for vancomycin resistance.

First, all the vancomycin-resistant CC5 staph bacteria have the same mutation in a gene called *DprA*. *DprA* appears to be involved in preventing horizontal gene transfer. A mutation in this gene might make it easier for the staph to take up DNA from other bacteria, such as *Enterococcus*.

Second, the CC5 strains lack a set of genes that encode an antibiotic that kills other bacteria. Perhaps this antibiotic normally kills *Enterococcus* near the staph, which would explain why horizontal transfer between the two species does not typically occur.

Finally, Gilmore and his team found that in place of that missing set of antibiotic genes, the vancomycin-resistant CC5 staph have a unique cluster of genes encoding proteins that confuse the human immune system. These proteins could make it easier for staph and other bacteria to grow in a wound because the host immune system would be less able to fight them off.

The lack of the antibiotic genes and the presence of new genes create the ideal conditions for a mixed infection, in which different species of pathogens mingle in a festering soup of contamination. Mixed infections are breeding grounds for antibiotic resistance because they are sites of gene flow among different organisms. In this explanation, the CC5 staph evolved via the three usual mechanisms—mutation, natural selection, and gene flow (via horizontal transfer)—to be more susceptible to take up the vancomycin-resistance allele from *Enterococcus*.

In addition to natural selection (and its intriguing permutation called *sexual selection*; see "Sex and Selection"), mutation, and gene flow,

# Sex and Selection

*Staphylococcus* and other bacteria replicate asexually by copying their DNA and dividing in two. But adding sex to the equation complicates reproduction. Another mechanism by which species evolve is called sexual selection. In **sexual selection**, nature selects a trait that increases an individual's chance of mating—even if that trait decreases the individual's chance of survival.

Sexual selection favors individuals that are good at getting mates, and it often helps explain differences between males and females in size, courtship behavior, and other traits. Species whose males and females are distinctly different in appearance, such as peacocks, lions, and ducks, are said to exhibit **sexual dimorphism**. In many species, the members of one sex—often females—

are choosy about whether to mate. In birds, for example, brightly colored males may perform elaborate displays in their attempts to woo a mate. In other species, males may attract attention by other means, such as calling vigorously. Females then select as their mates the males with the loudest calls.

Yet, some characteristics that increase an individual's chance of mating can *decrease* its chance of survival. For example, male túngara frogs perform a complex mating call that may or may not end in one or more "chucks." Females prefer to mate with males that emit chucks, but frog-eating bats use that same sound to help them locate their prey. As a result, a frog's attempt to locate a mate can end in disaster.

A male peacock performs an elaborate mating display to a potential mate.

Male túngara frogs face an ecological trade-off: the same type of call that is most successful at attracting females for mating also attracts predatory bats to dinner.

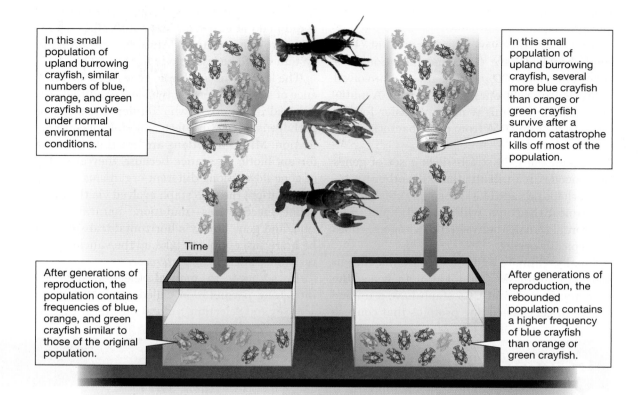

In this small population of upland burrowing crayfish, similar numbers of blue, orange, and green crayfish survive under normal environmental conditions.

In this small population of upland burrowing crayfish, several more blue crayfish than orange or green crayfish survive after a random catastrophe kills off most of the population.

Time

After generations of reproduction, the population contains frequencies of blue, orange, and green crayfish similar to those of the original population.

After generations of reproduction, the rebounded population contains a higher frequency of blue crayfish than orange or green crayfish.

Figure 12.14

## A genetic bottleneck is a type of genetic drift

Two small populations of upland burrowing crayfish are compared. The one on the right experiences a genetic bottleneck event. This is genetic drift—a change in the frequency of a trait that is not associated with natural selection. In fact, the blue crayfish could be less well adapted to the environment than the other crayfish are.

**Q1:** Why do you think a genetic bottleneck is more likely to occur in a small population than in a large population?

**Q2:** Genetic drift is often described as a "chance event." Give other examples of chance events that could cause a genetic bottleneck.

**Q3:** Which resulting population has more genetic diversity?

there is one other mechanism by which organisms evolve. Although alleles that code for beneficial traits like antibiotic resistance for bacteria are usually selected for and maintained in a population by the nonrandom action of natural selection, in some cases, chance events can cause allele frequencies to change from a parent generation to the next generation.

**Genetic drift** is a change in allele frequencies caused by random differences in survival and reproduction among the individuals in a population. Just by chance, some individuals in a generation may leave behind more descendants than other individuals do. In this case, the genes of the next generation will be the genes of the "lucky" individuals, not necessarily the "better" individuals. Genetic drift occurs in all populations, including mammals and bacteria, but it is more likely to cause evolution in a small population than in a large one.

Two ways in which strong genetic drift may occur include genetic bottlenecks and the founder effect. A **genetic bottleneck** is a drop in the size of a population, for at least one generation, that causes a loss of genetic variation (**Figure 12.14**). A genetic bottleneck can threaten the survival of a population.

In the 1970s, for example, the population of the endangered Florida panther plummeted because of hunting and habitat destruction. The species barely escaped extinction. At one point, experts believed that only six wild individuals in the whole species were still alive. This rapid population reduction created a genetic bottleneck in which an estimated half of the genetic variation within the species was lost, and severe inbreeding among the individuals that were left resulted in maladies including low sperm counts and abnormally shaped sperm in male panthers (**Figure 12.15**).

Thankfully, panther numbers have increased to about 80–100 individuals in recent years, in part because of captive breeding programs. Most

Figure 12.16

**Early Dutch colonists in South Africa**

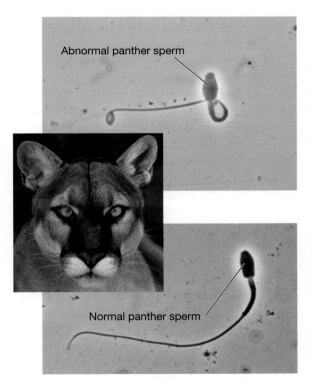

Figure 12.15

**Abnormally shaped sperm in the rare Florida panther**

Florida panthers have more abnormal sperm than do panthers from other populations—a possible effect of a genetic bottleneck promoting the increase of harmful alleles in the Florida panther population.

of the members of this rebounded population are incredibly similar genetically, because of the limited variety of alleles present in the original six individuals. In an effort to restore gene flow to the population, in 1995 researchers released eight female Texas cougars, close genetic cousins to the Florida panthers, into the panthers' habitat. Five of those females produced eight litters of kittens—more genetically diverse, stronger, and longer-lived than their fathers.

The **founder effect** occurs when a small group of individuals establishes a new population isolated from its original, larger population. For example, in South Africa a population of people called Afrikaners descended primarily from a few Dutch colonists (**Figure 12.16**). Today, the Afrikaner population has an unusually high frequency of the allele that causes Huntington disease because those original colonists by chance carried the allele when they settled in the area.

# After Vancomycin

When the initial case of VRSA popped up back in 2002, the patient's foot was carefully treated with an old topical drug that is rarely used today. The only other options were one or two new drugs that hadn't yet been released on the market. "There are still some antibiotics we can use for treatment, but we recommend those be used very responsibly in order to maintain their effectiveness," says Sievert. "You don't want to give them out when not necessary because you don't want to see resistance to them. You save them as the end of the line, so something is available if the commonly used drugs end up failing because of resistance." After 9 months of careful treatment, the patient healed and was able to keep her foot. Other cases of VRSA have similarly been resolved by the use of the newer antibiotics and the delicate care of doctors.

But what happens when even these drugs no longer work against staph? Pharmaceutical companies are developing few new antibiotics, says Frank DeLeo, acting chief of the

Laboratory of Human Bacterial Pathogenesis at the National Institute of Allergy and Infectious Diseases, and bacteria evolve more quickly than we can develop new antibiotics. It's a vicious cycle, he says. "As we continue to use antibiotics, populations of microbes will develop resistance to antibiotics." Recently, microbes appear to be winning the struggle. In July 2016, researchers reported the second U.S. appearance of a superbug resistant to colistin, an antibiotic of last resort. "The golden age of antibiotics appears to be coming to an end," the *Los Angeles Times* wrote.

VRSA has not yet become as widespread as MRSA. Scientists believe the reason may be that staph does not handle the resistance allele for vancomycin very well; the gene from *Enterococcus* is bulky and difficult to manage, so the staph individuals that receive the gene are often outcompeted by other staph individuals in the environment. "Though it allows them to survive, they don't do well with it. There's a [survival] cost to having this resistance," says Gilmore. In the absence of vancomycin, staph is better off without the resistance allele, so perhaps staph individuals lose it or don't pick it up when not in the presence of the drug. VRSA therefore remains "a rare, unstable organism in the environment that isn't very hardy, so it doesn't last and spread," says Sievert. "That's the good news."

But there's still trepidation that staph may adapt and get comfortable with this resistance gene, just as staph adapted to methicillin resistance, says Gilmore. If staph survives and reproduces more easily with the resistance gene, the gene will likely spread to staph throughout the host/human population. "We don't know if that's a possibility," says Gilmore. "My suspicion is that it is." (See "How Can *You* Make a Difference? Help Prevent Antibiotic Resistance!")

In fact, the thirteenth VRSA case, from a man in Delaware in July 2012, was not a CC5 strain, says Gilmore—suggesting that vancomycin resistance is spreading to other strains of staph. "It's disconcerting," he adds. It seems that other populations of staph may now also be evolving vancomycin resistance. "The main worry is that this would move into a strain that was highly transmissible in the community," adds DeLeo. "The potential is there."

## How Can *You* Make a Difference? Help Prevent Antibiotic Resistance!

The U.S. Centers for Disease Control and Prevention (CDC) recommends the following guidelines:

- Take antibiotics exactly as the doctor prescribes. Do not skip doses. Complete the prescribed course of treatment, even when you start feeling better.

- Only take antibiotics prescribed for you; do not share or use leftover antibiotics. Antibiotics treat specific types of infections. Taking the wrong medicine may delay treatment and allow bacteria to multiply.

- Do not save antibiotics for the next illness. Discard any leftover medication once the prescribed course of treatment is completed.

- Do not ask for antibiotics when your doctor thinks you do not need them. Remember that antibiotics have side effects. When your doctor says you don't need an antibiotic, taking one might do more harm than good.

- Prevent infections by practicing good hand hygiene and getting recommended vaccines.

SOURCE: http://www.cdc.gov/Features/AntibioticResistance.

# Race against Resistance

"Each year in the United States, more than 2 million people become infected with antibiotic-resistant bacteria, including methicillin-resistant *Staphylococcus aureus* (MRSA), resulting in at least 23,000 deaths. To combat antibiotic resistance, doctors need novel antibiotics, yet fewer and fewer new antibiotics come to pharmacies each year. In 2011, Congress passed the Generating Antibiotic Incentives Now (GAIN) Act to stimulate the development of new antibiotics, and it's working. So far, six new antibiotics have been approved through that program.

Assessment available in smartwork5

## Antibacterial drugs approved by the FDA

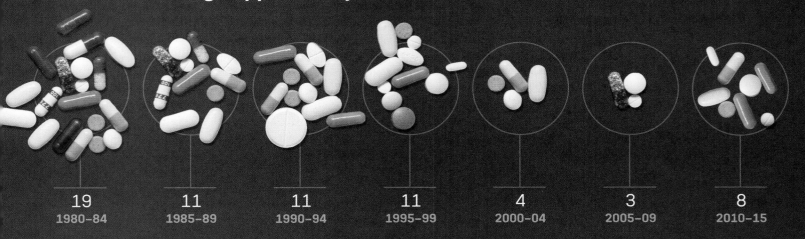

| 19 | 11 | 11 | 11 | 4 | 3 | 8 |
|----|----|----|----|----|----|----|
| 1980–84 | 1985–89 | 1990–94 | 1995–99 | 2000–04 | 2005–09 | 2010–15 |

## Time line of antibiotic resistance

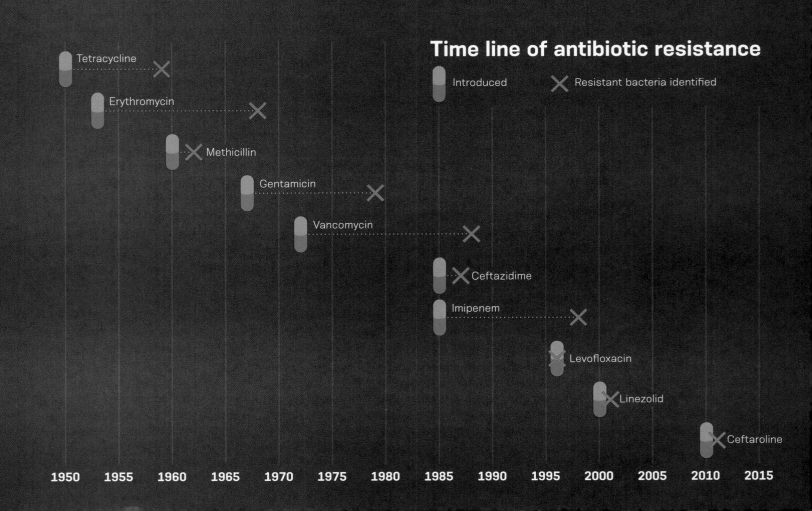

Introduced        ✕ Resistant bacteria identified

Tetracycline

Erythromycin

Methicillin

Gentamicin

Vancomycin

Ceftazidime

Imipenem

Levofloxacin

Linezolid

Ceftaroline

1950   1955   1960   1965   1970   1975   1980   1985   1990   1995   2000   2005   2010   2015

# REVIEWING THE SCIENCE

- Natural selection for inherited traits occurs in three common patterns: directional, stabilizing, and disruptive.

- In **directional selection**, individuals at one phenotypic extreme of a given genetic trait have an advantage over all others in the population.

- In **stabilizing selection**, individuals with intermediate phenotypes have an advantage over all others in the population.

- During **disruptive selection**, individuals with either extreme phenotype have an advantage over those with an intermediate phenotype.

- In **convergent evolution**, distantly related organisms (those without a recent common ancestor) evolve similar structures in response to similar environmental challenges.

- All mechanisms of evolution depend on the genetic variation provided by new alleles created by **mutation**.

- **Sexual selection** occurs when a trait increases an individual's chance of mating even if it *decreases* that individual's chance of survival.

- **Gene flow** is the exchange of alleles between separate populations.

- **Genetic drift** is a change in allele frequencies produced by *random* differences in survival and reproduction in a small population, and most dramatically occurs through one of two processes: a genetic bottleneck or the founder effect.

- A **genetic bottleneck** occurs when a drop in the size of a population leads to a loss of genetic variation in the new, rebounded population.

- The **founder effect** occurs when a few individuals from a large population establish a new population, leading to a loss of genetic variation in the new, isolated population.

# THE QUESTIONS

## The Basics

**1** The founder effect is a type of (**genetic drift / gene flow**) in which individuals in one small group of a large population (**establish a new distant population / are the only survivors**) and then reproduce.

**2** Unlike natural selection, _____ is not related to an individual's ability to survive and may result in offspring that are less well adapted to survive in a particular environment.

(a) genetic drift

(b) sexual selection

(c) directional selection

(d) convergent evolution

**3** Which of the following statements about convergent evolution is true?

(a) It demonstrates how similar environments can lead to different physical structures.

(b) It demonstrates how similar environments can lead to the same physical structures.

(c) It demonstrates that similarity of structures is due to descent from a common ancestor.

(d) It demonstrates that similarity of structures is due to random chance.

**4** Evolution is most accurately described as a change in allele frequencies in _____ over time.

(a) an individual

(b) a species

(c) a population

(d) a community

## Challenge Yourself

**5** In a population, which individuals are most likely to survive and reproduce?

(a) The individuals that are the most different from the others in the population.

(b) The individuals that are best adapted to the environment.

(c) The largest individuals in the group.

(d) The individuals that can catch the most prey.

**6** A study of a population of the goldenrod wildflower finds that large individuals consistently survive and reproduce at a higher rate than small or medium-sized individuals. Assuming size is an inherited trait, the most likely evolutionary mechanism at work here is

(a) disruptive selection.

(b) directional selection.

(c) stabilizing selection.

(d) natural selection, but it is not possible to tell whether it is disruptive, directional, or stabilizing.

**7** Explain how, because of sexual selection, an individual might be very successful at surviving (natural selection), but not pass on genes to the next generation.

## Try Something New

**8** Two large populations of the same species found in neighboring locations that have very different environments are observed to become genetically more similar over time. Which of the four main evolutionary mechanisms is the most likely cause of this trend? Justify your answer.

**9** The Tasmanian devil, a marsupial indigenous to the island of Tasmania (and formerly mainland Australia as well), experienced a population bottleneck in the late 1800s when farmers did their best to eradicate it. After it became a protected species, the population rebounded, but it is now experiencing a health crisis putting it at risk for disappearing again. Many current Tasmanian devil populations are plagued by a type of cancer called devil facial tumor disease, which occurs inside individual animals' mouths. Afflicted Tasmanian devils can actually pass their cancer cells from one animal to another during mating rituals that include vicious biting around the mouth.

Unlike the immune systems of other species, including humans, the Tasmanian devil's immune system does not reject the passed cells as foreign or nonself (as we reject a liver transplant from an unmatched donor), but accepts them as if they were their own cells. Why would a population bottleneck result in the inability of one devil's immune system to recognize another devil's cells as foreign?

**10** Global warming is causing more and more ice to melt each year at far-northern latitudes, exposing more bare ground than ever before. These vast areas of brown ground coloration make polar bears (which are white) much more conspicuous to their prey. Recently, an infant polar bear was born with brown fur. This polar bear survived to adulthood and has sired several offspring with brown fur. Which of the following is a plausible explanation of how the brown fur trait appeared in these polar bears?

(a) A polar bear realized it would be better to be brown in order to hide more effectively. It induced mutations to occur in its fur pigment gene, which resulted in a change in pigment from white to brown fur.

(b) One or more random mutations occurred in the fur pigment gene in an individual polar bear embryo, which resulted in a change in pigment from white to brown fur.

(c) Increased temperatures due to global warming caused targeted mutations in the fur pigment gene in an individual polar bear embryo, which resulted in a change in pigment from white to brown fur.

(d) A female polar bear realized it would be better for her offspring to be brown and therefore mated with a grizzly bear to achieve this result.

**11** In the garden shed belonging to one of this text's authors, stabilizing selection has occurred over the past 10 years in the house mouse, *Mus musculus*. Which of the following scenarios is an example of stabilizing selection?

(a) Small and medium-sized mice cannot reach the seed shelf in the shed and therefore are at a disadvantage for finding food, so they do not survive and reproduce as well as large mice do.

(b) Small mice cannot reach the seed shelf, and large mice are easily seen by hawks circling above. Medium-sized mice therefore survive and reproduce better than both small and large mice.

(c) Small mice can easily cross the yard to the vegetable garden, and large mice can easily reach the seed shelf. Medium-sized mice have trouble with the seed shelf and are seen by hawks in the yard. Small and large mice survive and reproduce much better than medium-sized mice.

(d) All of these are examples of stabilizing selection.

(e) None of these are examples of stabilizing selection.

## Leveling Up

**12** **What do *you* think?** One way to prevent a small population of a plant or animal species from going extinct is to deliberately introduce some individuals from a large population of the same species into the smaller population. In terms of the evolutionary mechanisms discussed in this chapter, what are the potential benefits and drawbacks of transferring individuals from one population to another? Do you think biologists and concerned citizens should take such actions?

**13** *Write Now* **biology: mechanisms of evolution** This assignment explores the mechanisms of evolution through five selected short stories from *Welcome to the Monkey House* by Kurt Vonnegut Jr. Answer the questions associated with each story.

**"Harrison Bergeron"**
What message is this story trying to send? Cite examples from the story and relate them to the mechanisms of evolution from this chapter.

**"Welcome to the Monkey House"**
Is this story an example of sexual selection? Why or why not? Cite examples from the story and from this chapter to support your thinking.

**"The Euphio Question"**
If technology could produce such an instrument, how would it affect the evolution of humans? What about the evolution of other species on Earth?

**"Unready to Wear"**
Relate this story to as many of the mechanisms of evolution from this chapter as you can. Cite examples from the story and the chapter to support your thinking.

**"Tomorrow and Tomorrow and Tomorrow"**
Do you think these types of drugs are a good or bad thing? Where would you draw the line on technology's ability to extend life? How would drugs like these affect the natural selection and evolution of humans? What about the evolution of other species on Earth?

**For more, visit digital.wwnorton.com/bionow2 for access to:**

# Fast Lizards, Slow Corals

*The rapid evolution of lizards and the slow growth of corals offer clues about how species are born.*

## After reading this chapter you should be able to:

- Diagram the relationships among adaptive traits, reproductive fitness, and evolution by natural selection.

- Defend the utility of the biological species concept, and discuss its shortcomings.

- Define speciation and the role of genetic divergence among populations in creating new species.

- Differentiate between allopatric and sympatric speciation.

- Describe the two kinds of reproductive isolating mechanisms and explain their role in speciation.

- Define and give an example of coevolution.

CHAPTER
13

ADAPTATION
AND SPECIES

In 1971, the United States was at war in Vietnam, a gallon of gas cost 40 cents, a computer engineer sent the first e-mail—and an Israeli biologist named Eviatar Nevo captured 10 lizards on Pod Kopište, a small, rocky island off the coast of Croatia. Each lizard was about the length of a pinky finger and as heavy as a nickel. There was nothing remarkable about them.

Nevo and his team released the captured lizards, a common Mediterranean species called *Podarcis sicula*, on a nearby island just 3 miles away, within sight of the original island but separated by a deep ocean gulf (**Figure 13.1**). Pod Mrčaru is a smaller, plant-covered island that was already inhabited by two other lizard species. Nevo was curious to find out what would happen when the three species began to compete for resources on Pod Mrčaru. But Nevo never got the chance to return to the island. Shortly after he and his team departed, unrest broke out across Croatia, and war held the region in a stranglehold throughout the 1980s and '90s.

Scientists did not return to Pod Mrčaru until 2004. While visiting the University of Antwerp in Belgium, biologist Duncan Irschick from the University of Massachusetts, Amherst, and his Belgian colleagues decided to investigate the mysterious island. One of the researchers, familiar with an obscure paper that Nevo had published in 1972, thought it would be interesting to see what, if anything, had happened to the lizards. Further, Irschick would be able to complete some research on lizard behavior that he had planned to do in Europe anyway. "We didn't know what we'd find," recalls Irschick. "We decided, 'Let's just go check.'" This time, the lizards on the island were quite remarkable.

Figure 13.1

**View of Pod Kopište from Pod Mrčaru**

The island Pod Kopište (top), the original home of the *Podarcis sicula* lizards (bottom), is only 3 miles from their new home on the island Pod Mrčaru.

# Leaping Lizards

Back in the 1970s, Nevo's ten *Podarcis sicula* lizards had been well adapted to the rocky, sparsely vegetated environment of Pod Kopište. They had specific **adaptive traits**, inherited characteristics that enabled them to survive and reproduce successfully on the island. Adaptive traits can be structural features, biochemical traits, or behaviors. In this case, the lizards were fast, with long legs that may have helped them catch the insects that made up most of their diet on the island. They were also territorial, fighting with other lizards over space and mating partners. These adaptive traits, among others, enabled the lizards to survive and reproduce better than competitors lacking those traits on Pod Kopište.

When transplanted to Pod Mrčaru, however, the lizards faced a new environment. Pod Kopište is large and rocky with few plants, and the lizards ate primarily insects; Pod Mrčaru is smaller yet has an abundant supply of plants, including leaves and stems of local shrubs and grasses (**Figure 13.2**). Either the lizards would adapt to their new surroundings, or they would die. The term **adaptation** is commonly applied to adaptive traits *or* the process of evolution through natural selection that brings about adaptive traits, as discussed in Chapter 12. Therefore, an adaptation can be a trait that is advantageous to an individual or a population or, in broader terms, it can be the evolutionary

Pod Kopište, the island from which the original ten *Podarcis sicula* lizards were taken, is rocky and sparsely vegetated.

Pod Mrčaru is a smaller island covered with lush vegetation.

**Figure 13.2**

**Pod Kopište's sparse vegetation and Pod Mrčaru's lush vegetation**

Longer, wider head

Thicker body

New digestive structure

**Figure 13.3**

***Podarcis sicula* lizard on Pod Mrčaru, 33 years after introduction of the species**

*process* of natural selection that enables a good match between a population of organisms and their environment. If the population of lizards was going to survive, it would need to adapt to its new environment.

When Irschick and his team landed on Pod Mrčaru in 2004, they set to work catching lizards. It wasn't hard; the island was swarming with them. Irschick could sit on a rock and simply pick up lizards as they ran by. There were several thousand lizards on the 7-acre island. When Irschick and his team looked closely, they realized that *all* the lizards were *P. sicula*. The ten transplants had wiped out the other two species of lizards. What's more, the adapted lizards looked strange compared to *P. sicula* lizards on other islands nearby (**Figure 13.3**). "They were these really big, chunky lizards," recalls Anthony Herrel, then a postdoctoral fellow in Irschick's lab and now a researcher at the French National Centre for Scientific Research in Paris. "They were unlike lizards on any of the other islands."

Twice a year for 3 years, the group returned to Pod Mrčaru to collect, weigh, and measure the lizards. They also took small pieces of the tails (which grow back) to test the lizards' DNA and compare it to the original population of *P. sicula* on Pod Kopište. Those DNA tests confirmed that the lizards on Pod Mrčaru were indeed descended from the original ten that Nevo had transported to the island. In addition to taking size and weight measurements, the team tested the strength of the lizards' bite. They also dissected two samples of dead lizards.

What the researchers found was unprecedented. In about 33 years, from the time Nevo stepped off the island to the time Irschick stepped on, the species had evolved dramatically. The descendant lizards' heads were larger and shaped differently, making the lizard's bite much stronger than that of the original lizards brought to the island. The descendant lizards also had a unique digestive-tract structure called a cecal valve, a set of muscles between the large and small intestine that slow down food digestion, enabling the lizards to better process the cellulose of plants. It's rare for lizards to have these structures, says Herrel. Only a few plant-eating lizard species, like iguanas, have cecal valves.

**DUNCAN IRSCHICK**

Duncan Irschick is a biologist at the University of Massachusetts, Amherst, who studies animal function and evolution, specializing in research on animal movement and gecko adhesion.

The new traits were the result of the lizards adapting to a different food source, says Irschick: on Pod Kopište, they ate insects, but on Pod Mrčaru, they began to feast on plants. Flushing the stomachs of a few lizards revealed that the lizards' diet was made up of two-thirds plant material. Over time, the lizards that had evolved a new head shape and digestive structures survived and reproduced better than those that had not, because these traits enabled them to take advantage of the new food sources on Pod Mrčaru.

The lizards evolved not only physical adaptations, but behavioral ones too. They became less territorial and less aggressive and were mating more often, probably because more space and food were available. Adaptations, whether behavioral, physical, or biochemical, have three important characteristics. First, they closely match an organism with its environment; in this case, the lizards evolved to match the ecosystem on Pod Mrčaru. Second, adaptations are often complex, such as the lizards' new gut structure. Finally, adaptations help the organism accomplish important functions, such as feeding and mating.

The lizards on Pod Mrčaru show that evolution by natural selection can improve the adaptive fit of organisms to their environment not only over long periods of time, as with the evolution of whales described in Chapter 11, but also over surprisingly short periods of time. In just 33 years, the *P. sicula* lizard population evolved both physical and behavioral traits that helped it flourish on the new island. "It was really, truly rapid evolution across multiple facets," says Irschick.

## What Makes a Species?

Despite how impressive the lizards' quick adaptation may be, natural selection does not always result in a perfect match between an organism and its environment. In many cases, animals fail to adapt successfully. In fact, scientists estimate that 99 percent of all species that have ever lived are now extinct. Every extinct species is a silent testament to a failure to adapt in the face of adversity.

It's tempting to label the descendant lizards as a new species, but Irschick is hesitant to do so without more testing. Most commonly, the term **species** is used to refer to members of a group that can mate with one another to produce fertile offspring. According to the **biological species concept**, a species is a group of natural populations that can interbreed to produce fertile offspring and cannot breed with other such groups; that is, they are **reproductively isolated** from other populations (**Figure 13.4**). Irschick and Herrel have yet to test whether the Pod Mrčaru lizards can still mate with their cousins back on Pod Kopište. According to the

Figure 13.4

**A male and a female South Pacific rattlesnake confirm that they are the same species**
After performing this mating ritual, these snakes successfully mated, producing viable offspring.

**Q1:** How do we know that these rattlesnakes are members of the same species?

**Q2:** How would you design an experiment to determine whether two populations of snakes are distinct species according to the biological species concept?

**Q3:** For which types of populations does the biological species concept *not* work as a way of determining how they are related?

### ANTHONY HERREL

Anthony Herrel worked as a research fellow in Duncan Irschick's lab before starting his own lab at the French National Centre for Scientific Research in Paris, where he studies the evolution of feeding and locomotion in vertebrates.

Figure 13.5

**One species or two?**
Though they look quite different, these tree frogs are genetically similar enough to be considered distinctly colored variations of the same species. Breeding these frogs with each other would determine their classification under the biological species concept.

Q1: List three kinds of information that scientists use to distinguish between species.

Q2: What differences can you observe between the individuals in the photos? Why are these differences not enough to confirm that they are from two different species?

Q3: How is genetic divergence among populations determined?

biological species concept, if individuals from the two populations of lizards can still breed and produce fertile offspring, they are not different species. This same idea holds true for all sexually reproducing organisms. (Incidentally, it's not obvious why sexual reproduction is so common; see "Why Sex?")

The definition of "species" is not a black-and-white issue; it is a frequently discussed, multi-faceted topic in evolution, and the biological species concept doesn't always apply. For example, not all species can be defined by their ability to interbreed, such as prokaryotes like bacteria, which reproduce asexually. To handle these cases and others, scientists may use biogeographic information, DNA sequence similarity, and **morphology**—the organisms' physical characteristics—to identify and distinguish species (**Figure 13.5**).

# Why Sex?

Sex is ubiquitous in the animal and plant kingdoms. An estimated 99 percent of multicellular eukaryotes are capable of sexual reproduction, which involves the joining of two haploid gametes produced through meiosis. Yet sex is very costly for individuals, so scientists have struggled to explain why it is so prevalent compared to asexual reproduction. And no, it's not because sex feels good; the first eukaryotes to engage in sex were single-celled protists some 2 billion years ago, long before animals developed neurons capable of giving an individual a sense of pleasure.

### Costs of Sex

1. Time and energy must be invested to find or attract a mate.

2. Parents pass on only 50 percent of their genetic material to offspring, as opposed to the 100 percent that is passed on through asexual reproduction.

3. Gene combinations that have benefited the parents may be shuffled and broken apart during meiosis and recombination.

### Possible Benefits of Sex

1. The genetic diversity created by sexual reproduction is critical for adaptation to new environments.

2. Sexual reproduction can help a population eliminate detrimental alleles and generate new beneficial alleles.

3. Rapid genetic change that occurs through sexual recombination can help a population evolve resistance to parasitic infections.

The lizards that adapted to the island of Pod Mrčaru are physically different enough from their cousins on Pod Kopište to be considered a different species, says Herrel, yet their DNA sequences are close to identical, so he believes they might still be able to interbreed. **Speciation** is the process by which one species splits to form two species or more. Fundamentally, speciation occurs because of **genetic divergence**, the accumulation of differences in the DNA sequences of genes in two or more populations of organisms over time; as a result, the populations become more and more genetically dissimilar.

The two populations of lizards may continue to diverge, becoming more and more different because of their **geographic isolation** from one another, and may eventually become two unarguably distinct species. This is one of the most common ways that new species form: individuals of a single population become geographically separated from one another. This process can begin when a newly formed geographic barrier, such as a river, a canyon, or a mountain chain, isolates two populations of a single species (**Figure 13.6**). Such geographic isolation can also occur when a few members of a species colonize a region that is difficult to reach, such as an island located far outside the usual geographic range of the species.

Geographically isolated populations are disconnected genetically; there is little or no gene flow between them. Without gene flow, the other mechanisms of evolution we saw in Chapter 12—mutation, genetic drift, and natural selection—can more easily cause populations to diverge from one another. If populations remain isolated long enough, they can evolve into new species. The formation of new species from geographically isolated populations is called **allopatric speciation** (*allo*, "other"; *patric*, "country"), as shown in **Figure 13.7**.

But what happens when there isn't a physical barrier between two populations? In the ocean, for example, plants and animals can drift

The Kaibab squirrel is confined to the North Rim of the Grand Canyon.

Abert's squirrel lives on the South Rim and other southern locations, all the way into Mexico

Figure 13.6

**The Grand Canyon is a geographic barrier for squirrels**
The Kaibab squirrel population became isolated from the Abert's squirrel population when the Colorado River cut the Grand Canyon—which is as deep as 6,000 feet in some places. With gene flow between them blocked, probably beginning about 5 million years ago, the two populations eventually accumulated enough genetic differences to become two distinct species.

**Q1:** What is the definition of gene flow? How was gene flow blocked between these species?

**Q2:** Name as many types of geographic barriers as you can. Which do you think would be the best at blocking gene flow?

**Q3:** Are geographic barriers universal for all species? If not, name a geographic barrier that might block gene flow for one species but not another.

around or swim extremely long distances. There are few physical barriers yet many separate, unique species. How do new species form when populations are free to mix and mingle?

# Caribbean Corals

Carlos Prada did not spend much of his time at the University of Puerto Rico in a classroom. Instead, he spent it in the ocean, studying corals, a type of marine invertebrate. On regular dives onto the Caribbean reef, the graduate student documented the morphology of a species of coral named *Eunicea flexuosa*, commonly called "sea fans." At 12 different locations along the Puerto Rican coast, Prada observed different morphologies of the coral at different depths. In shallow areas, less than 15 feet from the surface, the sea fans were wide, with a broad network of thick branches like a bush. In deeper areas, between 15 and 55 feet from the surface, the sea fans were much taller and spindly, resembling trees with networks of thin branches.

"They're supposed to be the same species, but they really looked different at different depths," says Prada. Curious about whether the morphologies were due to genetics or a result of the surrounding environment shaping the animals as they grew, Prada began

Figure 13.7

**Physical barriers can produce allopatric speciation by blocking gene flow**
Allopatric speciation can occur when populations are separated by a geographic barrier, such as a rising sea.

A single plant species is distributed over a broad geographic range.

Time

The sea level rises and isolates plant populations from one another. The populations may adapt to different environments on opposite sides of the barrier, indirectly causing genetic changes that reduce their ability to interbreed.

Time

When the barrier is removed, the plants recolonize the intervening area and mingle, but do not interbreed.

Range of overlap

**Q1:** What factors must be present for allopatric speciation to occur?

**Q2:** If a geographic barrier is removed and the two reunited populations intermingle and breed, what attributes must the offspring have in order for the two populations, according to the biological species concept, to be considered still the same species?

**Q3:** If the two populations in question 2 are determined to still be the same species, did allopatric speciation occur?

Shallow-water sea fan

Deep-water sea fan

Carlos Prada transplants coral to waters of different depths.

Figure 13.8

**Different corals at different depths**
These two corals were once considered the same species, *Eunicea flexuosa*, commonly called a "sea fan." (Source: Photos courtesy of Carlos Prada.)

carefully transplanting the corals, moving deep-water sea fans to shallow depths and vice versa (**Figure 13.8**). He found that when transplanted, the corals did change. The shallow-water sea fans became taller and more spindly when planted in deep water, and the deep-water sea fans became wider in shallow waters, but—critically—neither made a complete transition to the alternate shape. The lack of a total transformation by either form to the other suggested to Prada that the corals, while they likely share a common ancestor, are actually two species that have adapted to their respective water depths.

When Prada finished his graduate work in Puerto Rico, he e-mailed a professor at Louisiana State University who studied speciation in ocean animals. With wavy, bleached-blond hair, Michael Hellberg looks more like a California surfer than a professor, but this evolutionary biologist has long been fascinated with how one species splits to become two, especially in the ocean. "Say you have a new lake forming, and a species becomes isolated in the lake. Then it's pretty obvious there's not going to be a lot of interbreeding to fight against, and the species just adapts. To me, there's no mystery in that," says Hellberg. This would be an example of allopatric speciation. "I've always tried to target groups where species look closely related and where ranges of the species overlap. That makes things a lot harder."

Hellberg welcomed Prada into his crew, and the two set out to extend Prada's work to find out whether his idea—that coral evolve different adaptive traits at different depths in the ocean, leading to the formation of new species—was unique to coral reefs in Puerto Rico or could be observed in other areas around the Caribbean. With Hellberg's support, Prada traveled to the Bahamas, Panama, and Curaçao to observe and take samples from sea fan colonies.

As he waited for Prada to return home with the data, Hellberg remained skeptical of the idea of **ecological isolation**, that two closely related species in the same territory could be reproductively isolated by slight differences in habitat. But

**MICHAEL HELLBERG AND CARLOS PRADA**

Michael Hellberg (right) is an evolutionary biologist at Louisiana State University who studies how species evolve in marine environments. Carlos Prada (left) was a graduate student in Hellberg's lab, and is now a postdoctoral researcher at Penn State studying how organisms cope with changes in the environment.

as Prada sampled more and more locations, the evidence was convincing. Indeed, everywhere he looked, Prada saw the same thing he had observed in Puerto Rico: broad, leaflike coral at shallow depths, and tall, sticklike coral in deeper waters. Distance between the populations did not matter; some shallow-water and deep-water fans were close enough together that Prada could reach out and bend them to touch each other. But depth did matter. The corals were physically close enough to easily interbreed in the water, yet the species had somehow become specialized to two different depths (**Figure 13.9**).

Prada brought his coral samples back to the lab and performed tests to see how genetically similar the two groups of corals were. Prada and Hellberg found that all the shallow-water sea fans across the Caribbean were more closely related to each other than they were to any of the deep-water sea fans. The same held true for the deep-water sea fans: they were more closely related to each other than to any of the shallow-water fans. In the DNA data, Hellberg had seen some genetic exchange between the two populations in the past, yet that exchange of genetic materials had been limited across the two groups, and each species had bred almost exclusively among its own populations.

# Different Depths, Different Habitats

How did a few yards of depth produce habitats so different that the corals evolved different adaptations to each depth? The scientists are still investigating the differences between the shallow and deep habitats, but they have some hypotheses. One is that the coral have adapted their morphologies, and possibly their biochemistry, to suit different symbiotic algae that grow on them at different depths.

Symbiotic algae live on coral and, through photosynthesis, use sunlight to produce energy and organic compounds that coral use to maintain and grow calcium carbonate skeletons. In turn, the coral provide the algae with a sheltered place to live and produce carbon dioxide that the algae use during photosynthesis. (This cooperative relationship between species is known as mutualism; see Chapter 20.) But different species of algae have

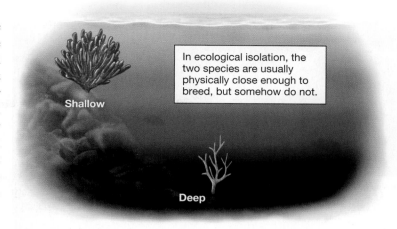

In ecological isolation, the two species are usually physically close enough to breed, but somehow do not.

Shallow

Deep

Figure 13.9

**Depth of water provides ecological isolation for sea fans**
The two depths shown here differ slightly in their quantity and quality of light, the force of the waves or current, the amount of sediment deposited on them, the number and type of predators, and the availability and type of food.

different light and nutrient requirements, and organisms 15 feet below the surface receive a lot of sunlight, while those at 30 feet receive less light, and light of different wavelengths. Prada found different types of algae living on the two coral populations (**Figure 13.10**), so it's possible that the sea fans evolved to be better hosts for the most successful algae at their depth.

In many cases in the ocean, as with coral and algae, the interaction between two species so strongly influences their survival that they have evolved in tandem—a phenomenon known as **coevolution**. The term "coevolution" encompasses a wide variety of ways in which an adaptation in one species evolves alongside a complementary adaptation in another species.

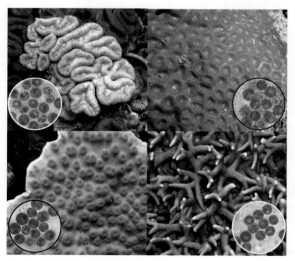

Figure 13.10

**Different species of corals and their resident algae**
Although all these algae (single-celled eukaryotic organisms) look the same under a microscope, they are actually different species with different light requirements and photosynthetic capacities.

Flowers that coevolved with hummingbirds produce nectar, are colored to attract the birds, and are shaped specifically to fit the hummingbirds' bills.

Figure 13.11

**Coevolution at its finest**

This hummingbird's bill fits perfectly into this flower for easy access to the hummingbird's favorite food, nectar. The hummingbird feasts and then distributes the flower's pollen by carrying the pollen along to its next meal. It's a win-win situation!

> **Q1:** Describe how coevolution, as with the hummingbird bill and hummingbird-pollinated flowers, is different from the kind of evolution described in Chapters 11 and 12.
>
> **Q2:** Is coevolution the same thing as convergent evolution, described in Chapter 12? Why or why not?
>
> **Q3:** Do you think one species' adapting over time to feed specifically and extremely successfully on another species is an example of coevolution? Why or why not?

# So Many Chromosomes

New plant species can form in a single generation as a result of **polyploidy**, a condition in which an individual gains an extra full set or two (or three) of chromosomes. Humans and most other eukaryotes are *diploid* (having two sets of chromosomes), but some organisms are *triploid* (three sets) or *tetraploid* (four sets), or have an even higher number of chromosomes. Polyploidy is invariably fatal in people, but in many plant species it is not lethal.

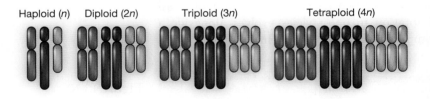

Haploid (*n*)    Diploid (2*n*)    Triploid (3*n*)    Tetraploid (4*n*)

Polyploidy can occur when two different species hybridize and produce an offspring with an odd number of chromosomes, because of improper alignment of the chromosomes during mitosis (see Chapter 6). This increase in chromosome sets can lead to reproductive isolation because the chromosome number in the gametes of the new polyploid individual no longer matches the number in the gametes of either of its parents.

Polyploidy has had a large effect on life on Earth: more than half of all plant species alive today are descended from species that originated by polyploidy. A few animal species also appear to have originated by polyploidy, including several species of lizards, salamanders, and fish.

Another example of coevolution is the relationship between hummingbirds and certain species of flowers (**Figure 13.11**).

Alternatively, or in addition to hosting specific algae, Hellberg hypothesizes that the shallow-water and deep-water sea fans may have adapted to the type and density of sedimentation found at different ocean depths. Or perhaps predators are more abundant at certain depths. Whatever the major habitat factors driving the corals' adaptation, natural selection led to the formation of two different coral populations tailored to the ocean depths at which they grow.

Today, Hellberg considers the two populations different species. "What makes this interesting is that you have populations that could potentially interbreed, yet they maintain their differences in the face of each other," says Hellberg. "For just about any biological question you want to ask, they are independent entities." The formation of new species in the *absence* of geographic isolation is called **sympatric speciation** (from *sym*, "together"; **Figure 13.12**). Sympatric speciation is a particularly important process in plants (see "So Many Chromosomes").

Still, a question lingers: Why don't the two coral species interbreed? They appear to be capable of it, because of evidence of some genetic

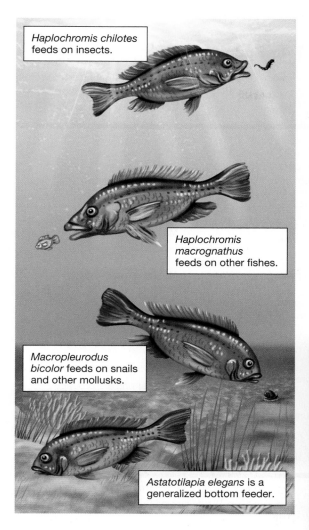

Haplochromis chilotes feeds on insects.

Haplochromis macrognathus feeds on other fishes.

Macropleurodus bicolor feeds on snails and other mollusks.

Astatotilapia elegans is a generalized bottom feeder.

**Figure 13.12**

**Sympatric speciation drives diversity among Lake Victoria cichlid species**

Scientists have described some 500 species of cichlid fishes in Lake Victoria. Genetic analyses indicate that they all descended from just two ancestor species over the past 100,000 years. These four species show some of the differences in feeding behavior and morphology.

**Q1:** What is the main difference between allopatric and sympatric speciation?

**Q2:** Name two events that must happen for both allopatric speciation and sympatric speciation to occur.

**Q3:** Do you think all of the 500 species in Lake Victoria arose through sympatric speciation? Why or why not?

mixing in the past, yet there is no evidence that interbreeding is common; only a few rare hybrids grow on the ocean floor between them. When two species are reproductively isolated from each other, we say that **reproductive barriers** exist between those species. Reproductive barriers are often divided into two categories: *prezygotic* and *postzygotic*.

Barriers that prevent a male gamete (such as a human sperm) and a female gamete (such as a human egg) from fusing to form a zygote are **prezygotic barriers**. Prezygotic barriers act *before* the zygote exists (**Figure 13.13**). Barriers that prevent zygotes from developing into

Blue-footed boobies point their beaks, wings, and tails upward in a mating dance called "sky pointing."

**Figure 13.13**

**The blue-footed booby courtship dance is a prezygotic, behavioral reproductive barrier**

This species of booby has a unique ritual dance that must be accurately completed before mating. Other booby species do not perform exactly the same dance and therefore do not mate with the blue-footed booby.

**Q1:** What does "prezygotic" mean?

**Q2:** How is the booby's ritual dance a prezygotic reproductive barrier?

**Q3:** What are some other prezygotic reproductive barriers besides a mating dance?

Table 13.1

## Reproductive Barriers That Isolate Two Species in the Same Geographic Region

| Type of Barrier | Description | Effect |
|---|---|---|
| **Prezygotic** | | |
| Ecological isolation | The two species breed in different portions of their habitat, in different seasons, or at different times of day. | Mating is prevented. |
| Behavioral isolation | The two species respond poorly to each other's courtship displays or other mating behaviors. | Mating is prevented. |
| Mechanical isolation | The two species are physically unable to mate. | Mating is prevented. |
| Gametic isolation | The gametes of the two species cannot fuse, or they survive poorly in the reproductive tract of the other species. | Fertilization is prevented. |
| **Postzygotic** | | |
| Zygote death | Zygotes fail to develop properly, and they die before birth. | No offspring are produced. |
| Hybrid sterility | Hybrids survive but are unable to produce viable offspring. | No offspring are produced. |
| Hybrid performance | Hybrids survive poorly or reproduce poorly. | Hybrids are not successful. |

healthy and fertile offspring are called **postzygotic barriers**, which act *after* the zygote is formed.

A wide variety of cellular, anatomical, physiological, and behavioral mechanisms generate pre- and postzygotic reproductive barriers, but they all have the same overall effect: little or no mating occurs, and therefore few or no alleles are exchanged between species (**Table 13.1**). Hellberg suspects that some prezygotic barriers exist between the two sea fan species—perhaps the gametes of the two species don't fuse successfully, as in some sea urchins (**Figure 13.14**)—but there may also be a postzygotic barrier such as generation time.

It seems likely that the reason hybrids between the two species do not succeed is the uniquely long generation time for these corals, says Prada. Sea fans don't reach reproductive age until they are 15–20 years old, and they continue reproducing until they are 60 years old or more. So, while coral gametes and larvae can and do disperse far from their parents and may interbreed or take root at incorrect depths, natural selection then has 15–20 years to winnow out the less successful offspring. If there are small differences in survival rates between the two species at a particular depth—if a deep-water coral fares slightly worse in a shallow area than a shallow-water coral does—those differences will become amplified over 15 years, and by the time the corals reach reproductive age it is likely that only the shallow-water species will still be alive in the shallow area.

New species form and adapt to their environments in many different ways, from island-bound lizards to depth-dependent corals. And from those varied beginnings comes the vast diversity of life that exists on our planet today, the result of billions of years of evolution. In

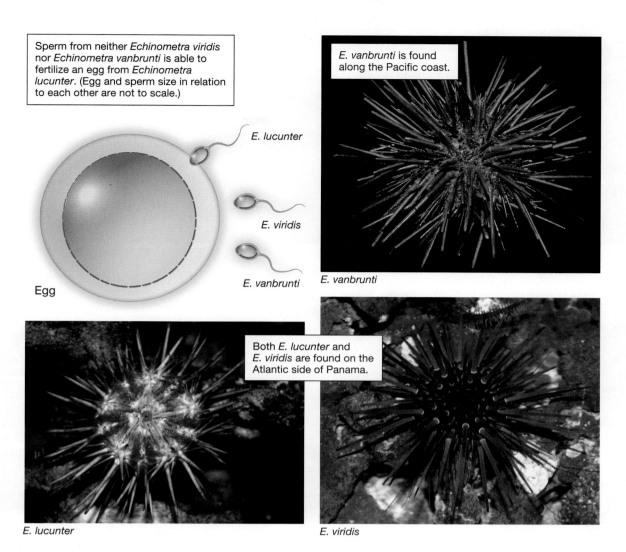

Sperm from neither *Echinometra viridis* nor *Echinometra vanbrunti* is able to fertilize an egg from *Echinometra lucunter*. (Egg and sperm size in relation to each other are not to scale.)

E. lucunter

E. viridis

E. vanbrunti

Egg

*E. vanbrunti* is found along the Pacific coast.

*E. vanbrunti*

Both *E. lucunter* and *E. viridis* are found on the Atlantic side of Panama.

*E. lucunter*

*E. viridis*

## Figure 13.14

**Gametic isolation among sea urchin species**

The sea urchin species found along both coasts of Panama illustrate gametic isolation as a reproductive barrier. At one point long ago, the Panama landmass was not continuous and did not separate the two oceans. These three species likely evolved through both allopatric and sympatric speciation from a recent common ancestor after the oceans were separated.

**Q1:** Which species is/are sympatric with *Echinometra lucunter*?

**Q2:** Which species is/are allopatric with *E. lucunter*?

**Q3:** If two individuals have incompatible gametes, what will be the result of a mating event between them?

the next section of this book, we explore the dramatic transitions in life-forms that have walked, crawled, and swum around our planet, from small dinosaurs with wings to the most successful, most cunning species ever to traverse the planet: humans.

# On the Diversity of Species

There are an estimated 8.74 million eukaryotic species on Earth, yet we are familiar with only a fraction of those. Scientists estimate that 86% of land species and 91% of aquatic species have not been discovered. Of the known species, insects top the chart with an estimated 1 million different species.

Assessment available in smartwork5

## Numbers of known species

**Fish**
31,200

**Crustaceans**
47,000

**Mollusks**
85,000

**Fungi**
99,000

**Arachnids**
102,200

**Insects**
1,000,000

**Plants**
310,100

**Mammals**
5,500

**Sponges**
6,000

**Amphibians**
6,500

**Reptiles**
8,700

**Jellyfish and polyps**
9,800

**Birds**
10,000

**Millipedes and centipedes**
16,100

**Segmented worms**
16,800

**Flatworms**
20,000

**Roundworms**
25,000

# REVIEWING THE SCIENCE

- **Adaptive traits,** or adaptations, are inherited characteristics that improve an individual's chances of surviving and reproducing in a specific environment.

- Natural selection leads to **adaptation**, genetic changes that improve the reproductive success of organisms in their environment over time. Adaptation does not produce "perfection."

- According to the **biological species concept**, a **species** is a group of populations that interbreed and can produce live and fertile offspring.

- The process of **speciation**, one species splitting into two or more species, is the by-product of **reproductive isolation** and **genetic divergence** between populations.

- **Allopatric speciation** occurs when populations of a species become **geographically isolated**, limiting gene flow and making genetic divergence more likely.

- Speciation that occurs between populations lacking geographic isolation is called **sympatric speciation**.

- In **ecological isolation**, two closely related species in the same area are reproductively isolated by slight differences in habitat.

- When adaptive traits of one species evolve in concert with adaptive traits in another species, **coevolution** has occurred.

- Reproductive isolating mechanisms can be **prezygotic barriers** (before zygote formation) or **postzygotic barriers** (after zygote formation).

# THE QUESTIONS

## The Basics

**1** When two populations are reproductively isolated, what else must occur for speciation to happen?

(a) gene flow

(b) genetic divergence

(c) coevolution

(d) convergent evolution

(e) none of the above

**2** Traits that are inherited and improve an individual's ability to survive and reproduce are called _____ traits.

(a) adaptive

(b) polymorphic

(c) biological

(d) sympatric

(e) allopatric

**3** Prezygotic isolating mechanisms prevent hybrid offspring from occurring between species because

(a) the resulting offspring are not fertile and cannot reproduce.

(b) the egg and sperm fuse and form a zygote, but it does not survive.

(c) the egg and sperm do not ever meet or, if they do, cannot fuse to form a zygote.

(d) all of the above

**4** Which of the following reproductive isolating mechanisms prevents mating because the two species are physically unable to mate?

(a) ecological isolation

(b) behavioral isolation

(c) mechanical isolation

(d) gametic isolation

(e) all of the above

**❺** Select the correct terms:
Speciation due to geographic separation and (**genetic divergence / gene flow**) of two populations is considered (**allopatric / sympatric**) speciation.

**❻** Select the correct phrases:
(**Prezygotic barriers / Postzygotic barriers**) are reproductive isolating mechanisms that occur after a zygote has been formed. An example would be (**an infertile hybrid / gametes that cannot fuse**).

## Challenge Yourself

**❼** Natural selection

(a) leads to species being better adapted to their environment.

(b) may lead to speciation if there is no gene flow between populations.

(c) may lead to genetic divergence among populations in differing environments.

(d) all of the above

(e) none of the above

**❽** Which of the following examples describe an adaptive trait? (Select all that apply.)

(a) A rainforest tree that is vulnerable to storm damage.

(b) A male bird that is more successful than others at attracting a female mate.

(c) A rabbit that is better camouflaged in its environment.

(d) A desert plant that is able to survive drought.

(e) A frog that is more noticeable to predators.

**❾** Place the following events in the order in which they are most likely to occur by numbering them from 1 to 5.

_____ a. Speciation occurs.

_____ b. Genetic divergence takes place.

_____ c. Geographic barrier arises.

_____ d. Gene flow is discontinued.

_____ e. Species adapts to a new environment.

## Try Something New

**❿** Distinct species that are able to interbreed in nature are said to "hybridize," and their offspring are called "hybrids." The gray oak and the Gambel oak can mate to produce fertile hybrids in regions where they co-occur. However, the gene flow in nature is sufficiently limited that, overall, the two species remain phenotypically distinct. If the hybrid offspring survive well and reproduce to the extent that there is a large population of hybrid individuals that breed between themselves but do not interbreed with either of the two original parent species (the gray and the Gambel), which of the following would you say most likely led to the new hybrid species?

(a) prezygotic reproductive barriers

(b) sympatric speciation

(c) allopatric speciation

(d) postzygotic reproductive barriers

(e) none of the above

**⓫** The four-eyed fish, *Anableps anableps*, really has only two eyes, which function as four, enabling the fish to see clearly through both air and water. *A. anableps* is a surface feeder, so the ability to see above water helps it locate prey such as insects. Its unique eyes also enable it to scan simultaneously for predators attacking from above (such as birds) or below (such as other fish). This species of fish _____ crafted by natural selection.

*Anableps anableps*

(a) is an example of coevolution

(b) has an evolutionary adaptation

(c) is an example of sympatric speciation

(d) is an example of allopatric speciation

**⓬** Give an example of a barrier that might bring about allopatric speciation. In addition to this barrier, what would be necessary for speciation to occur?

**13** Researchers from the Smithsonian Institution were startled to discover that the 3 species of *Starksia* blennies they had been studying in the Caribbean islands were really 10 different species. How could these researchers have thought that these 10 different reef fish species were only 3 species?

## Leveling Up

**14** *Write Now* **biology: adaptations** Select an organism (other than humans) that you find interesting. Research two adaptations of your organism and describe them in detail. Explain carefully why each of these features is considered an important adaptation for its species. Discuss how a change in the environment might change the usefulness of these adaptations.

**15** **Doing science** The eastern and western meadowlarks are very similar in morphology. Both are grassland species with some overlap in their range in the upper Midwest, and each sings a distinctly different song. Create a hypothesis as to whether these are two different species or two populations of the same species, and describe an experiment that would enable you to test your hypothesis. What result would support your hypothesis, and what result would cause you to reject your hypothesis?

**16** **What do *you* think?** According to the Defenders of Wildlife organization, there were 20–30 million American range bison in the old West. During the late 1800s the American range bison were hunted to the brink of extinction, leaving behind a bottleneck population of only 1,091 individuals. The population has since rebounded to about 500,000 bison. Unfortunately, almost all of these bison are the descendants of these few individuals crossbred with domestic cattle by ranchers. Scientists and conservationists want to genetically test bison to find those of pure bison origin to preserve the species. Only these, they argue, should be called American range bison and be allowed to roam free in the national parks as bison. They think hybrids should be confined to farms and ranches, should be called "beefalo" rather than bison, and should not be afforded the protection that pure bison currently have. What do you think? Should bison tainted with cattle genes be removed from free-range parks? Should the government spend scarce conservation monies on genetic testing and breeding efforts to preserve the pure bison population? Investigate conservation efforts and the costs of genetically testing and relocating bison to help you with your decision. Is speciation at the hands of human beings now part of evolution as we know it?

**For more, visit digital.wwnorton.com/bionow2 for access to:**

# The First Bird

*A fresh wave of fossils challenges the identity of the first dino-bird.*

----------------------------------------------------------------

**After reading this chapter you should be able to:**

- Identify critical events on a time line of life on Earth and explain their evolutionary importance.

- Interpret an evolutionary tree of a group of organisms.

- Explain how new scientific data may change an evolutionary tree without challenging our understanding of evolution.

- Classify a given species at each level of the Linnaean system.

- Give an example of an adaptation that enabled a group of organisms to transition from an aquatic existence to life on land.

- Describe the impact of prehistoric mass extinctions on Earth's biodiversity, and infer the likely impact of the current extinction rate on biodiversity.

249

**BIODIVERSITY**

I n 1861, quarry workers in Germany unearthed the fossil of a crow-sized bird. The 150-million-year-old preserved skeleton looked bizarre: it had feathers and clawed hands like a bird, but teeth and a long, bony tail like a reptile.

After close examination by paleontologists, the fossil was hailed as a transitional form between birds and reptiles and labeled the oldest known bird. Named *Archaeopteryx*, the feathered dinosaur became famous around the world as the first solid evidence that birds descended from dinosaurs (**Figure 14.1**). Charles Darwin, who proposed the theory of descent with modification, called the discovery "a grand case for me."

Over 100 years later, a young fossil hunter named Xu Xing bent over the dirt at a dig site in Liaoning province in northern China. It was Xu's first time in the field since finishing college. An excellent student, Xu had wanted to study economics at Peking University in Beijing, but at the time, students in China did not pick their majors, and Xu was assigned paleontology. Luckily, farmers in Liaoning had recently begun to uncover huge numbers of fossils, and research was booming.

So, as a grudging paleontologist, Xu began searching for dinosaurs in Liaoning. He had a particular interest in feathered dinosaurs. By the late twentieth century, the idea that birds descended from dinosaurs was no longer a hypothesis, but a well-established scientific theory backed by a mountain of data. And atop that mountain of data sat *Archaeopteryx*, the earliest known bird.

Little did Xu know, as he dug through the dirt that day, that he would soon be the one to knock *Archaeopteryx* from its perch.

# Dinosaurs and Domains

Although the *Archaeopteryx* fossil is old—150 million years old, in fact—much older fossils have been found.

Our solar system and Earth formed 4.6 billion years ago. Some of the oldest known rocks on Earth are 3.8 billion years old and contain carbon deposits that hint at life. Cell-like structures have been found in layered mounds of sedimentary rock called stromatolites that formed 3.7 billion years ago, and projections based on DNA analysis also support the idea that life appeared on Earth at that time.

The question of how life arose from nonlife is one of the greatest riddles in biology, but scientists have little doubt that all life on Earth is related. As noted in Chapter 1, all living organisms are united by a basic set of characteristics.

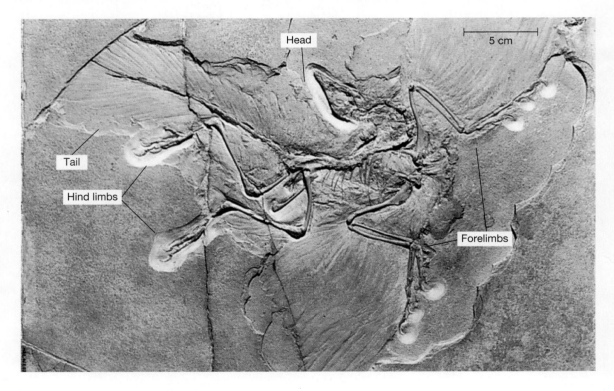

### Figure 14.1

**The original *Archaeopteryx* fossil**

There are 11 complete *Archaeopteryx* fossils, but this one from Germany was the first to be found. Note the faint impressions of feathers on the wings and tail.

Life shares this set of common properties because all living organisms descended from a common ancestor, known as the *universal ancestor*. This hypothetical ancestral cell is placed at the base of the tree of life. From that cell, all life emerged. Biological diversity, or **biodiversity**, embraces the variety of all the world's living things, as well as their interactions with each other and the ecosystems they inhabit. Biodiversity can be described at the level of genes, or species, or entire ecosystems.

In spite of intense worldwide interest, scientists do not know the exact number of species alive today. Most traditional estimates have fallen in the range of 3–30 million species, but in 2016, scientists combined microbial, plant, and animal data sets from around the world, scaled up, and estimated that Earth could contain nearly a trillion species. So far, about 1.5 million species have been collected, named, and placed into an evolutionary tree, meaning it is possible that only one ten-thousandth of 1 percent of species are known to us. In other words, 99.9999 percent of species are yet undiscovered.

Life throughout history and today is so diverse that biologists created a classification system to organize it into categories. The **domains** form the highest, most inclusive hierarchical level in the organization of life, describing the most basic and ancient divisions among living organisms. There are three domains of life (**Figure 14.2**):

• **Bacteria**, which includes familiar disease-causing bacteria such as *E. coli*

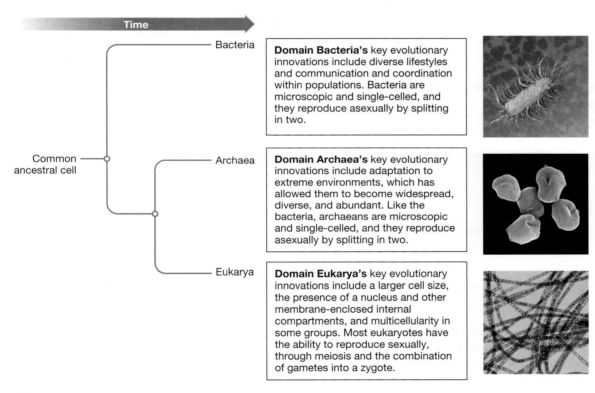

Figure 14.2

**Three domains of life**
This "tree of life" shows the relationships of the three domains of life.

Q1: Why is there a shared line from the universal ancestor for Archaea and Eukarya?

Q2: Where would birds be found within this figure? What about humans?

Q3: To which domain would you expect a disease-causing organism to belong? What if the organism was multicellular?

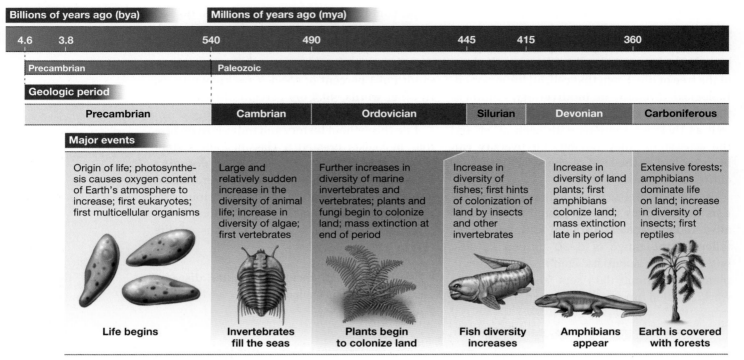

Figure 14.3

## The geologic timescale and major events in the history of life
The history of life can be divided into 12 major geologic time periods, beginning with the Precambrian (4.6 bya to 540 mya) and extending to the Quaternary (2.6 mya to the present). This time line is not drawn to scale; to do so would require extending the diagram off the book page to the left by more than 5 feet (1.5 meters). ▶▮

- **Archaea**, which consists of single-celled organisms best known for living in extremely harsh environments

- **Eukarya**, which includes all other living organisms, from amoebas to plants to fungi to animals

Humans, dinosaurs, and birds are all part of the Eukarya domain. They are *eukaryotes*. Bacteria and Archaea are two different domains—Archaea are more closely related and in some ways more similar to Eukarya than to Bacteria—yet because neither Bacteria nor Archaea are eukaryotes, the two have traditionally been lumped under a common label: *prokaryotes*.

Prokaryotes first appear in the fossil record at about 3.7 billion years ago (**Figure 14.3**), but the first eukaryotes did not evolve until over a billion years later. Luckily for us, and all other eukaryotes, roughly 2.8 billion years ago a group of bacteria evolved a type of photosynthesis that releases oxygen as a by-product. As a result, the oxygen concentration in the atmosphere increased over time, and about 2.1 billion years ago the first single-celled eukaryotes evolved. When the oxygen concentration reached its current level, by about 650 million years ago (mya), the evolution of larger, more complex multicellular organisms became possible, including fish, then land plants, then insects, amphibians, and reptiles. One group of reptiles, which would eventually dominate most

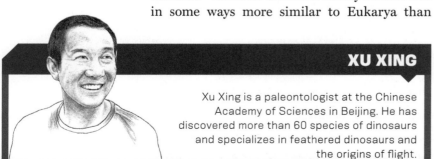

### XU XING

Xu Xing is a paleontologist at the Chinese Academy of Sciences in Beijing. He has discovered more than 60 species of dinosaurs and specializes in feathered dinosaurs and the origins of flight.

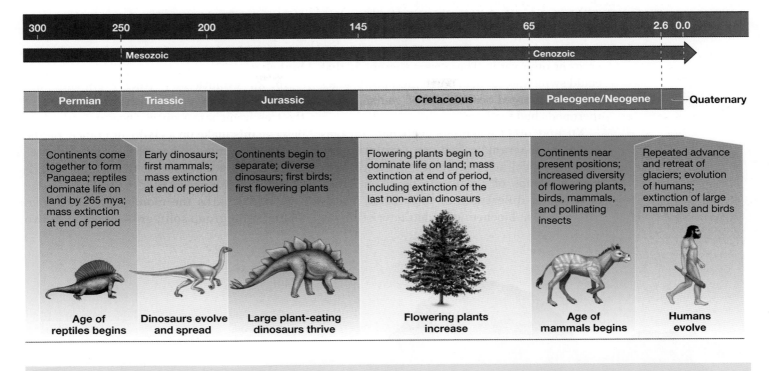

| 300 | 250 | 200 | 145 | 65 | 2.6 0.0 |
|---|---|---|---|---|---|

Mesozoic | Cenozoic

| Permian | Triassic | Jurassic | Cretaceous | Paleogene/Neogene | Quaternary |

Continents come together to form Pangaea; reptiles dominate life on land by 265 mya; mass extinction at end of period

Early dinosaurs; first mammals; mass extinction at end of period

Continents begin to separate; diverse dinosaurs; first birds; first flowering plants

Flowering plants begin to dominate life on land; mass extinction at end of period, including extinction of the last non-avian dinosaurs

Continents near present positions; increased diversity of flowering plants, birds, mammals, and pollinating insects

Repeated advance and retreat of glaciers; evolution of humans; extinction of large mammals and birds

**Age of reptiles begins** | **Dinosaurs evolve and spread** | **Large plant-eating dinosaurs thrive** | **Flowering plants increase** | **Age of mammals begins** | **Humans evolve**

**Q1:** During what geologic period did life on Earth begin?

**Q2:** How long ago did species begin to move from water to land? What period was this?

**Q3:** In what period would *Archaeopteryx* have been alive?

other species, was the dinosaurs. Dinosaurs first appeared about 230 mya, during the Triassic period, and they took over the planet.

# Feathered Friends

Xu Xing may not have wanted to be a paleontologist when he went to college, but by the time he graduated, he was hooked on dinosaurs. Over the next 20 years, Xu became one of the most productive researchers in his field. To date, he has discovered and named more than 60 extinct species—mostly dinosaurs, but also a reptile and a salamander. And the majority of those dinosaur fossils have feathers (**Figure 14.4**).

As scientists traced back the **lineage**, or line of descent, from birds to dinosaurs, it became clear that birds are most closely related to theropods—fast-moving dinosaurs that ran

Figure 14.4

**A feathered dinosaur tail, trapped in amber**

In 2016, scientists discovered a 99-million-year-old dinosaur tail, complete with its feathers, in a piece of amber.

on two legs and had hollow, thin-walled bones (as birds do). Theropods were a diverse group of dinosaurs (**Figure 14.5**). Most were carnivores, including insectivores, but a few were herbivores or omnivores. There were theropods that could swim and eat fish. Some theropod species boasted enlarged scales, and of course, many theropods had feathers.

Xu and other scientists map out lineages using a diagram called an **evolutionary tree**, a model of evolutionary relationships among groups of organisms based on similarities and differences in their DNA, physical features, biochemical characteristics, or

some combination of these. An evolutionary tree maps the relationships between ancestral groups and their descendants, and it clusters the most closely related groups on neighboring branches.

In an evolutionary tree, organisms are depicted as if they were leaves at the tips of the tree branches. A given ancestor and all its descendants make up a **clade**, or branch, on the evolutionary tree. *Archaeopteryx* and all subsequent animals that evolved from it are considered a clade (**Figure 14.6**).

A **node** marks the moment in time when an ancestral group split, or diverged, into two

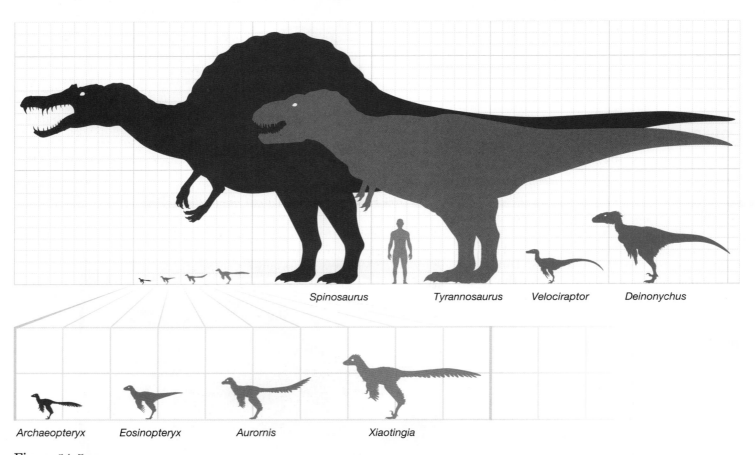

Spinosaurus      Tyrannosaurus   Velociraptor   Deinonychus

Archaeopteryx    Eosinopteryx    Aurornis       Xiaotingia

Figure 14.5

**Dinosaurs large and small**

Theropods ranged in size from tiny, like a chicken, to huge, like the group's most famous member, *Tyrannosaurus rex.*

**Q1:** In what ways were theropods the same as modern birds? Give at least two similarities.

**Q2:** In what ways did theropods differ from modern birds? Give at least two differences.

**Q3:** Birds are often referred to as "living dinosaurs." Is this accurate? Why or why not?

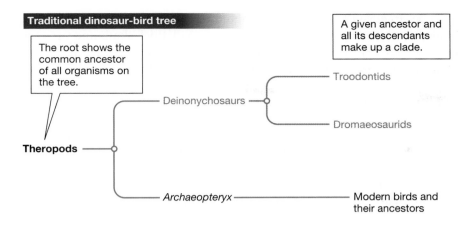

**Traditional dinosaur-bird tree**

The root shows the common ancestor of all organisms on the tree.

A given ancestor and all its descendants make up a clade.

Theropods — Deinonychosaurs — Troodontids / Dromaeosaurids

*Archaeopteryx* — Modern birds and their ancestors

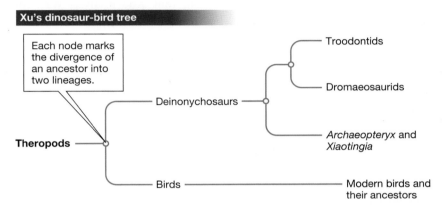

**Xu's dinosaur-bird tree**

Each node marks the divergence of an ancestor into two lineages.

Theropods — Deinonychosaurs — Troodontids / Dromaeosaurids

*Archaeopteryx* and *Xiaotingia*

Birds — Modern birds and their ancestors

**Figure 14.6**

**The evolutionary origins of birds**
(*Top*) The traditional evolutionary tree, showing *Archaeopteryx* as an early bird that split off from the deinonychosaurs (birdlike, carnivorous dinosaur). (*Bottom*) The evolutionary tree proposed by Xu after the discovery of *Xiaotingia*, with *Archaeopteryx* as a deinonychosaur rather than an early bird.

**Q1:** In the traditional tree, identify the node showing the common ancestor for early birds and dinosaurs.

**Q2:** What do both the traditional tree and Xu's tree suggest about troodontids and dromaeosaurids?

**Q3:** In both trees, identify the node for the common ancestor of *Archaeopteryx* and other birds. In what way are the nodes different in the two trees?

separate lineages. A node represents the **most recent common ancestor** of two lineages in question—that is, the most *immediate* ancestor that *both* lineages share. For over 100 years, researchers considered *Archaeopteryx* the most recent common ancestor of both birds and dinosaurs and thus placed it at the root of the avian clade—the first bird (**Figure 14.6**, top).

The first bird, that is, until Xu stumbled across a fossil that threw the field into controversy. In 2008, Xu visited the Shandong Tianyu Museum of Nature, a dinosaur museum in eastern China. There, he happened upon a unique fossil that had been collected by a farmer in Liaoning and sold through a dealer to the museum. The fossil, entombed in yellowish rock, shows a small, birdlike dinosaur seemingly craning its neck forward and spreading its short wings (**Figure 14.7**). "I saw it and said, 'Oh, this is an important species,'" recalls Xu. He asked the museum to let him study it.

As Xu examined the fossil, which he later named *Xiaotingia zhengi*, he wondered where it belonged on the dinosaur-bird evolutionary tree.

To find out, he analyzed *shared derived traits* of the fossil and similar early birds. **Shared derived traits** are unique features common to all members of a group that originated in the group's most recent common ancestor and then were passed down in the group (but not in groups that are not direct descendants of that ancestor). In this case, the original ancestor

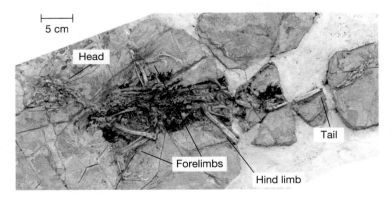

**Figure 14.7**

*Xiaotingia zhengi*: a controversial new leaf on the dinosaur-bird evolutionary tree in question was assumed to be *Archaeopteryx*, and the shared traits included feathers, clawed hands, and a long, bony tail.

By comparing *Xiaotingia* to *Archaeopteryx* and other related species, Xu created a new evolutionary tree of early birds (**Figure 14.6**, bottom). Suddenly, *Archaeopteryx* wasn't in the avian clade. Instead, *Archaeopteryx* and *Xiaotingia* were in a different clade, grouped with deinonychosaurs—the small, birdlike, carnivorous dinosaurs commonly called raptors. (The term "raptor" refers to carnivorous modern birds like hawks and owls, but it is also used informally to describe this group of dinosaurs.) *Archaeopteryx*, Xu believed, was not the first bird, but a raptor. It had feathers, but its descendants did not evolve into birds.

"It was a big change," says Xu. Yet, he adds, it wasn't entirely unexpected. In the last 20 years, more and more fossils of early avian species have been discovered—and the more Xu compared *Archaeopteryx*, discovered in Europe, to other early birds discovered in China, the less it looked like a bird and the more it looked like a raptor. Early birds have small, thick skulls and two toes on each foot; *Archaeopteryx*, on the other hand, has a long, almost pointy skull and three toes on its feet. "*Archaeopteryx* is just so different from other early birds," says Xu.

Xu published his revised evolutionary tree in 2011. It took the scientific world by storm. Some researchers embraced the idea: "Perhaps the time has come to finally accept that *Archaeopteryx* was just another small, feathered bird-like theropod fluttering around in the Jurassic," wrote one paleontologist in the journal *Nature*. Others disagreed, arguing that Xu's analysis was not convincing. For them, *Archaeopteryx* remained the first bird.

# The History of Life on Earth

Xu knew that one new fossil was but a single shred of evidence and therefore unlikely, by itself, to convince the scientific community. "The evidence is not very strong," he admits, "but it was a question we wanted to discuss and add more analyses to, including old and new fossils." The identity of the first bird is important because our understanding of how flight evolved is based on the classification of *Archaeopteryx*. If *Archaeopteryx* is not the species from which flight evolved, then our understanding of flight is wrong.

Xu's redesign of the early bird evolutionary tree raised another important question: Did flight evolve more than once? If *Archaeopteryx*, which looked capable of flight, was a bird, then flight probably evolved just once among dinosaurs, in the avian lineage. If *Archaeopteryx* was not a bird, and was instead simply a raptor that could fly, then flight evolved at least twice among dinosaurs, once in raptors and once in avians—but only avians went on to evolve into modern birds.

Biological classification helps us answer important evolutionary questions like these. In addition to recognizing three broad domains of life, biologists group the Eukarya into four distinct **kingdoms**, the second-highest level in the hierarchical classification of life (**Figure 14.8**).

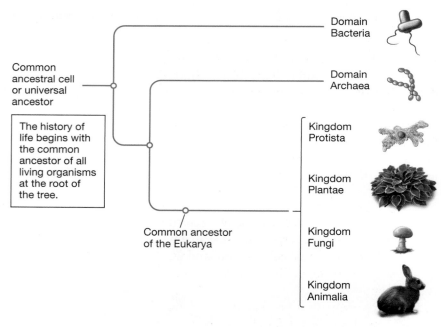

Figure 14.8

**The tree of life**

The domains Bacteria and Archaea do not have separate kingdoms within them. The domain Eukarya encompasses four kingdoms: Protista (protists, an artificial grouping that includes organisms such as amoebas and algae), Plantae (plants), Fungi (including yeasts and mushroom-producing species), and Animalia (animals).

These kingdoms are **Protista**, a diverse group that includes amoebas and algae; **Plantae**, which encompasses all plants; **Fungi**, which includes mushrooms, molds, and yeasts; and **Animalia**, which encompasses all animals, including dinosaurs, birds, and humans. The members of each of these kingdoms share evolutionary innovations that adapted the organisms to their environments, enabling them to live and reproduce successfully (**Figure 14.9**).

Below the level of kingdom, biological classification can get even more specific using the **Linnaean hierarchy**, a system of biological classification devised in the eighteenth century by a Swedish naturalist named Carolus Linnaeus. The smallest unit of classification in the Linnaean hierarchy—the **species**—reflects individuals that are the most related to each other. The most closely related species are grouped together to form a **genus** (plural "genera"). Using these two categories in the hierarchy, every species is given a unique, two-word Latin name, called its **scientific name**. The first word of the name identifies the genus to which the organism belongs; the second word defines the species. Comparing our own species name, *Homo sapiens*, to that of the Neanderthals, *Homo neanderthalensis*, shows that the two are classified as separate species but belong to the same genus, *Homo* (**Figure 14.10**).

In the Linnaean hierarchy, each species is placed in successively larger and more inclusive categories beyond genus. Closely related genera are grouped into a **family**. Closely related families are grouped into an **order**. Closely related orders are grouped into a **class**. Closely related classes are grouped into a **phylum** (plural "phyla"). And, you guessed it, closely related phyla are grouped together into a kingdom.

The earliest forms of life evolved in water. About 650 mya (during the Precambrian period), the number of organisms appearing in the fossil record increased. At that time, much of Earth was covered by shallow seas filled with small, mostly single-celled organisms floating freely in the water.

Then, about 540 mya, the world experienced an astonishing burst of evolutionary activity, with a dramatic increase in biodiversity. Most of the major living animal groups first appear in the fossil record during this time, popularly known as the **Cambrian explosion**. The Cambrian explosion changed the face of life on Earth from a world of relatively simple, slow-moving, soft-bodied scavengers and herbivores to a world dominated by large, fast predators. The presence of predators sped up the evolution of Cambrian herbivores, judging by the variety of scales and shells and other protective body coverings typical of many Cambrian, but not Precambrian, fossils (**Figure 14.11**).

The Cambrian explosion, however, occurred primarily in the oceans. Because life first evolved in water, the colonization of land by living organisms posed enormous challenges. Indeed, many of the functions basic to life, including support, movement, reproduction, and the regulation of heat, had to be handled very differently on land than in water. About 480 mya, near the beginning of the Ordovician period, plants were the first organisms to meet these challenges. These early terrestrial colonists were single-celled or had just a few cells.

Fungi are thought to have made their way onto land next, according to fossil evidence. For example, scientists have found fossils of terrestrial fungi that are 455–460 million years old. Reconstructing the evolutionary history of eukaryotes from DNA data, scientists estimate that the common ancestor of fungi and animals diverged from all other eukaryotes about 1.5 billion years ago, and fungi diverged from their closest cousins, the animals, about 10 million years after that.

Next, land plants evolved and diversified from the original green algae that made it to land. By 360 mya, at the end of the Devonian period, Earth was covered with plants. The first land animals likely emerged about 400 mya. Many of the early animal colonists on land were carnivores; others fed on living plants or decaying plant material.

The first vertebrates to colonize land were amphibians, the earliest fossils of which date to about 365 mya (see Chapter 17 for more on vertebrates). Early amphibians descended from lobe-finned fishes. Amphibians were the most abundant large organisms on land for about 100 million years. Then, in the late Permian period, reptiles took over as the most common vertebrate group. Reptiles were the first group

## Protista

The kingdom Protista is an artificial grouping defined by what members of this group are not: protists are not plants, animals, fungi, bacteria, or archaeans.

Although most protists are harmless, the best-known ones are pathogenic, like *Plasmodium vivax*, the protist that causes malaria.

Animal-like protists are consumers.

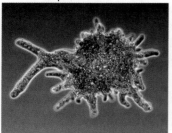

Fungus-like protists are decomposers.

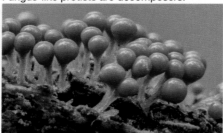

Plant-like protists are photosynthetic.

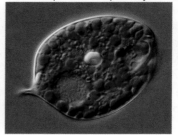

Most protists are single-celled and microscopic, and they can swim with the help of one or more flagella, or by waving a carpet of tiny hairs called cilia.

### Protista

Diplomonads, others | Euglenoids, others | Ciliates, others | Brown algae, diatoms, others | Forams, others | Red algae | Green algae

The evolutionary history of the protists remains unresolved.

Common ancestor of plants and the green algae

Common ancestor of the Eukarya

## Plantae

Plants are multicellular autotrophs and mostly terrestrial. Because plants are producers, they form the basis of essentially all food webs on land.

Bryophytes (mosses and liverworts)

Plants have a waxy covering, known as the *cuticle*, that covers their above-ground parts. A waxy cuticle holds in moisture—an important adaptation to life on land.

Ferns

Gymnosperms

Gymnosperms were the first plants to evolve pollen, a microscopic structure that contains sperm cells, which freed them from a dependence on water for fertilization. Gymnosperms were also the first to evolve seeds, which can be disseminated so they will not compete with the mother plant for sunlight, or for water and nutrients in the soil.

Angiosperms

Flowering plants, or angiosperms, are dominant and the most diverse group of plants on our planet. The keys to the success of angiosperms are the flower, a structure that evolved through modification of early plant reproductive organs, and the fruit, a fleshy ovary wall that protects and helps disperse the seeds inside it.

Figure 14.9

**The four kingdoms of the domain Eukarya**

**Q1:** What group of organisms shares the most recent common ancestor with plants?

Zygomycetes (molds)

Fungi digest organic material outside the body and absorb the molecules released as breakdown products.

Ascomycetes (sac fungi)

Basidiomycetes (club fungi)

Fungi play several roles in terrestrial ecosystems. Many are decomposers, acting as garbage processors and recyclers by speeding the return of the nutrients in dead and dying organisms to the ecosystem.

Fungi are similar to animals in that they store surplus food energy in the form of glycogen. Like some animals, such as insects and lobsters, fungi produce a tough material called *chitin* that strengthens and protects the body. Unlike animals, fungal cells have a protective cell wall that wraps around the plasma membrane and encases the cells.

| Plantae | Protista | Animalia | Fungi |
|---|---|---|---|

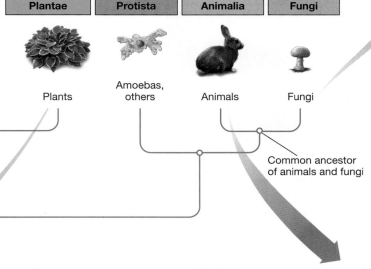

Plants

Amoebas, others

Animals

Fungi

Common ancestor of animals and fungi

Animals are multicellular ingestive heterotrophs, obtaining energy and carbon by ingesting food. All animals are consumers, and some are important decomposers in the ecosystems they inhabit.

Animal cells differ from those of plants and fungi in that they lack cell walls. Instead, many cells in the animal body are enveloped in, or attached to, a felt-like layer known as the *extracellular matrix*. An important evolutionary innovation of animals is the development of true tissues. Most animals have two or three main tissue layers that give rise to a structurally complex body.

Sponges

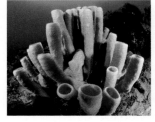

Cnidarians

Mollusks

The sponges are the most ancient animal lineage. Cnidarians, a group that includes jellyfish and corals, evolved next. The remaining animal phyla fall into two groups: the *protostomes* and *deuterostomes*, distinguished by different patterns of embryonic development. Protostomes comprise more than 20 separate subgroups, including mollusks (such as snails), annelids (segmented worms), and arthropods (including spiders and insects). Deuterostomes include echinoderms (sea stars and their relatives) and the chordates. The chordates are a large group that includes all animals with backbones, such as fish, birds, and humans.

Arthropods

Echinoderms

Chordates

Q2: Are fungi more closely related to plants or to animals? Does the answer surprise you? Why or why not?

Q3: If you were to create an evolutionary tree in which amoebas were included within the kingdom of organisms to which they were the most closely related (rather than with protists, where they are currently placed), where would you put them?

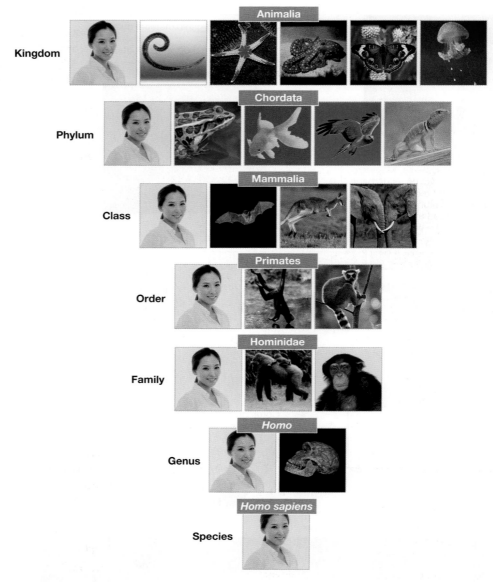

Figure 14.10

**The Linnaean hierarchy of classification**

Q1: Within which category are individuals most closely related to one another?

Q2: Within which category are individual species most distantly related?

Q3: Are individual species more closely related within the same order or within the same family?

of vertebrates that could reproduce without returning to open water, because they lay amniotic eggs that have a built-in food source and are protected from drying out by a hardened shell, compared to the jellylike sac that encloses the eggs of other vertebrates. The evolution of the amniotic egg was a major event in the history of life because it established a new evolutionary branch, the amniotes, which later included all reptiles, birds, and mammals.

And so, with the rise of reptiles 230 mya, the age of dinosaurs began.

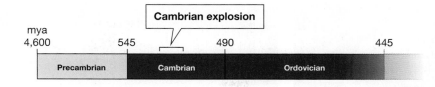

mya
4,600      545      Cambrian explosion     490              445

Precambrian      Cambrian      Ordovician

### Figure 14.11

**Cambrian biodiversity**
During the Cambrian period, animal diversity increased dramatically. The remains of many of these species have been found in Canada (the Burgess Shale), China (the Maotianshan Shale), and fossil beds in Greenland and Sweden. Some of the fossils look familiar, resembling sponges and brachiopods, but do not appear to be related to any living groups of animals.

# Tussling with Trees

In January 2013, a year and a half after Xu published his controversial paper about *Xiaotingia* and *Archaeopteryx*, his work received unexpected support. Pascal Godefroit, a paleontologist at the Royal Belgian Institute of Natural Sciences, reported the discovery of a feathered dinosaur called *Eosinopteryx brevipenna*. A commercial collector in northeastern China had dug up *Eosinopteryx* in the same area where *Xiaotingia* was discovered. The tiny 161-million-year-old dinosaur was preserved as a virtually complete skeleton, with its legs bent and arms out, as if about to jump (**Figure 14.12**).

When Godefroit and his team added *Eosinopteryx* to the evolutionary tree of feathered dinosaurs, they came to the same surprising conclusion that Xu had: *Archaeopteryx* was not a bird. Instead, *Archaeopteryx* was a raptor along with *Eosinopteryx* and *Xiaotingia*. All three species share traits that early birds did not have, such as arms longer than their legs, reduced tail plumage, and primitive feather development.

But science is a continuously changing process, and only 4 months later, Godefroit published data about another birdlike fossil that caused him to revise his hypothesis once again. This time it was a feathered dinosaur that Godefroit found collecting dust in the archives of a Chinese museum. The 18-inch-long fossil had small, sharp teeth and long forelimbs. Godefroit believed the dinosaur, which his team named *Aurornis xui* to honor Xu's work, probably couldn't fly, but instead used its wings to glide from tree to tree. But its other features, including the hip bones, were clearly shared by modern birds (**Figure 14.13**).

Using *Aurornis*, Godefroit constructed another evolutionary tree. This time he started from scratch, compiling data on almost a

### PASCAL GODEFROIT

Pascal Godefroit is a paleontologist at the Royal Belgian Institute of Natural Sciences in Brussels. He has discovered numerous feathered dinosaurs.

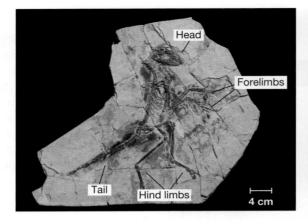

Figure 14.12

**Another feathered dinosaur, *Eosinopteryx brevipenna*, supports Xu's new tree**

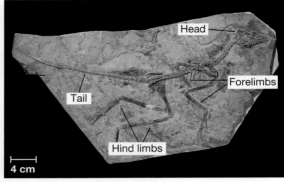

Figure 14.13

**A namesake fossil, *Aurornis xui*, illuminates the dinosaur-bird transition**

thousand characteristics of skeletons of 101 species of dinosaurs and birds. "It's very impressive," paleontologist Mike Lee at the South Australian Museum told *National Geographic*. "They considered more than twice as much anatomical information as even the best previous analyses."

Contrary to the tree that arose from his first study, Godefroit's second tree places *Archaeopteryx* back on its roost in the bird family, although no longer as the oldest bird. That place belongs to *Aurornis*, says Godefroit (**Figure 14.14**).

Still, the debate is not over. "Of course we need more evidence and more work," says Xu. "Many of these new species are a possible candidate for

Figure 14.14

**The early birds**

Godefroit's 2013 study places *Archaeopteryx* and *Xiaotingia* with birds rather than dinosaurs, as in the traditional dinosaur-bird evolutionary tree (see Figure 14.6, top). But it places *Aurornis* as the earliest known bird on the evolutionary tree.

**Q1:** Is *Xiaotingia* an earlier or later bird than *Archaeopteryx* in this tree?

**Q2:** If a future study, based on more fossils or new measurements, placed *Archaeopteryx* back with dinosaurs, would this suggest that birds are not related to dinosaurs? Why or why not?

**Q3:** If you were to create an evolutionary tree of modern birds, where would you expect to place the roadrunner (judging by its appearance in this figure) as compared to a house sparrow or pigeon?

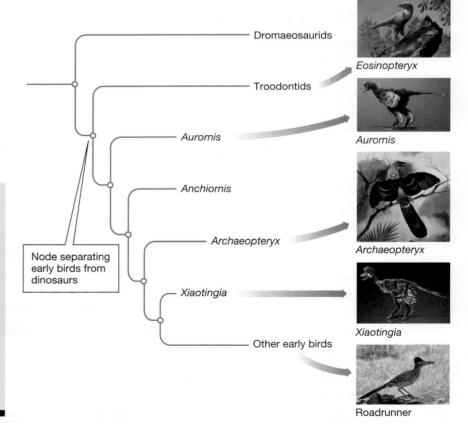

Dromaeosaurids

*Eosinopteryx*

Troodontids

*Aurornis*

*Aurornis*

Node separating early birds from dinosaurs

*Anchiornis*

*Archaeopteryx*

*Archaeopteryx*

*Xiaotingia*

*Xiaotingia*

Other early birds

Roadrunner

the earliest bird." There are likely to be more fossils that will shake up the bird tree, says Xu, but that's a good problem to have in science. "There are so many new species, it just makes it difficult for us all to agree."

Xu, Godefroit, and others are sure to continue to dig up dinosaur fossils for some time yet. Dinosaurs arose about 230 mya and dominated the planet from about 200 mya to about 65 mya. Then the majority of them went extinct, except for those that evolved into birds. As the fossil record shows, species have regularly gone extinct throughout the history of life. The rate at which this has happened—that is, the number of species that have gone extinct during a given period—has varied over time, from low to very high. At the upper end of this scale, the fossil record shows that there have been five **mass extinctions**, periods of time during which great numbers of species go extinct (**Figure 14.15**).

Although difficult to determine, the causes of the five mass extinctions are thought to include such factors as climate change, massive volcanic eruptions, changes in the composition of marine and atmospheric gases, and sea level changes. The Cretaceous extinction event occurred

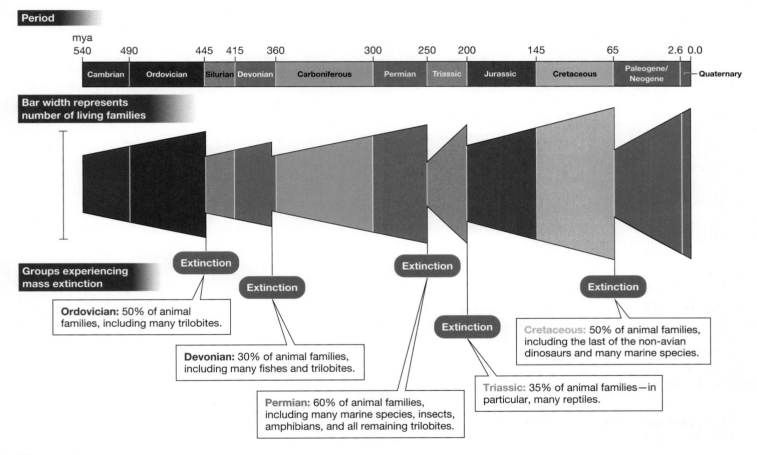

Figure 14.15

**Mass extinctions and biodiversity**

In addition to the marine and terrestrial animal groups shown here, plant groups were severely affected by the five mass extinctions that have occurred in Earth's history. After each extinction, life again diversified.

Q1:   What extinction event occurred about 200 mya? What animal groups were most affected by this event?

Q2:   Which of the mass extinctions appears to have removed the most animal groups? How long ago did this extinction occur?

Q3:   The best studied of the mass extinctions is the Cretaceous extinction. Why do you think it has been better studied than the other extinctions?

# The Sixth Extinction

On at least five occasions, mass extinctions have occurred across the globe, caused at different times by climate change, volcanic eruptions, and possible asteroids. Today, scientists agree we are in the midst of a sixth extinction, and this time, we, the human race, are the cause.

Assessment available in smartwork5

## Ghosts of species past

### Passenger Pigeon

The passenger pigeon, the most common bird in North America 200 years ago, was hunted to extinction in the nineteenth century. The last wild bird was shot in 1900, and the last captive bird died in 1914.

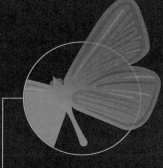

### Xerces Blue Butterfly

The Xerces blue butterfly once lived on sand dunes around San Francisco—until its habitat was destroyed by urban development. It was last seen in 1943.

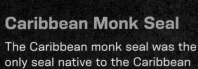

### Caribbean Monk Seal

The Caribbean monk seal was the only seal native to the Caribbean Sea and Gulf of Mexico. It was overhunted for oil, and its food sources were overfished. It was last sighted in 1952.

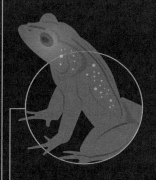

### Golden Toad

Once common in the cloud forests of Monteverde, Costa Rica, the golden toad has not been seen since 1989. Pollution and global warming likely contributed to its extinction.

## Number of species known to be extinct or extinct in the wild since 1500

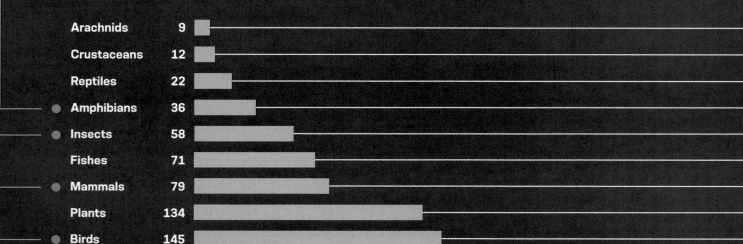

| Group | Number |
|---|---|
| Arachnids | 9 |
| Crustaceans | 12 |
| Reptiles | 22 |
| Amphibians | 36 |
| Insects | 58 |
| Fishes | 71 |
| Mammals | 79 |
| Plants | 134 |
| Birds | 145 |
| Mollusks | 324 |

about 65 mya and wiped out three-fourths of plant and animal species on Earth, including non-avian dinosaurs. Researchers suspect that a massive comet or asteroid slammed into the Gulf of Mexico, choking the skies around the planet with debris and decreasing the ability of plants to photosynthesize. As the plants died, so, too, did animals further up the food chain.

Mass extinctions affect the diversity of life in two main ways: First, entire groups of organisms perish, changing the history of life forever. Second, the extinction of one or more dominant groups of organisms can provide new opportunities for groups that previously were of relatively minor importance, thereby dramatically altering the course of evolution. When a group of organisms expands to take on new ecological roles and to form new species and higher taxonomic groups, that group is said to have undergone an **adaptive radiation**. Some of the great adaptive radiations in the history of life occurred after mass extinctions, such as when the mammals diversified after the extinction of the dinosaurs.

Today, one species of mammal, *Homo sapiens*, dominates life on land, and our impact on biodiversity is unprecedented. Because of human activities, the world is losing species at an alarming rate. In July 2016, researchers at the United Nations Environment Programme World Conservation Monitoring Centre in the United Kingdom announced that worldwide biodiversity had fallen below predetermined "safe" levels, thresholds below which ecological function is likely to be negatively affected. An estimated 58 percent of the world's land has lost more than 10 percent of its biodiversity, the researchers found, with grasslands and biodiversity hot spots such as the Amazon rainforest hit the hardest.

That is alarming news, because nonhuman organisms provide even the most basic requirements for human life. Plants and plantlike protists produce the oxygen we breathe and provide us with food. Whole ecosystems provide so-called "ecosystem services," environmental benefits that humans rely on. For example, coast redwood trees in northern California intercept fog, mist, and rain, channeling the water onto and into the ground.

Biologists today assert that we are on our way toward a new mass extinction, and the cause is clear: the activities of the ever-increasing number of humans living on, and exploiting, Earth.

# REVIEWING THE SCIENCE

- The first single-celled organisms resembled bacteria and probably evolved about 3.7 billion years ago.

- The most basic and ancient branches of the tree of life define three **domains**: **Bacteria**, **Archaea**, and **Eukarya**. All life-forms fall into one of these three domains. The Eukarya are further divided into four kingdoms: **Protista**, **Fungi**, **Plantae**, and **Animalia**. The variety of these life-forms, as well as their interactions with each other and the ecosystems they inhabit, is **biodiversity**.

- Scientists use **evolutionary trees** to model ancestor-descendant relationships among different organisms. The tips of branches represent existing groups of organisms, and each **node** represents the moment when an ancestor split into two descendant groups. A **clade** is an ancestral species and all its descendants.

- Closely related groups of organisms share distinctive features that originated in their **most recent common ancestor**. These **shared derived traits** are used to identify the **lineages** of a species.

- The **Linnaean hierarchy** places each **species** in successively larger and more inclusive categories. Closely related species are grouped together into a **genus**, related genera are grouped into a **family**, related families into an **order**, related orders into a **class**, related classes into a **phylum**, and finally, related phyla into a **kingdom**.

- Organisms are identified by their genus and species names, together referred to as their **scientific name**.

- The release of oxygen by photosynthetic bacteria caused oxygen concentrations in the atmosphere to increase. Rising oxygen concentrations made possible the evolution of single-celled eukaryotes about 2.1 billion years ago. Multicellular eukaryotes followed about 650 mya.

- Life in the oceans changed dramatically during the **Cambrian explosion**, when large predators and well-defended herbivores suddenly appear in the fossil record. The Cambrian explosion is an example of **adaptive radiation**, in which a group of organisms

take on new ecological roles and form new species and higher taxonomic groups.

- The land was first colonized by plants (about 480 mya), fungi (about 460 mya), and invertebrates (insects about 400 mya), which were followed later by vertebrates (about 365 mya). Each group evolved unique adaptations that enabled life to flourish on land.

- There have been five **mass extinctions** during the history of life on Earth. The extinction of a dominant group of organisms may provide new opportunities for other groups.

# THE QUESTIONS

## The Basics

**1** The first single-celled organisms on Earth probably evolved

(a) 3,700 years ago

(b) 3.7 million years ago

(c) 3.7 billion years ago

(d) We have no idea when the first life on Earth appeared.

**2** The production of _____ by prokaryotes increased its atmospheric concentration and enabled more complex forms of life to evolve.

(a) carbon dioxide

(b) oxygen

(c) nitrogen

(d) all of the above

**3** The term "Cambrian explosion" refers to

(a) an increase in biodiversity.

(b) a mass extinction.

(c) an atmospheric detonation due to increasing gas concentrations.

(d) an oceanic eruption caused by underwater volcanic activity.

**4** Which of the following terms most specifically describes what occurs when a group of organisms expands to take on new ecological roles, forming new species and higher taxonomic groups in the process?

(a) speciation

(b) mass extinction

(c) evolution

(d) adaptive radiation

**5** Match each term with its definition.

| | |
|---|---|
| CLADE | 1. A distinctive feature that originated in two groups' most recent common ancestor. |
| NODE | 2. An ancestor and all its descendants. |
| LINEAGE | 3. The point at which an ancestral group splits into two separate lineages. |

| | |
|---|---|
| EVOLUTIONARY TREE | 4. A diagram showing the evolutionary relationships among a related group of organisms. |
| SHARED DERIVED TRAIT | 5. The line of descent of a group of organisms. |

**6** Select the correct terms:
The domains Archaea and Bacteria are referred to as (**prokaryotes / eukaryotes**). The domain (**Eukarya / Prokarya**) includes four kingdoms. The kingdom (**Plantae / Fungi**) was first to make the transition to land. The kingdom (**Animalia / Plantae**) is most closely related to the kingdom Fungi.

**7** All of the following are considered possible causes of the five mass extinctions *except*

(a) climate change.

(b) change in the composition of atmospheric or marine gases.

(c) comet or asteroid strike.

(d) worldwide thunderstorms.

(e) volcanic eruptions.

## Challenge Yourself

**8** *Archaeopteryx*

(a) belongs to the domain Prokarya.

(b) belongs to the kingdom Protista.

(c) lived during the Precambrian period.

(d) was much larger than modern birds.

(e) is an early example of the evolution of birds.

**9** Identify the domain and kingdom for each of the following organisms.

(a) *Archaeopteryx*

(b) chanterelle mushrooms

(c) palm tree

(d) green algae

(e) you

**10** "King Philip came over for good soup" is a mnemonic to remember each level of the Linnaean hierarchy.

a. List each level of the Linnaean hierarchy, in the order indicated by the mnemonic.

b. What level of the Linnaean hierarchy is missing from the mnemonic?

c. Create your own mnemonic to remember the Linnaean hierarchy.

**11** Place the following evolutionary events in the correct order from earliest to most recent by numbering them from 1 to 5.

_____ a. Cambrian explosion

_____ b. origin of life on Earth

_____ c. plants' transition to land

_____ d. oxygen-rich environment created by bacteria

_____ e. evolution of birds from dinosaurs

**12** Fill in the tree below to show the evolutionary relationships among the domains and kingdoms of life.

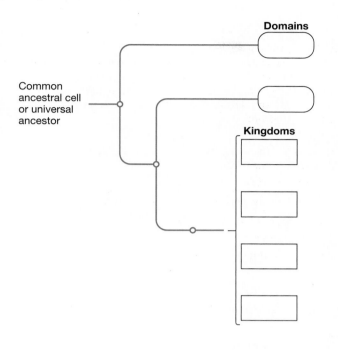

Domains

Common ancestral cell or universal ancestor

Kingdoms

## Try Something New

**13** In each of the following cases, which domain(s) or kingdom(s) might contain organisms with the traits described? (You may need to review Figure 14.9 to answer.)

a. motile, with flagella or cilia

b. All organisms within the domain or kingdom are single-celled.

c. found in extreme environments (for example, high temperature, low oxygen, high salt)

d. multicellular with organ systems

e. photosynthetic

**14** The traditional dinosaur-bird tree (Figure 14.6, top) can be restated as a hypothesis: "We hypothesize that *Archaeopteryx* is an early bird, and that birds split off from the closely related dinosaur groups Troodontidae and Dromaeosauridae." Restate Xu's tree (Figure 14.6, bottom) and Godefroit's tree (Figure 14.14) as hypotheses.

## Leveling Up

**15** *Write Now* **biology: early life on Earth** Read the *New York Times* article "World's Oldest Fossils Found in Greenland" (http://www.nytimes.com/2016/09/01/science/oldest-fossils-on-earth.html). Answer the following questions, writing one paragraph for each.

a. Summarize the findings of the study reported in the *Times* in your own words.

b. If the scientific community reached a consensus that these are *not* fossils, how would this conclusion affect our understanding of early life on Earth? Explain why such a conclusion would not be a failure of the scientific method, but rather a natural result of it.

c. Define the Late Heavy Bombardment (LHB). Describe how our understanding of the LHB has changed, and how this change has influenced our hypotheses about early life on Earth.

**16** **Is it science?** Some people became quite upset when scientists agreed that the dinosaur *Brontosaurus* was in fact *Apatosaurus*, that *Triceratops* was only a juvenile *Torosaurus*, and that Pluto was not a planet. They clung to their *Brontosaurus*, or *Triceratops*, or planet Pluto, dismissing the scientists' arguments by saying, "Scientists are always changing their minds." Explain how scientists' "changing their minds" is not a problem but instead a necessary and useful feature of science, using the science described in this chapter as an example.

**For more, visit digital.wwnorton.com/bionow2 for access to:**

# Navel Gazing

*Scientists discover hundreds of new species of microbes—burrowed deep in our belly buttons.*

## After reading this chapter you should be able to:

- Differentiate between bacteria and archaea, and explain why they are together referred to as prokaryotes.
- Explain why prokaryotes can reproduce more rapidly than eukaryotes and why rapid reproduction can be advantageous.
- Give examples of the environments in which archaeans are found.
- Illustrate through example the breadth of metabolic diversity found in bacteria.
- Explain the function of at least one cellular structure that is unique to prokaryotes.
- Explain how citizen science contributes to scientific research.

CHAPTER

15

BACTERIA
AND ARCHAEA

The gathering crowd jostled for position around the table. The scientists behind the table smiled as they handed out long, thin cotton swabs. Holly Menninger thanked them for the swab, then stared down at her shirt. She had been told what to do: lift your shirt, insert the swab into your belly button, swirl it around, place it into a test tube, seal the top. "Oh, the things we do for science," Holly thought. Then she swirled.

After returning her tube, Menninger—an entomologist by training and, at the time, coordinator of the New York Invasive Species Research Institute at Cornell University—struck up a conversation with the researchers manning the table, including Rob Dunn of North Carolina State University. Dunn, an applied ecologist and writer with an infectious passion for science, had initiated the project to answer a simple question: What lives on our skin? The belly button, as small and strange as it is, was an excellent place to look.

If you pick away the lint, the leftover schmutz in your belly button isn't dirt; it's alive. Our bodies are ecosystems, home to an estimated 39 trillion resident **microbes** (microscopic organisms), primarily bacteria. Considering that the average human body is made up of 30 trillion human cells, that's about a 1:1 ratio of microbial cells to human cells. (The old 10:1 ratio is a miscalculation that has unfortunately become enshrined in popular culture.) The **human microbiome**—the complete collection of microbes that live in and on our cells and bodies—affects human gut health, brains, and even body odor.

Microbes live in our guts, on our eyelashes, and yes, in our belly "holes," as Dunn calls them. In fact, the "belly hole" is an ideal place for a scientific study of resident microbes because it is a protected, moist patch of skin and one of the few areas that individuals don't regularly wash. Dunn knew that his students, if they could determine which microbes were swimming around in individual volunteers' navels, could then dig into the more burning question: *Why* do we each have the microbes we do?

"We know that which microbes you have on your skin influences your risk of infection, how attractive you are to other people, and how attractive you are to mosquitoes," says Dunn. "So this question of 'What determines which microbes are on your skin?' is super intriguing."

Of the three domains of life on Earth, Bacteria was the first to split off from the shared ancestor of the Archaea and Eukarya (**Figure 15.1**). Some fossil evidence places that split at about 3.48 billion years ago, yet a 2016 discovery in Greenland suggests that bacteria existed as early as 3.7 billion years ago, close to a period of time when Earth was being bombarded by asteroids (**Figure 15.2**).

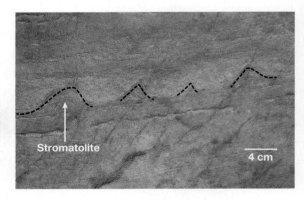

Figure 15.2

**Earth's first life left its mark in Greenland rocks**

In 2016, Australian scientists found stromatolites, the fossilized secretions of bacteria, in rocks exposed by receding glaciers. These rocks formed 3.7 million years ago in a part of the world that is now Greenland.

Figure 15.1

**Two of the three domains of life**

Bacteria and Archaea are two distinct domains. They share many characteristics that are not found in the Eukarya.

Archaea likely split from Eukarya much later, around 2.7 billion years ago. The microbes that make up Bacteria and Archaea display many small but significant differences in their DNA, plasma membrane structure, and metabolism. They also share several important characteristics and so they are traditionally lumped under a common label: *prokaryotes.*

**Prokaryotes** are single-celled organisms, with a single loop of DNA floating free in the cytoplasm of the cell (review Figure 4.8 for a comparison of prokaryotic and eukaryotic cells). Unlike cells found in the Eukarya, prokaryotes do not have membrane-enclosed organelles. Prokaryotic cells are not only simpler than eukaryotic cells; they are smaller—almost exclusively microscopic, and invisible to the naked eye because of their diminutive size. The simple structure and single loop of DNA enable prokaryotes to reproduce at a much more rapid rate than eukaryotes, doubling in number every 10–30 minutes (**Figure 15.3**).

Prokaryotes in both domains are widespread and extremely abundant, and they display an astonishing diversity in metabolism. The vast majority of life on Earth is single-celled and prokaryotic. Scientists estimate that the number of prokaryotes on Earth is about 5,000,000,000, 000,000,000,000,000,000,000 (5 nonillion, or $5 \times 10^{30}$). For example, prokaryotes—not fishes or algae—are the most abundant organisms in the open ocean, where they play a crucial role in the ecology of our biosphere.

The success of prokaryotes is due in part to how quickly a prokaryotic population reproduces, and also to the fact that they can live practically anywhere, including many places where few other forms of life can, such as deep, hot thermal vents or the acidic environment of the human intestine. Humans are saturated in microbes. We are each a zoo.

Menninger was so intrigued by Dunn's project that when a job became available in his lab a few months later, she immediately applied for and got the position. By then, the team had gathered about 60 swabs at local events, including the aforementioned conference, and now they had bigger ambitions: Dunn hired Menninger to spearhead a massive public outreach effort to acquire belly button swabs from all around the country. In **citizen science** projects such as this, the public participates in research by collecting

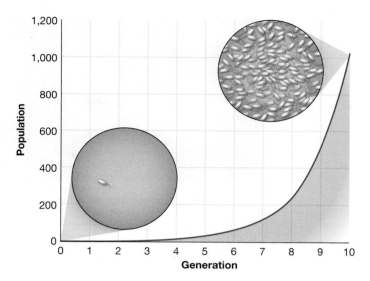

Figure 15.3

**Prokaryotes are capable of extremely rapid population growth**
An individual prokaryote is able to divide in two within 10–30 minutes. Those two can each divide in two in the same amount of time. This means that even a single bacterium or archaean can become a large population in a very short period of time.

**Q1:** If an individual prokaryote divides every 20 minutes, how many individuals will there be after an hour?

**Q2:** If the generation time is 20 minutes, how much time will have gone by when the final generation shown has doubled?

**Q3:** Many bacteria are able to reproduce more quickly in warmer conditions. What does this suggest to you about the importance of refrigerating foods?

and sometimes even analyzing data in cooperation with professional scientists (**Figure 15.4**).

"At the time, we were right at the forefront of an explosion in research, learning about how important the microorganisms that live on and in our bodies are to our health and well-being," says Menninger. "And this was a really great project to get people talking about the skin microbiome."

**ROB DUNN**

Rob Dunn is an applied ecologist at North Carolina State University. His fascination with species around us in our everyday lives has led him to research projects on microbes in our belly buttons, armpits, homes, and more.

Data collection

Citizen scientists from the Republic of the Congo are learning to map the forest with GPS.

Data analysis

A citizen scientist in North America uses an app to document plant budding dates.

Participation as an experimental subject

Sebastian Seung, a professor at Princeton University, works with citizen scientists around the world to map the brain through a video game.

Figure 15.4

### Citizen scientists work alongside professional scientists to increase our knowledge about the natural world

There are many ways you can contribute to science—by collecting data, analyzing data, or even being an experimental subject like the video game players who work with Sebastian Seung. And the advantages are not one-way: citizen scientists learn new things, often have fun, and report feeling a sense of purpose and contribution.

**Q1:** In which of the three ways did the navel microbiome participants contribute?

**Q2:** Which of the advantages listed above do you think the navel microbiome citizen scientists received?

**Q3:** Would you be willing to contribute to the navel microbiome project? Why or why not?

Dutch tradesman Antonie van Leeuwenhoek first opened our eyes to the microbial world back in 1668, when he used simple, handcrafted microscopes to discover bacteria swimming around in pond water. In the intervening 350 years, our knowledge of the microbial world has come a long way.

Today, we know that bacteria and archaeans make up more than two-thirds of the species found on Earth. In 2016, microbiologists at UC Berkeley published a revised evolutionary tree of life that, for the first time, included the vast diversity of bacteria and archaeans lurking in Earth's nooks and crannies. With the data used to generate that tree, they showed that life is clearly dominated by prokaryotes (**Figure 15.5**).

# Merry Microbes

In 2010, as a gag, an undergraduate student in Dunn's lab decided to make a Christmas card with streaks of microbes taken from people in the lab. She asked for belly button swabs from her coworkers, spread each sample on a petri dish, and grew the bacteria into colonies. As she grew bacteria from multiple individuals, it quickly became clear that there was more variety among the microbes growing on people in the lab than had been expected.

"Then it went from being a fun lab project to a serious question," says Dunn. The Christmas gag had raised an important question: Why do one person's microbes differ from another's? From his experience studying ecology, Dunn knew that microbes are critical in our lives. But neither Dunn nor others knew which factors determine which particular skin microbes a person has, or lacks.

So Dunn and his team began to study the locally collected swabs. An early part of the experiment engaged participants by visually depicting each person's microbial menagerie. The team plated each sample onto a petri dish and then took a picture of it for the microbes' owner. Although prokaryotes are single-celled organisms, some form colonies or long chains of identical cells, produced by repeated splitting begun by one original cell. This bacterial pattern of reproduction resulted in bright, dramatic patterns on the plates (**Figure 15.6**).

Under the microscope, the individual microbes were even more diverse and elaborate. Bacteria and Archaea cells are quite variable in shape, ranging from rods to spheres to spirals (**Figure 15.7**). Still, they all have a basic structural plan. Most bacteria and many archaeans have a protective cell wall that surrounds the plasma membrane. Some have an additional wrapping around that cell wall called a **capsule**. The capsule, made of slippery biomolecules, works like an invisibility cloak: it helps disease-causing bacteria evade the immune system that protects organisms like us from foreign invaders.

Dunn's team observed many bacteria whose surface was covered in short, hairlike projections called **pili** (singular "pilus"), a common bacterial

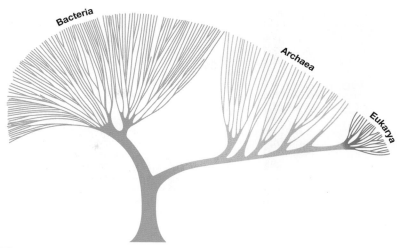

Figure 15.5

**The most recent tree of life**
Researchers sequenced DNA from individuals in 3,083 genera across all three domains. This information was combined by phylum, and the placement of each phylum in the tree was based on its relatedness to other phyla, as measured by DNA similarity.

**Q1:** Where in the figure would you place the first life found on Earth?

**Q2:** Identify where in the figure the Bacteria split off from the ancestor of Archaea and Eukarya.

**Q3:** The figure (and thus the study) demonstrates that Archaea and Eukarya are more closely related to each other than to Bacteria. How is that illustrated?

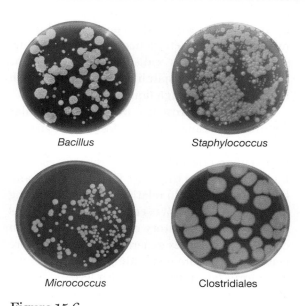

*Bacillus*          *Staphylococcus*

*Micrococcus*          Clostridiales

Figure 15.6

**Belly button bacterial biodiversity**
Each petri dish contains colonies of the bacteria collected from one person's belly button sample.

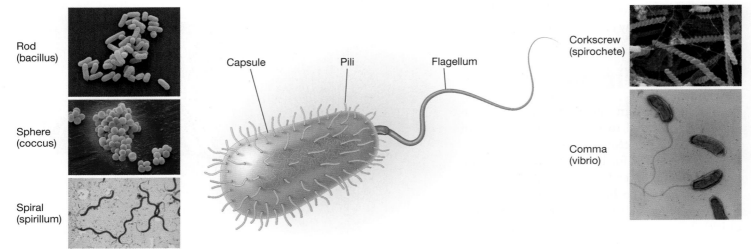

Rod
(bacillus)

Sphere
(coccus)

Spiral
(spirillum)

Capsule    Pili    Flagellum

Corkscrew
(spirochete)

Comma
(vibrio)

## Figure 15.7

## A simple structure, a diversity of forms

Bacteria and Archaea share a simple cell structure relative to eukaryotes, but they display a striking diversity of shapes and may have additional structures that perform special functions.

**Q1:** Which of these shapes do you think *Streptococcus* would take?

**Q2:** From the micrographs here, does it appear that all prokaryotes have a flagellum?

**Q3:** Which one of these shapes is most clearly capable of self-motility? Why?

feature. Bacteria use pili to link together to form bacterial mats, or to attach to surfaces in their environment, such as the cells of the human intestine. Some bacteria have one or more long, whiplike structures called **flagella** (singular "flagellum"), which spin like a propeller to push the bacterium through liquid.

Under the microscope, one obvious cellular feature is missing: prokaryotes do not have a true nucleus. They typically have much less DNA than eukaryotic cells have. Prokaryotes have less DNA because they have far fewer genes and because they contain relatively little non-coding DNA (DNA that is not used to construct proteins but may have regulatory or other functions). In contrast, eukaryotes generally have many more genes and far more non-coding DNA.

Dunn's lab partnered with Noah Fierer, an ecologist at the University of Colorado, to sequence the DNA from each belly button sample to determine the identity of the resident microbes, especially those that don't grow in a laboratory culture. To do so, lab members first had to access that DNA by mechanically crushing

the microbes and chemically dissolving their cell membranes with detergent, similar to how dish soap dissolves grease clinging to a dirty dish.

Next, they ran the sample through a silica strainer, called a column, to capture the DNA while allowing other debris to pass through. Once the DNA was isolated, the team amplified it using the polymerase chain reaction technique (see Figure 9.9) and then sequenced one short section of it, about 250 nucleotides of a well-studied gene called 16S ribosomal RNA (16S rRNA). Although 16S rRNA is found in all bacteria and archaeans, it is highly variable from one species to another, providing an identifying tag, like a fingerprint or barcode, to distinguish one species of microbe from another.

In 2012, the team published results from the initial set of swabs: among 92 belly buttons, the team identified 1,400 species of bacteria. "About 600 or so don't match up in obvious ways with known species, which is to say either they are new to science or we don't know them well enough," Dunn said in a media interview after the paper was published.

Archaea found in mineral hot springs are able to withstand extreme acidity *and* high temperatures.

Hydrogen-eating, methane-producing archaeans have been found in deep-sea thermal vents.

A new archaean species was recently found in an abandoned copper mine. Others have been found living in acidic drainage from mines.

Figure 15.8

## Archaeans are everywhere

Archaeans, and bacteria to a lesser extent, thrive in environments that humans are incapable of surviving in.

**Q1:** Which shape in Figure 15.7 corresponds to the archaeans from deep-sea thermal vents?

**Q2:** Why are many archaeans referred to as "extremophiles"?

**Q3:** Is there anywhere you think archaeans could *not* survive? Justify your answer.

They had discovered new species of bacteria playing house in our navels. Still, Dunn was most interested in finding patterns hiding in the data. The first pattern they found was that the microbes in belly buttons are somewhat predictable. Six species showed up in about 80 percent of people, and when present, those species tended to be most widespread. "If I go to a cocktail party, I can tell you which species will be most abundant in the room," says Menninger with a laugh. That distribution pattern reminded Dunn of rainforest ecology: in any given forest, the types of plants might vary, but an ecologist can depend on a certain few tree types dominating the landscape.

Likewise, infrequent species tended to be present only in small quantities in belly buttons. Some were familiar: *Bacillus subtilis*, known to cause foot odor, also lives in some belly buttons (where it likely causes the same smell). Others were quite rare: one participant hosted a species previously found only in the soil of Japan, yet the man had never been to Japan. Dunn found a species on himself known solely for eating pesticides. Why it was on his skin remains a mystery.

Amazingly, the team found more than just bacteria. On one particularly fragrant individual, who said he had not washed in years, they detected two species of archaeans. Although some bacteria thrive in unusual environments, the domain Archaea is well known for the extreme lifestyles of its members (**Figure 15.8**). Some are extreme **thermophiles** that live in geysers, hot springs, and hydrothermal vents—cracks in the seafloor that spew boiling water. The cells of most organisms cannot function at such high temperatures, but thermophiles have come up with evolutionary innovations that enable them to succeed where others cannot. Other archaeans, classified as **halophiles**, thrive in very salty, high-sodium environments where nothing else can live.

Yet archaeans have also been found in less exotic locations. In another citizen science project from Dunn and Fierer, students analyzed the microbial DNA swabs from home surfaces—kitchen counters, door frames, television screens, and more. Once again, they found several species of archaeans. "They are actually a lot more common than people tend to realize," says Menninger. "We're starting to have the molecular tools that allow us to find them." Dunn thinks the discovery is a bit ironic: "We spent so much time finding archaeans in the first place, in hard-to-reach, faraway places, and somebody could have just had some introspection with their own belly hole."

Yet back in the belly button, Dunn was unable to answer his main question about the bacteria: Why do individuals have the bacteria they do? The team looked at gender, ethnicity, how often participants washed their belly buttons, age, and more, yet they were unable to attribute any of the variation to biological or lifestyle factors.

So, as scientists tend to do, Dunn went looking for more samples. When Menninger joined the team, she initiated a program to have students and volunteers from around the country send in belly button swabs. Thanks to Menninger's outreach efforts, volunteers nationwide swabbed, swirled, and sealed. Soon the lab had over 600 samples to work with. That's a lot of microbes.

# Talk, but No Sex

Like the plants and animals in a rainforest, microbes in an ecosystem communicate with each other, reproduce, and more. Researchers used to think prokaryotes were the ultimate loners—self-sufficient and maintaining strict single-celled lifestyles. Then they discovered that bacteria physically interact with each other in many ways.

Some bacteria (and Archaea) have a unique system of cell-to-cell communication called **quorum sensing** that enables them to sense and respond to other bacteria in the area in accordance with the density of the population. Disease-causing bacteria, for instance, may begin to multiply rapidly upon sensing that their numbers are high enough to overwhelm the host organism's immune system. Other bacteria coordinate their behavior by forming tough aggregates called *biofilms* that are made up of the same or different species (**Figure 15.9**).

It's strange to think about, but all these activities are happening on our skin: "Everything you can imagine life doing, happens in you," says Dunn, including reproduction. Prokaryotes

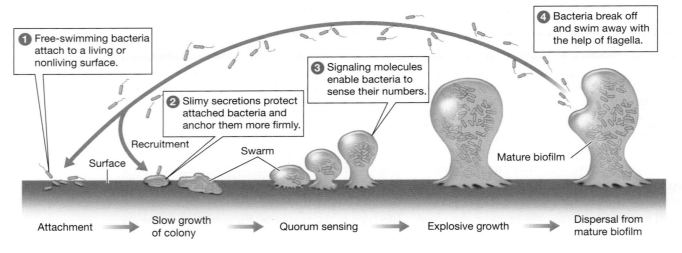

① Free-swimming bacteria attach to a living or nonliving surface.

② Slimy secretions protect attached bacteria and anchor them more firmly.

③ Signaling molecules enable bacteria to sense their numbers.

④ Bacteria break off and swim away with the help of flagella.

Surface

Recruitment

Swarm

Mature biofilm

| Attachment | Slow growth of colony | Quorum sensing | Explosive growth | Dispersal from mature biofilm |

Figure 15.9

**Quorum sensing enables coordinated behavior in bacteria**
With quorum sensing, a bacterial population can increase its virulence, reproductive rate, or antibiotic resistance under the appropriate conditions. Biofilms are produced via quorum sensing to protect bacteria from environmental hazards.

Q1:   How do individual bacteria know that they have a "quorum"?

Q2:   There is a well-known biofilm found in your mouth. What is it?

Q3:   Under what conditions might bacteria want to coordinate (via quorum sensing) to increase their reproductive rate?

typically reproduce by splitting in two in a process called **binary fission**, a form of asexual reproduction (see Figure 6.3). The DNA in the parent cell is copied before fission, and one copy is transferred to each of the resulting daughter cells. The genetic information in the daughter cells is virtually identical to that of the parent cell, as is invariably the case in asexual reproduction.

Although sexual reproduction has not been seen in prokaryotes, they are adept DNA pickpockets: microbes can capture bits of DNA from their environment or other bacteria and incorporate them into their own genetic material. The transfer of genetic material between microbes is known as *horizontal gene transfer* and it involves plasmids, loops of extra DNA in the cytoplasm of prokaryotes (see Figure 12.12). (Horizontal gene transfer has been seen in some eukaryotes as well. Bdelloid rotifers, for example, a type of microscopic freshwater animal, stole about 8 percent of the DNA in their genome from bacteria.) A bacterium can directly trade DNA with another bacterium through a process known as *bacterial conjugation*. Alternatively, when a bacterium dies, the cell may burst open and another bacterium may simply take up the released DNA.

Some types of bacteria—but no archaeans—can undergo **sporulation**, the formation of thick-walled dormant structures called spores. Spores are the bomb shelters of microbes: they can survive boiling and freezing, thereby allowing the microbes to hang out for a long period of time until the conditions are again favorable to reproduce. In many cases, even antibiotics can't kill bacterial spores.

# All Hands on Deck

Dunn, Menninger, and the team analyzed the additional samples. Still, they were unable to explain the differences among belly button microbes. They could not identify any factors affecting the number and types of species a person might host.

So they decided to take a step that more and more scientists are taking—they made all their data freely available online for other researchers to use (with all identifying information about the participants removed). "We asked other researchers to let us know what their insights are, because we're all in this together," says Menninger. "We need all hands on deck."

The gambit worked. Mathematician Sharon Berwick at the University of Maryland took a new approach to the data. Instead of focusing on characteristics of the participants, such as gender or location, she looked at the characteristics of the microbes and found that people tend to be dominated by one of two types of belly button bacteria: *aerobic* or *anaerobic*. **Aerobes** are prokaryotes that need oxygen gas to survive. *Micrococcus* species, which Dunn found in the navel, are aerobes; they need oxygen. Therefore, species of *Micrococcus* are unlikely to do well deep inside a belly button, says Dunn, but they appear to thrive on the surface.

Another type of navel resident, species of *Clostridia*, do not use oxygen. They are **anaerobes**, prokaryotes that survive without oxygen. In fact, some anaerobes may actually be poisoned by oxygen. Among the anaerobic archaeans are several species of **methanogens**, which feed on hydrogen and produce methane gas as a by-product of their metabolism. The ability to exist in both oxygen-rich and oxygen-free environments is another reason why prokaryotes can be found in most habitats.

Some prokaryotes can even switch between the two. One of the most common bacteria Dunn's team found in belly buttons was *Staphylococcus*. Although it is best known as a pathogen, staph on your skin is typically a good guy. On the skin, it is beneficial for your health, fighting off other pathogens that want to crowd in. Some species of staph, such as *Staphylococcus epidermidis*, are typically aerobes, but when oxygen is in short supply, they switch over to a special type of anaerobic metabolism known as **fermentation**, involving the breakdown of sugars (yes, the same process that is used to ferment beer and wine; to review fermentation, see Figure 5.12). In other words, as Dunn wrote in one description of *S. epidermidis*, "Right now you might be making a teeny tiny bit of navel wine."

Like staph, *Bacillus subtilis*, a rod-shaped, spore-forming bacterium, can grow in both aerobic and anaerobic conditions. *B. subtilis* is

**HOLLY MENNINGER**

Holly Menninger is the director of public science at North Carolina State University, where she works to engage the public in citizen science and science education.

one of the warrior clans in our belly buttons and other areas of the skin, producing antibiotics that kill off other bacteria and even foot fungi, says Dunn. Prokaryotes are either consumers, called **heterotrophs**, or producers, called **autotrophs** (**Figure 15.10**). Autotrophs make food on their own, but heterotrophs, like *B. subtilis*, obtain energy by taking it from other sources. Specifically, *B. subtilis* is a **chemoheterotroph**, an organism that consumes organic molecules to get energy (in the form of chemical bonds) and carbon (in the form of carbon-containing molecules). Some species of blue-green algae are **photoheterotrophs**, which acquire carbon from organic sources but their energy from sunlight.

Some autotrophs, called **photoautotrophs**, absorb the energy of sunlight and take in carbon dioxide to conduct photosynthesis. Others, called **chemoautotrophs**, get their energy from inorganic chemicals in their environment, including iron ore, hydrogen sulfide, and ammonia, instead of from sunlight.

In fact, the very first photoautotrophs on the planet were prokaryotes. One type of aquatic bacteria, the cyanobacteria, is believed to be responsible for changing Earth's chemistry by evolving photosynthesis and producing oxygen gas as a by-product. Oxygen gas accumulated in the air and water, and the levels rose from next to nothing to almost 10 percent about 2.1 billion years ago. At that time in the fossil record, eukaryotes appear, suggesting that the oxygen generated by cyanobacteria may have facilitated the evolution of eukaryotes, especially multicellular forms.

Today, prokaryotes continue to aid eukaryotes. Bacteria directly help plants through a process known as **nitrogen fixation**. Plants need nitrogen in the form of ammonia or nitrate, which they cannot make themselves. Bacteria, however, can take nitrogen gas from the air—our

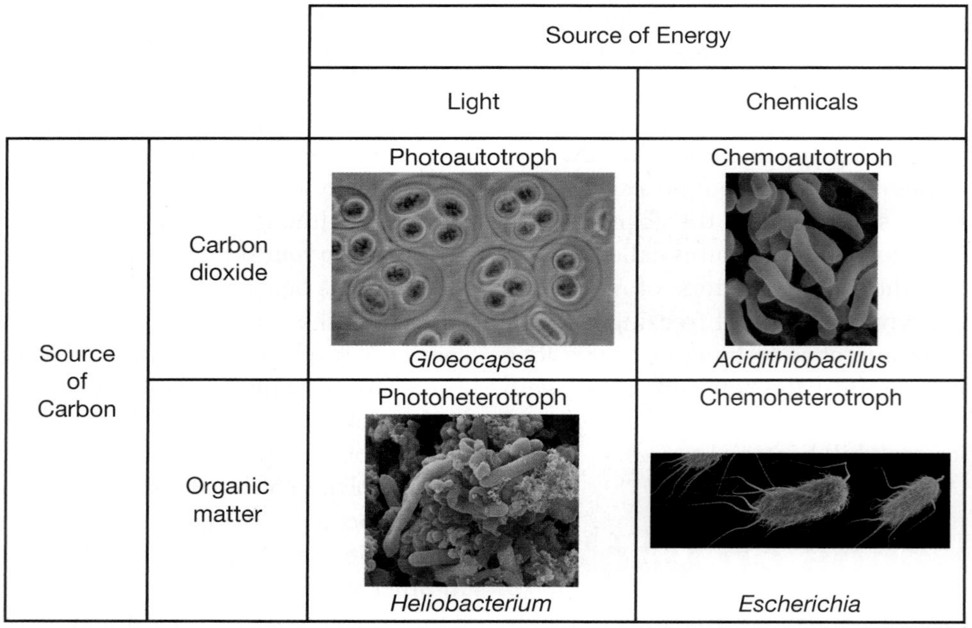

| | | Source of Energy | |
|---|---|---|---|
| | | Light | Chemicals |
| **Source of Carbon** | Carbon dioxide | Photoautotroph <br> *Gloeocapsa* | Chemoautotroph <br> *Acidithiobacillus* |
| | Organic matter | Photoheterotroph <br> *Heliobacterium* | Chemoheterotroph <br> *Escherichia* |

Figure 15.10

**How prokaryotes feed themselves**

Prokaryotes have many more ways to feed themselves than do eukaryotes. They are categorized by their energy source, and by the source of carbon they use.

**Q1:** What source of energy would you expect a cave-dwelling prokaryote to use?

**Q2:** In which of these categories would you place the bacteria responsible for nitrogen fixing? Why?

**Q3:** In which of these categories do decomposers belong? Explain your reasoning.

atmosphere is 78 percent nitrogen—and convert it to ammonia, making it available for plants.

Autotrophs aren't the only microbes that enable life on Earth. Many heterotrophic bacteria and archaeans are decomposers, consumers that extract nutrients from the remains of dead organisms and from waste products such as urine and feces. Decomposers play a crucial role in **nutrient cycling**: by breaking down dead organisms or waste products, decomposers release the chemical elements locked in the biological material and return them to the environment. Those released elements, such as potassium or nitrogen or phosphorus, are used then by autotrophs and eventually by heterotrophs as well.

## Healthy Balance

Since van Leeuwenhoek first looked through his microscope, people tended to regard microbes with suspicion and fear (**Figure 15.11**). We vilified

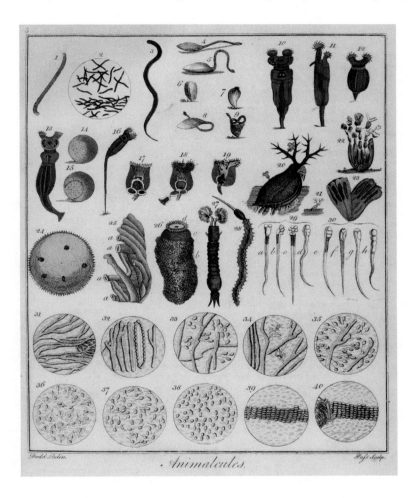

Figure 15.11

**The first illustrations of microbes**
Antonie van Leeuwenhoek's descriptions and illustrations of microscopic life led to his being considered the "father of microbiology."

**Q1:** From the prokaryotic structures shown in Figure 15.7, what shape would you assign to drawing number 8 in van Leeuwenhoek's illustration?

**Q2:** Which of the large prokaryote drawings has the coccus shape?

**Q3:** Do you think all of these "animalcules" drawn by van Leeuwenhoek are prokaryotes? Why or why not?

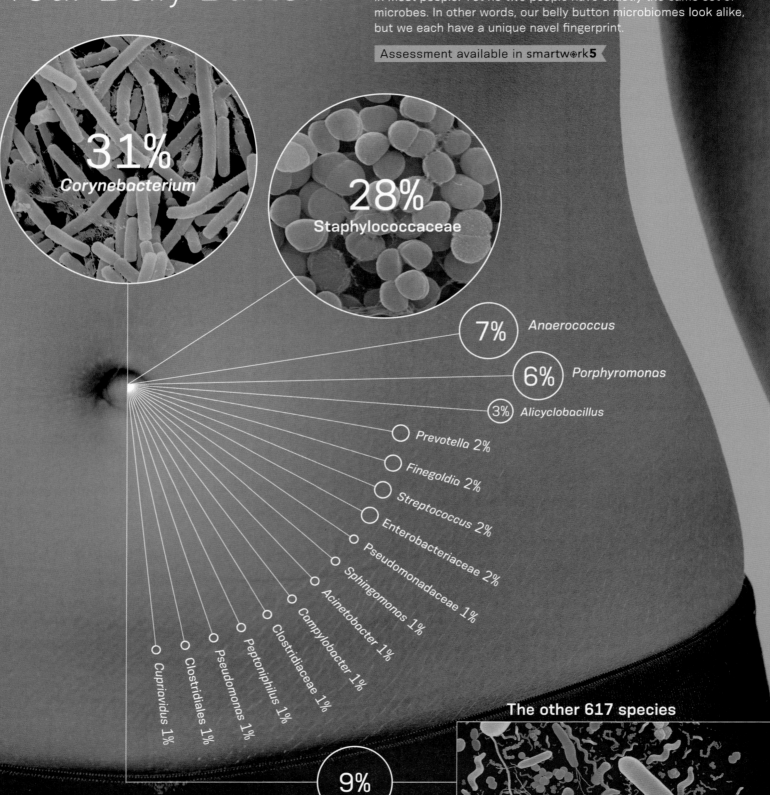

# The Bugs in Your Belly Button

There's no way around it: Your belly button is full of bugs. According to research from Rob Dunn's lab at North Carolina State University, the most abundant species in a belly button—those with the biggest populations, such as *Corynebacterium* and *Staphylococcus*—are found in most people. Yet no two people have exactly the same set of microbes. In other words, our belly button microbiomes look alike, but we each have a unique navel fingerprint.

Assessment available in smartwork**5**

**31%** *Corynebacterium*

**28%** Staphylococcaceae

7% *Anaerococcus*

6% *Porphyromonas*

3% *Alicyclobacillus*

*Prevotella* 2%

*Finegoldia* 2%

*Streptococcus* 2%

Enterobacteriaceae 2%

Pseudomonadaceae 1%

*Sphingomonas* 1%

*Acinetobacter* 1%

*Campylobacter* 1%

Clostridiaceae 1%

*Peptoniphilus* 1%

*Pseudomonas* 1%

Clostridiales 1%

*Cupriavidus* 1%

The other 617 species

9%

microbes as the bad guys that our bodies were constantly at war with. But now we recognize that our ecosystems wouldn't function without microbes: we would have no oxygen to breathe, no plants to eat. And microbes in our gut and on our skin, when in healthy balance, promote human health.

We emphasize a healthy balance because it is possible for communities in the human microbiome to shift out of balance—a phenomenon called *dysbiosis*, which may cause illness. Alternatively, pathogenic bacteria can make us sick as well. Although the great majority of bacteria are harmless and many are actually beneficial to humans, some cause mild to deadly disease. Interestingly, archaeans are not known to be pathogens of any organism.

The biodiversity of microbes on our skin actually helps keep pathogens away. When a dangerous microbe lands on your skin, even before it meets the immune system it meets other microbes. If the skin has a diversity of microbes, odds are "one of them has the ability to kick the butt" of the pathogen, says Dunn.

After the overwhelming response to the belly button project, the lab kicked off a new citizen science spectacular: "Armpit-pa-looza" (**Figure 15.12**). As with the belly button, the team began collecting swabs of microbes living in people's armpits. In a preliminary study, the team found that the use of antiperspirants or deodorants dramatically inhibits bacterial growth, affecting the composition of the microbial populations on the skin more strongly than any other factor. "It now looks

Figure 15.12

**Sampling armpits for science**
Citizen scientists volunteer to have samples taken of the prokaryotes living under their arms.

like one of the biggest things affecting skin microbes in general is the use of antiperspirant or anti-odor products," says Dunn.

Now the team is taking a look at microbes on the skin of dogs. Skin wounds on pet pooches actually heal about four times faster than human skin wounds do, says Dunn, and he suspects that skin microbes are involved. "We're starting to figure out if we can predict wound healing rates as a function of which microbes are there to start with," says Dunn. "I'm pretty excited about that."

# REVIEWING THE SCIENCE

- The non-Eukarya, or **prokaryotes**, fall into two domains: Bacteria and Archaea.

- All prokaryotes are **microbes**—microscopic, single-celled organisms—but the Bacteria and Archaea differ in significant ways, such as in their DNA, plasma membrane structure, and metabolism.

- Prokaryotes can reproduce extremely rapidly and are the most numerous life-forms on Earth. They also have the most widespread distribution.

- Many prokaryotes, particularly bacteria, have specialized structures. Bacteria with a **capsule** surrounding the cell wall can avoid detection by

organisms' immune systems. Short, hairlike **pili** help bacteria attach to surfaces and to each other. Whiplike **flagella** assist in locomotion.

- Some prokaryotes, particularly archaeans, thrive in extreme environments. **Thermophiles**, for example, live in extremely hot places, and **halophiles**, in very salty places.

- Prokaryotes typically reproduce by **binary fission**, and they may acquire DNA from each other or their environment via horizontal gene transfer.

- Some prokaryotes use **quorum sensing** to communicate with each other, and some undergo **sporulation** for protection.

- Prokaryotes exhibit unmatched diversity in methods of getting and using nutrients and energy. **Autotrophs** (producers) make their own food, whereas **heterotrophs** (consumers) obtain food from other sources. **Chemoautotrophs** and **chemoheterotrophs** use inorganic chemicals as their energy source; **photoautotrophs** and **photoheterotrophs** use sunlight.

- Prokaryotes perform key tasks in ecosystems, including photosynthesis, **nitrogen fixation** (providing nitrogen to plants), and **nutrient cycling** (decomposing dead organic matter).

- Although prokaryotes are useful to humanity in many ways, some cause deadly diseases.

contains some species that are pathogens. (**Prokaryotes / Eukaryotes**) have smaller, less complex cells and are able to reproduce at a more rapid rate. Some (**prokaryotes / eukaryotes**) are multicellular.

# THE QUESTIONS

## The Basics

**1** Prokaryotes include

(a) archaeans.

(b) bacteria.

(c) fungi.

(d) both a and b

(e) both b and c

**2** Some prokaryotes can

(a) break down chemicals for energy.

(b) use sunlight for energy.

(c) create their own energy.

(d) harvest energy from other organisms.

(e) all of the above

**3** Prokaryotes are extremely abundant because

(a) they can survive in a narrow range of environments.

(b) they have a single loop of DNA.

(c) they reproduce very rapidly.

(d) they are able to form biofilms.

(e) all of the above

**4** Quorum sensing

(a) is the transfer of plasmid DNA from one bacterium to another.

(b) enables bacteria to form biofilms.

(c) is the formation of thick-walled dormant structures, called spores, under conditions unfavorable for growth.

(d) enables bacteria to switch from cellular respiration to fermentation when they sense that oxygen levels are low.

**5** Select the correct terms:
Of the two types of prokaryotes, (**Archaea / Bacteria**) are more closely related to eukaryotes. The domain (**Archaea / Bacteria**)

## Challenge Yourself

**6** Which of the following can prokaryotes *not* do?

(a) communicate with each other about environmental conditions

(b) reproduce sexually

(c) live in extreme environments, including high salt, high temperature, and high atmospheric pressure

(d) share DNA with other prokaryotes

(e) double in population size two or more times an hour

**7** Which of the following descriptions is true *only* of prokaryotes?

(a) They reproduce through binary fission.

(b) Their genetic material is DNA.

(c) They are unicellular.

(d) They contain organelles within the cell.

(e) none of the above

**8** Place the following steps of the belly button research described in this chapter in the correct order by numbering them from 1 to 5.

_____ a. An observation was made that belly button bacteria seemed to differ across individuals.

_____ b. The DNA was sequenced.

_____ c. DNA was isolated from the samples.

_____ d. Volunteers had their navels swabbed to collect bacterial samples.

_____ e. The bacterial samples were grown in petri dishes.

## Try Something New

**9** Which of the following is citizen science?

(a) reporting when the first hummingbirds arrive at your feeders in the spring

(b) completing a Facebook quiz on your horoscope

(c) filling out your medical history at the doctor's office

(d) all of the above

(e) none of the above

**10** There are more genera of bacteria and archaeans in the world than of eukaryotes, and more individuals. However, the total mass of prokaryotes is thought to be approximately equal to that of eukaryotes. How can this be?

**11** Why are prokaryotes able to replicate so much more quickly than eukaryotes? (*Hint:* What cellular components must be copied before a cell splits in two?) Why is this difference in replication rate an important part of our vulnerability to bacterial pathogens?

## Leveling Up

**12** **Doing science** The research discussed in this chapter would not have been possible without the many citizen scientists who shared samples of their belly button microbes with the research team. Choose a citizen science project: Perform an Internet search on the term "citizen science project," or find regularly updated lists of projects at Scitable (http://www.scitable.com) and Your Wild Life (http://www.yourwildlife.org). Write a one-paragraph summary of the goals of the project you have chosen. Would you participate in this project? Why or why not? Describe a citizen science project that you *would* want to contribute to.

**13** **Looking at data** The table here is based on the published results of the study discussed in the chapter. It shows the number of bacterial phylotypes (kinds of bacteria) that were found on a particular percentage of people.

a. How many bacterial phylotypes were found on only 10 percent or fewer of the people sampled? How many were found on more than 90 percent of people?

b. How many phylotypes were found on more than half of the people sampled?

c. Describe in one or two sentences the frequency of phylotypes found in the study. Were most phylotypes rare or common?

| Phylotypes | Percentage of Human Samples Where Found |
|---|---|
| 2,188 | 1–10 |
| 97 | 11–20 |
| 31 | 21–30 |
| 12 | 31–40 |
| 17 | 41–50 |
| 11 | 51–60 |
| 4 | 61–70 |
| 2 | 71–80 |
| 2 | 81–90 |
| 4 | 91–100 |

**For more, visit digital.wwnorton.com/bionow2 for access to:**

# The Dirt on Black-Market Plants

*Nighttime poaching and illegal trade threaten to drive rare plants and fungi to extinction.*

## After reading this chapter you should be able to:

- Compare and contrast prokaryotes and eukaryotes.
- Explain what is distinctive about the kingdom Protista.
- Outline the key evolutionary innovations of plants.
- Connect structure and function in how fungi obtain energy from the environment.
- Identify key characteristics and give an example organism for each kingdom covered in this chapter.

CHAPTER

16

PLANTS, FUNGI, AND PROTISTS

Jacob Phelps spent a lot of his time in Thailand trying not to arouse suspicion. Walking among vendor stalls at wildlife markets in Bangkok, the tall, thin graduate student stopped occasionally to help a trader trim dead leaves off his plants, or to chat about the weather. Wherever he walked, plants surrounded him—hanging from the ceiling, stuffed into boxes, piled into mounds.

Phelps was careful about the questions he asked, because he was there to document illegal activity. The market traders had tens of thousands of plants for sale, most of which weren't supposed to be there. But when Phelps cautiously asked whether the plants were wild or rare, the traders were surprisingly relaxed. Concern for plant conservation, and enforcement of trading laws, is so limited that illegal sales occur openly at public markets across Southeast Asia.

Even Thailand's wildlife trade management authority at the time claimed that illegal trade in ornamental flowers was limited, "found in small case [*sic*] in some parties." But Phelps was familiar enough with plants to know that most of what he saw in the Thailand markets were wild, protected plants that were illegal to sell: the charismatic pink blossoms of *Dendrobium* orchids; the dense, fragrant plumes of *Rhynchostylis* flowers; the delicate, wavy petals of the coveted lady slipper orchid, *Paphiopedilum*. Plant enthusiasts prize such species for their beauty, fragrance, and rarity. Coveted wild orchids make up more than 80 percent of the plants traded at these unregulated markets (**Figure 16.1**).

As a graduate student at the National University of Singapore, Phelps received the blessing of his PhD supervisor, plant ecologist Ted Webb, to conduct an extensive survey of the illegal plant trade at the four largest plant markets in Thailand. It was the first time such a survey had been conducted of plant markets anywhere in Southeast Asia. In fact, the illegal trade of plants is often called the "invisible wildlife trade" because it is so rarely discussed or documented, in contrast to the widely publicized illegal trade of animals and animal products such as elephant ivory and rhinoceros horns.

Plants often take second-tier status behind animals—it's easier to get the public to care about a fuzzy baby tiger than a sprouting redwood tree—but their biology is no less wondrous. In Chapter 15 we explored the microscopic worlds of the Bacteria and Archaea domains of life. Here, we begin to meet the Eukarya, consisting of four kingdoms: Protista, Plantae, Fungi, and Animalia. Animalia will be covered in detail in Chapter 17.

The defining feature of the Eukarya is a true nucleus: instead of floating free in the cytoplasm, eukaryotic DNA is enclosed in two concentric layers of cell membranes that form a nuclear envelope. In addition to the nucleus, eukaryotes have a great variety of membrane-enclosed subcellular compartments, many of which are specialized for various tasks—such as sending messages, producing energy, or

Figure 16.1

**A plant market in Thailand**
Jacob Phelps visited plant markets like this one to conduct research on illegal trade in endangered orchid species.

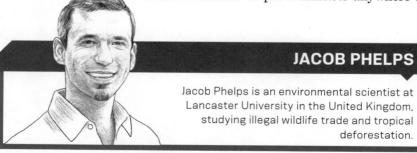

**JACOB PHELPS**

Jacob Phelps is an environmental scientist at Lancaster University in the United Kingdom, studying illegal wildlife trade and tropical deforestation.

cleaning up—so that the cell can function efficiently (review Figure 4.8 for a comparison of prokaryotic and eukaryotic cells, and Figure 4.9 for an overview of eukaryotic organelles). All of these organelles take up space, so the diameter of a eukaryotic cell is, on average, 10 times larger than that of a prokaryotic cell, and the eukaryotic cell volume is a thousandfold greater.

This compartmentalization of the cell interior enables eukaryotic cells to do things that most prokaryotic cells cannot do. For example, some eukaryotes engulf their prey and digest them internally. That's the way many single-celled eukaryotes, such as the blob-like amoebas, eat: they extend gooey cytoplasmic arms to engulf other cells whole, then digest their prey with an elaborate system of internal compartments for digestion, ridding the cell of waste and storing surplus food (**Figure 16.2**).

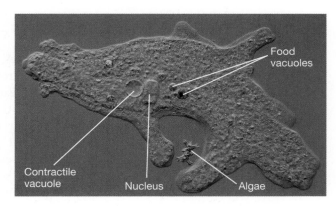

Figure 16.2

**Eukaryotes have a true nucleus and compartmentalized cells**
The more complex cell structure of eukaryotes enables them to perform functions that are impossible for prokaryotes. Intracellular compartments such as food vacuoles enable this amoeba to digest its food, a single-celled alga. The amoeba expels excess water with the help of its contractile vacuole, another type of intracellular organelle.

# Fungi Play Well with Others

Plants probably would not have been as successful on land if they had not entered into a mutualistic relationship with fungi almost immediately on their arrival. Today, the vast majority of plants in the wild have mutualistic fungi, known as **mycorrhizal fungi**, associated with their root systems. (The mutualistic relationships between plants and mycorrhizal fungi are called "mycorrhizae.") Truffles, morels, and chanterelles, all beloved by gourmets, are the reproductive structures of mycorrhizal fungi.

Mycorrhizal fungi form thick, spongy mats of mycelium on and in the roots of their plant hosts and also extend into the surroundings, sometimes permeating several acres of the soil around the root. Mycelia are thinner, more extensively branched, and in closer contact with the soil than even the thinnest branches on a plant root. As a result, a mycelial mat plumbs far more water and mineral nutrients, such as phosphorus and nitrogen, than the plant's root system could absorb on its own. In return for sharing absorbed water and mineral nutrients, the fungus obtains sugars that the plant manufactures through photosynthesis.

Mycorrhizal fungi assist in providing nutrients to orchid seeds, which are tiny and lack stored food. The embryo within a newly sprouted orchid seed could not survive without the mycorrhizal network linking the seedling to mature photosynthesizing plants, from which the seedling draws nourishment until it can photosynthesize on its own.

Fungal mutualisms are not found only with other eukaryotes. A **lichen** is a mutualistic association between

Plants that have no mycorrhizal fungi associated with their roots do not grow as well…

…as those that host the beneficial fungi.

a photosynthetic prokaryote and a fungus. The fungus receives sugars and other carbon compounds from its photosynthetic partner, usually a green alga or a cyanobacterium. In return, the fungus produces lichen acids, a mixture of chemicals that scientists believe may function to protect both the fungus and its partner from being eaten by predators.

Lichens grow very slowly and are often pioneers in barren environments. Lichen acids wear down a rocky surface, facilitating soil formation. Soil particles build up from the slow weathering of rock, and over time, other life-forms, including plants, gain a toehold in the newly made soil.

# Peculiar Protists

Amoebas are single-celled eukaryotes. Most single-celled eukaryotes are **protists**. Protista is a strange kingdom that is defined by exclusion instead of inclusion: although all protists are eukaryotes, they are grouped together in this kingdom simply because they are *not* plants, animals, or fungi (**Figure 16.3**). Researchers have proposed several classification schemes to split the protists into multiple kingdoms; however, no consensus exists on the best way to do so. For now, protists remain divided into two traditional, broad categories: the **protozoans**, which are nonphotosynthetic and motile (capable of moving); and the **algae**, which are photosynthetic and may or may not be motile.

Protists are diverse in size, shape, cellular organization, and mode of nutrition. Most protists are single-celled and microscopic, but some protists have evolved from free-living single cells into multicellular associations, such as slime molds and kelp, large seaweed such as kelp. Certain single-celled protists are bound by nothing more than a flexible plasma membrane, while others are covered in protective sheets, heavy coats, or other types of armor.

Most protists are motile and can swim with the help of one or more flagella, or by waving a carpet of tiny hairs called cilia. Others can crawl on a solid surface with the help of cellular projections called *pseudopodia* (false feet).

Many protists are heterotrophs and eat other organisms. Some of these heterotrophs function as decomposers, breaking down waste material and releasing nutrients into the environment to be taken up by producers and cycled back into the food chain. Other protists are nutritional

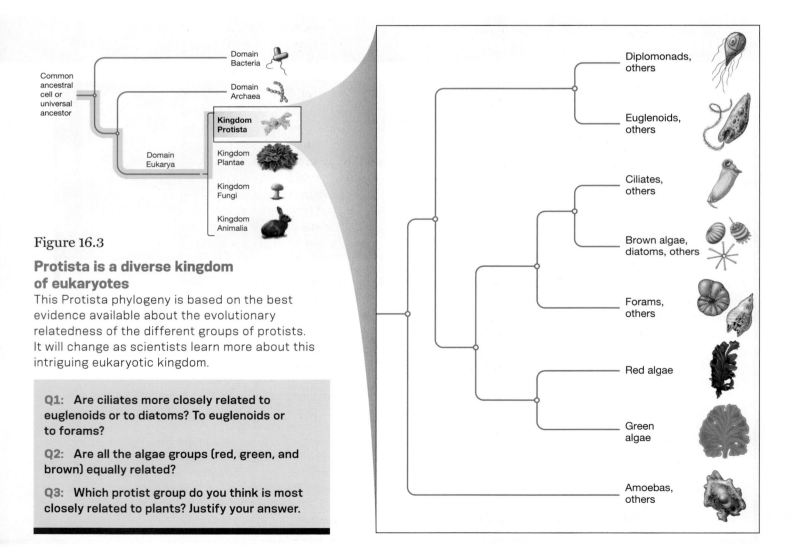

Figure 16.3

**Protista is a diverse kingdom of eukaryotes**

This Protista phylogeny is based on the best evidence available about the evolutionary relatedness of the different groups of protists. It will change as scientists learn more about this intriguing eukaryotic kingdom.

**Q1:** Are ciliates more closely related to euglenoids or to diatoms? To euglenoids or to forams?

**Q2:** Are all the algae groups (red, green, and brown) equally related?

**Q3:** Which protist group do you think is most closely related to plants? Justify your answer.

opportunists, or *mixotrophs*, organisms that use energy and carbon from a variety of sources to fuel their growth and reproduction.

Although most protists are harmless, many of the best-known protists are pathogens (disease-causing agents), such as *Plasmodium*, which is transmitted by mosquitoes and causes malaria, and *Toxoplasma gondii*, passed to humans through poorly cooked food or cat feces. Toxoplasmosis typically causes mild flu-like symptoms, but can be passed from mother to child in utero and cause more severe symptoms.

## Two Cells Are Better Than One

Multicellular forms evolved several times among different lineages of the eukaryotes. A multicellular organism is a well-integrated assemblage of genetically identical cells, in which different groups of cells perform distinctly specialized functions. This functional compartmentalization makes it easier for multicellular eukaryotes to sense and respond to the external environment, through the use of complex structures such as leaves, fruit, eyes and wings.

Multicellularity also enables an individual organism to grow large, which can be advantageous for evading potential predators. A bigger individual can also gather resources from its environment more effectively than can a smaller individual. Having more resources, such as light or food, may translate into producing more surviving offspring, the ultimate measure of biological success.

To make those offspring, eukaryotes reproduce via either asexual or sexual reproduction. Asexual reproduction is common among species in this domain and generates genetically identical offspring. Protists, for example, can split into two in a process similar to binary fission in prokaryotes. Many plants reproduce asexually by fragmenting into pieces, each piece developing into a new individual.

But it is through the second type of reproduction that eukaryotes have made an indelible mark on our planet. By combining genetic information from two parents, sexual reproduction produces offspring that are genetically different from each other and from both parents. Although their life cycle is distinctly different from that of animals, plants similarly produce embryos: the fusion of egg and sperm produces a single cell, called a zygote, which then divides to produce a multicellular structure called an embryo. Sexual reproduction is one means by which natural populations become genetically diverse; it is why there are an estimated 400,000 species of plants on Earth.

Through these unique traits—a true nucleus, cellular compartmentalization, multicellularity, and sexual reproduction—eukaryotes have evolved into amazingly diverse and dynamic species. Of these species, Phelps was particularly concerned about orchids. In addition to spending time in Bangkok, the budding ecologist devoted months traveling along Thailand's borders with Laos and Myanmar, wandering hours upon hours through wildlife markets. Over time, he built local networks that allowed him to interview more than 150 plant harvesters and intermediaries, including market traders, online traders, nursery owners, and more.

Orchids are, on paper, one of the most heavily protected families, representing almost 75 percent of all species, plant or animal, for which international trade is regulated. Artificially propagated orchids can be traded legally, but wild orchid species in Southeast Asia are protected, says Phelps: "None should be collected from the wild, unless you have specialized permissions." Yet, as he came to discover, there is no enforcement of those protections in Thailand or across the region. Orchids and other ornamental and medicinal plants were not only being traded, but also poached from nearby countries, threatening the very existence of certain species and the overall biodiversity of the region.

## Green-Fingered Thieves

Under the dark of night, the thieves worked swiftly. Armed with shovels, they snuck past the closed gift shop and made a beeline for their quarry. They knew exactly what to take.

The next morning, the staff at the Quail Botanical Gardens in Encinitas, California, discovered the theft. In a frantic effort, they spread details of the heist to local newspapers, television, and online community message boards. Their quick action paid off; the loot became too hot. Within days, an anonymous tip led authorities to a rural road where the thieves

Figure 16.4

**Cycads: ancient, rare, and hard to move**
Several workers struggle to move a cycad. In spite of their size, many cycads have been stolen from public and private gardens, and then sold on the black market.

The thieves had attempted to steal a group of rare African cycads, an ancient type of plant that once lived alongside dinosaurs in the Jurassic period (**Figure 16.4**). As with the Thai orchids, horticulturalists around the world highly prize cycads as ornamental plants. Today, a rare, mature cycad can fetch $20,000 or more on the international black market. That's right— enough to fund a year of college. There are about 300 species of cycads, most threatened with extinction.

Ironically, 4 of the 21 cycads stolen from the Quail Botanical Gardens (now the San Diego Botanic Garden) were part of a rescue program; the plants had been illegally brought into the United States, where they were seized upon import by authorities, and the botanical garden had taken over their care. Some of the plants survived the ordeal and were replanted. But they weren't the only cycads stolen that year; a nursery in San Diego lost almost 40 large cycads to thieves, and many homes in the Long Beach area had plants swiped right out of their front yards.

had dumped their haul in a sorry-looking pile: 21 stocky plants, specimens with thick trunks covered in woody scales and a crown of long, green palm-like leaves, like a pineapple.

**Plants** are multicellular autotrophs that are mostly terrestrial (**Figure 16.5**). Like the green

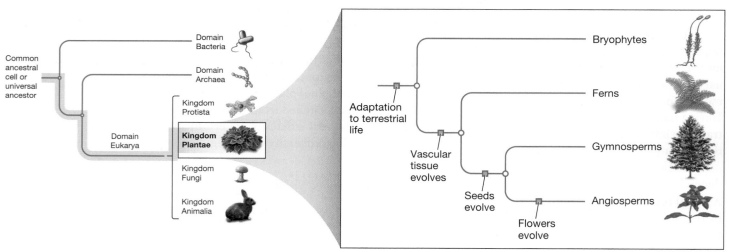

Figure 16.5

**Plants are eukaryotic autotrophs**
Plants are photosynthetic, making their own food from sunlight, and are almost exclusively found on land. They are at the base of almost all terrestrial food webs.

**Q1:** What evolutionary innovation separates all land plants from their aquatic ancestors?

**Q2:** How do ferns differ from bryophytes? Do they share this difference with other plant groups? (You will need to read ahead to answer this question.)

**Q3:** What group(s) might a plant with seeds belong to? What about a plant with flowers?

algae from which they evolved, plants use chloroplasts in order to photosynthesize. Most photosynthesis in plants takes place in their leaves, which typically have a broad, flat surface—a design that maximizes light interception. Because plants are producers, they form the basis of essentially all food webs on land. They may not be as cute, fuzzy, or exciting as animals, but animals wouldn't exist without them: nearly all organisms on land ultimately depend on plants for food, either directly by eating plants or indirectly by eating other organisms that eat plants.

In addition to being food, plants are valuable for many other reasons. Many organisms live on or in plants, or in soils largely made up of decomposed plants. By soaking up rainwater in their roots and other tissues, plants prevent runoff and erosion that can contaminate streams. Plants (and algae) also recycle carbon dioxide and produce the oxygen we breathe.

Many of these traits are adaptations to the challenges of living on land (**Figure 16.6**). The greatest of those challenges is how to obtain and conserve water. The first plants, the ancestors of

**Algae (aquatic)**

- Photosynthesis occurs and $CO_2$ is absorbed throughout the organism.

- Water and minerals are absorbed by the whole organism.

- Water supports the whole organism.

- Dehydration is not an issue in an aquatic environment.

**Plants (terrestrial)**

- Photosynthesis occurs and $CO_2$ is absorbed primarily in the leaves. Air enters leaf cells through minute openings in the cuticle. Many plants have more elaborate pores (*stomata*) that open and close to regulate the flow of gases into and out of the leaf.

- Roots absorb water and minerals from the soil.

- Roots anchor and support the plant within the ground. Lignin and vascular tissues help support the plant aboveground.

- The waxy cuticle holds in moisture to keep plant tissues from drying out, even when exposed to sunlight and air throughout the day.

Figure 16.6

**Moving to land brought unique challenges to plants**
Terrestrial plants evolved in response to challenges that their aquatic ancestors had not faced, which resulted in adaptations designed to slow dehydration, provide support and anchoring, and enable photosynthesis and nutrient uptake.

Q1: In what ways are terrestrial plants and their aquatic ancestors the same? Give at least two similarities.

Q2: In what ways do terrestrial plants and their aquatic ancestors differ? Give at least two differences.

Q3: Would you predict that aquatic plants (which have secondarily evolved to live in water) would be more like plants in a rainforest or more like desert plants? Explain your reasoning.

present-day **bryophytes** (liverworts and mosses), grew as ground-hugging carpets of greenery. These simple plants had relatively thin bodies, often just a few cells thick, and absorbed water through a wicking action. However, absorption by direct contact cannot transport fluids effectively in a plant that rises a foot or more aboveground, so more complex plants, including cycads, have a network of tissues, called the **vascular system**, that is made up of tubelike structures specialized for transporting fluids. Roots, found in most plants, also have an extensive vascular system. The first land plant with a vascular system was the ancestor of present-day **ferns**.

Like green algae from the protist kingdom, plant cells have strong but flexible cell walls composed of the substance known as **cellulose**, which gives structural strength to the cells, including those of the low-growing bryophytes. Beginning with the ferns, plants evolved yet another type of strengthening material, **lignin**, that allowed them to grow even taller. One of the strongest materials in nature, lignin links together cellulose fibers in the cell wall to create a rigid network. Lignin is the reason that some cycads in Japan can grow over 20 feet (7 meters) tall, though it takes them about 50–100 years to achieve such heights.

**Figure 16.7**

**A flower attracts many pollinators**

Pollination in the angiosperms is more efficient than in the wind-pollinated gymnosperms. Flowers attract animal pollinators that carry pollen to and from individual plants of the same species.

One reason rare cycads are so prized is that it can take a long time to raise and grow them. Cycads are **gymnosperms**, the first plants to evolve **pollen**, a microscopic structure containing sperm cells that can be lofted into the air in massive quantities. The evolution of pollen freed gymnosperms from a dependence on water for fertilization: instead of having to transport sperm cells through water, they could do it through air.

Gymnosperms were also the first plants to evolve the **seed**, which consists of the plant embryo and a supply of stored food, all encased in a protective covering. The embryo uses this stored food to grow until it is able to make its own food via photosynthesis. Seeds also provide embryos with protection from drying and from attack by predators. Unfortunately, they don't provide protection from poachers. Today, the San Diego Botanic Garden keeps not only its prized cycads, but also their valuable seeds, in a greenhouse under lock and key.

# Searching for Flowers

Botanical gardens also go to extreme measures to keep flowering plants, or **angiosperms**, safe for long-term conservation. Orchids, as we have seen, are the most common illegally traded angiosperms. Others include *Galanthus*, or "snowdrop" plants, with delicate white flowers. In 2012, a single bulb of a rare variety of snowdrop fetched $945 at auction.

Compared to gymnosperms, angiosperms are a relatively recent development in the history of life. With about 250,000 species today, angiosperms are the most dominant and diverse group of plants on our planet. Nearly all agricultural crops are flowering plants. These plants also provide humans with materials such as cotton and pharmaceuticals.

Angiosperms' key evolutionary innovation is the flower, a structure that evolved through modification of the cone-like reproductive organs of gymnosperms. **Flowers** are structures that enhance sexual reproduction in angiosperms by bringing male gametes (sperm cells) to the female gametes (egg cells) in highly efficient ways, by attracting animal pollinators through scent, shape, and color (**Figure 16.7**).

Gymnosperms produce "naked" seeds that sit bare, unwrapped in any additional layers, on the modified leaf (the scale in a pine cone). In angiosperms, the modified leaf evolved into the ovary wall, which consists of tissue layers that enclose and protect the egg-bearing structures, or **ovules**. After fertilization, the ovules develop into seeds, and the ovary wall that enclosed them becomes the fruit wall.

At the Thailand plant markets, Phelps had trouble identifying orchid species until they flowered, so he went back again and again—four times per year to four different markets—looking for newly opened flowers, listing species he recognized, taking photos of others, and even occasionally asking for a flower off a plant he did not recognize, which he quickly stored in a vial of alcohol to take back for identification. In the end, Phelps gathered evidence of 348 orchid species in 93 genera, representing 13–22 percent of the area's known orchid flora, and tens of thousands of individual plants, including several new species.

Phelps's results were shockingly different from those published in preexisting government reports on plant trades among Thailand, Laos, and Myanmar. The Convention on International Trade in Endangered Species of Wild Fauna and Flora, or CITES, is an international treaty of over 175 countries that monitors and regulates international trade of plants and animals. Member countries are required to produce permits for wildlife trade of species protected under the agreement, including all wild orchids. These permits are used to guarantee that plants are legally harvested in a sustainable way that does not endanger either the species or the environment. "CITES is about conservation and sustainable use," says Anne St. John, a biologist with the U.S. Fish and Wildlife Service's Division of Management Authority, which implements CITES in the United States. "The goal is to ensure that these species are around for our grandchildren. That includes not only tigers and elephants, but also bigleaf mahogany and Brazilian rosewood."

And the orchids of Southeast Asia. Over nine years, Laos reported permits for the export of just 20 wild-collected orchids into Thailand; Myanmar reported none. Yet during just one day with a single market trader at the border between Laos and Thailand, Phelps documented that the woman sold at least 168 plants of eight different genera. In just one day, she sold eight times more plants than the government reported as sold over 9 years. "It's totally anecdotal, but incredibly illustrative of the problem," says Phelps. "This trade is completely unacknowledged. It's an open secret."

Other countries are working hard to crack down on the plant black market. American ginseng, a short leafy plant with a tan, gnarled root commonly used in Chinese herbal medicine, is the largest CITES-regulated plant export of the United States (**Figure 16.8**). A pound of quality, dried ginseng can sell for up to $900, so some people try to bypass CITES permits, harvesting plants that are too young (legally

Figure 16.8

**Ginseng plants carpet a forest floor**
After pollination, the ginseng flowers develop a bright-red seed head that helps ginseng hunters (inset) find it in the filtered light of the forest.

**Q1:** What feature(s) of the ginseng plant tell you that it is not a bryophyte?

**Q2:** What feature(s) of the ginseng plant tell you that it is not a fern or gymnosperm?

**Q3:** Because of the CITES classification of ginseng, you are not allowed to sell plants younger than 5 years even if they grew on your own land. Do you agree with that law? Why or why not?

traded roots must be 5 years or older), out of season, or poached from federal lands.

"As you would expect, species that are high value and in high demand are the subject of illegal trade," says St. John. The illegal trade of ginseng has even spawned a reality show on the History television channel: *Appalachian Outlaws*. To combat the criminal activity, some states have taken to spraying ginseng plants on state and federal lands with paint, so if a marked plant shows up in a batch of ginseng to be exported to China, authorities will know where it came from. Those caught selling protected plants face fines and jail time, and violating CITES is a federal crime.

# Truffle Trouble

The illegal wildlife trade is a problem not just for animals and plants, but also for **fungi**, none of which have been listed as CITES-protected species so far. Fungi are absorptive heterotrophs:

they digest organic material outside the body and absorb the breakdown products. The majority of fungal species fall into three main groups: **zygomycetes**, which contains many species of molds; **ascomycetes**, a diverse group informally known as sac fungi; and the more familiar **basidiomycetes**, or club fungi (**Figure 16.9**). Each of these groups differs in—and is named for—its unique reproductive structures.

Fungi have properties in common with both plants and animals. Like plant cells, all fungal cells have a protective cell wall that wraps around the plasma membrane and encases the cell. However, fungi are similar to animals in that they store surplus food energy in the form of glycogen.

In 2012, two bandits broke through security gates and stole an estimated $60,000 of fungi from a locked warehouse. The loot was truffles, the fruiting body of a particular group of ascomycetes. Certain white European truffles can sell for as much as $3,600 per pound, making them the most expensive food in the world. That high price also makes them a target for thieves, especially since

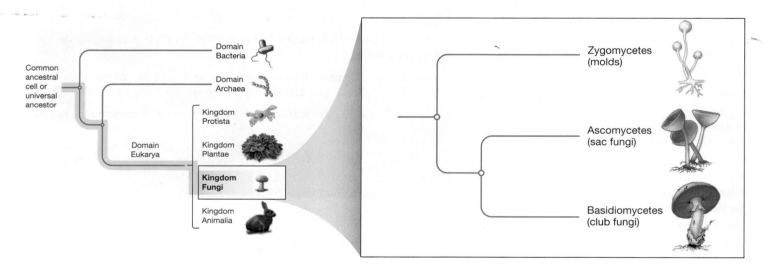

Figure 16.9

**Fungi are eukaryotic absorptive heterotrophs**
Fungi must take their food from other organisms, and they do this in a unique way. Instead of ingesting their food, as most heterotrophs do, fungi use chemicals to break down their food outside of their bodies and then absorb the nutrients.

Q1: What group of fungi most resembles the mushrooms you buy in a grocery store?

Q2: Are sac fungi more closely related to molds or to club fungi?

Q3: How do we know that fungi are eukaryotes rather than prokaryotes?

Figure 16.10

**A truffle hunter and his dog search for a fungus beloved by gourmets**

While farmers continue attempts to grow truffles commercially, their success has been limited. For now, we must rely on truffle hunters and their dogs (or pigs, which can also be trained to sniff out truffles) for this fungal delicacy.

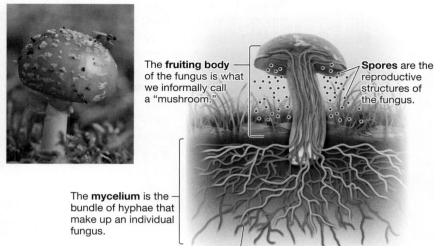

The **fruiting body** of the fungus is what we informally call a "mushroom."

**Spores** are the reproductive structures of the fungus.

The **mycelium** is the bundle of hyphae that make up an individual fungus.

A **hypha** is a single fungal thread.

Figure 16.11

**A fungus lives both belowground and aboveground**

The main body of a fungus is unseen, belowground. To reproduce, the fungus generates a fruiting body that is usually aboveground and releases spores. Spores travel by wind and develop into new fungi.

Q1: Why is it important that the fruiting body is aboveground?

Q2: What part of a fungus is the mushroom that you can buy in the grocery store?

Q3: Write a sentence in your own words that uses the terms "mycelium," "fruiting body," and "spore" correctly.

white truffles cannot be cultivated in greenhouses as other fungi can.

There's danger in the forests too. Truffle hunters use dogs and pigs to sniff out wild truffles (**Figure 16.10**), and some have reported competitors planting spiked traps or poisoned meatballs in forests to eliminate trained dogs. Other hounds have been stolen from their owners, never to be seen again.

That's a lot of drama for some fungi. Most fungi are multicellular, but there are some single-celled species collectively known as **yeasts**. Many of us are familiar with yeasts thanks to two important products they produce: alcohol and carbon dioxide, crucial to the rising of bread, the brewing of beer, and the fermenting of wine.

Fungi's key evolutionary innovation is their body form, which is well suited for absorptive heterotrophy (**Figure 16.11**). They are made up of a network of fine, colorless, branching, hairlike threads called **hyphae** (singular "hypha"), which absorb nutrients from the environment. The entire bundle of hyphae, composing the main body of the fungus, is called the **mycelium** (plural "mycelia").

Because of this unique body form, many fungi are decomposers and consume nonliving organic material. As they eat, these fungi release back into the environment inorganic chemicals that had previously been trapped in the bodies of dead organisms. Once these chemicals are back in the environment, plants and algae scoop them up and use them to manufacture food. Fungi interact with plants in other important ways too: the vast majority of wild plants have mutualistic fungi in their root systems that help the plants

# Food Banks

Over millennia, farmers have saved seeds from crops that were easiest to grow, process, or store, and have used those seeds and their offspring year after year. The results of this domestication process are staple crops we rely on today: corn, soybeans, wheat. Yet to feed a growing human population, we're going to have to look beyond what we're familiar with, to the smorgasbord of edible plants that pepper the planet.

Assessment available in smartwork**5**

## 30,000
terrestrial plants are known to be edible.

## 7,000
are cultivated or collected by humans for food.

## 30
crops provide 95% of the food energy taken in by the world's human population.

Rice   Maize   Wheat   Millet   Sorghum

## 5
cereal crops provide 60% of the food energy taken in by the world's human population.

absorb nutrients from the soil (see "Fungi Play Well with Others" on page 287).

Despite that assistance, fungi and plants are not always close friends. Fungi are the most significant parasite of plants: they are responsible for two-thirds of all plant diseases, causing more crop damage than bacteria, viruses, and insect pests combined.

Like many plants (and animals and protists), fungi can reproduce both asexually and sexually. Some species appear to multiply only asexually, and most multicellular fungi can reproduce asexually through fragmentation—that is, by simply breaking off from the mother colony. When fungi do reproduce sexually, they do not have distinct male and female individuals. Instead, a sexually reproducing mycelium belongs to one of two (or more!) mating types. Each mating type can mate successfully with only one of the other types. After mating, a **fruiting body** is formed that may be large enough to be readily observed.

Fungal fruiting bodies release offspring as sexual spores. A **spore** is a reproductive structure that can survive for long periods of time in a dormant state and will sprout under favorable conditions to produce the body of the organism. Spores released from a fruiting body that is raised up in the air are better able to catch a ride on wind currents or to attract animals that can carry them far and wide. Many fungi also produce asexual spores.

# Fighting for the Future

The illegal sale of plants and fungi could drive species to extinction. In 2015, for example, the International Union for Conservation of Nature (IUCN), a global environmental authority, announced that 31 percent of cactus species—renowned for their unique forms and beautiful flowers—are in danger of extinction, and that the greatest threat to these plants is illegal trade (**Figure 16.12**).

During his studies, Phelps observed traders selling plants that are highly endangered in the wild. He published results from his thesis in 2015 in the journal *Biological Conservation,* pleading with scientific and policy communities for a greater focus on these plants and calling

Figure 16.12

**The greatest threat to endangered cacti is illegal trade**
Like orchids and cycads, the main threat to most species of cacti is illegal trade. Collectors pay high prices to smugglers who are willing to risk fines and prison time to meet the demand.

the illegal plant trade "a major conservation challenge that has been almost completely overlooked."

"It's hard to get people to care about plants, and nobody can care if we don't have data to show it's a problem," says Phelps, who is now trying to engage governments and experts to raise awareness and initiate policies that will better regulate plant trade. At the CITES conference in September 2016, country delegates increased protections for a large number of plant species, including ponytail palms, which are in demand as houseplants, and all rosewoods, whose wood is used to manufacture fine musical instruments.

"All these plants are part of a larger ecosystem," says St. John. "They serve important purposes in temperature regulation, in creating oxygen, and more. We need to recognize that plants are an integral part of a healthy ecosystem and key to the diversity of life."

**ANNE ST. JOHN**

Anne St. John is a biologist with the Division of Management Authority for the U.S. Fish and Wildlife Service.

# REVIEWING THE SCIENCE

- The domain Eukarya is traditionally divided into four major kingdoms: Protista, Plantae, Fungi, and Animalia. Eukaryotes possess a true nucleus; they have complex subcellular compartments, which enable larger cell size. Sexual reproduction and multicellularity are key evolutionary innovations of the Eukarya.

- Kingdom Protista, the **protists**, lumps together many evolutionarily distinct lineages under one highly diverse grouping. Most protists are single-celled and microscopic.

- Protists are traditionally divided into two categories: **protozoans**, which are nonphotosynthetic and motile; and **algae**, which are photosynthetic and may or may not be motile.

- **Plants** are descended from green algae and have evolved numerous evolutionary innovations to adapt to life on land. All plants photosynthesize and use **cellulose** to strengthen their cell walls.

- The first plants, the ancestors of present-day **bryophytes** (liverworts and mosses), had relatively thin bodies and absorbed water through a wicking action.

- **Ferns** were the earliest plant group to grow larger and taller than the bryophytes. This extended growth required both cellulose and **lignin** to provide structure to plant bodies, as well as a network of fluid-transporting tissues called the **vascular system**.

- **Pollen** and **seeds** first evolved among the **gymnosperms**, the plant group that includes cycads.

- **Angiosperms** evolved **flowers**, and they enclose their seeds in the fruit. Many angiosperms recruit animals to deliver pollen and also to disperse their seeds.

- The **fungi** are distinguished by their mode of nutrition: they acquire their nutrients by absorption, digesting their food outside of their bodies. To accomplish this, their bodies consist of a **mycelium** composed of hairlike threads called **hyphae**.

- Fungal reproduction may be sexual or asexual. During reproduction, a **fruiting body** is formed that releases **spores** into the environment. Much as seeds do, these spores then develop into new individuals.

- There are at least three main groups of fungi—**zygomycetes**, **ascomycetes**, and **basidiomycetes**—and each group is characterized by distinctive reproductive structures (fruiting bodies).

- Most plant roots in natural habitats form close associations with beneficial fungi, called **mycorrhizal fungi**. A **lichen** is a mutually beneficial association between a fungus and a photosynthetic microbe, usually a green alga or a cyanobacterium.

# THE QUESTIONS

## The Basics

**1** Which of the following descriptions is *not* true of all eukaryotes?
(a) They are multicellular.
(b) They have cellular organelles.
(c) They have a larger cell size than prokaryotes have.
(d) They have a true nucleus.
(e) All of the above are characteristics of all eukaryotes.

**2** Protists
(a) are the largest prokaryotes.
(b) are all single-celled.
(c) include plants and fungi.
(d) are an unnatural grouping, placed together for convenience.
(e) are all photosynthetic.

**3** Fungi
(a) reproduce only sexually.
(b) reproduce only asexually.
(c) may have multiple mating types within a single species.
(d) are more closely related to plants than to animals.
(e) are more closely related to protists than to animals.

**4** Which of these evolutionary innovations enabled larger cell size?
(a) autotrophic mode of nutrition
(b) multicellularity
(c) sexual reproduction
(d) subcellular compartmentalization
(e) all of the above

**5** Select the correct terms:
Green algae belong in the kingdom (**Plantae / Protista**). They are (**aquatic / terrestrial**) and are (**autotrophs / heterotrophs**).

## Challenge Yourself

**6** Place the following adaptations in the correct order from earliest to most recent by numbering them from 1 to 5.

____ a. vascular tissue

____ b. flowers

____ c. seeds

____ d. multicellularity

____ e. movement to land

**7** Which of the following groups contains only multicellular species?

(a) algae

(b) protists

(c) eukaryotes

(d) fungi

(e) angiosperms

**8** Of the kingdoms covered in this chapter, which, if any, are composed of only autotrophic species? Which, if any, have only heterotrophs?

## Try Something New

**9** Symbiosis is a long-term and intimate association between two different types of organisms. A symbiotic organism may live on or inside another species. For each of the following symbiotic relationships, (1) define the relationship in one to three sentences, (2) identify the domain—and for Eukarya, the kingdom—of each partner in the relationship, and (3) discuss whether the relationship evolved as a mutualism (both benefit), commensalism (one benefits, while the other is not affected), or parasitism (one benefits to the detriment of the other). You may need to read more than your textbook to answer this question.

a. mycorrhizae

b. lichens

c. hermit crabs/shells

d. malaria

**10** Which plant groups produce pollen, and what is the adaptive value of pollen?

**11** Plants are not the only organisms susceptible to fungal infections. Give an example of a fungal infection to which humans are susceptible.

## Leveling Up

**12** **Life choices** If fungi are more closely related to animals than to plants, should vegetarians and vegans refrain from eating fungi? Why or why not?

**13** *Write Now* **biology: is a mass extinction under way?** The International Union for Conservation of Nature maintains what it calls its Red List, which identifies the world's threatened species. To be defined as such, a species must face a high to extremely high risk of extinction in the wild. The 2016 Red List contains almost 25,000 species threatened with extinction out of the approximately 79,000 species assessed.

Because this assessment accounts for only a small percentage of the world's 1.7 million described species, the total number of species threatened with extinction worldwide may actually be much larger. For example, only 6,051 of 1 million described insect species have been assessed in terms of their survival risk.

The Red List is based on an easy-to-understand system for categorizing extinction risk. It is also objective, yielding consistent results when used by different people. These two attributes have earned the Red List international recognition as an effective tool for assessing extinction risk.

a. If the threatened species listed by the IUCN do become extinct and the percentages of species under threat in other taxonomic groups turn out to be similar to those listed, then the percentages of species that will go extinct will approach the proportions lost in some of the previous mass extinctions. Does this mean that a mass extinction is under way? Why or why not? Explain your reasoning, using information about mass extinctions from Chapter 14 (in particular, Figure 14.15) to support your argument.

b. What are the causes of the current high extinction rates? Compare the causes of the current situation with those of previous mass extinction events. In what ways are they similar, and in what ways do they differ?

c. Why are some groups in more danger of extinction than others? Choose one group that has a large proportion of species threatened with extinction, and find out more about it (you can begin at http://www.iucnredlist.org):

- What kinds of habitat are they found in?
- Does a particular aspect of their ecology—for example, feeding or reproduction—make them more vulnerable to extinction?
- Is anything being done to protect them and/or their habitat?
- Has there been an increase or decrease in the number of species within the group that have been identified as vulnerable to extinction?

**For more, visit digital.wwnorton.com/bionow2 for access to:**

# Neanderthal Sex

*The relationship between modern humans and Neanderthals just got a whole lot spicier.*

## After reading this chapter you should be able to:

- Identify the key characteristics of animals, and give an example organism for each group within the chordates.

- Explain the significance of symmetry and segmentation for some animal groups.

- Compare and contrast the three types of mammals.

- Interpret an evolutionary tree of the primates.

- Illustrate the inheritance of mitochondrial DNA as compared to nuclear DNA.

- Describe the evidence suggesting that prehistoric hominin species reproduced with our direct ancestors.

# CHAPTER
## 17

## ANIMALS AND HUMAN EVOLUTION

In popular culture, Neanderthals are ugly, hairy cavemen with big brains but no wits. The first Neanderthal bones were discovered in Germany in 1856, and since then we've cultivated an image of our closest extinct human relatives as hulking brutes who communicated by grunting, walked like chimps, and hit each other over the head with clubs. Yet, over the years, paleontologists have dug up fossilized vocal bones, sophisticated tools, and other evidence suggesting that Neanderthals were a fairly advanced group—and not as different from our own species as we once believed.

The big revelation came in 2010, when scientists sequenced the Neanderthal genome and compared it to the modern human genome: Humans have some Neanderthal DNA. Our two species may have been a whole lot closer than we thought. "There could have been interbreeding between modern humans and Neanderthals," says Silvana Condemi, a researcher at the National Center for Scientific Research (CNRS) at Aix-Marseille University in France. "We can imagine they not only exchanged culture, but exchanged genes." That's right: mounting evidence suggests that Neanderthals and modern humans had sex.

This story begins where many stories end— with a pile of bones. In the early years of the twenty-first century, Laura Longo, a curator at the Civic Natural History Museum of Verona, Italy, decided to take a second look at a group of fossils excavated from a rock shelter called Riparo Mezzena. Riparo Mezzena is nestled in the Lessini Mountains in northern Italy, a wide-open landscape speckled with large rocks and evergreen trees. The region is snowy and silent in the winter but green and thriving in the summer, when paleontologists come to work.

In the 1950s, paleontologists had carefully collected fossils of Neanderthals that lived around Riparo Mezzena about 35,000 years ago, late in the history of Neanderthals. But the fossils had sat at the museum, untouched, for over 50 years (**Figure 17.1**). Now, Longo believed, the fossils could help answer a hotly debated question about human history: How closely did modern humans and Neanderthals interact?

# Animal Kingdom

Modern humans and Neanderthals lived in some of the same areas of Europe at the same times. Because of this proximity, some paleontologists hypothesize that as modern humans expanded their territory, Neanderthals were

Figure 17.1

**Fossil remains**
A jawbone like this one was found at the Riparo Mezzena rock shelter in the Lessini Mountains in Italy. The individual it belonged to lived between 40,000 and 30,000 years ago.

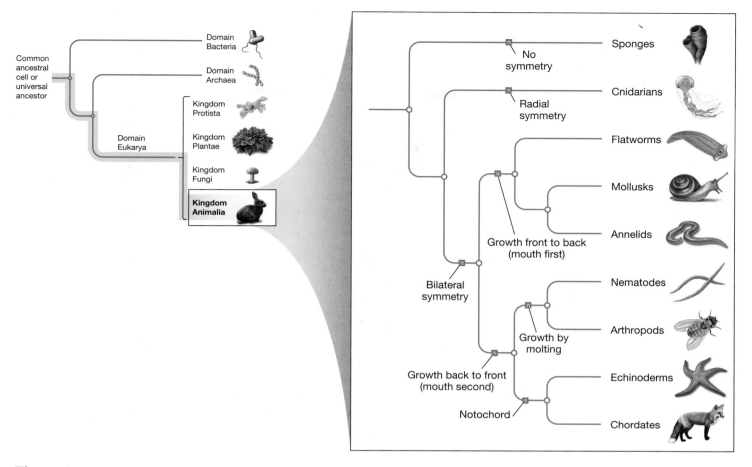

Figure 17.2

## Animals are heterotrophic eukaryotes

Animals ingest their food, which can be organisms from any of the other domains and kingdoms, and then digest it internally.

**Q1:** Are mollusks more closely related to flatworms or to annelids? Explain your reasoning.

**Q2:** If you found an animal with no symmetry, to what group would it belong? Give an example of an animal that is radially symmetrical, and one that is bilaterally symmetrical.

**Q3:** How do we know that a sponge is an animal, and not a plant or an alga?

quickly driven to extinction, and therefore the two did not live side by side. Others claim the opposite: that Neanderthals, *Homo neanderthalensis*, were slowly incorporated into the population of newly incoming humans, *Homo sapiens*. But to know how humans evolved, it is important first to realize where we came from.

All species in the *Homo* genus, of which *H. sapiens* is the only one alive today, are in the Animalia kingdom. **Animals** are multicellular ingestive heterotrophs; that is, we obtain energy and carbon by ingesting food into our bodies and digesting it internally. Animals first evolved

some 700 million years ago, descended from flagellated protists, single-celled organisms with whiplike tails that thrived in wet habitats and began to use oxygen to break down food into energy. From those humble origins, countless animal species evolved—beginning with sponges and including mollusks such as snails and clams, the annelids (segmented worms), and arthropods like crustaceans, spiders, and insects (**Figure 17.2**).

**Chordates** make up a large phylum that includes all animals with backbones, such as fishes, birds, and mammals (see **Figure 17.3** and

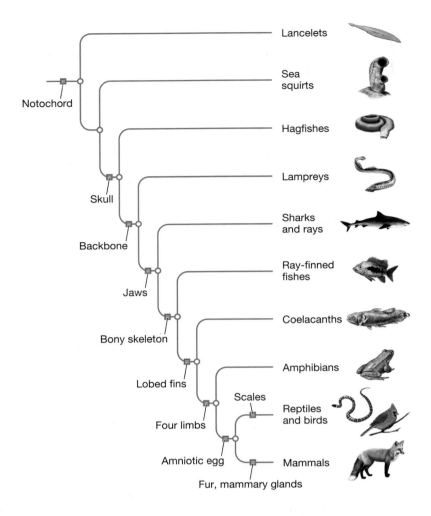

## Figure 17.3

### Not all chordates have a backbone

As with all eukaryotes except the protists, the evolutionary relatedness of the chordates is very well known. Not all chordates have a backbone, but all have a notochord, which is the precursor to part of the vertebrate backbone.

**Q1:** Do amphibians have amniotic eggs?

**Q2:** What group of animals has jaws but not a bony skeleton?

**Q3:** When people talk about animals, they are sometimes referring only to mammals. How would you explain to them their error?

"Get a Backbone!" on page 305). The phylum also includes several subgroups of less familiar animals, such as sea squirts and lancelets, that have a nerve cord along the back of the body but no backbone. Chordates that do possess a backbone are known as **vertebrates**. All other phyla of animals are informally lumped together as invertebrates. But keep in mind that "invertebrates" is not an evolutionarily meaningful label; it is an artificial grouping (like that of protists) of animals with varied evolutionary histories and different sets of evolutionary adaptations.

Body plan is another key factor that distinguishes one group of animals from another. All animals except sponges, the most ancient of the animal lineages, have a distinct body symmetry. The asymmetric sponges aside, animals can be divided into two main groups: those with radial

# Get a Backbone!

Within the kingdom Animalia, humans belong to the phylum of chordates and the subphylum of vertebrates. All chordates have a dorsal **notochord**, a flexible rod along the center of the body that is critical for development. In vertebrates, the dorsal notochord has evolved to become the cushioning discs between **vertebrae** (singular "vertebra"), which are strong, hollow sections of the backbone, or vertebral column. Vertebrates include fishes, amphibians (frogs and salamanders), reptiles (snakes, lizards, turtles, and crocodiles), birds, and mammals.

The jawless fishes were the first vertebrates to evolve. Their skeletons—including the backbone—were made from a strong but flexible tissue called **cartilage**. Only a few groups of jawless fishes have survived to the present day, most notably the lampreys. The next great leap in vertebrate evolution was hinged jaws, which enabled predators to grab and swallow prey efficiently. The evolution of teeth made jaws even more effective because teeth enabled animals to seize and tear food.

Another major step in the evolution of vertebrates was the replacement of the cartilage-based skeleton with a denser tissue strengthened by calcium salts: bone. Although the descendants of cartilaginous fishes—sharks, skates, and rays—are still with us today, bony fishes are far more diversified and widespread in both marine and freshwater environments. With more than 30,000 species, bony fishes are the most diverse vertebrates today.

The advent of lungs was a crucial milestone in the transition of vertebrates onto land. Amphibians made this transition only partially; they can live on land but must return to the water to lay eggs and breed. The several thousand species of amphibians include frogs and salamanders.

Reptiles were the first vertebrates to head into drier environments, and they evolved a number of adaptive traits to deal with the risk of dehydration. These adaptations included skin covered in waterproof scales, a water-conserving excretory system, and the amniotic egg with its calcium-rich protective shell, which retards moisture loss while allowing the entry of life-giving oxygen and the release of waste carbon dioxide for the developing embryo. Reptiles dominated Earth during the age of the dinosaurs, and the dinosaurs' descendants (as we saw in Chapter 14) remain with us today in the form of birds. Like mammals, birds are warm-blooded, but they have feathers instead of fur for insulation. At least 10,000 different species of birds are living today.

---

symmetry and those with bilateral symmetry. The body of an animal with **radial symmetry** (**Figure 17.4**, left)—including cnidarians like jellyfish, sea anemones, and corals—can be sliced symmetrically along any number of vertical planes that pass through the center of the animal, like cutting a pie. Radial symmetry gives an animal sweeping, 360-degree access to its environment. The animal can snare food drifting in from any direction of the compass, and it can also sense and respond to danger from any side.

Animals with **bilateral symmetry** (**Figure 17.4**, right), on the other hand, can be divided by just one plane passing vertically from the top to the bottom of the animal into two halves that mirror each other. Bilateral animals, which include all chordates, have distinct right and left sides, with near-identical body parts on each side. The symmetrical arrangement of body parts on either side of a central body facilitates movement in bilateral animals. The paired arrangement of limbs or fins, for example, enables quick and efficient movement on land

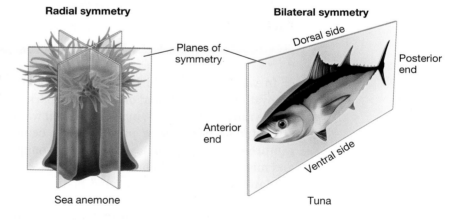

**Figure 17.4**

**Body symmetry in animals**

All animals other than sponges have symmetrical bodies that enable them to better sense and respond to the world around them.

> **Q1:** Is a sea star radially or bilaterally symmetrical?
>
> **Q2:** What advantage might a bilaterally symmetrical animal have over one that is radially symmetrical, and vice versa?
>
> **Q3:** What kind of symmetry do you (a human) have?

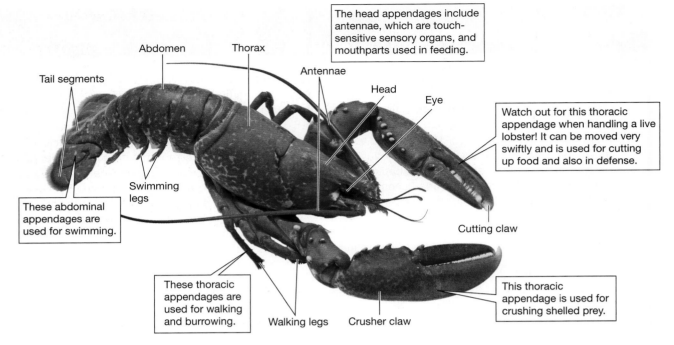

Tail segments

Abdomen

Thorax

Antennae

Head

Eye

The head appendages include antennae, which are touch-sensitive sensory organs, and mouthparts used in feeding.

Watch out for this thoracic appendage when handling a live lobster! It can be moved very swiftly and is used for cutting up food and also in defense.

Swimming legs

These abdominal appendages are used for swimming.

These thoracic appendages are used for walking and burrowing.

Walking legs

Crusher claw

Cutting claw

This thoracic appendage is used for crushing shelled prey.

Figure 17.5

**Segmentation in animals**

Segmentation, a body plan in which segments repeat, enabled the evolution of diverse uses of appendages, shown here in a lobster.

> **Q1:** List all of the lobster's thoracic appendages.
>
> **Q2:** Which of the lobster's appendages is most important for sensing the environment?
>
> **Q3:** What body segments do you (a human) have?

or in water. Locomotion is a key evolutionary innovation of animals and one that has sparked a wide range of behaviors, including varied ways of capturing prey, eating prey, avoiding being captured, attracting mates, caring for young, and migrating to new habitats.

Many animals have segmented bodies; that is, their body plan consists of repeated units known as segments (**Figure 17.5**). Specialized body parts, known as appendages, often originate in pairs from specific segments of the body, such as fins or limbs. Over evolutionary time, the segments and the appendages that spring from them have evolved diverse form and function, enabling the animal body to adapt to new habitats or acquire new modes of life.

The evolution of just the posterior segments of arthropods illustrates how evolution can take a basic body plan and modify it to produce many variations over time: The last segment in the body of arthropods has evolved into the delicate abdomen of the butterfly, the piercing abdomen of the wasp, and the delicious tail of the lobster. The front appendage of vertebrates has evolved as an arm in humans, a wing in birds, a flipper in whales, an almost nonexistent nub in snakes, and a front leg in salamanders and lizards.

# Mammals R Us

To get back to our egocentric focus—that is, humans—we are part of the kingdom Animalia and the class Mammalia. Humans share specific features with all other mammals, including body hair, sweat glands, and milk produced by mammary glands. And at the risk of tooting our own horn, **mammals** have been a highly successful class of animals, with over 5,000 species living in a variety of habitats. This is largely thanks to

Eutherian

Marsupial

Monotreme

Figure 17.6

**Three kinds of mammals**

All mammals have milk-producing mammary glands for suckling their young. Eutherians like the polar bear give birth to well-developed young. Kangaroos are marsupials, and give birth to immature young that finish developing in a pouch. Monotremes like the platypus lay eggs instead of giving birth to live young.

**Q1:** The Virginia opossum, or possum, is the only North American marsupial. Describe how its young are born and develop.

**Q2:** If you were a monotreme, would you still produce milk and nurse your young?

**Q3:** What kind of mammal is a cow? How about a human?

the extinction of dinosaurs; if dinosaurs still roamed Earth, it is likely that mammals would be nothing but dinner. Lucky for us, mammals have replaced dinosaurs as the top predators in most terrestrial habitats, and they thrive in both marine and freshwater environments. Only one type of mammal can fly—bats—although a few others can glide through the air.

Mammals can be divided into three broad categories, all of which feed their offspring with milk (**Figure 17.6**). More than 95 percent of mammals alive today are **eutherians**, including humans. A unifying characteristic of eutherians is that the offspring are nourished inside the mother's body through a special organ called the *placenta* and are therefore born in a relatively well-developed state. **Marsupials**, the second category of mammals, have a simple placenta, resulting in offspring born early, which then complete development in an external pocket or pouch. Marsupials are found mainly in Australia and New Zealand, with a few species in the Americas.

Making up the third category, **monotremes** are egg-laying mammals that lack a placenta altogether. The only living species of monotremes are just one platypus species and several echidna species, all confined to Australia and New Guinea.

Moving now from class to order, humans and Neanderthals are in the order of the **primates** (**Figure 17.7**). Like all other primates, we have flexible shoulder and elbow joints, five functional fingers and toes, thumbs that are **opposable** (that is, they can be placed opposite each of the other four fingers), flat nails (instead of claws), and brains that are large in relation to our body size.

Within the primates we are members of the ape family, the **hominids**. We are not just closely related to apes; we *are* apes. As such, we share many characteristics with other apes, especially chimpanzees, including the use of tools, a capacity for symbolic language, and the performance of deliberate acts of deception. But we are part of a distinct branch of apes called **hominins**—the "human" branch of the ape family that includes

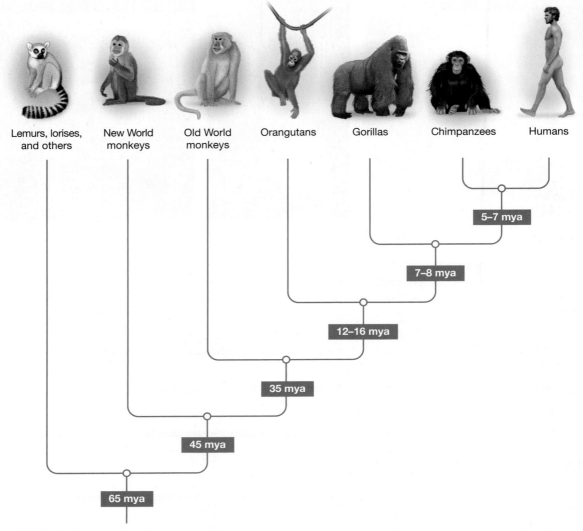

Lemurs, lorises, and others | New World monkeys | Old World monkeys | Orangutans | Gorillas | Chimpanzees | Humans

5–7 mya

7–8 mya

12–16 mya

35 mya

45 mya

65 mya

Figure 17.7

### The primates include lemurs, monkeys, and apes

Genetic analyses and a series of spectacular fossil discoveries have led scientists to propose that the human lineage diverged from that of chimpanzees about 5–7 million years ago (mya). Similar evidence suggests that the evolutionary lineage leading to humans diverged from the lineage leading to gorillas about 7–8 mya, and from the lineage leading to orangutans about 12–16 mya.

Q1: According to this evolutionary tree, which primate group is most closely related to humans?

Q2: According to this evolutionary tree, which primate group is most distantly related to humans?

Q3: What characteristics are common to all the primates, including humans?

our extinct relatives, such as Neanderthals. The members of the hominin lineage have one or more humanlike features—for example, thick tooth enamel or upright posture—that set them apart from the other apes, like gorillas and chimpanzees.

## Rise of the Apes

A major step in hominin evolution, and the main feature that distinguishes hominins from other hominids, was the shift from moving on four legs to being **bipedal**, walking upright on

# Hereditary Heirlooms

Because of a shared common ancestor 1.6 billion years ago, humans share DNA with all animals, plants, and fungi. But how much? Take a look to see how much genetic material you have in common with other organisms.

Assessment available in smartwork5

## Percentage of genes shared with humans

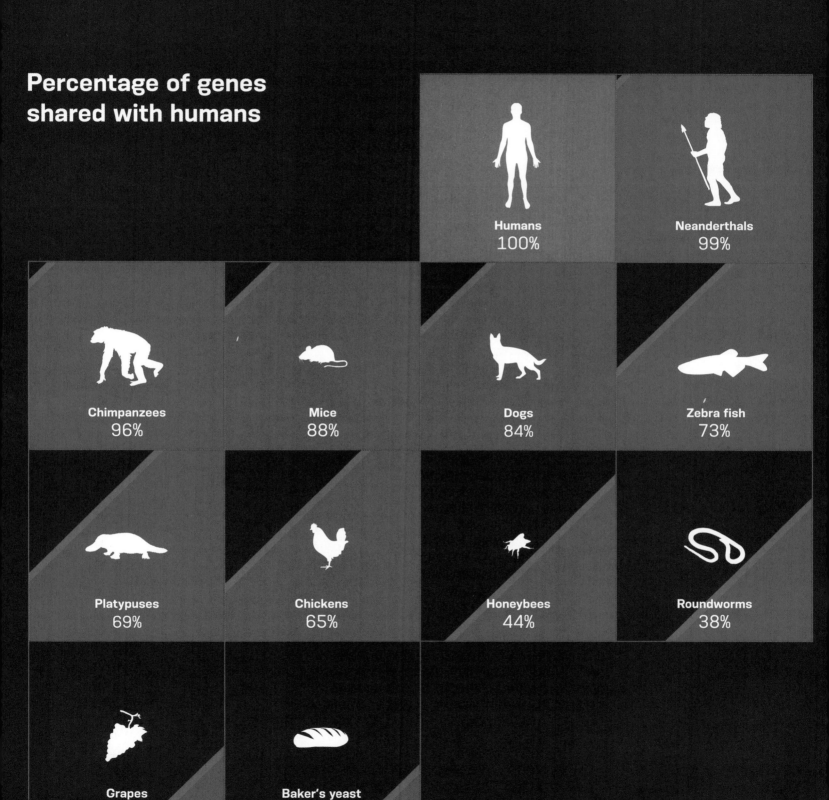

Humans
100%

Neanderthals
99%

Chimpanzees
96%

Mice
88%

Dogs
84%

Zebra fish
73%

Platypuses
69%

Chickens
65%

Honeybees
44%

Roundworms
38%

Grapes
24%

Baker's yeast
18%

BIODIVERSITY

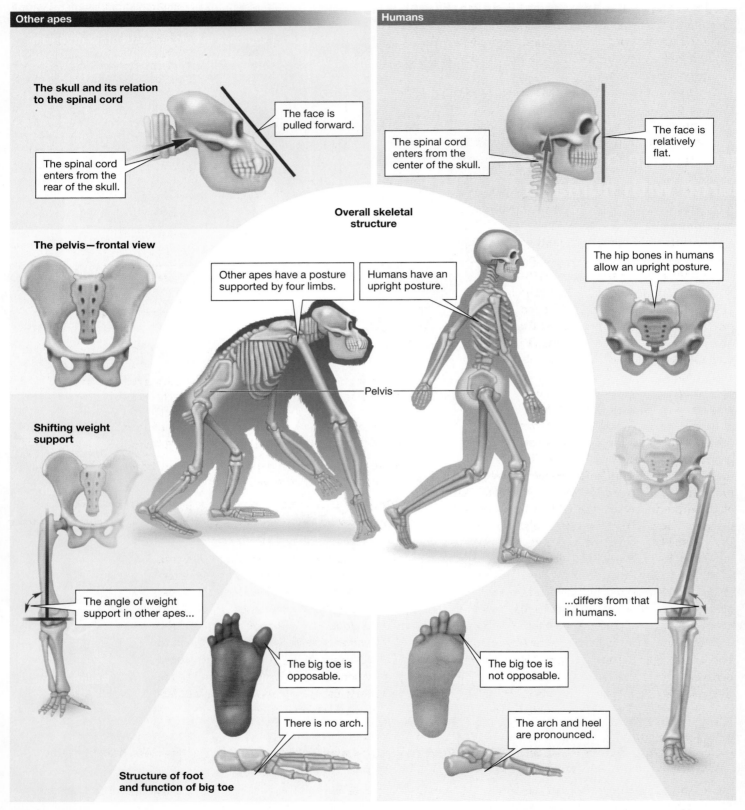

**Other apes**

**The skull and its relation to the spinal cord**

The face is pulled forward.

The spinal cord enters from the rear of the skull.

**The pelvis—frontal view**

**Shifting weight support**

The angle of weight support in other apes...

**Structure of foot and function of big toe**

The big toe is opposable.

There is no arch.

**Humans**

The spinal cord enters from the center of the skull.

The face is relatively flat.

**Overall skeletal structure**

Other apes have a posture supported by four limbs.

Humans have an upright posture.

The hip bones in humans allow an upright posture.

Pelvis

...differs from that in humans.

The big toe is not opposable.

The arch and heel are pronounced.

Figure 17.8

**Evolutionary differences between humans and other apes**

The switch to walking upright required a drastic reorganization of primate anatomy, especially of the hip bones.

two legs (**Figure 17.8**). Many skeletal changes accompanied the switch to walking upright, including the loss of opposable toes, as you will notice if you try touching your little toe with the big toe on the same foot.

The loss of opposable toes that accompanied walking upright would have been a handicap in trees, since opposable toes help grasp branches during climbing. It is therefore likely that bipedalism was an adaptation for living on the ground. Walking on two feet freed the hands to carry food, tools, and weapons, and it also elevated the head, enabling the walker to see farther and over more things.

The shift to life on the ground was probably not sudden or complete. The skeletal structure of some of the oldest fossil hominins (3–3.5 million years old) indicates that they walked upright. However, foot bones and fossilized footprints show that the hominins living at that time still had partially opposable big toes (**Figure 17.9**). Perhaps they still occasionally climbed trees.

The earliest known hominin is *Sahelanthropus tchadensis*, identified from a 6- to 7-million-year-old skull discovered in 2002. Other early hominins include *Ardipithecus ramidus*, who lived 4.4 million years ago, and several *Australopithecus* species that are 3–4.2 million years old, including the first full-time walker with the first modern foot: *Australopithecus afarensis*. All of these hominins are thought to have walked upright. Their brains were still relatively small (less than 400 cubic centimeters in volume), and their skulls and teeth were more similar to those of other apes than to those of humans (**Figure 17.10**). A typical modern human has a brain volume of about 1,400 cubic centimeters, about the same volume as a 1.5-liter soda bottle.

Within the hominin branch is the *Homo* genus. The fossils identified at Riparo Mezzena were believed to be *Homo neanderthalensis* bones, but they had never been closely studied. So, Longo, the curator from Verona, assembled a team of

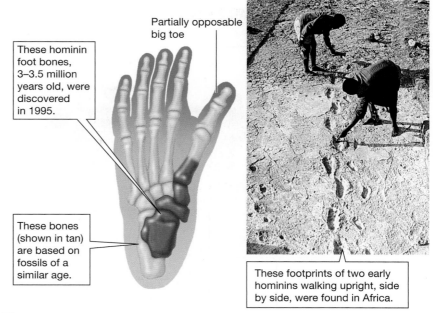

Partially opposable big toe

These hominin foot bones, 3–3.5 million years old, were discovered in 1995.

These bones (shown in tan) are based on fossils of a similar age.

These footprints of two early hominins walking upright, side by side, were found in Africa.

Figure 17.9

**Early hominins had an upright stance and partially opposable big toes**
Fossilized foot bones show that some hominins living between 3.5 and 3 million years ago walked upright but had partially opposable big toes.

**Q1:** What other reason besides continuing to use trees might explain why early hominins had partially opposable big toes?

**Q2:** In what way does the pattern of footprints in this figure suggest that the print makers were walking upright?

**Q3:** Why do you think we no longer have partially opposable big toes?

researchers to analyze them, including Condemi, the anthropologist. Condemi had long been interested in the movement of Neanderthals across Europe and how their populations overlapped with modern *Homo sapiens* populations, and she wanted to compare the Riparo Mezzena fossils to fossils from Neanderthal groups that had been dug up elsewhere around Europe, hoping to see what the fossils might say about how Neanderthals and modern humans interacted.

**Q1:** Through natural selection, deleterious traits will tend to disappear from a population over time. Which traits might have been deleterious for ground-dwelling early hominins?

**Q2:** Through natural selection, advantageous traits will tend to persist in a population over time. Which traits might have been advantageous for ground-dwelling early hominins?

**Q3:** Adaptations to upright walking also mean that human females have more difficulty giving birth than do females of other species. What adaptation would you predict has had the greatest impact on this result?

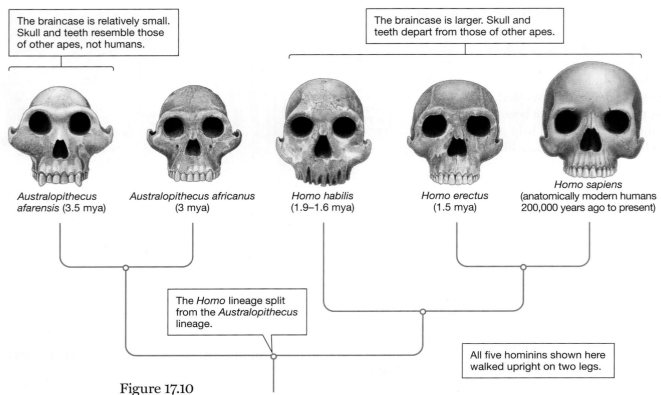

The braincase is relatively small. Skull and teeth resemble those of other apes, not humans.

The braincase is larger. Skull and teeth depart from those of other apes.

*Australopithecus afarensis* (3.5 mya)

*Australopithecus africanus* (3 mya)

*Homo habilis* (1.9–1.6 mya)

*Homo erectus* (1.5 mya)

*Homo sapiens* (anatomically modern humans 200,000 years ago to present)

The *Homo* lineage split from the *Australopithecus* lineage.

All five hominins shown here walked upright on two legs.

Figure 17.10

**A gallery of hominin skulls**

This tree shows the evolutionary relationships and the skulls of five hominin species. A complete evolutionary tree of hominins would be "bushier," with multiple side branches emerging at different times.

**Q1:** Where would the Neanderthal species branch be on this tree?

**Q2:** How would the Neanderthal skull differ from the *Homo erectus* skull?

**Q3:** How would the Neanderthal skull differ from the *Homo sapiens* skull?

# Hominins United

"For years, it was a very simple story: Neanderthals either disappeared quickly when modern humans came or they integrated with humans," says Condemi. But she suspected that the dynamic was more complex. "In some regions, Neanderthals disappeared very quickly. In other regions, we have evidence the two [species] could have overlapped." Researchers, including Condemi, had often suggested that during the overlap, Neanderthals and humans interbred. But there was little physical evidence.

Even DNA evidence initially suggested that there was no interbreeding. DNA from the mitochondria of Neanderthals (isolated first from a single Neanderthal fossil in 1997 and then from

four Neanderthal fossils in 2004) was compared to modern human mitochondrial DNA, and the tests showed that there was no genetic overlap between the species. Mitochondrial DNA (mtDNA) is unique because it is passed down virtually unchanged from mother to child, so it can be tracked from one generation, or one species, to another (**Figure 17.11**). But modern *Homo sapiens* did not have Neanderthal mitochondrial DNA, so it appeared there was no interbreeding, at least not involving female Neanderthals.

That mitochondrial work was performed by Svante Pääbo and researchers at the Max Planck Institute for Evolutionary Anthropology in Leipzig, Germany. Pääbo is one of the founders of the effort to use genetics to study early

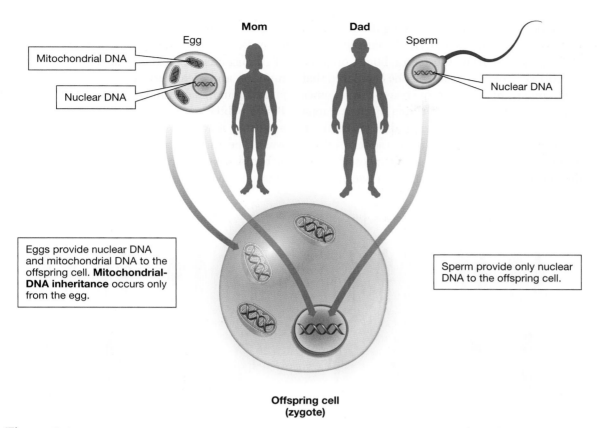

**Mom** **Dad**

Egg

Mitochondrial DNA

Nuclear DNA

Sperm

Nuclear DNA

Eggs provide nuclear DNA and mitochondrial DNA to the offspring cell. **Mitochondrial-DNA inheritance** occurs only from the egg.

Sperm provide only nuclear DNA to the offspring cell.

**Offspring cell (zygote)**

Figure 17.11

## Mitochondrial DNA comes only from your mom (not from your dad)

In addition to the chromosomes in the nucleus (nuclear DNA), cells also have DNA in their mitochondria, called mitochondrial DNA. Mitochondria and their resident DNA all come from Mom.

**Q1:** Why does mitochondrial DNA come only from your mother?

**Q2:** If a Neanderthal-human hybrid was born to a human mother and a Neanderthal father, could you tell by mitochondrial-DNA sequencing?

**Q3:** If a Neanderthal-human hybrid was born to a human father and a Neanderthal mother, could you tell by mitochondrial-DNA sequencing?

humans and other ancient populations, and he has pioneered numerous techniques to extract delicate DNA from even the tiniest slivers of fossilized bones. At the time that he began his work, it was easier to find and extract mitochondrial DNA because cells contain hundreds of copies of this type of DNA, and only one copy of nuclear DNA. In addition, mitochondrial DNA can be isolated from cells and tissues that aren't so well preserved and from damaged DNA. Whole-genome sequencing, by contrast, requires well-preserved cells or tissues with fully intact nuclear DNA. But despite his initial findings that modern humans and Neanderthals did not share

mtDNA, Pääbo still thought there might be room for some small contribution. Being a diligent scientist, he decided to look even deeper—this time at the whole Neanderthal genome.

### SVANTE PÄÄBO

Svante Pääbo is a Swedish geneticist who directs the Department of Genetics at the Max Planck Institute for Evolutionary Anthropology in Leipzig, Germany. He specializes in using genetics to study early humans and other ancient populations.

Pääbo spent 4 years sequencing the 1.5 billion base pairs in the Neanderthal genome, using DNA extracted from the femur bones of three 38,000-year-old females. Then he compared that long, composite genome sequence to the genomes of five living humans from China, France, Papua New Guinea, southern Africa, and western Africa.

According to the results, all modern ethnic groups, other than Africans, carry traces of Neanderthal DNA in their genomes—between 1 and 4 percent. Most of us have a little Neanderthal in us. But whether that nuclear DNA is the result of thousands of sexual encounters between humans and Neanderthals or a few one-night stands remains unknown (**Figure 17.12**). When Pääbo published his work in 2010, he and others admitted it was possible that the shared DNA wasn't necessarily a product of interbreeding. It could have been a remnant of DNA from a common shared ancestor.

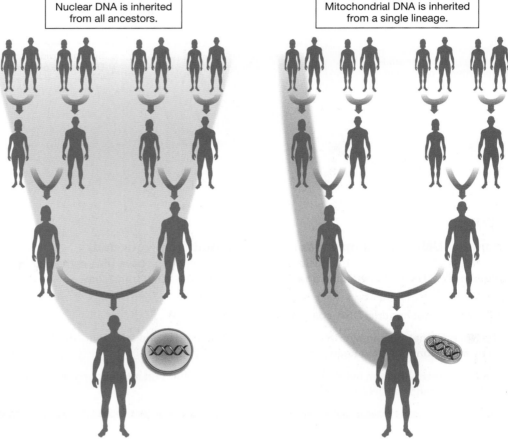

Figure 17.12

**Nuclear-DNA inheritance**

By sequencing the bases of nuclear DNA, scientists can determine how related an individual is to its ancestors, both male and female (see Chapter 11). Mitochondrial-DNA sequencing can determine only how related an individual is to the female ancestors on its mother's side.

**Q1:** If a human-Neanderthal hybrid was born to a human mother and a Neanderthal father, could you tell by whole-genome DNA sequencing that it was a hybrid?

**Q2:** If a human-Neanderthal hybrid was born to a Neanderthal mother and a human father, could you tell by whole-genome DNA sequencing that it was a hybrid?

**Q3:** Under what circumstances are scientists able to do whole-genome sequencing, and when are they restricted to mitochondrial-DNA sequencing?

Who would that ancestor have been? The oldest *Homo* fossil fragments were found in Ethiopia in 2015 and date between 2.80 and 2.75 million years ago (mya), suggesting that the earliest members of the genus *Homo* originated in Africa 2–3 mya. More complete early *Homo* fossils exist from the period 1.9–1.6 mya; these fossils have been given the species name *Homo habilis*. The oldest *H. habilis* fossils resemble those of *Australopithecus africanus*, yet more recent *H. habilis* fossils show a more rounded skull and a face that isn't pulled as far forward. *H. habilis* fossils therefore provide an excellent record of the evolutionary shift from ancestral hominins (in *Australopithecus*) to more recent species, such as *Homo erectus*, the most likely candidate for a shared common ancestor of Neanderthals and modern humans (**Figure 17.13**).

Figure 17.13

**Meet the folks**

Cartoon representations of seven hominin species depicting their average height in centimeters and their presumed features. *Homo sapiens* is represented as a 6-foot-tall male for reference.

Q1: Are you surprised by the interpretations of the hominins in this picture? Why or why not?

Q2: Describe the main differences that distinguish the hominin species.

Q3: From what you've learned about these species, do you think these representations are accurate? How can you find more information about each species to help you answer this question?

Taller and more robust than *H. habilis*, *H. erectus* also had a larger brain and a skull more like that of modern humans. It is likely that by 500,000 years ago, *H. erectus* could use, but not necessarily make, fire. In addition, *H. erectus* probably hunted large animals, as suggested by a remarkable 2010 discovery in Germany of three 400,000-year-old spears, each about 2 meters long and designed for throwing with a forward center of gravity (like a modern javelin). *H. erectus*, or one of the other *Homo* ancestors, migrated from Africa about 2 million years ago. From there, this ancestor species spread around the Middle East and into Asia. *Homo* fossils dating from the period 1.9–1.7 mya have been found in the central Asian republic of Georgia, in China, and in Indonesia.

Overall, current research on *H. habilis*, *H. erectus*, and other early *Homo* species indicates that there were more species of *Homo* than was once thought, and that several of these species existed in the same places and times. More research and evidence will be necessary before general agreement is reached regarding the exact number of early *Homo* species and their evolutionary relationships.

So, was the Neanderthal DNA found in modern human genomes simply a remnant of a common ancestor? In 2012, Pääbo's team and others were able to determine the age of the pieces of Neanderthal DNA in the human genome. They found that the DNA had been introduced into our genome between 90,000 and 40,000 years ago, around the same time that modern humans spread out of Africa and met the Neanderthals. A remnant of DNA from a common ancestor would have been 10 times older.

## All in the Family

The fossil record indicates that the first *Homo sapiens*, called archaic *H. sapiens*, originated between 400,000 and 300,000 years ago.

## Uniquely Human?

Yes, the frontal lobe of the brain is unique to human beings, and it enables us to reason like no other animal on our planet. But what about the rest of the attributes that we so commonly consider unique to us? Humans pride themselves on their intelligence and deep emotional connections to others, but are these really only human traits?

- **Language.** Researchers once believed language was an exclusively human trait. We now know that chimpanzees in the wild use sign language, with approximately 70 different signs for distinct words. Meanwhile, other primates, birds, whales, and bats have distinct, learned vocalizations that they use to communicate.

- **Memory.** Some have suggested humans alone possess the ability to store memories. But dogs easily learn and remember many commands, while crows can learn and remember shapes better than human adults can, and they can use causal reasoning, not trial and error, to unlock doors and find hidden objects.

- **Social culture.** Once thought to be strictly human, social culture is a learned trait that chimpanzees, Japanese macaques, and killer whales pass on throughout their populations. Tool use by dolphins, elephants, and octopi varies in its specifics from population to population—a sure sign of learned behavior.

- **Emotions.** Our emotions make us human, right? Others in the animal kingdom have been documented expressing empathy (elephants), grief (dolphins, elephants), jealousy (apes), curiosity (cats, lizards), altruism (apes), and gratitude (whales). Apes have been seen laughing at a clumsy fellow ape and using deception to outwit a family member.

- **Self-Awareness.** The ability to recognize oneself in the mirror, or show self-awareness, was once thought to be ours alone. As it turns out, all the apes, some gibbons, elephants, magpies, and some whales pass the mirror test of self-awareness.

- **Morality.** Finally, what about a sense of morality or an understanding of social norms? Monkeys and rats will not accept offered food if, in doing so, a fellow member of their species receives an electric shock.

To be sure, there *is* something unique about humans that lies at the intersection of all these abilities. However, our expanding knowledge of animal behavior can't help but make us feel more closely connected to the other species with whom we share this planet.

**Figure 17.14**

**Anatomically modern humans evolved in Africa**

The earliest known archaeological specimens of modern humans (*Homo sapiens*) come from Africa. The dates provided give the age of the earliest evidence that anatomically modern humans lived in different regions of the world. These dates are continually challenged by new fossil evidence that scientists must then work to confirm.

**Q1:** What evidence suggests that Neanderthals never lived in Africa?

**Q2:** How does the hypothesized origin of modern humans (*Homo sapiens*) differ from the hypothesized origin of Neanderthals (*Homo neanderthalensis*)?

**Q3:** What species of hominins other than the Neanderthals may have commingled with modern humans?

Archaic *H. sapiens* bore features intermediate between those of *H. erectus* and those of "anatomically modern" *H. sapiens*—our species—which arose some 195,000–200,000 years ago. These ancestors of anatomically modern humans developed new tools and new ways of making tools, used new foods, and built complex shelters. (But humans are not the only organisms to do many of these things; see "Uniquely Human?" on page 316.)

Early populations of archaic humans eventually gave rise to both the Neanderthals (who lived from 300,000 to 28,000 years ago) and us—that is, anatomically modern humans. There is some debate as to whether Neanderthals are simply an odd form of archaic *H. sapiens*, or their own distinct species. That question has yet to be resolved.

According to the out-of-Africa hypothesis, anatomically modern humans first evolved in Africa about 195,000–200,000 years ago from a unique population of archaic *H. sapiens*, and then spread into other continents to live alongside other hominins (**Figure 17.14**). Evidence from the fossil record indicates that anatomically modern humans overlapped in time with *H. erectus* and Neanderthal populations, yet remained distinct from them. Neanderthals and modern humans coexisted in western Asia for about 80,000 years, and in Europe for some 10,000 years, until modern humans completely replaced all other *Homo* populations.

But what happened in that intervening time? Were modern humans and Neanderthals friendly neighbors, or were the latter quickly wiped out by the former?

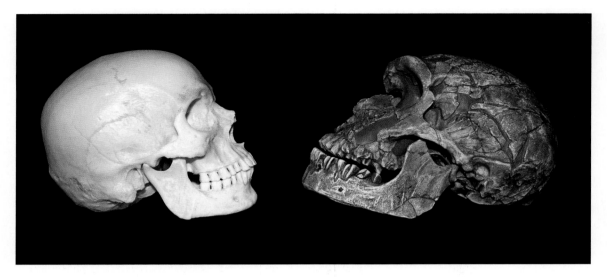

Figure 17.15

**Neanderthal and human skulls**
The skull on the right is from a Neanderthal; the one on the left, from a modern human. Notice the differences in the eyebrow ridge, forehead, and lower jaw.

**Q1:** Describe the difference you observe between the modern human skull's chin and the Neanderthal skull's lower jaw.

**Q2:** What other differences do you observe between the two skulls?

**Q3:** Why would you expect a hybrid of Neanderthals and modern humans to have intermediate features?

With a previous team, Condemi had used fossil evidence from southern Italy to determine that modern humans arrived on the Italian peninsula between 45,000 and 43,000 years ago, *before* the disappearance of Neanderthals. So the two populations likely made contact in

**SILVANA CONDEMI**

Silvana Condemi is an anthropologist and research director at the National Center for Scientific Research (CNRS) at Aix-Marseille University in France. She studies the movement of Neanderthals across Europe and how their populations overlapped with modern humans.

Italy. During this period, it's possible that "there was a kind of interbreeding," says Condemi. But if the two species interbred and had children, what did those children look like? And why hadn't Condemi's team found their bones?

Then Condemi examined the Riparo Mezzena bones. One caught her attention. It was a jawbone, a mandible, from a late Neanderthal living in Italy at the same time that modern humans had already made their way into Europe (see **Figure 17.1**). But the jawbone didn't look like a Neanderthal's, which has no chin. Instead, the face of the Riparo Mezzena individual, when reconstructed with three-dimensional imaging, had an intermediate jaw, something between no chin and a strongly projected chin. Because chins are a feature unique to modern humans (**Figure 17.15**),

Figure 17.16

**Neanderthal-human hybrid**
Paleontological reconstruction of a Neanderthal-human hybrid.

the jaw appeared to be something of a hybrid between a Neanderthal and a modern human. "This, in my view, could only be a sign of interbreeding," says Condemi.

To back up her hypothesis, Condemi's team analyzed the fossil's DNA. The fossil had Neanderthal mitochondrial DNA, confirming that at least the individual's mother was a Neanderthal. From the DNA and imaging evidence, Condemi and her team concluded that it was the child of a "female Neanderthal who mated with a male *Homo sapiens*" (**Figure 17.16**). What's more, she added, this evidence supports the idea of a slow transition from Neanderthals to anatomically modern humans, in which the two species intermingled in both culture and sex, rather than the abrupt extinction of Neanderthals when modern humans arrived.

It is unlikely that Condemi's finding will be the last word on Neanderthal-human interbreeding. In 2016, the first in-depth genetic analysis of a Neanderthal Y chromosome revealed mutations in genes on the Neanderthal chromosome that would have been incompatible with comparable genes in the modern human Y chromosome. In other words, these Y chromosome mutations could have discouraged or prevented the formation of a hybrid Neanderthal–modern human species.

An even bigger question lingers for future scientists: While modern humans continued to develop their culture and populate the planet, Neanderthals went extinct. Why? Perhaps another pile of bones will someday reveal the answer.

# REVIEWING THE SCIENCE

- **Animals** are multicellular ingestive heterotrophs. The **chordates** include all animals with a dorsal nerve cord. Chordates with a backbone are **vertebrates**, and all other phyla of animals are informally designated as invertebrates.

- Sponges lack distinct symmetry; some invertebrates have **radial symmetry**; all other animals, including all chordates, have **bilateral symmetry**.

- Many animal groups have a body plan consisting of repeated units known as **segments. Appendages,** body parts with specialized functions, develop in pairs from particular segments of the body. The evolution of segmentation and appendages has allowed animals to exploit new habitat and food sources in novel ways.

- **Mammals** are grouped into three categories: **Eutherians** are placental and born in a well-developed state. **Marsupials** have a simple placenta and finish development in a pouch. **Monotremes** lack a placenta and lay eggs.

- Humans are mammals (class), and more specifically **primates** (order), having flexible shoulder and elbow joints, five functional fingers and toes, **opposable** thumbs, flat nails (instead of claws), and large brains.

- **Hominins** are characterized by **bipedalism**, the ability to walk upright on two legs. They are a branch of the apes, the **hominids** (family).

- **Mitochondrial-DNA inheritance** occurs exclusively through the maternal egg; the sperm contributes essentially no mitochondria and none of the mitochondrial DNA. **Nuclear-DNA inheritance** occurs equally through both the eggs and sperm; all individuals have half maternal and half paternal nuclear DNA.

# THE QUESTIONS

## The Basics

**1** Which of the following statements is *not* true?

(a) A single evolutionary line led from *Ardipithecus ramidus* to modern humans.

(b) Some hominid traits evolved more rapidly than others.

(c) Brain size increased greatly from early hominids to *Homo sapiens*.

(d) Toolmaking technology has improved greatly over the past 300,000 years.

**2** The out-of-Africa hypothesis states that

(a) all new species of hominins arose in Africa and then migrated to the rest of the world.

(b) many new species of hominins arose outside of Africa and then migrated back to Africa later.

(c) all hominin speciation events occurred outside of Africa.

(d) all speciation events occurred in Africa, but only *Homo sapiens* distributed itself across the globe.

**3** Which of the following features do sponges lack that some other animals have?

(a) segmented body plan

(b) symmetrical body plan

(c) backbone

(d) all of the above

**4** Select the correct terms:

(**Mitochondrial DNA / Nuclear DNA**) is passed on only from the maternal line. (**Mitochondrial DNA / Nuclear DNA**) is inherited from both the mother and the father.

**5** For each of the following cases, identify whether the animal group has bilateral symmetry (B), radial symmetry (R), or no distinct symmetry (N).

_____a. sponges

_____b. cnidarians

_____c. arthropods

_____d. chordates

_____e. primates

## Challenge Yourself

**6** _____ specimens have features that are intermediate between those of *Australopithecus africanus* and *Homo erectus* and provide an amazing record of the evolutionary shift from ancestral hominin characteristics seen in *Australopithecus* fossils to more recent ones seen in *H. erectus* fossils.

(a) *Homo sapiens*

(b) *Homo neanderthalensis*

(c) *Homo habilis*

(d) *Ardipithecus ramidus*

**7** Briefly describe the key differences that distinguish monotreme, marsupial, and eutherian mammals.

**8** Place the following hominids in the correct order from earliest to most recent by numbering them from 1 to 5.

_____ a. archaic *Homo sapiens*

_____ b. *Australopithecus afarensis*

_____ c. modern *Homo sapiens*

_____ d. *Homo habilis*

_____ e. *Homo erectus*

## Try Something New

**9** You visit your local museum of natural history and come upon an exhibit showcasing hominin fossils that date back 300,000–400,000 years ago. You notice that these fossils have features intermediate between those of *Homo erectus* and those of "anatomically modern" *Homo sapiens*. Who were these fossils?

(a) archaic *Homo erectus*

(b) archaic *Homo sapiens*

(c) *Homo neanderthalensis*

(d) *Homo habilis*

**10** In 2004, scientists discovered the fossilized remains of an extinct species they named *Tiktaalik roseae*. The fossil appears to be a transitional species between (aquatic) fishes and (terrestrial) four-legged amphibians. In what environments do you predict this animal lived?

**11** Give an example animal for some of the key adaptations of chordates: (a) backbone, (b) skull, (c) amniotic egg, (d) bony skeleton. Choose a different animal for each adaptation.

**12** How does the fact that all ethnic groups *except* Africans contain some Neanderthal DNA (1–4 percent of their DNA) support the out-of-Africa hypothesis for the origin of modern humans (*Homo sapiens*)?

## Leveling Up

**13** *Write Now* **biology: if we were not alone** Fossil evidence indicates that in the relatively recent past (about 30,000 years ago), anatomically modern humans, or *Homo sapiens*, may have shared the planet with at least three other distinct hominins: *H. erectus*, *H. neanderthalensis*, and *H. floresiensis*. If one or more of these species were alive today, how would their existence affect the world as we know it?

**14** **Is it science?** Watch the original 1968 movie *Planet of the Apes*. Document as many scientific problems with the movie as you can. Which of the apes' adaptations would be biologically possible, and which ones would be impossible, from your understanding of apes on Earth today? If this species did evolve to have the adaptive traits of *Homo*, would its members still be called apes?

**For more, visit digital.wwnorton.com/bionow2 for access to:**

# Amazon on Fire

*Can the world's largest rainforest survive a vicious cycle of wildfires and climate change?*

**After reading this chapter you should be able to:**

- Define the biosphere and the role that humans play in it.
- Articulate the difference between a biotic factor and an abiotic factor in the study of ecology.
- Distinguish between climate and weather.
- Describe the greenhouse effect.
- Compare and contrast the water cycle and the carbon cycle.
- Explain how global warming contributes to climate change.
- List the consequences of climate change, and describe one consequence in detail.

# CHAPTER
# 18

## GENERAL
## PRINCIPLES OF
## ECOLOGY

human experience. We _____
and we use fire in our ca_____
tion, on a daily basis," says Ba_____
have an imperfect relationship w_____
vulnerable to fire; we don't comp_____
this tool."

Balch decided to try to control it j_____
to learn about fire's impact on the_____
By performing a planned burn in a_____
area, her team could collect data that_____
obtained by studying an accidental _____
fire: inventories of the plants in the a_____
and after a fire, a census of the anim_____
area, and other measurements (Fig_____
With a crew of other researchers, Bal_____
trated an experiment using three 0_____
kilometer plots of forest. One control _____
never be burned, one experimental_____
burned once every 3 years, and _____

_____ment,
_____includes
_____organisms)
_____igure 18.3).
_____natural world
_____to change our
_____lt, perhaps even

_____n the Amazon in
_____t the Yale School of
_____dies. She is an
_____w organ-
_____d affect

Flames rise from the forest floor, licking at Jennifer Balch's heels. Balch walks carefully ahead of the heat, stoking it, encouraging it. She tips a large metal container, dripping flaming kerosene onto another pile of dead branches on the ground (**Figure 18.1**). The forest burns behind her.

Balch steps back to look at her work—the destruction of a small patch of the Amazon rainforest. Balch, an ecologist at the University of Colorado Boulder, is in the southern part of the Amazon basin in Brazil, in the state called Mato Grosso. There, on a small square kilometer and a half of land belonging to a soybean farmer, Balch is experimenting to see what happens when the Amazon burns.

The Amazon is the largest rainforest in the world and a critical part of our planet's biosphere. The **biosphere** consists of all of Earth's organisms, plus the physical space we all inhabit. It includes inorganic chemicals like water, our nitrogen-rich atmosphere, every living organism, and more (**Figure 18.2**). Put more simply, the biosphere is the integration of all the environments on Earth. It is crucial to our survival and well-being because humans depend on the biosphere for food and raw materials.

Within the biosphere, the Amazon is home to more than half of the world's millions of species of plants and animals. It also contains one-fifth of the world's freshwater. Yet the Amazon is under threat: since 1960, the human population of the Brazilian Amazon region has increased from 6 million to 25 million people, leading to a dramatic expansion of agriculture, including major deforestation as humans cut down trees to make room for roads, cattle, and fields. The forest cover in the Amazon has declined by 20 percent during this period, and the grasslands and pastures that have taken its place—including soybean fields like the one Balch works at—are threatening the rainforest that is left.

### JENNIFER BALCH

Jennifer Balch is an ecologist at the University of Colorado Boulder who studies how fire disturbance affects ecosystems. In 2004, she initiated a one-of-a-kind experimental burn study in the Amazon to see how wildfires affect the tropical rainforest.

Figure 18.1

**Jennifer Balch ignites a controlled burn in an experimental plot in the rainforest**

It's traditional for farmers in Brazil to clear land for agriculture by setting fires to burn away trees and brush. In addition, fire is a useful tool for clearing out old grasses to encourage the growth of new grasses for cattle to feed on. Yet when these fires are not kept under control, they can spread to nearby rainforest and destroy it. Historically, this has not been a major concern, because the Amazon is wet and humid, so forest fires burned slowly, extinguishing themselves before spreading too far. Yet a dramatic shift in the planet's climate is making the Amazon drier and more vulnerable to wildfires.

Our planet is warming. As a result of global climate change, the temperature in Earth's atmosphere is increasing, causing less rainfall and more drought in the Amazon, leaving the forests primed for fire. And they've begun to burn. Even worse, that loss of trees adds to the warming, resulting in a vicious cycle in which the Amazon becomes hotter and drier, burns again, and contributes to further climate change.

Figure 18.2

**The biosphere is Earth and all of its inhabitants**

This is a view of Earth from space. The atmosphere, Earth's surface, and all of the organisms make up the biosphere.

Abiotic factors: rocks, water, air

Biotic factors: plants, animals, microbes

Figure 18.3

**Amazon rainforest ecology**

Ecology is the study of how living organisms (biotic factors) interact with other organisms, and how they all interact with their nonliving (abiotic) environment.

**Q1:** List as many biotic and abiotic factors in this photograph as you can.

**Q2:** Is the forest part of the biotic or abiotic environment? Explain.

**Q3:** Is the river part of the biotic or abiotic environment? Explain.

Balch and other scientists have been investigating whether it might be possible to intervene in that cycle to slow or stop the loss of the Amazon rainforest. Their latest experiment, burning plots of Amazon rainforest to see what happens afterward, is an experiment in ecology. **Ecology** is the scientific study of interactions between organisms and their environment, where the environment of an organism includes both **biotic** factors (other living organisms) and **abiotic** (nonliving) factors (**Figure 18.3**). Ecology helps us understand the natural world we live in, but humans continue to change our biosphere in ways that are difficult, perhaps even impossible, to fix.

# Hot and Dry

Balch began setting fires in the Amazon in 2004 as a graduate student at the Yale School of Forestry and Environmental Studies. She is an ecologist, a scientist fascinated by how organisms and environments interact with and affect each other.

Balch has always been interested in the link between humans and fire. "Fire is integral to the human experience. We play with fire as kids, and we use fire in our cars, through combustion, on a daily basis," says Balch. "But humans have an imperfect relationship with fire. We are vulnerable to fire; we don't completely control this tool."

Balch decided to try to control it just enough to learn about fire's impact on the Amazon. By performing a planned burn in a restricted area, her team could collect data that cannot be obtained by studying an accidental or escaped fire: inventories of the plants in the area before and after a fire, a census of the animals in the area, and other measurements (**Figure 18.4**). With a crew of other researchers, Balch orchestrated an experiment using three 0.5-square-kilometer plots of forest. One control plot would never be burned, one experimental plot would be burned once every 3 years, and a second experimental plot would be burned once every year.

Figure 18.4

**Scientists take measurements before a planned burn in the Amazon**

To prevent fires in their two experimental plots from spreading, each plot was protected by a firebreak—a perimeter around the plot, several meters wide, cleared of all plants, sticks, grasses, and debris. Once the plots were prepared, it was time to set the fires. The team walked along 10 kilometers of trail cutting through the plots, dripping flaming kerosene from special tanks and watching the forest burn.

The experiment began not a moment too soon. In 2005, the Amazon experienced an extreme drought. Experts called it a "hundred-year drought," meaning that such a drought is expected to occur, on average, only once in a century. But just five years later, in 2010, there was another hundred-year drought, this one even more widespread and severe. And in both cases, extensive wildfires followed the drought, destroying over

**MICHAEL COE**

Michael Coe is a hydrologist and leader of the Amazon Program at the Woods Hole Research Center in Massachusetts. He investigates how humans affect the water and energy balance in tropical South America.

30,000 square miles of rainforest. In 2005 alone, fires increased 20-fold in the year following the drought. "It was a huge increase," says Michael Coe, a researcher at the Woods Hole Research Center in Massachusetts and one of Balch's collaborators. "And if humans are increasing drought frequency, that's going to be a big problem."

Humans, researchers suspect, are increasing the number and intensity of droughts in the Amazon through climate change. It's important to distinguish between "climate" and "weather." **Weather** refers to short-term atmospheric conditions in a limited geographic area, such as today's temperature, precipitation, wind, humidity, and cloud cover. **Climate** describes the prevailing weather of a region over relatively long periods of time (30 years or more). Organisms are more strongly influenced by climate than by any other feature of their environment. On land, for example, features of climate such as temperature and precipitation determine whether a particular region is desert, grassland, or tropical forest.

**Climate change**, then, is a large-scale and long-term *alteration* in Earth's climate, and it includes such phenomena as global warming, change in rainfall patterns, and increased frequency of violent storms. Although Earth has gone through many changes in its average climate over its 4.6-billion-year history, the speed of the change that has taken place in the past 100 years is without precedent in the climate record. Climate change in recent history has been caused to a large extent by human actions, and its consequences are likely to be negative for people and ecosystems around the world (**Figure 18.5**).

"There are a lot of indications that climate change causes increased drought events, causing increased fire events," says Balch. "It's hard to predict the future, but that's definitely the trend."

## A Warmer World

The terms "global warming" and "climate change" are related but not synonymous. **Global warming** is a significant increase in the average surface temperature of Earth over decades or more. Temperature on Earth is generally determined by the angle at which sunlight strikes the planet. Sunlight strikes Earth most directly at the equator, but at a more slanted angle near the North and

**Trift Glacier, Switzerland**
1948    2002    2006

Figure 18.5

## Consequences of climate change

Extreme weather and new rainfall patterns caused by climate change have led to increased flooding in some areas and drought in others, bringing about ecosystem and habitat destruction, and species extinction. The increase in global temperatures has led to melting of glaciers and polar ice, causing sea level rise, increase in ocean temperatures, and coral reef bleaching and death. Effects on our food sources include increased crop failure and depletion of important fisheries.

**Q1:** Name two ways in which climate change affects the frequency and severity of floods.

**Q2:** How has climate change caused a rise in sea level?

**Q3:** Give an example of an environmental effect of climate change in your state or region.

South Poles (**Figure 18.6**). Where sunlight hits the atmosphere more directly, its solar energy is more concentrated. For this reason, more solar energy reaches the equator, making it and neighboring tropical regions much warmer than the poles. Because Earth is tilted on its axis as it revolves around the sun on its annual orbit, sunlight hits different areas of the planet at different angles over the course of the year. This is the cause of our seasons: the Northern Hemisphere experiences summer when Earth is tilted toward the sun, while the Southern Hemisphere simultaneously experiences winter because it is tilted away.

Global warming, however, is not dependent on how sunlight strikes the Earth. Instead, it is caused by an increase in **greenhouse gases**. Some gases in Earth's atmosphere, such as carbon dioxide ($CO_2$), water vapor ($H_2O$), methane ($CH_4$), and nitrous oxide ($N_2O$), absorb heat that radiates away from Earth's surface, preventing it from being released into space. These gases are called greenhouse gases because they

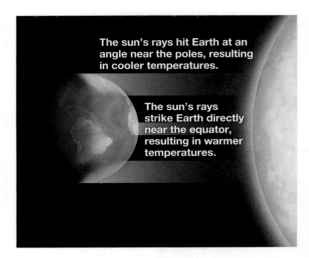

The sun's rays hit Earth at an angle near the poles, resulting in cooler temperatures.

The sun's rays strike Earth directly near the equator, resulting in warmer temperatures.

Figure 18.6

**Sunlight strikes Earth most directly at the equator**

The angle at which the rays of the sun strike Earth determines how much energy or heat reaches Earth's surface. The more direct the rays are when they strike Earth, the more heat they deliver.

> **Q1:** Why is it colder at the poles than at the equator?
>
> **Q2:** Why is it warmer at the equator than at the poles?
>
> **Q3:** During part of the year the Northern Hemisphere is tilted at a more direct angle to the sun than the Southern Hemisphere is, and for the other part of the year the opposite is true. How does this tilt explain temperature differences in summer and winter?

function much as the walls of a greenhouse or the windows of a car do: they let in sunlight and trap heat in a process known as the **greenhouse effect** (**Figure 18.7**).

Greenhouse gases are not inherently bad; in fact, they have existed in Earth's atmosphere for more than 4 billion years, and they play an important role in maintaining temperatures that are warm enough for life to thrive on Earth. Yet human activities, primarily the burning of fossil fuels, have released an excess of greenhouse gases into the atmosphere, especially the infamous king of greenhouse gases: **carbon dioxide** ($CO_2$).

Scientists have estimated atmospheric $CO_2$ levels for both the recent and the relatively distant past, up to hundreds of thousands of years ago, by measuring $CO_2$ concentrations in air bubbles trapped in ice. This evidence shows a near-perfect historical correlation between $CO_2$ levels and the surface temperature on Earth. Yet during the last 200 years, levels of atmospheric $CO_2$ have risen greatly—from roughly 280 to 400 parts per million (**Figure 18.8**). Measurements from ice bubbles show that this rate of increase is greater than even the most sudden increase that occurred naturally during the past 420,000 years. Carbon dioxide levels are now higher than those estimated for any time during that period.

The cause of the increasing levels of atmospheric $CO_2$ has been linked to the burning of fossil fuels like coal and oil, which releases $CO_2$ into the air. About 75 percent of the current yearly increase in atmospheric $CO_2$ is due to the burning of fossil fuels. Logging and burning of forests are responsible for most of the remaining 25 percent of the increase, but industrial processes also make a significant contribution. "Humans have a giant impact on the environment," says Coe. "With more than seven billion people on the planet, individual decisions add up to big global changes." (For more on the environmental impact of personal choices, see "How Big Is Your Ecological Footprint?" on page 332.)

All this is to say that human burning of fossil fuels has led to an increase in carbon dioxide in our atmosphere. That carbon dioxide acts as a greenhouse gas and traps additional heat as it leaves the atmosphere, causing temperatures on Earth to rise. Since the early twentieth century, Earth's mean surface temperature has increased by about 1.4°F (0.8°C), and it is estimated to rise another 2°F–11.5°F (1.1°C–6.4°C) in the future (**Figure 18.9**). As of the date this book went to publication, 2016 was the hottest year ever recorded; 2015 was previously the hottest year ever; 2014 was the winner before that. We keep beating our own record.

Our globe is warming. And the effects of that warming on the biosphere are now evident, especially in the Amazon, where trees dominate the landscape. Trees act as a layer connecting the atmosphere above them and the ground beneath them, absorbing water from the ground and releasing oxygen into the air. Any change in tree cover can change the local

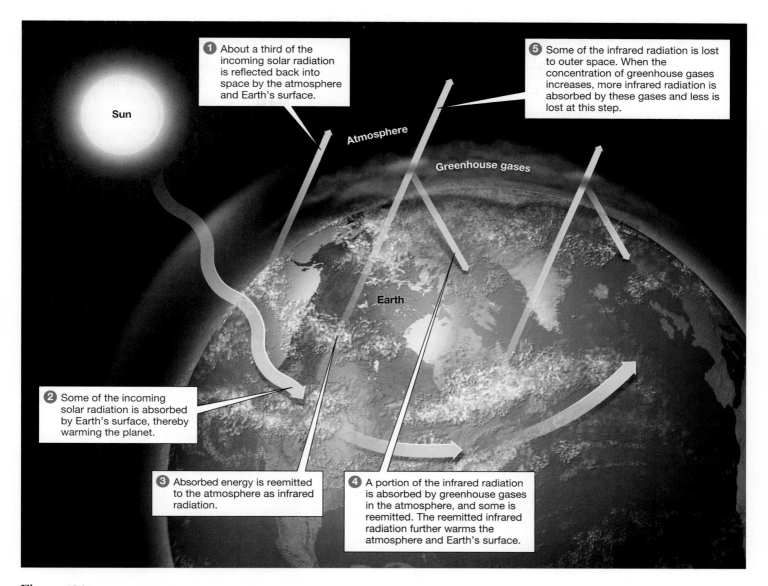

**Figure 18.7**

**How greenhouse gases warm the surface of Earth**

Carbon dioxide ($CO_2$), water vapor ($H_2O$), methane ($CH_4$), and nitrous oxide ($N_2O$) are known as greenhouse gases because they absorb and trap heat that would otherwise radiate away from Earth. ▶▌

**Q1:** How much of the incoming solar energy is reflected back to outer space?

**Q2:** What kind of energy is reemitted to the atmosphere after being absorbed by Earth's surface?

**Q3:** How are greenhouse gases like a blanket on your bed at night?

climate, and increasing wildfires are devastating that tree cover.

"The big question is, What's the limit for these tropical forests? How much fire can they withstand before they convert to something else?" says Balch. That something else would be grasslands, also called savanna. If major parts of the Amazon rainforest were to convert to savanna, the climate would be significantly affected, and there would be no easy way—perhaps no way at all—to revert that land back to rainforest.

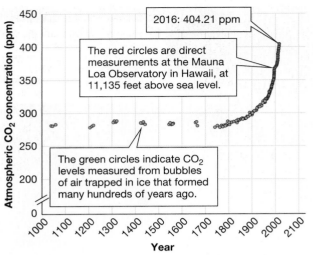

Figure 18.8

## Atmospheric CO₂ levels are rising rapidly

Atmospheric $CO_2$ levels (measured in parts per million, or ppm) have increased greatly in the past 200 years.

> **Q1:** What measurements do the green circles represent?
>
> **Q2:** What measurements do the red circles represent?
>
> **Q3:** For approximately how many years has the Mauna Loa Observatory been recording $CO_2$ levels?

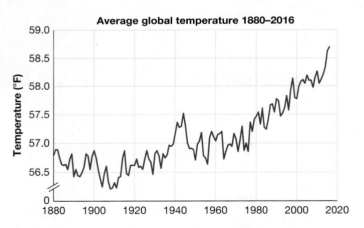

Figure 18.9

## Global temperatures are on the rise

Average global temperature has increased greatly over the 140 years since it has been recorded.

> **Q1:** In what years were global temperatures the lowest?
>
> **Q2:** In what years were global temperatures the highest?
>
> **Q3:** What trend is apparent in this graph of actual global temperatures?

# Fire and Water

During her first burn in 2004, Balch was surprised to see that the plants in the two experimental plots were fairly resistant to fire. The Amazon's tall, dense tree canopy creates humid air below it, protecting the forest from most fires, which tend to burn slowly under such conditions. For this reason, Balch's experimental plots initially suffered little damage.

That wasn't true in the following years. Remember that the Amazon saw a major drought in 2005, and droughts cause tree death, allowing more sunlight to reach the forest floor and dry the leaves and shrubs there, priming the forest for fire (**Figure 18.10**). During the planned burns after the drought in 2005, the two experimental plots experienced major damage. The majority of the trees on the plots died, and grasses took their place—particularly at the forest edge, where it was hottest and driest.

In both the 1-year and the 3-year burned plots, grasses invaded from adjacent pasturelands. This experimental result suggests that wildfires do indeed push the Amazon toward a drier,

Figure 18.10

## Before and after the Amazon drought of 2005

The drought devastated the forest to this extent over much of the forest's range.

# Forest Devastation

Agriculture is the main driver of tropical deforestation, and Brazil clears more land, by far, than any other country in the world. Since 1990, Brazil has cleared over 42 million hectares of forest, an area the size of California. Unless drastic measures are taken, deforestation will continue as farmers and companies clear land to meet rising global food demands.

Assessment available in smartwork5

## Causes of deforestation
(Latin America)

1%
3% 2%
27%
67%

- Agriculture (commercial)
- Agriculture (local/subsistence)
- Infrastructure (e.g., roads)
- Mining
- Urban expansion

## Drivers of forest degradation over time
(Latin America)

8% 3%
16%
73%

- Timber logging
- Uncontrolled fires
- Wood for fuel and charcoal
- Livestock grazing in forest

## Top 5 countries that cleared the most forest between 2000 and 2005

Deforested area, in thousands of hectares

| Brazil | Indonesia | Sudan | Myanmar | Zambia |
|--------|-----------|-------|---------|--------|
| 15,515 | 9,357 | 2,945 | 2,332 | 2,224 |

# How Big Is Your Ecological Footprint?

An action or process is **sustainable** if it can be continued indefinitely without causing serious damage to the environment. The current human impact on the biosphere is *not* sustainable.

Each of us can help build a more sustainable society. We can advocate legislation that fosters less destructive and more efficient use of natural resources, patronize businesses that take measures to lessen their negative impact on the planet, support sustainable agriculture, and modify our own lifestyles. For example, we can increase our use of renewable energy and energy-efficient appliances; reduce all unnecessary use of fossil fuels (for instance, by biking to work or using public transportation); buy seafood from sustainable fisheries; use "green" building materials; and reduce, reuse, and recycle waste. Experts estimate that more than 200 million women around the world wish to limit their family size but have no access to family planning. Those of us who live in developed countries can support aid efforts that provide education, health care, and family-planning services in developing countries.

One measure of sustainability is an **ecological footprint**, which is the area of biologically productive land and water that an individual or a population requires to produce the resources it consumes and to absorb the waste it produces. Scientists compute an ecological footprint using standardized mathematical procedures and express it in *global hectares* (gha). One gha is equivalent to one hectare (2.47 acres) of *biologically productive* space. Approximately one-fourth of Earth's surface is considered biologically productive; this definition excludes areas such as glaciers, deserts, and the open ocean.

According to recent estimates, the ecological footprint of the average person in the world is 2.7 gha, which is about 60 percent higher than the 1.7 gha that would be needed to support each of the world's 7.4 billion people in a sustainable manner. An ecological footprint can also be expressed in **Earth equivalents**, the number of planet Earths needed to provide the resources we use and absorb the wastes we produce. Currently, the global population uses 1.6 Earth equivalents each year (as shown at the bottom of the figure on the facing page).

Overall, such estimates suggest that, since the late 1970s, people have been using resources faster than they can be replenished—a pattern of resource use that, by definition, is not sustainable. As the world population grows, the amount of biologically productive land available per person continues to decline, increasing the speed at which Earth's resources are consumed.

The per capita consumption of Earth's resources by different countries is most directly related to energy demand, affluence, and a technology-driven lifestyle. As people in populous countries such as China and India become wealthier, their ecological footprints are growing rapidly.

What is *your* ecological footprint? If you are a typical American college student, your footprint is probably close to the U.S. average of 8.2 gha. It would take nearly five planet Earths to support the human population if everyone on Earth enjoyed the same lifestyle that you do (see the top row of the accompanying figure). Your ecological footprint depends on four main types of resource use:

1. *Carbon footprint*, or energy use

2. *Food footprint*, or the land and energy and water it takes to grow what you eat and drink

3. *Built-up land footprint*, which includes the building infrastructure (from schools to malls) that supports your lifestyle

4. *Goods-and-services footprint*, which includes your use of everything from home appliances to paper products

If you drive a gas guzzler, live in a large suburban house, routinely eat higher up on the food chain (more beef than grains or veggies/fruits), and do not recycle much, your footprint is likely to be higher than that of a person who uses public transportation, shares an apartment, eats mostly plant-based foods, and sends relatively little to the local landfill. Most of us can significantly reduce our ecological footprint with little reduction in our quality of life, while bestowing an outsized benefit on our planet.

---

grassier savanna-like ecosystem (**Figure 18.11**). "This could be the trigger for changing forest to savanna," says Balch. "The message is that repeated disturbances, such as multiple fires or drought and fire, cause forest to reach its tipping point, to shift into grassland."

While Balch and her team observed changes in plants after fire, ecologist Michael Coe investigated how fire affects its nemesis: water. "The forest is an incredibly important part of the **hydrologic cycle**," the circulation of water from the land to the sky and back again, says Coe (**Figure 18.12**).

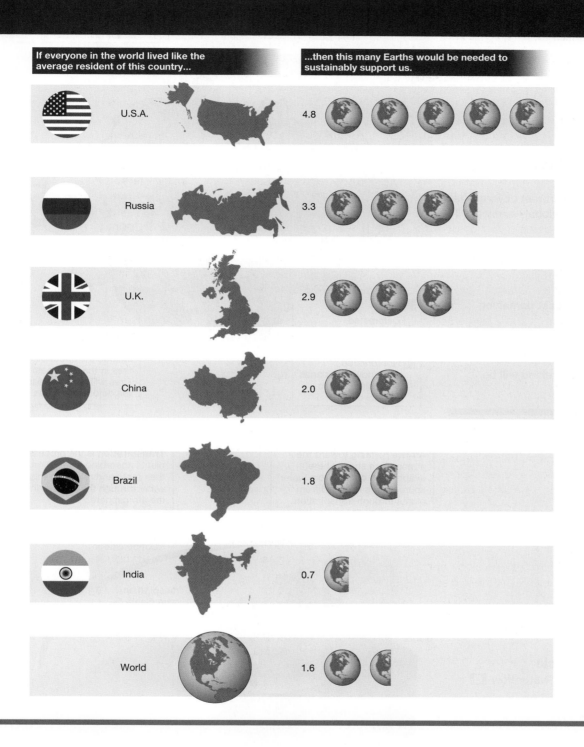

| | | |
| --- | --- | --- |
| U.S.A. | 4.8 | |
| Russia | 3.3 | |
| U.K. | 2.9 | |
| China | 2.0 | |
| Brazil | 1.8 | |
| India | 0.7 | |
| World | 1.6 | |

Near the equator, direct sunlight causes water to evaporate from Earth's surface. The warm, moist air rises because heat causes it to expand, making it less dense and lighter than air that has not been heated. But as it rises, the warm, moist air begins to cool. Because cool air cannot hold as much water as warm air can, much of the moisture from a cooling airmass is "wrung out" and falls as rain.

For this reason, most tropical regions, including the Amazon, receive ample rainfall. Earth has four giant **convection cells** in which warm,

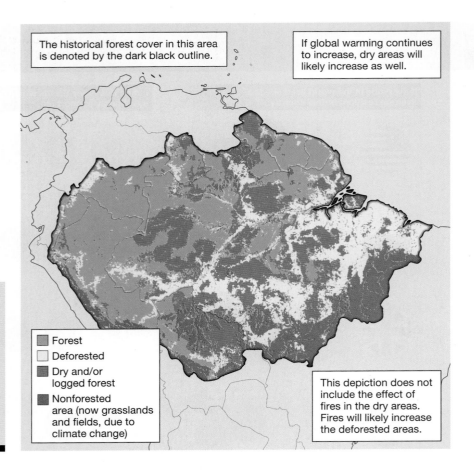

The historical forest cover in this area is denoted by the dark black outline.

If global warming continues to increase, dry areas will likely increase as well.

**Legend:**
- Forest
- Deforested
- Dry and/or logged forest
- Nonforested area (now grasslands and fields, due to climate change)

This depiction does not include the effect of fires in the dry areas. Fires will likely increase the deforested areas.

## Figure 18.11

### A projection of the Amazon rainforest in 2030

This map of future Amazon rainforest coverage is based on the assumption that global warming will persist but not increase.

**Q1:** Where will fire most seriously affect the Amazon rainforest?

**Q2:** Where will fire be the least damaging to the Amazon rainforest?

**Q3:** This map does not include an increase in pasturelands for grazing animals. Do you think more or less pastureland will be needed in 2030? Explain.

Arrows pointing toward the atmosphere show where evaporation occurs from water sources or is released from plants through transpiration to form humidity or clouds.

**Transpiration** is the process of plants absorbing water through their roots and releasing this water through their leaves into the atmosphere.

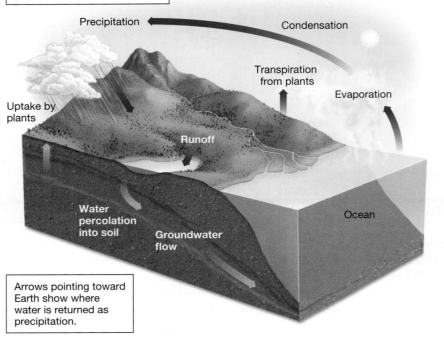

Precipitation | Condensation

Transpiration from plants

Evaporation

Uptake by plants

Runoff

Water percolation into soil

Groundwater flow

Ocean

Arrows pointing toward Earth show where water is returned as precipitation.

## Figure 18.12

### The hydrologic (water) cycle

Arrows indicate the direction of water flow.

**Q1:** What is transpiration?

**Q2:** Why is transpiration important to the water cycle?

**Q3:** If there are fewer plants and therefore less transpiration in a given area, what will happen to the humidity or cloud cover in this area?

moist air rises and cools, releasing moisture as rain or snow depending on temperature, and then sinks back to the ground as dry air (**Figure 18.13**). These convection cells, in combination with the angle of sunlight striking the Earth, play a large role in the creation of regional environments on Earth, such as rainforests and deserts.

In the Amazon, trees are very important players in the water cycle, adds Coe. "The trees pull water out of the soil and evaporate it into the atmosphere in the process of photosynthesizing (through transpiration), so they're the mediators between the rainfall and the streams," he says. "Burning the trees greatly reduces the amount of water getting back into the atmosphere." (See Chapter 5 for more on photosynthesis.)

Each year of Balch's study, Coe joined Balch at the test site. There, his team dug 10-meter-deep soil pits—long, dark caverns in which they inserted instruments to measure the moisture content of the soil. In a healthy forest ecosystem, trees absorb a lot of the water in the soil, leaving it nice and dry, with only minimal water runoff into streams. This is what Coe observed in the control plot where nothing was burned. "A healthy forest uses up almost all the water [in the soil]," says Coe.

But in the other two plots—burned every year or every 3 years—he found that the soil in the pits was wet to the touch. When the forest burned, trees died, so nothing absorbed the moisture from the soil. Consequently, nearby streams were overflowing with water—up to four times the volume of water seen in healthy forests. "That's not a good thing," says Coe. "We're circumventing the natural cycle. Instead of this water going back into the atmosphere, creating more rain and driving vegetation, it's flushing the water out of the system."

And on the burned plots, the invasive grasses that took the place of the trees have very shallow roots, absorb less moisture from the ground, and evaporate less water into the air. In this way, deforestation of large areas—whether through unintentional wildfires or the intentional cutting down of trees—results in less rainfall and hotter temperatures. "If you deforest enough of it, you're going to really decrease the rainfall over a broad swath of this region," says Coe.

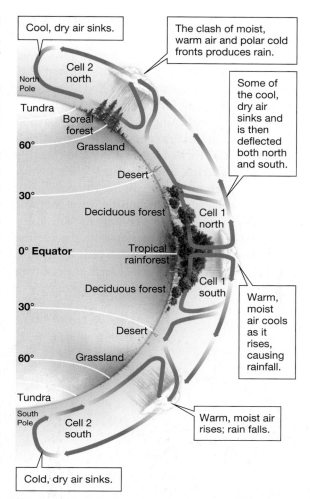

Figure 18.13

**Earth has four giant convection cells**
Two giant convection cells are located in the Northern Hemisphere and two in the Southern Hemisphere.

**Q1:** How do the patterns of rainfall in the Northern and Southern Hemispheres compare?

**Q2:** How do the patterns in the kinds of environments shown in the Northern and Southern Hemispheres compare?

**Q3:** What happens at the equator to make this region so wet?

# The Carbon Games

In addition to measuring the hydrologic cycle and the expansion of grasses into the region, the teams led by Balch and Coe measured how much carbon was released into the atmosphere when the forests burned.

Carbon, in the form of $CO_2$ gas, makes up only about 0.04 percent of Earth's atmosphere, although that percentage has been creeping upward every year for the last 200 years and causing global warming, as we've seen. Carbon is also found in Earth's crust, where carbon-rich sediments and rocks formed from the remains of ancient marine and terrestrial organisms. Carbon is present in every living thing.

Living cells are built mostly from organic molecules—molecules that contain carbon atoms bonded to hydrogen atoms. After oxygen, carbon is the most abundant element in cells by weight; every one of the large biomolecules in an organism has a backbone of carbon atoms. Living organisms, in both aquatic and terrestrial ecosystems, acquire carbon mostly through photosynthesis. Aquatic producers, such as photosynthetic bacteria and algae, absorb dissolved $CO_2$ and convert it into organic molecules using sunlight as a source of energy. Plants, the most important producers in terrestrial ecosystems, absorb $CO_2$ from the atmosphere and transform it into food with the help of sunlight and water. This is how a plant builds itself—its leaves, stems and branches, flowers, and other structures.

The transfer of carbon within biotic communities and their physical surroundings, the abiotic world, is known as the global **carbon cycle** (**Figure 18.14**). (For more on nutrient

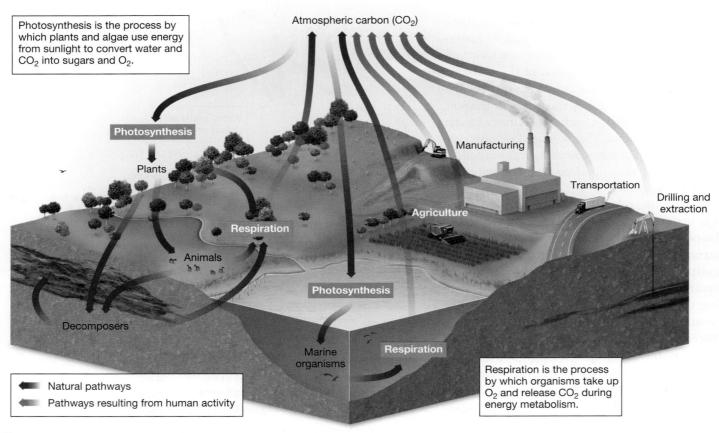

**Figure 18.14**

**The carbon cycle**
Arrows indicate the direction of carbon flow.

**Q1:** What are three ways that carbon is released into the atmosphere?

**Q2:** Are all of the pathways you listed for question 1 affected by human activity?

**Q3:** What are two biotic reservoirs of carbon?

cycling, see Chapter 21.) One way that carbon is transferred between the biotic and abiotic worlds is through combustion—the burning of carbon-rich materials, living or not.

Some of the organic matter from ancient organisms has been transformed by geologic processes into deposits of fossil fuels such as petroleum, coal, and natural gas. When we extract these fossil fuels and burn them to meet our energy needs, the carbon that was locked in these deposits for hundreds of millions of years is released into the atmosphere as carbon dioxide.

Plants also release carbon back into the atmosphere when they are burned. To assess the amount of carbon released by a burning rainforest, Balch and her team measured the amount of leaf litter and branches—the biomass—on the forest floor both before and after a burn, as well as the number of trees affected. Half of that biomass is carbon, so half of the difference between the biomass before and after the burn is the amount of carbon that was released into the atmosphere. It was a lot.

The very first burn, for example, released "about 20 tons of carbon per hectare," says Balch. That's approximately equal to the carbon emitted by a passenger car driven 864,000 miles (burning about 40,000 gallons of gasoline). And burns have another, indirect, detrimental effect that Balch wasn't able to directly measure: they destroy trees that normally absorb carbon dioxide from the air. In a normal year, trees in the Amazon absorb approximately 1.5 billion tons of carbon dioxide, helping to slow climate change. In this way, they act as a **carbon sink**, a natural or artificial reservoir that absorbs more carbon than it releases.

But during the 2005 drought, as trees died and rotted, the rainforest stopped absorbing more carbon dioxide than it gave off, and it actually released carbon *into* the atmosphere. In 2005, 5 billion tons of carbon were released. Instead of acting as a carbon sink, the Amazon became a **carbon source**, a reservoir that releases more carbon than it absorbs (**Figure 18.15**). And with increasing temperatures, decreasing rainfall, and increasing wildfires, ecologists fear that the Amazon could transition more and more into a carbon source instead of a carbon sink. If that happens, our planet will lose one of its greatest buffers against future climate change.

Figure 18.15

**Carbon sinks and sources**

Q1: How does a carbon source contribute to global warming?

Q2: How does a carbon sink protect against global warming?

Q3: How can trees act as both a source and a sink?

# Waiting and Watching

Balch and her crew have stopped burning the plots and have begun to record what happens to them as they recover. In addition to making direct observations of which plants grow back and how their growth affects the water and carbon cycles, the team is conducting new experiments on the burned plots to see whether it is possible to help prevent the invasion of grasses and encourage a return to rainforest.

In one area, the scientists are planning to plant different types of trees to see whether particular species can successfully reestablish themselves in a mat of grasses and retake that land. Another experiment, also still in the planning phase, will test different amounts of watering to see whether the lack of precipitation is preventing the growth of new trees. If any strategies work well on the small plots, perhaps they can be adapted to save larger areas of rainforest affected by fire.

"We're hoping we find ways to help the forest recover," says Balch. Still, the researchers worry about the future of the region. "Have we crossed a threshold where this part of the forest isn't going to grow back as rainforest?" asks Coe, worry in his voice. If that is true for the small experimental plots, it bodes danger for large regions of the Amazon that border agricultural land and are threatened by wildfires.

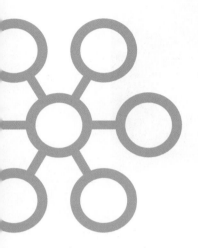

# REVIEWING THE SCIENCE

- **Ecology** is the study of interactions between organisms and their environment. All ecological interactions occur in the **biosphere**, which consists of all living organisms on Earth (**biotic** factors), together with the physical environment they inhabit (**abiotic** factors).

- **Weather** consists of the short-term atmospheric conditions in a given area, such as temperature, precipitation, wind, humidity, and cloud cover. **Climate**, the prevailing weather of a specific region over relatively long periods of time, has a major effect on the biosphere. Climate is determined by incoming solar radiation, global movements of air and water, and major features of Earth's surface.

- **Climate change** is a large-scale and long-term alteration in Earth's climate, and it includes such phenomena as more extreme temperatures, change in rainfall patterns, and increased frequency of violent storms.

- Carbon dioxide acts as a **greenhouse gas** and contributes to the **greenhouse effect** by absorbing heat released at Earth's surface, rather than allowing it to leave the atmosphere. The greenhouse effect has led to a rapidly rising average global temperature, a phenomenon known as **global warming**.

- The concentration of **carbon dioxide** ($CO_2$) gas in the atmosphere is increasing at a dramatic rate because of an increase in **carbon sources**, such as the release of $CO_2$ through the burning of fossil fuels, and a loss of **carbon sinks**, such as the absorption of $CO_2$ through photosynthesis in large forests.

- Both the **hydrologic cycle** and the **carbon cycle** are affected by global warming, which disrupts the natural cycling of water and carbon molecules in the biosphere.

- The area of the biosphere required to produce the resources and to absorb the waste produced by an individual or population is known as the **ecological footprint**. The ecological footprint is often expressed in **Earth equivalents**, the number of planet Earths needed to provide the resources and absorb the waste of an individual or population.

# THE QUESTIONS

## The Basics

**1** The biosphere is

(a) all organisms on Earth, together with their physical environments.

(b) crucial to human survival and well-being.

(c) a source of food and raw materials for human society.

(d) a web of interconnected ecosystems.

(e) all of the above

**2** Greenhouse gases function by

(a) blocking sunlight but letting out heat from Earth to outer space.

(b) absorbing heat that radiates from Earth that would otherwise escape to outer space.

(c) absorbing heat that radiates from the sun toward Earth.

(d) releasing heat that radiates from Earth into outer space.

**3** Which of the following is an abiotic component of the biosphere?

(a) algae

(b) insects

(c) lichen

(d) water

(e) none of the above

**4** Place the following components of the hydrologic cycle in the correct order by numbering them from 1 to 5, beginning with precipitation.

_____ a. precipitation

_____ b. uptake by plants

_____ c. transpiration

_____ d. water percolation into soil

_____ e. condensation

**5** Select the correct terms:

(Climate / Weather) is the short-term atmospheric conditions in an area, while (climate / weather) is the average atmospheric conditions in a given area over a long period of time. (Climate change / Global warming) consists of the long-term and large-scale changes in atmospheric conditions, which have been brought about by (climate change / global warming), an increase in the average global temperature.

## Challenge Yourself

**6** The carbon cycle and the water (hydrologic) cycle are similar in that

(a) both cycle molecules between abiotic and biotic components of the environment.

(b) photosynthesis is a critical process in both.

(c) both are involved in global warming and climate change.

(d) all of the above

(e) none of the above

**7** An ecologist is studying an area of forested land that experiences a forest fire every 5 years. In the years between forest fires, the area of study absorbs 300 million tons of $CO_2$. If a forest fire releases 200 million tons of $CO_2$ in this area, does this forested area act as a carbon sink or a carbon source? Why?

**8** Place the following events in the correct order of their onset by numbering them from 1 to 5, beginning with the earliest.

_____ a. Climate change, including more extreme weather

_____ b. Increase in human population and activities

_____ c. Increase in release of $CO_2$ and other greenhouse gases into the atmosphere

_____ d. Increase in calls to control the production and release of greenhouse gases

_____ e. Global warming

## Try Something New

**9** Which of the following behaviors could *decrease* the production of greenhouse gases, and thus perhaps curb global warming?

(a) using ethanol or biodiesel rather than fossil fuels to power autos

(b) using coal rather than gas to warm homes

(c) using solar or wind power rather than fossil fuels to produce electricity

(d) all of the above

(e) none of the above

**10** Reviewing Figure 18.13, identify the type of environment at your latitude, and explain how it is influenced by Earth's convection cells.

**11** Discuss briefly (in one to two sentences) how the more extreme weather caused by global warming may affect where you live.

## Leveling Up

**12** **Looking at data** Review Figures 18.8 and 18.9. Describe the change in atmospheric $CO_2$ levels from 1880 to the present, and the change in average global temperature for the same time period. What were the $CO_2$ level and the average temperature in 1880? in 1960? in 2016? Is there a relationship between the two variables? How would you explain the pattern you see here?

**13** **Life choices** You can estimate your impact on the planet by using one of the many "footprint calculators" on the Internet. Take three online quizzes available from organizations such as the Global Footprint Network, the Nature Conservancy, and the U.S. Environmental Protection Agency. Each site calculates your ecological footprint a little differently. Compare and contrast the different sites you used. Which one do you think is the most accurate? Why? Which one do you think is the most superficial? Why? Write a list of ways you can decrease your ecological footprint, and try to adhere to your new lifestyle!

**14** *Write Now* **biology: climate denial** Watch the documentary *Frontline's Climate of Doubt* (2012), produced by PBS. Write an essay reflecting on what you learned from the film and from this chapter about global warming. Use your knowledge of your ecological footprint to reflect on how important it is for all citizens to be aware of this global crisis.

**For more, visit digital.wwnorton.com/bionow2 for access to:**

# Zika-Busting Mosquitoes

*Genetically engineered insects help stop the spread of dengue, malaria, and Zika.*

- - - - - - - - - - - - - - - - - - - - - - - - - - - - - - - - - - - - - - - - - - -

## After reading this chapter you should be able to:

- Articulate the difference between population size and population density.
- Interpret graphical population data to determine whether the population exhibits logistic or exponential growth.
- Explain the concept of carrying capacity, and give examples of factors that can increase or decrease carrying capacity for a population.
- Differentiate between density-dependent and density-independent changes in population size.
- Illustrate population cycles of predators and prey.

CHAPTER

19

GROWTH OF
POPULATIONS

t is impossible to know when an infected mosquito bit the first person in Miami-Dade County, Florida. But on Friday, July 29, 2016, health officials confirmed that a woman and three men in a small area north of downtown Miami had been infected with the Zika virus through the bite of a local mosquito. It was the moment everyone had been dreading: the first known cases of the virus being transmitted by mosquitoes in the United States. "As we have anticipated, Zika is now here," Tom Frieden, the director of the U.S. Centers for Disease Control and Prevention (CDC), told reporters that afternoon (**Figure 19.1**).

By July 2016, Zika had become a household name. Zika fever, the illness that results from infection with Zika virus, is a mild sickness that typically lasts less than a week and involves fever, red eyes, headache, and a possible rash. But Zika has another, much more dire effect on health. A year before the Miami outbreak, doctors in Brazil had begun noticing an increasing number of infants born with a serious birth defect known as microcephaly, in which the head and brain are smaller than expected; they have not developed properly. Microcephaly can cause many problems in growing infants, including seizures, speech and intellectual delays, feeding and movement problems, and hearing and vision loss. Brazil's new cases of microcephaly were soon linked to a rapidly spreading virus passed through the bite of a mosquito: Zika.

The scientific evidence linking Zika virus and microcephaly is strong: infected mothers pass the virus to their fetuses, where it may stunt brain growth (**Figure 19.2**). Studies suggest that infants exposed to the Zika virus in utero are more than 50 times more likely to be born with microcephaly than those who are not exposed.

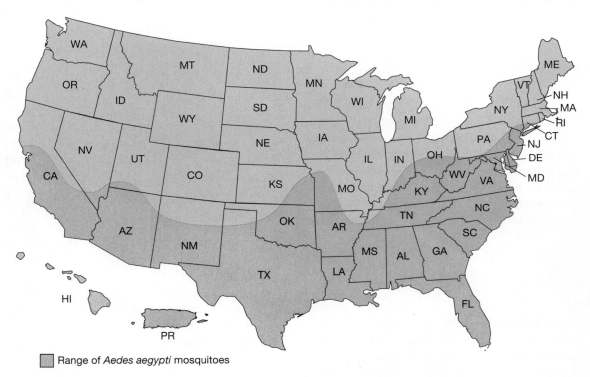

Range of *Aedes aegypti* mosquitoes

Figure 19.1

**Range of the mosquito species most likely to carry the Zika virus**

*Aedes aegypti* mosquitoes spread viruses like Zika, dengue, and chikungunya more often than other mosquito species do.

Q1: What parts of the United States are within the range of the mosquito that carries Zika?

Q2: What areas are *not* within the mosquito's range? Why do you think that is?

Q3: Find your own state, and where you are in the state. Are you at risk of contracting Zika?

In February 2016, the World Health Organization declared the spread of Zika virus to be a "public health emergency of international concern."

Since the outbreak in Brazil, governments around the world have begun looking for ways to stop Zika, and the local transmission in Florida galvanized U.S. officials and scientists. "It's scary," says Matthew DeGennaro, a biologist who studies mosquito genetics and behavior at Florida International University in Miami. "But people in other countries face these things all the time, and not much is being done about it. It is possible this [outbreak] will focus Americans' attention on mosquito-borne illness."

The human species has a long history of battling mosquito-borne viruses, including dengue virus and West Nile virus. Along the way, we've learned that when a vaccine is lacking, the best way to stop the virus is to stop the insect. Yet despite efforts to eradicate mosquito populations around the world, the pests continue to roar back no matter what we throw at them. Mosquitoes quickly develop resistance to common insecticides, for example, and in at least one case they showed an adapted behavior, attacking farmers outdoors in the early morning rather than indoors during the night when the farmers were protected by mosquito nets.

But now there's a new weapon in the arsenal against mosquito-borne diseases. Taking a technological approach to the problem, scientists have begun to target mosquito populations by using the insects themselves as weapons. It's a new kind of war—mosquito versus mosquito.

# Population Control

One month after the first four reported cases of Zika transmission in Florida, officials captured Zika-infected mosquitoes at seven different locations on Miami Beach—the smoking gun to show that local disease transmission was indeed occurring. By October 2016, mosquitoes bearing Zika virus had spread another 3 miles north.

Zika is spread by females of both the *Aedes aegypti* species and the less common *Aedes albopictus* species, although the latter is less likely to bite humans because it breeds in rural areas and feeds on other animals in addition to humans. These two types of mosquitoes also

Figure 19.2

**A Brazilian mother holds her microcephalic baby**
The incidence of microcephaly in newborns has skyrocketed, and the increase has been linked to Zika infection of the mother during pregnancy.

serve as carriers, or *vectors*, for dengue, chikungunya, and other viruses. (Malaria, which is caused by parasitic protists, is transmitted among humans by female mosquitoes of the genus *Anopheles*.) Active primarily during the day, a female mosquito feeds on human blood via a tubelike mouthpart (which male mosquitoes lack) that pierces the skin of a host. During that moment, virus particles in the mosquito's saliva can be transferred to the person's skin. Once on the skin, the virus is able to replicate in skin cells and then can spread to the lymph nodes and bloodstream.

In addition to being transmitted by mosquitoes, Zika can spread from person to person via body fluids: blood, tears, semen, and saliva. In some cases, the virus can remain in those fluids for months after

**MATTHEW DEGENNARO**

Matthew DeGennaro is a neurogeneticist at Florida International University in Miami, studying mosquito genetics and behavior. As a Florida resident, he witnessed the outbreak of Zika in Miami and advocates the use of genetically modified mosquitoes to control mosquito populations and stop the spread of disease.

infection (**Figure 19.3**). Slowing the spread of Zika, therefore, involves the screening of donated blood and the proper use of condoms during sex. But true prevention ultimately requires decreasing the incidence of mosquito bites, which can be accomplished by getting rid of the mosquitoes.

Scientists have taken many approaches to trying to reduce the mosquitoes' **population size**, the total number of individuals in the population. A **population** is a group of organisms of the same species in a defined area. Population size tends to change over time—sometimes increasing, sometimes decreasing. Whether the size of a population increases or decreases depends on the number of births and deaths in the population, as well as on the number of individuals that enter (immigrate) or

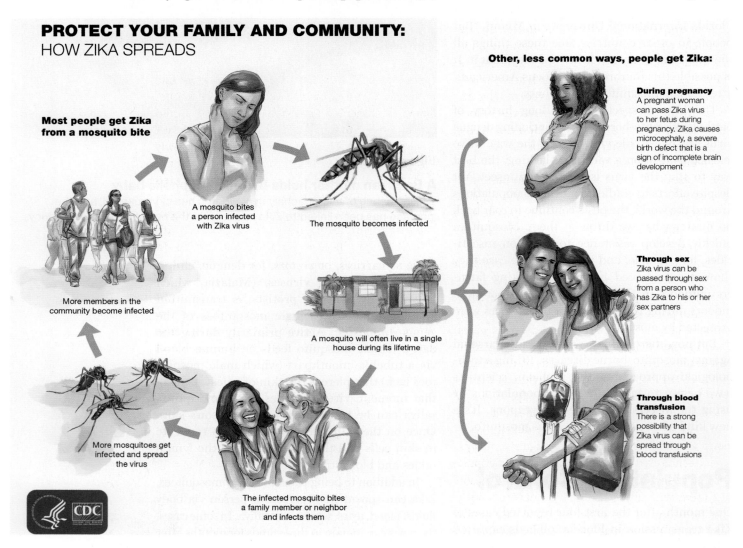

Figure 19.3

**How Zika is transmitted**
The CDC produces posters such as this for public health issues including Zika, sexually transmitted diseases, and the flu.

**Q1:** What is the main way by which someone is infected with the Zika virus?

**Q2:** Judging by the poster and your knowledge of mosquito behavior, what can you do to decrease your risk of being infected with the Zika virus?

**Q3:** Besides the transmission methods shown on the poster, what are some other ways you could become infected with the Zika virus?

leave (emigrate) the population. Birth and immigration increase population size; death and emigration reduce it. Environmental factors also have a strong impact on population size. Mosquito populations, for example, increase in warmer, wetter weather when the conditions are ripe for reproduction.

To target mosquito populations, scientists rely on **population ecology**, the study of the size and structure of populations and how they change over time and space. Population ecology can help us determine where mosquitoes live, eat, and breed. Mosquitoes congregate and lay their eggs at water sources, and many prefer urban environments with trash receptacles and concentrated groups of people. At these locations we find the highest **population density**, or number of individuals per unit of area.

To calculate population density, total population size is divided by the corresponding area of interest. For example, in an urban neighborhood of Rio de Janeiro, Brazil, where people were being regularly infected with dengue virus, scientists captured mosquitoes in mosquito traps and calculated the neighborhood's population of female mosquitoes to be 3,505 pregnant females. The neighborhood has an area of 911 acres, so the female mosquito population density was 3.85 mosquitoes per acre (3,505 mosquitoes divided by 911 acres). Keep in mind that population density is often difficult to measure, because it depends on an accurate count of the population size. Individuals may be hard to detect, may move between populations, or may inhabit a complex, hard-to-define area.

Scientists can track population density to see how well eradication efforts work against mosquitoes. As mentioned earlier, one of the most popular ways of reducing mosquito populations is poisoning them with insecticides. But that's not ideal, scientists agree. "Insecticides affect the entire insect population usually, and that leads to overall imbalances in the environment," says DeGennaro. "So putting a chemical into the environment is something you need to do very carefully."

An alternative approach, thanks to advanced genetic technologies, is to tweak the genes of the mosquitoes to sabotage their reproduction. If the mosquitoes can't reproduce, population size will plummet, and fewer mosquitoes means fewer people being bitten and infected. This gene tweaking is called **genetic modification**, or **GM**—altering the genes of an organism for a specific purpose. It is also referred to as genetic engineering, or GE. In contrast to insecticides, GM mosquitoes can directly target just one insect species, says DeGennaro. "You're only affecting one species of mosquito, the species that is the primary vector for Zika, dengue, and more."

Early field tests of GM mosquitoes in Brazil appeared to stop the spread of dengue. But could they do the same for Zika?

## Rapid Spread

Before it hit Florida, Zika virus struck Brazil like a tsunami. In a span of 3 months in early 2015, Brazil documented nearly 7,000 cases of mild illness similar to dengue fever, though no one yet suspected Zika. That's the problem with a newly emerging infectious disease: no one is looking for it.

Then, in May, a national laboratory reported that Zika virus was circulating around the country. Two months later, Brazilian health workers began reporting neurological disorders associated with Zika infection, including brain inflammation. In October came the worst news of all: reports of an unusual spike in newborns with microcephaly.

That dramatic spread and link to microcephaly put Zika on the map, but it isn't actually a new virus. Zika virus—a round particle with a dense core packed with RNA (**Figure 19.4**)—was first

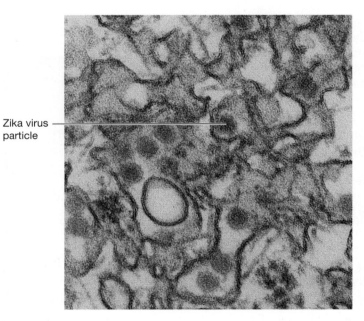

Zika virus particle

Figure 19.4

**Zika is a spherical RNA virus**

isolated in 1947 in the Zika Forest of Uganda, for which it is named. The first human cases of Zika were detected in Uganda in 1952, and the little-studied virus subsequently spread across central Asia, the South Pacific, and into South and Central America and the Caribbean.

Humans are no strangers to mosquito-borne diseases (see "Mosquito-Borne Diseases"). While Zika and other diseases were once isolated to specific areas of the globe, global travel and human population growth, along with climate change, have increased mosquitoes' ability to proliferate and spread disease on a grand scale. In the second half of the twentieth century, our planet saw the fastest rate of human population growth in its history. **Population doubling time**—the time it takes a population to double in size—is a good measure of how fast a population is growing. For example, the doubling time for the U.S. human population at our current reproductive rate is about 100 years. In stark contrast, the population doubling time for mosquitoes can be as little as 30 days in an optimal environment.

## Reaching Capacity

Now back to mosquitoes, and how to get rid of them. Scientists argue strongly that understanding mosquito population ecology is a prerequisite for eliminating mosquito-borne diseases. For example, many mosquitoes lay their eggs directly on the surface of water. That knowledge led to one of the most important public health efforts to reduce mosquito-borne diseases: the removal of pools of standing water. Eliminating these breeding sites makes it possible to reduce the **carrying capacity** of the environment, the maximum population size that can be sustained in a given area.

In most species, the growth rate of a population decreases as the population size nears the carrying capacity, because resources such as food and water begin to run out. If there are only a few locations of standing water in a given area, for example, a limited number of eggs can be laid in that area, thereby limiting the size of the population. Any limiting resource needed for survival, such as habitat, food, or water, will determine an environment's carrying capacity for a specific population. Because different species have unique needs, the same environment can have different carrying capacities for different resident species. For the same reason, two different environments in which the same species lives will have different carrying capacities. At the carrying capacity, the population growth rate is zero.

If a population has no constraints on its resources, it will experience **exponential growth**, which occurs when a population increases by a constant proportion over a constant time interval, such as one year. Exponential growth is represented by a **J-shaped growth curve** (**Figure 19.5**). A population that approaches its carrying capacity because of constrained resources will experience **logistic growth**, in which the population grows nearly exponentially at first but then stabilizes at the maximum population size that can be supported indefinitely by the environment. Logistic growth is represented by an **S-shaped growth curve**.

## Mosquito-Borne Diseases

Mosquitoes cause more human suffering than any other organism (with the exception of our own species). Here are just a few of the deadly or debilitating diseases they carry:

- *Dengue virus* is a huge problem for the population of Brazil. Spread by the same mosquitoes as Zika, dengue also has no vaccine or cure, and it causes a disease chillingly known as "breakbone fever." Symptoms include continuous joint pain, vomiting, high fever, and severe rashes. In 2015, Brazil chalked up a record 1.6 million cases and 863 deaths to dengue—an 82.5 percent increase over 2014.

- *Malaria*, an ancient disease that has been traced as far back as 2700 BCE, affects an estimated 40 percent of the world's population. Hundreds of thousands of deaths are still reported annually from malaria.

- *Yellow fever*, against which there is a vaccine, still leads to an estimated 30,000 deaths per year.

- *West Nile virus*, which emerged in the 1930s, can invade the nervous system and cause death.

- *Chikungunya* is found mainly in Africa and Asia, but cases have been reported in Florida. The main symptoms are fever, rash, and joint pain that can last for years. Only one in 1,000 cases results in death.

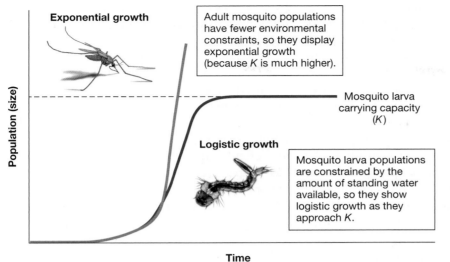

**Exponential growth**

Adult mosquito populations have fewer environmental constraints, so they display exponential growth (because *K* is much higher).

Mosquito larva carrying capacity (*K*)

**Logistic growth**

Mosquito larva populations are constrained by the amount of standing water available, so they show logistic growth as they approach *K*.

Population (size)

Time

Figure 19.5

**Populations can experience exponential growth or logistic growth**
A population that is not constrained by resources or by the environment can grow exponentially, whereas one that is constrained by a set carrying capacity will show logistic growth. 

**Q1:** Which form of population growth displays a J-shaped curve?

**Q2:** Which form of population growth displays an S-shaped curve?

**Q3:** Describe a situation in which a population initially shows exponential growth and later shows logistic growth.

Little research has been done on growth curves for mosquito populations, primarily because the insects are small and airborne, making them hard to track. But we do know how the human population has changed: Over the last 500 years, Earth's human population exhibited both logistic and exponential growth. At the end of the last ice age, in approximately 10,000 BCE, there were only 5 million people on Earth. With the advent of agriculture in about 8000 BCE, the world population began to rise logistically, until about 200 years ago. Then, alongside the use of fossil fuels and the industrial revolution, human population growth exploded exponentially. Modern populations have continued to show exponential growth to the detriment of the environment. Current estimates of the carrying capacity of Earth range from 2 billion to over 1,000 billion people, with the majority of studies insisting that 9–10 billion people is the maximum number that Earth can support. At current population growth rates, we will reach this number by the year 2050 (**Figure 19.6**).

# Seeking Change

With human populations increasing in number and concentrating in urban areas, we can expect the continued spread of mosquito-borne illnesses. Thankfully, recent results from field

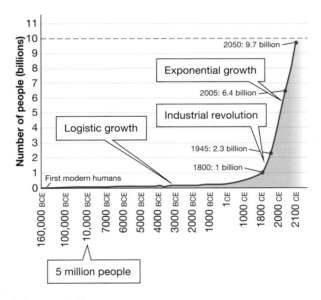

Figure 19.6

**Curves of logistic and exponential growth in the world human population**
The dashed line indicates the United Nations' estimated carrying capacity of Earth.

**Q1:** According to this graph, approximately when did exponential growth begin?

**Q2:** What milestone corresponds to the transition from logistic to exponential population growth?

**Q3:** What is the UN's projected carrying capacity of Earth, and when will we reach it?

tests of GM mosquitoes in Brazil suggest that we may have a new way to stop that spread.

Starting in 2002, British company Oxitec began producing a GM line of the *Aedes aegypti* mosquito. The company inserts a single gene into a line of lab-bred mosquitoes. This gene, designed to work only in insect cells, produces a protein that prevents the mosquito from transitioning between two of its life stages: larva and adult. In the lab, the mosquitoes are exposed to an antibiotic (tetracycline) that neutralizes the protein so that the mosquitoes can grow to be adults. Then, only males are released into the wild, because the males don't bite and so will not contribute to the spread of disease. Once free, the male GM mosquitoes mate with wild females and produce offspring with the lethal gene—and

no access to antibiotics—so the offspring die off before becoming adults and therefore do not reproduce. In this way the size of the population is reduced (**Figure 19.7**).

Populations—mosquito, human, or otherwise—can change in a density-dependent or density-independent manner. **Density-dependent population change** occurs when birth and death rates change as the population density changes. The number of offspring produced and the death rate are often density-dependent. Food shortages, lack of space, and habitat deterioration—all these factors influence a population more strongly as it increases in density (**Figure 19.8**).

In addition, when a population has many individuals, disease spreads more rapidly (because individuals tend to encounter one another more

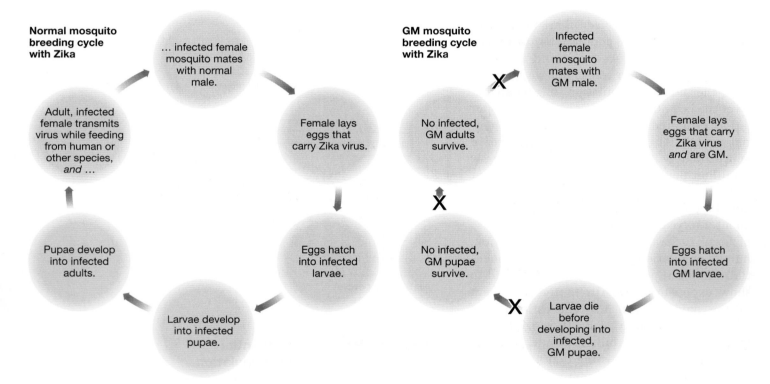

Figure 19.7

**How mosquitoes are genetically modified to stop the Zika virus**
Genetically modified (GM) male mosquitoes have been created and released in Brazil and Florida, where they mate with wild females and produce offspring that cannot grow to adulthood.

Q1: List all the stages of the mosquito life cycle.

Q2: Which life cycle stage is vulnerable to the GM treatment?

Q3: Why are only male GM mosquitoes released into the wild, rather than both males and females?

The number of seeds produced per plant drops dramatically under increasingly crowded conditions.

Figure 19.8

## Overcrowded conditions result in density-dependent population change

Overcrowding affects many species. The plantain, shown here, is a small, herbaceous plant that has decreased reproduction in crowded conditions.

**Q1:** What factors may be limiting growth and reproduction in the plantain's crowded conditions?

**Q2:** Why are overcrowded conditions considered density-dependent population changes?

**Q3:** Relate this example of overcrowded conditions to the human population growth shown in Figure 19.6. How do you think the situations are similar? How are they different?

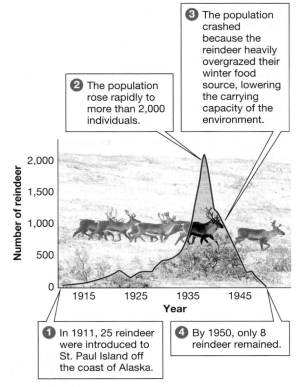

**3** The population crashed because the reindeer heavily overgrazed their winter food source, lowering the carrying capacity of the environment.

**2** The population rose rapidly to more than 2,000 individuals.

**1** In 1911, 25 reindeer were introduced to St. Paul Island off the coast of Alaska.

**4** By 1950, only 8 reindeer remained.

Figure 19.9

## Habitat destruction results in density-dependent population change

**Q1:** In what year did the reindeer's numbers begin to rise exponentially?

**Q2:** In what years was the reindeer's population growth logistic?

**Q3:** How do you predict the graph (population size) would change if someone had begun bringing in supplemental food for the reindeer in 1940? Draw a sketch of your prediction.

often), and predators may pose a greater risk (because most predators prefer to hunt abundant sources of food). Disease and predators obviously increase the death rate. These changes are also density-dependent. A 2014 study found that the mosquito *Culex pipiens*, which spreads the West Nile virus, undergoes strong density-dependent population changes arising from competition for resources within the watery habitats where it lays its eggs. In other words, larvae compete with each other for food. But once they become adults, the population is no longer density-dependent, because the mosquitoes become airborne and have wider access to food.

For some organisms, if a population exceeds the carrying capacity of its environment by depleting its resources, it may damage that environment so badly that the carrying capacity is lowered for a long time. A drop in the carrying capacity means that the habitat cannot support as many individuals as it once could. Such habitat deterioration may result in widespread starvation and death, causing the population to decrease rapidly (**Figure 19.9**).

Not all population changes are due to density. **Density-independent population change** occurs

ECOLOGY

when populations are held in check by factors unrelated to population density. Density-independent factors can prevent populations from reaching high densities in the first place. Year-to-year variation in weather, for example, may cause conditions to be suitable for rapid population growth. On the other hand, poor weather conditions may reduce the growth of a population directly (by freezing the eggs of a mosquito, for example) or indirectly (such as by decreasing the number of food plants available to an animal).

Other natural disturbances, such as fires and floods, can also limit the growth of populations in a density-independent way. Finally, the effects of environmental pollutants, such as the insecticide DDT, are density-independent; such pollutants can threaten natural populations with extinction (**Figure 19.10**).

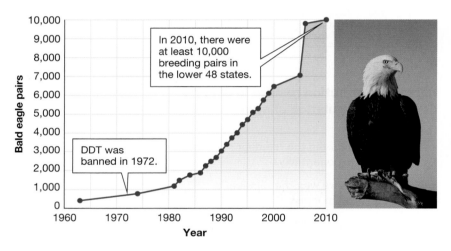

Figure 19.10

**Banning the use of the pesticide DDT removed a density-independent population limit**

DDT poisoning was directly responsible for declining eagle populations by the middle of the twentieth century. By the early 1960s, population counts indicated that only 417 breeding pairs of bald eagles remained in the lower 48 states—a huge drop from the estimated 100,000 breeding pairs present in 1800. Bald eagle populations increased dramatically after DDT was banned.

**Q1:** In what year did the bald eagle population rise to more than 2,000 breeding pairs?

**Q2:** Give some examples of possible density-dependent limits on bald eagle populations.

**Q3:** Is the population growth of bald eagles more like logistic or exponential growth? Explain why you think so.

Populations of many species rise and fall unpredictably over time. These **irregular fluctuations** in population size are far more common in nature than a smooth rise to a stable population size. In the 1950s, for example, Brazil mounted a massive antimosquito campaign to combat yellow fever—a disease also transmitted by *Aedes aegypti*—that included spraying insecticides and encouraging citizens to get rid of standing water. The success of the program led officials to declare in 1958 that *Aedes aegypti* had been eradicated. If only that had been true. By the 1970s the mosquitoes had come roaring back, breeding and spreading like crazy—an unwelcome irregular fluctuation. And now the pests are resistant to many chemical attacks.

Populations can also exhibit **cyclical fluctuations**, predictable patterns that occur with seasonal changes in temperature and precipitation or when at least one of two species is strongly influenced by the other. The Canadian lynx, for example, depends on the snowshoe hare for food, so lynx populations increase when hare populations rise, and they decrease when hare populations drop. In this example, the population cycles are also density-dependent population changes because each population is affected by the other's numbers. They cycle together in response to each other's density (**Figure 19.11**).

# Friendly Fight

The Oxitec GM mosquitoes—trademarked "Friendly" *Aedes aegypti*—were first released in Brazil in February 2011, under the direction of biochemist Margareth Capurro of the University of São Paulo. In a densely populated suburb in northeastern Brazil, the released mosquitoes reduced the wild population of *Aedes aegypti* by 85 percent, says Capurro. In 2013, a similar test in the village of Mandacaru resulted in a 96 percent reduction of the wild mosquito population in the area after only 6 months.

"For the first time, we demonstrated that transgenic mosquitoes can work," says Capurro, who is now developing alternate GM mosquito lines in her own laboratory, attempting to make the GM insects even more potent against their wild counterparts. For example, Capurro's lab is trying out different genetic mutations to create

## Figure 19.11

**Populations of two species may increase and decrease together**
The Canadian lynx depends on the snowshoe hare for food, so the number of lynx is strongly influenced by the number of hare.

**Q1:** During which years did the hare likely have the greatest food supply?

**Q2:** Besides the number of hare, what other factors might contribute to the number of lynx?

**Q3:** Can you draw an average carrying-capacity line on these graphs? Why or why not?

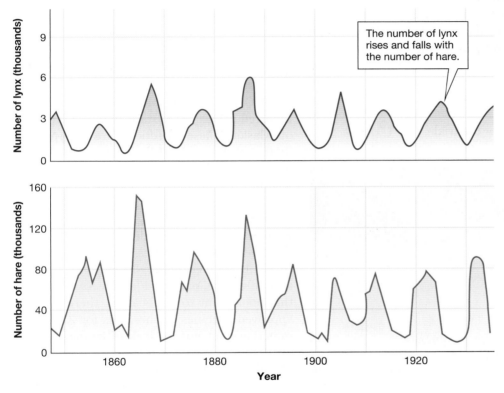

The number of lynx rises and falls with the number of hare.

a line of mosquitoes that does not require the use of an antibiotic to be kept alive in the laboratory, which would be less expensive. "We are finishing the lines now. They are very promising," says Capurro. Other genetic modification techniques are being attempted in labs elsewhere to make *Anopheles* incapable of carrying the malaria parasite, for example.

More recently, Oxitec released Friendly *Aedes aegypti* in a neighborhood of 5,000 residents in the city of Piracicaba, Brazil, to combat the spread of dengue virus. The company released 3–4 million mosquitoes per month over a period of 10 months. The experiment reduced the number of wild *A. aegypti* larvae in the area by 82 percent, and the neighborhood experienced a

### MARGARETH CAPURRO

Margareth Capurro is a biochemist at the University of São Paulo in Brazil. Starting in 2011, she led two field releases of Oxitec's genetically modified mosquitoes in Brazil, and today she continues to explore the best way to use GM mosquitoes to control mosquito-borne viruses.

91 percent reduction in cases of dengue, according to local officials.

In August 2016, shortly after the first cases of Zika infection from local mosquitoes were reported, the U.S. Food and Drug Administration (FDA) approved a field test of the Oxitec mosquitoes in Key Haven, Florida. Critics argue that the approach is too expensive, because mosquitoes have to be bred and released multiple times. There are also concerns that the engineered gene won't correctly insert in the genome or will become somehow deactivated in the wild. Release of such mosquitoes would simply exacerbate the problem, adding reproducing mosquitoes into the environment. After examining such claims, the FDA concluded that the proposed field trial in Florida "would not result in significant impacts on the environment." In a 3–2 vote in November 2016, the Florida Keys Mosquito Control District approved a trial of genetically modified mosquitoes, yet local opposition continues, and the district continues to seek a location for the trial.

Figure 19.12

**Some Florida residents oppose the release of GM mosquitoes in their community**
Misinformation about genetically modified organisms, including mosquitoes, has caused concerns about their use.

Some reject the plan solely on the grounds that it involves genetic modification (**Figure 19.12**). "The public fears genetic engineering. Nearly all politicians don't understand it," said bioethicist Arthur Caplan at the New York University School of Medicine, as quoted in a 2016 article in the *Atlantic*. "I don't think the issue is economic. It is ignorance, distrust, fear of the unknown." Recent polls suggest that the tide may be changing: three independent polls found that between 53 and 78 percent of Americans support the release of GM mosquitoes in the United States. "I think it will work well," says DeGennaro. "It's time we use modern methods here in Florida like they are using in Brazil to reduce mosquito populations."

## Just the Beginning?

As of October 2016, Zika infections had been reported in every U.S. state, with the exception of Alaska. These infections are due primarily to travel outside of the country. In Florida, however, local infections continue; as of this writing, 214 people have locally acquired the virus since 2015.

The CDC continues to warn pregnant women away from traveling to any area with Zika, which now includes dozens of countries; three U.S. territories; Brownsville, Texas; and, of course, parts of Florida. As of the date this book went to publication, there were no approved vaccines or therapies for Zika virus infection—but not for lack of trying. In June 2016, California-based Inovio Pharmaceuticals dosed the first humans with a prospective Zika vaccine. Even if the trials go well, however, it could take years for a vaccine to reach the market.

It remains too soon to tell whether GM mosquitoes will stop the spread of Zika. Success will depend on population ecology: on mosquito populations, the human population, and how much virus is circulating among both. But history suggests that in the fight against mosquitoes, it's time to try something new.

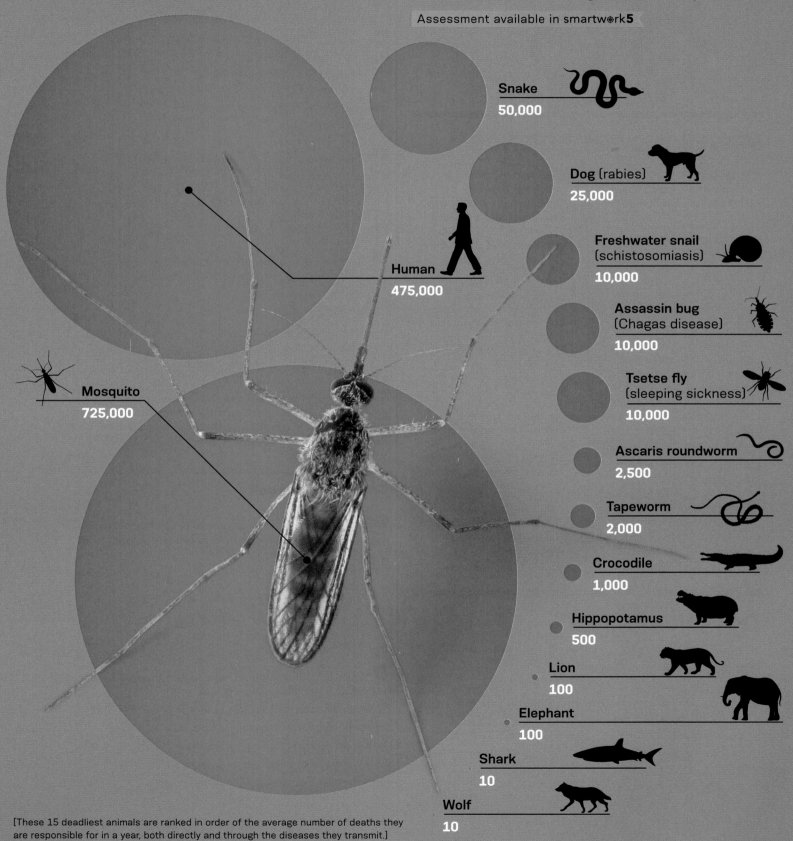

# World's Deadliest Animals

You may have heard that humans are the deadliest animals on the planet. It's true that we, as a species, kill hundreds of thousands of humans. But there's one family of animals that has us beat: mosquitoes. Many species of these small, pesky insects transmit harmful infections, including Zika fever, malaria, West Nile disease, dengue fever, and many more.

Assessment available in smartwork**5**

Mosquito
**725,000**

Human
**475,000**

Snake
**50,000**

Dog (rabies)
**25,000**

Freshwater snail
(schistosomiasis)
**10,000**

Assassin bug
(Chagas disease)
**10,000**

Tsetse fly
(sleeping sickness)
**10,000**

Ascaris roundworm
**2,500**

Tapeworm
**2,000**

Crocodile
**1,000**

Hippopotamus
**500**

Lion
**100**

Elephant
**100**

Shark
**10**

Wolf
**10**

[These 15 deadliest animals are ranked in order of the average number of deaths they are responsible for in a year, both directly and through the diseases they transmit.]

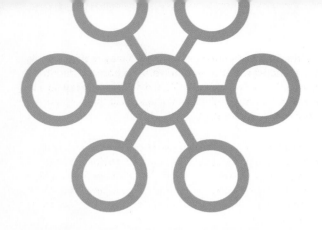

# REVIEWING THE SCIENCE

- A **population** is a group of individuals of a single species located within a particular area.

- **Population size** is the total number of individuals in a population. **Population density** is the population size divided by the area covered by that population.

- Environmental factors such as lack of space, food shortages, predators, disease, and habitat deterioration limit populations. These factors affect the **carrying capacity**, the number of individuals that can live in an environment indefinitely.

- In **logistic growth**, a population grows exponentially at first but then stabilizes after reaching the carrying capacity. It is associated with an **S-shaped growth curve** when graphed.

- In **exponential growth**, a population increases by a constant proportion from one generation to the next generation. It is associated with a **J-shaped growth curve** when graphed.

- **Density-dependent population change** occurs when birth and death rates are affected by population density, which is the case when many individuals occupy the same space and therefore compete for resources.

- **Density-independent population change** occurs when population size is affected by factors that have nothing to do with the density of the population.

- Populations may rise and fall over time, exhibiting **irregular fluctuations** or **cyclical fluctuations**.

# THE QUESTIONS

## The Basics

**1** A group of interacting individuals of a single species located within a particular area is called

(a) a biosphere.

(b) an ecosystem.

(c) a community.

(d) a population.

**2** A population that is growing exponentially increases

(a) by the same number of individuals in each generation.

(b) by a constant proportion in each generation.

(c) in some years and decreases in other years.

(d) none of the above

**3** In a population with an S-shaped (logistic) growth curve, after an initial period of rapid increase the number of individuals

(a) continues to increase exponentially.

(b) drops rapidly.

(c) remains near the carrying capacity.

(d) cycles regularly.

**4** The growth of populations can be limited by

(a) natural disturbances.

(b) weather.

(c) food shortages.

(d) all of the above

**5** The maximum number of individuals in a population that can be supported indefinitely by the population's environment is called the

(a) carrying capacity.

(b) J-shaped curve.

(c) sustainable size.

(d) exponential size.

**6** Select the correct terms:

(**Density-independent / Density-dependent**) factors limit the growth of populations more strongly at high densities than at low densities. One example of a (**density-independent / density-dependent**) factor is a natural disaster.

## Challenge Yourself

**7** A population of plants has a density of 12 plants per square meter and covers an area of 100 square meters. What is the population size?

(a) 120

(b) 1,200

(c) 12

(d) 0.12

(e) none of the above

**8** Which of the following would *not* cause the carrying capacity of a mosquito population to change?

(a) lower availability of standing water in which to lay eggs

(b) fewer animals for female mosquitoes to feed on

(c) warmer weather that increases overwintering survivorship

(d) All of the above would change carrying capacity.

(e) None of the above would change carrying capacity.

**9** Place the following elements in the strategy to control mosquito populations in the correct order from earliest to latest by numbering them from 1 to 5.

_____ a. Scientists genetically modify male mosquitoes in the lab.

_____ b. Mosquito offspring produced by the matings do not survive to adulthood.

_____ c. GM male mosquitoes mate with normal wild females.

_____ d. The mosquito population decreases dramatically.

_____ e. Scientists release GM male mosquitoes into the wild.

## Try Something New

**10** Draw a diagram that illustrates the possible fluctuating population cycles of a predator, such as the mountain lion, and its prey species, such as deer. The prey regularly experiences population crashes when it overgrazes its food source. Label your axes.

**11** An artificial pond at a college campus was populated with several mating pairs of pond slider turtles. After many years of slow population growth, campus administrators installed a coin-operated food dispenser at the pond's edge. Students and community members bought and fed the turtles an enormous amount of food each day, causing the population of turtles in the pond to increase exponentially. How did the installation of the food dispenser affect the carrying capacity of the turtle population at this pond?

**12** Suppose that population ecologists at the college in question 11 determined that the number of turtles in the pond had increased from 6 individuals to 24 individuals in just the first year after installation of the coin-operated food dispenser. What is the population doubling time for the turtles in this population?

(a) 6 years

(b) 6 months

(c) 1 year

(d) 1 month

## Leveling Up

**13** **What do *you* think?** Some Florida residents do not want to allow genetically modified (GM) mosquitoes to be released, even though they have been demonstrated to significantly reduce mosquito-borne illnesses in Brazil.

a. Should Floridians be allowed to decide not to have GM mosquitoes released in their neighborhoods? Should this question be decided by vote? Explain your reasoning.

b. Currently, the Florida Department of Health may require people to remove standing water from their yards in order to decrease breeding habitat for mosquitoes. Should people be allowed not to comply with this order, given that the yard is their private property? Explain your reasoning.

**14** **Life choices** Do you feel that people should choose to limit the number of children they have, given the increasing human population? Alternatively, should people focus on decreasing their own ecological footprint—given that, for example, one American uses the same amount of resources as do four Chinese citizens? Explain your reasoning.

**15** **Is it science?** Your best friend found a Reddit thread stating that GM mosquitoes do not suppress the population size of Zika-bearing mosquitoes, and instead claimed that Zika was *caused* by the release of GM mosquitoes. Putting aside the idea that Zika was caused by GM mosquitoes, which is refuted by multiple studies as discussed on many legitimate websites, you decide to simply show your friend some data comparing mosquito population sizes in areas with and without GM mosquitoes. In your own words, write a paragraph that describes the results shown in the figure here and explains to your friend how they relate to the scientific claims he read.

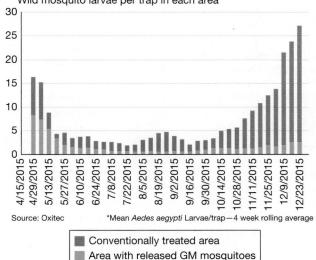

**Mosquito larvae population numbers in areas with and without released GM mosquitoes**

Wild mosquito larvae per trap in each area*

Source: Oxitec          *Mean *Aedes aegypti* Larvae/trap—4 week rolling average

■ Conventionally treated area
■ Area with released GM mosquitoes

**For more, visit digital.wwnorton.com/bionow2 for access to:**

# Of Wolves and Trees

*The extermination of wolves in Yellowstone National Park had unforeseen effects on the park's ecosystem. Can the return of this top predator restore order?*

## After reading this chapter you should be able to:

- Explain the concept of an ecological community and how it relates to ecological diversity.
- Articulate the role of relative species abundance and species richness in defining ecological diversity.
- Distinguish between a food chain and a food web.
- Illustrate how the removal of a keystone species disrupts an ecological community.
- Describe the four major kinds of species interactions, and give an example of each.
- Compare and contrast primary and secondary succession.

CHAPTER
20

COMMUNITIES
OF ORGANISMS

obert Beschta will always remember the day he visited Yellowstone National Park's Lamar Valley in 1996. Beschta, a hydrologist, was there to observe the Lamar River, which winds through the valley's lush lowland of grass and sage (**Figure 20.1**). But on that day, as he walked toward the waterway, he noticed something odd: all around the valley, there were not many trees. The few tall, white aspens he saw looked haggard, their bark eaten away. And there were no young saplings to be seen.

Beschta had studied forestry and rivers for decades and knew what a healthy valley was supposed to look like. This was not it. Beschta approached the river and observed that its banks were also devoid of trees. The leafy green cottonwoods and wide willows that had once arched gracefully from the riverbank over the water were absent. And with no tree roots to hold the soil in place, the riverbanks themselves were jagged and eroding. "I was dumbstruck," recalls Beschta. Something unprecedented was happening in Yellowstone.

Beschta returned to Oregon State University (OSU), where he worked, and described his observations in a seminar. He showed pictures of aspens with their bark stripped away and empty riverbanks where saplings should have been growing. In the audience, William Ripple sat up a little straighter. Also a scientist at OSU, Ripple studied forest ecology. He was particularly interested in aspen trees, which grow as tall as 70 feet and live up to 150 years. From what Beschta was saying, aspens were no longer growing in Yellowstone. Ripple wanted to know why.

"It was a scientific mystery as to what was the cause of the decline," he recalls. In that moment, listening to Beschta, Ripple knew exactly what his next research project would be: to document the extent of the aspen decline and determine why it was occurring.

Within a year, Ripple and one of his graduate students, Eric Larsen, traveled to the Lamar Valley. There, they drilled small holes into the centers of aspens and removed from each a plug of wood about the diameter of a pencil. Then they counted the growth rings in each plug—one for every year the tree has been in existence—to determine the age of the tree. They found that most of the aspens had begun to grow *prior* to 1920. After 1920, almost no new trees had begun growing.

"We started scratching our heads at that point," says Ripple. He, Beschta, and Larsen began to brainstorm reasons why the trees might have stopped regenerating. They looked for environmental changes in the 1920s that could have done it: a fire that killed off saplings, or a change in climate that reduced the trees' ability to reproduce. But nothing lined up—until an ordinary moment gave Ripple an extraordinary idea.

Standing in a gift shop in Grand Teton National Park, just south of Yellowstone, Ripple looked up at a poster on the wall. It featured a grove of tall, white aspen trees in the winter. In the middle of the trees, its paws covered in snow, stood a large gray wolf. "That was an 'aha' moment," says Ripple. "I thought, 'Maybe the wolf protects the aspen.'"

## ROBERT BESCHTA

Robert Beschta is a professor emeritus at Oregon State University. He studies water processes in forest and rangeland ecosystems.

Figure 20.1

**The Lamar River flows through Yellowstone National Park**

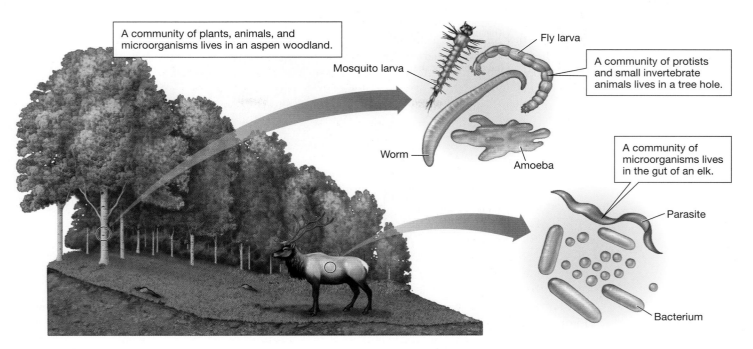

**Figure 20.2**

**Ecological communities come in all sizes**

Smaller communities can be nested within a larger community. This aspen woodland community contains the smaller communities of a temporary pool of water in a tree hole and an elk's gut, among others.

**Q1:** List another species that is part of this community.

**Q2:** Of which community could this aspen woodland be a smaller part?

**Q3:** Which other small communities could be found within this larger community?

The idea of wolves protecting aspens initially seems nonsensical. How would meat-eating predators protect trees? They wouldn't—at least not directly. Ripple surmised that wolves in Yellowstone might have had an indirect effect on the community. As an ecologist, Ripple had long studied **ecological communities**, associations of species that live in the same area. Communities vary in size and complexity, from a small group of microorganisms in a temporary pool of water to the whole of Yellowstone Park, home to an estimated 322 species of birds, 67 species of mammals, 1,349 species of plants, and an uncounted number of insects (**Figure 20.2**).

Ripple knew that an ecological community is characterized by the diversity of species that live there, and that diversity is governed by two things: the **relative species abundance** (how common one species is when compared to another) and the **species richness** (the total number of different species that live in the community; **Figure 20.3**). Ripple also knew that communities are subject to constant change, and that something must have changed in Yellowstone.

Ecological communities change naturally as a result of interactions between and among species, and as a result of interactions between species and their physical environment. Ripple knew that both the relative abundance and the

**WILLIAM RIPPLE**

William Ripple is director of the Trophic Cascades Program at Oregon State University. There, he leads a research project investigating how gray wolves affect other species in the Yellowstone ecosystem.

This community has higher relative species abundance for white-barked trees than the community shown below.

This community has higher species richness than the community shown above.

## Figure 20.3

### Two measures of species diversity

The diversity of an ecological community is determined by relative species abundance and also species richness. High relative species abundance means that one species dominates the community (so the community is less diverse), and high species richness means that many species are present in the community (so it is more diverse).

**Q1:** If relative species abundance increased in the first community, how would the figure look different from now?

**Q2:** If species richness decreased in the second community, how would the figure look different from now?

**Q3:** How do relative species abundance and species richness define the species diversity of a forest community?

richness of species had changed in Yellowstone during the previous century. The relative species abundance had changed because aspen trees and willows were in decline, elk and coyote numbers had increased, and bison and beaver populations had decreased. Many of the common inhabitants of Yellowstone had changed in abundance; therefore, the relative species abundance of this community had changed.

The species richness had also changed because early in the twentieth century, a *keystone species*, the gray wolf, had gone missing. The total number of species in the community had gone down by only one, but that seemingly small change had a big impact. A **keystone species** is a species that has a disproportionately large effect on a community, relative to the species' abundance. There are few wolves compared to, say, rabbits, yet the wolves have a stronger effect on the community. Keystone species in other communities include prairie dogs in the American western plains, hummingbird pollinators in the Sonoran Desert, and sea stars in intertidal waters (**Figure 20.4**).

Keystone species are often recognized only when they go missing and their disappearance results in dramatic changes to the rest of the community. And that is exactly what happened in Yellowstone in the 1920s: the mighty gray wolf disappeared. Or, to be accurate, the mighty gray wolf was exterminated.

**How a predator maintains diversity**

A *Pisaster* sea star feeding on a mussel. Without this keystone species, the mussels crowd out other species in their community.

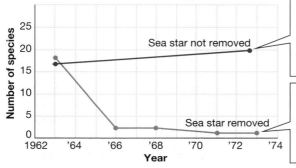

**Loss of a keystone species reduces diversity**

Sea star not removed

Sea star removed

Sea stars completely eliminated mussels in submerged areas of this marine community, enabling other intertidal species to thrive there.

When the sea star *Pisaster* was removed from a community experimentally, the number of species dropped from 18 to 1, a mussel.

Figure 20.4

**The star of the community**

The sea star *Pisaster ochraceus* is an example of a keystone species. In a classic experiment conducted along the Pacific coast of Washington State in 1963, sea stars were removed from one site while an adjacent site was left undisturbed.

**Q1:** How many species were left in 1966 in the community where sea stars were *not* removed?

**Q2:** How many species were left in 1966 in the community where sea stars *were* removed?

**Q3:** How do your answers to questions 1 and 2 demonstrate the importance of a keystone species for the maintenance of diversity in a community?

# A Key Loss

In the early 1900s, ranchers and homesteaders killed wolves all across the United States and eliminated them in many eastern states. Then, in 1915, the U.S. government began subsidizing wolf extermination programs all over the country, and a systematic slaughter began. Under the national program, states paid bounties of up to $150 for individual wolf pelts. Wolves, feared and hated by private landowners for killing livestock, were trapped, shot, and skinned. The extermination happened quickly: the last known wolf den in Yellowstone was destroyed in 1926. At least 136 wolves, maybe more, were killed during the eradication campaign in Yellowstone. Once it was done, the park was wolf-free.

Seven decades later, staring at the poster of a wolf standing among the aspens, Ripple wondered whether the loss of that keystone species had contributed to the decline of aspen trees. He immediately looked up the historical records to see whether the timing matched. Lo and behold, the last wolf had been killed at

about the same time the aspens stopped regenerating, in the mid-1920s. Suddenly, it seemed obvious why the aspens had declined: wolves kill elk and elk eat aspens—three species in the same food chain.

A **food chain** is a simple list of who eats whom. In scientific terms, it is the direct path by which nutrients are transferred through the community. A **food web**, on the other hand, is a more complex diagram of all the food chains in a single ecosystem and how they interact and overlap (**Figure 20.5**).

In the wolf-elk-aspen food chain, aspen are the **producers**, the organisms at the bottom of the food chain that use energy from the sun to produce their own food through photosynthesis. In Yellowstone and on land all over Earth, photosynthetic plants like trees, grasses, and shrubs are the major producers (**Figure 20.6**). In aquatic biomes, photosynthetic plankton are the major producers, as we'll see in Chapter 21.

Further up the food chain are **consumers**, organisms that obtain energy by eating all or parts of other organisms or their remains. Elk and wolves are both consumers: elk eat aspens, and wolves eat elk. In the Yellowstone food chain, elk are **primary consumers**: they eat producers. Wolves are **secondary consumers** because they eat primary consumers. This sequence of organisms eating organisms can continue: a bird that eats a spider that ate a beetle that ate a plant is a **tertiary consumer**; a killer whale that eats a leopard seal that ate a sea bass that ate a krill that ate a phytoplankton is a **quaternary consumer**. We will explore the flow of energy up the food chain in Chapter 21.

The more they discussed it, the more Ripple, Larsen, and Beschta believed that the loss of wolves in Yellowstone had allowed the elk population to flourish and eat so many young trees that the aspen population could not regenerate. "We developed a hypothesis that maybe the killing of wolves actually affected the reproduction of aspen trees," says Ripple. Any change in species diversity will have a ripple effect (no pun intended) throughout the community, and the wolves of Yellowstone were no exception. Ripple and Larsen published their hypothesis in 2000, suggesting that the loss of wolves had led to increased elk

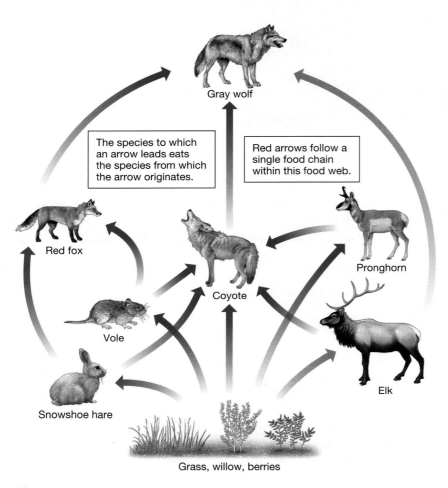

The species to which an arrow leads eats the species from which the arrow originates.

Red arrows follow a single food chain within this food web.

Gray wolf
Red fox
Coyote
Pronghorn
Vole
Elk
Snowshoe hare
Grass, willow, berries

Figure 20.5

**Food webs show how energy moves through a community**

Food webs are composed of many food chains that show one species eating another.

Q1: What species eats the coyote?

Q2: What species does the coyote eat?

Q3: What do you think would happen to a community that lost its coyotes?

In tropical rainforests, the abundant producers (plants) store chemical energy harvested from sunlight, which in turn is readily available to consumers.

In deserts, the sparseness of plant life means that relatively little chemical energy is available for consumers.

Figure 20.6

**Producers are the energy base of a food chain**

All communities and ecosystems have producers, but they vary in abundance, depending on the circumstances.

**Q1:** Where do producers acquire the energy they need to perform their function in the food chain?

**Q2:** Why are producers necessary for life on Earth?

**Q3:** Given the lower abundance of producers in desert regions, compared to tropical rainforests, what would you predict about the abundance of consumers in the two environments?

populations and altered elk movements and feeding patterns. In other words, with wolves gone, elk were free to find and eat young aspen shoots whenever they wanted, with no fear of wolves.

In 2001, Beschta returned to the Lamar Valley and collected data on another species of tree, the cottonwood, which can live more than 200 years. He documented the same trend as the aspens showed: in the 1920s, cottonwoods suddenly stopped generating new, young trees. In fact, since the 1970s, not a single new cottonwood had been established. "This was dramatic," recalls Beschta. "It's a big deal when you can't have a single cottonwood in this large valley make it to a mature tree."

It is possible for a consumer to eat a species to extinction, so if elk populations had continued to grow unchecked, they could have reduced the aspen and cottonwood populations to zero, disrupting the ecological community permanently. But before that could happen, something dramatic occurred in Yellowstone: humans brought the wolves back.

# A Second Ripple Effect

In 1973, wolves became the first animals to be protected under the Endangered Species Act. It was the dawn of the modern conservation movement, and the idea of returning wolves to Yellowstone grew in popularity. It took some time, but eventually, lawmakers agreed to the plan, and between 1995 and 1997, 41 wild wolves were captured in Canada and released in Yellowstone. It didn't take long for the wolf populations to recover: by 2007, an estimated 170 wolves lived in and around Yellowstone; today that estimate is 528. As of January 2016, at least 98 wolves in 10 packs were living within the park limits (**Figure 20.7**).

The loss, and then return, of the wolves had significant impacts on **species interactions** in the park. There are four central ways in which species in a community interact: *mutualism*, *commensalism*, *predation*, and *competition*.

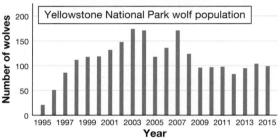

Figure 20.7

**Wolves today in Yellowstone**

**Q1:** When were wolves first reintroduced to Yellowstone?

**Q2:** What was the highest number of wolves observed in Yellowstone? In what year did that occur?

**Q3:** Why did it take a few years after the reintroduction of wolves for the aspen population to increase, as well as the beavers and bears?

The classification is based on whether the interaction is beneficial, harmful, or neutral to each species involved. These interactions affect where organisms live and how large their populations grow. Species interactions also drive natural selection and evolution, thereby changing the composition of communities over short and long periods of time.

With wolves back in the park, Ripple suddenly had a way to test his hypothesis about aspen decline. If the loss of wolves was responsible for aspen loss, then the return of the wolves should incite a revival of aspens (and other woody plants,

including cottonwoods). But he needed a way to quantify that change. "It's a scientifically difficult task to connect a wolf to a plant. Obviously wolves don't consume plants, so instead we had to connect the dots. There are data that wolves affect elk, and other data that elk affect plants," says Ripple.

Ripple and Beschta went into the field in 2006 and again in 2010 to take measurements. In addition to recording the ages of the trees, they looked for and documented signs of elk feeding on aspens—such as scars where branches or buds had been bitten off—and measured the heights of young aspens. Beschta did the same with the cottonwoods. They were eager to see which species interactions would occur now that wolves were back.

The first type of species interaction, **mutualism**, occurs when two species interact and both benefit. For example, Yellowstone is home to 4,600 bison, the largest land mammals in North America. Bison have a mutualistic relationship with the black-billed magpie. Pests such as ticks burrow into a bison's short, dense hair to suck the beast's blood, but hungry little magpies perch on top of the bison and eat those ticks. Thus, both the bison and the magpie benefit from close interaction with one another. Mutualism is common and important in ecosystems all over Earth: many species receive benefits from, and provide benefits to, other species, as the clownfish-anemone partnership in **Figure 20.8** illustrates. These benefits increase the survival and reproduction of both interacting species.

When they aren't perched atop bison, black-billed magpies can be found in large nests atop deciduous or evergreen trees, where they reproduce once a year. These trees, another member of the community, share a commensal relationship with the magpies. **Commensalism** happens when one partner benefits while the other is neither helped nor harmed—in this case, the magpie benefits from having a safe place to lay eggs, and the interaction has no effect on the tree. Another example of a commensal relationship is barnacles living on a whale (**Figure 20.9**).

As you might have guessed, not all species interactions are as pleasant as those among bison, birds, and trees. In two other types of interactions, at least one of the two species is harmed: *competition* (which we will return to in a moment) and *predation*.

The clownfish gains protection from predators by hiding within the sea anemone's stinging tentacles. The clownfish is not harmed, because a thick mucus covers its body.

The anemone is protected from grazing predators by the clownfish and absorbs the fish's nutrient-rich excrement.

**Figure 20.8**

### Mutualism: friends in need
Both the clownfish and the sea anemone benefit from their relationship.

**Q1:** What might happen to an anemone without a resident clownfish?

**Q2:** What would happen to a clownfish that did not produce a mucous coating?

**Q3:** In the example of mutualism given in the text, what would happen if ticks were no longer able to feed on bison?

In **predation**, one species benefits and the other is harmed, and **predators** are defined as consumers that eat part or all of other organisms. A **parasite** is a kind of predator that lives in or on the organism it harms, its **host**. An important group of parasites is pathogens, which cause disease in their hosts. The bacteria that cause strep throat, tuberculosis, and pneumonia are pathogens, for example. Many organisms have evolved mechanisms to avoid being hosts, such as immune systems to help fight off parasitic diseases and infections.

Other types of predators are distinguished by what they eat. Elk are **herbivores**, animals that eat plants. Yellowstone elk feed on the shoots, saplings, and new branches of woody plants like aspen, cottonwood, and willow, especially in winter, when other plants are scarce. Wolves are **carnivores**, animals (and, in rare cases, plants) that kill other animals for food. Yellowstone wolves predominantly eat elk, especially in winter, but they also eat deer and any small mammals they can catch, notably beaver. Other animals, such as raccoons and coyotes, eat both animals and plants, so we call them **omnivores**. Both raccoons and coyotes are also **scavengers**, animals that eat dead or dying plants and animals. Of the species that are scavengers, **decomposers**, such as fungi, dissolve their food; and **detritivores**, including worms and millipedes, mechanically break apart and consume their food.

The barnacles on a whale's snout enjoy a continuous stream of nutrient-rich water flowing across them from which they can feed. The whale is neither harmed nor helped by these stowaways.

**Figure 20.9**

### Commensalism: a whale of a ride!
This gray whale's snout is covered in barnacles.

**Q1:** How do barnacles benefit from living on a whale?

**Q2:** Do you think a whale could avoid being colonized by barnacles? Why or why not?

**Q3:** Explain why detritivores are considered commensal to the organisms they consume.

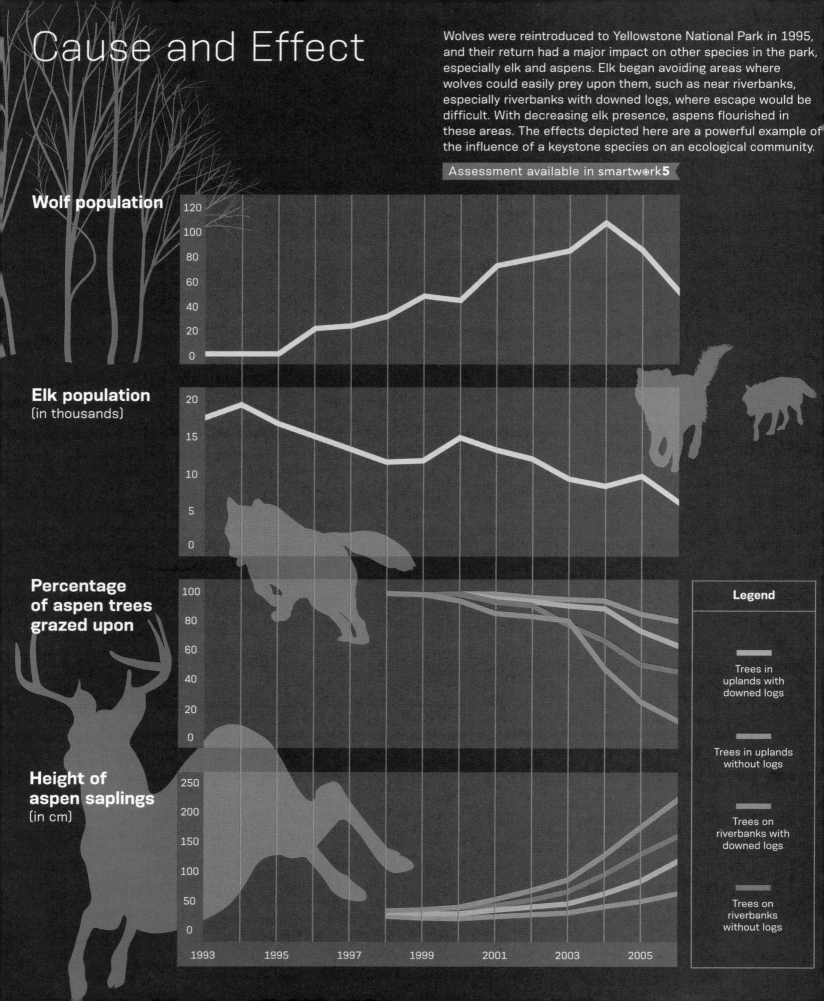

# Cause and Effect

Wolves were reintroduced to Yellowstone National Park in 1995, and their return had a major impact on other species in the park, especially elk and aspens. Elk began avoiding areas where wolves could easily prey upon them, such as near riverbanks, especially riverbanks with downed logs, where escape would be difficult. With decreasing elk presence, aspens flourished in these areas. The effects depicted here are a powerful example of the influence of a keystone species on an ecological community.

Assessment available in smartwork5

**Wolf population**

120
100
80
60
40
20
0

**Elk population**
(in thousands)

20
15
10
5
0

**Percentage of aspen trees grazed upon**

100
80
60
40
20
0

**Height of aspen saplings**
(in cm)

250
200
150
100
50
0

1993    1995    1997    1999    2001    2003    2005

**Legend**

Trees in uplands with downed logs

Trees in uplands without logs

Trees on riverbanks with downed logs

Trees on riverbanks without logs

This cheetah is a predator whose many prey species include the African hare.

The parasite *Hippobosca longipennis*, the louse fly, lives off carnivores like the cheetah by sucking their blood, but does not kill its host.

Figure 20.10

**Predators can also be prey**
While the cheetah is what we traditionally think of as a predator, it can also be prey to parasites, and eventually to decomposers and detritivores.

**Q1:** What kind of predator is the cheetah in the figure?

**Q2:** What kind of predator is the louse in the figure?

**Q3:** What kind of predator are the elk that graze on aspen tree saplings?

The animals eaten by predators are called **prey**. All of the animal residents of Yellowstone (and most plants) are eaten by other species, except for grizzly bears, mountain lions, eagles, and gray wolves. These four animals are all at the top of the food chain. They are top predators, although they are themselves prey to parasites (**Figure 20.10**).

# Back in the Park

Now that wolves, a top predator in the community, were back in Yellowstone, how would their prey, the elk, react? And how would that reaction affect the elk's food, the trees?

Using the plant measurements taken in the park and comparing those measurements to historical data, Beschta and Ripple found that between 1998 and 2010, as the wolf population in the park grew, the elk population decreased and therefore fewer aspens were eaten. In 1998, essentially 100 percent of the young aspen plants were being preyed upon, but by 2010, only 18–24 percent were being eaten. In addition, average aspen heights increased for all areas that the scientists observed. Cottonwoods experienced the same revival. In the 1970s, cottonwoods had entirely stopped adding new young saplings, but by 2012, some 4,660 young cottonwoods had grown to over 2 meters high.

Together, the aspen and cottonwood data sets convinced Ripple and Beschta that the reintroduction of wolves was responsible for a cascade of species interactions leading to the restoration of aspen and cottonwood populations. "With

ECOLOGY

**Warning coloration**

The bright colors of the poison dart frog warn potential predators of the deadly chemicals contained in its tissues.

**Mimicry**

The viceroy butterfly (left) mimics the color and pattern of the monarch butterfly (right), which contains toxic compounds.

**Camouflage**

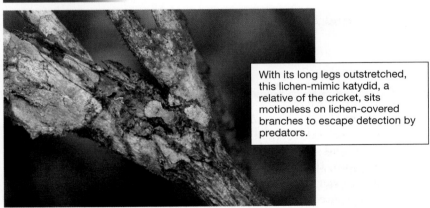

With its long legs outstretched, this lichen-mimic katydid, a relative of the cricket, sits motionless on lichen-covered branches to escape detection by predators.

Figure 20.11

**Adaptive coloration responses to predation**

Prey species have adapted many elaborate strategies to avoid being eaten by predators, including warning coloration, mimicry, and camouflage.

Q1: How do predators know that brightly colored prey are usually toxic?

Q2: Do you think mimicry works if the toxic species is in low abundance? Why or why not?

Q3: Why is camouflage considered an adaptive response to predation?

wolves back, young, woody plants are doing better and growing taller," says Ripple. "Plant communities are beginning to recover." Different species are growing and spreading at different rates of recovery, he adds, "but there's enough new growth that we suggest it is in support of our basic hypothesis, that the presence or absence of the top predator—the wolf in this case—makes a difference in these plant populations."

Since Ripple and Beschta's discovery, the scientific community has been debating two potential reasons why the plants are flourishing with the return of the wolves. The most straightforward possibility is that the elk population has decreased: wolves kill elk, so there are fewer elk to consume the plants. Yet some of the tree populations seemed to recover faster than the drop in the elk populations would suggest. So a second possibility is that the presence of wolves led to a change in elk behavior called a *fear effect*. Often, the presence of a predator in an area can affect the behavior of its prey. In this case, it is possible that elk stopped grazing in areas where they could easily be seen by wolves, such as along the banks of the Lamar River.

"These two mechanisms, the population density and fear behavior, are difficult to tease apart, and we're working on that," says Ripple. "Many today believe it [the change in plant populations in Yellowstone] is due to a combination of the two." Beschta agrees: "In my opinion, they've both been going on."

# Safety in Numbers and Colors

Elk may avoid lingering at streams as a way to evade their predators, but other prey have far more elaborate strategies to avoid being consumed. The poison dart frog, for example, is among the most toxic animals on Earth, and it evolved bright colors as **warning coloration** to alert potential predators to the dangerous chemicals in its tissues (**Figure 20.11**, top). Such warning coloration can be highly effective. Young blue jays, for example, quickly learn not to eat brightly colored monarch butterflies, which contain chemicals that cause nausea and, at high doses, death.

The success of goshawk attacks on wood pigeons decreases greatly when there are many pigeons in a flock.

Although a single musk ox may be vulnerable to predators such as wolves, a group forming a circle makes a difficult target.

Figure 20.12

**Safety in numbers**

Animals that live in groups are better able to warn each other and sometimes fend off attacking predators.

Q1: What percentage of pigeons are caught when they are alone and not in a flock?

Q2: For wood pigeons, what is the minimum number of individuals that provides protection from goshawks?

Q3: Why do you think a group of musk oxen versus a lone musk ox would be safer from a pack of wolves?

Then there are species that, though not poisonous, evolved coloration to make them look as if they were. Through **mimicry**, the viceroy butterfly, which is not poisonous, imitates the color and pattern of the monarch butterfly (**Figure 20.11**, middle). That "borrowed" coloration scares away blue jays and other birds that may have felt sick the last time they ate a monarch. Another mechanism to avoid being eaten is **camouflage**, any type of coloration or appearance that makes an organism hard to find or hard to catch (**Figure 20.11**, bottom). Finally, many prey, from musk oxen to wood pigeons, have evolved a different strategy to avoid becoming dinner: living together. By group living, these animals are able to act together to warn each other when a predator is about to attack and even to repel attacks as a united front (**Figure 20.12**).

In predation, the predator benefits. And each of the other two types of species interactions we've discussed—mutualism and commensalism—benefits at least one of the species involved. But in one final type of species interaction, **competition**, no one benefits. Instead, both interacting species are negatively affected.

Competition most often occurs when two species share an important but limited resource, such as food or space. In Yellowstone, both beavers and elk eat woody plants. For both species, woody plants are part of the **ecological niche**, the set of conditions and resources that a population needs in order to survive and reproduce in its habitat. Because the niches of beavers and elk overlap, these species compete. When two (or three or more) species compete, each has a negative effect on the other because one is using resources that the other then cannot access. (If resources are abundant, however, there may be no competition between species, even if their niches overlap.)

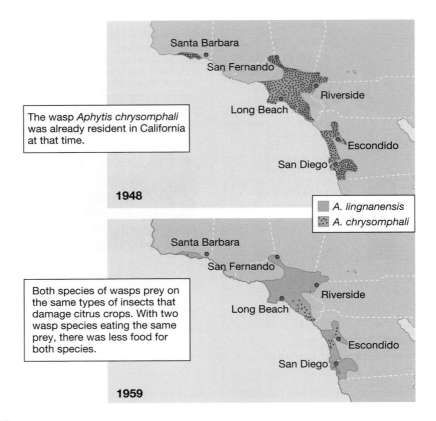

The wasp *Aphytis chrysomphali* was already resident in California at that time.

**1948**

A. lingnanensis
A. chrysomphali

Both species of wasps prey on the same types of insects that damage citrus crops. With two wasp species eating the same prey, there was less food for both species.

**1959**

Figure 20.13

**Exploitative competition: a new species moves in**

Two different species of wasps—*Aphytis lingnanensis* and *Aphytis chrysomphali*—feed on the same resource but do not directly compete for access to it.

**Q1:** Which species appears to be the superior competitor?

**Q2:** Why is this example considered exploitative competition?

**Q3:** What would you predict if these species had undergone competitive exclusion? (You will need to read the next paragraph to answer this question.)

There are two main categories of competition: *exploitative* and *interference*. In **exploitative competition**, species compete indirectly for shared resources, such as food. In this case, each species reduces the amount of the resource that is available for the other species, but they do not directly interact or come in contact with each other (**Figure 20.13**). When wolves returned and elk populations declined, beavers had less competition for food, especially willow trees, and the number of beaver colonies in Yellowstone rose from 1 in 1996 to 12 in 2009. By 2015, Yellowstone had an estimated 100 beaver colonies.

Elk and bears also interact through exploitative competition. Elk eat the leaves and branches on shrubs, resulting in a decrease in the number of berries the shrubs produce. That's not good for grizzly bears, which love to eat berries. Knowing of this relationship, Ripple hypothesized that a decrease in elk would result in an increase in berry-producing shrubs, and that bears would be eating more berries. To test this hypothesis, Ripple, Beschta, and others spent 2 years analyzing grizzly scat that had been collected in the park. They compared the percentage of fruit in current scat to that of scat data that had been collected and saved before 1995, prior to the wolf return. Over a 19-year period, they found that the percentage of berries in the grizzly diet went up as elk populations went down. This is an example of the **competitive exclusion principle**, which predicts that different species that use the same resource can coexist only if one of the species adapts to using other resources. In this case, grizzlies adapted to the presence of elk by finding food sources other than berries, and in the absence of elk they increased their consumption of berries.

Organisms also compete through **interference competition**, in which one organism directly excludes another from the use of a resource. In Yellowstone, for example, bears and wolves often fight over the carcass of an animal (**Figure 20.14**). In interference competition, one individual physically interferes with another individual that is trying to obtain a resource.

# A Community Restored

Today, the return of the wolf is having a clear and significant impact on the ecological community of Yellowstone, from the rebirth of aspens and cottonwoods to growing beaver populations. These are signs of **succession**, the process by which the species in a community change over time. "We're on a very important upward trend," says Beschta.

All ecological communities change over time, sometimes because of human intervention, as in Yellowstone, but also because of natural changes in species composition—the number of individuals in a population often changes as the seasons change, for example—and natural disturbances such as fires, floods, and windstorms. In addition, communities

Figure 20.14

**Interference competition can be dramatic**

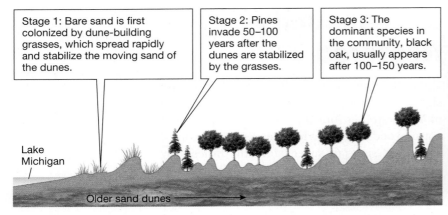

Stage 1: Bare sand is first colonized by dune-building grasses, which spread rapidly and stabilize the moving sand of the dunes.

Stage 2: Pines invade 50–100 years after the dunes are stabilized by the grasses.

Stage 3: The dominant species in the community, black oak, usually appears after 100–150 years.

Lake Michigan

Older sand dunes

Stage 4: Climax communities of black oak have lasted up to 12,000 years.

Figure 20.15

**Primary succession: from nothing to climax community**
Sand becomes woodland near Lake Michigan.

**Q1:** What species represents the first colonizers of the sand dunes?

**Q2:** What species is the intermediate species, and how does it become the dominant species?

**Q3:** What species is the mature, climax community species, and how does it become the dominant species?

can broadly change by the slow loss or gain of populations of species over long periods of time through natural selection.

In nature, ecologists have observed two major types of succession: primary and secondary. **Primary succession** occurs in newly created habitats, such as an island that just emerged from the sea, the soil left behind a retreating glacier, or new sand dunes (**Figure 20.15**). A new habitat begins with no species. Often the first species to colonize an area alters the habitat in ways that enable later-arriving species to thrive. A specific type of flowering plant may grow on a new island, for example, and then a species of bee that feeds on that flower subsequently joins the habitat.

**Secondary succession** occurs after a disturbance within a community—such as the loss of a keystone species like the wolves in Yellowstone or the plant loss caused by a forest fire—reduces the number of species in that community. During secondary succession, communities usually regain the successional state that existed before the disturbance. This type of succession does not take as long as primary succession, because some species still exist in the community.

Luckily, communities can and do bounce back from disturbances, but the time required to regain a previous state varies from years to decades to centuries. Yellowstone is currently experiencing a secondary succession as wolf, aspen, and beaver populations slowly return. It will likely still be a while before the park returns to its previous state, says Ripple. "We were 70 years without wolves, and now we're less than 20 years since wolves have returned, so this is going to take time." But he and others are hopeful that Yellowstone will return to its status as a **climax community**, a mature community whose species composition remains stable over long periods of time (**Figure 20.16**).

Yet even as Yellowstone's recovery is under way, ecological communities around the planet are being threatened by the loss of other keystone species, especially large carnivores. In 2013, Ripple and colleagues analyzed 31 carnivore species around the globe, including leopards, lions, cougars, and sea otters. They found that more than 75 percent of those 31 species are declining, and that 17 of them now occupy less than half of their former ranges. In 2016, they found that bushmeat hunting, in particular, is

In 1988, a large fire destroyed a portion of the mature lodgepole-pine forest in Yellowstone National Park.

By 1992, the lodgepole-pine forest regrowth was gaining momentum.

An example of a mature, climax community lodgepole-pine forest.

Figure 20.16

**Secondary succession: from disturbance to climax community**
Lodgepole-pine forest has been slowly but steadily regrowing in Yellowstone National Park.

**Q1:** What other types of disturbances could you imagine destroying a forest?

**Q2:** How is secondary succession different from primary succession?

**Q3:** What is a climax community?

driving many mammal species to extinction, including 113 species in Southeast Asia. Because of all the species interactions discussed here, it is clear that those changes will have major effects on ecological communities. "Humans are affecting predators around the globe in a major way," says Ripple. "It's a worldwide issue."

# REVIEWING THE SCIENCE

- An **ecological community** can be characterized by its species composition, or diversity. This diversity has two components: **relative species abundance** (the number of individuals of each species that exist in the community) and **species richness** (the total number of different species that live in the community).

- **Keystone species** have a disproportionately large effect, relative to their own abundance, on the richness and abundances of the other species in a community. The removal or disappearance of these keystone species results in dramatic changes to the rest of the community.

- A **food chain** is a single direct line of who eats whom among species in a community. A **food web** depicts how overlapping food chains of a community are connected.

- **Producers**, organisms found at the bottom of a food chain that use light energy to produce their own food, are eaten by **consumers**.

- Consumers are classified as **carnivores**, **herbivores**, or **omnivores**, depending on whether they eat animals or plants or both, respectively.

- Dead and dying organisms are prey for many kinds of consumers. **Scavengers** are omnivores or carnivores that hunt for dead and dying prey. **Decomposers** dissolve their food, and **detritivores** mechanically break apart and consume their food.

- **Species interactions** in a community can be beneficial, harmful, or without benefit or harm to each of the interacting species.

  - In **mutualism**, both species benefit.

  - In **commensalism**, one species benefits at no cost to the other.

  - In **predation**, one species benefits and the other is harmed. **Parasites** are predators that live in or on their **hosts**.

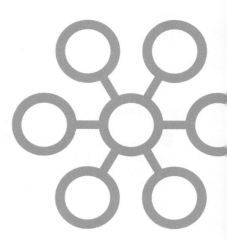

- In **competition**, both species are harmed. Competition occurs when **ecological niches** overlap and includes **exploitative competition** and **interference competition**. The **competitive exclusion principle** predicts that different species that use the same resource can coexist only if one of the species adapts to using other resources.

- **Succession** establishes new communities (**primary succession**) and replaces disturbed communities (**secondary succession**). In stable environments without disturbances, called mature communities or **climax communities**, species composition remains stable over long periods of time.

# THE QUESTIONS

## The Basics

**1** A single sequence of feeding relationships describing who eats whom in a community is a

(a) life history.

(b) keystone relationship.

(c) food web.

(d) food chain.

**2** The process of species replacement over time in a community is called

(a) global climate change.

(b) succession.

(c) competition.

(d) community change.

**3** Organisms that can produce their own food from an external source of energy without having to eat other organisms are called

(a) suppliers.

(b) consumers.

(c) producers.

(d) keystone species.

**4** A low-abundance species that has a large effect on the composition of an ecological community, especially when removed from that community, is called a

(a) predator.

(b) herbivore.

(c) keystone species.

(d) dominant species.

**5** Select the correct terms:

A cheetah eats an antelope that ate some grass. The cheetah is a (secondary consumer / primary consumer), while the antelope is a (secondary consumer / primary consumer). The grass is a (tertiary consumer / producer).

## Challenge Yourself

**6** Wolves are considered a keystone species in Yellowstone because

(a) their removal in the early twentieth century caused many changes to the Yellowstone ecological community.

(b) their reintroduction in the late twentieth century caused many changes to the Yellowstone ecological community.

(c) when they were removed, elk populations increased, leading to increased competition with beavers and bears, which then declined.

(d) when they were reintroduced, aspen populations began to increase because of decreased predation from elk.

(e) all of the above

**7** Link each species interaction with an example of the interaction.

| | |
|---|---|
| MUTUALISM | 1. Elk graze on aspen. |
| COMMENSALISM | 2. Bison allow magpies to perch on them; the birds eat ticks they find on the bison's bodies. |
| PREDATION | 3. Birds nest in aspen trees. |
| COMPETITION | 4. Beavers and elk eat the same trees. |

**8** Place the following elements of the scientists' study of the relationship between aspen populations and wolf populations in the correct order from earliest to latest by numbering them from 1 to 5.

_____ a. Beschta hypothesized that the decimation of wolf populations in the early twentieth century allowed elk populations to grow, thus increasing grazing of the aspen by elk.

_____ b. Beschta observed there were few trees in the river valley.

_____ c. Ripple heard Beschta describe his observations in a talk.

_____ d. Ripple and his graduate student collected data showing that no new trees had grown in the valley since the 1920s.

_____ e. With the reintroduction of wolves into the park, elk populations declined and aspen populations rebounded.

**9** Which ecological community would be more diverse: one with high relative species abundance or one with low relative species abundance? Which community would be more diverse: one with high species diversity or one with low species diversity?

## Try Something New

**10** When a female cat comes into heat and is ready to mate, she urinates more frequently and in a large number of places. Male cats from the neighborhood congregate near urine deposits and fight with each other for the female's attention and breeding rights. In what type of interaction are the male cats engaging?

(a) commensalism

(b) predation

(c) interference competition

(d) exploitative competition

(e) mutualism

**11** Rabbits can eat many plants, but they prefer some plants over others. Assume that the rabbits in a grassland community containing many plant species prefer to eat a species of grass that happens to be a superior competitor. If the rabbits are removed from the region, predict whether relative species abundance will increase or decrease, and whether species richess will increase or decrease.

**12** Identify whether each of the following is an example of primary succession (P) or secondary succession (S) in Rocky Mountain National Park.

_____ a. A mountain slope was cleared of evergreen trees and is now sprouting aspen trees.

_____ b. A mountainside with mature evergreen trees has dominated the landscape for generations.

_____ c. An area of soil, sand, and rocks remains after a dam burst and flooded the area.

_____ d. Lichens and mosses grow on bare rock at upper elevations because of an increase in average yearly temperatures.

**13** Analyze the food web shown in the accompanying figure and answer the following questions:

a. Which species do *not* have a predator shown on the food web?

b. Which species shown on the food web performs photosynthesis (captures light energy to make its own food)?

c. Which species have only one predator and only one prey shown on the web?

d. Which species has the most predators shown on the web?

## Leveling Up

**14** **Doing science** Citizen science is an amazing way for anyone to get involved in scientific research. Search the Internet for citizen science projects relevant to this chapter's topics, using keywords like "parasite," "predator," "prey," and "group living," to name a few. Examples include Project Monarch Health (http://www.monarchparasites.org), in which volunteers sample wild monarch butterflies for a protozoan parasite to track its prevalence across North America. Participate in a project and, in writing, reflect on what you learned.

**15** *Write Now* **biology: saving a species** Write a letter to your congressperson in which you use the concepts learned in this chapter to explain how wolves (or a different species near you) are beneficial to our wilderness ecosystems. Explain what role the species plays in the environment, what has caused it to be endangered or at risk now, what the likely effect will be if it goes extinct (locally or globally), and what could be done to ensure its survival.

---

**For more, visit digital.wwnorton.com/bionow2 for access to:**

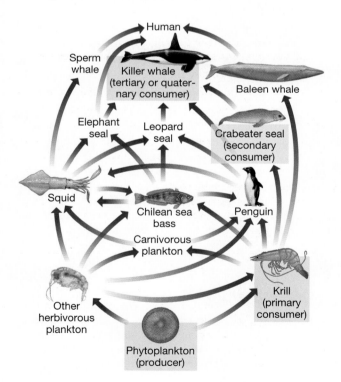

# Here and Gone

*Researchers discover an alarming decline of plankton in the ocean.*

- - - - - - - - - - - - - - - - - - - - - - - - - - - - - - - - - - - - - -

## After reading this chapter you should be able to:

- ◆ Define an ecosystem and explain how it differs from an ecological community.

- ◆ Compare and contrast nutrient and energy cycles.

- ◆ Place the members of a specified ecosystem into an energy pyramid at the appropriate trophic level.

- ◆ Identify where on Earth a specified biome is found, and compare its relative productivity to other biomes.

- ◆ Define net primary productivity (NPP) and explain its significance to studies of ecosystems.

# CHAPTER

# 21

## ECOSYSTEMS

As a teenager, Daniel Boyce made extra cash working as a deckhand on fishing boats, so he was no stranger to the ocean. Eventually, Boyce turned that interest in marine environments into a career, and in 2007 he joined the lab of marine biologist Boris Worm at Dalhousie University in Nova Scotia, Canada.

Several years earlier, Worm had discovered that the industrialized fishing boom that began in the 1950s had decimated predatory fish communities. Boyce decided to follow up on his mentor's work by studying how that loss of predatory fish reverberated down the food chain. Had the disappearance of ocean predators affected plankton? Plankton, a diverse group of free-floating organisms, are the base of the ocean food web, supporting virtually all marine animals. Although they are tiny, they play a mighty role in marine environments.

When Boyce proposed this study, little did he know that it would uncover a profound shift in oceans around the world—a finding so shocking that just suggesting it would plunge him and his collaborators into a heated public controversy.

# Going Green

Boyce's initial goal, similar to that of William Ripple's work studying wolves in Yellowstone National Park (see Chapter 20), was to show how the loss of a top predator affects the environment in which it lives. In Ripple's story, a variety of species living in Yellowstone interacted to form what we call an *ecological community*. A group of communities interacting with one another and with the physical environment they share is an **ecosystem** (**Figure 21.1**). To say it another way, an ecosystem is characterized by interactions of organisms in the biotic (living) world with each other, as well as with the abiotic (nonliving) world.

An ecosystem may be small or large; a puddle teeming with protists is an ecosystem, as is the Atlantic Ocean. And smaller ecosystems can be nested inside larger, more complex ecosystems. This variety means that ecosystems do not always have sharply defined physical boundaries. Instead, ecologists often define an ecosystem by the

Figure 21.1

## Ecosystems in Nova Scotia, Canada

Overfishing has decimated fish populations in Nova Scotia, affecting the entire ecosystem in which these fish reside. Both biotic and abiotic elements of the ecosystem have been affected.

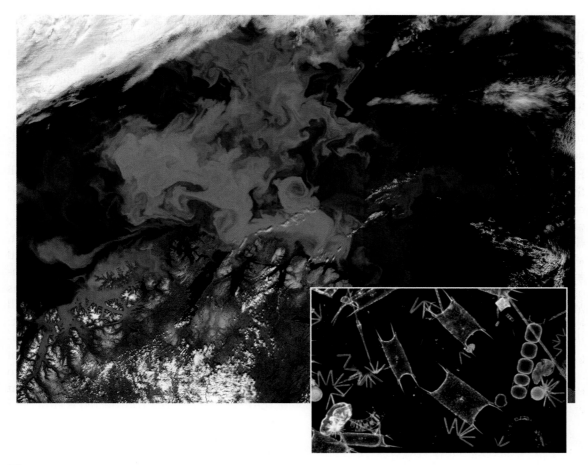

Figure 21.2

**Phytoplankton bloom**
The turquoise area in this aerial photo is a phytoplankton bloom occurring off the coast of Norway. When a population of phytoplankton (inset) increases rapidly, it discolors the water in which it resides.

distinctive ways in which it functions, especially the means by which *energy* and *nutrients* are acquired and distributed by the biotic community.

The activity of primary producers, in particular, profoundly influences the characteristics of an ecosystem. Ecologists often describe an ecosystem according to the types of producers it contains and the consumers that the producers support. A duckweed-covered pond, a tallgrass prairie, and a beech-maple woodland are all examples of ecosystems that can be defined by the specific types of producers that capture and supply energy to consumers.

To see how overfishing was affecting food chains in the ocean, Boyce first looked to the primary producers of the ocean ecosystem: phytoplankton. These small, floating microalgae come in a fantastic array of shapes and sizes, from smooth orbs to segmented spirals to pointy crescents (**Figure 21.2**). Phytoplankton are primarily microscopic, but in large groups they form the green color often seen in water. The more phytoplankton in the water, the greener the water is; the less phytoplankton, the bluer it is.

Phytoplankton are green because they are photosynthetic: they convert light energy from the sun into chemical energy using chlorophyll, the green pigment critical to the process

**DANIEL BOYCE AND BORIS WORM**

Daniel Boyce (left) is a postdoctoral researcher at Queen's University in Ontario, Canada. As a graduate student with Boris Worm (right), he studied the amount of phytoplankton biomass in the ocean. Boris Worm is a marine biologist at Dalhousie University in Nova Scotia, Canada, studying global marine biodiversity.

of photosynthesis. Because they are photosynthetic, these water-living organisms inhabit the top layer of water in the ocean (and almost every body of freshwater as well), a location that gives them easy access to sunlight. Thanks to their ability to photosynthesize using sunlight, phytoplankton are primary producers and the central means through which energy enters the ocean ecosystem. If overfishing was affecting phytoplankton levels, Boyce worried, the whole ocean ecosystem could be in danger.

# Bottom of the Pyramid

Energy and nutrients flow through ecosystems in distinctive patterns. First, let's consider the path of energy. Producers, like phytoplankton, capture

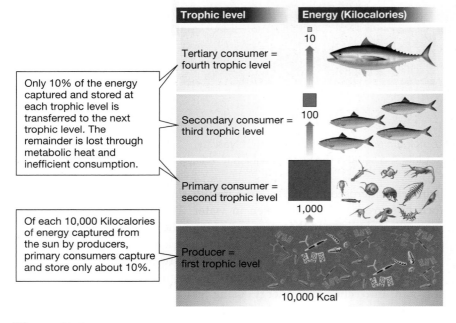

Only 10% of the energy captured and stored at each trophic level is transferred to the next trophic level. The remainder is lost through metabolic heat and inefficient consumption.

Of each 10,000 Kilocalories of energy captured from the sun by producers, primary consumers capture and store only about 10%.

## Figure 21.3

### Energy pyramid

The levels of the energy pyramid correspond to steps in a food chain.

**Q1:** What percentage of the original 10,000 Kilocalories is available to a shark that might eat the tuna in this figure?

**Q2:** What trophic level and term would describe a predator of tuna?

**Q3:** Give an example of a primary consumer in a terrestrial environment.

energy from the sun and transform it into fuel energy. That fuel is passed up the food chain as one organism eats another. An **energy pyramid** represents the amount of energy available to organisms in an ecosystem. Each level of the pyramid corresponds to a step in a food chain and is called a **trophic level**. In the ocean, for instance, phytoplankton are on the first trophic level. Zooplankton, larger, multicellular plankton that feed on phytoplankton, are the second trophic level. Small fishes such as herring are the third level, and large fishes such as tuna are the fourth (**Figure 21.3**).

At each trophic level, a portion of the energy captured by producers is lost as **metabolic heat**, the heat released as a by-product of chemical reactions within a cell, especially during cellular respiration. Organisms lose a lot of energy as metabolic heat, as revealed by the fact that a small room crowded with people rapidly becomes hot; that warmth is the result of metabolic heat leaving our bodies. On average, roughly 10 percent of the energy at one trophic level is transferred to the next trophic level. The remaining 90 percent of the energy that is not transferred is either not consumed (for example, when we eat an apple, we eat only a small part of the apple tree), is not taken up by the consumer's body (for example, we cannot digest the cellulose contained in the apple), or is lost as metabolic heat.

Because of this steady loss of heat, energy flows in *only one direction* through ecosystems. It enters Earth's ecosystems from the sun (in most cases) and leaves them as metabolic heat. Every unit of energy captured by producers is eventually lost from the biotic world as heat. Therefore, energy cannot be recycled within an ecosystem. It travels up an energy pyramid, never down.

In contrast, **nutrients**—chemical elements required by living organisms—*are* recycled and reused within and across ecosystems. Although Earth receives a constant stream of light energy from the sun, our planet does not acquire more nutrients on a daily basis; rather, a constant and finite pool of nutrients cycles through the land, water, and air. If nutrients were not cycled between organisms and the physical environment, life on Earth would not exist.

Nutrients pass through the abiotic world, from rocks and mineral deposits into soil, water, and air, and then on to the biotic world via absorption by producers. Once in the biotic world, they are cycled among consumers for varying lengths

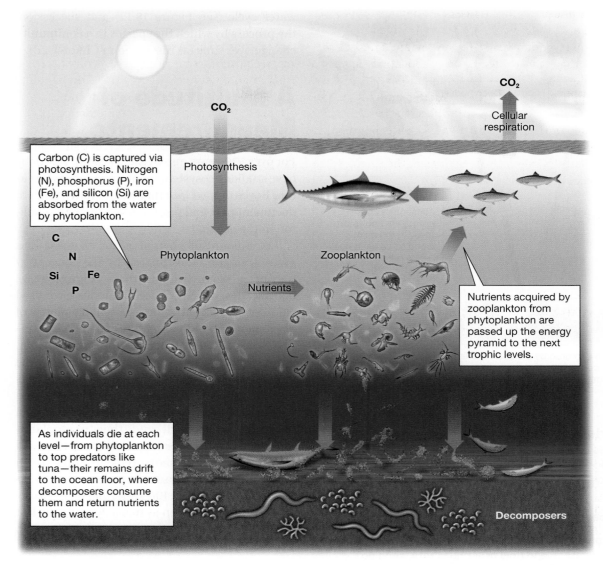

**Figure 21.4**

## Nutrients cycle within and beyond the ocean ecosystem

In the ocean, phytoplankton absorb nutrients from the abiotic world. Within the biotic world, these nutrients are cycled among consumers, beginning with zooplankton and moving up through trophic levels. The nutrients then return to the abiotic world when dead or dying organisms are broken down by decomposers into their constituent elements.

**Q1:** Which organisms are the producers in this ecosystem?

**Q2:** How do nutrients flow from the abiotic to the biotic components of the ecosystem?

**Q3:** How do nutrients flow from the biotic to the abiotic components of the ecosystem?

of time. Phytoplankton, for example, require the nutrients nitrogen, phosphorus, iron, and silicon for growth. When zooplankton eat phytoplankton, they take up those nutrients, and so on up the food chain (**Figure 21.4**).

Nutrients are eventually returned to the abiotic world when **decomposers** break down the dead bodies of other organisms. In some ecosystems, decomposers break down 80 percent of the *biomass*, or biological material,

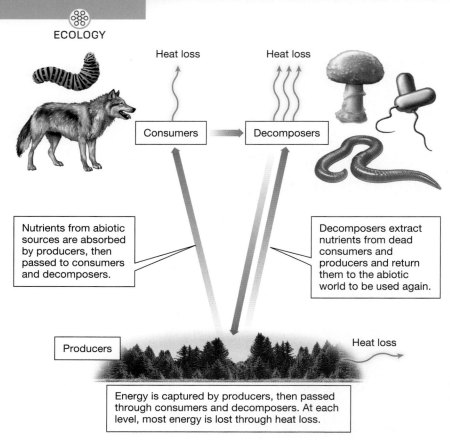

Figure 21.5

### Energy flow and nutrient cycling

Unlike energy, which moves up and out of ecosystems, nutrients are constantly cycled between the abiotic and biotic worlds. Important nutrients for the biotic world include carbon (C), potassium (K), phosphorus (P), and nitrogen (N).

**Q1:** How is a decomposer different from a more typical consumer?

**Q2:** What is the difference between how carbon is brought into the biotic portion of the ecosystem, and how other nutrients, such as phosphorus, are brought in?

**Q3:** Describe all the points at which heat is lost in this figure.

made by producers. Without decomposers, nutrients could not be repeatedly reused, and life would cease because all essential nutrients would remain locked up in the bodies of dead organisms. In this way, decomposers are the "cleaners" of an ecosystem. Bacteria and fungi are important decomposers in the ocean, as are hagfishes, worms, and others.

Ecologists and earth scientists use the term "nutrient cycle" to describe the passage of a chemical element through an ecosystem (**Figure 21.5**). The nutrient cycle and the flow of energy are two of four processes that link the biotic and abiotic worlds in an ecosystem. These **ecosystem processes** also include the

water cycle (see Figure 18.12) and succession, the process by which the species in a community change over time (as discussed in Chapter 20).

# A Multitude of Measurements

For over 100 years, researchers around the globe have studied ecosystems containing phytoplankton. Boyce tapped into that wealth of research to document past and present levels of phytoplankton in the oceans.

The amount of phytoplankton biomass in a given area can be estimated by the concentration of chlorophyll found there (**Figure 21.6**). For decades, nearly all ocean studies have used chlorophyll concentration as a reliable metric of phytoplankton biomass. Chlorophyll concentration is measured by detecting the color of water. Water takes on deeper shades of green as the amount of chlorophyll increases. When there is no chlorophyll, water appears clear.

Ideally, Boyce would have used satellite data to detect chlorophyll and thus phytoplankton concentrations, since satellites today take high-resolution color measurements of the ocean surface. Yet Boyce planned to review phytoplankton levels over the past 100 years, and high-quality satellite data have been available for only the last decade. He needed another source of data.

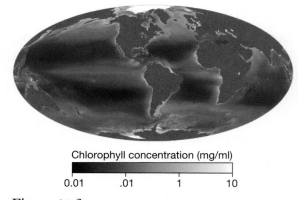

Figure 21.6

### Average chlorophyll concentration in the oceans

Phytoplankton are most abundant in high latitudes, along coastlines and continental shelves, and along the equator in the Pacific and Atlantic Oceans (yellow)—but are scarce in remote oceans (dark blue).

In a first-of-its-kind analysis, Boyce, together with his adviser Boris Worm and the oceanographer Marlon Lewis, combined two types of chlorophyll measurements. The first type, dating all the way back to 1899, were recorded with nothing more than a rope and a disk.

In 1865, the pope asked priest and astronomer Pietro Angelo Secchi to measure the clarity of water in the Mediterranean Sea for the purposes of the papal navy. Secchi designed one of the simplest measurement devices ever used: a dinner plate–sized disk painted with black and white stripes attached to a rope. The disk is lowered into water until the white stripes disappear (as they become obscured by chlorophyll from phytoplankton), and the depth at that point is recorded (**Figure 21.7**; see also the chapter-opening photo). Chlorophyll concentrations derived from Secchi disk measurements have been corroborated by satellite data, so scientists know they are reliable.

In addition to gathering Secchi disk data, scientists at sea use lab tools to directly measure the quantity of chlorophyll in the water (as opposed to observing its color and relating that to chlorophyll concentration). Boyce found hundreds of thousands of these direct chlorophyll measurements online in open-source databases. "There's been a huge increase in the amount of publicly available oceanographic data out there," says Boyce.

But to use the data, Boyce first had to separate the wheat from the chaff. "With any big database, there are bound to be measurements that are entered incorrectly, for whatever reason," says Boyce. He, Worm, and Lewis ruled out measurements that were inappropriate for their study, such as those taken where the ocean floor was less than 25 meters deep, because changes of water transparency in those cases could be caused by sediment or runoff from landmasses nearby rather than by phytoplankton.

The team analyzed each data set separately—Secchi disk measurements and direct chlorophyll measurements—and then together. To combine the two, they converted all the Secchi measurements into the same units as those used for direct chlorophyll concentrations. In total, the blended data set included 445,237 chlorophyll measurements collected between 1899 and 2008.

What they found made them pause. With those two different methods of analysis, they identified a significant decline in phytoplankton levels—a whopping 60–80 percent—in Earth's oceans where data were available during the

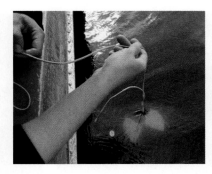

Figure 21.7

**Secchi disks indirectly measure chlorophyll concentrations**
The Secchi disk is lowered into water until its white stripes become obscured by the chlorophyll in phytoplankton.

last century. Overall, phytoplankton appeared to have declined by about 1 percent of the global average each year.

One percent sounds like a small number, but that amount every year just since 1950 translates into a staggering total phytoplankton decrease of 50 percent in the world's oceans.

# The Precious 1 Percent

A 50 percent loss in the main producer in any ecosystem is a worrisome number, but especially with respect to phytoplankton. Phytoplankton support fisheries, produce half the oxygen we breathe, and take in carbon dioxide from the atmosphere, which helps offset the greenhouse effect and global warming.

An ocean with less phytoplankton will function differently because ecosystems depend on **energy capture**, the trapping and storing of solar energy by the producers at the base of the ecosystem's energy pyramid. Herbivores, carnivores, and detritivores all depend indirectly on energy capture. If an ecosystem has an abundance of producers, it can often support more consumers at higher trophic levels. In a tropical forest, for example, an abundance of plants capture energy from the sun, and the forest teems with life. On the flip side, relatively little energy is captured in an environment with few producers. In tundra or desert regions, for example, less food is available, and fewer animals can live there. These significant differences have prompted ecologists to categorize large areas of Earth's surface into distinct regions, called **biomes**, that are defined by their unique climatic and ecological features (**Figure 21.8**).

Assessing the overall amount of energy captured by producers is important in determining how an ecosystem works, because energy

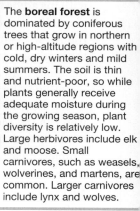

The **boreal forest** is dominated by coniferous trees that grow in northern or high-altitude regions with cold, dry winters and mild summers. The soil is thin and nutrient-poor, so while plants generally receive adequate moisture during the growing season, plant diversity is relatively low. Large herbivores include elk and moose. Small carnivores, such as weasels, wolverines, and martens, are common. Larger carnivores include lynx and wolves.

The **tundra** biome is found at the poles and on mountaintops. Trees are absent or scarce because of the short growing season. The vegetation is dominated by low-growing flowering plants, and the boggy landscape is covered in mosses and lichens, important food sources of herbivores. Rodents provide food for carnivores like foxes and wolves. Bears and musk oxen are among the few large mammals.

The **chaparral** is a shrubland biome dominated by dense growths of drought-resistant plants in regions with cool, rainy winters and hot, dry summers. These conditions make the chaparral exceptionally susceptible to wildfires. Common vegetation in the California chaparral includes scrub oak, pines, mountain mahogany, manzanita, and the chemise bush. Small mammals such as jackrabbits and gophers are common, and there are many species of lizards and snakes.

**Terrestrial biomes:**

- Tundra
- Boreal forest
- Temperate forest
- Grassland
- Chaparral
- Desert
- Tropical forest

Ecosystems within the **freshwater biome** are heavily influenced by the terrestrial biomes that they border or through which their water flows. Lakes are landlocked bodies of standing freshwater. Rivers are bodies of freshwater that move continuously in a single direction. Wetlands are characterized by standing water that is shallow enough for rooted plants to emerge above the water surface. A bog is a freshwater wetland with stagnant, oxygen-poor water, and low productivity and species diversity. In contrast, grassy marshes and tree-filled swamps are highly productive wetlands with a high diversity of organisms.

The **tropical forest** biome is characterized by warm temperatures, about 12 hours of daylight each day, and either seasonally heavy or year-round rains, which tend to leach nutrients from the soil. Soils in this biome further tend to be nutrient-poor because a large percentage of nutrients are locked up in the living tissues (biomass) of organisms. Tropical rainforests, which may receive in excess of 80 inches (200 cm) of rain annually, are some of the most productive ecosystems on Earth, with a rich diversity of organisms.

The **grassland** biome cannot sustain vigorous tree growth, but its moisture levels are not as low as in deserts. Grasslands are found in both temperate and tropical latitudes and are dominated by grasses and herbaceous plants such as coneflower and shooting star. Scattered trees are found in some, such as the tropical grasslands known as the savanna. Burrowing rodents like voles and prairie dogs may aerate the soil, thereby improving growing conditions. Many grasslands have been converted to agriculture.

The **temperate forest** biome is dominated by trees and shrubs adapted to relatively rich soil, snowy winters, and moist, warm summers. These forests display greater species diversity than do the tundra and boreal forest biomes: oak, maple, hickory, beech, and elm are common. Herbivores include squirrels, rabbits, deer, raccoons, and beavers, while bobcats, mountain lions, and bears make up the carnivores. Amphibians and reptiles are common.

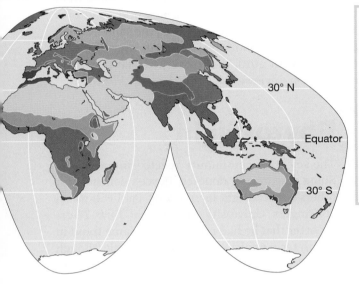

30° N

Equator

30° S

Because air in the **desert** biome lacks moisture, it does not retain heat well. As a result, temperatures can be above 113°F (45°C) in the daytime and then plunge to near freezing at night. Desert plants have small leaves to minimize water loss. Succulents, such as cacti, store water in their fleshy stems or leaves. Most animals in the desert are nocturnal, hiding in burrows during the heat of the day and emerging at night to feed.

**Estuaries** are the shallowest but most productive of the aquatic biomes. They are tidal ecosystems where rivers flow into the ocean. They have a constant ebb and flow of fresh and salt water, and organisms must be able to tolerate daily changes in saltwater concentrations. The plentiful light, the abundant supply of nutrients delivered by the river system, and the regular stirring of nutrient-rich sediments by water flow create a rich and diverse community of photosynthesizers. Grasses and sedges are the dominant vegetation in most estuaries.

The **marine biome**, characterized by salt water, is the largest biome on our planet. The *coastal region* stretches from the shoreline to the edge of the continental shelf and is highly productive because of the availability of nutrients and oxygen. A majority of Earth's marine species live in the coastal region. The intertidal zone, closest to the shore, is a challenging environment where organisms such as seaweeds, worms, crabs, sea stars, sea anemones, and mussels are submerged and exposed to dry air on a twice-daily basis. The relatively nutrient-poor *open ocean* begins about 40 miles offshore and is much less productive than coastal waters.

## Figure 21.8

### Amazing biomes

Biomes do not begin and end abruptly, but rather transition into one another. Terrestrial biomes are categorized by temperature, precipitation, and altitude; aquatic biomes are determined by proximity to shorelines.

capture influences the amount of food available to other organisms. The **net primary productivity** (**NPP**) of an ecosystem is the energy, acquired through photosynthesis over a particular time period, that is available for the growth and reproduction of producers. NPP is the amount of energy captured by photosynthetic organisms minus the amount they expend on cellular respiration and other maintenance processes. NPP is typically determined by estimating the amount of carbon captured during photosynthesis. This can be done by measuring the amount of new biomass produced by the photosynthetic organisms in a given area during a specified period of time.

According to scientists' estimates, the NPP of all producers on Earth exceeds 100 billion tons of carbon biomass per year. Roughly half of this productivity comes from phytoplankton in the ocean. Therefore, phytoplankton capture approximately 50 billion tons of carbon per year. So if Boyce's calculations are right, a loss of 1 percent of that biomass is 500 million tons of organic matter lost from the oceans *each year*. That's a lot of biomass to lose.

Net primary productivity relies on four things: sunlight, water, temperature, and the availability of nutrients. The most productive ecosystems on land are tropical forests; the least productive are deserts and tundra (including some mountaintop communities). The most productive ecosystems in water are estuaries—regions where rivers empty into the sea—because nutrients drained off the land stimulate the growth and reproduction of phytoplankton and other producers, which in turn nourish large populations of consumers. The least productive aquatic biome is the deep ocean, where sunlight does not penetrate.

Despite similarities between the NPP requirements on land and in the ocean, the global pattern of NPP differs between the two. On land, the NPP is highest at the equator and decreases toward the poles. But in the ocean, the general pattern relates not to latitude but to distance from shore: the productivity of marine ecosystems is typically high in ocean regions close to land and relatively low in the open ocean (**Figure 21.9**). This is because nutrients needed by aquatic photosynthetic organisms are in better supply near land, thanks to delivery from streams and rivers. Wetlands such as swamps and marshes, which trap soil sediments rich in nutrients and organic matter, can be so productive that they match the productivity levels of tropical forests.

A loss of 500 million tons of phytoplankton each year could potentially affect the ocean's NPP and ocean life. "Almost all biological life in the ocean depends on phytoplankton. A reduction in the biomass of phytoplankton will result in less secondary production in the oceans," says Boyce. That means fewer sharks, whales, fishes, eels—you name it.

The team's discovery was shocking, to say the least. No one else had documented a global decline in phytoplankton before. So, to be confident in their results, Boyce and Worm did several more rounds of data analysis, checking again and again to make sure they were using the right numbers in ways that correctly represented what was happening in the natural world. And over and over, they came back with the same results: global phytoplankton declined over the last century. In 2010, they published that finding in the peer-reviewed scientific journal *Nature*.

## Phyto-Fight

The scientific community reacted immediately. Some researchers doubted Boyce's conclusions; others were outright incredulous. Paul Falkowski at Rutgers University told a *New York Times* reporter that he had not found the same trend in a long-term analysis of the North Pacific (though Boyce contends that their trends were very similar), and another team had actually seen an increase in phytoplankton starting around 1978 in the central North Pacific. Falkowski called Boyce's paper "provocative" but said he would "wait another several years" to see whether satellite data would back up the finding.

Then, in 2011, three separate research teams went so far as to publish formal critiques of the work. One suggested that the declining trend was an error resulting from the use of two different types of measurements: Secchi disk readings and direct chlorophyll measurements. A second team echoed that idea, reanalyzed the data in a way that showed an increase in phytoplankton, and then bluntly concluded, "Our results indicate that much, if not all, of the century-long decline reported by [Boyce] is attributable to this [sampling bias] and not to a global decrease in phytoplankton biomass."

The third team noted that Boyce's finding conflicted with eight decades of data on phytoplankton biomass collected by a large project called the Continuous Plankton Recorder (CPR)

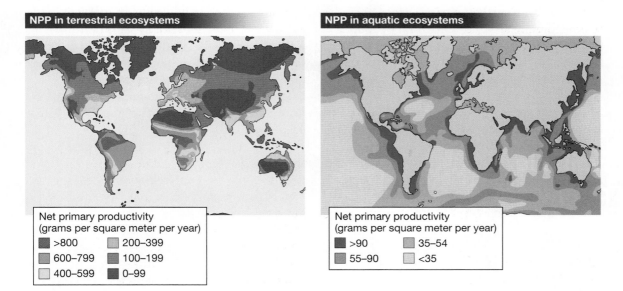

**NPP in terrestrial ecosystems**

**NPP in aquatic ecosystems**

Net primary productivity
(grams per square meter per year)
- \>800
- 600–799
- 400–599
- 200–399
- 100–199
- 0–99

Net primary productivity
(grams per square meter per year)
- \>90
- 55–90
- 35–54
- \<35

Figure 21.9

### Global variation in net primary productivity

NPP can be measured as the number of grams of new biomass made by producers each year in a square meter of each biome's area. NPP varies greatly across both terrestrial and aquatic ecosystems.

**Q1:** Which terrestrial biome has the lowest NPP? Which aquatic biome?

**Q2:** Where are the most productive terrestrial biomes located?

**Q3:** Give a possible reason for your answer to question 2.

survey, which monitors the Northeast Atlantic Ocean. The CPR survey, started in 1931, employs a unique instrument pulled through the ocean by commercial fishing vessels to collect millions of samples of plankton. The CPR survey found that over the last 20–50 years, phytoplankton biomass increased in the Northeast Atlantic, says Abigail McQuatters-Gollop, a former researcher at the Sir Alister Hardy Foundation for Ocean Science, which operates the survey.

In response, Boyce, Worm, and Lewis didn't get angry; they got focused. After reading the critiques, the three researchers went back to the data. First they applied a correction factor suggested by the critics, in the hopes of removing any bias between the two types of data. "We did that, and the trends remained similar," says Boyce. Next, they again estimated changes over time individually for the two data sources. "That didn't change the trends either," says Boyce. After that, they incorporated additional suggestions by their peers and created a new, expanded database of chlorophyll measurements to work from. Finally, they reestimated changes in chlorophyll

using this new database and their revised analysis methods, but the phytoplankton still seemed to be declining, independently of the type of data or how the data were analyzed. The three researchers published their reanalysis in a series of three papers in 2011, 2012, and 2014, demonstrating the same decline again and again.

But the additional work has not silenced the critics. "It's still pretty hotly debated," admits Boyce. "The story is not over by any means." In 2011, in fact, Worm traveled to an international plankton conference where he, McQuatters-Gollop, and others debated the topic in front of a live audience. "It was an amicable meeting that generated loads of discussion," says

### ABIGAIL MCQUATTERS-GOLLOP

Abigail McQuatters-Gollop was a science and policy researcher at the Sir Alister Hardy Foundation for Ocean Science, home of the Continuous Plankton Recorder survey, from 2010 to 2015. She is now a lecturer in marine conservation at Plymouth University in the United Kingdom.

# Productive Plants

Data on net primary productivity (NPP)—the total energy available in an ecosystem for the growth and reproduction of primary producers—is sparse for whole biomes, but researchers have been able to estimate the NPP of those listed here using information about different vegetation types and carbon sources in each area.

Assessment available in smartwork5

## Total energy available from primary producers

(in grams per square meter per year)

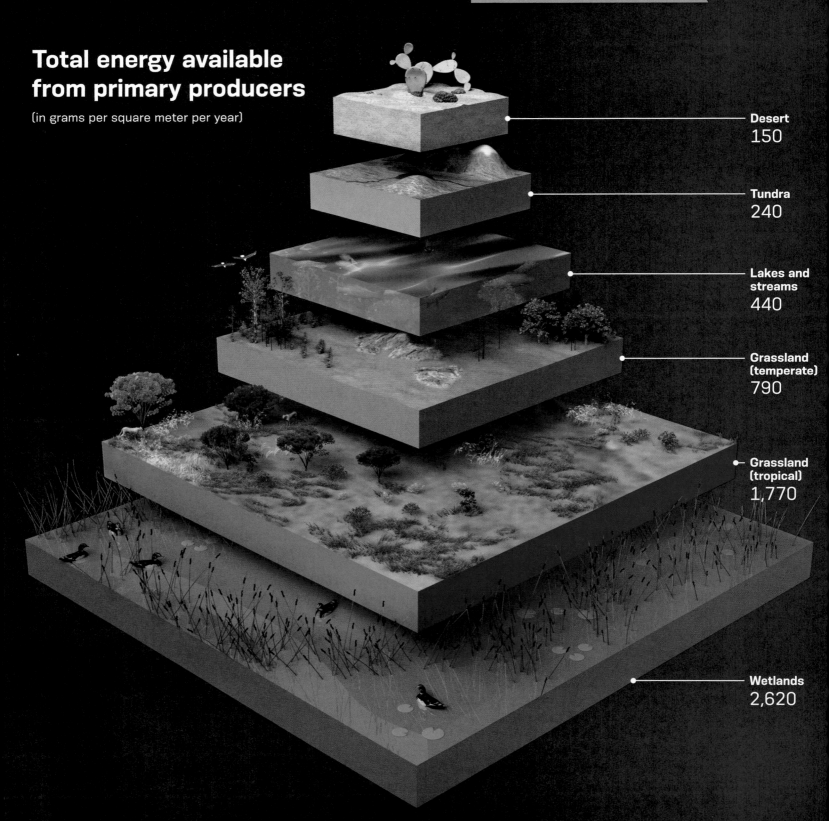

Desert
150

Tundra
240

Lakes and streams
440

Grassland (temperate)
790

Grassland (tropical)
1,770

Wetlands
2,620

McQuatters-Gollop. Still, they could not agree on whether phytoplankton populations have increased or decreased in the ocean. At the meeting, the researchers agreed that the best thing to do was to combine as many data sets as possible.

# Heating Up

In 2015, the long-awaited satellite data came in—and they weren't good. Ocean color measurements from two NASA satellites led scientists to conclude that diatoms, the largest type of phytoplankton, had declined more than 1 percent per year from 1998 to 2012 globally, with major losses in the North Pacific, North Indian, and Equatorial Indian Oceans.

It looks like Boyce, Worm, and Lewis may be right, but the important question of why a decline is occurring remains. As part of his research, Boyce investigated possible causes of the downward trend. In one study, he compared changes in sea surface temperature to changes in chlorophyll levels and noted a strong correlation: over the last 100 years, chlorophyll concentrations declined *and* ocean temperatures increased, in line with increases in global warning. This correlation has been closely followed in recent years (**Figure 21.10**).

But correlation does not prove causation, and much more work needs to be done to support the idea that global warming caused the decline. It is a logical hypothesis, however, because as the planet warms, water in the oceans mixes less, limiting the nutrients delivered to the surface from decomposers in the deep sea. As a result, phytoplankton do not receive the nutrients they need for growth and reproduction. In 2014, Boyce published experimental data supporting the hypothesis. Working with marine scientists in Germany, he found that warming the water in a controlled, experimental ocean water enclosure led to reduced phytoplankton biomass.

A 2014 study by a large, international group of researchers predicted that an increase in ocean temperature due to global warming would cause phytoplankton biomass to decrease by 6 percent by the end of this century. Consequences of continued phytoplankton decline could include altering the carbon cycle between the ocean and the atmosphere, changing heat distribution in the ocean, and causing a decrease in the supply of food in the ocean. Whatever the case, if phytoplankton populations are decreasing, says Boyce—and he is sure they are—there will be profound effects, which have only just begun.

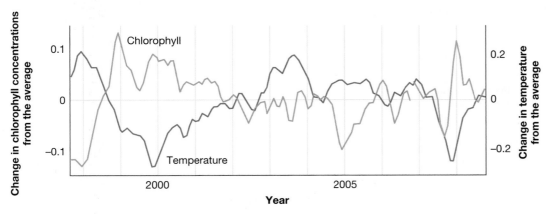

Figure 21.10

**As ocean temperature increases, chlorophyll decreases**
Between late 1997 and mid-2008, satellites observed that warmer-than-average temperatures (red line) were correlated with below-average chlorophyll concentrations (green line).

**Q1:** In what years were chlorophyll levels the highest?

**Q2:** In what years were the temperature changes from the average the greatest?

**Q3:** How do you predict this graph will look 10 years from now?

# REVIEWING THE SCIENCE

- An **ecosystem** consists of communities of organisms and the physical environment in which those communities live. It is the sum of all biotic elements interacting with all abiotic elements. Energy, materials, and organisms can move from one ecosystem to another.

- An **energy pyramid** represents the amount of energy available to each **trophic level** of a food chain in an ecosystem.

- Energy enters an ecosystem when producers capture it from an external source, such as the sun. A portion of the energy captured by producers is lost as **metabolic heat** at each trophic level. As a result, energy flows in only one direction through ecosystems.

- **Nutrients** are the chemical elements required by living organisms. Unlike energy, nutrients are recycled and reused within and across ecosystems. Earth has a fixed amount of nutrients.

- **Decomposers** break down the dead bodies of other organisms, both consumers and producers, releasing the nutrients in the bodies of dead organisms back to the physical environment.

- Four **ecosystem processes** link the biotic and abiotic worlds in an ecosystem: nutrient cycling, energy flow, water cycling, and succession.

- Ecosystems depend on **energy capture**, the trapping of solar energy by producers via photosynthesis, and the storage of that energy as chemical compounds in their bodies.

- Earth is categorized into 10 major **biomes**, regions defined by their climatic and ecological features.

- **Net primary productivity (NPP)** is the energy acquired through photosynthesis that is available for growth and reproduction to producers in an ecosystem. NPP is estimated by the amount of biomass produced in a given area during a specified period of time.

# THE QUESTIONS

## The Basics

**1** The movement of nutrients between organisms and the physical environment is called

(a) nutrient cycling.

(b) ecosystem services.

(c) net primary productivity.

(d) decomposition.

**2** How much energy is transferred up the energy pyramid from one trophic level to the next?

(a) 90%

(b) 50%

(c) 10%

(d) 10%–50%

**3** Which organisms are considered the "recyclers" of our planet?

(a) consumers

(b) producers

(c) phytoplankton

(d) decomposers

**4** The terrestrial biome that receives the most consistent year-round rainfall is

(a) wetland.

(b) boreal forest.

(c) tropical forest.

(d) chaparral.

**5** Link each term with the correct definition.

| | |
|---|---|
| BIOME | 1. The energy acquired through photosynthesis by producers of an ecosystem. |
| ECOSYSTEM | 2. A group of communities interacting with one another and with the physical environment they share. |
| NET PRIMARY PRODUCTIVITY | 3. A large, distinct region defined by its unique climatic and ecological features. |
| ECOLOGICAL COMMUNITY | 4. The populations of different species that live and interact with one another in a particular place. |

## Challenge Yourself

**6** Select the correct terms:
The biome characterized by shrubs and nonwoody plants that grow in regions with cool, rainy winters and hot, dry summers is (**tundra / chaparral**). Another biome with few trees, but in this case dominated by grasses and nonwoody plants, is (**grassland / tundra**). The most productive aquatic biome is (**freshwater / estuaries**).

**7** Which of the following is a component of an ecosystem but *not* of an ecological community?

(a) a producer

(b) water

(c) a secondary consumer

(d) a primary consumer

**8** Place the following elements of the scientists' study of oceanic phytoplankton levels in the correct order from earliest to latest by numbering them from 1 to 5.

_____ a. Boyce analyzed data from multiple sources showing that ocean phytoplankton levels have declined dramatically over the last 100 years.

_____ b. Worm demonstrated that industrialized fishing had decimated predatory fish communities.

_____ c. Secchi designed a device to measure phytoplankton levels in water.

_____ d. Boyce's finding of declining phytoplankton levels was challenged by fellow scientists, because of his methodology.

_____ e. Boyce, a graduate student in Worm's lab, decided to study how the loss of predatory fish reverberated through the oceanic ecosystem—in particular, its phytoplankton populations.

**9** How is an ecosystem different from an ecological community?

## Try Something New

**10** In the energy pyramid shown here, an owl, a cardinal, and a grasshopper are the fourth, third, and second trophic levels, respectively.

a. If each grasshopper passes 100 Kilocalories to the cardinal when eaten, how many grasshoppers would the cardinal have to eat to obtain 10,000 Kilocalories?

b. How many Kilocalories of grass are required to produce 10,000 Calories' worth of grasshopper?

c. Where does the lost energy go?

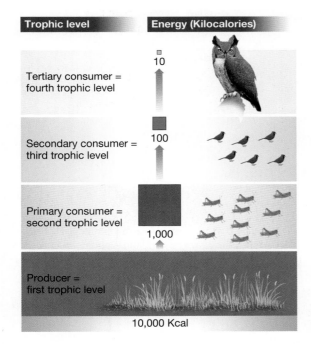

| Trophic level | Energy (Kilocalories) |
|---|---|
| Tertiary consumer = fourth trophic level | 10 |
| Secondary consumer = third trophic level | 100 |
| Primary consumer = second trophic level | 1,000 |
| Producer = first trophic level | 10,000 Kcal |

**11** Is the water cycle (see Figure 18.12) more similar to the movement of energy or to the movement of nutrients through an ecosystem? Justify your answer.

**12** While sloshing around in the swampy wetlands of a nearby forest preserve, a small child notices a rotting tree branch covered in fungi. What are these organisms doing on this tree branch?

(a) These are producers acquiring energy through photosynthesis.

(b) These are consumers acquiring energy from the wood and releasing nutrients back to the earth.

(c) These are decomposers acquiring energy from the wood and releasing nutrients back to the earth.

(d) These are decomposers releasing nutrients back to the earth, but acquiring no energy for themselves in the process.

## Leveling Up

**13** *Write Now* **biology: human-caused biome shifts** The location of Earth's different biomes depends on climate and altitude, for the most part. However, human activities play a role in the conversion of one biome to another, as has been seen many times in history. Research one major change in a biome category based on human activity, and describe how and why this change happened. Speculate on how this change could have been specifically avoided. (*Note:* Do not analyze a change via deforestation to agricultural land, since agricultural land is not a natural biome. *Hint:* Take a look at Easter Island as one example.)

**14** **Doing science** Join forces with millions of others by classifying phytoplankton on your computer. Do an Internet search for "citizen science phytoplankton" and sign in as a citizen scientist, complete the tutorial, and start helping researchers quantify the phytoplankton in our oceans.

**15** **What do *you* think?** Some people think the current U.S. Endangered Species Act should be replaced with a law designed to protect ecosystems, not species. The intent of such a law would be to focus conservation efforts on what its advocates think really matters in nature: whole ecosystems. Given how ecosystems are defined, do you think it would be easy or hard to determine the boundaries of what should and should not be protected if such a law was enacted? Give reasons for your answer.

**For more, visit digital.wwnorton.com/bionow2 for access to:**

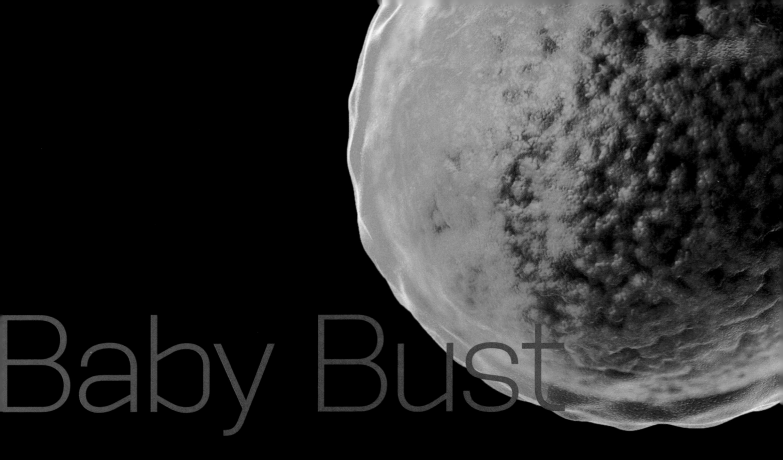

# Baby Bust

*Facing dwindling births, Denmark searches
to resolve problems of infertility.*

**After reading this chapter you should be able to:**

- Distinguish among tissues, organs, and organ systems.
- Explain the importance of homeostasis for life.
- Diagram the steps of gamete formation in males and females.
- Create a flowchart of the steps of fertilization in humans.
- Describe the stages of prenatal development in humans.
- Explain the role of a given hormone in the reproductive system.

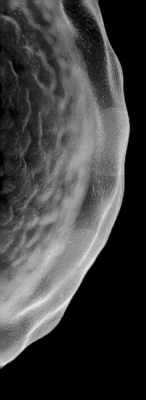

CHAPTER
22

HOMEOSTASIS,
REPRODUCTION,
AND
DEVELOPMENT

A woman slides a bra strap off her naked shoulder. In Danish, a seductive man's voice asks, "Can sex save Denmark's future?"

The commercial, aired on Danish television, then switches to a view of an empty playground. Produced by a travel agency, the ad encourages responsible Danes to book a romantic holiday with the company's "ovulation discount" and try to conceive while on vacation. It concludes with a large banner proclaiming, "Do It for Denmark" (**Figure 22.1**).

Birth rates reached a 27-year low in the year 2014 in Denmark, a Nordic country in northern Europe. When asked, most Danish couples said they would like to have two or three children, yet the present fertility rate is only 1.69 children per family—not high enough to maintain Denmark's current population. Infertility is now considered an epidemic in the country. In fact, one in 10 children in Denmark is conceived using reproductive technologies. "We see more and more couples needing to get assisted fertility treatment," Bjarne Christensen, secretary general of Sex and Society, Denmark's leading family-planning association, told *Bloomberg News*. "We see a lot of people who don't succeed in having children."

Commercials like the one described here make a patriotic, if whimsical, appeal to Danish citizens to have more children, but experts debate whether falling fertility rates are a cause for concern or for celebration. Some applaud the decrease as a way to slow global population growth and reduce human consumption of limited natural resources. But others in countries facing declining fertility rates—including Denmark, France, Singapore, and recently the United States—worry about the burdens that will be placed on a generation smaller than the one before it (**Figure 22.2**). Government officials fear that this trend will produce a smaller workforce and fewer young people to care for retirees, reducing their countries' economic strength.

To counteract falling birth rates, some governments have begun taking profertility stances, including providing free postnatal care and subsidized day care. In Denmark, Sex and Society, a nongovernmental organization, provides sex education materials for most of Denmark's schools, and it recently unveiled a new series of lesson plans entitled "This Is How You Have Children!" Instead of focusing on contraception and how to avoid becoming pregnant, these new classes educate students about what fertility is, how aging affects fertility, and when may be the best times to have children.

"We have for many years addressed the very important issues of how to avoid becoming

Figure 22.1

**A print advertisement in the "Do It for Denmark" campaign**

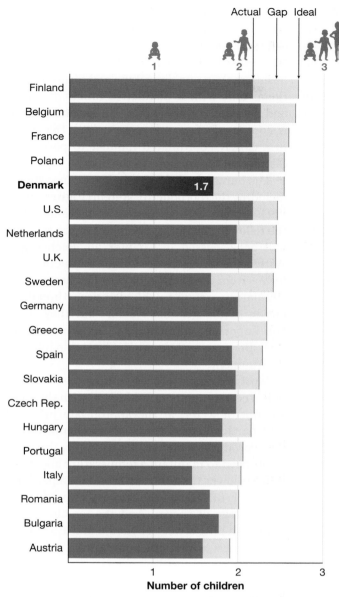

Actual Gap Ideal

Finland
Belgium
France
Poland
**Denmark** 1.7
U.S.
Netherlands
U.K.
Sweden
Germany
Greece
Spain
Slovakia
Czech Rep.
Hungary
Portugal
Italy
Romania
Bulgaria
Austria

1          2          3
**Number of children**

Sources: European data from 2011 Eurobarometer, U.S. data from 2006 and 2008 General Social Survey

## Figure 22.2

### Women in Europe and the United States want more children than they have

American and European women aged 40–54 were asked, "What do you think is the ideal number of children for a family to have?" Their answers were consistently higher than actual birth rates.

**Q1:** In which of the countries shown in the graph did women want the most children, on average? The least?

**Q2:** Which country has the largest gap between the number of children women would like to have and the number they actually have?

**Q3:** Which country has the highest average number of children? The lowest?

pregnant, how to avoid sexual diseases, how they have a right to their own bodies, but we totally forgot to tell the kids that we cannot have children forever," Søren Ziebe, head of Copenhagen University Hospital's fertility clinic, told reporters. "There is a biological limit."

## Seeking Stability

That biological limit may be the cause of Denmark's falling birth rate. Population studies show that, on average, couples are waiting longer to have children, and conception can be more difficult after age 30.

But other causes are also possible. Laboratories are investigating whether chemicals or other environmental factors might be affecting the human **reproductive system**, the parts of the body responsible for producing offspring.

Our bodies are highly efficient and well-coordinated communities of 200 different specialized cell types. **Anatomy** is the study of the structures that make up a complex multicellular body, and **physiology** is the science of the functions of anatomical structures. Through physiological research, scientists hope to determine why fertility rates are falling in Denmark, and whether anything can be done about it.

Their first step has been to examine the reproductive system, which consists of cells, tissues, and organs. **Tissues** are made up of cells that act in an integrated manner to perform a common set of functions (**Figure 22.3**). A tissue

may be composed of just one cell type, or it may contain multiple cell types; in either case, the cells that compose a tissue cooperate to perform the distinctive functions of that tissue. An **organ** has more than one tissue type and forms a functional unit with a distinctive function, shape, and location in the body. An **organ system** is composed of two or more organs that work in a closely coordinated manner to perform a distinct set of functions in the body.

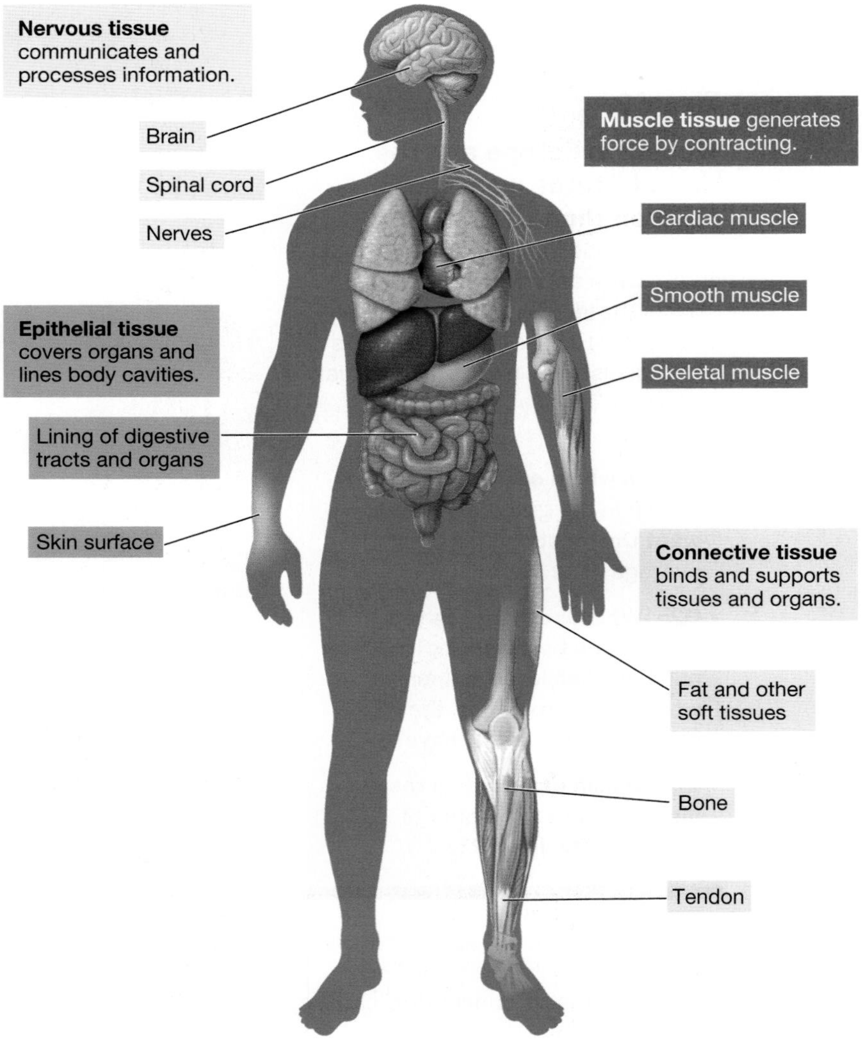

The human body has 11 major organ systems, including the reproductive system (**Figure 22.4**). Each system will be covered in more detail in the following chapters, and plant organ systems will be discussed in Chapter 26. Each organ system and its organs are unique in function and form—from the beating, blood-filled heart of the circulatory system to the electrical, threadlike nerves of the nervous system—but they share one important commonality: for proper function, an organ system requires a stable internal environment.

Most biological processes take place within only a certain temperature range, with the right amount of water, at the appropriate pH, and at a particular concentration of chemicals. In the reproductive system, for example, the male testes need to maintain a temperature about 1–2 degrees lower than the usual body temperature of 98.6°F (37°C) for the successful production and storage of sperm. And the female vagina requires a pH between about 3.8 and 4.5 to keep out bacteria while still maintaining a healthy environment for fertility.

These environments are regulated through **homeostasis**, the process of maintaining a relatively constant internal state despite changes in the external environment. Homeostasis enables an organism to continually sense its internal state and rapidly adjust. In this way, despite large fluctuations in the outside world, homeostasis maintains the internal conditions best suited for life processes.

Homeostasis occurs via **homeostatic pathways**, sequences of steps that reestablish homeostasis if there is any departure from the genetically determined normal state (also called the **set point**). Homeostatic pathways continually monitor the physical and chemical characteristics of the internal environment and trigger regulatory processes within the body if this monitoring system detects any deviation from the set point.

**Nervous tissue** communicates and processes information.

Brain

Spinal cord

Nerves

**Muscle tissue** generates force by contracting.

Cardiac muscle

Smooth muscle

Skeletal muscle

**Epithelial tissue** covers organs and lines body cavities.

Lining of digestive tracts and organs

Skin surface

**Connective tissue** binds and supports tissues and organs.

Fat and other soft tissues

Bone

Tendon

**Figure 22.3**

**Tissues in animals**

The different types of tissues found in the animal body can be placed into four broad categories: **epithelial**, **connective**, **muscle**, and **nervous tissue**.

Q1: Which tissue type is the primary component of skin?

Q2: Which tissue type is the primary component of bone?

Q3: Which tissue types does the hand contain?

The **urinary system** removes excess fluid from the body, along with waste products and toxins.

In the **digestive system**, large molecules of food are broken down in the mouth, stomach, and small intestine, and nutrients are absorbed in the small and large intestines.

The **circulatory system** diffuses oxygen from the lungs to the heart, which then pumps oxygen-rich blood to the rest of the body through a closed network of vessels.

The **respiratory system** brings in oxygen and expels carbon dioxide through the lungs.

The **endocrine system** works closely with the nervous system to regulate all other organ systems. It consists of a number of glands and secretory tissue.

The **integumentary system** is the largest organ system in the human body, covering and protecting the surface of the body.

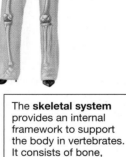

The **nervous system** is a key player in sensing the external world and the body's internal state, and it communicates with all of the other organ systems.

The **skeletal system** provides an internal framework to support the body in vertebrates. It consists of bone, cartilage, and ligaments.

The **muscular system** produces the force that moves structures within the body. It works closely with the skeletal system.

The **immune system** defends the body from invaders such as viruses, bacteria, and fungi.

The **reproductive system** generates gametes and, depending on the animal group, may also support fertilization and prenatal development.

## Figure 22.4

**Organ systems**

The 11 major organ systems of the human body work in an integrated manner.

**Q1:** Which organ system defends the body from infectious diseases such as the common cold or flu?

**Q2:** Which organ systems transport oxygen to cells?

**Q3:** Which organ systems regulate the activities of the other organ systems?

Homeostatic pathways depend on **feedback loops**. A **negative feedback** loop turns off or reduces the output of a process. For example, if a person drinks a large milk shake, the level of glucose in the blood rises. In response, cells in the pancreas produce insulin, which allows glucose to enter cells. So, blood glucose concentration declines. The regulation of body temperature also relies on negative feedback loops (**Figure 22.5**).

A **positive feedback** loop, on the other hand, increases the output of a process until an endpoint is reached. Blood clotting triggered by a broken blood vessel, for example, is a positive feedback loop: the process of clotting releases chemicals that lead to accelerated clotting, continuing until the clot plugs the break in the blood vessel wall. In this feedback loop, the chemicals increase the amount of clotting.

Overall, in homeostatic pathways negative feedback loops are more common than positive feedback loops. Examples of homeostatic pathways include *thermoregulation* (the control

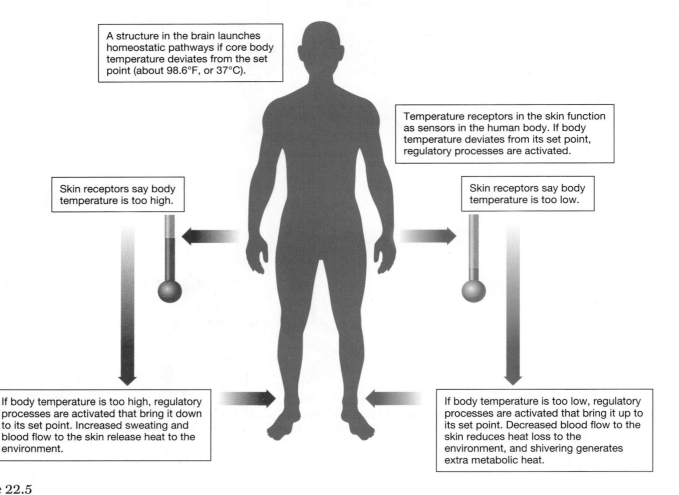

A structure in the brain launches homeostatic pathways if core body temperature deviates from the set point (about 98.6°F, or 37°C).

Temperature receptors in the skin function as sensors in the human body. If body temperature deviates from its set point, regulatory processes are activated.

Skin receptors say body temperature is too high.

Skin receptors say body temperature is too low.

If body temperature is too high, regulatory processes are activated that bring it down to its set point. Increased sweating and blood flow to the skin release heat to the environment.

If body temperature is too low, regulatory processes are activated that bring it up to its set point. Decreased blood flow to the skin reduces heat loss to the environment, and shivering generates extra metabolic heat.

Figure 22.5

## Homeostasis maintains stable internal conditions

Because of homeostatic pathways, large fluctuations in external conditions produce little or no change in the overall state within the animal body. For example, even when the outside temperature is very hot or cold, the human body's internal temperature stays within the narrow range required for survival. ▶▤

Q1: Why is it important to maintain a stable body temperature?

Q2: Which organ system is involved as a temperature sensor? (See Figure 22.4 for an overview of organ systems.)

Q3: Give another example of a homeostatic pathway in humans.

of heat gain and loss) and *osmoregulation* (the control of internal water content and solute concentration by an organism). Homeostasis is critical to the maintenance of our organ systems, including the reproductive system—which brings us back to doing it in Denmark.

# All in the Timing

Danish researchers have long suspected that the primary reason for declining fertility rates is that women are waiting longer to have children. In the 1970s, on average, a Danish woman gave birth to her first child at 24 years old. But in 2014, that average age was 29, with more and more women waiting until they were over 35 to have a first child. And the older a woman gets, the more difficult it is for her to conceive.

Humans reproduce through **sexual reproduction**, in which haploid gametes from a male (the **sperm**) and a female (the **egg**, or ovum) combine to form a diploid **zygote**, which develops into a multicellular individual that is genetically unique and different from either parent. Here, we will review primarily human sexual reproduction, but keep in mind that sex in other animals is more variable than our human perspective might lead us to expect (**Figure 22.6**). In particular, recall that some animals can also reproduce via **asexual reproduction**, in which cells from only one individual produce the offspring, so all of the offspring's genes come from that parent. (For more on the benefits and pitfalls of sexual and asexual reproduction, see "Why Sex?" on page 235 of Chapter 13.)

Human eggs develop through a series of cell divisions called **oogenesis**, the production of mature eggs capable of being fertilized (**Figure 22.7**, left). Oogenesis begins before birth, when germ line cells multiply and develop into immature, diploid egg cells called **primary oocytes**. At birth, the ovaries of a female already contain her entire lifetime supply of primary oocytes, about 1–2 million cells. These cells remain in a state of suspended development until the production of hormones at puberty stimulates one, or occasionally two, of them to mature each month in preparation for ovulation. By the time a female reaches puberty, at about

Sea stars can reproduce by breaking off an arm, which then regenerates into a new individual. Although some animals rely exclusively on asexual reproduction, most asexually reproducing species switch between sexual and asexual reproduction, depending on environmental conditions.

In frogs and other species, females release their eggs, which males then cover with sperm. External fertilization is common among aquatic animals; internal fertilization is more common among land animals.

Clownfish begin life as males, but the largest fish in a group will change to female. Other species begin as females and may then switch to males, while still other species are both male and female at the same time. Individuals that produce both functional testes and functional ovaries—and are therefore both male and female—are called **hermaphrodites**.

Figure 22.6

**Animals display a rich variety of reproductive systems**
Some animals reproduce by cloning themselves through asexual reproduction (left). Others fertilize gametes externally (middle), while still others are both male and female, either simultaneously or sequentially (right).

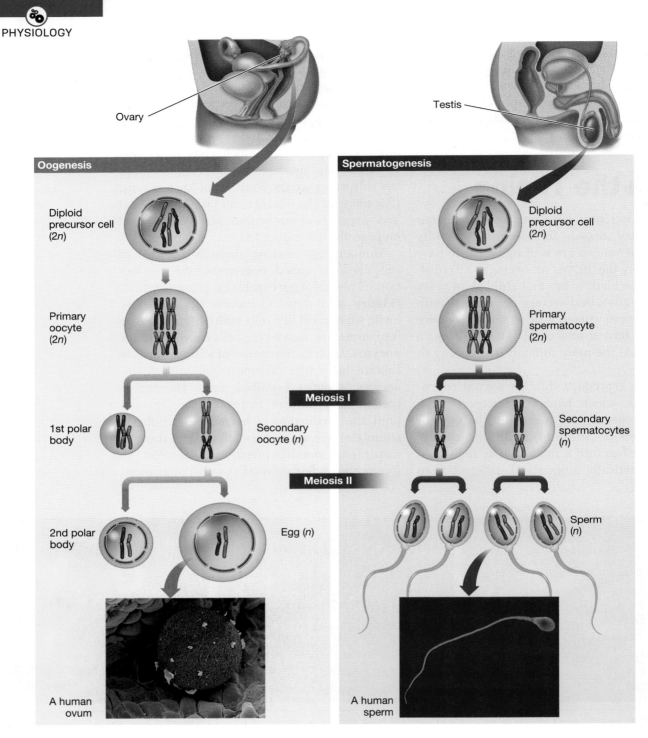

Ovary

Testis

**Oogenesis**

Diploid precursor cell (2*n*)

Primary oocyte (2*n*)

1st polar body

Secondary oocyte (*n*)

2nd polar body

Egg (*n*)

A human ovum

**Spermatogenesis**

Diploid precursor cell (2*n*)

Primary spermatocyte (2*n*)

**Meiosis I**

Secondary spermatocytes (*n*)

**Meiosis II**

Sperm (*n*)

A human sperm

### Figure 22.7

**Sexual reproduction requires the production of haploid gametes**

Oogenesis produces haploid eggs (ova), and spermatogenesis produces haploid sperm.

**Q1:** Identify one way in which oogenesis and spermatogenesis are the same.

**Q2:** How many eggs are produced from each precursor cell? How many sperm?

**Q3:** How much time elapses between the appearance of a precursor cell and the formation of an egg? How does this process differ for a sperm? (You will need to read ahead to answer this question.)

10–12 years of age, approximately 400,000 viable primary oocytes remain—still more than she will use in her lifetime.

Human females do not produce mature eggs continuously. Instead, individual eggs mature and are released in a hormone-driven sequence of events known as the **menstrual cycle** (**Figure 22.8**). The menstrual cycle averages about 28 days, but cycle lengths from 21 to 35 days are considered normal.

A woman has more primary oocytes than she will use in her lifetime, but evidence suggests that those eggs decline in quality as a woman ages—a conclusion supported by the increased risk of birth defects in children born to older mothers. In addition, if eggs from younger women are implanted into women over 40, the pregnancy rate equals the rate associated with the younger women who donated the eggs. In other words, young eggs result in young pregnancy rates.

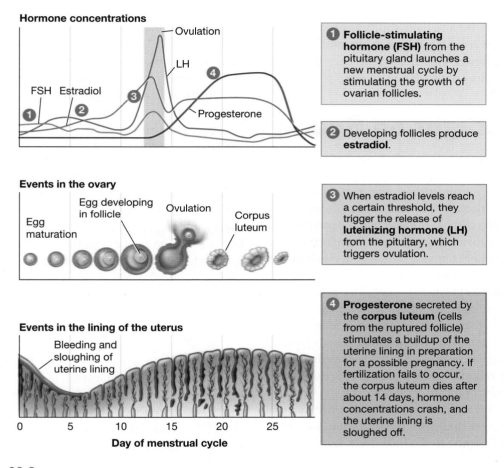

① **Follicle-stimulating hormone (FSH)** from the pituitary gland launches a new menstrual cycle by stimulating the growth of ovarian follicles.

② Developing follicles produce **estradiol.**

③ When estradiol levels reach a certain threshold, they trigger the release of **luteinizing hormone (LH)** from the pituitary, which triggers ovulation.

④ **Progesterone** secreted by the **corpus luteum** (cells from the ruptured follicle) stimulates a buildup of the uterine lining in preparation for a possible pregnancy. If fertilization fails to occur, the corpus luteum dies after about 14 days, hormone concentrations crash, and the uterine lining is sloughed off.

Figure 22.8

## The human menstrual cycle

A menstrual cycle begins with the first day of bleeding, which marks the end of the previous cycle. Over the next few weeks, a succession of hormones stimulates the release of an egg and signals the uterine lining to grow and thicken in preparation for a potential pregnancy. If pregnancy does not occur, hormone levels plummet and the lining is sloughed off as menstrual flow, ending that menstrual cycle.

Q1: Which hormones important for the menstrual cycle are produced in the pituitary gland?

Q2: Which hormone is involved in producing the uterine lining?

Q3: How is the egg follicle involved in producing hormones?

When a woman passes 40 years of age, she shows a clear drop in her ability to produce normal eggs and bear children. Human females reach menopause—the end of their reproductive lives—around the age of 50, although this is highly variable among women.

A decline in egg quality is one suggested cause of Denmark's fertility problem: women today are waiting longer to have children, so their eggs are older and their fertility has decreased. Age also affects the fertility of men, although males do not undergo the clearly identifiable menopause that is characteristic of females. Instead, their sex drive and their ability to produce sperm slowly decrease as they age. In addition, men produce fewer sperm as they age, decreasing their chances of fertilizing an egg.

# Spotlight on Sperm

While parents' increasing age at conception likely contributes to Denmark's fertility problem, researchers suspect that other factors are also at play—particularly, the quality of male sperm. The production of male gametes and the production of female gametes differ in several important ways. The supply of a female's primary oocytes is limited, and once a primary oocyte develops into a mature ovum, it is lost from the supply. In contrast, sperm precursor cells in males constantly replenish the pool of sperm. It is also noteworthy that in a normal menstrual cycle, a human female produces only one mature egg cell—in other words, only one egg per month—while a male produces hundreds of millions of sperm cells every day.

Another difference is that ova are typically much larger than sperm. The human egg is visible (just barely) to the naked eye, but individual sperm can be seen only under a microscope. Sperm contain little substance beyond chromosomes and the cellular machinery needed to move up the female reproductive tract, attach to an egg, and propel the sperm's chromosomes into the egg's cytoplasm. Sperm are simple

packages with valuable information inside. An ovum, on the other hand, is a plump, complex cell full of organelles.

The first reports of falling male sperm counts came in 1992 from Niels Skakkebæk at the University of Copenhagen. Looking at 61 different papers describing the semen quality of almost 15,000 men—both the density and volume of sperm in semen—Skakkebæk and his team found that sperm counts around the world had dropped by 50 percent from about 1940 to 1990. In a peer-reviewed paper they concluded, "As male fertility is to some extent correlated with sperm count, the results may reflect an overall reduction in male fertility." In other words, a decrease in the quality of semen could be lowering male fertility.

In human males, meiosis occurs inside structures called *seminiferous tubules*: twisty, spaghetti-like tubes that permeate the **testes**. In response to male hormones that surge at the onset of puberty, diploid germ line cells in the tubules start dividing to form sperm in a sequence of steps known as **spermatogenesis** (**Figure 22.7**, right). The average man produces about 300 million sperm each day. Surplus sperm that accumulate over time are degraded and reabsorbed by the cells that line the tubules.

Skakkebæk's 1992 study of declining semen quality sparked a lot of interest. Hundreds of studies followed, as researchers sought the cause for the decline. Many suspected an environmental cause, such as a toxin.

"We all wanted to study it in more detail," says Tina Kold Jensen, a researcher at the University of Southern Denmark. But instead of looking at past studies, Jensen and her collaborators determined that it was necessary to track these trends in real time. "We decided if we wanted to get closer to the truth, we had to collect our own data."

Jensen partnered with Skakkebæk and Niels Jørgensen at Copenhagen University Hospital. Starting around 1995, the scientists began recruiting young male volunteers at government-required physical exams, asking each to provide a sperm sample. By 2010 the team had amassed sperm from over 5,000 volunteers. When analyzed, the data seemed to contradict Skakkebæk's initial finding: over the 15 years of the study, semen quality did not decline; in fact, it improved slightly. Sperm concentrations rose from 43 million per milliliter in 1996–2000 to 48 million per milliliter in 2006–10.

**TINA KOLD JENSEN**

Tina Kold Jensen studies links between the environment and human reproduction as a professor of environmental medicine at the University of Southern Denmark.

# Preventing Pregnancy

The Food and Drug Administration approved the first birth control pill for sale in the United States in 1960. It became immediately popular as a way for women to control their fertility, and has been said to have contributed to a sharp increase in college attendance and graduation rates for women. Here is an overview of some of today's most popular forms of birth control, with data on just how effective each form is.

Assessment available in smartwork5

## Pregnancies per 100 women in a year

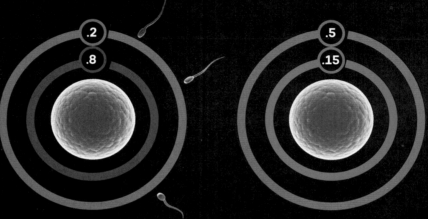

**Intrauterine Device**
T-shaped plastic device must be positioned inside the uterus by a health care professional.

○ *Hormonal*   .2

○ *Nonhormonal*   .8

**Female Sterilization**
The oviducts are sealed with clamps or by other surgical means.   .5

**Male Sterilization**
The tubes that carry sperm are sealed surgically by a health care professional.   .15

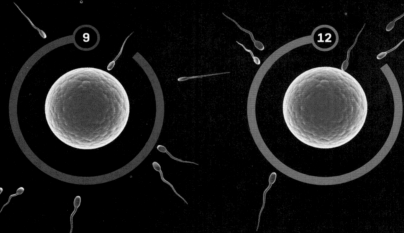

**The Pill**
Hormones suppress ovulation, preventing the release of an egg, by mimicking the hormone levels of pregnancy.   9

**Diaphragm**
Dome-shaped latex cup filled with spermicide, inserted before intercourse, covers the cervix and keeps sperm out of the uterus.   12

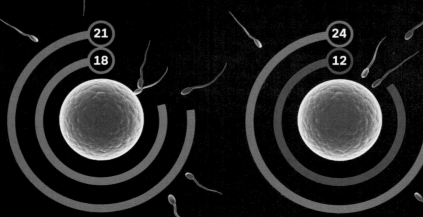

**Female Condom**
Plastic pouch, inserted before intercourse, lines the vagina to prevent sperm from entering.   21

**Male Condom**
Plastic or latex pouch covers the penis to keep sperm from entering the vagina.   18

**Sponge**
A sponge containing spermicide is inserted deep in the vagina prior to intercourse.

○ *Women who have given birth before*   24

○ *Women who have never given birth*   12

Still, the rise was not enough to suggest that fertility improved during that period. "Although we see a slight rise, only 23 percent of the young men had optimal semen quality," said Jørgensen in a statement when the study was released. "In fact, the semen quality of 27 percent of the men was so poor, it will probably take these men longer to make their partner pregnant," he added. "Furthermore, for 15 percent of the men, semen quality was so poor, they are likely to need fertility treatment in order to conceive."

# Driven by Hormones

Jørgensen and Jensen have gone on to study the factors that might have caused such low levels of semen quality. In May of 2014, their team found that 98 percent of 308 young men had detectable urinary levels of bisphenol A (BPA), a chemical found in plastics that disrupts the body's endocrine system (see Chapter 6 for more on BPA). Men with higher BPA levels, they discovered, also had higher levels of testosterone and other hormones. That correlation suggests that BPA could be affecting hormone feedback loops, but additional research is needed to identify possible mechanisms for such an effect.

Hormones regulate nearly all aspects of reproduction in animals, from mating behaviors to the development and birth of offspring. The emergence of sex-specific characteristics in the fetus and the maturation of reproductive organs during puberty are examples of long-term aspects of reproduction controlled by hormones. Hormones also regulate the recurring stimulation of sperm production in males and the monthly cycle of menstruation in females (see **Figure 22.8** for the role of hormones in the menstrual cycle, and Chapter 25 for more on hormones in general).

Testes and ovaries produce three major types of hormones: estrogens, progestogens, and androgens. Both males and females produce all three, but in different ratios; for example, males have more androgens than estrogens, and females have more estrogens than androgens.

- **Estrogens** play a role in determining female characteristics such as wide hips, a voice that is pitched higher than that of males, and the development of breast tissues. The primary estrogen is **estradiol**.

- **Progestogens** have a number of functions in the female body, including thickening the lining of the uterus and increasing its blood supply to create a suitable environment for a developing fetus. **Progesterone** is the most important of the progestogens.

- **Androgens** stimulate cells to develop characteristics of maleness, such as beard growth and the production of sperm. The primary androgen is **testosterone**. Together with another closely related androgen, testosterone directs the development of internal reproductive structures such as the sperm ducts and prostate gland. A third androgen directs the development of external structures such as the penis.

In addition to uncovering the link between BPA and hormone levels, the Danish research team recently found evidence that regular alcohol consumption may affect semen quality, possibly by changing testosterone levels. In particular, large amounts of alcohol significantly lowered semen quality: men who consumed 40 or more drinks in a week had a 33 percent reduction in sperm as compared with men who drank just 1–5 drinks per week. In their conclusion, the authors went so far as to warn young men to "avoid habitual alcohol intake."

# "Do It for Denmark"

Jørgensen, Jensen, and others continue to seek the causes for declining semen quality, while demographers track social reasons for decreasing fertility rates, such as the increasing age of mothers. "Of course there are a lot of factors involved, including social factors," says Jensen. "But it's not all due to social factors. When we talk about fertility rates, we need to think about biology as well."

In the meantime, as we saw earlier, Danish officials are working to encourage young people to reproduce. After years of teaching how pregnancy can be prevented through abstinence—refraining from sexual intercourse—and the use of **contraceptives** such as the Pill, Danish teachers will now also teach how fertilization occurs (**Figure 22.9**).

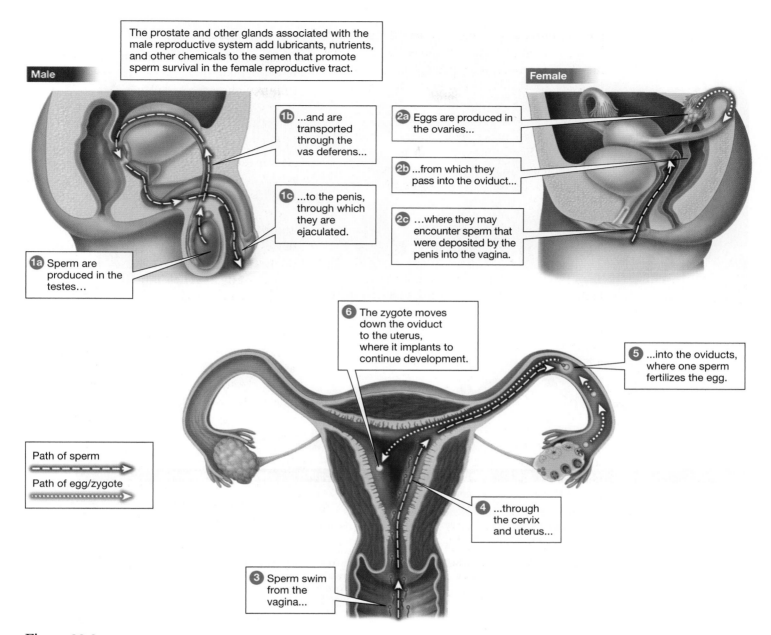

The prostate and other glands associated with the male reproductive system add lubricants, nutrients, and other chemicals to the semen that promote sperm survival in the female reproductive tract.

**Male**

**1b** ...and are transported through the vas deferens...

**1c** ...to the penis, through which they are ejaculated.

**1a** Sperm are produced in the testes...

**Female**

**2a** Eggs are produced in the ovaries...

**2b** ...from which they pass into the oviduct...

**2c** ...where they may encounter sperm that were deposited by the penis into the vagina.

**6** The zygote moves down the oviduct to the uterus, where it implants to continue development.

**5** ...into the oviducts, where one sperm fertilizes the egg.

Path of sperm

Path of egg/zygote

**4** ...through the cervix and uterus...

**3** Sperm swim from the vagina...

Figure 22.9

**Fertilization takes place in the oviduct**

Fertilization results in a zygote that can develop in the sheltered environment of the uterus.

**Q1:** If an egg is released but no sperm enter the oviduct, what is likely to occur?

**Q2:** If sperm enter the oviduct but no egg is present, what is likely to occur?

**Q3:** Sperm can live for up to 5 days inside a woman's body. Furthermore, eggs may be released at any point in the menstrual cycle, although midcycle ovulation is the norm. If you are trying *not* to become pregnant, when is it safe to have unprotected intercourse?

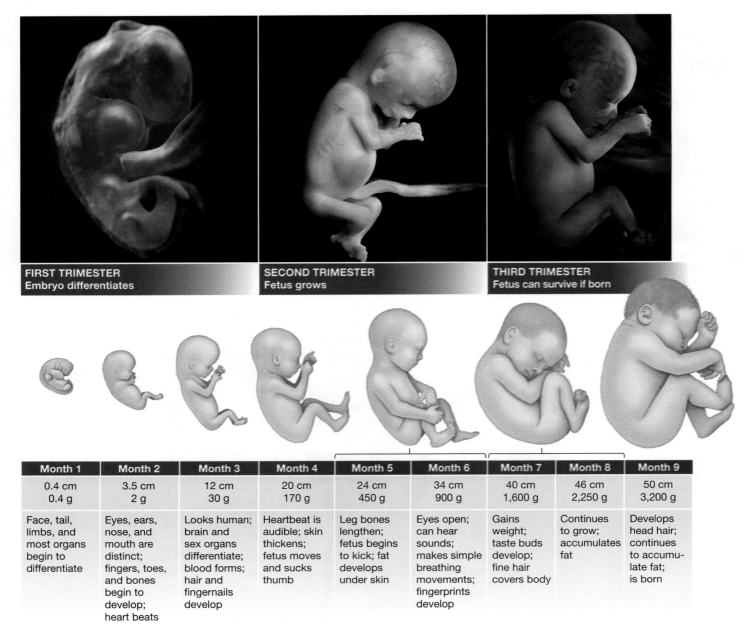

**FIRST TRIMESTER**
Embryo differentiates

**SECOND TRIMESTER**
Fetus grows

**THIRD TRIMESTER**
Fetus can survive if born

| Month 1 | Month 2 | Month 3 | Month 4 | Month 5 | Month 6 | Month 7 | Month 8 | Month 9 |
|---|---|---|---|---|---|---|---|---|
| 0.4 cm 0.4 g | 3.5 cm 2 g | 12 cm 30 g | 20 cm 170 g | 24 cm 450 g | 34 cm 900 g | 40 cm 1,600 g | 46 cm 2,250 g | 50 cm 3,200 g |
| Face, tail, limbs, and most organs begin to differentiate | Eyes, ears, nose, and mouth are distinct; fingers, toes, and bones begin to develop; heart beats | Looks human; brain and sex organs differentiate; blood forms; hair and fingernails develop | Heartbeat is audible; skin thickens; fetus moves and sucks thumb | Leg bones lengthen; fetus begins to kick; fat develops under skin | Eyes open; can hear sounds; makes simple breathing movements; fingerprints develop | Gains weight; taste buds develop; fine hair covers body | Continues to grow; accumulates fat | Develops head hair; continues to accumu-late fat; is born |

Figure 22.10

**Nine months in the womb**
Although it is a critical step in human reproduction, fertilization marks only the beginning of a 9-month-long period of development within the mother's uterus.

Q1: Place these terms in the correct order of development: embryo, fetus, infant, zygote.

Q2: In what trimester is the fetus most likely to survive outside its mother's body?

Q3: The first trimester is the most sensitive time for exposure to mutagens. Why might that be?

A woman releases an egg from her ovaries, or *ovulates*, about once every 28 days. The egg moves down the **oviduct**, or *fallopian tube*, a 4-inch-long tube that connects the **ovary** to the **uterus**. If the woman has unprotected sexual intercourse, the man's **penis** ejaculates almost 300 million sperm into her **vagina**. The sperm swim from the vagina into the uterus through an opening called the **cervix**, and then up the oviduct in response to a chemical signal released by the ovary.

Only a few hundred sperm manage to reach the egg in the oviduct, and only one of those lucky sperm fertilizes the egg to create a zygote. Although both parents contribute equally to the genetic material of the zygote, its organelles and other cellular machinery come almost entirely from the female.

Human development in the uterus averages about 38 weeks and is divided into three stages known as **trimesters**, each about 3 months long (**Figure 22.10**). During the first trimester, the zygote develops from a single cell into an **embryo** possessing all the main tissue types. All organ systems are established by the third month, and the developing individual is now known as a **fetus**. It is during these first 3 months that most birth defects occur, when the fetus's organs are initially formed, many of them severe enough to cause miscarriage. Birth defects, the leading cause of infant mortality in the United States, are structural changes that affect how the body forms or works. Physicians know the cause of some birth defects, such as fetal alcohol syndrome resulting from heavy drinking by the mother during pregnancy. However, the causes of most birth defects remain unknown, though they are likely due to a mix of genetic and environmental factors.

During the next 3 months, the second trimester, the organs develop and the fetus increases in size. By the start of the third trimester, fetal development has progressed to the point that, with the help of modern technology, the fetus has reasonably good odds of surviving outside the mother's body. It gains a good deal of weight during the third trimester, and its circulatory and respiratory systems prepare for living in a gaseous atmosphere rather than the watery world of the amniotic fluid.

By the end of the third trimester, the fetus is ready for its sudden transition from the uterus to the outside world: childbirth (**Figure 22.11**). The last few weeks of pregnancy are marked by hormonal changes. Specifically, higher estrogen levels in the mother's blood make the muscles of her uterus more sensitive to **oxytocin**, a hormone secreted by the fetus and, later in the birth process, by the mother's pituitary gland. Oxytocin stimulates the uterine muscles and causes the placenta to secrete prostaglandins, which reinforce the contractions. Labor begins when the muscles of the uterus begin to contract in response to these hormones.

In a positive feedback loop, more contractions cause the production of more oxytocin, and the strength of the contractions increases as more oxytocin is produced. The cervix begins to open, and the increasingly strong contractions eventually push the fetus out of the mother's body. At this point the positive feedback ends, and the contractions subside as oxytocin levels decrease. The placenta, often referred to as the "afterbirth," is expelled during the last stage of childbirth.

At birth, a baby becomes physically independent of its mother. It no longer obtains its oxygen and nutrients directly from her blood, and it must eat, breathe, and maintain homeostasis on its own. Development does not end when an animal is born. Humans spend about a quarter of their lives reaching full adult size. Most of this growth occurs during childhood, before sexual maturity.

Amazingly, the "Do It for Denmark" ad blitz appeared to work, at least temporarily. In the summer of 2016, 9 months after the commercials premiered, the country was preparing for a baby boom, with 1,200 more babies due to be born in that summer than in the previous summer. But in most developed countries, including the United States, all signs point to birth rates continuing to fall. In August of 2016, the U.S. government reported that American fertility had fallen

**NIELS JØRGENSEN**

Niels Jørgensen is a member of the Department of Growth and Reproduction at Copenhagen University Hospital in Denmark. He studies male infertility.

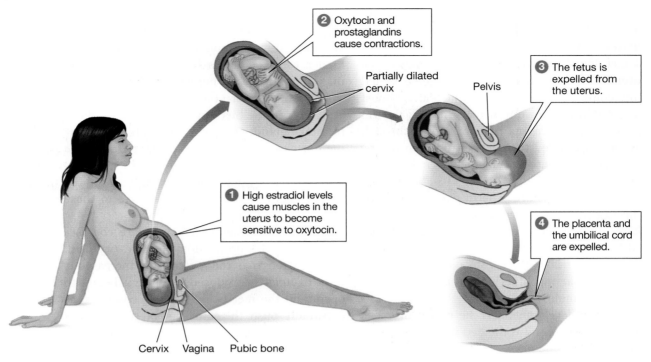

Figure 22.11

### Childbirth is orchestrated by hormones

Childbirth occurs in stages, driven by the hormone oxytocin. Oxytocin signals uterine muscles to contract. The contractions become stronger as the amount of oxytocin increases. The mother's cervix opens, and the fetus is eventually expelled from the uterus, followed by the placenta soon after.

**Q1:** What is the role of estradiol in childbirth?

**Q2:** Explain how the involvement of hormones in childbirth is an example of a positive feedback loop.

**Q3:** If a woman has been pregnant for more than 40 weeks, her doctor might give her an injection of oxytocin to precipitate labor. How would that bring about labor?

to a record low of just 1.86 births per woman, putting us on the same page as Denmark, with a birth rate below the amount needed to keep the country's population from falling. As of yet, however, the United States has not adopted policies designed to encourage women to have more children—though the idea is not far-fetched. Is "Do It for America" in our future?

300 million sperm, only one of which can fertilize the egg.

- During the first **trimester** of human development in the **uterus**, cells of the **embryo** rapidly differentiate into the various organs and structures present at birth. From the ninth week of development on, the developing human is called a **fetus**. During the second and third trimesters, the fetus grows rapidly.

- Childbirth occurs in stages. The hormone **oxytocin** signals uterine muscles to contract. The contractions become stronger as positive feedback increases the amount of oxytocin produced.

# REVIEWING THE SCIENCE

- Cells that work in an integrated manner to perform a common set of functions constitute a **tissue**. Four main types of tissues are found in vertebrates: **epithelial**, **connective**, **muscle**, and **nervous**. An **organ** is made up of more than one tissue type and forms a functional unit with a distinctive function, shape, and location in the body. An **organ system** is composed of two or more organs that work in a closely coordinated manner to perform a distinct set of functions.

- **Homeostasis** is the process of monitoring the internal environment of an organism. **Homeostatic pathways** have two basic features: sensors that monitor the internal environment, and regulatory processes that attempt to restore the normal internal state when deviations from optimal conditions are detected.

- Many homeostatic pathways are controlled by **feedback loops**. In **negative feedback** loops, the results of a process cause that process to slow down or stop. In **positive feedback** loops, the results of a process cause it to speed up.

- **Sexual reproduction** involves the organs of the **reproductive system** and the union of male and female gametes (**sperm** and **eggs**). Most animals produce offspring in this way, although some can individually produce genetically identical offspring via **asexual reproduction**.

- **Oogenesis** and **spermatogenesis** are the production of eggs and sperm, respectively. Fertilization fuses the haploid sperm and haploid egg to produce a diploid **zygote**. Males produce sperm in **testes**, and females produce eggs in **ovaries**.

- Approximately monthly, one egg released from a woman's ovary moves into the **oviduct**, where it can be fertilized. During sexual intercourse, the man's **penis** releases into the woman's **vagina** nearly

# THE QUESTIONS

## The Basics

**1** Denmark's birth rate is dropping because

(a) women are now older when they begin a family.

(b) men are now older when they begin a family.

(c) exposure to chemicals is decreasing the sperm count in men.

(d) all of the above

**2** Tissues

(a) are composed of cells that work in an integrated manner.

(b) have a distinctive shape and location in the body.

(c) are composed of multiple organs.

(d) are composed of only one cell type.

**3** Homeostasis does *not* maintain

(a) cellular pH.

(b) body temperature.

(c) environmental temperature.

(d) blood oxygen levels.

**4** Which of the following stimulate cells to develop the characteristics of maleness?

(a) estrogens

(b) spermatogens

(c) progestogens

(d) androgens

**5** What is the typical order of events to produce an embryo?

(a) gamete development, ovum release, intercourse, fertilization, implantation

(b) gamete development, ovum release, fertilization, implantation, intercourse

(c) gamete development, ovum release, intercourse, implantation, fertilization

(d) ovum release, gamete development, intercourse, fertilization, implantation

**6** Link each term with the correct definition.

**SPERMATOGENESIS**   1. Cells from only one parent produce offspring.

**OOGENESIS**   2. The process of producing sperm.

**SEXUAL REPRODUCTION**   3. The process of producing eggs (ova).

**ASEXUAL REPRODUCTION**   4. Gametes from two parents combine to produce offspring.

## Challenge Yourself

**7** Which of the following is an example of a positive feedback loop?

(a) the interaction of oxytocin and prostaglandins during labor

(b) the interaction of estradiol and progesterone during the menstrual cycle

(c) the interaction of glucose and insulin during eating

(d) both a and b

(e) both a and c

**8** Sea stars (also called starfish) sometimes break off one of their "arms," which then develops into a separate sea star. Is this an example of asexual or sexual reproduction?

**9** Select the correct terms:
Homeostasis maintains a constant internal state through (**homeopathic / homeostatic**) pathways that trigger regulatory processes when there is movement away from the body's (**set point / initiator**). For example, body temperature homeostasis is maintained through a (**negative / positive**) feedback loop—a process known as (**thermoregulation / osmoregulation**).

**10** Place the following events of the human menstrual cycle, beginning with the final day of bleeding, in the correct order by numbering them from 1 to 5.

_____ a. The corpus luteum produces progesterone for about 14 days.

_____ b. Follicle-stimulating hormone (FSH) stimulates the growth of egg follicles.

_____ c. The egg is released from the follicle (now corpus luteum).

_____ d. The uterine lining sloughs off (menstruation occurs) with decreased progesterone.

_____ e. Estrogen levels increase and trigger luteinizing hormone (LH).

## Try Something New

**11** By producing leptin, fat cells signal to the body that it has sufficient energy stores. Higher leptin levels then decrease the feeling of being hungry. With fewer fat cells, less leptin is produced, and hunger levels are higher. This is an example of

(a) oogenesis.

(b) an organ system.

(c) spermatogenesis.

(d) homeostasis.

(e) contraception.

**12** The actual effectiveness of birth control is often significantly lower than the theoretical effectiveness, or "effectiveness if used as directed." For example, there is a 9% annual pregnancy rate of women on the Pill, although its effectiveness is stated as over 99%. Similarly, although the condom is considered 98% effective, the associated annual pregnancy rate is between 10% and 18%. Would you expect a difference between theoretical and actual effectiveness for sterilization (female tubal ligation and male vasectomy)? If so, what would be the cause of that difference?

**13** In maintaining homeostasis, small animals face more challenges than do large animals. Large animals tend to exchange water, solutes, and heat with their environment more slowly than small animals do, because a large animal has a larger volume relative to its surface area than a smaller animal has. The ratio between these two quantities—surface area and volume—determines how quickly or slowly an animal can gain or lose water, solutes, or heat. When the ratio of surface area to volume is relatively high, gains and losses are rapid; when the ratio is relatively low, gains and losses happen more slowly. Explain how this ratio indicates the need to be extra careful to protect newborns from temperature extremes. What else does the ratio of surface area to volume suggest that we need to consider when caring for a newborn?

## Leveling Up

**14** **Looking at data** To answer the following questions, refer to the accompanying graph, which shows data collected for the National Survey of Family Growth by the U.S. CDC (Centers for Disease Control and Prevention).

a. What does the y-axis show?

b. How many children, on average, does a woman with a bachelor's degree (or higher) have?

c. How many children, on average, does a man with a high school diploma or GED have?

d. Describe in your own words what the graph shows about the relationship between education and reproduction in the United States.

e. State a scientific hypothesis that might explain the relationship you described in part (d).

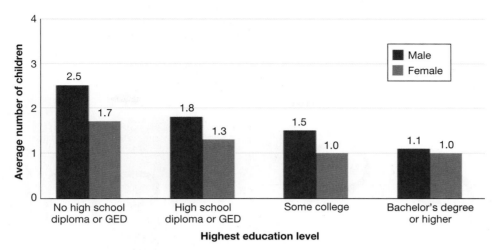

NOTE: GED is General Educational Development high school equivalency diploma.

**⑮** *Write Now* **biology: birth control and fertility rates** You are the aide to a U.S. senator, who has just sent you an e-mail including an attached article from a constituent in her (and your) home state. The constituent is concerned that people are having fewer children than in the past and thinks that the Senate should pass a law making it illegal to sell birth control to married couples. The senator asks you to write a two-page "white paper" (500 words) that she will use to respond to the e-mail and possibly also to propose legislation. The senator is extremely busy, and she neglected to take a biology class in college, so she is relying on you to give her a clear, concise, and accurate summary of the issue raised by the e-mail, as well as of possible actions to take.

Write a position paper addressing the following points (using about half a page, or 125 words, on each point). Begin by reading the article that was sent to the senator: http://national .deseretnews.com/article/1522/the-potential-impact-of-falling -fertility-rates-on-the-economy-and-culture.html. You may use other resources, including your textbook.

a. Summarize the main points of the article, defining terms (for example, "fertility rate," "replacement rate") as needed.

b. Find the fertility rate for your state, and compare it to the U.S. fertility rate. How do the two rates differ, and what is your best hypothesis for why they differ? Do the points made in the article hold for your state?

c. What are some potential challenges to the legislation proposed in the e-mail? Is there alternative legislation—or another action—that might have the same effect with fewer challenges?

d. In your final paragraph, advise the senator on how she should respond to the letter and whether she should propose the legislation recommended in the e-mail. Also provide an alternative recommendation, which does not have to involve increasing reproduction. Justify your opinion with data and logic.

---

**For more, visit digital.wwnorton.com/bionow2 for access to:**

# The Sunshine Vitamin

*Is vitamin D the new supernutrient?*

---

## After reading this chapter you should be able to:

◆ Explain the roles of the major classes of nutrients in maintaining a healthy body.

◆ For a given vitamin or mineral, describe its main functions and dietary sources, and the possible effects of a deficiency.

◆ List the components of the integumentary system, and the layers of skin.

◆ Relate the structure to the function of one component of the digestive system.

◆ Demonstrate how flexible and rigid elements of a joint work together to produce controlled motion.

◆ Compare and contrast skeletal, smooth, and cardiac muscle.

◆ Compare and contrast cartilage and bone.

CHAPTER

23

DIGESTIVE,
MUSCULAR,
AND SKELETAL
SYSTEMS

The babies' bones were not growing well. As a PhD student at McMaster University in Ontario, Canada, Hope Weiler was observing infants in a local hospital's neonatal intensive care unit, or NICU. They were being given dexamethasone, an anti-inflammatory steroid.

Prematurely born infants, or "preemies," often have chronic lung disease and cannot breathe on their own (**Figure 23.1**). To help these babies breathe, doctors commonly treated them with many doses of dexamethasone over the course of a month. But Weiler observed that the medication, while saving the infants' lives, also appeared to interfere with bone formation. Infants treated with dexamethasone had smaller heads, thinner bones, and shorter stature than those who didn't receive the medication.

"We wanted to know more about why that was happening," says Weiler, now the Canada Research Chair in Nutrition, Development, and Aging and a professor at McGill University in Quebec.

### HOPE WEILER

Hope Weiler is an associate professor of nutrition, development, and aging at McGill University in Quebec, Canada, where she studies the effects of nutrition on bone health.

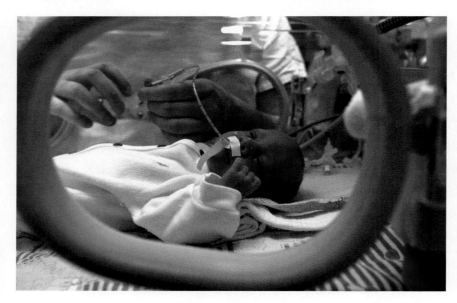

Figure 23.1

**A premature baby receiving oxygen via a nose tube**

Determined to find ways to help preemies grow, Weiler started a study tracking the bone mass of infants. During the study, she discovered that many newborn infants had low vitamin D levels in their cord blood.

**Vitamins** are small, organic nutrients needed by our bodies, but only in tiny amounts (**Figure 23.2**). They participate in a great variety of essential metabolic processes, such as helping blood cells form and maintaining brain function. Some vitamins bind to enzymes to speed up chemical reactions within a cell. Some act as a delivery service, supplying chemical groups needed in important metabolic reactions. Others act as signaling molecules. And some are even believed to work as antioxidants, substances that protect body tissues from destructive chemicals known as free radicals.

Most vitamins have multiple functions in the animal body, and vitamin D is no exception. Vitamin D, a fat-soluble vitamin, is required to absorb calcium from food and therefore is critical for bone growth and bone remodeling. Most vitamin D is naturally made by our bodies when ultraviolet (UV) rays are absorbed through the skin—an organ system in its own right, the *integumentary system* (see "The Skin We're In" on page 416).

Vitamins are a type of *nutrient*. **Nutrients** are components of foods that an organism needs to survive and grow. These can be micronutrients (vitamins and minerals) or macronutrients. Macronutrients are large organic molecules classified into three main categories: carbohydrates, lipids, and proteins. These biomolecules serve as sources of energy and furnish the body with chemical building blocks such as sugars, fatty acids, and amino acids. Although an adult human can synthesize some of the 20 amino acids needed to make proteins, we must get 9 of them, called **essential amino acids**, from food. Another type of macronutrient is dietary fiber, which does not contribute amino acids or energy to the body but is critical for survival because it affects how other nutrients are absorbed in the gut.

**Minerals** are inorganic chemicals that have critical biological functions. Carbon, hydrogen, oxygen, and nitrogen make up about 93 percent of the animal body, so, by convention, these four elements are excluded from the category of dietary minerals. But more than 20 other elements, such as fluoride, sodium, and

**Vitamin C** is abundant in fruits and vegetables, and it assists in the maintenance of teeth, bones, and other tissues. Deficiency of this vitamin leads to scurvy, in which teeth and bones degenerate.

Fish is the richest source of **vitamin D**; fortified foods (such as milk, soy milk, and breakfast cereals) are important sources for most people. A deficiency in vitamin D leads to poor formation of bones and teeth.

Folic acid, a **B vitamin**, is abundant in green vegetables, legumes, and whole grains. $B_{12}$, another B vitamin, is scarce in plant foods but abundant in milk, meat, fish, and poultry. It is important for the maintenance of teeth, bones, and other tissues.

**Vitamin E** is abundant in nuts, vegetable oils, whole grains, and egg yolk. It protects lipids in cell membranes and other cell components.

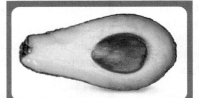

Leafy green vegetables and some fruits (e.g., avocado and kiwi) are rich in **vitamin K**, which is also manufactured by intestinal bacteria. A deficiency can cause prolonged bleeding and slow wound healing.

Carotene is responsible for the color of yellow and orange fruits and vegetables. It is converted into **vitamin A** within our bodies. Vitamin A aids in production of the visual pigment needed for good eyesight and is used in making bone.

## Figure 23.2

### Vitamins needed in the human diet

Humans need nine **water-soluble vitamins** (C and eight different B vitamins) and four **fat-soluble vitamins** (A, D, E, and K) from their diet. Because water-soluble vitamins are easily excreted in urine, they tend not to accumulate in body tissues, which means we must obtain these vitamins from food on a regular basis. Fat-soluble vitamins are not excreted as readily and tend to accumulate in body fat, so excessive consumption can lead to overdosing.

**Q1:** Which vitamins described in the figure are important for healthy bones?

**Q2:** Which vitamins are you more likely to overaccumulate?

**Q3:** In your own diet, are there any vitamins that you may not be eating enough of?

iodine, are essential for the normal function of most animals, and these are classified as dietary minerals. Calcium is the most abundant dietary mineral in the body; it makes up a large proportion of your bones.

Low levels of vitamin D have been identified not only in premature infants, but also in patients with ailments such as osteoporosis (a progressive bone disease), schizophrenia, erectile dysfunction, and more. If lack of vitamin D does contribute to these disorders, then supplementation with the vitamin should be protective. Many supplement manufacturers, in fact, now tout D as a "supervitamin" and "star supplement"

# The Skin We're In

The **integumentary system** is the largest organ system in the human body, accounting for almost 15 percent of our weight. It covers the body and protects it from environmental hazards such as extreme temperatures and dangerous pathogens. It also prevents water loss and protects the body from physical damage. Sensory receptors are embedded in the skin (see Chapter 24 for more on the nervous system), and skin is the site of vitamin D synthesis.

The integumentary system consists of the skin and structures embedded in the skin, such as hair and nails in humans, or feathers, hooves, and scales in other vertebrates. The skin is made up of three layers. Moving inward from the outermost layer, the layers are the epidermis, dermis, and hypodermis.

As the figure shows, the skin contains multiple tissue types. The epidermis is an example of epithelial tissue. The nerve endings are examples of nervous tissue. The arrector pili muscle is composed of smooth muscle tissue. Much of the dermis is made up of connective tissue. And adipose tissue dominates in the hypodermis, a thick insulating sheet under the other layers of the skin.

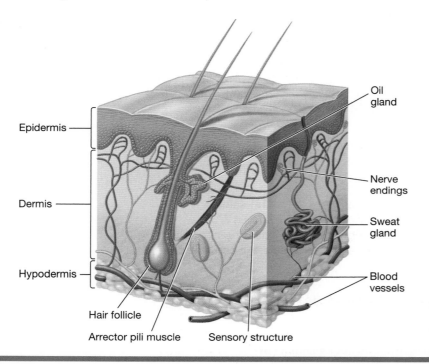

that will protect individuals from a wide range of diseases.

But has science shown that claim to be true? Just because low vitamin D and disease are correlated does not mean that one causes the other. So, scientists have long wondered whether low vitamin D actually causes disease or is involved in some other way.

# You Are What You Eat

In her experiment studying premature infants, Weiler hypothesized that low vitamin D levels were an indirect result of treatment with dexamethasone.

Dexamethasone is a powerful steroid that helps a premature baby's lungs work better, but it has a side effect: it interferes with calcium absorption in the gut. When the drug is present, the intestines do not absorb calcium correctly. To counteract this effect, the babies' bodies began trying to find ways to compensate—by using up their vitamin D.

Vitamin D promotes calcium absorption in the intestines and in bones. The babies were using up all their vitamin D in an effort to absorb more calcium in the gut, so they didn't have enough left circulating in their blood to build bones.

Absorbing vitamins is one important function of the **digestive system**, which also processes food and eliminates unusable waste. The digestive system of most animals consists of a long, hollow passageway, known as the digestive tract, and a number of accessory organs, such as the pancreas and liver (**Figure 23.3**).

Eating, or **ingestion**, is the first stage in the processing of food by the digestive system. Digestion, the mechanical and chemical breakdown of food, begins almost immediately after ingestion in many species.

During ingestion, a bite of food—say, a spoonful of cornflakes—is deposited in the **oral cavity**, the mouth. There, an array of different types of teeth, which are shaped to cut, crush, or grind food into smaller pieces, begin to break apart the cornflakes. Many small pieces of food provide a greater surface area for digestive enzymes to work on than do fewer large pieces.

The muscular tongue mixes the crushed cereal particles with saliva. **Saliva** contains enzymes that start to break down starches—or any carbohydrate—into sugars. If you chew a piece of bread long enough, for example, its starches are digested to sugar and it will begin to taste sweet. Saliva is also important for turning the crunchy cereal into a moist mass that can slip easily down the throat.

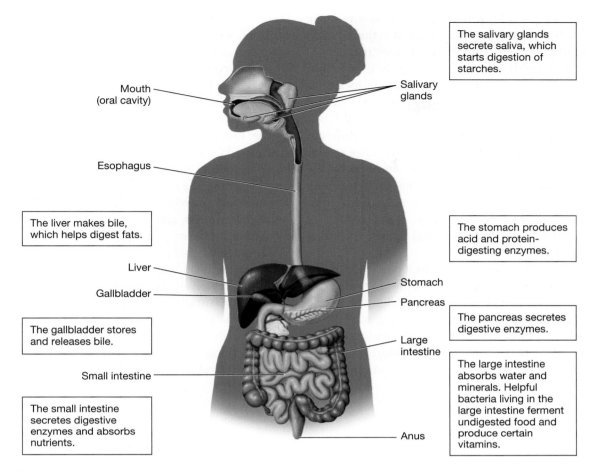

**Figure 23.3**

**The digestive system converts food into absorbable nutrients**
As food moves through the digestive system, it is broken down into small molecules that can be absorbed by the lining of the intestine.

The salivary glands secrete saliva, which starts digestion of starches.

The liver makes bile, which helps digest fats.

The gallbladder stores and releases bile.

The small intestine secretes digestive enzymes and absorbs nutrients.

The stomach produces acid and protein-digesting enzymes.

The pancreas secretes digestive enzymes.

The large intestine absorbs water and minerals. Helpful bacteria living in the large intestine ferment undigested food and produce certain vitamins.

Mouth (oral cavity)
Salivary glands
Esophagus
Liver
Gallbladder
Stomach
Pancreas
Large intestine
Small intestine
Anus

**Q1:** List, in order, the structures of the digestive system that a piece of swallowed food would pass through, beginning with the mouth.

**Q2:** What part of the digestive system hosts bacteria that produce vitamins?

**Q3:** What is the shared function of the liver and gallbladder?

The tongue assists in pushing the now-moist cereal into the throat, or **pharynx**, where the back of the mouth and the nasal cavity come together. The pharynx is the common entryway for both the air tube (the trachea) and the food tube (the **esophagus**). That is why, on occasion, you may cough up food or liquid that "went down the wrong tube"; it accidentally went down the trachea instead of the esophagus.

Normally, the pharynx is very good at separating air and food. When the mushy bite of cereal makes contact with the wall of the pharynx, it stimulates nerves that launch the **swallowing reflex**, in which a flap of tissue, called the epiglottis, seals off the entry into the trachea. The cereal is then pushed into the esophagus.

Waves of muscular contractions carry the cereal down the esophagus and into the **stomach**. Protein digestion begins in the stomach, which secretes acid and enzymes that break down complex protein molecules. Muscles in the wall of the stomach alternately contract and relax to mix the food particles with the acid and enzymes. The resulting watery mixture is stored in the stomach until it can move into the small intestine.

The **small intestine** is a highly coiled, thin tube about 3–4 centimeters in diameter. If straightened, the small intestine would extend about 20 feet (6 meters). The upper and lower regions of the small intestine serve different functions. The upper region, which lies nearer to the stomach, uses enzymes secreted by the **pancreas** and by the intestine itself to break down large molecules into simpler forms that the body can absorb. Here, the digestion of proteins, carbohydrates, and lipids, including fats, is completed.

The digestion of fats poses a particular problem because fats are not soluble in water, yet they need to be broken down and made to mix with the watery contents of the digestive tract. **Bile** is a fluid that helps digest fats by creating a coating that enables the fat globules to interact with water molecules to partially dissolve

them. The large globules break down into tiny droplets, which offer a larger work surface for lipid-degrading enzymes. Bile is produced by the **liver**, an organ that serves a multitude of functions. Some of the bile made by the liver is stored in the **gallbladder**, which dispenses the bile into the small intestine as needed.

The lower region of the small intestine is specialized for **absorption**, the uptake of mineral ions—including calcium—and small molecules by cells lining the cavity of the digestive tract. These small molecules include broken-down sugars, fatty acids, and amino acids. The innermost lining of the small intestine presents a large surface area for that process (**Figure 23.4**).

Vitamins, including D, are also absorbed in the lower region of the small intestine. Only a few foods naturally contain vitamin D, including fatty fishes such as salmon and tuna, and egg yolks. Most of our ingested vitamin D comes from fortified foods—foods to which a vitamin or mineral is added. Milk, for example, is

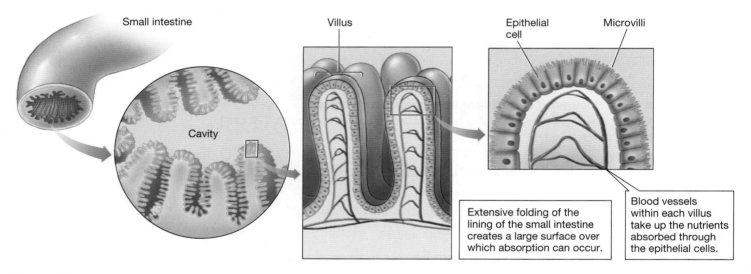

Small intestine     Villus     Epithelial cell     Microvilli

Cavity

Extensive folding of the lining of the small intestine creates a large surface over which absorption can occur.

Blood vessels within each villus take up the nutrients absorbed through the epithelial cells.

## Figure 23.4

### The small intestine is specialized for absorption

Nutrients are absorbed in the small intestine by large numbers of fingerlike projections called **villi**. Each villus is about 1 millimeter long, with a surface consisting of cells specialized for nutrient absorption. The plasma membrane of each of these cells also has many tiny projections, called microvilli. This complex folding of the intestinal lining produces almost 300 square meters of surface area for absorption.

Q1:   Why is a larger surface area important for absorption?

Q2:   In what way are the epithelial cells lining the villi modified to increase absorption?

Q3:   Where is the food coming from when it enters the small intestine? Where does it go next?

often fortified with vitamin D, voluntarily in the United States and by law in Canada. Both countries mandate that all baby formula be fortified with vitamin D. Breakfast cereals often contain added vitamin D as well.

Most of the nutrients absorbed by the digestive tract are sent to the bloodstream, which eventually delivers them to every cell in the body. Whether from the skin or the gut, vitamin D is sent to the liver and then to the kidney (**Figure 23.5**). In each organ, the nutrient is chemically modified to become an active form of the vitamin that our cells can use.

Our original bite of cereal contains very few nutrients by the time it arrives in the final segment of the digestive tract. This residual matter is prepared for **elimination**—the removal from the body of solid waste, consisting mostly of indigestible material and bacteria that inhabit the digestive tract—during

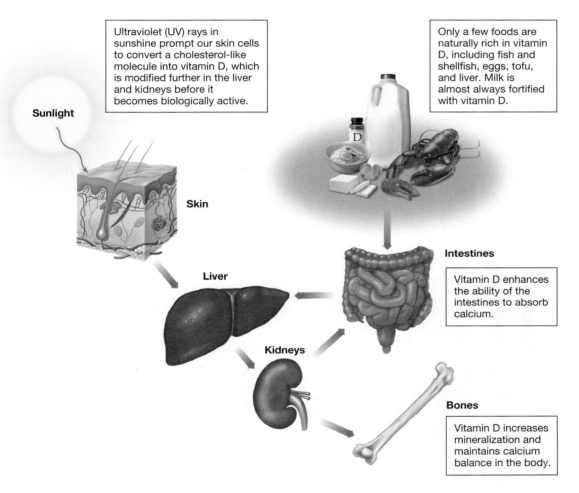

Ultraviolet (UV) rays in sunshine prompt our skin cells to convert a cholesterol-like molecule into vitamin D, which is modified further in the liver and kidneys before it becomes biologically active.

Only a few foods are naturally rich in vitamin D, including fish and shellfish, eggs, tofu, and liver. Milk is almost always fortified with vitamin D.

Sunlight

Skin

Liver

Kidneys

Intestines

Vitamin D enhances the ability of the intestines to absorb calcium.

Bones

Vitamin D increases mineralization and maintains calcium balance in the body.

Figure 23.5

**Sources of vitamin D**

Vitamin D is the only vitamin that we can manufacture entirely within our tissues, yet many Americans get inadequate amounts of vitamin D.

**Q1:** What organs of the digestive system are involved in processing vitamin D?

**Q2:** Do you get the majority of the vitamin D you need from the sun, from natural foods, or from dietary supplements?

**Q3:** It is true that you can increase your vitamin D levels by visiting a tanning booth (as described later in the chapter). Why is increasing vitamin D in this way considered a bad idea?

passage through the **large intestine**, or **colon**. The colon absorbs almost all remaining minerals and water from the waste. Then, large numbers of bacteria living in the colon break down the remaining waste to squeeze out the very last nutrients that the body can absorb. These bacteria also produce certain vitamins that are absorbed into the body from the colon. The waste, or **feces**, leaves the body through the **anus**, a muscle-lined opening.

# Building Bones

Weiler's study became one of many that, over the years, have confirmed the cause-and-effect role of vitamin D on bone health in the **skeletal system**: with low levels of vitamin D, the babies did not grow a healthy skeleton. Today, dexamethasone is no longer used to treat

premature babies, having been replaced with a safer drug.

Like most other vertebrates, humans have a bony internal skeleton that supports the body, gives it shape, and protects soft tissues and organs (**Figure 23.6**). The **axial skeleton**, made up of 80 bones, supports and protects the long axis of the body. It includes the skull, the ribs, and a long, bony spinal column. Although the axial skeleton plays a role in movement, its primary purpose is to protect vital organs. The 126 bones of the arms, legs, and pelvis make up the **appendicular skeleton** ("appendicular" means "relating to an appendage or limb"). These bones have more to do with motion than with protection.

Another important, though often passed over, part of our skeleton is **cartilage**, a dense tissue that combines strength with flexibility. In the human skeleton, cartilage gives form to the nose, the ears, and part of the rib cage. In addition, cartilage is found at nearly every point in the body where two bones would otherwise come into direct contact (see **Figure 23.6**); it creates a smooth surface that prevents the two bony surfaces from grinding against each other.

Cartilage does contain cells, but it consists primarily of nonliving, extracellular material—bundles of **collagen**, a tough but pliable protein

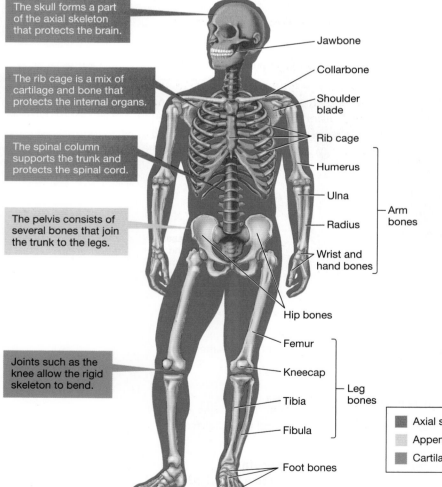

The skull forms a part of the axial skeleton that protects the brain.

The rib cage is a mix of cartilage and bone that protects the internal organs.

The spinal column supports the trunk and protects the spinal cord.

The pelvis consists of several bones that join the trunk to the legs.

Joints such as the knee allow the rigid skeleton to bend.

Jawbone
Collarbone
Shoulder blade
Rib cage
Humerus
Ulna
Radius
} Arm bones
Wrist and hand bones
Hip bones
Femur
Kneecap
Tibia
Fibula
} Leg bones
Foot bones

■ Axial skeleton
■ Appendicular skeleton
■ Cartilage

## Figure 23.6

### The human skeleton
The axial human skeleton protects vital organs; the appendicular skeleton facilitates movement.

**Q1:** The collarbone is part of which skeleton: axial or appendicular?

**Q2:** Which parts of the skeleton are made of cartilage?

**Q3:** Which skeleton—axial or appendicular—protects the central nervous system (the brain and spinal cord)?

found in a great variety of tissues, including skin, blood vessels, bones, teeth, and the lens of the eye.

Sharks belong to a small class of vertebrates that do not have a skeleton made of bones; their skeleton consists entirely of cartilage and connective tissue. Keep in mind, too, that not all animals have their skeleton on the inside. While humans and other vertebrates have an internal skeleton, or **endoskeleton**, many other animals, such as lobsters and insects, have an **exoskeleton**, an external skeleton that surrounds and encloses the soft tissues it supports (**Figure 23.7**).

Like cartilage, much of bone is made up of nonliving material. Still, it is a living tissue that has a blood and nerve supply. Specialized bone cells, called **osteocytes**, surround themselves with a hard, nonliving mineral matrix composed largely of calcium and phosphate. Vitamin D maintains calcium and phosphate concentrations in the body that are necessary to form that matrix. Although they are just single cells, osteocytes can live as long as the organism whose skeleton they belong to.

Bones are made of two major types of bone tissue (**Figure 23.8**): **Compact bone** forms the hard, white outer region. **Spongy bone**, honeycombed with numerous tiny cavities, lies inside the compact bone and is most abundant at the knobby ends of our long bones. Long bones and some others, such as the ribs and breastbone, have a hollow interior, which makes them light but strong. The cavities inside hollow bones contain **marrow**, a tissue that, depending on the type of bone, stores fat or produces blood cells.

Vitamin D, as previously noted, promotes calcium absorption in the gut and maintains calcium and phosphate concentrations. It is also needed for calcium absorption into bones, and thus is essential to bone growth and remodeling. Without it, bones can become thin, brittle, or misshapen—symptoms of a condition called rickets.

Rickets causes bones to soften, especially near joints, the junctions in the skeletal system that let the skeleton move in specific ways (**Figure 23.9**). Walking, for example, requires movement at the hips and knees, as well as at other joints. The lower jaw connects to the skull at a joint so that it can move relative to the rest of the skull, enabling us to chew and talk.

Figure 23.7

**A newly molted cicada emerges from its exoskeleton**
Exoskeletons provide a protective armor for many animals and also protect terrestrial invertebrates from excessive moisture loss. The rigidity of an exoskeleton means that immature animals that outgrow their exoskeletons must shed them periodically—a process known as molting.

Joints are held together by collagen-rich ligaments and tendons. **Ligaments** are specialized, flexible bands of connective tissue that join bone to bone, while **tendons** connect muscle to bone.

Wherever two moving parts rub against each other, as in a joint, wear can erode bone, and friction can waste energy. So, joints are lined with

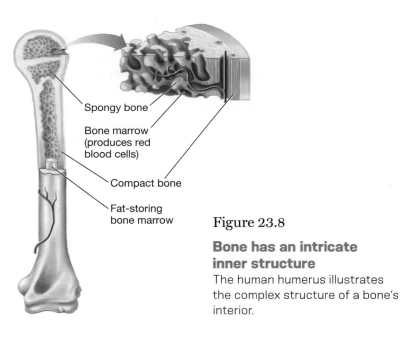

Spongy bone
Bone marrow (produces red blood cells)
Compact bone
Fat-storing bone marrow

Figure 23.8

**Bone has an intricate inner structure**
The human humerus illustrates the complex structure of a bone's interior.

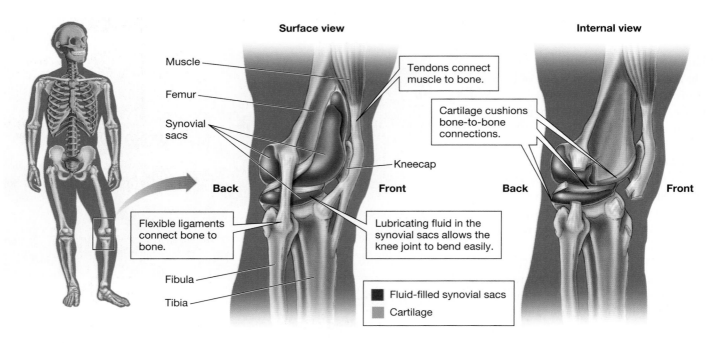

**Surface view**

Muscle

Femur

Synovial sacs

Tendons connect muscle to bone.

Cartilage cushions bone-to-bone connections.

Kneecap

**Back**　　　　**Front**

Flexible ligaments connect bone to bone.

Lubricating fluid in the synovial sacs allows the knee joint to bend easily.

Fibula

Tibia

**Internal view**

**Back**　　　　**Front**

■ Fluid-filled synovial sacs
■ Cartilage

Figure 23.9

**The human knee illustrates how rigidity and flexibility together enable movement**
Although our knees differ in detail from other joints in the body, they provide a good general model of the way flexible and rigid materials are combined in a joint to allow controlled motion.

**Q1:** What is the function of the synovial sacs?

**Q2:** Compare and contrast ligaments and tendons.

**Q3:** Knee injuries are some of the most common sports injuries. Why do you think that is?

sheets of tissue (synovial membranes) that form cavities called **synovial sacs**. The space inside each synovial sac is filled with a lubricating fluid that reduces friction between the two bony surfaces.

Altogether, these five components—bone, cartilage, ligaments, tendons, and synovial sacs—work to move a joint safely and with precise control.

# Show of Strength

There is now no doubt that vitamin D is vital for bone health, but over decades, scientists have also been amassing data on the role of vitamin D in nonskeletal tissues. One of these is muscle tissue, which forms the body's **muscular system**.

Muscle tissue is unique to animals. The skeleton and its **joints** are the framework for motion, while muscles provide the power necessary for movement. Muscle tissue possesses a crucial property:

it can contract and relax. When we walk, run, or jump, we are using *voluntary* contractions.

But even when we are still, we're using *involuntary* muscles: the heart is pumping blood, the lungs are pumping air, and food is being moved along the digestive tract by muscular contractions. Involuntary muscles do their work without our having to think about them (**Figure 23.10**). Two specialized types of muscle engage in involuntary contraction: cardiac muscle and smooth muscle. The vertebrate heart is the only organ that contains **cardiac muscle. Smooth muscle** is found in the digestive tract, the walls of blood vessels, the respiratory tract, the uterus, and the urinary bladder. Some smooth muscles contract rapidly, while other types of smooth muscle can sustain a contraction for hours.

In November 2014, researchers in Belgium compiled all the available data on the role of vitamin D in **skeletal muscle**, the muscles involved in voluntary contractions. Skeletal muscle consists of many bundles of muscle fibers. A **muscle fiber** is a long, narrow cell that can span the length of an entire muscle because it is made up of several muscle cells that fused together during development (**Figure 23.11**). Each muscle fiber is packed with cylindrical structures containing proteins that contract by bracing against each other. Each such cylinder is known as a **myofibril**. Myofibrils are organized into series of contractile units called **sarcomeres**.

The simultaneous contraction of sarcomeres, usually taking no more than a tenth of a second, produces the contraction of a whole muscle. Sarcomeres are visible as bands when seen through a microscope. At extremely high magnification, additional details of their structure become evident (**Figure 23.12**). Two kinds of protein filaments are arranged in a specific manner inside each sarcomere: each **actin filament** is made up of molecules of the protein actin, and each **myosin filament** consists of the protein myosin. Under a microscope, each end of a sarcomere appears as a dark line called the **Z disc**. The two Z discs that mark the ends of the sarcomere contain a large protein that provides anchor points for actin filaments. Between the free ends of the actin filaments, attached to the middle of the sarcomere, lie thicker myosin filaments. The sliding of the myosin filaments against the actin filaments, which requires calcium, enables the sarcomere to contract.

To assess the impact of vitamin D supplementation on skeletal muscle health, the Belgian authors identified 30 randomized, controlled trials involving a total of 5,615 individuals. In each trial, an experimental group of people was given vitamin D supplements while a control group was not.

Looking at the data from all of those trials, the researchers found a "small but significant positive effect of vitamin D supplementation on global muscle strength" and saw that supplementation seemed to be more effective in people aged 65 and older than in younger people.

We've now seen that scientific evidence connects vitamin D to bone and muscle health. Yet some studies correlate low vitamin D levels with almost every other tissue and ailment under the sun, from heart disease to prostate cancer to

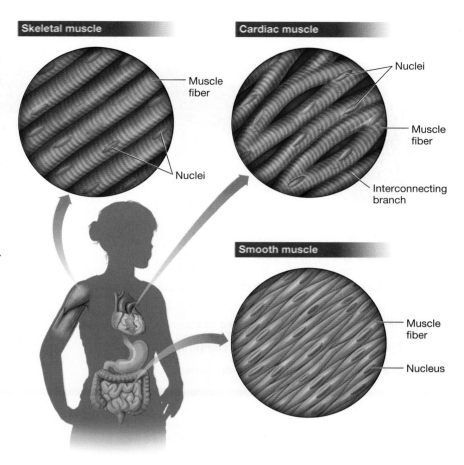

Figure 23.10

### Specialized types of muscles for different types of movement

Skeletal muscle has a distinctive banded appearance, brought about by the sarcomeres that make it up (see Figure 23.12). Cardiac muscle is also banded, and its muscle fibers are branched, which helps produce the coordinated contractions known as heartbeats. Because it lacks sarcomeres, smooth muscle has no visible bands.

**Q1:** Which types of muscles can you voluntarily contract?

**Q2:** Do the muscles in the heart contract voluntarily or involuntarily?

**Q3:** You do not have to think about breathing (otherwise, sleeping would be dangerous!), but you *can* increase or decrease your rate of breathing. Are the muscles involved in breathing, then, voluntary or involuntary?

dementia. But, as we'll keep repeating, correlation isn't necessarily causation: a recent study has cast a large shadow of doubt over whether lack of vitamin D is actually the *cause* of these nonskeletal disorders.

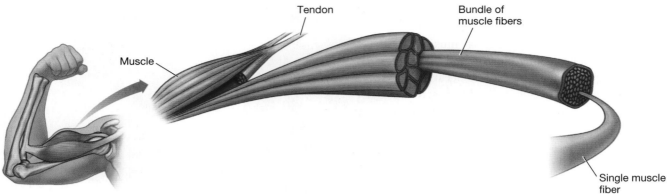

Tendon

Bundle of muscle fibers

Muscle

Single muscle fiber

Figure 23.11

## The muscles that move you

Both ends of a skeletal muscle are anchored by tendons to nearby support structures, such as bones. A muscle contains bundles of muscle fibers, each one running the length of the entire muscle.

**Q1:** How is skeletal muscle attached to bones in the skeleton?

**Q2:** Do skeletal muscles contract voluntarily or involuntarily?

**Q3:** Is a muscle fiber one cell or multiple cells?

# Beyond Bone

To one skin cancer researcher in France, the flurry of studies linking vitamin D to nonskeletal diseases was surprising. Epidemiologist Philippe Autier, vice president of population research at the International Prevention Research Institute (iPRI) in Lyon, France, had been following the vitamin D story from early on. He had grown skeptical of the health claims—that vitamin D could prevent a wide range of diseases—when the tanning industry began touting vitamin D as a benefit of tanning booths while conveniently glossing over the risks of skin cancer from the beds' UV rays (**Figure 23.13**).

Shortly after the tanning industry started promoting the health benefits of vitamin D, numerous studies began to emerge about the vitamin's role in a variety of nonskeletal diseases. "We

### PHILIPPE AUTIER

Vitamin D₃

HO

Philippe Autier is an epidemiologist at the International Prevention Research Institute in Lyon, France. He researches cancer, specifically the role of UV light in skin cancer.

were surprised by the fact that there were so many relationships between low vitamin D and diseases of all sorts, from brain disease to lung disease to infectious disease to cancer," says Autier. As he began reading the primary literature, he found that for virtually every disease, some correlation between vitamin D and the disease existed, and usually it followed a particular pattern.

"It was almost always that low vitamin D was associated with a greater risk of disease, and high vitamin D was protective," says Autier. But such conclusions were made from observational studies. Much rarer were randomized clinical trials, in which one group of people was given vitamin D supplements, another was not, and their likelihood of developing a particular ailment was compared. In those studies, Autier noticed, higher levels of vitamin D did not seem to protect the individuals taking supplements.

People began asking Autier whether they should take a vitamin D supplement to protect themselves. "The question was legitimate," says Autier, "so we had to address it."

With colleagues at iPRI, Autier analyzed data from 290 observational studies and 172 randomized trials examining the effects of vitamin D on nonskeletal disorders. They spent 6 years gathering and then analyzing the data from those

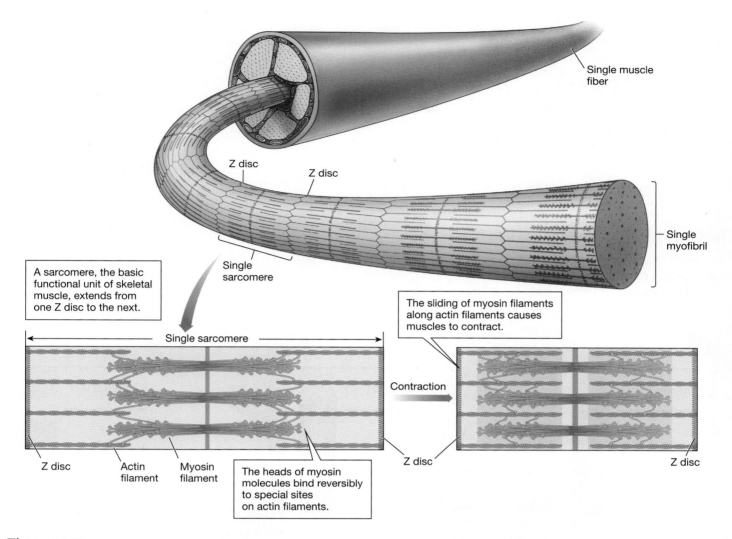

Single muscle fiber

Z disc

Z disc

Single myofibril

A sarcomere, the basic functional unit of skeletal muscle, extends from one Z disc to the next.

Single sarcomere

The sliding of myosin filaments along actin filaments causes muscles to contract.

Single sarcomere

Contraction

Z disc    Actin filament    Myosin filament

The heads of myosin molecules bind reversibly to special sites on actin filaments.

Z disc

Z disc

Z disc

## Figure 23.12

### The microscopic structure of muscle

Each muscle fiber contains myofibrils made up of sarcomeres. Muscle contraction depends on the movement of actin and myosin filaments within each sarcomere. At each end of a sarcomere is a Z disc; each of the two Z discs contains a large protein that provides anchor points for the actin filaments.

**Q1:** List muscle structures from smallest to largest, beginning with sarcomeres.

**Q2:** What are the components of the sarcomere?

**Q3:** Across animal species, the microscopic structure of muscles is the same. Why, then, are there differences in strength among animals?

studies. "We had a huge database with many servers," says Autier. "You should have seen my office." From the data deluge, they found that high doses of vitamin D did not prevent any of the disorders they looked at. Not one.

Autier's team concluded, therefore, that low vitamin D levels are not a cause of ill health.

So why, then, do so many sick patients have low levels of vitamin D? One hypothesis is that diseases (especially cancer) are often associated with inflammation, and inflammation can reduce vitamin D concentrations in the body.

Three months after Autier's paper came out, a second, independent research team from

# Nutritional Needs

Vitamins and minerals are essential nutrients that our bodies require in small amounts. Supplement manufacturers sell a range of different multivitamins, but the broad consensus from nutrition experts is that nothing substitutes for a healthy diet. Check out the data below to see how much of each nutrient you should be ingesting in a day, and some of the best foods in which to find them.

Assessment available in smartwork5

- Men's multivitamin
- RDA male
- Women's multivitamin
- RDA female

The recommended daily allowance (RDA) is the average daily dietary nutrient intake level that is sufficient to meet the nutrient requirements of nearly all healthy individuals (97%–98%) in a particular life stage and gender group.

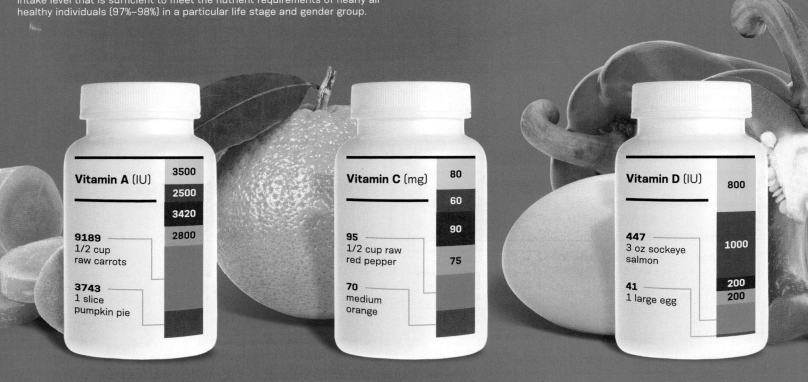

**Vitamin A (IU)**
3500
2500
3420
2800
9189 — 1/2 cup raw carrots
3743 — 1 slice pumpkin pie

**Vitamin C (mg)**
80
60
90
75
95 — 1/2 cup raw red pepper
70 — medium orange

**Vitamin D (IU)**
800
1000
200
200
447 — 3 oz sockeye salmon
41 — 1 large egg

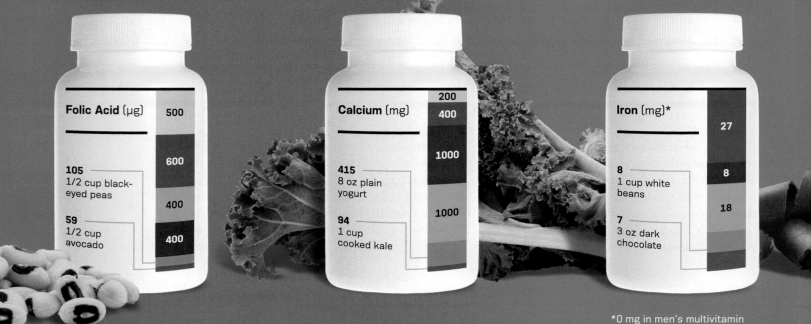

**Folic Acid (µg)**
500
600
400
400
105 — 1/2 cup black-eyed peas
59 — 1/2 cup avocado

**Calcium (mg)**
200
400
1000
1000
415 — 8 oz plain yogurt
94 — 1 cup cooked kale

**Iron (mg)***
27
8
18
8 — 1 cup white beans
7 — 3 oz dark chocolate

*0 mg in men's multivitamin

the University of Auckland in New Zealand published another meta-analysis that came to a similar conclusion: vitamin D supplementation does not reduce the risk of nonskeletal diseases.

All in all, these findings suggest that vitamin D supplements may not protect against nonskeletal disorders, though Autier emphasizes that their findings do not apply to pregnant women and young children, for whom vitamin D is critical during development.

Autier would be happy to see more large, randomized trials to confirm his findings. Weiler agrees. "Vitamin D doesn't have the same long history of people focusing on it, like heart disease," she says. "We know some relationships are there, but we haven't much looked at cause and effect."

In the meantime, consumers should not be nervous about taking vitamin D. The current recommended daily allowance from the Institute of Medicine, an impartial, nonprofit health and science policy institute, is 600 international units (IUs) of vitamin D per day to maintain health. (Seniors and breast-fed infants may need more.) Taking vitamin D for long periods of time in doses higher than 4,000 units per day is potentially unsafe, according to the National Institutes of Health, as it may cause excessively high levels of calcium in the blood. When blood levels of calcium are too high, the risk of heart attack and stroke rise dramatically.

Figure 23.13

**Tanning beds are a risky source of vitamin D**
Exposure to ultraviolet rays, either outdoors from the sun or indoors from tanning beds or lamps, greatly increases the risk of skin cancer. It can also cause eye cancer, cataracts, and premature wrinkling. About 400,000 Americans a year develop skin cancer as a result of indoor tanning.

Autier encourages other researchers to keep studying vitamin D. The story is not over. "It's quite good that we continue," he says, "because vitamin D levels could tell us a lot about the mechanisms of many diseases."

# REVIEWING THE SCIENCE

- Animals rely on **nutrients** for chemical building blocks and energy. Macronutrients are carbohydrates, proteins, and lipids.

- Micronutrients include vitamins and minerals. **Vitamins** are organic compounds obtained from food that regulate metabolic processes in the animal body. **Minerals** are inorganic molecules needed by the body in small amounts.

- The **integumentary system** is the largest organ system in the human body, covering and protecting the surface of the body. It consists of skin and the structures embedded in the skin, such as hair and nails.

- The **digestive system** is a tubular passageway that, in conjunction with accessory organs, processes ingested food.

  - After **ingestion**, food is broken down into smaller pieces in the **oral cavity** through the grinding action of teeth.

- **Saliva** moistens the food and begins the chemical breakdown of starch.

- Food passes from the oral cavity to the **pharynx**, down the **esophagus**, and into the acidic environment of the **stomach**, where protein **digestion** begins.

- Partially digested food moves from the stomach into the upper region of the **small intestine**, where enzymes secreted by the **pancreas** and the intestine complete digestion of the food. The **liver** produces **bile** (stored and delivered by the **gallbladder**), which helps digest fats.

- In the lower region of the small intestine, digested nutrients are absorbed into the body. The lining of the small intestine is highly folded and bears fingerlike projections (**villi**) that present a large surface area for absorbing nutrients.

- From the small intestine, any unabsorbed material moves into the **colon**, where remaining water and minerals are absorbed. Here, bacteria break down the waste and release nutrients that the body can absorb.

- The **axial skeleton** supports and protects vital organs along the long axis of the body. The **appendicular skeleton** is composed of the bones in the arms, legs, and pelvis, and is primarily involved with movement.

- Most bones are made of **spongy bone** surrounded by harder **compact bone**. Hollow bones contain **marrow**.

- **Ligaments** connect bone to bone. **Tendons** connect muscle to bone. Tendons and ligaments are made of **collagen**.

- Muscles provide the power necessary for movement. Muscles consist of **muscle fibers**, each of which is packed with **myofibrils**. Myofibrils contain repeating units, called **sarcomeres**, that contract when actin and myosin filaments slide past one another.

- Movements are created by specialized muscle types. **Cardiac muscle** is found in the heart, where its contractions pump blood. **Smooth muscle**, which is found in the digestive tract and blood vessels, contracts in waves. **Skeletal muscle** is under conscious control and has a banded appearance.

# THE QUESTIONS

## The Basics

**1** Which nutrient is most important for energy *storage*?

(a) proteins

(b) carbohydrates

(c) lipids

(d) all of the above

**2** Which of the following is true of vitamin D?

(a) A deficiency can interfere with the formation of bones and teeth.

(b) Sunlight is the only source.

(c) Leafy green vegetables are an excellent source.

(d) It is a water-soluble vitamin.

**3** Vitamin K is

(a) a necessary dietary mineral.

(b) manufactured by intestinal bacteria.

(c) important for good eyesight.

(d) a water-soluble vitamin.

**4** Which of the following is *not* true of the small intestine?

(a) It is specialized for absorption.

(b) It absorbs nutrients through villi and microvilli.

(c) It secretes digestive enzymes.

(d) It is specialized for elimination of waste.

**5** Use these terms correctly in the following sentence: adipose, dermis, epithelial. _____ tissue is found in the hypodermis and insulates us from temperature extremes. One level up, the _____ is composed mainly of connective tissue. And protecting us from the external environment is the epidermis, made up of _____ tissue.

**6** Link each term with the best definition.

| | |
|---|---|
| SKELETAL MUSCLE | 1. Muscle that is consciously controlled. |
| SMOOTH MUSCLE | 2. Type of muscle found in the heart. |
| CARDIAC MUSCLE | 3. Type of muscle used for walking and running. |
| VOLUNTARY MUSCLE | 4. Type of muscle found in the digestive system. |
| INVOLUNTARY MUSCLE | 5. Muscle that works without conscious control. |

## Challenge Yourself

**7** Which of the following are important for controlled movement of the knee?

(a) the axial and appendicular skeletons

(b) the ulna and the radius

(c) rigidity and flexibility

(d) the femur and the tibia

**8** Select the correct terms:

Within a (**muscle / tendon**), sarcomeres shorten when their (**actin / myosin**) filaments slide over (**actin / myosin**) filaments joined to (**Z / X**) discs on either end of the (**muscle fiber / sarcomere**).

**9** Place the following events of food movement through the digestive system in the correct order by numbering them from 1 to 5.

_____ a. Nutrients are absorbed through cells of the villi.

_____ b. Food is broken into smaller pieces through chewing.

_____ c. Digestive enzymes are released by accessory organs.

_____ d. Bacteria help digest food and produce some vitamins before sending wastes out of the body.

_____ e. Acids break down proteins for further digestion.

## Try Something New

**10** Which of the following is *not* involved in supporting movement of a joint?

(a) bone

(b) smooth muscle

(c) ligaments

(d) synovial sacs

(e) All of these are involved in joint movement.

**11** Our bones are constantly changing in response to how we live. Physical activity builds stronger bones. The bones in a pitcher's throwing arm, for example, are stronger and have larger ridges to which muscles can attach than do the same bones in the other arm. Inactivity leads to weaker bones because special osteocytes step up the rate at which they remove tissue from the bone when the skeleton is not under physical stress; such weakening is seen, for example, in a person confined to bed by an injury or illness. Which of the following activities would be the least effective in increasing bone strength? Explain why.

(a) walking

(b) dancing

(c) weight lifting

(d) swimming

(e) running/jogging

**12** A January 2013 Tumblr post that went viral on the Internet read as follows:

> eating is so bad*** i mean you put something in a cavity where you smash and destroy it with 32 protruding bones and then a meat tentacle pushes it into a vat of acid and after a few hours later you absorb its essence and transform it in[to] energy just wow.

Identify each element of the digestive system discussed in the post. Do you think this is an accurate description of digestion?

## Leveling Up

**13** **Looking at data** The recommended daily allowance (RDA) for nutrients is the average daily amount necessary to meet the nutrient requirements of most healthy adults (97%–98%). There are two kinds of necessary mineral nutrients: macrominerals, which your body needs in larger amounts; and trace minerals, which your body needs in smaller amounts. Macrominerals include calcium and phosphorus. Trace minerals include iron, copper, iodine, zinc, and fluoride.

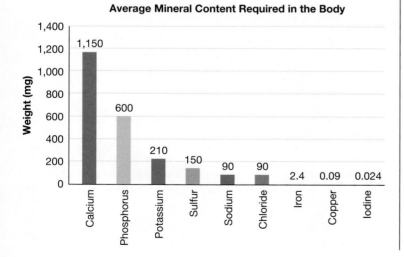

**Average Mineral Content Required in the Body**

a. According to the accompanying graph, what is the dividing line between macrominerals and trace minerals?

b. What does the *y*-axis show?

c. Describe in your own words what the graph shows.

d. How much potassium is needed by an average person?

e. Is the amount of potassium you stated in part (d) the same amount that we need to eat daily?

**14** **Life choices** Nutritionists agree that we can obtain all our necessary nutrients in optimal amounts from natural sources, provided we eat a well-balanced diet that includes a variety of foods. Recognizing that relatively few people meet this ideal in reality, some physicians advise their patients to take a multivitamin and mineral supplement as "added insurance." The use of this kind of dietary supplement raises the specter of toxicity from excessive intake, particularly of the fat-soluble vitamins. Most manufacturers, but by no means all, attempt to avoid high levels of the fat-soluble vitamins in their vitamin preparations, but the burden is largely on us to use supplements wisely, because this industry is not highly regulated by the government. Review the information provided on dietary supplements by the National Institutes of Health Office of Dietary Supplements (http://ods.od.nih.gov /HealthInformation/DS_WhatYouNeedToKnow.aspx) and the U.S. Food and Drug Administration, or FDA (http://www.fda.gov/Food /DietarySupplements/UsingDietarySupplements/ucm110567.htm). Take notes so that you can answer the following questions.

a. What aspects of the sale of dietary supplements does the government regulate?

b. Are you confident that this regulation is sufficient to make supplements safe to use as labeled? Why or why not?

c. List three recommendations that you found informative or helpful. Will they change your use of dietary supplements? Explain your reasoning.

d. Do you think a mobile app for dietary supplements would be helpful to have? Why or why not?

e. Identify one dietary supplement that you feel would be helpful for you to take. Will you take it? Why or why not?

**For more, visit digital.wwnorton.com/bionow2 for access to:**

# Body (Re)Building

*Could engineered human tissues, brought to life in the lab, replace failing organs in people?*

**After reading this chapter you should be able to:**

- Create a flowchart depicting the movement of blood through a cardiovascular system.
- Describe the function of the different components of blood, and compare the three types of blood vessels.
- Diagram the elements of the respiratory system, and show how air moves through it.
- Explain how gases are exchanged in the lungs, and why.
- Describe how the structure of a nephron relates to its function in the urinary system.
- Distinguish between the central and peripheral nervous systems.
- Identify the sensory systems active in humans.

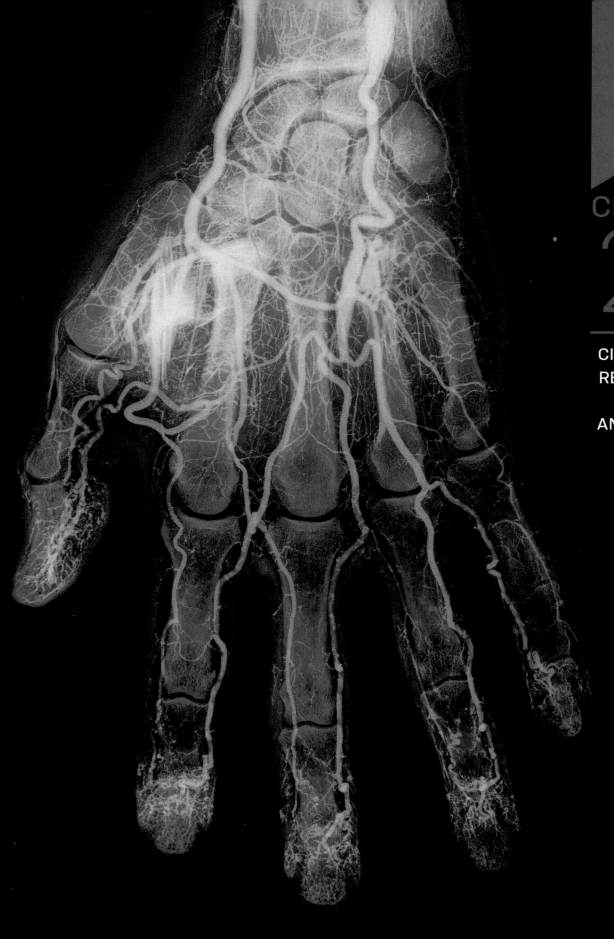

CHAPTER

24

CIRCULATORY,
RESPIRATORY,
URINARY,
AND NERVOUS
SYSTEMS

Standing over the operating table, covered head to toe in teal scrubs, physician Jeffrey Lawson lifts a long, white tube out of a bath of clear liquid. Carefully, slowly, he threads the tube through the unconscious patient's upper arm and stitches each side of the tube to an exposed blood vessel.

Bioengineer Laura Niklason stands to the side of the operating table, watching closely (**Figure 24.1**). She and Lawson created the tube being implanted in the patient. Once the operation is complete, Lawson steps back, and Niklason gives him a hug. "Congratulations to you," says Niklason happily. "We're saving the world!"

On June 5, 2013, Lawson, Niklason, and their team transplanted a laboratory-grown blood vessel into a human. It was the first such procedure in the United States and a major feat for the field of *tissue engineering*, the effort to grow or regenerate tissues or organs using engineering materials and principles. Other engineered tissues implanted into humans have included nerve grafts, bladders, and windpipes.

Lawson and Niklason imagine a future in which any type of organ can be constructed in the lab. In an ideal world, "we'll be able to grow all sorts of tissues for patients, so a surgeon can literally reach up on a shelf, pull down a tissue graft, and implant it in a patient," Niklason said in 2013, shortly after the first blood vessel transplant. "That's really going to be a revolution—being able to grow replacement parts for patients, that they don't have to wait for."

Organ shortage is a major concern in health care (**Figure 24.2**). According to the U.S. government, the number of people waiting for an organ would fill a football stadium—twice. Every day, an average of 18 people die waiting for transplants of kidneys, hearts, livers, lungs, and more.

One possible solution to this organ crisis is the Frankenstein-like approach of tissue engineering: building organs in the lab. It sounds far-fetched, but research teams have already begun to engineer complicated organs such as kidneys, lungs, even whole hearts. These are just a few of the organs that make up the major organ systems in our bodies, including the circulatory, respiratory, urinary, and nervous systems—each one critical to the healthy functioning of our bodies (see Figure 22.4 for an overview).

That healthy functioning, including growth and homeostasis (see Figure 22.5), depends on the internal transfer of transport proteins (see Figure 4.4), waste products, and other substances in the body. Distances inside the bodies of most multicellular organisms are far too great for diffusion to be an effective means of distributing these materials, so elaborate organ systems exist to transport them around the body. The **circulatory system** moves oxygen from the lungs to the heart, which then pumps oxygen-rich blood to the rest of the body. The **respiratory system** brings in oxygen and expels carbon dioxide to support cellular respiration. The **urinary system** removes excess fluid from the body, along with waste products, toxins, and other water-soluble substances that are not needed.

These three systems are interconnected in the body: Carbon dioxide collected by the circulatory system is delivered to the respiratory system for exchange with the outside environment. The circulatory system brings substances dissolved in blood to the urinary system for discharge into the environment. And a unique high-speed communication system called the **nervous system** coordinates the many muscles involved in the functioning of organs, including the heart, lungs, and bladder. The nervous system directs the rapid contractions of muscles, and processes

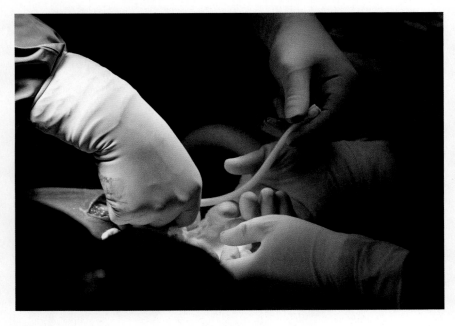

Figure 24.1

**A bioengineered blood vessel**

In 2013, a team of doctors at Duke University was the first in the United States to implant a bioengineered blood vessel into a patient.

information received by the senses, such as touch, sound, and sight. In this way, the nervous system enables an animal to detect food, find a mate, avoid predators, and respond to extremes of heat and cold.

These organ systems, along with all the other organ systems introduced in Chapter 22, govern how your body operates. If one of your organs fails, a whole system can shut down. That's why researchers are exploring how to build new organs, writing a recipe for flesh and blood.

# Emergency Meeting

Niklason and Lawson met at Duke University Medical Center in 1999 while performing surgery together. Waiting in the operating room to move their patient, the two struck up a conversation. Within moments, they discovered they shared strikingly similar interests: Lawson was a surgeon who spent his days suturing blood vessels in the hospital and growing different types of blood cells in his lab, and Niklason was an anesthesiologist and bioengineer who had recently opened her own lab to build blood vessel tubes. Both were dreaming of the same goal—to create a blood vessel from scratch—and they had complementary skills. "It made us a perfect partnership," says Lawson.

That first conversation bloomed into a friendship and working partnership. Together, Lawson and Niklason started building prototypes of blood vessels, using his knowledge of blood cells and her knowledge of the engineering forces at work in blood vessels. **Blood vessels** are a critical part of the **cardiovascular system**, a closed circulatory system consisting of a muscular *heart*, a complex network of *blood vessels* that collectively form a closed loop, and *blood* that circulates through the heart and blood vessels. Almost all of our cells lie within 0.03 millimeter of a blood vessel with which they exchange materials by diffusion. Carrying

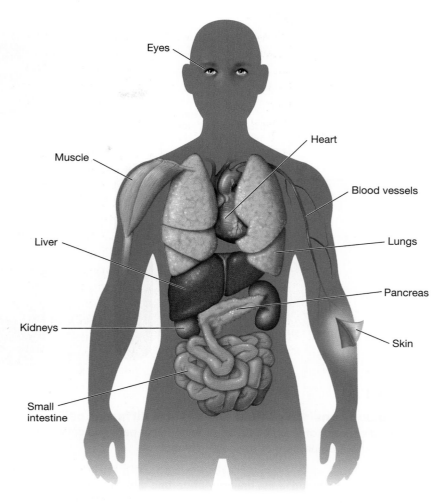

Figure 24.2

**Organs needed for donation**

blood so close to all the trillions of cells in our bodies requires an extensive network of vessels.

Lawson and Niklason built blood vessels created of various materials and cell types and then implanted them in rats to see whether blood would successfully flow through the tubes. **Blood** is composed of cells and cell fragments that float

**JEFFREY LAWSON AND LAURA NIKLASON**

Jeffrey Lawson is a physician-scientist at Duke University Medical Center in North Carolina who studies novel therapies for vascular surgery, including bioengineered blood vessels. Laura Niklason is a doctor and professor of anesthesiology and biomedical engineering at Yale University in Connecticut. Her research includes cardiac, lung, and blood vessel bioengineering.

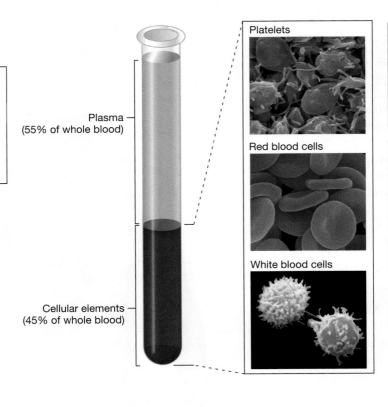

Blood plasma is 92% water and contains dissolved gases, ions, and molecules that are critical for homeostasis, as nutrients, or as signaling molecules. Much of the carbon dioxide carried in the blood is dissolved in plasma.

Plasma
(55% of whole blood)

Cellular elements
(45% of whole blood)

Platelets

Red blood cells

White blood cells

Platelets are small cell fragments. They can clump together to help stop the loss of blood if a blood vessel is damaged. Platelets release substances that stimulate plasma proteins to create a meshwork of protein strands, platelets, and blood cells to collectively form a blood clot.

A mature red blood cell has no nucleus, and its cytoplasm is packed with oxygen-binding proteins called hemoglobin. Each hemoglobin molecule can carry up to four oxygen molecules. Because each human blood cell contains about 280 million hemoglobin molecules, a single one of these cells can bind over a billion molecules of oxygen.

Several different kinds of white blood cells help defend the body from invading organisms.

**Figure 24.3**

**Human blood consists of fluid and several different cell types**

Whole blood consists of plasma and different kinds of cells and cell fragments, three of which are shown here. Red blood cells account for about 95 percent of the cells in blood.

**Q1:** Where in the blood is the majority of carbon dioxide carried?

**Q2:** Where in the blood is the majority of oxygen carried?

**Q3:** What would happen if your red blood cells carried a mutation that made the hemoglobin less effective at binding to oxygen (as in sickle-cell disease)?

in a fluid known as **plasma** (**Figure 24.3**). Blood plasma itself has a low capacity for transporting dissolved oxygen, but **red blood cells** in the plasma carry significant amounts of oxygen, greatly increasing the oxygen-carrying capacity of blood.

In 2005, the duo hit upon a technique that seemed to work. They collected smooth muscle cells from the blood vessels of organ donors and grew those muscle cells on a biodegradable frame shaped like a blood vessel. The cells worked like little machines, churning out proteins that formed a three-dimensional scaffold of connective tissue called the *extracellular matrix*. Then the original, biodegradable frame dissolved, leaving behind a sturdy tube of cells and extracellular matrix.

There was still a catch. The immune system of a human body rejects cells from another person (see Chapter 25 for more on the immune system), so Lawson and Niklason had to wash away the original donor muscle cells to leave behind just the tubular extracellular matrix. "It's like we kept the mortar but the bricks all washed away," says Lawson. They tested their creation in rats. It worked. They had made a functioning blood vessel.

The vessels were first tested in humans in Poland in December 2012. The researchers found that once an engineered tube is implanted into a patient, the patient's own blood and muscle cells take up residence in the tube, filling in the cracks. "What started off as our structure now becomes your tissue," says Lawson. "And it all seals together so it doesn't leak."

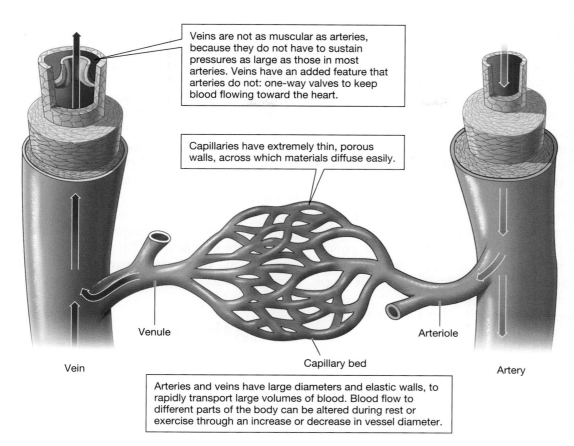

Veins are not as muscular as arteries, because they do not have to sustain pressures as large as those in most arteries. Veins have an added feature that arteries do not: one-way valves to keep blood flowing toward the heart.

Capillaries have extremely thin, porous walls, across which materials diffuse easily.

Arteries and veins have large diameters and elastic walls, to rapidly transport large volumes of blood. Blood flow to different parts of the body can be altered during rest or exercise through an increase or decrease in vessel diameter.

Venule

Arteriole

Vein

Capillary bed

Artery

Figure 24.4

## Arteries, veins, and capillaries

Arteries carry blood away from the heart. Veins carry blood toward the heart. Arterioles and venules are smaller arteries and veins, respectively. The narrowest arteries and veins connect with each other in a fine network known as a capillary bed.

**Q1:** Why are arteries more muscular than veins?

**Q2:** What structural feature(s) of capillaries enable easier diffusion into and out of surrounding tissues?

**Q3:** Why do you think capillaries are not typically transplanted?

A critical feature of the engineered vessel is that it has the structural strength of a normal blood vessel in the body, so it can withstand the force of the blood pulsing through it. The human body has three major kinds of blood vessels through which blood flows: arteries, veins, and capillaries (**Figure 24.4**). **Arteries** are large vessels (0.1–10 mm in diameter) that transport blood away from the heart. **Veins** (0.1–2 mm) are large vessels that carry blood back to the heart. **Capillaries**, the smallest vessels, at 0.005–0.01 mm, exchange materials by diffusion with nearby cells.

The large vessels—arteries and veins—are built for mass transport of blood. Currently, Niklason and Lawson have built blood vessels with diameters of 6 mm (about the width of a pencil) and 3 mm. But they also expect to be able to make and transplant larger vessels, such as the aorta, the main artery of the body. Capillaries, however, are very small and are not typically transplanted. They are built for slower movement of blood, and their large surface area facilitates the exchange of materials with surrounding cells.

Niklason and Lawson's blood vessels are not the only tissues in the cardiovascular system that researchers are trying to engineer. Scientists are attempting to grow the **heart**, a muscular organ the size of a fist in humans that works as the body's circulatory pump (**Figure 24.5**). Like the hearts of all other mammals, the human heart is divided into four chambers that form two distinct pumping units, which are independent but coordinated.

The left atrium receives oxygenated blood from the lungs and pumps it to the left ventricle, which pumps it through the **systemic circuit** to cells performing cellular respiration. The right atrium receives blood low in oxygen and laden with $CO_2$ returning from the systemic circuit, and pumps it to the right ventricle, which pumps it through the **pulmonary circuit** for gas exchange in the lungs.

Together, these chambers pump some 7,000 liters of blood per day—about 1,850 gallons. **Heart rate** is the number of times a heart beats per minute; an average resting (not exercising) human heart beats about 60–100 times per minute. The force of blood pushing through blood vessels is called **blood pressure**. As the heart contracts to push blood out, blood pressure is at its highest, and when the heart relaxes after each contraction, blood pressure is at its lowest. The first pressure is referred to as **systole** and is the top number in a blood pressure reading. The second, **diastole**, is the bottom number in a blood pressure reading. The human circulatory system adjusts the heart rate and patterns of blood distribution according to the body's needs. Blood pressure is at its highest as it leaves the heart, and by the time it reaches the veins it is at its lowest.

The heart might seem like too complex an organ to replicate by means of tissue engineering techniques, but researchers have demonstrated some success using a tactic similar to the one Niklason and Lawson took with blood vessels, starting with an entire organ as a scaffold (such as a pig or rat heart, of which there is a ready supply, or a heart from an organ donor). They use detergents to strip away all the original cells that would cause an immune reaction, then repopulate the heart with cells that are a better match for the patient. Using this technique, researchers have created pumping rat hearts in a dish, though these bioengineered hearts are still too primitive to work when transplanted into an animal.

Since the initial implant of Niklason and Lawson's blood vessel, 60 more implants of bioengineered blood vessels have been performed in the United States. And a company that Niklason cofounded and Lawson now serves as chief medical officer, is testing the blood vessels in human studies. Of the first 20 patients in Poland who received engineered blood vessels, none of the transplants became infected, and additional implants in a total of 60 patients saw no evidence of immune rejection.

"We've got a tube that works to put into people, so now we have the groundwork to make things more complicated than a blood vessel," says Lawson. "There is still an unlimited amount of science to do."

# Breathe In, Breathe Out

In 2010, after she had moved from Duke to Yale University, Niklason expanded her research program from creating simple blood vessels to attempting to build a lung. **Lungs** are the main organs of the respiratory system, which carries air from the nose (or mouth) to the lungs through a series of tubular passageways. These airways allow air to move between the external environment and the inside of the body—specifically, the gas exchange surfaces in the lungs.

The process of taking air into the lungs (inhaling) and expelling air from them (exhaling) is called **breathing** (**Figure 24.6**). The air we inhale is about 21 percent oxygen and contains little carbon dioxide and water vapor. The air we exhale has less oxygen (15 percent) and contains about 4 percent each of carbon dioxide and water vapor. That's because these gases are exchanged at the surface of the cells that line our lungs: Oxygen is removed from the inhaled air and sent to the bloodstream, while carbon dioxide and water vapor are removed from the bloodstream and added to the air that is exhaled. Remember that our bodies need oxygen to obtain energy via cellular respiration (see Figure 5.11).

At Yale, Niklason's team bioengineered a rat lung and implanted it in a rat. There, it briefly supported gas exchange for the animal before filling up with fluid, suggesting that more work

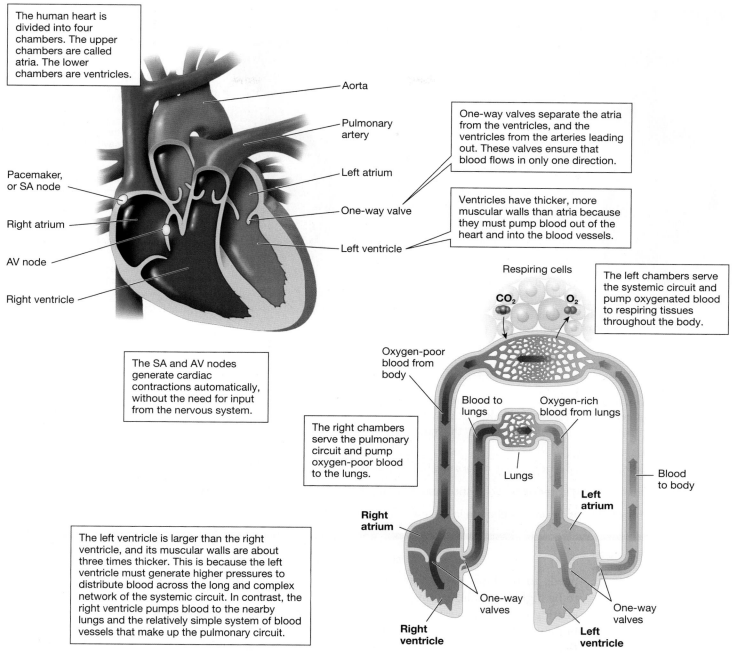

The human heart is divided into four chambers. The upper chambers are called atria. The lower chambers are ventricles.

Aorta

Pulmonary artery

Left atrium

One-way valves separate the atria from the ventricles, and the ventricles from the arteries leading out. These valves ensure that blood flows in only one direction.

Pacemaker, or SA node

Right atrium

One-way valve

Ventricles have thicker, more muscular walls than atria because they must pump blood out of the heart and into the blood vessels.

AV node

Left ventricle

Right ventricle

Respiring cells

The left chambers serve the systemic circuit and pump oxygenated blood to respiring tissues throughout the body.

$CO_2$   $O_2$

The SA and AV nodes generate cardiac contractions automatically, without the need for input from the nervous system.

Oxygen-poor blood from body

Blood to lungs

Oxygen-rich blood from lungs

The right chambers serve the pulmonary circuit and pump oxygen-poor blood to the lungs.

Lungs

Blood to body

**Right atrium**

**Left atrium**

The left ventricle is larger than the right ventricle, and its muscular walls are about three times thicker. This is because the left ventricle must generate higher pressures to distribute blood across the long and complex network of the systemic circuit. In contrast, the right ventricle pumps blood to the nearby lungs and the relatively simple system of blood vessels that make up the pulmonary circuit.

One-way valves

One-way valves

**Right ventricle**

**Left ventricle**

## Figure 24.5

### The human heart

The right and left sides of the heart function as two separate pumps, although the two upper chambers, or **atria** (singular "atrium"), contract in unison, as do the two lower chambers, the **ventricles**. This unified contraction begins with a signal from the **pacemaker**, or **sinoatrial (SA) node**. The signal causes both atria to contract; it also causes the **atrioventricular (AV) node** to pass the signal on to the ventricles about a tenth of a second later. The short delay allows the atria to empty completely. Note that the heart is shown from the front of the body, so the left atrium is on the right side of the diagram, and so forth.

**Q1:** Beginning with the left atrium, list the locations of a drop of blood as it moves through the circulatory system.

**Q2:** Why is the left ventricle larger than the right ventricle, and why are its walls thicker and more muscular?

**Q3:** Some people have an artificial pacemaker implanted in their heart. What is its function?

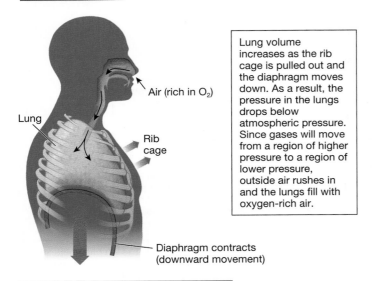

**Inhalation**

Air (rich in O₂)

Lung

Rib cage

Lung volume increases as the rib cage is pulled out and the diaphragm moves down. As a result, the pressure in the lungs drops below atmospheric pressure. Since gases will move from a region of higher pressure to a region of lower pressure, outside air rushes in and the lungs fill with oxygen-rich air.

Diaphragm contracts (downward movement)

The diaphragm is a thick sheet of muscle that forms the floor of the chest cavity.

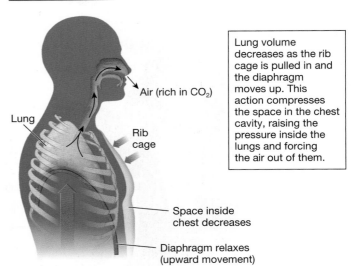

**Exhalation**

Air (rich in CO₂)

Lung

Rib cage

Lung volume decreases as the rib cage is pulled in and the diaphragm moves up. This action compresses the space in the chest cavity, raising the pressure inside the lungs and forcing the air out of them.

Space inside chest decreases

Diaphragm relaxes (upward movement)

## Figure 24.6

### Breathing

Breathing involves two main steps: inhalation, when air is pulled into the lungs, and exhalation, when air is pushed out of the lungs. Most of the time, breathing is controlled automatically by sensory systems located in the heart and brain. If we choose, we can also control our breathing with a system of muscles, the most important of which are the rib muscles and the diaphragm.

**Q1:** When is air richer in oxygen—as it enters the body or as it exits the body?

**Q2:** The figure shows air entering through the nose. Where else can air enter the respiratory system?

**Q3:** Explain how the body creates a change in air pressure during breathing.

needs to be done. Any engineered lung needs to be highly reliable, as lungs are vital to transporting oxygen throughout the body. Although humans can live for more than a week without food and for a few days without water, a mere 4 minutes without oxygen results in irreversible brain damage. And lung tissue is notoriously bad at repairing itself, so lung transplants are often the only option in the case of lung damage from disease or trauma.

The respiratory system can be divided into two parts (**Figure 24.7**). The **upper respiratory system** includes airways in the nose, mouth, and throat. When we inhale, air enters through the nostrils and moves into each nasal cavity. Next the air enters the throat, or **pharynx**, an area where the back of the mouth and the two nasal cavities join together into a single passageway.

From the pharynx, air moves into the **larynx**, or voice box, which forms the entryway to the windpipe, or **trachea**.

The trachea is the start of the **lower respiratory system**. Within the chest, the trachea branches into two smaller tubes called **bronchi** (singular "bronchus"). Each bronchus leads to one of the paired lungs, the organs where gases are exchanged. Together, the trachea, bronchi, and lungs make up the lower respiratory system.

Inside the lungs is where the respiratory system gets intricate: The bronchi divide into **bronchioles**, a series of branching, ever-smaller tubes. The tiniest bronchioles open into the **alveoli** (singular "alveolus"), small clusters of sacs that resemble a bunch of grapes. Gases are exchanged across the moist surface

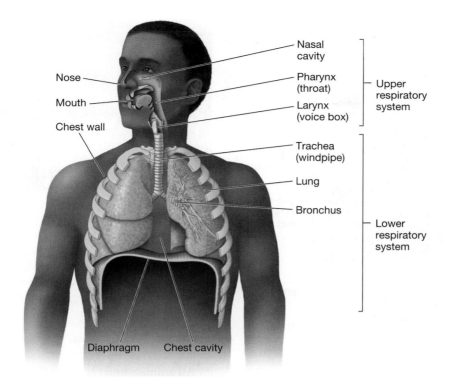

Figure 24.7

**The human respiratory system**
The respiratory system consists of the upper respiratory system (nose, mouth, throat, and larynx) and the lower respiratory system (the trachea, bronchi, and lungs).

> **Q1:** Beginning with the nose, list the locations of a molecule of oxygen as it moves through the respiratory system.
>
> **Q2:** From the lung, where would the oxygen molecule move?
>
> **Q3:** How might an upper respiratory infection like the common cold affect your respiratory system and thus your breathing?

of the thin layer of cells that line each alveolar sac, surrounded by capillaries (**Figure 24.8**). Niklason continues to try to achieve this gas exchange in an engineered lung. So far, her rat lung transplants have managed this exchange for only about 2 hours before failing.

## Waste Not

At Massachusetts General Hospital in Boston, Harald Ott, a pioneer in the field of organ bioengineering, has engineered whole rat, pig, and human lungs by cleaning donated organs and repopulating them with new cells (**Figure 24.9**).

These organs, however, have yet to be successfully transplanted.

In 2013, Ott's lab used the scaffolding technique to engineer rat kidneys, which were then transplanted into rats. In vertebrates, **kidneys** are a set of paired organs that maintain water

**HARALD OTT**

Harald Ott is a thoracic surgeon at Massachusetts General Hospital and heads a laboratory focused on organ engineering and regeneration at Harvard Medical School.

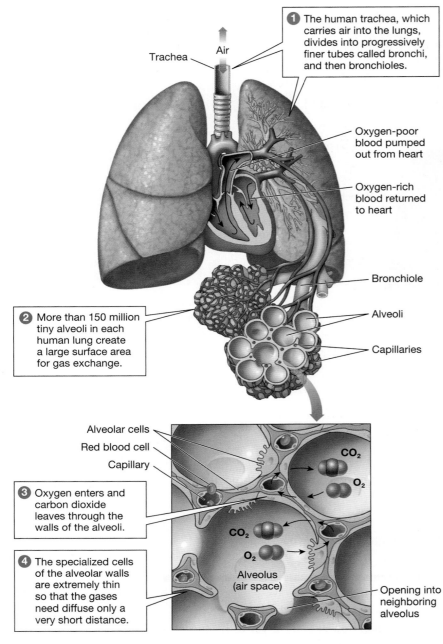

**1** The human trachea, which carries air into the lungs, divides into progressively finer tubes called bronchi, and then bronchioles.

Trachea — Air

Oxygen-poor blood pumped out from heart

Oxygen-rich blood returned to heart

Bronchiole

Alveoli

Capillaries

**2** More than 150 million tiny alveoli in each human lung create a large surface area for gas exchange.

Alveolar cells
Red blood cell
Capillary

**3** Oxygen enters and carbon dioxide leaves through the walls of the alveoli.

**4** The specialized cells of the alveolar walls are extremely thin so that the gases need diffuse only a very short distance.

$CO_2$
$O_2$
$CO_2$
$O_2$
Alveolus (air space)

Opening into neighboring alveolus

**Figure 24.8**

### Gas exchange in the alveoli

The structure of the lungs speeds the diffusion of oxygen and carbon dioxide into and out of the body by providing a large surface area for gas exchange.

**Q1:** Why is a large surface area important for gas exchange?

**Q2:** Does carbon dioxide move into or out of the alveoli? Into or out of capillaries at the surface of the alveoli?

**Q3:** When a person has pneumonia, the alveoli may fill with fluids. Why would this be a problem?

**Figure 24.9**

### A bioengineered rat lung

Researchers at Massachusetts General Hospital have bioengineered the lungs of rats, pigs, and humans. None have yet been capable of functioning within a living individual.

and solute homeostasis, serving as key components of the urinary system.

All animals must regulate the concentrations of solutes, such as sodium and calcium, in their body fluids, but terrestrial animals face the additional challenges of conserving water and retaining vital solutes. The solute composition of an animal is affected by metabolic activities within the body: as cells metabolize biomolecules, they use up chemicals dissolved in body fluids and produce new ones. This process results in waste products that must somehow be removed from the body.

Kidneys filter and regulate the composition of the blood as it moves through them. When blood leaves the kidneys and returns to the circulatory system, it is cleansed of metabolic wastes and carries water and solutes in normal amounts. The volume of blood leaving the kidneys is slightly smaller than the entering volume because some water is lost to make **urine**, the waste-carrying solution that is expelled from the body.

The blood-cleansing work of the kidneys—**filtration**—is performed by the kidney's basic functional unit, the **nephron** (**Figure 24.10**). Each human kidney has about a million of these tiny filtration units, tightly integrated with surrounding capillaries. When the kidneys fail, an individual can no longer filter waste. Approximately 100,000 individuals in the United States currently await kidney transplantation. As they wait, these patients receive dialysis three times per week, during which a machine filters their blood for them. This lifesaving technology

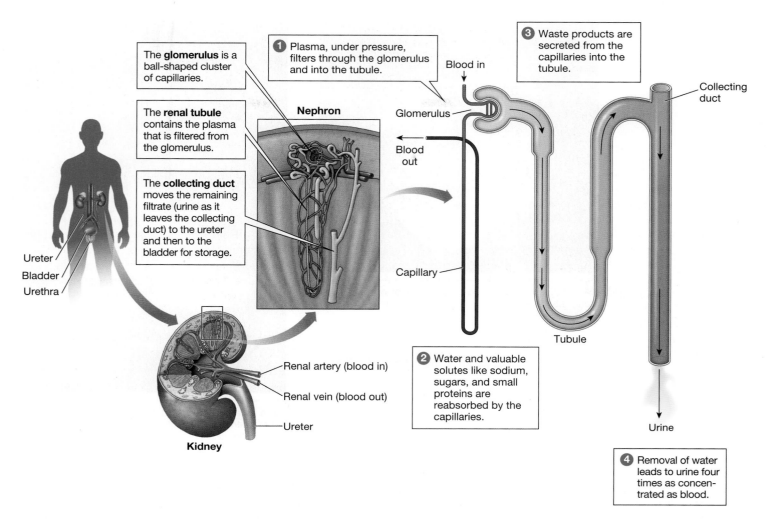

**Figure 24.10**

## The kidney, a critical component of the human urinary system

The kidney regulates internal water content, balances solute concentrations, and removes toxic wastes. All of this work occurs within the nephron.

**Q1:** What is released from the kidney, and where does it go after release?

**Q2:** What is the difference between reabsorption and secretion?

**Q3:** Alcohol suppresses the kidney's ability to reabsorb water. What common consequence of drinking alcohol is related to this fact, and how might you alleviate this problem?

---

requires a significant time commitment, and dialysis patients have a high risk of infection.

An engineered kidney would provide a permanent solution for these patients. But an engineered kidney would need to do more than just remove waste from the blood. Filtration is only one of three parts of the kidney's job. A second important function is the **reabsorption** of water and valuable solutes such as sodium, chloride, and sugars before they leave the kidneys. A third function of the kidney is **secretion**: the kidney actively transports excess quantities of substances such as potassium and hydrogen ions, and some medications and toxins, from the blood into the liquid passing through the kidney.

The concentrated fluid that results from the combination of filtration, reabsorption, and secretion is urine. Urine from the many

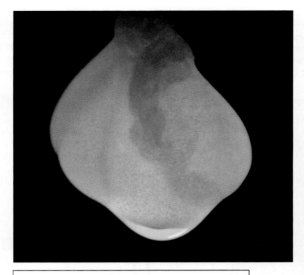

A rat kidney with all its cells removed, leaving only the extracellular matrix made of collagen.

## Figure 24.11

**A bioengineered kidney: before and after**
The Ott lab at Massachusetts General Hospital bioengineered rat kidneys that function when transplanted. Before implanting, the kidney is flushed of its donor cells and repopulated (seeded) with recipient cells so that it will not be rejected by the recipient's immune system when transplanted.

The collagen "scaffold" has now been reseeded with cells.

collecting ducts in a kidney drains into a long tube, the *ureter*, which delivers the fluid to the urinary bladder for storage. In urination, the bladder empties through a tube called the *urethra*. In Ott's lab in Boston, the bioengineered rat kidneys successfully produced urine when transplanted into rats (**Figure 24.11**).

# Coming to Your Senses

To date, the kidney is the most complex organ re-created in the lab and successfully transplanted into an animal. Meanwhile, bioengineering research in a different organ system—the nervous system—has already yielded astonishing improvements in patients' lives, including the life of navy corpsman Edward Bonfiglio Jr.

At 5:00 a.m. on August 27, 2009, Edward Bonfiglio awoke to the sound of the phone ringing. It was his son, Edward Jr., a member of the U.S. Navy serving on active duty in Afghanistan. "Dad?" Edward said, "I got shot. Don't tell Mom."

During a routine foot patrol, Edward's unit had been ambushed, and he'd been shot in his left leg. The bullet hit his sciatic nerve, a long, thick bundle of nerve fibers that runs from the lower back down the leg. Edward immediately lost all feeling and function below his left knee. Doctors found that the bullet had sliced a 5-centimeter gap in the sciatic nerve. Without that nerve intact, the leg would not function.

After Edward arrived home in the United States, doctors gave him two options: either amputate the leg, or repair the damaged nerve with a new kind of nerve graft, a bioengineered tube that could potentially reconnect the two ends of his severed nerve and bring feeling back to his leg. Edward chose the latter—to try to repair the nerve in his leg, part of his peripheral nervous system.

The nervous system of vertebrates is a communication system that transmits signals among various parts of the body. It can be divided into

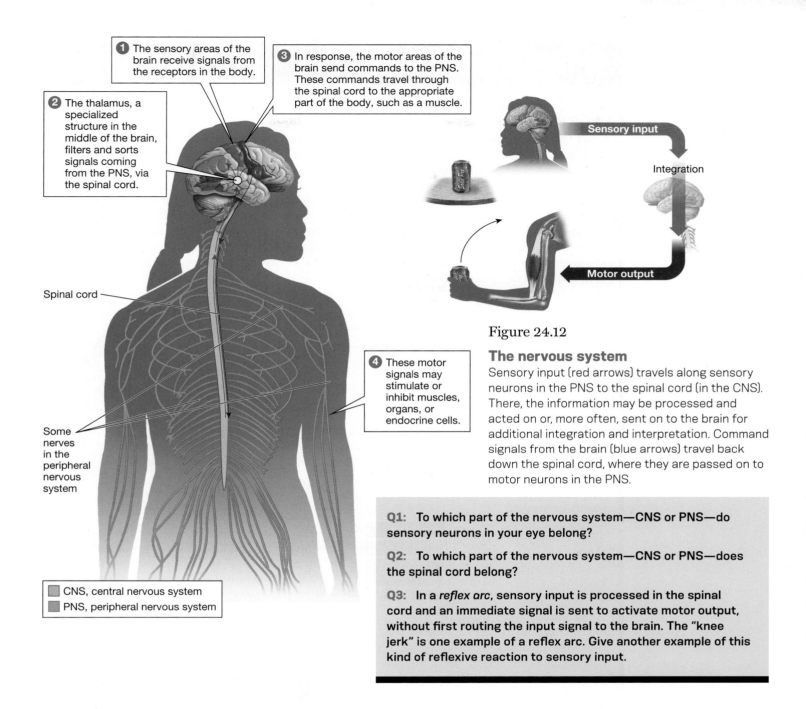

**1** The sensory areas of the brain receive signals from the receptors in the body.

**2** The thalamus, a specialized structure in the middle of the brain, filters and sorts signals coming from the PNS, via the spinal cord.

**3** In response, the motor areas of the brain send commands to the PNS. These commands travel through the spinal cord to the appropriate part of the body, such as a muscle.

Spinal cord

**4** These motor signals may stimulate or inhibit muscles, organs, or endocrine cells.

Some nerves in the peripheral nervous system

Sensory input

Integration

Motor output

CNS, central nervous system

PNS, peripheral nervous system

Figure 24.12

**The nervous system**

Sensory input (red arrows) travels along sensory neurons in the PNS to the spinal cord (in the CNS). There, the information may be processed and acted on or, more often, sent on to the brain for additional integration and interpretation. Command signals from the brain (blue arrows) travel back down the spinal cord, where they are passed on to motor neurons in the PNS.

**Q1:** To which part of the nervous system—CNS or PNS—do sensory neurons in your eye belong?

**Q2:** To which part of the nervous system—CNS or PNS—does the spinal cord belong?

**Q3:** In a *reflex arc*, sensory input is processed in the spinal cord and an immediate signal is sent to activate motor output, without first routing the input signal to the brain. The "knee jerk" is one example of a reflex arc. Give another example of this kind of reflexive reaction to sensory input.

two main units: the **central nervous system**, or **CNS**, and the **peripheral nervous system**, or **PNS** (**Figure 24.12**). The CNS consists of the brain and spinal cord. The **brain** has a large capacity for processing diverse types of sensory information, and it controls and coordinates nerve signals throughout the body (see "What's in Your Head?" on page 444). The **spinal cord** is a thick central nerve cord that is continuous with the brain, acting as a filter between the brain and sensory neurons.

The PNS consists of **sensory organs**—such as the eyes and ears—plus the nerves (except for the retinal, optic, and olfactory nerves, which are part of the CNS), including the nerve in Edward Bonfiglio's leg. The PNS converts stimuli from the sensory organs into **sensory input**—signals that are received, transmitted, and processed by the CNS. Animals are constantly bombarded with sensory input from their external and internal environments. Five main classes of sensory receptors in humans receive this input; some

# What's in Your Head?

The human brain is mind-bogglingly complex. With an estimated 100 billion nerve cells, it is the epicenter of the nervous system. The brain is the organ that most distinctly sets humans apart from other species, giving us our capacity to reason, feel, and remember. The brain is made up of *gray matter* (the cell bodies of neurons) and *white matter* (the branching network of neuronal projections—the dendrites and axons, winding tendrils that connect one neuron to many other neurons).

The human brain weighs about 3 pounds and has three main sections: *forebrain*, *midbrain*, and *hindbrain* (Figure 1). The forebrain contains the *thalamus*, a central switchboard that processes and directs incoming sensory information, and the *cerebrum*, whose outer layer (the *cerebral cortex*) handles most of the brain's actual information processing. The forebrain also contains the *hypothalamus*, *hippocampus*, and *amygdala*, structures that make up the *limbic system*, our emotional control center. The limbic system regulates emotions and motivations, and participates in memory formation.

The midbrain and hindbrain together make up the *brain stem*. The midbrain coordinates sensory information from the thalamus with the peripheral nervous system for simple physical movement. The main structure in the midbrain is the *reticular formation*, which regulates consciousness. Finally, the hindbrain, evolutionarily the oldest part of the brain, controls the most basic functions, including balance, heart rate, and breathing. The hindbrain consists of the *medulla oblongata*, *pons*, and *cerebellum*.

Neuroscience is an extremely active area of research. Scanning technologies such as diffusion MRI, which tracks the diffusion of water through white matter, and functional MRI, which detects brain activity by measuring changes in blood flow, are helping us map the brain at an unprecedented level of detail. Figure 2 is a scan from a diffusion MRI (artificially colored) depicting the elegant structure of white-matter fibers twisting through the brain.

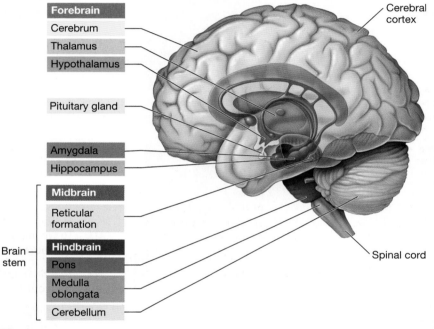

Forebrain
Cerebrum
Thalamus
Hypothalamus
Pituitary gland
Amygdala
Hippocampus
Midbrain
Reticular formation
Brain stem
Hindbrain
Pons
Medulla oblongata
Cerebellum

Cerebral cortex
Spinal cord

**Figure 1**

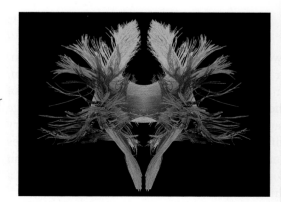

**Figure 2**

## Table 24.1

# Different Ways to Sense the World

| Receptor Type | Stimulus | Sense(s) |
|---|---|---|
| Chemoreceptors | Chemicals | Taste, smell |
| Photoreceptors | Light | Vision |
| Mechanoreceptors | Physical changes | Touch, hearing, proprioception (body position), balance |
| Thermoreceptors | Moderate heat and cold | Thermoreception (gradations of heat and cold) |
| Pain receptors | Injury, noxious chemicals, chemical and physical irritants | Pain, itch |
| Electroreceptors* | Electrical fields (especially those generated by muscle contractions of other animals) | Electrical sense |
| Magnetoreceptors* | Magnetic fields | Magnetic sense |

*Electroreceptors and magnetoreceptors are found in many animals, including most vertebrates. They are not known to be active in humans, however.

additional categories, not known to be active in humans, are found in other animals (**Table 24.1**).

Through sensory input, the PNS gathers information from the external and internal environments and sends it to the CNS. For example, imagine placing your hand near a hot stove; the heat input to your skin is transmitted as a signal from the PNS to the CNS. The CNS integrates and processes the information and generates a signal in response, dictating a particular action, or **motor output**, such as moving your hand away from the stove. The PNS then relays that signal to the body part that will complete the action: the hand moves away from the stove.

All this action in the nervous system is conducted by specialized cells called **neurons** that transmit signals from one part of the body to another in a fraction of a second (**Figure 24.13**). The structure of a neuron reflects its unique function, and different types of neurons respond to different types of stimuli.

Neurons in the PNS are fragile and can be damaged easily. For instance, if you've ever hit your "funny bone," you've actually caused trauma to your ulnar nerve at the elbow, resulting in numbness and tingling. A more severe blow—such as trauma from a broken bone, or a bullet wound as in Edward's case—can cause loss of motor or sensory function.

Unlike lungs, kidneys, or the heart, nerves do have the ability to repair themselves, but they need help. "If you have an injury to a nerve, we can get those nerves to regenerate, but they need a pathway, like a sidewalk, to migrate on," says Christine Schmidt, a biomedical engineer at the University of Florida. As a postdoctoral fellow, Schmidt started looking for biomaterials to act as sidewalks to guide severed neurons to migrate.

The traditional approach to nerve repair is to remove a piece of nerve from somewhere else in a patient's body—such as a nerve in the leg that receives sensory input from the top of the foot—and stitch it into the damaged nerve to restore function. However, this process requires surgery to take the nerve from the leg, and feeling on the top of the foot is lost. And in some cases, using a person's own nerves is not an option. Edward's gap was 5 centimeters wide—far too big for a nerve from another part of his body to work.

## CHRISTINE SCHMIDT

Christine Schmidt heads the biomedical engineering department at the University of Florida. She develops biomaterials to guide nerves to regenerate.

PHYSIOLOGY

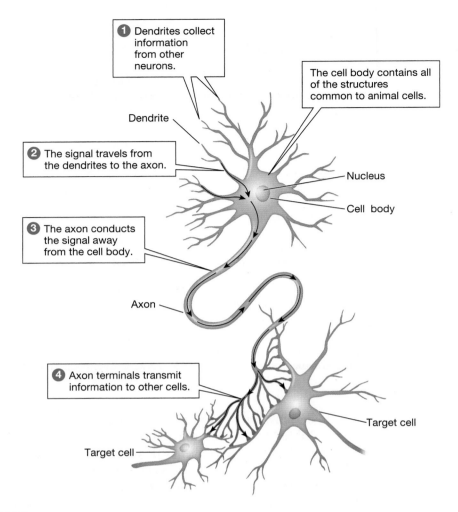

**①** Dendrites collect information from other neurons.

The cell body contains all of the structures common to animal cells.

Dendrite

**②** The signal travels from the dendrites to the axon.

Nucleus

Cell body

**③** The axon conducts the signal away from the cell body.

Axon

**④** Axon terminals transmit information to other cells.

Target cell

Target cell

Figure 24.13

**Neurons transmit signals from one cell to another**
A neuron receives information from other cells, including other neurons, through one or more **dendrites**. The neuron pictured here has a single long **axon** with branched endings. Axons carry signals away from the cell body and to another cell. Two target cells are shown here.

**Q1:** How do neurons look different from other cells you've learned about in this book?

**Q2:** What is the function of these differences?

**Q3:** Some people are born without the capacity to feel pain. Although it might initially sound nice not to feel pain, it is actually quite dangerous. Describe a situation in which it would be dangerous not to feel pain.

Another option is a synthetic graft, a hollow plastic tube inserted to provide a way for nerve cells to grow from one location to another. Yet these tubes can be used only in very small places and, like artificial blood vessels, can be rejected by the immune system.

"There was really a need for grafts that could provide better guidance for neurons, without using one's own nerves," says Schmidt. Schmidt's lab obtained nerves from animal cadavers and spent 3 years testing different methods of washing the nerves to remove cells that could cause immune rejection. "It was like a cooking experiment," says Schmidt—trying to find the right combination of chemicals and physical forces, such as rinsing and swirling the nerves, to retain

# Have a Heart

Over 100,000 people in the United States are right now waiting for an organ transplant. For many of them, the phone call that a suitable donor organ is available may never come. On average, 21 people die each day in the U.S. while awaiting a transplant. Consider registering to be an organ donor: By donating your organs after you die, you could save or improve as many as 50 lives.

**Assessment available in smartwork5**

## Organs needed vs. organs transplanted

Waiting-list registrations and completed transplants for 2013

| Heart | |
|---|---|
| 3,700 needed | 2,531 transplanted |

| Lung | |
|---|---|
| 1,600 needed | 1,923 transplanted |

| Liver | |
|---|---|
| 19,000 needed | 6,455 transplanted |

| Pancreas | |
|---|---|
| 1,200 needed | 256 transplanted |

| Kidney | |
|---|---|
| 108,000 needed | 16,895 transplanted |

Figure 24.14

**Using bioengineering to repair a damaged nerve**
Edward Bonfiglio chose experimental surgery over amputation, and today he can walk and run without assistance.

the nerve's architecture but remove all the original cells. Her team perfected the process in 2004 and published the results.

Right away, a company involved in nerve repair called and asked Schmidt about using her washing process on human tissue. Together, she and the company tested the process on human cadaver tissues. After a successful demonstration in those tissues, the company soon began selling off-the-shelf nerve grafts for use in hospitals. In late 2009, Edward Bonfiglio had one of these grafts implanted into his leg (**Figure 24.14**). Within months of his surgery, he wiggled his toes. It was "one of the greatest moments I had in my entire life," he later said.

There's plenty more bioengineering to be done in the nervous system, says Schmidt. For instance, no nerve grafts can currently repair injuries to the spinal cord. But she's not giving up anytime soon. "It's rewarding," says Schmidt. "It's pretty neat to take something all the way from the bench to impacting patients."

# REVIEWING THE SCIENCE

- Humans and other vertebrates have a **cardiovascular system**, a closed **circulatory system** with a chambered heart that pumps blood through a complex network of **blood vessels**.

- **Blood** is composed of cells and cell fragments that float in a fluid known as **plasma**. **Red blood cells** are the main oxygen carriers in the blood.

- Large vessels are built for mass transport of blood. **Arteries** are large vessels that transport blood away from the heart. **Veins** are large vessels that carry blood back to the heart. **Capillaries**, the smallest vessels, facilitate the exchange of materials with surrounding cells.

- The human cardiovascular system has two main circuits. In the **pulmonary circuit**, oxygen-deficient blood is pumped to the lungs. In the **systemic circuit**, oxygenated blood returning from the lungs is pumped out to body tissues.

- The mammalian **heart** is composed of four chambers that make up two separate muscular pumps, each composed of an **atrium** and a **ventricle**. The left atrium and ventricle pump blood to the body; the right atrium and ventricle pump blood to the lungs.

- The human **respiratory system** carries air from the nose (or mouth) to the **lungs**, eventually reaching

clusters of tiny sacs in the lungs called **alveoli**. Gas exchange takes place in the alveoli, where oxygen diffuses into the blood and carbon dioxide diffuses out of it.

- Inhalation and exhalation are controlled by the contraction of muscles, especially those of the diaphragm and the rib cage.

- In the **urinary system** of many animals, including humans, **kidneys** regulate body water and solute concentrations. The kidney's basic unit, the **nephron**, performs three functions: **filtration**, **reabsorption**, and **secretion**. The resulting concentrated solution, **urine**, is carried by ducts to the bladder and excreted from the body.

- The vertebrate **nervous system** is divided into the **central nervous system** (**CNS**), consisting of the **brain** and **spinal cord**, and the **peripheral nervous system** (**PNS**), consisting of the sensory organs and all the remaining nerves.

- **Sensory organs** convert environmental stimuli into nerve impulses that are carried by sensory **neurons** to the CNS. All human senses rely on **sensory input** from just five types of sensory receptors.

- The CNS integrates sensory information and sends an output signal, often **motor output**, through the PNS to the appropriate body part.

# THE QUESTIONS

## The Basics

**1** Blood plasma transports
(a) waste products.
(b) water.
(c) solutes.
(d) all of the above

**2** In humans, where is the gas exchange surface located?
(a) pharynx
(b) bronchus
(c) alveolus
(d) bronchiole

**3** Which blood vessels carry blood back toward the heart?
(a) veins
(b) arteries
(c) ventricles
(d) capillaries

**4** In a neuron, the _____ conducts signals to other cells.
(a) dendrite
(b) axon
(c) nucleus
(d) cell body

**5** Which of the following functions is *not* performed by the nephron?
(a) filtration
(b) reabsorption
(c) secretion
(d) deletion

**6** Identify each of the following items as belonging to either the central nervous system (C) or the peripheral nervous system (P).
____ a. brain
____ b. sensory organs
____ c. spinal column
____ d. temperature-sensing nerves in the skin

**7** Link each term with the correct definition.

| | |
|---|---|
| **SA NODE** | 1. Sends blood to the right ventricle. |
| **LEFT VENTRICLE** | 2. Pumps blood to the systemic circuit. |
| **RIGHT ATRIUM** | 3. Serves as the heart's pacemaker. |
| **PULMONARY CIRCUIT** | 4. Pumps oxygen-poor blood to the lungs. |

## Challenge Yourself

**8** Which of the following sensory receptors is *not* active in humans?
(a) pain receptor
(b) electroreceptor
(c) mechanoreceptor
(d) chemoreceptor

**9** Select the correct terms:
Kidneys conduct their primary work of filtering the blood through their basic functional unit, the (**neuron / nephron**). Besides filtration, the kidney must (**reabsorb / secrete**) water and important solutes, and (**reabsorb / secrete**) toxins and overabundant substances. Urine drains from the kidneys into the bladder via the (**urethra / ureter**) and empties from the bladder via the (**urethra / ureter**).

**10** Beginning with sensory input from a mechanoreceptor in your toe, place the response events of your nervous system in the correct order by numbering them from 1 to 5.
____ a. Command signals travel from the brain to the spinal cord.
____ b. Input is sent to the brain.
____ c. A sensory signal is sent to the spinal cord.
____ d. The muscle responds to the signal.
____ e. Processing occurs in the brain.

**11** How do veins and capillaries differ in both structure and function?

## Try Something New

**12** Which of the following consequences would most likely result from kidney failure?
(a) Heart rate would be uncontrolled.
(b) Toxin levels in the bloodstream would increase.
(c) Incorrect sensory signals would be sent to the brain.
(d) Oxygen levels would fluctuate dramatically.
(e) None of these would occur because of kidney failure.

**13** If an organism has a greater concentration of $CO_2$ in its lungs than in its blood, will there be (a) net transport of $CO_2$ from the lung air space to the alveolar capillaries, or (b) net transport of $CO_2$ from the alveolar capillaries to the lung air space? Explain the reasoning for your answer.

**14** The number of times our hearts beat per minute is referred to as heart rate. Each heartbeat lasts a little less than 1 second and consists of a series of events called the *cardiac cycle*. The blood pressure measured in a doctor's office reflects the pressure in the arteries leading to the body from the left ventricle. A blood pressure reading of 120/80 (systole/diastole), for example, means that contraction of the left ventricle generates 120 millimeters of mercury (mm Hg) of pressure in the arteries, followed by a drop to 80 mm Hg when the ventricles relax and refill.

a. Blood vessels become less flexible with age. How might this change affect the body's ability to respond to environmental changes with changes in blood pressure?

b. How would you predict that heart rate changes with exercise in the short term, and in the long term?

c. "White coat hypertension" refers to the higher blood pressure displayed at the doctor's office than in normal daily life. How would you explain this phenomenon?

## Leveling Up

**15** **What do *you* think?** In the highly competitive world of endurance sports such as marathon running, cross-country skiing, and bicycle racing, any means of improving performance offers a significant advantage. In recent years, many cyclists have admitted to injecting the hormone erythropoietin (EPO), also known as blood doping. Made naturally in the kidneys, EPO increases the oxygen-carrying capacity of blood by stimulating red blood cell production. In addition to increasing the production of red blood cells, EPO stimulates the growth of capillaries that carry oxygen to tissues.

A synthetic form of EPO developed by drug companies is used to treat patients with anemia, kidney damage, and malaria. Athletes engaging in "blood doping" subject themselves to many health risks. An excess of red blood cells can make the blood so thick that it clots or fails to flow easily through the heart. Nearly two dozen endurance athletes are thought to have died of heart attacks caused by doping with EPO. Because EPO is a naturally occurring hormone, identifying synthetic EPO in the blood or urine samples of athletes has been difficult.

a. What is EPO, and why does it offer a performance advantage in sports, especially endurance events such as cycling and rowing? How could taking EPO kill a person?

b. Some cyclists increase their red blood cell counts by training at high altitudes. The low oxygen content of mountain air triggers the natural release of EPO. Other athletes have accomplished the same thing by spending time in special low-oxygen tents. Do you think either of these approaches is more acceptable than injecting EPO or cells engineered to express EPO? Where would you draw the line, and why?

**16** **Life choices** As of 2016, almost half of American adults have identified themselves as organ donors in the event of their death. Unfortunately, many more people are in need of an organ than can

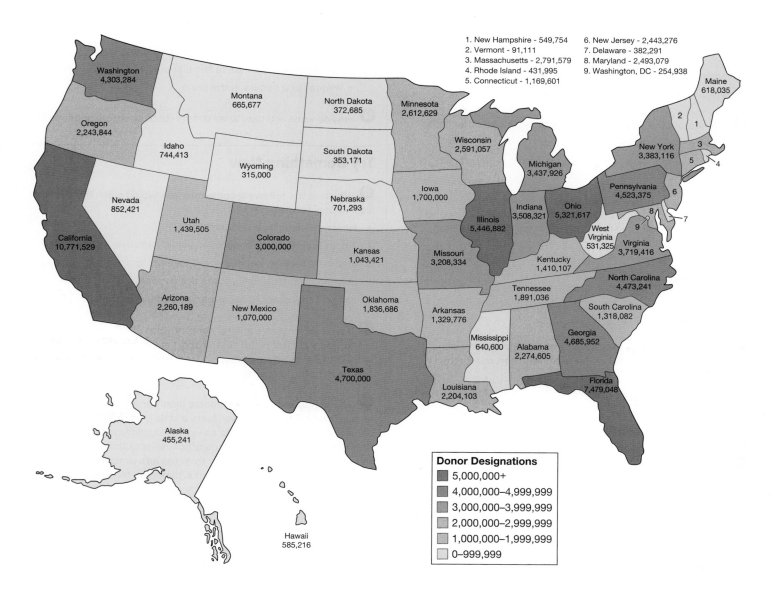

| | |
|---|---|
| 1. New Hampshire - 549,754 | 6. New Jersey - 2,443,276 |
| 2. Vermont - 91,111 | 7. Delaware - 382,291 |
| 3. Massachusetts - 2,791,579 | 8. Maryland - 2,493,079 |
| 4. Rhode Island - 431,995 | 9. Washington, DC - 254,938 |
| 5. Connecticut - 1,169,601 | |

**Donor Designations**
- 5,000,000+
- 4,000,000–4,999,999
- 3,000,000–3,999,999
- 2,000,000–2,999,999
- 1,000,000–1,999,999
- 0–999,999

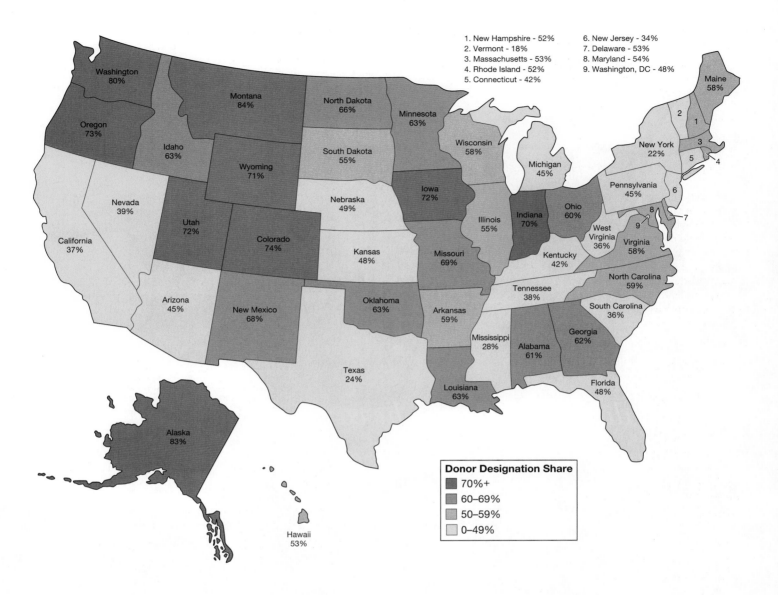

1. New Hampshire - 52%
2. Vermont - 18%
3. Massachusetts - 53%
4. Rhode Island - 52%
5. Connecticut - 42%
6. New Jersey - 34%
7. Delaware - 53%
8. Maryland - 54%
9. Washington, DC - 48%

Washington 80%
Montana 84%
North Dakota 66%
Minnesota 63%
Maine 58%
Oregon 73%
Idaho 63%
Wyoming 71%
South Dakota 55%
Wisconsin 58%
Michigan 45%
New York 22%
Nevada 39%
Utah 72%
Colorado 74%
Nebraska 49%
Iowa 72%
Illinois 55%
Indiana 70%
Ohio 60%
Pennsylvania 45%
West Virginia 36%
Virginia 58%
California 37%
Kansas 48%
Missouri 69%
Kentucky 42%
North Carolina 59%
Arizona 45%
New Mexico 68%
Oklahoma 63%
Arkansas 59%
Tennessee 38%
South Carolina 36%
Mississippi 28%
Alabama 61%
Georgia 62%
Texas 24%
Louisiana 63%
Florida 48%
Alaska 83%
Hawaii 53%

**Donor Designation Share**
- 70%+
- 60–69%
- 50–59%
- 0–49%

be helped, and this number continues to rise. Every day, about 80 people receive organ transplants but another 18 people die, waiting for a transplant that never came. Refer to the two maps above to answer the first three questions here.

a. Which state has the highest number of organ donors? The lowest number?

b. Which state has the highest percentage of organ donors? The lowest?

c. Where does your state fall, in terms of the percentage of organ donors? Does this surprise you? Why or why not?

d. Are you an organ donor? Why or why not?

e. Suggest one way that a state (yours or another) could increase the number of its residents who identify themselves as organ donors.

**For more, visit digital.wwnorton.com/bionow2 for access to:**

# Testing the Iceman

*A Dutch daredevil claims he can fend off disease with his mind. Two skeptical scientists take the case.*

- Explain how cells communicate with each other via the endocrine system.
- Describe different ways a hormone can act on a target cell.
- Identify the immune system's first, second, and third lines of defense.
- Compare and contrast the role of white blood cells in the innate and adaptive immune systems.
- Diagram the processes of inflammation and blood clotting.
- Distinguish between a primary and a secondary adaptive immune response.
- Create a flowchart depicting the sequence of events as a vertebrate immune system responds to a pathogen.

CHAPTER

25

ENDOCRINE
AND IMMUNE
SYSTEMS

The scantily clad young men lie on the ground, looking up toward the sky. With sunglasses on and hands propped behind their heads, they look as if they're tanning at the beach—but there are no piña coladas or warm sand here. Instead, these 18 men, wearing only swim trunks, are lying on cold, white snow in the mountains of Poland. And lying with them is the Iceman.

Wim Hof, a Dutch daredevil known as the "Iceman," who holds numerous world records for cold exposure, breathes deeply, leading the youths in an exercise. Over 4 days, he will train them to tolerate extreme cold. During his rigorous program, they will swim in near-freezing water every day and climb a snow-covered mountain in just shorts (**Figure 25.1**). Hof claims that exposure to the cold, combined with meditation and breathing exercises, will enable the men to fend off illness and disease.

Matthijs Kox stands to the side of the Iceman's trainees, taking notes. Kox, a researcher in intensive-care medicine at Radboud University Medical Center in the Netherlands, first met Hof in 2010, when the Iceman was visiting another laboratory at the university. A team in the physiology department was measuring Hof's ability to regulate his core temperature while standing in an ice bath (**Figure 25.2**). The scientists were surprised to find that rather than decreasing as expected, Hof's core temperature actually increased, and his metabolism climbed. While standing in the ice bath talking to his examiners, Hof mentioned that he could also consciously modulate his autonomic nervous system and immune system.

It was an unbelievable claim. The autonomic nervous system operates body functions that humans cannot voluntarily control, such as heartbeat and blood pressure. The **immune system**—a remarkable defense system that protects us against most infectious agents—has also long been known to be involuntary.

Figure 25.1

**"Iceman" Wim Hof trains volunteers under extreme conditions**

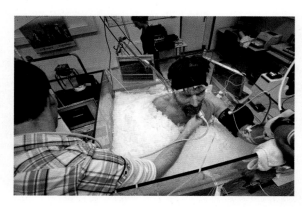

Figure 25.2

**Hof's vital signs are monitored while he is immersed in ice**

But Hof had a history of doing the unbelievable. He claimed the Guinness World Record for longest ice bath by staying immersed in ice for 1 hour, 52 minutes, and 42 seconds. He climbed part of Mount Everest wearing nothing but shorts. He ran a marathon through the snow at 4°F (20°C), again wearing only shorts.

Hof's testers in the physiology unit told him that a Radboud University professor named Peter Pickkers had a way to measure a person's immune response. So, Hof hoofed it to Pickkers's office, shook his hand, and said, "I can modulate my immune system. I heard you can measure it. Will you measure mine?"

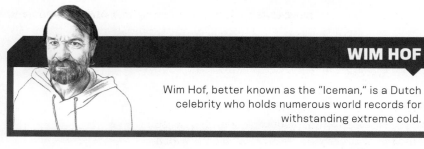

**WIM HOF**

Wim Hof, better known as the "Iceman," is a Dutch celebrity who holds numerous world records for withstanding extreme cold.

# Hormonal Changes

Pickkers was skeptical of Hof's claim, which had a whiff of pseudoscience. But Hof was an interesting character, so Pickkers went online and watched videos of his feats. "There were remarkable things I did not know of—things that, if you had asked me beforehand, I would have said, 'That's not possible. It's not possible to run half a marathon barefoot in the snow,'" says Pickkers. "But he did that."

Pickkers raised the idea of testing Hof to Kox, who was one of Pickkers's PhD students at the time, studying how the brain and immune system interact. Pickkers and Kox discussed the possibility at length, and they decided to give Hof a chance to document his claim. But they would do it while adhering strictly to the principles of the scientific process. "You can imagine some people wondered what we were doing with this guy," says Kox. "So we really focused on doing this in a very sound, precise manner, with no doubt about the scientific integrity of the project."

Hof claimed that the regimen for consciously controlling his immune system required three components: cold exposure, meditation, and breathing exercises. So, the team tested Hof's blood before and after an 80-minute full-body ice bath while Hof performed breathing and meditation exercises. Each time the scientists took blood, they went back to the lab and exposed the blood cells to molecules of endotoxin, a substance found in the cell walls of bacteria that activates an immune response in the human body. They wanted to see how Hof's immune cells in the blood would react to the endotoxin. After the regimen of ice, breathing, and meditation, Hof's cells had a far more subdued immune system response, showing very low levels of proteins associated with activation of the immune system, compared to similar cells before the regimen. The cause of that subdued immune response was unclear, but the researchers suspected that stress hormones played a role.

**Hormones** are signaling molecules produced by certain cells that tell other cells what to do under specific situations or at certain times in the life cycle of the individual. Hormones are produced by specialized secretory cells of the **endocrine system** (**Figure 25.3**).

Secretory cells are often organized into discrete organs called **endocrine glands**. Major

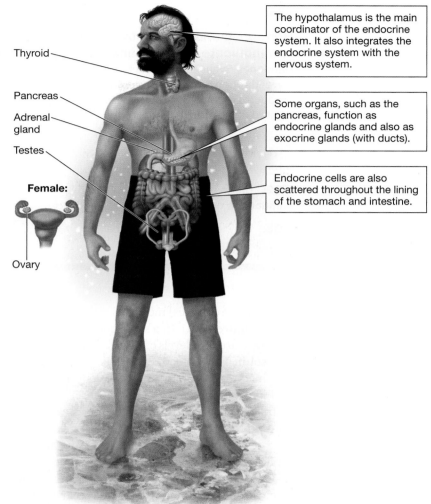

Thyroid

Pancreas

Adrenal gland

Testes

**Female:**

Ovary

The hypothalamus is the main coordinator of the endocrine system. It also integrates the endocrine system with the nervous system.

Some organs, such as the pancreas, function as endocrine glands and also as exocrine glands (with ducts).

Endocrine cells are also scattered throughout the lining of the stomach and intestine.

Figure 25.3

## The endocrine system is composed of hormone-secreting cells

The endocrine system consists of cells organized into ductless glands, plus scattered endocrine cells embedded in other tissues or organs. These cells all release hormones directly into the circulatory system.

**Q1:** What organ coordinates the endocrine system?

**Q2:** How does an endocrine gland differ from an exocrine gland?

**Q3:** How do male and female endocrine systems differ?

endocrine glands are located throughout the human body. Unlike exocrine glands, such as tear ducts, endocrine glands do not have ducts or tubes that deliver secretions from the gland to the site of action. Instead, endocrine glands release hormones into body fluids such as blood, which

Figure 25.4

### Hormones enable cells to communicate with one another

Hormones released by endocrine cells travel through the circulatory system to produce a response in target cells often located at a distance in the body.

Q1: How do hormones travel to target cells?

Q2: Distinguish between an endocrine cell and a target cell.

Q3: Why is a hormone called a signaling molecule?

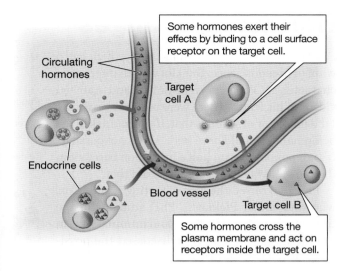

Some hormones exert their effects by binding to a cell surface receptor on the target cell.

Circulating hormones

Target cell A

Endocrine cells

Blood vessel

Target cell B

Some hormones cross the plasma membrane and act on receptors inside the target cell.

carries these chemical messengers throughout the body (**Figure 25.4**). In a subsequent experiment, Kox measured the levels of a stress hormone called cortisol in Hof's blood. After the ice, breathing, and meditation regimen, Hof's blood contained far higher levels of cortisol than before.

In the human body, most hormones can travel only as fast as the blood moves, which means they take several seconds or more to arrive at their target cells. Hormones coordinate functions that take place over timescales ranging from seconds (such as quickly increasing one's heartbeat in reaction to fear) to months (such as preparing a uterus to contract during the birthing process; see Figure 22.11).

Typically, hormones become greatly diluted after they are released into the circulatory system. They must therefore be able to exercise their effects at very low concentrations, as vitamins do. Hormones are effective in small amounts because they bind to their targets with great specificity. Cortisol, for example, binds to

a very specific receptor present on the surface of almost every cell in the body.

Cortisol, adrenaline (also called epinephrine), and noradrenaline (norepinephrine) are three hormones produced by the **adrenal glands**, a pair of endocrine glands that sit atop the kidneys. The release of these hormones launches a number of rapid physiological responses, including boosting blood glucose levels to increase energy levels in times of stress.

When a single hormone molecule binds to its receptor, it sets in motion a chain of events that may ultimately activate thousands of protein molecules in the target cell (**Figure 25.5**). When cortisol binds to its receptor, it initiates a pathway that results in the regulation of genes involved in development, metabolism, and immune response. This signal amplification—from a single hormone molecule to the activation of many proteins and genes—means that just a few hormone molecules can have a substantial impact on target cells. Through its effects on many cells, a hormone can exert a profound influence on the body as a whole.

Some hormone-secreting cells are not organized into distinct glands like the adrenal glands, but are instead embedded as single cells or clusters of cells within other specialized tissues and organs. For example, the main role of the kidneys is to filter blood, yet some cells in the kidneys produce hormones that stimulate red blood cell production. Altogether, the endocrine glands and the endocrine cells embedded in other organs, such as the kidneys, make up the endocrine system.

### PETER PICKKERS AND MATTHIJS KOX

Peter Pickkers (left) is a professor of experimental intensive-care medicine who studies the innate immune system. Matthijs Kox (right) is a researcher in intensive-care medicine. Both work at Radboud University Medical Center in the Netherlands.

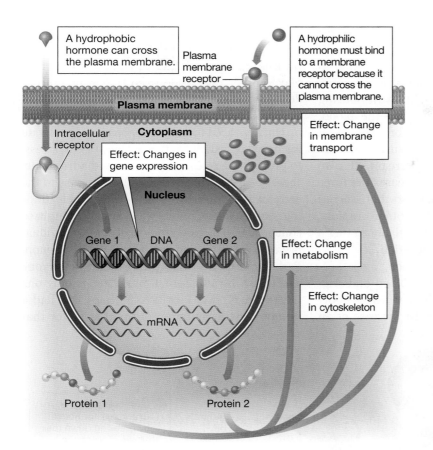

A hydrophobic hormone can cross the plasma membrane.

Plasma membrane receptor

A hydrophilic hormone must bind to a membrane receptor because it cannot cross the plasma membrane.

Plasma membrane

Intracellular receptor

Cytoplasm

Effect: Changes in gene expression

Effect: Change in membrane transport

Nucleus

Gene 1 DNA Gene 2

Effect: Change in metabolism

mRNA

Effect: Change in cytoskeleton

Protein 1 Protein 2

Figure 25.5

## Hormonal signals are amplified within the cell

Hormones are effective at low concentrations because of their specificity and because tiny amounts of a hormone can generate a large internal signal within a target cell.

**Q1:** Describe the two ways that a hormone outside a cell can exert its effect on a cell.

**Q2:** Within the cell, how does a hormone bring about a change in cell activity?

**Q3:** It takes very little of the hormone cortisol to have large effects throughout the body. Explain why.

# Brain-Body Connection

Testing Hof's cells alone wasn't enough for Kox and Pickkers. They wanted to measure his entire body's immune response, and wanted to know if the breathing and meditation techniques made any difference. After the ice bath, they asked Hof to perform his meditation and breathing techniques while they injected him directly with the endotoxin. In previous experiments, healthy volunteers injected with endotoxin had experienced fever, headaches, and shivering, accompanied by high levels of signaling proteins, called **cytokines**, that immune system cells use to communicate when an invader is present.

At various times before and after the injection, Kox measured Hof's blood levels for hormones and cytokines. Kox then compared Hof's results to those of a control group of 112 healthy volunteers who had previously taken the same test. To the scientists' surprise, as soon as Hof began practicing his breathing techniques,

his cortisol and adrenaline levels skyrocketed. And unlike the other volunteers, Hof reported almost no flu-like symptoms. Topping it off, the concentration of cytokines in his blood—indicative of an immune response—was less than half that of the control group. Hof appeared to have suppressed his immune system—voluntarily.

How was that possible? Kox considered the possibilities. First, a tiny region at the base of the vertebrate brain, the **hypothalamus**, coordinates the endocrine system and integrates it with the nervous system (refer back to **Figure 25.3**). The hypothalamus contains neurons, which interact with the brain, and endocrine cells, which produce hormones. It is a literal brain-body connection.

One well-known part of that connection involves the adrenal glands. In response to stress messages from the brain, the adrenal glands release adrenaline into the blood. If a man sees a rattlesnake in front of him, for example, he is likely to jump back or at least freeze in place, his heart racing. This quick response is due to the connection between the nervous system and the adrenal glands (**Figure 25.6**): The nervous system processes visual information (*Snake!*) and transmits an alarm signal to the adrenal glands within a fraction of a second. The adrenal glands kick in right away, pouring adrenaline and noradrenaline into the blood. Adrenaline stimulates glycogen breakdown in liver and skeletal muscle cells, causing glucose to be released into the bloodstream. It also speeds up the heartbeat and increases the force with which the heart contracts, so that glucose is delivered throughout the body more rapidly. In this way, glucose becomes available to fuel a rapid response to a stressful situation.

Within just a few seconds, then, these hormones increase the pumping of blood and trigger the release of glucose, all of which support the next move: fight or flight. In the case of an encounter with a snake, that may mean either arming oneself with a stout stick or running away. It turns out, however, that adrenaline and cortisol play another role, aside from triggering glucose delivery: research has shown that the hormones also subdue the activity of immune system cells.

Hof appeared to be able to consciously activate his nervous system (in the absence of real stress) to prompt the release of cortisol and adrenaline, thereby suppressing his immune response to the endotoxin voluntarily. It was a "remarkable" finding, says Pickkers, but he wasn't ready to jump to conclusions. The case study of a single individual is weak scientific evidence for any phenomenon. Perhaps Hof was simply an outlier: "Everyone can play a little baseball, but there is only one Derek Jeter," says Pickkers. Perhaps Hof had a unique genetic mutation or another factor that

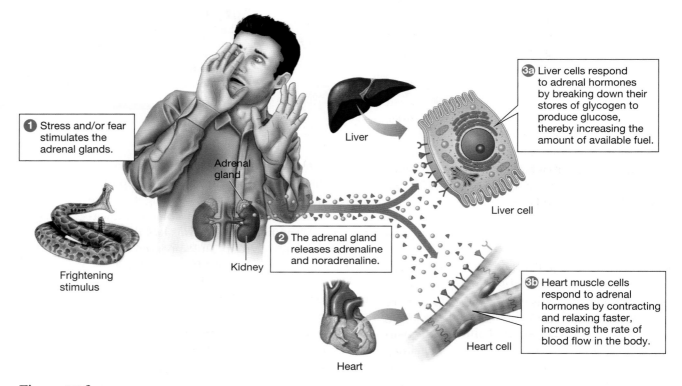

**1** Stress and/or fear stimulates the adrenal glands.

Adrenal gland

Kidney

Frightening stimulus

**2** The adrenal gland releases adrenaline and noradrenaline.

Liver

**3a** Liver cells respond to adrenal hormones by breaking down their stores of glycogen to produce glucose, thereby increasing the amount of available fuel.

Liver cell

**3b** Heart muscle cells respond to adrenal hormones by contracting and relaxing faster, increasing the rate of blood flow in the body.

Heart cell

Heart

Figure 25.6

**Adrenal hormones produce a rapid response to stress or fear**

The adrenal glands produce adrenaline (epinephrine) and noradrenaline (norepinephrine), which trigger the rapid release and delivery of stored energy.

**Q1:** Describe an event (other than the one illustrated in the figure) that might cause the release of adrenaline.

**Q2:** What organs does adrenaline affect?

**Q3:** What do you think would happen if your adrenal glands were constantly releasing adrenaline?

enabled him to control his autonomic nervous and immune systems.

But Hof claimed that he was not an outlier, that he could teach his technique to anyone. "I'm sure everybody is able to do this," Hof told Pickkers. Pickkers challenged him to demonstrate that he could do this. For scientific validation, Hof needed to teach his method to a group of healthy volunteers so that Pickkers could then compare that group's immune responses to those of an untrained control group of volunteers. In this controlled way, Hof might produce stronger scientific evidence for his claim.

# Innate Defenders

If Hof was right—if it really was possible to voluntarily control the immune system—the discovery would do more than change our understanding of the immune system; it would offer hope to people with autoimmune diseases, individuals in whom the immune system is overactive.

When healthy, the immune system protects animals from most infectious agents, called **pathogens**. Human pathogens include viruses, bacteria, and protists, as well as some fungi and multicellular animals such as parasitic worms. A well-known example is human immunodeficiency virus or HIV, the virus responsible for AIDS (see "What Makes HIV so Deadly?" on page 460).

Pathogens infect animals only if they can find a way into the body. An animal's first line of defense against pathogens is its **external defenses**, which reduce the likelihood that a harmful organism or virus will gain access to internal tissues. Linings that separate the "outside" from the "inside" of the body—the skin and the linings of the lungs, for example—act as a physical barrier to keep out most pathogens. Other external defenses include chemical agents (such as enzymes) and chemical environments (such as acidic conditions) that keep invaders from attaching to or growing on body surfaces (**Figure 25.7**).

Although external defenses do a good job of keeping out most pathogens, the body is still vulnerable. Mucous membranes, in the nasal cavity and other parts of the body, are a common point of entry for pathogens. Wounds, in the form of cuts, abrasions, and punctures, are common, and many pathogens will take advantage of breaks in the skin to gain entry to their hosts.

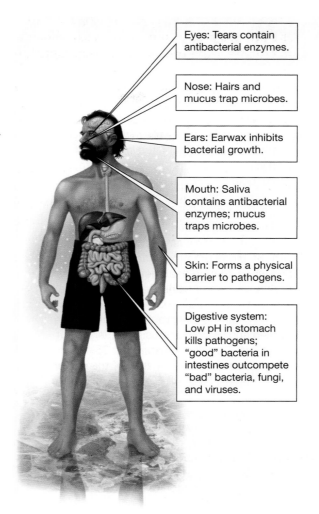

Eyes: Tears contain antibacterial enzymes.

Nose: Hairs and mucus trap microbes.

Ears: Earwax inhibits bacterial growth.

Mouth: Saliva contains antibacterial enzymes; mucus traps microbes.

Skin: Forms a physical barrier to pathogens.

Digestive system: Low pH in stomach kills pathogens; "good" bacteria in intestines outcompete "bad" bacteria, fungi, and viruses.

Figure 25.7

**The immune system's first line of defense is to prevent the entry of pathogens**
Our skin and the linings of our respiratory and digestive systems form physical and chemical barriers against pathogens.

**Q1:** What is the main physical barrier that animals use to keep out pathogens?

**Q2:** Give an example of a chemical defense within the digestive system.

**Q3:** Explain why rubbing your eyes and nose during flu and cold season is not recommended.

Once inside, pathogens confront a second line of defense: the cells and defensive proteins of the **innate immune system**. To mount an internal defense that kills, disables, or isolates invading

# What Makes HIV So Deadly?

In the early 1980s, doctors in the United States began to notice that gay men were dying of a variety of rare diseases, including a skin cancer called Kaposi's sarcoma, an unusual kind of pneumonia, and other infections that most people ordinarily shake off. By the mid-1980s it was clear that patients with the syndrome—named acquired immunodeficiency syndrome, or AIDS—had broken immune systems, the result of infection by a virus called the human immunodeficiency virus, or HIV.

In North America and Europe, the number of new cases rapidly increased, claiming the lives of tens of thousands of people each year. Initially, most new cases were limited to gay men, intravenous drug users, and people who had received blood transfusions. The common denominator was contact with the blood or body fluids of others: couples during sex, drug users when sharing used needles, and surgical and hemophilia patients who received blood transfusions contaminated with HIV.

In time, safe-sex education and clean-needle programs reduced the rates of infections among gay men and blood transfusion patients, but the virus spread to other populations. Globally, an estimated 35 million people have died of AIDS-related illnesses since the start of the epidemic. In just one year, 2015, 1.1 million people died of AIDS, and 36.7 million were living with HIV.

Inside the bloodstream, HIV enters immune system cells and reproduces inside them, eventually killing so many immune cells that the body's defenses are crippled. In the short term, remaining immune system cells track down HIV-infected cells and destroy them. Because the immune cells do such a good job of killing HIV-infected cells in the blood, most people with HIV have about a decade of normal health before they become ill, even without any treatment.

Over time, however, the HIV particles in the body evolve. As HIV evolves, the immune system cells no longer recognize and kill the virus. The population of HIV particles increases and begins destroying immune system cells faster than they can multiply. The body no longer has the immune system cells it needs to fight off infections by bacteria, yeasts, and other viruses. Once the immune system collapses, a person is vulnerable to almost any infection.

So far, there is no effective vaccine or cure for HIV. But a variety of new drugs enable people with AIDS to live years longer with fewer symptoms than they used to. "HIV cocktails," as the standard mixture of therapeutic drugs are called, prevent the viral genetic material from replicating and prevent the virus from merging with plasma membranes and entering cells. But the drugs can cost hundreds or thousands of dollars a month. Because treatment is so costly, only one in five AIDS patients in Africa and Asia receives effective treatment. For now, the best way to slow the spread of the disease remains safe-sex education, the free availability of condoms, and clean-needle programs.

pathogens, the body first must recognize that an invader is present. Although a person is not consciously aware of it, a healthy body can distinguish foreign invaders (nonself) from its own cells (self). If the internal defenders fail to tell self from nonself, they mistakenly attack the body's own cells, leading to autoimmune diseases such as rheumatoid arthritis (in which immune cells attack the lining of the membranes that surround joints) or type 1 diabetes (in which immune cells attack the pancreas, which makes insulin).

If there was a way, as Hof claims, to subdue the immune system at will, people with auto-immune diseases might have another avenue, aside from expensive drug therapies, by which to control their rebellious immune systems.

## Team Effort

Despite Hof's personal achievement, Pickkers and Kox didn't really think he would be able to teach others to voluntarily control their innate immune systems. "We thought it would be a negative result," says Pickkers. If nothing else, Hof had been performing his technique for 30 years, so even if he could teach it, Pickkers doubted he could teach a novice enough to influence his or her own immune system in just a few days. Hof disagreed, arguing that a short training regimen would be sufficient to impart the ability. Kox didn't think most of the volunteers would even make it through the training.

It was no easy study for the participants. Over 4 days, 18 healthy, young, male volunteers were taken into the mountains of Poland and exposed to the cold in various ways: standing in the snow barefoot for 30 minutes, lying in the snow bare-chested for 20 minutes, swimming in ice-cold water each day for several minutes, hiking up a snowy mountain in nothing but shorts and shoes. Hof also taught them his meditation and breathing techniques, including deep inhalations and exhalations.

Contrary to Kox's expectation, all 18 participants completed the training, and 12 of them were then randomly selected to come back to the lab for the final part of the experiment: exposure to the endotoxin, to test whether they could consciously regulate their immune response. Their results would then be compared to those of 12 healthy controls who had also been exposed to the endotoxin but had not received Hof's training. Now the scientists would finally find out whether Hof could teach someone to control the activity of the innate immune system.

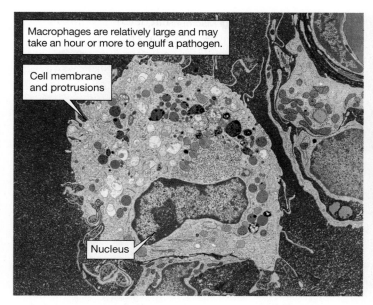

Macrophages are relatively large and may take an hour or more to engulf a pathogen.

Cell membrane and protrusions

Nucleus

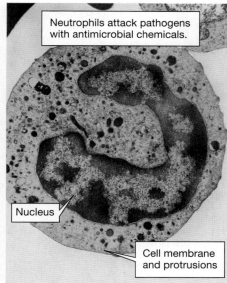

Neutrophils attack pathogens with antimicrobial chemicals.

Nucleus

Cell membrane and protrusions

Figure 25.8

## Phagocytes destroy pathogens by engulfing them

Phagocytes are a kind of white blood cell, a family of defense cells found in body fluids including blood, where they intercept invading pathogens. Two different kinds of phagocytes—a macrophage and a neutrophil—are seen in these colorized transmission electron micrographs (TEMs).

**Q1:** Place these terms in order from most to least inclusive: neutrophil, white blood cell, phagocyte, innate immune system.

**Q2:** Compare and contrast macrophages and neutrophils.

**Q3:** Why would it be a problem if your innate immune system identified the insulin-producing cells in your pancreas as "nonself"?

The innate immune system reacts to cells or molecules that do not belong in the body by activating defense cells and proteins to eliminate the unwelcome guests. A suite of pathogen-recognizing cells called **phagocytes**, a type of white blood cell, mark and destroy foreign invaders by engulfing and digesting them (**Figure 25.8**).

This immune response is said to be innate (inherent) because the necessary components are constantly at the ready for deployment against an invading pathogen. The innate response can be local, occurring at the point of entry, or global, involving the whole body. Like the external defense system, innate immunity is indiscriminate as to which foreign invaders it repels, so it is considered a **nonspecific response**. The innate immune system is an ancient defense mechanism found in both invertebrates and vertebrates.

In addition to defending against invaders, the innate immune system plays two other critical roles. First, it responds to tissue damage from a pathogen invasion or wound by mounting an immediate and coordinated sequence of events known as **inflammation** (**Figure 25.9**). Cytokines are a clear marker of inflammation, and thus a good way to measure the action of the immune system. A second role of the innate immune system is clotting blood to close a wound. Sealing an open wound reduces blood loss and restores the integrity of external defense barriers (**Figure 25.10**).

To test the innate immune response of the 24 study participants (the 12 trained volunteers and the 12 untrained controls), the scientists injected each participant with endotoxin and monitored them for 6 hours. Hof visited his trainees during the experiment, coaching them through his breathing techniques.

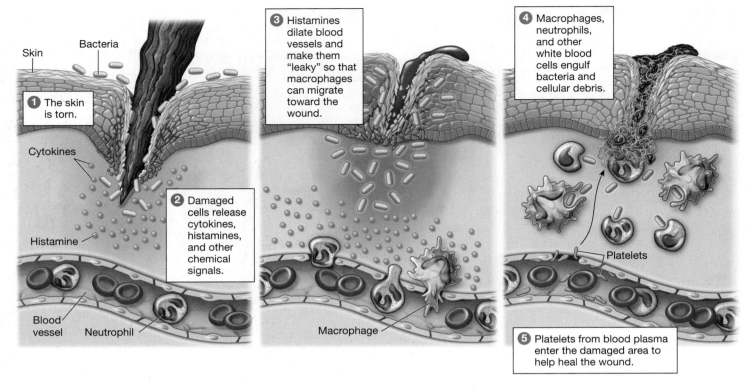

Skin  Bacteria

**1** The skin is torn.

Cytokines

Histamine

Blood vessel  Neutrophil

**2** Damaged cells release cytokines, histamines, and other chemical signals.

**3** Histamines dilate blood vessels and make them "leaky" so that macrophages can migrate toward the wound.

Macrophage

**4** Macrophages, neutrophils, and other white blood cells engulf bacteria and cellular debris.

Platelets

**5** Platelets from blood plasma enter the damaged area to help heal the wound.

## Figure 25.9

### The inflammatory response acts against invading pathogens

Inflammation occurs when the innate immune system swings into action after cellular damage is detected, cleaning up damaged tissues and preventing the spread of pathogens. Inflammation can occur anywhere inside the body. Here we see an inflammatory response following a puncture wound to the skin. ▶️

**Q1:** What is the role of white blood cells in inflammation?

**Q2:** What would happen if histamines were not produced during inflammation?

**Q3:** Why is inflammation called a "nonspecific" immune response?

## Figure 25.10

### Blood clots help prevent pathogens that may be present in a wound from spreading

Sticky cell fragments, or **platelets** (shown here in light blue), and clotting proteins (yellow) form a gel-like mesh that traps blood cells, creating a blood clot that seals broken skin. Clotting can begin as quickly as 15 seconds after tissue damage occurs. Growth of new tissue eventually repairs the wound more permanently.

**Q1:** Why is blood clotting an important immune response?

**Q2:** How are the inflammatory response and blood clotting similar?

**Q3:** Some people have a genetic disorder in which their blood cannot clot. Why would this be a problem?

The results were clear: after being injected, and while performing the breathing techniques, the trainees showed higher adrenaline levels than the controls—higher even than the adrenaline produced by a person's first bungee jump. "They produced more adrenaline just lying in bed than somebody standing in front of an abyss going to jump in fear for the first time," says Hof. "That means direct control of your hormone system, and your hormones have a direct relationship to the immune system."

In addition, the trainees had fewer flu-like symptoms and lower fevers, and their cytokines—the signaling proteins of the immune system and markers of inflammation—were at less than half the level of the control group. "We were very surprised," says Kox. "It was impressive that these guys could do all this cold exposure training, but I still thought the chances were slim they'd be able to modulate their immune systems. But the results were so convincing."

While Hof's trainees' results may point the way to identifying new, less expensive treatments for autoimmune diseases, don't try this at home! We have an innate immune system for a reason—to protect against pathogens—so voluntarily subduing it without the guidance of a doctor is not a good idea.

# Adapting to the Enemy

Kox and Pickkers's study did not address the human immune system's third line of defense. In contrast to the broad responses of external defenses and the innate immune system, the more complex **adaptive immune system** is tailored against specific invaders. There is no evidence, yet, that Wim Hof is able to control his adaptive immune response.

Adaptive immunity goes beyond simply recognizing something as nonself. Instead, specialized defense cells are trained to recognize only one strain of pathogen and to activate a **specific response**. The adaptive immune system is based in the lymphatic system, which is itself an important part of both the immune system and the circulatory system. It relies on two main weapon systems: antibody-mediated immunity and cell-mediated immunity (**Figure 25.11**).

**Antibody-mediated immunity** uses powerful Y-shaped proteins called **antibodies** to recognize and attack invaders. Antibodies recognize nonself markers—**antigens**—on the pathogen and mark the pathogen for destruction. **B cells**, specialized lymphocytes created and matured in the bone marrow, produce thousands of antibodies per second aimed specifically at the pathogen that has been recognized.

**Cell-mediated immunity** recognizes cells that have been infected by a pathogen such as a virus, as well as cancer cells (for more on cancer, see "Cancer: Uncontrolled Cell Division" on page 101 of Chapter 6). **T cells**, lymphocytes created in the bone marrow and matured in the thymus, recognize markers on the surface of a cell that has been infected. The T cells then kill the infected cells so that thay cannot spread the disease to other cells.

Compared to innate immunity, adaptive immunity is slow to mobilize. However, it is the most sophisticated and effective of animal defense systems because of the amazing selectivity with which the adaptive immune system attacks a particular invader.

Adaptive immunity occurs in two stages. The very first time a person is exposed to a particular pathogen, the **primary immune response** is activated. This response takes time—more than 2 weeks sometimes—to reach full steam. Because of that slow start, and because pathogens multiply so rapidly, people infected with an aggressive pathogen for the first time can sometimes lose the race, becoming ill and dying. Therefore, any pathogen that is new to humans is particularly dangerous, and the nonspecific response of innate immunity may be more effective in combating it.

A distinctive feature of the adaptive immune system is **immune memory**, the capacity of this defense system to remember a first encounter with a specific pathogen and to mobilize a speedy and targeted response to future infection by the same strain. This "memory" is what enables us to become immune to attacks by the same strain after we suffer the disease a first time. Once you've had measles, for example, you never get sick from the measles virus again, because the adaptive immune system recognizes the virus and quickly eradicates it the next time. Keep in mind that this means you are not born with immune memory. Each individual must

## Lymphatic ducts and associated organs

Adenoids

Tonsils

Thymus

Lymph nodes

Lymphocytes move around the body through the lymphatic and circulatory systems.

They develop in the thymus…

Spleen

Lymphatic ducts

Appendix

…and in the marrow of the bones.

## Origin and maturation of lymphocytes

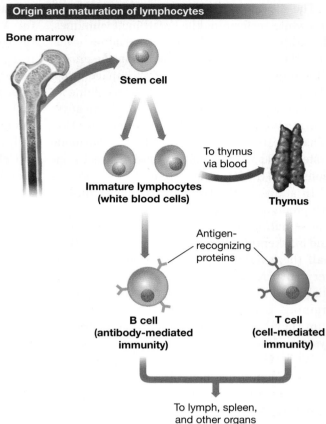

Bone marrow

Stem cell

Immature lymphocytes (white blood cells)

To thymus via blood

Thymus

Antigen-recognizing proteins

B cell (antibody-mediated immunity)

T cell (cell-mediated immunity)

To lymph, spleen, and other organs

## Figure 25.11

### Adaptive immunity resides in the lymphatic system

(Left) The **lymphatic system** consists of lymphatic ducts, lymph nodes, and associated organs. (Right) **Lymphocytes** originate from stem cells in bone marrow. B cells mature in the bone marrow; T cells mature in the thymus. Lymphocytes circulate in the lymphatic and circulatory systems and accumulate in lymph nodes and other organs, such as the spleen, appendix, and tonsils.

**Q1:** Why are B and T cells so named?

**Q2:** In what way is this immune system "adaptive"?

**Q3:** Why is the adaptive immune response considered the third layer of the immune system?

build one over time by being exposed to various pathogens.

Because of immune memory, the adaptive immune system produces a faster, more dramatic response to pathogens when it encounters them a second time. The second encounter, when the adaptive immune system is poised and ready to respond, is the **secondary immune response**.

We acquire immunity in two ways: either actively or passively. **Active immunity** to a particular pathogen develops when our own bodies produce antibodies against that pathogen; they are not received from an outside source.

This happens naturally when we're exposed to certain pathogens, such as the measles virus. We can also acquire active immunity to certain diseases through vaccination, which introduces antigens to the body in a harmless form.

**Passive immunity** develops when antibodies that were not made by our own bodies are received. A human fetus acquires antibodies from exchanges between its blood and its mother's blood. This antibody sharing continues after birth: mother's milk is rich in antibodies because the mother's immune system has encountered many antigens and made many antibodies in her lifetime. Thanks

# Driven by Hormones

Hormones regulate not only the classical fight-or-flight response, but also the sleep and wake cycles of all people. These signaling molecules in the blood also become elevated or depressed during periods of stress (cramming for a test, anyone?) and exercise. What are your levels of melatonin, adrenaline, and cortisol right now?

Assessment available in smartwork5

## Hormones in the human body

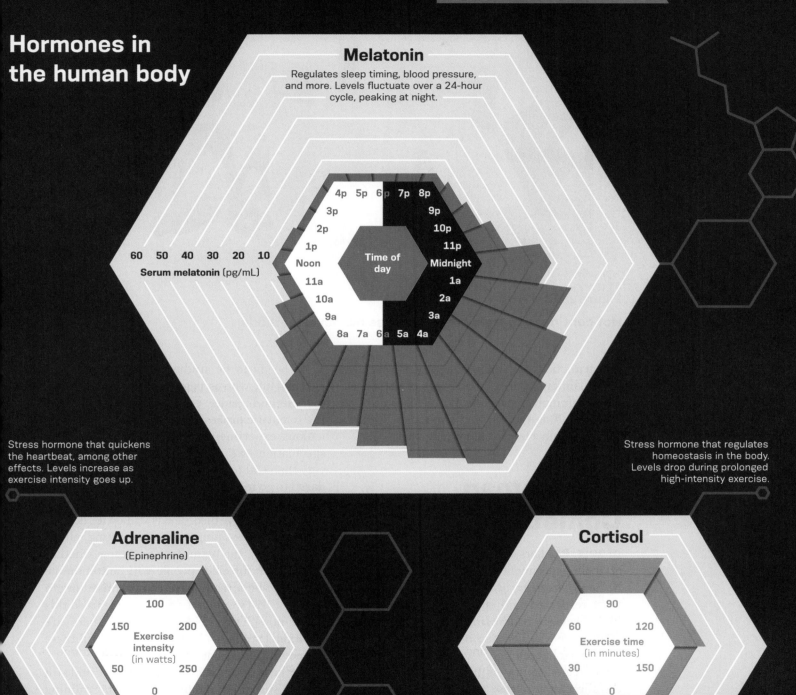

### Melatonin
Regulates sleep timing, blood pressure, and more. Levels fluctuate over a 24-hour cycle, peaking at night.

**Serum melatonin (pg/mL)**
60 50 40 30 20 10

Time of day

4p 5p 6p 7p 8p
3p 9p
2p 10p
1p 11p
Noon Midnight
11a 1a
10a 2a
9a 3a
8a 7a 6a 5a 4a

Stress hormone that quickens the heartbeat, among other effects. Levels increase as exercise intensity goes up.

Stress hormone that regulates homeostasis in the body. Levels drop during prolonged high-intensity exercise.

### Adrenaline
(Epinephrine)

**Exercise intensity (in watts)**
100
150   200
50    250
0
200
400
600
800
1000

**Plasma epinephrine (pg/mL)**

### Cortisol

**Exercise time (in minutes)**
90
60    120
30    150
0
4
8
12

**Cortisol (µg/dL)**

Figure 25.12

**Hof continues to train recruits to use his novel methods**

to that antibody-rich milk, a nursing baby receives passive immunity to a broad range of pathogens. Passive immunity produces no memory cells, so it wears off as the received antibodies degrade, usually within a few weeks or months.

# The Iceman Cometh

In 2014, Pickkers and Kox published the results of their experiment in the *Proceedings of the National Academy of Sciences*, one of the world's most respected and cited peer-reviewed journals. They are conducting follow-up studies to determine whether one or more of the three parts of Hof's technique—cold exposure, breathing, and meditation—is primarily responsible for the adrenaline release and subsequent immune suppression, and exactly how these effects come about. Kox suspects that the breathing techniques are the main factor, since Hof's breathing appears to trigger the release of hormones, but he cannot yet be sure.

"It needs to be studied a whole lot more," agrees Hof, who is eager to continue putting his method under the magnifying glass of the scientific process (**Figure 25.12**). "By meticulous experiments and measurements—not speculation—we want to show this works. I'm very thankful to the professors at Radboud who dared to go into this."

Whether Hof's technique can help individuals with autoimmune disorders is still up for debate. Although the training worked for young, physically fit men, the team does not know whether it will work for older people with autoimmune diseases who already have compromised organ systems. "We would not advise people to do this [right now]," says Pickkers, before there are additional studies. "We have to be careful there are no unwanted side effects or risks."

But he is optimistic about the future and still sounds surprised about how well the training worked. "We confirmed that, indeed, using the techniques of Wim Hof, humans are able to modulate their autonomic nervous system and influence their immune response," says Pickkers. "It is remarkable."

more times, creating immune memory.

- **Active immunity** can be acquired through natural exposure to a pathogen or through a vaccine. **Passive immunity** comes from receiving antibodies that were not made by our own bodies, such as when a fetus acquires antibodies from its mother.

# REVIEWING THE SCIENCE

- A **hormone** is a signaling molecule distributed through the body by the circulatory system. Because hormones move only as quickly as the blood moves, they tend to coordinate functions that are slower and longer-lasting than those under the influence of the nervous system.

- A single hormone may affect many different kinds of target cells, potentially triggering a different response in each. Hormones act on target cells either by moving through the plasma membrane to the cell's interior or by acting on receptors embedded in the plasma membrane.

- The **endocrine system** is made up of the glands and specialized cells that produce hormones. The **hypothalamus**, located in the forebrain, coordinates the endocrine system and integrates it with the nervous system. The **adrenal glands**, located in the kidneys, produce hormones responsible for the fight-or-flight response.

- The vertebrate **immune system** possesses three layers of defenses against **pathogens**.

  - The first layer consists of **external defenses**: physical and chemical barriers, including the skin and the linings of the respiratory and digestive systems.

- The second line of defense is the **innate immune system**. Several types of blood cells and molecules produce the **nonspecific responses** of the innate immune system, including **phagocytes** such as macrophages and neutrophils, which engulf and destroy pathogens. Tissue damage stimulates **inflammation** and blood clotting.

- The third line of defense is the **adaptive immune system**, providing long-term defenses in the form of **specific responses** to pathogens and parasites. These responses are mediated by powerful proteins called **antibodies**, or by cells.

- The **lymphatic system** provides the primary sites for adaptive immunity. White blood cells called **lymphocytes** confer specific immunity. Immature lymphocytes differentiate into **B cells** in the bone marrow and **T cells** in the thymus. Each lymphocyte has special membrane proteins that bind to only a specific **antigen** of a specific pathogen.

- The **primary immune response** from the adaptive immune system is relatively slow and mild. The **secondary immune response** is a faster, stronger response to a pathogen that has been encountered one or

# THE QUESTIONS

## The Basics

**1** Hormones are

(a) secretory cells.

(b) endocrine glands.

(c) signaling molecules.

(d) target cells.

**2** Which of the following is *not* true of hormones?

(a) They are distributed through body fluids.

(b) They must be present in large amounts to be effective.

(c) They are produced by specialized cells.

(d) They act on target cells.

**3** Adrenaline

(a) is produced in the adrenal glands.

(b) increases the amount of glucose in the bloodstream.

(c) suppresses immune system activity.

(d) all of the above

**4** The _____ is the immune system's second line of defense against pathogens.

(a) innate immune system

(b) adaptive immune system

(c) combination of physical and chemical barriers to pathogen entry

(d) all of the above

**5** Which of the following is/are *not* a part of the innate immune system?

(a) phagocytes

(b) antibodies

(c) inflammation

(d) clotting

**6** Link each term with the correct definition.

| | |
|---|---|
| **ADAPTIVE IMMUNE RESPONSE** | 1. The glands and specialized cells that produce hormones. |
| **ENDOCRINE SYSTEM** | 2. The blood cells and molecules that provide a nonspecific response to pathogens. |
| **INNATE IMMUNE RESPONSE** | 3. The region of the brain that coordinates the endocrine system and integrates it with the nervous system. |
| **HYPOTHALAMUS** | 4. Long-term defense against pathogens centered in the lymphatic system. |

## Challenge Yourself

**7** Identify whether each of the following is characteristic of either antibody-mediated (A) or cell-mediated (C) immunity.

___ a. The immune response relies on Y-shaped proteins to identify pathogens.

___ b. B cells produce proteins specific to a pathogen.

___ c. Lymphocytes that matured in the thymus identify infected cells.

___ d. Antigens on the pathogen allow it to be identified as nonself.

___ e. Infected cells are destroyed so that an infection cannot spread to other cells.

**8** Select the correct terms:

The first time you are exposed to a pathogen, the (**primary / secondary**) immune response is activated. The (**primary / secondary**) immune response to a pathogen is stronger and more rapid. You acquire (**active / passive**) immunity to a pathogen when your own body creates the antibodies against that pathogen. (**Active / Passive**) immunity comes from the antibodies produced by another person, such as your mother when you were in utero or nursing. The immunity conferred by vaccines is an example of (**active / passive**) immunity.

**9** Beginning with a perceived threat (for example, a spider), identify the correct order of events in the stress response by numbering them from 1 to 5.

___ a. Target cells amplify the hormonal signal to produce a response.

___ b. The liver breaks down glycogen to glucose, and the heart increases its rate and the force of its contractions.

___ c. Adrenaline reaches target cells in the liver and heart.

___ d. The hypothalamus signals the adrenal glands that a threat is present.

___ e. The adrenal glands release adrenaline into the bloodstream.

## Try Something New

**10** Wim Hof and his trainees had increased levels of the stress hormone adrenaline and decreased immune function during the experiments described in this chapter. How might these changes negatively affect the endocrine and immune systems over the long term?

**11** Which of the following is *not* true of a B cell?

(a) It is a kind of lymphocyte.

(b) It is produced in the bone marrow.

(c) It matures in the thymus.

(d) It is part of the adaptive immune response.

(e) All of the above are true of B cells.

**12** Increased body temperature (fever) is part of the body's innate immune response. Fever is uncomfortable and can be dangerous if very high. It is often treated with over-the-counter medicines like acetaminophen, ibuprofen, naproxen, or aspirin. What are possible negative effects of this treatment?

## Leveling Up

**13** **Life choices** While clotting is an important component of the innate immune response, it can also be dangerous. For example, a blood clot could block an artery to the heart or brain, leading to a heart attack or stroke. Aspirin reduces blood clotting by interfering with the body's production of a lipid called thromboxane A2. This lipid normally helps platelets clump together (see Figure 25.10), so aspirin, by inhibiting its production, reduces clotting and "thins the blood." Some doctors may prescribe a daily dose of aspirin for patients who are at risk of heart attack or stroke. Review the U.S. Preventive Services Task Force 2016 recommendations on daily aspirin therapy (https://www.uspreventiveservicestaskforce .org/Page/Document/RecommendationStatementFinal/aspirin-to -prevent-cardiovascular-disease-and-cancer), and then answer the following questions.

a. Do you fall into one of the categories for which daily aspirin therapy is recommended? If yes, which one? If no, is there an aspirin therapy category that you think you'll be in eventually?

b. How strong is the evidence supporting aspirin therapy in the category you identified in the previous question, if any? (See the "Grade" column in the task force recommendations.)

c. With this information in hand, do you plan to take aspirin daily at some point in your life?

d. Will you speak with your doctor before taking aspirin daily? Why or why not?

e. Do you know anyone who has had a heart attack or stroke? Do they take aspirin daily?

**14 Doing science** Pickkers and Kox say that their next experiment will attempt to determine whether one or more of Hof's techniques—cold exposure, meditation, and breathing exercises—is primarily responsible for the increased adrenaline release and decreased immune response. Kox predicts that the breathing exercises will prove to be most important. (You can view a video of cold-exposure training at https://www.youtube.com/watch?v=ziXm9oWJm6A.) Imagine that it is your responsibility to design the next experiment for Pickkers and Kox. Include answers to the following questions in the description of your experimental design.

a. What are your experimental hypotheses?

b. Give at least one prediction for each hypothesis.

c. Identify your control group and treatment group(s). How many subjects will be in each group? Justify your sample size.

d. Give a detailed description of the treatment for each group.

**For more, visit digital.wwnorton.com/bionow2 for access to:**

# Amber Waves of Grain

*We've been growing the same domesticated crops for thousands of years. To survive the future, we're going to need new ones.*

**After reading this chapter you should be able to:**

- Identify the structure and explain the function of plant tissues, organs, and organ systems.
- Give an example of how plants use chemical means to survive, grow, and reproduce.
- Compare and contrast how plants and animals grow.
- Explain how a plant obtains a given nutrient.
- Diagram the alternation of generations in a plant life cycle.
- Explain the variety of ways in which plants are pollinated, and how this relates to the structure of flowers.
- Describe how plants disperse their seeds and how fruits have evolved to assist this effort.

CHAPTER

26

PLANT
PHYSIOLOGY

Lee DeHaan wanders through a field of grain, golden knee-high stalks brushing against his faded jeans. Tall, with an angular face and goatee, DeHaan looks the part of a farmer, with a floppy hat and tanned arms to boot. But though he grew up on a corn farm in Minnesota, DeHaan is now an *agronomist*, a researcher studying the science of producing and using plants for food, energy, land preservation, and more. In these fields around his office at the Land Institute in Salina, Kansas, DeHaan is domesticating a new crop with the potential to transform agriculture—a grain bred from humble prairie grass (**Figure 26.1**).

The way we plant and harvest crops is unsustainable. Nearly 70 percent of the freshwater used by humans is put toward irrigation. Farmland is losing productivity because of deforestation, overgrazing, and poor agricultural practices that lead to erosion and pollution. Staple crop production worldwide has leveled off. All this does not bode well when our population is projected to grow by another 3 billion people over the next half century.

We need new crops. The common crops grown in the United States today—corn, wheat, and soybeans—were domesticated by our ancestors thousands of years ago to produce high yields and be easily harvested and replanted. Yet these crops are failing to meet our sustainability needs. They require large amounts of water, fertilizer, and pesticides, and they are vulnerable to weather changes, pests, and diseases. They are inefficient and delicate; we need food crops that are hardy and resilient.

Unfortunately, humans stopped domesticating new crop plants long ago. So, despite Earth's rich diversity of plants—over 300,000 species, including more than 50,000 species of edible plants—we rely on just 30 of them to provide 95 percent of our food (see "Food Banks" on page 296 in Chapter 16).

More than 250,000 of Earth's plant species are *flowering plants* (also called *angiosperms*). Worldwide, people get over 80 percent of their calories from flowering plants such as grasses (wheat, rice, corn), legumes (peas, beans, peanuts), potatoes, and sweet potatoes. But as climate change affects agriculture—increasing temperatures and severe weather, and causing more disease—and as the global population grows, we're going to need tough, plentiful crops. That's no easy requirement to meet.

# Perfecting Plants

In 2001, as a young and ambitious plant breeder, DeHaan joined a research team at the Land Institute, a 600-acre research center devoted to developing sustainable alternatives in agriculture. At the time, he was the young scientist on the team, so his bosses handed him an ambitious long-term project: to create a new type of grain by domesticating a wild grass that grows year after year.

Plants can be grouped into three categories on the basis of their life cycle: annuals, biennials, and perennials.

- **Annuals** complete their entire life cycle in one year. In flowering plants, an annual has one year to grow from a seed into a mature plant, produce flowers, and make the seeds that will start the next generation. Annual crops must be replanted every year.

- **Biennials** grow and mature for a year but do not initially reproduce. Reproduction takes place in the second year of growth. After the second year, biennial crops must be replanted.

Figure 26.1

**Harvest at the Land Institute**
Researchers gather ripe stalks of a prairie grass that holds hope for feeding our growing population.

- **Perennials** live three years or more, and sometimes for hundreds or even thousands of years. Perennials, which once made up much of the natural grasslands that dominated Earth, are alive year-round and are efficient at nutrient cycling and water management.

Wheat, corn, rice, soybeans—these are all annuals. Our ancestors saw the advantages of breeding annuals: compared with perennials, annuals produce more seed, and replanting them every season speeds the process of domestication. But perennials have advantages too, and it was those advantages that DeHaan wanted to tap into by breeding a perennial as a food crop. First, perennials do not waste energy by regrowing **roots** each year, as annuals do. Instead, they grow long, deep roots that anchor in the soil and remain there year after year. These roots enable perennials to absorb water and nutrients more efficiently than annuals do. The roots also outcompete weeds, so less weed killer is needed to grow perennials. Deep roots also hold carbon in the soil, acting as a *carbon sink*.

Flowering plants can also be classified as either monocots or dicots (**Figure 26.2**).

## LEE DEHAAN

Lee DeHaan is an agronomist at the Land Institute in Salina, Kansas, where he leads a new crop domestication program.

Grain-producing plants such as wheat, rice, and corn are **monocots**, with fibrous, branching roots. Other plants, including soybeans and oak trees, are **dicots**, which grow straight, thick taproots.

Roots are part of the plant body, which is relatively simple in its organization, compared to the bodies of vertebrate animals (**Figure 26.3**). Plant bodies are made of three basic tissue types: dermal, ground, and vascular.

- **Dermal tissues**, which form the outermost layer of the plant, protect the plant from the outside environment and control the flow of materials into and out of the plant.

- **Ground tissues**, which form the intermediate layer, make up the bulk of the plant body and perform a wide range of functions, including physical support, wound repair, and photosynthesis.

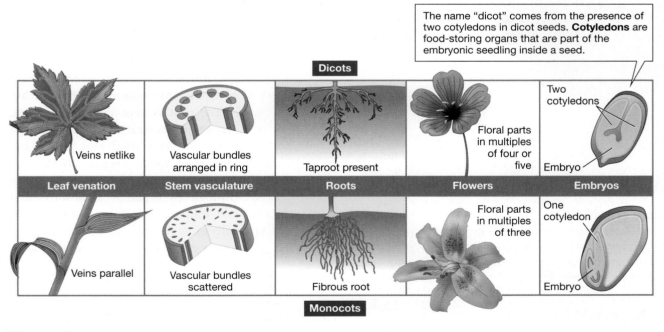

### Figure 26.2

**Monocots and dicots**

Flowering plants have traditionally been classified into two main groups—*dicots* and *monocots*—on the basis of their external form and internal structure. The dicots are the larger of these two informal categories and encompass about 175,000 species, including beans, squash, oak trees, and roses. Monocots include all the grasses, members of the lily family, palm trees, and banana plants.

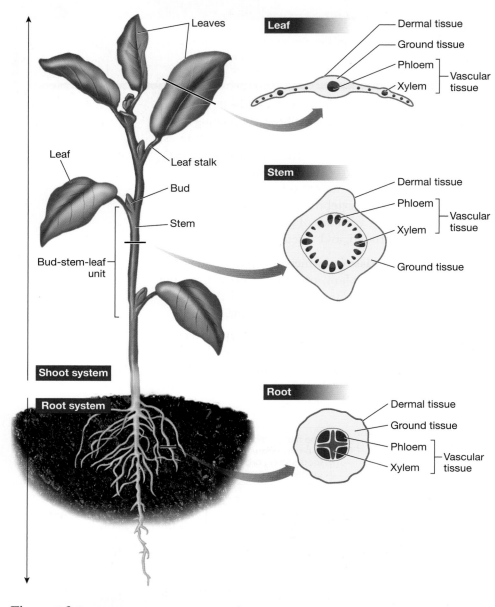

Figure 26.3

## How plants are organized

Plants have three basic tissue types (dermal, ground, and vascular) which make up three types of organs (roots, stems, and leaves). Belowground, plants grow by extending old roots and producing new lateral roots. Aboveground, plants grow by adding new bud-stem-leaf units. ▶

**Q1:** What is the function of the vascular tissue?

**Q2:** A plant organ is green if the cells within it contain chloroplasts. In the figure, which plant organ does not contain chloroplasts, and why do you think that is?

**Q3:** Which tissue type has chloroplast-containing cells? Why?

- **Vascular tissues**, which are found in or near the center of the plant body, contain stacks of long cells forming continuous tubes that run throughout the plant body, linking all organs of the root and shoot systems. The vascular tissue known as **phloem** ships sugars from the leaves, where they are produced, to living cells in every part of the plant; **xylem**, in contrast, transports water and minerals, absorbed from the soil, upward from the roots and outward to the leaves.

DeHaan's attempt to domesticate intermediate wheatgrass (*Thinopyrum intermedium*)—a wild, flowering perennial plant—started as a side project. Widely used for hay and pasture, intermediate wheatgrass grows wild across the western United States and Canada. It has tall, thin shoots that grow green in the fall and turn golden brown in the spring and summer, and long, deep roots that stretch belowground.

The body of a flowering plant, whether intermediate wheatgrass, corn, or roses, can be divided into those two basic organ systems: the belowground *root system* and the aboveground *shoot system*. These two organ systems are specialized for life in two very different environments: roots in soil, shoots in air.

The **root system** anchors the plant, absorbs water and nutrients from the soil, transports food and water, and may store food (**Figure 26.4**). *Root hairs* greatly increase the surface area through which plants can absorb water and mineral nutrients. Annuals typically grow short roots. Perennials grow longer roots, which last year after year, depending on the species.

Roots are one of plants' three basic organs. The other two are stems and leaves, which form the **shoot system** (**Figure 26.5**). **Stems** provide the plant with structural support, transport food and water, and hold leaves up to intercept light. Although cells in the stems of many plants can perform photosynthesis, most sugars are produced by

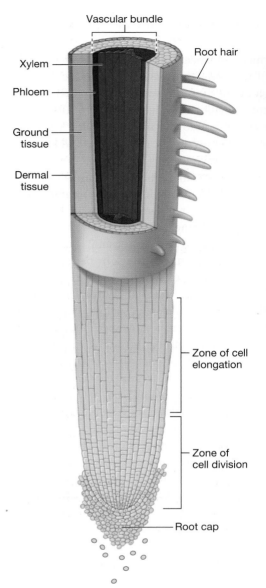

Figure 26.4

## The root system

Roots produce numerous outgrowths from dermal cells called *root hairs*, which aid in water and nutrient absorption. Plant roots have a region of active cell division, protected by the root cap, and a region of cell elongation, in which cells increase in size and complete their development.

**Q1:** How do root hairs increase the amount of water and nutrients that a plant can absorb?

**Q2:** How do roots make a plant more stable?

**Q3:** Given question 2, which do you predict would be more stable in harsh weather conditions: annuals or perennials?

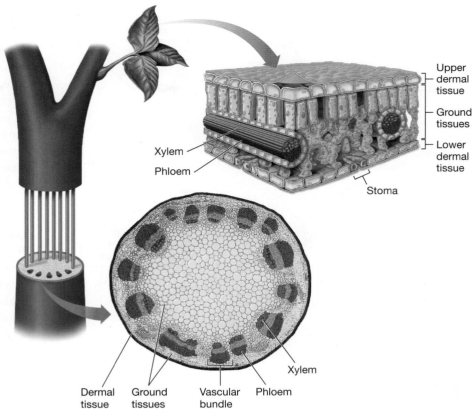

Figure 26.5

## The shoot system

Leaves produce the majority of the plant's food through photosynthesis. Stems provide structural support and may perform a limited amount of photosynthesis.

**Q1:** Which part of the shoot system—the bud, stem, or leaf—produces flowers?

**Q2:** Find the stoma in the figure. How is its location on the leaf important for its function?

**Q3:** What nutrients does phloem move from the leaves to other parts of the plant?

photosynthesis in the **leaves**. Leaves provide a broad, sunlight-capturing surface to maximize energy capture. Wheat and intermediate wheatgrass have long, pointy leaves that grow from the stems. At the tip of each shoot, and the base of many leaves, is a bud. Under the right conditions, buds produce new shoots or flowers.

**Flowers** house the structures that produce male and female gametes and, in many species, also facilitate the delivery of the sperm-bearing pollen to the female reproductive organs

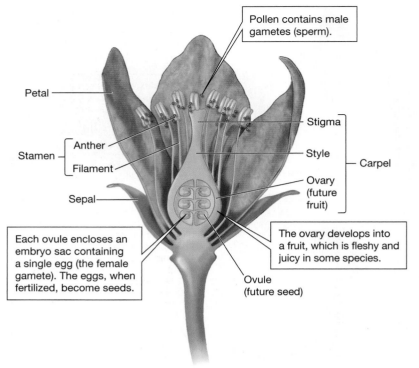

Pollen contains male gametes (sperm).

Petal

Stamen — Anther / Filament

Sepal

Stigma

Style

Carpel

Ovary (future fruit)

Each ovule encloses an embryo sac containing a single egg (the female gamete). The eggs, when fertilized, become seeds.

The ovary develops into a fruit, which is fleshy and juicy in some species.

Ovule (future seed)

Figure 26.6

**Four whorls make a flower**

The various parts of a flower are arranged in concentric rings, or whorls. From the outermost whorl inward, a typical flower consists of four whorls: sepals, petals, stamens, and carpels. (All of the petals in a flower are collectively known as the *corolla*. The collective term for all of the sepals is the *calyx*.)

(**Figure 26.6**). The ovary of the flower, following fertilization, develops into fruit, which helps disperse seeds in a highly effective ways.

The outer layer of leaves is made up of dermal tissues, which include pores known as **stomata** (singular "stoma") that control gas exchange. Plants open stomata to let in carbon dioxide needed for photosynthesis. When open, stomata release oxygen and lose water through evaporation. Most plants open their stomata in the daytime and close them at night, conserving water when photosynthesis is not an option. A plant experiencing water stress because of an inadequate water supply will close its stomata to conserve water, no matter the time of day. Intermediate wheatgrass and other deep-rooted perennials are less likely to experience water stress, because of their extensive underground network of roots to extract water from the ground. Yet they do often become dormant and close their stomata during the hot summer.

# Breeding Begins

Intermediate wheatgrass first came to the attention of agronomists in 1983, when the Rodale Institute, a Pennsylvania research facility that studies organic agriculture, evaluated close to 100 species of perennial grasses for traits including seed size, flavor, and harvestability. Intermediate wheatgrass emerged the winner: it is hardy and currently not susceptible to any major pest or disease.

Disease resistance is important because crops are at the mercy of pathogens. About 20 years ago, for example, a fungal disease called "Fusarium head blight" wiped out Minnesota's barley crop. Because plants are under attack by pathogens and predators, they have developed a rich variety of mechanisms to deter attacks (**Figure 26.7**, left). In addition to physical defenses such as thorns, many plants contain chemical substances that are toxic to herbivores. Nicotine, the addictive chemical in cigarette smoke, protects tobacco plants from insect predators. Caffeine protects the leaves and seeds of coffee and tea plants from potential predators. Plants also circulate and secrete antimicrobial chemicals to kill pathogens that could infect them.

In addition to using chemicals to protect themselves, plants produce chemicals called **hormones** to coordinate internal activities necessary for growth and reproduction (**Figure 26.7**, right). Hormones are active at very low concentrations (see Chapter 25 for more on hormones).

After intermediate wheatgrass proved to be resistant to pests and produced better seeds than other perennial grasses do, the Rodale Institute began collecting strains of the plant from all around the world. Over 12 years, researchers bred the strains, selected the best 20, and then identified the top 14 individual plants.

At the Land Institute, DeHaan received seed from the 14 Rodale plants and began breeding them, tinkering with the possibilities of intermediate wheatgrass. In his first year, DeHaan planted some 3,000 plants in an effort to create a large group of diverse individuals. Then, he watched the grass grow.

Plants can increase in length by **primary growth**, in which the plant's shoots and roots lengthen. They can also grow in thickness, called **secondary growth**. Secondary growth includes the thickening of both stems and roots.

Unlike most animals, which grow early in their lives and then level off in a pattern

## Plants have three lines of defense against the external environment

**1** A tough outer surface protects the plant from water loss and heat and cold stress, and serves to block pathogens from entry. Thorns and other defenses keep predators away.

**2** Plants deploy nonspecific chemical defenses, which involve a range of broadly targeted antipathogen and antiherbivore chemicals, including hormones that signal other parts of the plant to manufacture chemical weapons and have them "battle ready."

**3** Plants also have specific chemical defenses. For example, they have a number of genes that respond to complementary genes in a pathogen. If a specific pathogen is detected, protective defenses including toxic chemicals are unleashed.

## Five main hormones control the internal environment of a plant

**1** **Auxins** are necessary for cell division and the formation of organs such as roots. They are involved in *phototropism*, the growth of shoots toward the light, and in *gravitropism*, the growth of roots toward the ground and the growth of shoots away from the ground.

**2** **Cytokinins** are necessary for cell division, and they promote shoot formation. The levels of both cytokinins and auxins decline sharply just before plants drop their leaves in the fall season.

**3** **Gibberellins** bring about stem growth through both cell elongation and cell division. They also stimulate seeds to germinate.

**4** **Abscisic acid (ABA)** mediates adaptive responses to drought, cold, heat, and other stresses.

**5** **Ethylene** stimulates the ripening of some fruits, including apples, bananas, avocadoes, and tomatoes. Ethylene activates enzymes that convert starches into sugars, resulting in a sweeter fruit.

**Figure 26.7**

**Plants use chemicals to survive, grow, and reproduce**
Plants use chemicals to protect themselves from both living and nonliving threats in their environment. And like animals, plants use hormones to coordinate internal activities necessary for growth and reproduction.

**Q1:** Give an example of how plants use chemicals to defend themselves.

**Q2:** Of the plant hormones introduced in this figure, which promote growth?

**Q3:** What is the first line of plant defense? Is this a chemical defense?

termed **determinate growth**, most plants grow throughout their lives—a pattern called **indeterminate growth**. Also unlike most animals, plants are able to grow through **modular growth**, the repeated addition of "modules" of budstem-leaf units aboveground or new lateral roots belowground.

Indeterminate and modular growth habits give plants great flexibility to respond to changing environmental conditions, such as high or low levels of sunlight, water, or nutrients. Plants tend to add many new parts when conditions are favorable and few new parts when conditions are not. This flexibility also enables plants to replace damaged tissues and organs. In fact, plants are so flexible in their development that most living cells in the adult plant body can generate whole new plants.

When it was time to harvest the intermediate wheatgrass, DeHaan and his team went around their field and recorded traits of the plants, such as their heights and how early they flowered. Then the team plucked the heads off the plants and placed them in bar-coded bags. Back in the lab, technicians analyzed the seeds from each plant, recording size and weight—the larger the seed, the more grain can be produced for food—and noting how easily the seeds shed their outer husks, because the less sticky a grain's husk is, the easier it is to process that grain.

DeHaan's first experimental crop grew plants of many sizes and traits. "It's like scratching off lottery tickets," DeHaan told National Public Radio in 2009. "Maybe there's something amazing in there. We'll see." As he recorded traits, DeHaan found that most of his best performers—those with larger seeds and deeper roots—came from a few families, so DeHaan went back into the field and uprooted those specific plants. These were brought into the warm greenhouse for the winter and bred with each other. Their offspring were then planted the following fall, and the process repeated.

But the more DeHaan bred these particular plants, the more he restricted their gene pool, creating a genetic bottleneck by excluding genes that had been in the original population of 3,000 plants. He began to worry that he might have lost genes needed to improve certain traits. One particular trait he struggled to identify and retain in his crop was the ability to mature early in the season.

Many plants, especially species native to temperate regions, perceive the seasons by sensing the length of the day. This is possible because day length varies with the season: days are shorter in winter and longer in summer. This sensing of the duration of light and dark in a 24-hour cycle is known as **photoperiodism**. Plants use day length to sense when conditions are favorable for flowering and seed germination. The dormancy of buds through fall and winter, as well as their regrowth in spring, is also influenced by photoperiodism, as well as changes in temperature and precipitation. DeHaan wanted to identify wheatgrass that flowered early in the spring, so from the U.S. Department of Agriculture (USDA) he obtained hundreds of wild collections of wheatgrass that mature early, and those that flowered early he crossed with his own plants.

By 2010, DeHaan had something good: he had doubled both the seed yield and the weight of seeds of his intermediate wheatgrass. "It was succeeding a lot faster than we thought," says DeHaan. "It was really easy to grow." Unfortunately, "intermediate wheatgrass" doesn't sound like a tasty grain one might like in cereal or bread. So DeHaan and his team renamed the newly domesticated grain "Kernza," a combination of the word "kernel" and the name of a native tribe of the region, Kanza (which also inspired the state name "Kansas").

Kernza® perennial grain is different from wild intermediate wheatgrass in several ways. First, the seeds are larger. "When I started working with it, the typical seed weighed 3.5 milligrams," DeHaan said in 2010. "Now, our best seeds are 10 milligrams." That's progress, but there's more to be done: an average wheat seed weighs 35 milligrams (**Figure 26.8**).

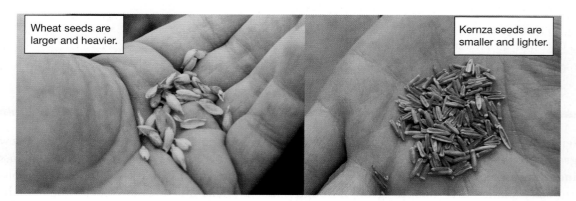

Figure 26.8

**Wheat and Kernza seeds, side by side**
Kernza growers hope to eventually breed plants with seeds as large as those of wheat. They still have a long way to go.

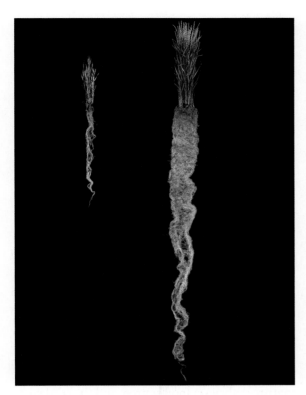

Figure 26.9

**Kernza roots grow deeper than wheat roots**

Kernza's longer roots (right) make the plant better able to collect water from surrounding soil, and to store carbon and nitrogen that would otherwise be lost to the air or waterways.

Then there are the roots. Kernza's roots grow down to 10 feet (**Figure 26.9**). The deep roots make the plant very efficient at sucking water out of the soil. That makes it more resistant to climate change than annual crops are, and the plant has already been shown to fare better in drought conditions than traditional wheat does. Furthermore, the long roots hold soil together, preventing erosion, which carries away fertile soil, fertilizers, and pesticides with it. Major crop-producing countries all around the world lose tens of billions of tons of topsoil every year to erosion.

Kernza's long roots have additional benefits. A recent study found that within 2 years of planting, Kernza beat annual wheat at accessing groundwater, accumulating carbon in the soil, and absorbing nitrogen fertilizer. To grow, plants need $CO_2$, water, and mineral nutrients, especially nitrogen, phosphorus, and potassium. Most of the dry weight of a plant comes from $CO_2$ absorbed from the air and converted into carbohydrates by photosynthesis. Plants require **macronutrients** (nitrogen, phosphorus, potassium, calcium, sulfur, and magnesium) in relatively large amounts and **micronutrients** (including iron, zinc, and copper) in small amounts. Carbon, oxygen, and hydrogen are obtained from air or water; the rest of the nutrients that plants need must be obtained from soil. In agriculture, farmers add most of the nutrients to the soil as fertilizer, but fertilizer can run off into water, polluting it. Perennial plants make better use of fertilizer than do annuals, by efficiently retaining and recycling it. In the same study mentioned earlier, Kernza reduced the amount of nitrogen leached into the nearby ecosystem by up to 99 percent compared with wheat.

# Crop Collaboration

Large seeds and deep roots are good traits, but they aren't enough to make Kernza a successful crop. Other specialty grain crops yield about 1,000 pounds per acre. In Kansas, Kernza was yielding less than 300 pounds per acre. So DeHaan sought help. Around 2007, DeHaan took his Kernza plants north to the University of Minnesota, where he had once been a student. He began working with agronomist Donald Wyse and other wheat breeders at the university. Planting in Minnesota turned out to be a boon: Kernza had larger yields there because intermediate wheatgrass is a cool-season grass and performs better in a cooler place.

**DONALD WYSE**

Donald Wyse develops and promotes winter annual and perennial crops at the University of Minnesota, where he is a professor of agronomy and plant genetics.

Wyse was an ideal collaborator. He had spent years breeding perennial and winter annual crops for agriculture in Minnesota, including perennial flaxseed and perennial sunflower. One of the great values of a perennial, in addition to its having deep roots that prevent erosion and requiring less fertilizer and pesticide, says Wyse, is that its roots and ground tissues are active during a longer period of the year, absorbing more solar energy and thus increasing an area's net primary productivity (see Chapter 21 for a discussion of NPP). This, he says, is "high-efficiency agriculture."

In Minnesota, for example, the roots of corn and soybeans are active for just 2½ months a year, usually after the heaviest rains. That means all the sunlight from the other 9½ months of the year is wasted. But perennials can perform photosynthesis from the moment the snow melts in the spring until the first snowfall of the winter, says Wyse. With perennials, he says, "we're harvesting more energy."

In 2011, DeHaan, Wyse, and the Minnesota team planted a field of more than 2,000 intermediate-wheatgrass plants from 69 families and then measured traits like biomass, grain yield, and seed shape. They've continued this process every year, sometimes pairing complementary strengths and weaknesses. In the summer of 2014, they planted 14,000 seedlings.

The domestication continues, and it is not easy. DeHaan estimates it will be another 10 years before Kernza is ready to compete economically with wheat. One of the challenges of breeding intermediate wheatgrass has to do with how it reproduces. Many plants, such as dandelions and poplar trees, can reproduce asexually, when a parent plant forms a genetically identical clone. Some crops, including potatoes, apples, and grapes, are also propagated as clones. But staple crops, including grains, reproduce sexually.

The overall principle of sexual reproduction in plants is similar to that in animals: A haploid male gamete (*sperm*) fuses with a haploid female gamete (*egg*) to give rise to a diploid cell, the *zygote*, which undergoes mitosis to create a multicellular diploid *embryo*. In time, the embryo develops into an individual offspring, which represents the next generation (**Figure 26.10**).

Plant and animal life cycles differ in one key respect: In animals, meiosis creates gametes and nothing but gametes. In plants, meiosis generates haploid cells called **spores**. Each spore undergoes mitotic divisions to create a haploid, multicellular **gametophyte**. Specialized cells in the gametophyte differentiate to produce sperm or egg cells.

In an animal, gametes are the only haploid cells, and animals have no such thing as a gametophyte. However, gametophytes are part of the life cycle of every plant. In plants like mosses, the gametophytes are a major multicellular phase of the life cycle. In flowering plants like Kernza, gametophytes are reduced to several cells contained in flowers.

Once fertilized, the zygote develops into an embryo inside the ovule, which is contained within the ovary, at the base of the flower. The outer layers of the ovule harden into a protective seed coat. Each seed contains the ingredients for growing a young plant of the next generation: a mature embryo, a food source, and the seed coat.

The ovary surrounding the seed forms a *fruit*, yet another plant organ. Once a fruit is formed, the embryo enters dormancy, and the seed is dispersed from its parent (**Figure 26.11**). Seeds are often dispersed by wind, by attaching to the fur of animals, or by being eaten (and excreted) by animals. Dormancy ends when the embryo is stimulated by favorable conditions to start growing again. Then the seed germinates, and a seedling grows into a plant that will mature and produce flowers and fruits, the beginning of another generation. This adult plant is a **sporophyte**, analogous to an individual animal in that both are multicellular, diploid organisms.

Annual wheat plants *self-pollinate*: they reproduce by fertilizing themselves with their own pollen to create a zygote before their flowers even open. But intermediate wheatgrass, and many other plants, do not self-pollinate. They require a mate. Given that plants cannot travel to find mates, how does sperm-containing pollen from one plant reach the eggs of another plant?

In some species, like grasses and pine trees, pollen is transported by wind. Wind can carry pollen long distances, but most pollen blown by wind lands in inhospitable places (such as parking lots or lakes), not on the flower of another plant of the same species. Many flowering plants

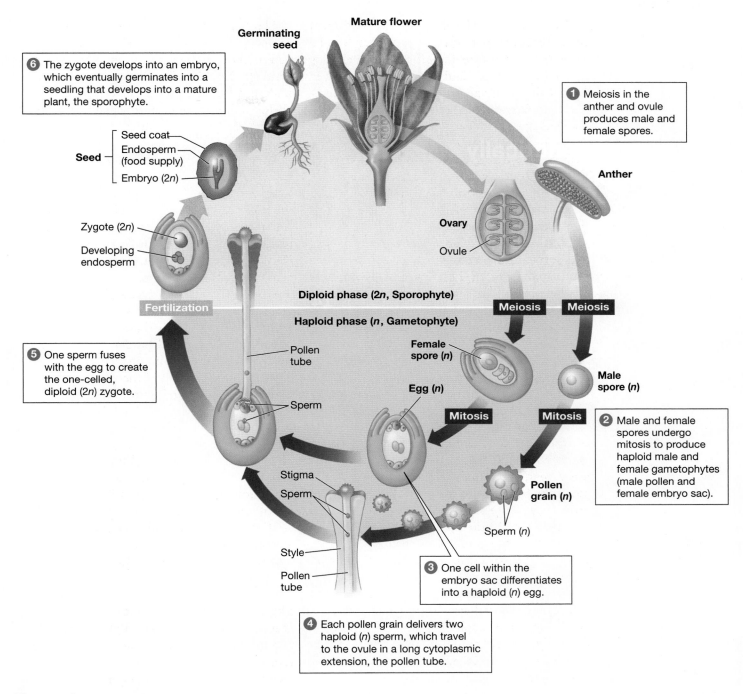

**Mature flower**

**Germinating seed**

**6** The zygote develops into an embryo, which eventually germinates into a seedling that develops into a mature plant, the sporophyte.

**1** Meiosis in the anther and ovule produces male and female spores.

**Anther**

Seed
- Seed coat
- Endosperm (food supply)
- Embryo (2n)

**Ovary**

Ovule

Zygote (2n)

Developing endosperm

**Diploid phase (2n, Sporophyte)**

**Fertilization**

**Haploid phase (n, Gametophyte)**

Meiosis    Meiosis

**5** One sperm fuses with the egg to create the one-celled, diploid (2n) zygote.

Pollen tube

**Female spore (n)**

**Male spore (n)**

**Egg (n)**

Sperm

Mitosis    Mitosis

**2** Male and female spores undergo mitosis to produce haploid male and female gametophytes (male pollen and female embryo sac).

Stigma

Sperm

**Pollen grain (n)**

Sperm (n)

Style

Pollen tube

**3** One cell within the embryo sac differentiates into a haploid (n) egg.

**4** Each pollen grain delivers two haploid (n) sperm, which travel to the ovule in a long cytoplasmic extension, the pollen tube.

**Figure 26.10**

**From generation to generation**

The life cycle of flowering plants is marked by **alternation of generations**. Haploid (n) stages of the life cycle are shown in purple; diploid (2n) stages, in orange.

**Q1:** Are eggs and sperm haploid or diploid? Are they part of the sporophyte or gametophyte generation?

**Q2:** Why is the plant life cycle, but not the animal life cycle, referred to as an "alternation of generations"?

**Q3:** How does asexual reproduction differ from the plant life cycle diagrammed here?

# GM Crops Take Root

Since 1996, the total land area used to grow [genetically]
modified (GM) crops has expanded dramatic[ally.]
of these crops, including the varieties depict[ed here,]
are enhanced with the insertion of a gene th[at confers]
insect resistance or herbicide tolerance. Bro[ad scientific]
consensus has concluded that food derived [from GM]
crops poses no greater risk to health than fo[od from]
nonmodified crops.

Assessment available in smartwork**5**

## Global area of genetically modified crops

Area shown in million hectares

Soybean    Maize    Cotton    Canola

**14**
**3**
**3**
**8**
1998

**26**
**3**
**1**

**36**
**7**
**3**
**12**
2002

**48**
**9**
**5**
**19**
2004

**58**
**6**
**25**

**65**
**15**
**7**
**37**
2008

**74**
**20**
**8**
**46**
2010

**80**
**9**
**56**

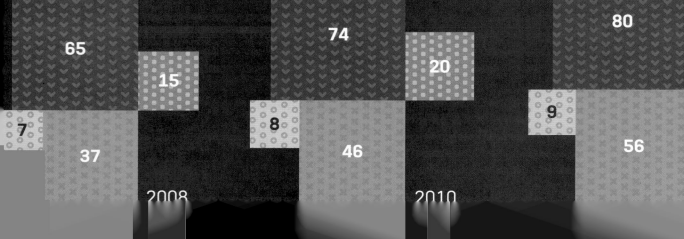

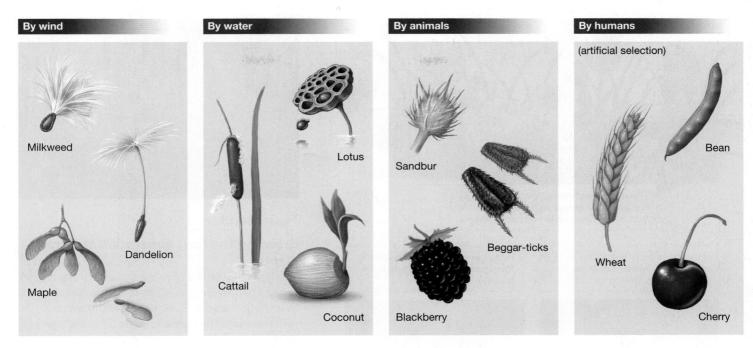

Figure 26.11

**Plants spread their seeds in a variety of ways**

Many plants spread their seeds via the wind or water, but others use modified seeds and fruit to attract animals or attach to their fur. Humans have artificially selected the fruit and seeds of many species to make them more edible.

**Q1:** What kinds of seeds would you expect to be contained within a sugary fruit: seeds spread by water or by animals? Why?

**Q2:** What kinds of seeds would you expect to weigh less: those spread by air or those artificially selected by humans? Why?

**Q3:** How does the height of a tree affect its ability to spread its seeds widely?

sidestep this problem by having their pollen transported by animals such as insects, birds, and even mammals (**Figure 26.12**). Plants attract these **pollinators** with brightly colored, sweet-smelling flowers filled with sugary nectar. It is a mutualistic relationship, in which both species benefit.

DeHaan and his team need to mate specific plants together in order to breed for specific traits. So, instead of relying on wind or pollinators, they put a bag over the flowering stalks of two plants they want to breed so that pollen from one plant passes directly to the second plant in the bag, and no others (**Figure 26.13**). "We call them 'plant condoms,'" says DeHaan with a laugh. "The bags keep crosses we don't want to happen from happening."

And because an intermediate-wheatgrass plant can breed with only another plant—rather than with itself, as annual wheat can do—the offspring are genetically diverse; no child is *exactly* like its parent. In fact, a child can be very different from its parents, because there is a great deal of variation in the genome of intermediate wheatgrass. Each gene may have six or more versions, or alleles, in a single plant, says DeHaan. So, when breeding new plants, the team can never match the exact genotype and phenotype of a parent but rather must constantly juggle traits to try to capture the right set of traits in a single line of plants. For example, DeHaan is currently working to grow shorter plants, because taller plants tend to tip over, which makes them difficult to harvest.

The distinctive colors, shapes, and smells of flowers often attract highly specific pollinators that are efficient pollen dispersal agents.

Honeybees carry the pollen that dusts their bodies from flower to flower as they search for food.

Birds have good color vision and favor red flowers with long floral tubes.

Figure 26.12

### Bribing animals to do the work

Pollinators provide stationary plants with a way of transporting sperm to eggs. The spectacular colors, shapes, and odors of flowers, in combination with food rewards such as nectar, lure pollinating animals into visiting several flowers of the same species, incidentally transferring pollen in the process.

**Q1:** Many crops, including apples, peppers, and tomatoes, depend on insect pollinators. What would happen if no pollinator visited a plant?

**Q2:** Although they look very different, flower petals are actually modified leaves. Do flowers still perform the main function of leaves? Explain your reasoning.

**Q3:** Some flowers look like an insect. How might this resemblance attract pollinators?

## Perennial Pancakes

Early on, DeHaan estimated it would take some 50 years for a new plant to be domesticated to match the yield of wheat, but now USDA and University of Minnesota researchers are

Figure 26.13

### Breeding the next generation of Kernza plants

The flowering stalks of Kernza plants are enclosed in bags to prevent the pollen from spreading to any other plants. Scientists identify plants with the traits they want to pass on to the next generation and selectively breed them in this way.

**Q1:** What traits are scientists selecting for in each new generation of Kernza?

**Q2:** What would happen if the Kernza plants were allowed to freely pollinate, rather than being selectively pollinated by hand?

**Q3:** After choosing two plants to breed, researchers put the plant heads together under a bag as shown in the photo. Why do they do this before the plants begin to pollinate?

supplementing his hands-on breeding work with gene sequencing efforts. These scientists have been associating genetic markers with particular traits, such as seed size. Once they have enough traits mapped on the genome, they will no longer have to wait for a plant to mature to observe its phenotype; instead they will simply sample the DNA of young seedlings and predict the plant's traits on the basis of genetic markers.

"Within a very short period of time, we'll be able to identify key genes in the domestication process, and at a very low cost," says Wyse. With these tools, Wyse thinks the breeders won't take too long to match the yield of wheat. "It may only take half a decade to see dramatic improvements in these new crops," says Wyse. "We think crops of this type are going to have great value."

But there's more to a food crop than how well it grows and can be harvested. It also has to taste good. DeHaan has used Kernza flour to make cookies, cakes, scones, bread, and more (**Figure 26.14**). He and other Land Institute staff have enjoyed them, though at least one food scientist says the flour will need some flavor improvement to go mainstream. A team at the University of Minnesota is currently experimenting with Kernza's food traits, including taste, texture, and ability to rise.

Both beer brewers and sustainable food companies have expressed an interest in using the new grain. DeHaan and Wyse also imagine that the crop by-product could be used as hay or to produce biofuel. "We'd love this to be a dual-use crop," says DeHaan. Farmers, for example, could harvest the grain for a food company and then use the "residue" plant material in the field, typically the leftover bottom of the stalk, to make biofuel. "Those two things together could result in an economically viable crop," he says.

Figure 26.14

**Kernza flour**
Scientists are finding uses for Kernza in traditional baking, and also using it to make alcohol, animal feedstock, and hopefully in the future, even biofuel.

# REVIEWING THE SCIENCE

- Plant bodies are made of three basic tissue types: **Dermal tissues** control the flow of materials into and out of the plant and protect the plant from attack. **Ground tissues** make up the bulk of the plant body and participate in support, wound repair, and photosynthesis. **Phloem** and **xylem** are the two types of **vascular tissue**, transporting sugars and water, respectively, through the entire plant.

- The plant body contains two basic organ systems: the belowground **root system** and the aboveground **shoot system**. Plant organs include roots, stems, leaves, flowers, and fruits. **Roots** absorb water and mineral nutrients, anchor the plant, and store food. **Leaves** produce the majority of the plant's food through photosynthesis. **Stems** support the plant and may perform a limited amount of photosynthesis.

- Plants defend themselves against the external environment with a tough outer covering, physical weapons such as thorns, and an arsenal of toxic chemicals.

- Like animals, plants use **hormones**, chemicals produced at low concentrations, to coordinate growth and reproduction.

- Plants have **indeterminate growth**: they can grow throughout their lives. Plants grow by adding repeating bud-stem-leaf units aboveground and root units below; a pattern called **modular growth**.

- Aboveground, plants increase in length by **primary growth**, the lengthening of shoots and roots. Plants increase in thickness by **secondary growth**, the thickening of stems and roots.

- To grow, plants need $CO_2$, water, and mineral nutrients. **Macronutrients** (especially nitrogen, phosphorus, and potassium) are required in large quantities.

- Many plants can reproduce asexually, but most can also reproduce sexually. Meiosis in plants produces a single-celled **spore**, which divides through mitosis to create a haploid, multicellular **gametophyte**. Cells in the gametophyte differentiate into egg or sperm. The fertilization of the egg by sperm generates a diploid zygote, which gives rise to the diploid, multicellular **sporophyte**. This **alternation of generations** (gametophyte and sporophyte) is a hallmark of plant life cycles.

- Fertilization produces a zygote, which divides and develops into an embryo. The embryo is located within the ovule, which hardens into a protective seed coat. Each seed contains the ingredients for growing a young plant of the next generation: a mature embryo, a food source, and the seed coat.

- In flowering plants, male and female reproductive parts are contained in flowers. Flowers attract animal **pollinators**, which provide immobile plants with a way of transporting sperm-containing pollen to eggs.

# THE QUESTIONS

## The Basics

**1** Water is transported through the plant body by

(a) xylem.

(b) phloem.

(c) stomata.

(d) root hairs.

(e) flowers.

**2** Which of the following is a chemical defense strategy used by plants?

(a) maintaining a tough dermal layer

(b) storing poisons in leaves

(c) signaling plant shoots to grow toward the light

(d) stimulating the ripening of fruit

(e) none of the above

**3** Which plant tissue controls the flow of materials into and out of the plant and protects the plant from attack?

(a) vascular tissue

(b) ground tissue

(c) dermal tissue

(d) phloem

(e) xylem

**4** What kind of seed is best for being dispersed by wind?

(a) dandelion

(b) apple

(c) coconut

(d) Kernza

(e) a seed held within a bur

**5** Alternation of generations

(a) involves a haploid gametophyte.

(b) involves a diploid sporophyte.

(c) is central to plant life cycles.

(d) all of the above

(e) none of the above

**6** Link each term with the correct definition.

| | |
|---|---|
| **MODULAR GROWTH** | 1. Increases in length by the division of cells located at the tip of each stem and each root. |
| **INDETERMINATE GROWN** | 2. Grows throughout its life. |
| **PRIMARY GROWTH** | 3. Grows in size by repeatedly adding the same basic bud-stem-leaf unit. |
| **SECONDARY GROWTH** | 4. Increases in thickness. |

## Challenge Yourself

**7** Which of the following is *not* a plant organ?

(a) stem

(b) fruit

(c) vascular bundles

(d) root

(e) flower

**8** Which *micro*nutrient do plants need to absorb from the environment to survive and grow?

(a) carbon

(b) sugar

(c) potassium

(d) nitrogen

(e) zinc

**9** Select the correct terms:

(**Annual / Biennial**) plants complete their life cycle in one year, whereas (**biennial / perennial**) plants live for three or more years. Wheat is a(n) (**annual / perennial**), whereas Kernza is a(n) (**annual / perennial**). Kernza is also a (**monocot / dicot**), which has a (**fibrous root / taproot**).

**10** Beginning with the stage following spores, place the following steps of the plant life cycle in the correct order by numbering them from 1 to 4.

____ a. Zygote develops into an embryo.

____ b. Gametophyte develops.

____ c. Seed germinates.

____ d. Egg is fertilized.

## Try Something New

**11** Some plants are pollinated by bats, which are active at night. What would you predict about how bat-pollinated flowers differ from bird-pollinated flowers? (*Hint:* Remember that birds are active during the day.)

**12** Ethylene is a plant hormone that causes fruits to ripen by converting starches into sugars. It also signals flowers to open, seeds to germinate, and leaves to shed. You may have observed that placing a ripe banana or apple next to another fruit causes it to ripen more quickly. Furthermore, if you place fruit in a paper bag (rather than exposed on the counter), it ripens more quickly. How would you explain this?

**13** Several of the plants domesticated by humans have highly modified organs. For example, potatoes are modified stems, and carrots and sweet potatoes are modified roots. From which plant organ do you think onions and tomatoes are modified? Why?

## Leveling Up

**14** **Looking at data** The debate over genetically modified (GM) crops and genetically modified organisms (GMOs) has been impassioned but not always informed. Genetically modified crops are those with DNA altered through biotechnology, in order to increase productivity or resistance to disease.

More than 93 percent of the soybeans harvested in the United States, fed almost exclusively to domesticated animals, are genetically modified for pesticide and herbicide resistance. Examine the accompanying graph, which shows the yield in bushels per acre of soybeans from 1980 to 2010. The red circle is the yield for organic soybeans in 2007, which is based on information from 1,331 organic farms.

a. What does the *x*-axis represent? The *y*-axis? What does the data line on the graph represent?

b. Describe the trend in soybean productivity from 1980 to 2010.

c. The yield for organic soybeans in 2007 is only 66 percent of the average yield for that year. What year of conventional soybean production does this production match?

d. Critics say that the widespread growth and consumption of GM crops may have unexpected health and environmental costs. Proponents argue that genetic modification is the only way to feed a rapidly growing population. What do you think?

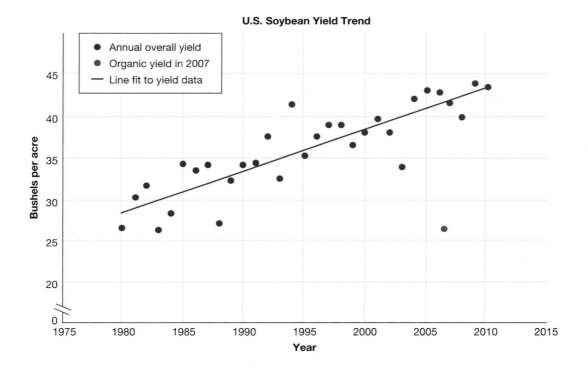

**U.S. Soybean Yield Trend**

- ● Annual overall yield
- ● Organic yield in 2007
- — Line fit to yield data

**15** **Doing science** Although you are an expert on Kernza® perennial grain, you have been asked to contribute your scientific expertise to developing another perennial crop, *Silphium integrifolium* (sometimes referred to as a "perennial sunflower"), that could replace annual crops grown as sources of cooking oil.

a. Read the Land Institute's overview of its research on *Silphium*: http://landinstitute.org/our-work/perennial-crops/silphium.

b. The *Silphium* researchers say that they aim to identify critical ways in which *Silphium* responds to changes in the environment each year, but that they will need to follow up such observations with experiments. Are the scientists conducting observational or experimental research on *Silphium*? Explain your answer.

c. Propose a hypothesis about an environmental variable that you believe might increase oil production in *Silphium*, and then identify a testable prediction from that hypothesis. Design an experiment to test the prediction, identifying your independent and dependent variables, the control group, and the treatment conditions. Create a graph to show the results you expect to find (1) if your hypothesis is supported and (2) if your hypothesis is not supported.

**For more, visit digital.wwnorton.com/bionow2 for access to:**

# Answers

## CHAPTER 1

### END-OF-CHAPTER ANSWERS

1. a, c, e. Viruses meet no criteria except for evolving as groups. The diamond meets no criteria.

2. b

3. observation, hypothesis, predictions

4. (a) 1, (b) 4, (c) 5, (d) 3, (e) 6, (f) 2, (g) 7

5. (a) organ, (b) organism, (c) population, (d) ecosystem, (e) organ system, (f) community

6. Observation: Some bats have white noses.
   Hypothesis: Bats with white noses are infected with a fungus.
   Experiment: Inject bats with a fungicide and observe whether they are less likely to develop white-nose syndrome than are bats who are given sham injections.

7. b

8. c

9. (a) reproducing autonomously, (b) responding to the environment, (c) obtaining energy from the environment, (d) evolving as a group, (e) consisting of one or more cells

10. c

11. (a) hypothesis, (b) result, (c) experiment, (d) observation, (e) result

12. (a) No; a question is not a hypothesis. (b) Yes; this is a plausible and falsifiable explanation. (c) Yes; this is a plausible and falsifiable explanation. (d) No; a question is not a hypothesis. (e) No; proposing the existence of a "mysterious cloud" does not generate testable predictions.

### ANSWERS TO FIGURE QUESTIONS

#### Figure 1.1

**Q1:** What were the original observation and question of the scientists studying the sick bats?

**A1:** They observed many dead bats and many bats with white noses. They questioned whether the high death rates were in some way related to the white noses.

**Q2:** At what point in the scientific method would a scientist decide on the methods she should use to test her hypothesis?

**A2:** After predictions are generated from a hypothesis, and before experiments are run to test those predictions.

**Q3:** How might you explain the scientific method to someone who complains that "scientists are always changing their minds; how can we trust what they say?"

**A3:** Science is a process, and nothing is ever "proved" in science, so we have to expect that our best understanding of nature will change as science proceeds.

#### Figure 1.2

**Q1:** Which step(s) in the scientific method does this photograph illustrate?

**A1:** Observation, and testing the predictions of the hypothesis through observational study (either descriptive or analytical).

**Q2:** What types of environmental data might the researchers have collected?

**A2:** Examples of environmental data would include temperature and humidity readings within the cave, as well as soil and air samples to see whether fungal spores are present.

**Q3:** Why do you think the researchers are wearing protective gear?

**A3:** To avoid coming into contact with pathogens.

#### Figure 1.4

**Q1:** State the hypothesis that this advertisement is claiming was scientifically tested.

**A1:** Lucky Strike cigarettes taste milder than other leading brands of cigarettes.

**Q2:** State a prediction that comes from this hypothesis. Is it testable? Why or why not?

**A2:** Prediction: Subjects who smoke a Lucky Strike cigarette will say it is milder than cigarettes from other brands that they are asked to smoke. Yes, this prediction is testable, because it can be measured and repeated.

**Q3:** Explain in your own words why the hypothesis cannot be "proved."

**A3:** Nothing in science can be proved. In this case, only a small number of subjects can be tested, and there might always be people (who weren't tested) who think another brand is milder.

## Figure 1.5

**Q1:** Give a possible hypothesis that could be tested by weighing the bats.

**A1:** Hypothesis: Bats with WNS weigh less than uninfected bats.

**Q2:** State the hypothesis being tested in the photo on the bottom right.

**A2:** Hypothesis: Healthy bats injected with a fungicide will have lower rates of infection by WNS than will bats that are sham injected (no fungicide).

**Q3:** Explain in your own words why an experimental study is the only way to show a cause-effect relationship.

**A3:** One possible explanation: If you find a relationship between two variables in an observational study, you can't know which one causes the other, or even whether a third variable is causing both to occur. In an experimental study, it is possible to manipulate one variable and see whether that manipulation causes a second variable to change.

## Figure 1.6

**Q1:** Which is the control group in this experiment, and what are the three treatment groups?

**A1:** Control group: Housed alone, with no exposure to WNS. Treatment groups: Housed in air contact but not physical contact with infected bats; housed in physical contact with infected bats; with fungus applied to wings.

**Q2:** What is the hypothesis being tested in this experiment?

**A2:** Hypothesis: WNS is caused by contact with the fungus *Geomyces destructans*.

**Q3:** In one or two sentences, state the conclusions you can draw from the experiment. Was the hypothesis supported? Why or why not?

**A3:** Bats that come into physical contact with *Geomyces destructans* are highly likely to develop WNS, whereas those exposed only through the air do not develop WNS. The hypothesis was supported in part (physical versus air contact).

## Figure 1.7

**Q1:** What is the control group in this experiment, and what are the two treatment groups?

**A1:** Control group: Sham injection. Treatment groups: Injected with *Geomyces destructans* from North America; injected with *G. destructans* from Europe.

**Q2:** At day 40, approximately how many individuals were alive in each treatment group? At day 80? At day 100?

**A2:** Day 40: 100% of all groups survived. Day 80: 100% of control and North American *G. destructans* (Gd) groups survived; 14 of 18 in the European Gd group survived. Day 100: 100% of control group survived; 13 of 18 in the North American Gd group survived; only 2 of 18 in the European Gd group survived (although the European Gd experiment was stopped at about day 90).

**Q3:** In one or two sentences, state the conclusions you can draw from the experiment. Was the hypothesis supported? Why or why not?

**A3:** Conclusion: *Geomyces destructans* causes WNS and leads to higher mortality. The hypothesis was supported. The study found that WNS was, in fact, caused by the fungus.

## Figure 1.8

**Q1:** Give one *fact* about bats that you learned from this chapter.

**A1:** One possible fact: Bats can develop fungal infections.

**Q2:** What is another example of evidence for the *germ theory of disease*? (*Hint:* Think about human diseases.)

**A2:** One example would be strep throat, caused by *Streptococcus* bacteria and cured by antibiotics.

**Q3:** Explain in your own words the difference between a fact and a hypothesis, and between a hypothesis and a theory.

**A3:** Hypotheses are not as certain and are more complex than facts; they are simpler and less well documented than theories.

## Figure 1.9

**Q1:** Give examples of other kinds of organs that mammals such as bats have. (*Hint:* Think of the organs in your own body.)

**A1:** Examples include kidney, liver, heart, lungs.

**Q2:** Are bats in California part of the community of bats in upstate New York, if they are of the same species? Why or why not?

**A2:** No, they are not, because they do not interact with each other.

**Q3:** Is the soil in a cave where bats live a part of the bats' population, community, or ecosystem? Explain your reasoning.

**A3:** Soil is part of the ecosystem; populations and communities are composed only of living things, and soil is part of the physical environment.

# CHAPTER 2

## END-OF-CHAPTER ANSWERS

1. b

2. c

3. d

4. scientific literacy: 2, basic research: 5, applied research: 1, secondary literature: 4, primary literature: 3

5. credentials, bias, secondary literature

6. (a) 1, (b) 6, (c) 3, (d) 2, (e) 4, (f) 5

7. a

8. (a) primary, (b) neither, (c) neither, (d) secondary

9. c

10. (a) Pseudoscience. Dr. Oz has relevant credentials but, from the success of his TV show, appears to have an agenda. In addition, the idea of a "fat-burning" dietary aid does not align with current scientific understanding.

    (b) Pseudoscience. The idea of a scientific conspiracy is a red flag, as is the rejection of scientific consensus. And though the

*Wall Street Journal* is a respected newspaper, it tends to have a conservative bias, especially since being bought by Rupert Murdoch.

(c) Real science. Practicing scientists have reported experimental findings to their peers, but they do not yet appear to have published in a peer-reviewed journal, which will need to be the next step.

(d) Pseudoscience. Astrologers are not scientists. There is no scientific evidence that date of birth has any effect on personality.

(e) Real science. The journal *Diabetes* is a peer-reviewed and well-established scientific journal. We do not know, however, the design of the study or the sample size.

11.  (a) The first graph shows the incidence of pertussis (whooping cough) cases in the United States from 1922 to 2012, with the inset showing a zoomed-in view of 1990–2012. The $x$-axis shows the year from which the pertussis cases were counted; the $y$-axis shows the counts. Any point on the line shows the total number of cases reported in the corresponding year on the $x$-axis. The general trend of the graph is a decrease in cases of pertussis over time.

(b) The reason for the increase in cases in the first decade of the twenty-first century is probably that parents were choosing not to have their children vaccinated.

(c) The second graph shows the incidence of pertussis (whooping cough) cases in the United States by age group from 1990 to 2012. The $x$-axis shows again the year from which the pertussis cases were counted, and the $y$-axis shows the counts. This graph's $y$-axis shows not the total number of cases, but rather the number of cases per 100,000 individuals in the population of that particular age group. One point on each line shows the number of cases per 100,000 individuals in the population of that particular age group reported in the corresponding year on the $x$-axis. The general trend is an increase in cases of pertussis by age group over time. The lines differ by age group. The younger the age group, the greater the increase in pertussis cases per year.

(d) The incidence of pertussis cases per 100,000 individuals in the population of children under 1 year old is very high, and in people over the age of 20 it is very low. Increase over time is much greater in children under 1 year old and is very minimal in people over the age of 20.

(e) It appears to be very risky not to vaccinate. I would vaccinate!

12.  (1) Evaluate the credentials of the group members recommending the vaccine (in this case it would be members of the ACIP, FDA, and CDC).

(2) Assess any possible biases of the people in step 1.

(3) Read the secondary literature (for example, recent articles in *Popular Science* or the *New York Times*) for an overview of the issue.

(4) Read recent primary literature that was cited in the secondary literature (for example, articles in *JAMA* or the *New England Journal of Medicine* or *Vaccine*).

(5) Review papers from the primary literature for credentials and biases of the authors, good research design, reasonable sample size, and conclusions.

## ANSWERS TO FIGURE QUESTIONS

### Figure 2.1

**Q1:** Describe in one sentence how a vaccine creates immunity to a virus.

**A1:** A vaccine causes the immune system to produce antibodies in response to an inactive or harmless virus. Those antibodies then immediately identify and attack any live virus (of that type) that enters the body.

**Q2:** Why is it impossible to become infected with a virus from a vaccine composed of viral proteins?

**A2:** Proteins don't replicate; DNA or RNA is needed for a virus to replicate.

**Q3:** Natural immunity occurs without a vaccine, just by exposure to a particular stimulus, like the chicken pox virus. Explain why people don't get chicken pox twice.

**A3:** Your body creates antibodies when you are exposed to, for example, chicken pox virus. If you are exposed a second time, those antibodies immediately identify and attack the chicken pox virus (just as in question 1 above).

### Figure 2.2

**Q1:** Before vaccinations, which diseases had the highest and lowest mortality rates? What are these mortality rates?

**A1:** Highest: Tetanus, 90%. Lowest: Pertussis, 5%.

**Q2:** After vaccinations, which diseases have the highest and lowest mortality rates? What are these mortality rates?

**A2:** Highest: Tetanus, 14%. Lowest: Polio, diphtheria, and smallpox, 0%.

**Q3:** If there were a sudden outbreak of whooping cough at a university where pertussis vaccinations were not required and no one was protected, how many students would die? What is the probability or chance that you would die if infected?

**A3:** At California State University, Northridge, for example, which has 40,000 students total, 40 students would die—a 0.1% chance of death for any individual student.

### Figure 2.5

**Q1:** Why are we less confident of scientific claims made over social media?

**A1:** Claims made over social media are not subject to peer review before being "published." Therefore, people can—and do—make ridiculous scientific claims in social media.

**Q2:** Where would you place a blog in this figure? Would it matter whether or not it was written by a practicing scientist? Explain your reasoning.

**A2:** A blog could be listed either under social media or under secondary literature, depending on the quality of the content. In the case of a science blog written by a practicing scientist, the choice would depend on the subject matter, but it would likely be listed under secondary literature. (Counterargument: The physicist who thinks global climate change is a scientific conspiracy.)

**Q3:** Give an example of when you would rely on secondary literature to evaluate a scientific claim and an example of when you would go to the primary literature. What is the basis of that decision?

**A3:** You might go to the secondary literature only for something that is not life-threatening—for example, what kind of exercise to do, or whether to turn down the thermostat in your bedroom. Primary literature is challenging reading, so it is usually left for life-critical choices—for example, whether to vaccinate your children, or possibly whether to eat a vegetarian or vegan diet.

## Figure 2.6

**Q1:** How much did organic food sales grow during the period covered in the graph? How much did the incidence of autism grow?

**A1:** Organic food sales grew from $5 billion to $25 billion (a fivefold increase). Autism diagnoses grew from about 50,000 individuals to over 250,000 (also a fivefold increase).

**Q2:** Why might both organic food sales and autism prevalence have increased during this time period? A Reddit user in the original discussion thread suggested that both might be affected by increasing wealth in the United States. How might increased wealth affect these variables?

**A2:** People with more disposable income are able to spend more on food (hence the rise in organic food sales), and they are also better able to take their children in for advanced medical care (possibly the reason for increased identification of autism disorder).

**Q3:** In what way has the vaccine-autism debate been confused by people misinterpreting correlation as causation?

**A3:** Because vaccination was increasing at the same time that autism rates were rising (correlation), people suggested that the former caused the latter (causation). In addition, the time at which children are typically vaccinated is about the same age at which autism symptoms typically appear.

## Figure 2.7

**Q1:** State the hypothesis of the people who believe that vaccines cause autism. What is an alternative hypothesis?

**A1:** Hypothesis: Vaccination harms a child's immune system and stimulates the development of autism. One alternative hypothesis: Autism is caused by a genetic predisposition, and symptoms begin to show up in the second year of a child's life.

**Q2:** What part(s) of the figure show where Wakefield's study failed to meet the standards of the scientific method?

**A2:** Wakefield's study really falls apart in steps 4–6. It was not carefully designed and definitely not reproducible. The sample studied was not of sufficient size and did not include a random control group. The conclusions did not follow from the analysis of the experimental results, and the study was published without adequate review from scientists in the field.

**Q3:** Why does only one arrow point to "real science," whereas multiple arrows point to pseudoscience?

**A3:** All of the criteria listed in the figure must be met for a study to meet the expectations of the scientific method.

## Figure 2.8

**Q1:** Why is it important to know the education and expertise of a person making a scientific claim?

**A1:** The opinion of a person who doesn't really understand the science behind a claim is not valid.

**Q2:** List at least five possible biases that people making scientific claims might have.

**A2:** (1) They could make money if you buy a product related to the claim. (2) They might win a lawsuit if the judge believes the claim. (3) They could become famous if people accept the claim. (4) Their religious beliefs might be supported if the claim is true. (5) Their political beliefs might be supported if the claim is true.

**Q3:** Describe a situation in which you might not dismiss the scientific claim of a person who did not have appropriate credentials, or who had a bias toward the claim.

**A3:** Paul Offit's case provides an example: He made a great deal of money by selling a rotavirus vaccine he created. However, his expertise is extremely strong, and he does not have an ongoing financial interest in supporting vaccination.

## Figure 2.10

**Q1:** Why do vaccine manufacturers begin with tests on animals or cell lines before moving on to adult human subjects?

**A1:** They need to be sure of the safety of the vaccine before exposing people to it.

**Q2:** What ongoing testing and reporting are vaccines subjected to?

**A2:** Manufacturers test all vaccine lots, the FDA regularly inspects manufacturing facilities, the Advisory Committee on Immunization Practices (ACIP) and the CDC director review all test results before approving vaccines, and vaccine safety is continually monitored through the Vaccine Adverse Event Recording System (VAERS) and the Vaccine Safety Datalink (VSD).

**Q3:** What do ACIP, FDA, and CDC stand for, and what is the role of each in evaluating vaccines?

**A3:** ACIP = Advisory Committee on Immunization Practices; FDA = Food and Drug Administration; CDC = Centers for Disease Control and Prevention. All are involved in the initial approval and ongoing monitoring of vaccines.

## Figure 2.11

**Q1:** What happens to an immunized person when a disease spreads through a population? (*Hint:* In the graphic, follow an immunized individual before and after a disease spreads.)

**A1:** The immunized person does not contract the disease.

**Q2:** Explain why a disease is less likely to spread to vulnerable members of a population if most people are immunized.

**A2:** "Herd immunity" means that fewer people contract the disease, and therefore vulnerable people are less likely to be exposed to a contagious person.

**Q3:** How does vaccination help an individual person? How does it help that person's community?

**A3:** An individual who is vaccinated is much less likely to become ill and is therefore less likely to pass on a disease to others in the community.

# CHAPTER 3

## END-OF-CHAPTER ANSWERS

1. a

2. d

3. ion: 3, matter: 6, solution: 9, element: 5, chemical compound: 8, molecule: 4, isotope: 7, polymer: 2, atom: 1

4. a

5. polymers, sugar, nucleotides, are not, carbon

6. a, c, e

7. Carbon can be the basis of more complex molecules than can hydrogen or oxygen because carbon can form four bonds (versus one for hydrogen and two for oxygen).

8. Six of the amino acids produced contained sulfur; these could not have formed in the absence of a sulfur-containing reactant.

9. b

10. b

11. (a) oven, (b) coffee maker, (c) neither, (d) oven, (e) coffee maker

12. The nonpolar end of the detergent will bond to the oil in the salad dressing while the polar end bonds to the water molecules, lifting the oil into the wash water. Vinegar is a polar molecule, so it will dissolve in the wash water; you don't need detergent to remove vinegar.

## ANSWERS TO FIGURE QUESTIONS

### Figure 3.1

**Q1:** How many protons, electrons, and neutrons does the hydrogen atom shown here have? What are the atomic number and the atomic mass number of the hydrogen atom?

**A1:** 1 proton, 1 electron, 0 neutrons. The atomic number (number of protons) and the atomic mass number (number of protons plus neutrons) are both 1.

**Q2:** What are the atomic number and the atomic mass number of the carbon isotope shown?

**A2:** The atomic number is 6 (6 protons); the atomic mass number, 12 (6 protons plus 6 neutrons).

**Q3:** Nitrogen-11 is an isotope of nitrogen that has 7 protons and 4 neutrons. What are the atomic number and atomic mass number of nitrogen-11?

**A3:** The atomic number is 7 (7 protons); the atomic mass number, 11 (7 protons plus 4 neutrons).

### Figure 3.2

**Q1:** Before the experiment was run, the apparatus was sterilized and then carefully sealed. Why was this an important thing to do?

**A1:** Having a spotlessly clean apparatus was important to ensure that any amino acids were produced by the experimental conditions, not from contamination.

**Q2:** Why is inclusion of methane in the gas flasks an essential part of the hypothesis that complex organic molecules were formed in the early atmosphere of Earth? (*Hint:* What makes a molecule organic?)

**A2:** Amino acids contain carbon, and methane (a simple organic molecule) is the only chemical in the mixture that contains carbon.

**Q3:** Answer this question after reading about Miller's "steam injection" experiments: Where was the steam injected in the experimental apparatus?

**A3:** The steam was injected straight into the gas chamber, allowing the gases to interact with the water (and the other gases) more directly.

### Figure 3.4

**Q1:** Where are the covalent bonds in this figure?

**A1:** The covalent bonds are located at the electrons shared between the oxygen atom and the two hydrogen atoms.

**Q2:** This figure shows a water molecule ($H_2O$). A hydrogen molecule ($H_2$) consists of two hydrogen nuclei that share two electrons. Draw a simple diagram of a hydrogen molecule indicating the positions of the two electrons.

**A2:** The electrons are equidistant from the two hydrogen nuclei because there is no difference in the strength of their attraction to electrons.

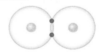

**Q3:** When table salt (sodium chloride, NaCl) dissolves in water, it separates into a sodium ion ($Na^+$) and a chloride ion ($Cl^-$). Which portion of a water molecule would attract the sodium ion, and which portion would attract the chloride ion?

**A3:** The $Na^+$ ions are attracted to the partial negative charges on the oxygen atoms, and the $Cl^-$ ions are attracted to the partial positive charges on the hydrogen atoms.

### Figure 3.5

**Q1:** Describe what will happen to the molecules of olive oil if you shake the bottle and then leave it alone for an hour. What about the molecules of vinegar?

**A1:** When you shake the bottle, the olive oil will form small beads of oil. As the bottle rests, these beads of oil will float back to the top of the bottle and coalesce into a layer of oil; that is, the bottle will return to its starting condition. Shaking the bottle will distribute the vinegar molecules through the water, and they will remain in solution.

**Q2:** What would happen if you added another fat to the bottle, such as bacon grease, and shook it?

**A2:** Bacon grease is a fat and, like the olive oil, would form small beads of fat that would float back to the top of the bottle and coalesce into a layer of grease.

**Q3:** Given how sugar behaves when it is mixed into coffee or tea, would you predict that it is hydrophobic or hydrophilic?

**A3:** Sugar dissolves in coffee and tea; therefore, it is hydrophilic.

## Figure 3.6

**Q1:** Identify where in the picture water can be seen in its liquid, solid, and gas states.

**A1:** The water in the hot springs is in a liquid state. The snow and ice are solid water. Water vapor (steam) in the air is a gas.

**Q2:** In the gas state, water molecules move too rapidly and are too far apart to form hydrogen bonds. Compare the volumes occupied by an equal number of water molecules in the liquid, solid, and gas states.

**A2:** The same number of molecules will occupy more volume in both ice (solid) and water vapor (gas) than in liquid water. The volume of ice is defined (9% greater than the volume in water) because the water molecules form an ordered array. The volume of water vapor depends on its temperature.

**Q3:** Explain in your own words how ice floats on water.

**A3:** Water molecules are farther apart in ice than they are in liquid water. Because the same number of water molecules occupies a greater volume in ice than in water, the density of ice is lower than that of water, and ice floats.

## Figure 3.7

**Q1:** Suppose you were going to repeat Miller's experiments. How would you decide how much of each gas to include in the chamber?

**A1:** Miller wanted to replicate the atmosphere of early Earth, so he based his mixtures on estimates of the proportions of the gases in the atmosphere at that time.

**Q2:** Why did the addition of steam to the gases in Miller's second set of experiments increase the yield of amino acids?

**A2:** The steam (hot water vapor) added heat energy to the mixture, and the water enabled more kinds of chemical reactions.

**Q3:** Miller used electrical energy in his experiment. What other forms of energy were present in the early atmosphere of Earth that could have led to the formation of complex molecules?

**A3:** Heat and ultraviolet light were present in early Earth's atmosphere.

## Figure 3.8

**Q1:** How did the NASA scientists find the fragments of the meteorite that exploded over eastern Africa?

**A1:** The fragments from the asteroid were dark and stood out against the light-colored sand in Sudan.

**Q2:** What piece of evidence suggests that amino acids found in the meteorite fragments originated in outer space?

**A2:** Some of the amino acids from the meteorite fragments have a "right-handed" orientation, whereas all naturally occurring amino acids on Earth are in a "left-handed" orientation.

**Q3:** Speculate on the significance of finding extraterrestrial amino acids.

**A3:** The existence of extraterrestrial amino acids suggests that at least some of the building blocks of life are found in outer space.

## Figure 3.9

**Q1:** Which has a higher concentration of free hydrogen ions: vinegar, pH 2.8; or milk, pH 6.5?

**A1:** A *high* concentration of free hydrogen ions corresponds to a *low* pH. Thus, vinegar has a higher concentration of free hydrogen ions than milk has.

**Q2:** What happens to the concentration of free hydrogen ions in your stomach when you drink a glass of milk?

**A2:** The concentration of free hydrogen ions decreases (that is, the pH increases).

**Q3:** Black coffee has a pH of 5. Does adding coffee to water (pH 7) increase or decrease the concentration of free hydrogen ions in the liquid?

**A3:** The concentration of free hydrogen ions increases.

## Figure 3.10

**Q1:** In methane gas, how many electrons is each hydrogen atom sharing? How many is the carbon atom sharing?

**A1:** Each hydrogen is sharing one electron with carbon. Carbon is sharing four electrons.

**Q2:** In carbon dioxide, how many electrons is each oxygen atom sharing? How many is the carbon atom sharing?

**A2:** Each oxygen is sharing two electrons with carbon. Carbon is sharing four electrons.

**Q3:** Draw a molecule of formaldehyde ($CH_2O$). How many electrons is the oxygen atom sharing with the carbon atom? How many is the carbon atom sharing with the oxygen atom and with each hydrogen atom?

**A3:** Oxygen is sharing two electrons with carbon. Carbon is sharing two electrons with oxygen and one electron with each of the hydrogens.

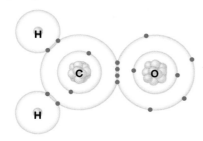

# CHAPTER 4

## END-OF-CHAPTER ANSWERS

1. c

2. d

3. receptor-mediated endocytosis: 2, phagocytosis: 1, pinocytosis: 4, exocytosis: 3

4. chloroplast: 7, Golgi apparatus: 4, lysosome: 5, mitochondrion: 6, nucleus: 1, rough endoplasmic reticulum: 2, smooth endoplasmic reticulum: 3

5.

| Component | Prokaryotes | Eukaryotes Animals | Eukaryotes Plants |
|---|---|---|---|
| Plasma membrane | X | X | X |
| Cellulose cell wall | | | X |
| Nucleus | | X | X |
| Endoplasmic reticulum | | X | X |
| Golgi apparatus | | X | X |
| Ribosomes | | X | X |
| Cytoskeleton | | X | X |
| Mitochondria | | X | X |
| Chloroplasts | | | X |

6. a

7. The dashed curve represents simple diffusion because as the concentration of solute increases, the rate of transport increases in a directly proportional way. The solid curve represents facilitated diffusion because it reaches the transport protein saturation line but cannot exceed it, since facilitated diffusion relies on these proteins to send the solutes across the membrane. If the transport proteins are fully in use, adding a higher concentration of solute will not increase the rate of transport.

8. b

9. right side, more, fewer

10. c

11. (a) isotonic, (b) neither gain nor lose, (c) equal to, (d) hypertonic, (e) lose, (f) higher than, (g) hypotonic, (h) gain, (i) lower than

## ANSWERS TO FIGURE QUESTIONS

### Figure 4.2

**Q1:** What was the purpose of inserting the gene that codes for blue pigment into the synthetic DNA?
**A1:** The blue color identified the cells that contained the DNA from *M. mycoides*.
**Q2:** What part of the transformed bacterium is synthetic?
**A2:** Only the DNA is synthetic; all of the structural components of the cells are from the *M. capricolum* cells into which the synthetic DNA was inserted.
**Q3:** Did this experiment create life?
**A3:** Although some articles in the popular press refer to the synthetic bacterium as a new life-form, it is better described as "repackaged life." The DNA is synthetic, but all the intracellular components that enable the DNA to function were already present in the cell.

### Figure 4.3

**Q1:** Why is it important that the phosphate head of a phospholipid is hydrophilic?
**A1:** The fact that the phosphate head is attracted to water (hydrophilic) and also to other phosphate heads means that a bilayer will form.
**Q2:** What essential component of a cell do liposomes lack, and why is that omission important?
**A2:** Liposomes lack genetic material (DNA), so the characteristics of a liposome are not transmitted to its descendants.
**Q3:** Could the tendency of phospholipid bilayers to spontaneously form spheres have played a role in the origin of life? (*Hint:* Refer to "The Characteristics of Living Organisms" on page 6 of Chapter 1.)
**A3:** Yes. Once phospholipids formed (how that happened is still an open question), they would have formed liposomes spontaneously, trapping substances in their interiors.

### Figure 4.4

**Q1:** In what ways is the plasma membrane a barrier, and in what ways is it a gatekeeper?
**A1:** It is a barrier in that it keeps out many molecules. It is a gatekeeper in that it selectively allows in other molecules.

**Q2:** Why can't ions (such as Na⁺) cross the plasma membrane without the help of a transport protein?

**A2:** The electrical charges on ions make them hydrophilic, so they cannot cross the lipid portion of the bilayer, which is hydrophobic, without the help of a transport protein.

**Q3:** If no energy were available to the cell, what forms of transport would not be able to occur? What forms of transport could occur? (*Hint:* Look ahead at Figures 4.5 and 4.6.)

**A3:** No form of active transport is possible without the input of energy. Any form of passive transport, including diffusion and osmosis, could occur even in the absence of an energy source.

## Figure 4.5

**Q1:** Is the dye at equilibrium in any of these glasses? Describe how the first glass will look when the dye is at equilibrium with the water.

**A1:** Yes, the dye is at equilibrium in the third glass. If the first glass reaches equilibrium, it will look like the third glass.

**Q2:** Will diffusion mix the molecules of dye evenly through the water, or is it necessary to shake the container to get a uniform mixture?

**A2:** Diffusion is sufficient to mix the dye thoroughly, but it is a very slow process and you would probably get tired of waiting.

**Q3:** Will diffusion mix the dye faster in hot water than in cold water? Why or why not? (*Hint:* Review the discussion of the behavior of water molecules at different temperatures in Chapter 3.)

**A3:** Diffusion is faster at higher temperatures because the water molecules have more energy, form and break hydrogen bonds at a higher rate, and hence move around more rapidly.

## Figure 4.6

**Q1:** What would the second diagram look like if the pores in the semipermeable membrane were too small to allow water molecules to pass through?

**A1:** The second diagram would look the same as the first diagram, since neither the sugar molecules nor the water molecules would be able to pass through the membrane.

**Q2:** What would the second diagram look like if the pores were large enough to let both water molecules and sugar molecules through?

**A2:** If the pores were large enough for sugar molecules to pass through, some sugar molecules would diffuse down the concentration gradient of sugar from the left side to the right while water molecules were diffusing from right to left down the concentration gradient of water. At equilibrium, the concentration of sugar and water would be the same on both sides of the membrane, and the depth of the solution would also be the same on both sides.

**Q3:** The fluid in an IV bag is isotonic to blood. What change would you see in the red blood cells of a patient if a bag of hypertonic solution was used in error?

**A3:** A hypertonic solution would be more dilute than blood, so an IV of hypertonic solution would dilute the blood—that is, increase the concentration of water in the blood. Now the red blood cells would have a lower concentration of water than the blood surrounding them, so water would move by osmosis into the red blood cells, causing them to swell.

## Figure 4.7

**Q1:** If endocytosis itself is nonspecific, how does receptor-mediated endocytosis bring only certain molecules into a cell?

**A1:** The receptor protein embedded in the plasma membrane attracts and holds only specific molecules. When they attach, the plasma membrane bulges inward to engulf the molecule.

**Q2:** What sorts of molecules could be moved by endocytosis or exocytosis, but not by diffusion?

**A2:** Endocytosis and exocytosis can move molecules that are too large to pass through the plasma membrane through diffusion.

**Q3:** How does the fluid that enters a cell via pinocytosis differ from the fluid that enters by osmosis?

**A3:** The fluid that fills the inward bulge of a cell during pinocytosis may contain molecules that are in solution in the fluid outside the cell, whereas osmosis allows only water molecules to enter a cell.

## Figure 4.8

**Q1:** What structures do prokaryotic and eukaryotic cells have in common?

**A1:** A plasma membrane, ribosomes, and DNA.

**Q2:** What cellular processes occur in both prokaryotic and eukaryotic cells?

**A2:** Both prokaryotic and eukaryotic cells regulate their internal concentration and the movement of substances in and out. Both require a supply of energy and carry out metabolic processes. Both respond to changes in their internal state and to conditions in the environment around them.

**Q3:** Both plants and animals are eukaryotes, but there are differences in their cellular structure. What are those differences?

**A3:** Plants have a cell wall and chloroplasts, while animals do not. Most animal cells also do not contain vacuoles.

# CHAPTER 5

## END-OF-CHAPTER ANSWERS

1. c

2. a

3. c

4. b

5. d

6. opposite, catabolism, produces

7.  (See figure below)

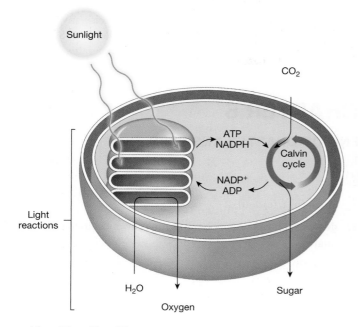

Sunlight

CO₂

Calvin cycle

ATP
NADPH

NADP⁺
ADP

Light reactions

H₂O

Oxygen

Sugar

8.  (a) 3, (b) 2, (c) 1, (d) 4

9.  The Calvin cycle cannot be maintained indefinitely, because it requires a source of energy to produce ATP. This energy source (glucose) is created from sunlight energy during the light reactions.

10. e

11. (a) The reaction without an enzyme requires more energy to proceed, because its line on the graph shows a higher peak on the $y$-axis representing how much energy is needed.

    (b) This reaction is catabolic because there is less energy in the products than in the reactants, as shown by the line's lower position on the $y$-axis at the end of the reaction. Catabolic reactions release energy; therefore, their products have less energy than their reactants.

12. c

13. If metabolic processes occur at a higher rate than can be sustained by cellular respiration, then it is possible to die from lack of usable energy.

14. a

## ANSWERS TO FIGURE QUESTIONS

### Figure 5.2

**Q1:** Why is photosynthesis called "primary production"?

**A1:** Because photosynthesis converts two inorganic molecules (carbon dioxide and water) into an organic molecule (glucose).

**Q2:** How does animal life depend on photosynthesis?

**A2:** Only photosynthetic organisms can create organic molecules from inorganic molecules, and animals cannot carry out photosynthesis. (There are a very few exceptions to that generalization.)

**Q3:** Explain how photosynthesis and cellular respiration are "complementary" processes.

**A3:** They use each other's products and reactants.

### Figure 5.3

**Q1:** What source of energy would plants use for anabolic reactions? Would an animal use the same kind of energy?

**A1:** Plants would use light energy from the sun for anabolism. Animals, in contrast, would use energy from food or glucose.

**Q2:** What source of energy would plants release in catabolic reactions? Would an animal release the same kind of energy?

**A2:** All living things release energy carriers, usually ATP, during catabolic reactions.

**Q3:** Create a mnemonic or jingle that helps you remember the difference between anabolism and catabolism.

**A3:** This question is just for fun and to aid in learning. Here's one possible answer: "Anabolism synthesizes; catabolism cannibalizes" (breaks down molecules).

### Figure 5.4

**Q1:** Define ATP in your own words.

**A1:** Something like this: "ATP is a molecule that cells use to store and deliver energy."

**Q2:** How is ATP involved in anabolism and catabolism? (*Hint:* Review Figure 5.3.)

**A2:** Anabolism requires energy, which is provided most often by ATP. Catabolism releases energy, and would be involved in creating ATP from ADP.

**Q3:** Arsenic disrupts ATP production. Why would this characteristic cause it to be a potent poison?

**A3:** If ATP stopped being produced, our cells could no longer function and we would quickly die.

### Figure 5.6

**Q1:** Is chlorophyll found only within chloroplasts?

**A1:** No. In bacteria it is embedded in membranes within the cell.

**Q2:** What could be an advantage of concentrating chlorophyll molecules in the membranes of chloroplasts?

**A2:** The location of the chlorophyll molecules maximizes the efficiency of the metabolic pathway of photosynthesis.

**Q3:** What is the advantage of having multiple chloroplasts per cell?

**A3:** Each chloroplast provides membranes needed for more and more embedded chlorophyll molecules.

### Figure 5.7

**Q1:** What is the source of the carbon dioxide used for photosynthesis?

**A1:** Atmospheric $CO_2$.

**Q2:** Which products of the light reactions of photosynthesis does the Calvin cycle use?

**A2:** ATP and NADPH.

**Q3:** What are the two major products of photosynthesis?

**A3:** Glucose and oxygen.

## Figure 5.8

**Q1:** Why is it important that enzymes are not permanently altered when they bind with substrate molecules?

**A1:** The fact that they are not permanently altered means that they can continue to perform their function without having to be present in great numbers or needing to be continually produced.

**Q2:** How would a higher temperature or higher salt concentration make it more difficult for an enzyme to function effectively?

**A2:** Both would change the shape of the enzyme so that it would not be able to bind to its specific molecules.

**Q3:** If a cell was unable to produce a particular enzyme necessary for a metabolic pathway, describe how the absence of that enzyme would affect the cell.

**A3:** The metabolic pathway would either proceed very slowly or, possibly, come to a complete halt, so the cell would not be able to function.

## Figure 5.11

**Q1:** What are the products of cellular respiration?

**A1:** Carbon dioxide and water.

**Q2:** Considering the inputs and products of each process, why is cellular respiration considered the reciprocal process to photosynthesis?

**A2:** Photosynthesis uses carbon dioxide and produces oxygen, whereas cellular respiration does the opposite, using oxygen and producing carbon dioxide.

**Q3:** Which of the three stages of cellular respiration—glycolysis, the Krebs cycle, or oxidative phosphorylation—could organisms have used 4 billion years ago, before photosynthesis by cyanobacteria released oxygen into the atmosphere?

**A3:** Glycolysis, because it does not depend on oxygen. As the text says, "Glycolysis was probably the earliest means of producing ATP from food molecules, and it is still the primary means of energy production in many prokaryotes."

## Figure 5.12

**Q1:** Which product released by fermentation accounts for the bubbles in beer?

**A1:** Carbon dioxide ($CO_2$).

**Q2:** Bakers of yeast breads also rely on fermentation, allowing bread to "rise" before baking. Describe what is occurring with the yeast as the bread rises.

**A2:** During bread rising, yeasts are performing metabolic functions—and growing and reproducing—using energy created through fermentation. The fermentation produces $CO_2$ as a waste product, which is trapped within the bread dough and causes it to rise.

**Q3:** Explain in your own words why lactic acid builds up in your muscles during strenuous physical activity.

**A3:** Something like this: "During strenuous exercise, my muscles can't get enough oxygen to produce all the needed ATP. So, glycolysis produces ATP anaerobically, which generates lactic acid as a by-product."

# CHAPTER 6

### END-OF-CHAPTER ANSWERS

1. b

2. a

3. (a) 4, (b) 1, (c) 2, (d) 3

4. Meiosis, binary fission, homologous chromosomes, sister chromatids

5. (a) 1, (b) 2, (c) 5, (d) 3, (e) 4

6. a

7. b

8. b

9. b

10. S phase. Chromosomes replicate during S phase; at some point during S phase, there will be the original DNA strands and half of the newly synthesized DNA strands. In all other phases, there is either the same amount of DNA as in the $G_1$ phase, or twice as much DNA as in $G_1$.

11. (a) 54, (b) 27, (c) 27

12. c

13. The $G_1$ checkpoint ensures that the cell is ready to divide—for example, that it is large enough and has enough energy to produce two normal daughter cells. The $G_2$ checkpoint ensures that the cell's DNA has been replicated and packed into pairs of sister chromatids. Bypassing the $G_1$ checkpoint could allow cells to divide before they're ready; bypassing the $G_2$ checkpoint could lead to the production of daughter cells with defective chromosomes.

### ANSWERS TO FIGURE QUESTIONS

## Figure 6.2

**Q1:** When is DNA replicated during the cell cycle?

**A1:** In interphase.

**Q2:** When in the cell cycle does DNA separate into the two genetically identical daughter cells?

**A2:** Late in the mitotic phase of cell division.

**Q3:** If a cell is not destined to separate into daughter cells, what phase does it enter? Is this part of the cell cycle?

**A3:** It enters the $G_0$ phase, which is not part of the cell cycle.

## Figure 6.3

**Q1:** Name one similarity between cell division in prokaryotes and cell division in eukaryotes.

**A1:** Both prokaryotes and eukaryotes replicate DNA, separate the DNA to opposite poles of the cell, and then physically split the cell into two parts.

**Q2:** Why is binary fission referred to as "asexual reproduction"?

**A2:** Because it does not involve sharing DNA between individuals, as in sexual reproduction.

**Q3:** Name one difference between cell division in prokaryotes and cell division in eukaryotes.

**A3:** Prokaryotic cells are smaller and simpler than eukaryotic cells, so cell division happens more rapidly and is less complex; for example, prokaryotes do not have organelles, so they do not have to be divided up among daughter cells.

## Figure 6.4

**Q1:** Do all cells in an organism enter each stage of mitosis at the same time? (*Hint:* See image of onion root tip at far left in the figure.)

**A1:** No.

**Q2:** What happens between the end of interphase and early prophase that changes the appearance of the chromosomes?

**A2:** The DNA within the chromosomes is condensed to prepare for division.

**Q3:** Explain in your own words the role of the mitotic spindle in mitosis.

**A3:** The mitotic spindle is responsible for accurately separating chromatids into daughter cells.

## Figure 6.5

**Q1:** Why is it important for a chromosome to be copied before mitosis?

**A1:** There need to be two copies of each chromosome, so that each daughter cell has identical genetic material.

**Q2:** Are sister chromatids attached at the centromere considered to be one or two chromosomes?

**A2:** Sister chromatids are, when attached, one chromosome. When split apart, they are two separate but identical chromosomes.

**Q3:** Why is the chromosome's DNA tightly packed for mitosis and cytokinesis? (*Hint:* Think about what would happen if it were unpackaged, as during interphase.)

**A3:** Tightly packed chromosomes are more easily divided into daughter cells.

## Figure 6.6

**Q1:** What could happen if the cell's checkpoints are disabled?

**A1:** The cell cycle could occur more quickly, because it will not stop to ensure that the checkpoint conditions are met.

**Q2:** What is the advantage of stopping the cell cycle if the cell's DNA is damaged?

**A2:** If the nutrient supply is inadequate, the daughter cells cannot grow. By stopping the cell cycle, the parent cell can wait until the nutrient supply increases.

**Q3:** Which part of the cell cycle may have been influenced in Soto and Sonnenschein's breast tumor cell experiments?

**A3:** The $G_0$ or $G_1$ checkpoint.

## Figure 6.7

**Q1:** Is a zygote haploid or diploid?

**A1:** Diploid.

**Q2:** What cellular process creates a baby from a zygote?

**A2:** Mitosis.

**Q3:** If a mother or father was exposed to BPA prior to conceiving a child, how might that explain potential birth defects in the fetus?

**A3:** If the parent's BPA exposure caused mutation in sex cells (eggs or sperm), that could lead to a birth defect in the fetus.

## Figure 6.8

**Q1:** Is a daughter cell haploid or diploid after the first meiotic division? How about after the second meiotic division?

**A1:** After meiosis I, a daughter cell is haploid (has one of each homologous chromosome). After meiosis II, each daughter cell is still haploid (has one of each homologous chromosome), and sister chromatids have split into separate daughter cells.

**Q2:** What is the difference between homologous chromosomes and sister chromatids?

**A2:** Sister chromatids are identical DNA molecules, replicated from a single DNA molecule, that remain bound to each other. They exist in a cell only from the S phase until anaphase of mitosis, or anaphase II of meiosis. A homologous chromosome pair consists of the two copies—one maternal, the other paternal—of the same type of chromosome. The pair is present at all times in diploid cells, but a haploid cell has just the paternal or the maternal copy.

**Q3:** If the skin cells of house cats contain 38 homologous pairs of chromosomes, how many chromosomes are present in the egg cells they produce?

**A3:** 38.

## Figure 6.9

**Q1:** Why is the term "crossing-over" appropriate for the exchange of DNA segments between homologous chromosomes?

**A1:** Segments of DNA physically "cross over" between homologous chromosomes.

**Q2:** At what stage of meiosis (I or II) does crossing-over occur?

**A2:** Meiosis I.

**Q3:** What would be the effect of crossing-over between two sister chromatids?

**A3:** There would be no effect, because sister chromatids are genetically identical.

**Figure 6.10**

**Q1:** During meiosis, does random assortment occur before or after crossing-over?

**A1:** After.

**Q2:** What would be the effect on genetic diversity if homologous chromosomes did not randomly separate into the daughter cells during meiosis?

**A2:** Genetic diversity would decrease.

**Q3:** With two pairs of homologous chromosomes, four kinds of gametes can be produced. How many kinds of gametes can be produced with three pairs of homologous chromosomes? What does this suggest for the 23 homologous pairs of chromosomes in human cells?

**A3:** Three pairs of homologous chromosomes can produce eight kinds of gametes. Therefore, 23 homologous pairs could produce huge variation in gametes. To be more precise, it could produce $2^{23}$ (8,388,608) different combinations of chromosomes in gametes.

# CHAPTER 7

## END-OF-CHAPTER ANSWERS

1.  genotype: 4, phenotype: 5, heterozygote: 1, homozygote: 2, dominant: 6, recessive: 3

2.  gene, alleles

3.  meiosis, segregation, independent assortment

4.  (a) M, (b) C, (c) C, (d) C, (e) M

5.  e

6.  e

7.  incomplete dominance

8.  d

9.  pleiotropic

10. b, c, e

11. The first-generation result suggests that the round-shape allele is dominant to the oval-shape allele. The next cross should be to breed offspring with themselves to create an $F_2$ generation, and the proportion of ovals would be 1 in 4, or 25 percent (3:1 round to oval).

12. Orange color must be dominant to black color. You could breed their offspring to test the hypothesis.

13. (a) *llff* (homozygous recessive for both long hair and no furnishings)
    (b) *LlFf, LLFF, LlFF, LLFf*
    (c) 9:3:3:1 short furnished to short unfurnished to long furnished to long unfurnished.

## ANSWERS TO FIGURE QUESTIONS

### Figure 7.2

**Q1:** What is the physical structure of a gene?

**A1:** A gene is a strand of DNA within a chromosome.

**Q2:** How many copies of each gene are found in the diploid cells in a woman's body?

**A2:** Two.

**Q3:** With 46 chromosomes in a human diploid cell, how many chromosomes come from the person's mother and how many from the father?

**A3:** Each parent contributes half of the chromosomes, so 23 chromosomes from the mother and 23 from the father.

### Figure 7.3

**Q1:** Which might you observe directly: the genotype or the phenotype?

**A1:** The phenotype may be directly observable.

**Q2:** Which poodle could be heterozygous: the one with the black coat or the one with the brown coat?

**A2:** Only the black-coated poodle may be heterozygous.

**Q3:** Can you identify with certainty the genotype of a black poodle? A brown poodle?

**A3:** No for a black poodle; yes for a brown poodle.

### Figure 7.4

**Q1:** What would you predict about the color of the $F_1$ plants' flowers?

**A1:** They should all be purple.

**Q2:** Why was it important that Mendel begin with pea plants that he knew bred true for flower color? Why couldn't he simply cross a purple-flowered plant and a white-flowered plant?

**A2:** The purple-flowered plant might be heterozygous *or* homozygous.

**Q3:** Over the years, Mendel experimented with more than 30,000 pea plants. Why did Mendel collect data on so many plants? Why didn't he study just one cross? *Hint:* Read "What Are the Odds?" on page 124 before answering.

**A3:** With data from more plants, there is a better chance that the results will accurately reflect reality.

### Figure 7.5

**Q1:** Why did Mendel's entire $F_1$ generation look the same?

**A1:** All of the $F_1$ plants were heterozygous, so they all had a purple phenotype.

**Q2:** The phenotype ratio in the $F_2$ generation is 3:1 purple-to-white flowers. What is the genotype ratio?

**A2:** 1:2:1.

**Q3:** Draw a Punnett square for a genetic cross of two heterozygous, black-coated dogs. What are the phenotype and genotype ratios of their offspring?

**A3:** Phenotype ratio is 3:1 black to brown. Genotype ratio is 1:2:1 *BB* to *Bb* to *bb*.

## Figure 7.6

**Q1:** List all the possible offspring genotypes and phenotypes.

**A1:** Round, yellow (dominant dominant): *RRYY, RrYY, RRYy, RrYy.*
Round, green (dominant recessive): *RRyy, Rryy.*
Wrinkled, yellow (recessive dominant): *rrYY, rrYy.*
Wrinkled, green (recessive recessive): *rryy.*

**Q2:** What is the offspring phenotype ratio?

**A2:** 9:3:3:1.

**Q3:** Complete a Punnett square for a genetic cross of two true-breeding Portuguese water dogs—one with a black, wavy coat (homozygous dominant, *BBWW*) and one with a brown, curly coat (homozygous recessive, *bbww*). What is the phenotype ratio of their offspring ($F_1$)? Now fill out another Punnett square, crossing two of the offspring. What is the phenotype ratio of the $F_2$ generation?

**A3:** $F_1$-generation phenotype ratio is 3:1 black wavy to brown curly. $F_2$-generation phenotype ratio is 9:3:3:1 black wavy to black curly to brown wavy to brown curly.

## Figure 7.7

**Q1:** Boxers are far more inbred than poodles. Why does that inbreeding make the former a better target for genetic studies of disease than the latter?

**A1:** Since they are more inbred, boxers are more likely than poodles to be homozygous for traits of interest.

**Q2:** Explain why a geneticist interested in finding a gene linked to cancer would want to look at the DNA of senior golden retrievers with *and* without cancer.

**A2:** To know whether there was a genetic difference between goldens with and without cancer, the DNA of cancer-free dogs would have to be known.

**Q3:** Obsessive-compulsive disorder (OCD) in humans is characterized by obsessive thoughts and compulsive behavior, such as pacing. Canine compulsive disorder (CCD) is characterized by compulsive behavior such as "flank sucking," sometimes seen in Doberman pinschers. Would you predict that the medications given to humans with OCD would decrease compulsive behaviors in CCD dogs? Why or why not?

**A3:** One would predict that they might, since dogs and humans share many genes and therefore OCD and CCD may share a common genetic basis and be treatable by the same means.

## Figure 7.8

**Q1:** What are the genotypes of a large and a small dog?

**A1:** *LL* (large) and *ll* (small).

**Q2:** Is it possible to have a heterozygous large dog? Explain why or why not.

**A2:** No, because *Ll* is a medium-sized dog.

**Q3:** Crossing a Great Dane and a Chihuahua is likely to be unsuccessful, even though they are members of the same species (and thus have compatible sperm and egg). Why is that? What are some potential risks of such a cross?

**A3:** It would be difficult for a Great Dane and a Chihuahua to mate, because of the size difference. One risk would be that if a female Chihuahua were to become pregnant this way, the pups would be too large for her to carry safely to term.

## Figure 7.9

**Q1:** What are the possible genotypes (at both genes) of the black dog? The yellow dog? The brown dog?

**A1:** Black dog: *BBEE, BbEE, BBEe, BbEe.*
Yellow dog: *BBee, Bbee, bbee.*
Brown dog: *bbEe, bbEE.*

**Q2:** Draw a Punnett square showing possible matings between the black dog (assuming it is heterozygous at both genes) and the yellow dog (assuming it is heterozygous at the *B* gene). List all the possible phenotypes of their offspring. (See Figure 7.6 for an example of a Punnett square made with two traits.)

**A2:** There are three possible offspring phenotypes: black, yellow, and brown.

**Q3:** If you wanted the most variable litter possible, what colors of Labrador retrievers would you cross?

**A3:** If they were true-breeding (homozygous), you would cross yellow and brown.

## Figure 7.10

**Q1:** The gene that brings about the pale Siamese body fur is also responsible in part for the typical blue eyes of the species. What is the term for this type of inheritance?

**A1:** Pleiotropy.

**Q2:** Siamese kittens that weigh more tend to have darker fur on their bodies. Why might this be?

**A2:** It could be that larger kittens have lower core temperatures, thus allowing more melanin to be produced on their bodies.

**Q3:** The Siamese cat pictured is called a "seal point" because it has seal-colored (dark brown) extremities. Some Siamese cats show the same color pattern, but the dark areas are of a lighter color or even a different shade—for example, lilac point, red point, blue point. What results would you predict if the experiments described in the text (shaving the cat and then increasing or decreasing temperature) were conducted on cats with these color patterns?

**A3:** The results should be the same, since even though the color of the melanin has changed, presumably the mechanism for laying down the melanin (based on temperature) has not.

# CHAPTER 8

## END-OF-CHAPTER ANSWERS

1. chromosomes, genes, loci, alleles

2. d

3. gene therapy: 3, in vitro fertilization: 5, preimplantation genetic diagnosis: 4, chorionic villus sampling: 1, amniocentesis: 2

4. The sex chromosomes are the two in the final pair. Since there is one large sex chromosome and one small one, the individual is a male.

5.

|     | H   | h   |
| --- | --- | --- |
| H   | HH  | Hh  |
| h   | Hh  | hh  |

6. d

7. e

8. c

9. b, d, e

10. a (where allele *T* represents stumpy tails and allele *t* represents long tails)

11. (a) from their mother, (b) no, (c) yes, (d) two types, (e) no, no, yes, female only

12. (a) $X^gX^g$, (b) $X^GX^g$, (c) $X^gY$, (d) $X^gY$, (e) $X^GX^g$

13. (a) *gg*, (b) *Gg*, (c) *Gg*, (d) *gg*

## ANSWERS TO FIGURE QUESTIONS

### Figure 8.2

**Q1:** Which two children in this pedigree have cystic fibrosis? How do you know?

**A1:** Individuals III-2 and III-3 have cystic fibrosis. They are depicted by filled symbols.

**Q2:** Does either parent of these two children have cystic fibrosis? If so, which one(s)? How do you know?

**A2:** Neither parent II-1 nor II-2 has cystic fibrosis. They are both depicted by open symbols.

**Q3:** Do any of the grandparents of these two children have cystic fibrosis? If so, which one(s)? How do you know?

**A3:** Yes. I-2, the paternal grandmother, has cystic fibrosis. She is depicted by a filled symbol.

### Figure 8.3

**Q1:** How many male and how many female descendants (individuals that did not join the family by marriage) does generation IV of Aldrich's pedigree contain?

**A1:** 8 males and 4 females.

**Q2:** What proportions of the male and female descendants in generation IV were affected by the disorder?

**A2:** 50% (4/8, or 0.50) of males, and 0% (0/4, or 0.00) of females.

**Q3:** Why did Aldrich hypothesize that the disease was X-linked? (You will need to read ahead to answer this question.)

**A3:** The disease was observed only in males, but their mothers' male relatives were also affected.

### Figure 8.4

**Q1:** Is this the karyotype of a male or a female?

**A1:** Male.

**Q2:** How would the karyotype of a person with Down syndrome differ from this karyotype?

**A2:** There would be three copies of chromosome 21.

**Q3:** The size of a chromosome correlates roughly with the number of genes residing on it. Why are an extra copy of chromosome 21 and a missing Y chromosome two of the least damaging chromosomal abnormalities?

**A3:** Chromosome 21 and the Y chromosome are two of the smallest chromosomes, so fewer genes would be affected by there being too many or too few of them (compared to a larger chromosome).

### Figure 8.5

**Q1:** What are the odds that a given egg cell will contain an X chromosome? A Y chromosome? What are those odds for a sperm cell?

**A1:** An egg cell has a 100% chance of containing an X chromosome and a 0% chance of containing a Y chromosome. For the sperm cell, the odds are 50/50 for both cases.

**Q2:** If a couple has two daughters, does that mean their next two children are more likely to be sons? Explain your reasoning. (*Hint:* Refer back to "What Are the Odds?" on page 124.)

**A2:** No, the probability for each event is independent of prior and future events.

**Q3:** Sisters share the same X chromosome inherited from their father, but they may inherit different X chromosomes from their mother. What is the probability that brothers share the same Y chromosome? What is the probability that brothers share the same X chromosome?

**A3:** Brothers have a 100% probability of sharing the same Y chromosome, and a 50% probability of sharing the same X chromosome.

### Figure 8.6

**Q1:** Why are changes in chromosome *number* almost always more severe than changes in chromosome structure?

**A1:** There are so many genes on an individual chromosome that deleting or adding an entire chromosome has massive effects on an individual.

**Q2:** In which part of meiosis would you predict that chromosomal abnormalities are produced? (Refer back to Chapter 6 if necessary.)

**A2:** During metaphase, when paired chromosomes are separated.

**Q3:** Create a mnemonic to help remember the four kinds of structural changes (for example, Doctors Improve Treatment Daily).

**A3:** Answers will vary.

## Figure 8.7

**Q1:** How do we know whether two chromosomes are homologous?

**A1:** If the chromosomes carry alleles of the same genes and align during cell division, they are homologous.

**Q2:** In one sentence, explain how the terms "gene," "locus," and "chromosome" are related.

**A2:** A gene is found at a particular location—a locus—on a chromosome.

**Q3:** If hair color were determined by a single gene, what would be an example of the gene's alleles?

**A3:** *B* for brown hair, *b* for blonde hair.

## Figure 8.9

**Q1:** Which of the children specified in this Punnett square represents Felix? What is his genotype?

**A1:** Felix is the "affected son," with genotype $X^aY$.

**Q2:** Explain why Felix is neither homozygous nor heterozygous for the *WAS* gene.

**A2:** The *WAS* gene is found on the X chromosome, and Felix carries only one X chromosome because he is a male. Only with two chromosomes are homozygous and heterozygous genotypes possible.

**Q3:** Create a Punnett square to illustrate the offspring that could result if Felix had children with a noncarrier woman. What is the probability that a son would have WAS? What is the probability that a daughter would be a carrier of WAS?

**A3:** A son would have a 0% probability of having WAS. A daughter would have a 100% probability of being a carrier.

## Figure 8.10

**Q1:** Which chromosome contains the gene for cystic fibrosis? For Tay-Sachs disease? For sickle-cell disease?

**A1:** Cystic fibrosis: chromosome 7. Tay-Sachs disease: chromosome 15. Sickle-cell disease: chromosome 11.

**Q2:** No known genetic disorders are encoded on the Y chromosome. Why do you think this is?

**A2:** The Y chromosome has very few genes; further, any disorder would always be expressed and therefore selected against.

**Q3:** In your own words, explain why most single-gene disorders are recessive rather than dominant.

**A3:** Dominant, single-gene disorders experience heavier selection than recessive disorders because they are always expressed (there are no carriers).

## Figure 8.12

**Q1:** Which of the children in this Punnett square represents Zoe? What is her genotype?

**A1:** Zoe is the "affected child," with genotype *aa*.

**Q2:** If Zoe's parents had another child, what is the probability that the child would have cystic fibrosis? That the child would be a CF carrier?

**A2:** The child would have a 25% (¼) probability of having CF, and a 50% (½) probability of being a carrier.

**Q3:** If Zoe is able to have a child of her own someday, and the other parent is not a carrier of cystic fibrosis (he would likely be tested before they chose to have children), what is the probability that the child would have cystic fibrosis? That the child would be a carrier?

**A3:** The child would have a 0% probability of having CF, and a 100% probability of being a carrier.

## Figure 8.13

**Q1:** What is the probability that a child with one parent who has an autosomal dominant disorder will inherit the disease?

**A1:** 50%.

**Q2:** Why are there no carriers with a dominant genetic disorder?

**A2:** Anyone with the gene would express the disorder.

**Q3:** Because dominant genetic disorders are rare, it is extremely rare for both parents to have the condition (genotype *Aa*). Draw a Punnett square with two *Aa* parents. What proportion of the offspring would have the disorder? What proportion would be normal?

**A3:** 75% of the offspring would have the disorder; 25% would be normal.

## Figure 8.14

**Q1:** Which gene was missing or damaged in Felix's case? From what chromosome would a healthy copy be taken?

**A1:** The gene missing or damaged in Felix was the *WAS* gene. A healthy copy of the gene would be taken from an X chromosome.

**Q2:** Why did Dr. Klein's group first conduct gene therapy on mice rather than on humans? What are the advantages and limitations of this approach?

**A2:** In case of unforeseen problems or dangers, a human would not be hurt during the early trials. The limitation is that mice are not humans, so the results are not directly transferable.

**Q3:** If Felix has children of his own someday, will they run the risk of inheriting his disorder, or has gene therapy removed that possibility? Explain your reasoning.

**A3:** Gene therapy does not replace the damaged gene; it simply enables a missing gene product to be made. Felix remains the same genetically, so there will be a risk of his children inheriting the disorder.

# CHAPTER 9

## END-OF-CHAPTER ANSWERS

1. c

2. d

3. a

4. d

5. nucleotide: 4, base pair: 1, DNA molecule: 3, base: 2

6. (See figure below)

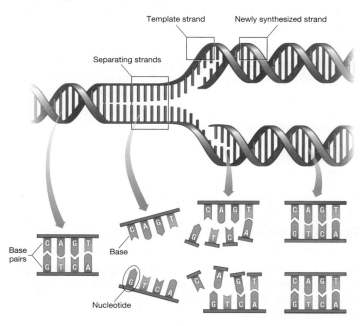

7. PCR, CRISPR

8. c

9. (a) 1, (b) 5, (c) 3, (d) 2, (e) 4

10. (a) D, (b) S, (c) I

11. ATGCAAATCCTGG
    TACGTTTAGGACC

12. (a) 20% T, (b) 30% G, (c) 30% C

13. It is not possible to delete the PERV sequences in each cell of the adult pig. Deleting the PERVs in the DNA of pig embryos ensures that none of the resulting cells will contain PERV sequences.

14. Yes, it matters. Noncoding DNA sequences still perform important functions and can easily be disrupted by mutations. Regulatory regions rely on specific nucleotide sequences, and they could become nonfunctional if those sequences are changed.

## ANSWERS TO FIGURE QUESTIONS

### Figure 9.2

**Q1:** Name one reason why, for a potential transplant, matching the size of organs shown would be important.

**A1:** The pig organs must fit into the space where the human organ was removed. Organs that are too big will not fit. Organs that are too small will not function at the level necessary to sustain life.

**Q2:** Name a tissue transplant for which size matching would not be as important.

**A2:** Tissue from the cornea of the eye need not be exactly the same size.

**Q3:** Name a tissue transplant for which size matching would not be important at all.

**A3:** Bone marrow will replenish in any organisms regardless of size matching.

### Figure 9.3

**Q1:** Name two base pairs.

**A1:** Adenine-thymine (A-T) and cytosine-guanine (C-G).

**Q2:** Why is the DNA structure referred to as a "ladder"? What part of the DNA represents the rungs of the ladder? What part represents the sides?

**A2:** The structure looks like a ladder. The rungs are hydrogen bonds, and the sides are the bases.

**Q3:** Is the hydrogen bond that holds the base pairs together a strong or weak chemical bond? (*Hint:* Refer to Chapter 3 to review chemical bonds, if necessary.)

**A3:** Weak.

### Figure 9.5

**Q1:** If all genes are composed of just four nucleotides, how can different genes carry different types of information?

**A1:** Each gene is composed of different numbers of nucleotides in different arrangements, thus allowing for many different types of information to be conveyed.

**Q2:** Would you expect to see more variation in the sequence of DNA bases between two members of the same species (such as humans) or between two individuals of different species (for example, humans and chickens)? Explain your reasoning.

**A2:** More variation would be expected between two individuals of different species. Within a species, individuals have the same genes but different alleles; across species, there are different genes *and* different alleles.

**Q3:** Do different alleles of a gene have the same DNA sequence or different DNA sequences?

**A3:** Different.

## Figure 9.6

**Q1:** What common mechanism is employed by the guide RNA to find its target DNA sequence?

**A1:** Complementary base-pairing between the guide RNA and one strand of the target DNA.

**Q2:** How many strands of DNA must Cas9 cut to be effective?

**A2:** Both strands must be cut.

**Q3:** Does Cas9 also cause the deletion of DNA from the genome?

**A3:** No. Normal DNA repair proteins cause the deletion of DNA from the genome.

## Figure 9.7

**Q1:** What are the structures that result from the first level of coiling around proteins called?

**A1:** Nucleosomes.

**Q2:** What makes up a "bead" and what makes up the "string" in the beads-on-a-string structure of DNA?

**A2:** The beads are the nucleosomes; the string is the double-stranded DNA linking them together.

**Q3:** What is the name for the structure that is more compact than the beads-on-a-string structure but less compact than an actual chromosome?

**A3:** Chromatin fiber.

## Figure 9.8

**Q1:** Which of the DNA strands here are the template strands? Why are they called "template" strands?

**A1:** The strands from the original double helix are the template strands. They are called this because new DNA is built using the information from them as a template.

**Q2:** What must be broken before replication can begin?

**A2:** Hydrogen bonds between base pairs.

**Q3:** In your own words, explain why replication is described as "semiconservative."

**A3:** It is semiconservative because each new double helix of DNA is composed of one "conserved" strand from the original DNA molecule and one new strand.

## Figure 9.9

**Q1:** PCR replicates DNA many times to increase the amount available for analysis. Why is this process called "amplification"?

**A1:** Because PCR substantially increases, or "amplifies," the quantity of DNA.

**Q2:** During the PCR cycle, what causes the DNA strands to separate?

**A2:** Heat.

**Q3:** Identify a difference between how PCR and DNA replication are accomplished.

**A3:** PCR is the targeted amplification of a specific region of DNA, while DNA replication is the duplication of an entire chromosome.

## Figure 9.10

**Q1:** Summarize how DNA repair works and why the repair mechanisms are essential for the normal function of cells and whole organisms.

**A1:** An error is detected and tagged, then the damaged section of DNA is removed and replaced. Without DNA repair, mutations would persist and potentially result in cell death.

**Q2:** Is DNA repair 100 percent effective?

**A2:** No.

**Q3:** What would happen to an organism if its DNA repair became less effective?

**A3:** Its cells would have more trouble operating properly as the DNA's genetic instructions became less accurate.

## Figure 9.11

**Q1:** What are the three types of point mutations?

**A1:** Substitution, insertion, and deletion.

**Q2:** Sickle-cell disease is an autosomal recessive genetic disorder. How many mutated hemoglobin alleles do people with sickle-cell disease have?

**A2:** Two.

**Q3:** Because of improved treatments, individuals with sickle-cell disease are now living into their forties, fifties, or longer. How might this extension of life span affect the prevalence of sickle-cell disease in the population?

**A3:** Sickle-cell disease will increase because people will be able to survive long enough to reproduce and pass on the sickle-cell allele.

## Figure 9.12

**Q1:** Name a step in this process that is similar to the original CRISPR method that removed the PERVs from the pig genome. (*Hint:* Review Figure 9.6.)

**A1:** CRISPR is employed to remove a gene from the genome.

**Q2:** Name a step in this process that is not included in the original CRISPR method that removed the PERVs from the pig genome.

**A2:** Human stem cells are added to the pig embryos.

**Q3:** What parts of this process would scientists need to change in order to develop several different human organs in a single pig?

**A3:** Scientists would need to use CRISPR to target only the genes that are required for the development of each organ in the pig. Those organs would then be replaced by human organs via use of the same human stem cells added to create a single human kidney in the pig.

# CHAPTER 10

## END-OF-CHAPTER ANSWERS

1. gene expression: 2, gene regulation: 3, transcription: 1, translation: 4

2. (a) tRNA, (b) rRNA, (c) mRNA, (d) tRNA, (e) rRNA

3. redundancy, ambiguity

4. (See figure below)

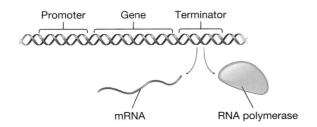

5. (a) 2, (b) 4, (c) 6, (d) 8, (e) 1, (f) 9, (g) 5, (h) 3, (i) 7

6. a, b, c, d

7. (a) asparagine, (b) stop codon, (c) isoleucine, (d) glycine, (e) proline

8. (a) CGU, CGC, CGA, CGG, AGA, or AGG; (b) GCU, GCC, GCA, or GCG; (c) AUG; (d) GGU, GGC, GGA, or GGG

9. b

10. e

11. At control point 2 (transcription), there might be an error in up-regulation. At control point 3 (breakdown of mRNA) there might be an error in down-regulation.

12. a

13. They are similar in that both involve building one molecule from another. They differ in that gene expression is DNA to RNA, while DNA replication is RNA to protein.

## ANSWERS TO FIGURE QUESTIONS

### Figure 10.3

**Q1:** Which is the faster way to produce vaccines: biopharming with plants or creating vaccines in eggs? Why is this important?

**A1:** Biopharming in plants is faster, making it possible to bring greater quantities of vaccine to the public to stem an outbreak before it becomes an epidemic.

**Q2:** How much cheaper is biopharming with plants than creating vaccines in eggs? Why is this important?

**A2:** Biopharming with plants is $364 million cheaper. The lower cost means that the vaccines will be marketed to the public at a much lower price. In addition, more insurance companies will cover the lower costs, and more individuals will receive the vaccine.

**Q3:** Why must tobacco-derived vaccines, or any new medications for that matter, be approved by the FDA?

**A3:** The FDA is responsible for ensuring the safety of drugs (and food). FDA approval is necessary because vaccines are drugs.

### Figure 10.5

**Q1:** In which of the step(s) illustrated here does DNA replication occur? In which step(s) does gene expression occur?

**A1:** Replication: steps 4 and 5. Expression: step 6.

**Q2:** Why do vaccine producers not simply replicate the entire viral genome, instead isolating the gene for one protein and replicating only that gene?

**A2:** Two reasons: It would take more time to replicate the entire genome, and more important, the added time would allow the vaccine to replicate and act like the infectious virus, causing disease rather than preventing it.

**Q3:** What role do the bacteria play in this process? Why are they needed?

**A3:** Because they can produce proteins much more quickly than eukaryotes can, the bacteria rapidly generate the proteins needed for the vaccine.

### Figure 10.6

**Q1:** Why is only one strand of DNA used as a template?

**A1:** Because the other strand would code for the *opposite* mRNA.

**Q2:** If a mutation occurred within the promoter or terminator region, do you think it would affect the mRNA transcribed? Why or why not?

**A2:** Yes, because it wouldn't be clear where the gene started or stopped, so the mRNA might not be transcribed at all, or it would grow too long.

**Q3:** The template strand of part of a gene has the base sequence TGAGAAGACCAGGGTTGT. What is the sequence of RNA transcribed from this DNA, assuming that RNA polymerase travels from left to right on this strand?

**A3:** ACUCUUCUGGUCCCAACA.

### Figure 10.7

**Q1:** In your own words, define RNA splicing. When during gene expression does it occur?

**A1:** RNA splicing removes the introns from the RNA transcript. It occurs after transcription, but before the mRNA transcript is translated into a protein.

**Q2:** What do you predict would happen if the introns were not removed from RNA before translation? Why would it be a problem if the introns were not removed?

**A2:** If the introns were not removed, they would be translated, and the protein would be much larger and presumably nonfunctional.

**Q3:** Where is the mRNA destined to go once it has been transported out of the nucleus?

**A3:** Into the cytoplasm.

## Figure 10.8

**Q1:** Which amino acid always begins an amino acid chain? Which codon and anticodon are associated with that amino acid?

**A1:** Methionine always begins the chain. The associated codon and anticodon are AUG and UAC, respectively.

**Q2:** Each of the codons for stopping translation binds to a tRNA molecule that does not carry an amino acid. How would the binding of a stop codon cause the completed amino acid chain to be released?

**A2:** Because there is no amino acid, the growing protein becomes detached from the mRNA.

**Q3:** Given the partial mRNA sequence that you specified in Figure 10.6's question 3 as being transcribed from the DNA template strand, what is the amino acid sequence that would be translated?

**A3:** The partial mRNA sequence ACUCUUCUGGUCCCAACA yields the amino acid sequence threonine-leucine-leucine-valine-proline-threonine.

## Figure 10.9

**Q1:** How many codons code for isoleucine? For tryptophan? For leucine?

**A1:** Isoleucine: 3. Tryptophan: 1. Leucine: 6.

**Q2:** What codons are associated with asparagine? With serine?

**A2:** Asparagine: AAU, AAC. Serine: AGU, AGC.

**Q3:** From the partial mRNA sequence that you specified in Figure 10.6's question 3 as being transcribed from the DNA template strand, remove only the first A. What amino acid sequence would be translated as a result of this change? How does that sequence compare to the amino acid sequence you translated from the original mRNA sequence? *Bonus:* What kind of mutation is this? (*Hint:* See Chapter 9.)

**A3:** The partial mRNA sequence CUCUUCUGGUCCCAACA yields the amino acid sequence leucine-phenylalanine-tryptophan-serine-glutamine. There is one less amino acid, and each of the amino acids is different. *Bonus:* Deletion.

## Figure 10.10

**Q1:** Why is an insertion or a deletion in a gene more likely to alter the protein product than a substitution, such as A for C, would?

**A1:** Because an insertion or deletion causes a "frameshift," so every single amino acid from that point on is likely to be different, as opposed to a substitution, where, at most, one amino acid changes.

**Q2:** Which would you expect to have more impact on an organism: a point mutation as shown here, or the insertion or deletion of a whole chromosome (discussed in Chapter 8)?

**A2:** The loss or addition of an entire chromosome, with all the genes on it, is likely to have far more impact on an organism than a mutation within a single gene.

**Q3:** Which mechanisms in a cell prevent mutations? (*Hint:* Refer back to Chapter 6 if needed.)

**A3:** Checkpoints in the cell cycle prevent (or at least repair) mutations.

## Figure 10.11

**Q1:** As illustrated here, at what control point is transcription regulated?

**A1:** Control point 2.

**Q2:** What is a possible advantage of regulating gene expression before transcription, versus after?

**A2:** The cell does not waste time and energy producing mRNA transcripts that it will not use.

**Q3:** If you wanted to up-regulate production of the hemagglutinin protein in a tobacco plant carrying the hemagglutinin gene, at which control point(s) would that be possible? Justify your reasoning.

**A3:** Control points 1–4 all could have an impact on the amount of hemagglutinin being produced by a cell. The levels of DNA compaction, transcription, mRNA degradation, and translation all work together to increase or decrease the production of hemagglutinin. Modifying any one of these or any in combination can have a large impact on production.

# CHAPTER 11

## END-OF-CHAPTER ANSWERS

1. c

2. b

3. c

4. biogeography: 4, fossil record: 1, DNA sequence similarity: 2, embryonic similarity: 5, homologous traits: 3

5. d

6. b

7. adaptation, natural selection

8. In artificial selection, humans choose which organisms survive and reproduce. In natural selection, the environment determines which organisms survive and reproduce, by selecting for individuals with more beneficial inherited traits and against individuals with less useful ones.

9. d

10. species Y

11. d

12. All (or most) of the current continents.

## ANSWERS TO FIGURE QUESTIONS

## Figure 11.3

**Q1:** What is selective breeding, and how does it work?

**A1:** Selective breeding is the process by which humans allow only individuals with certain inherited characteristics to breed. Generation after generation, the selective breeder chooses which individuals mate and pass their traits to offspring.

**Q2:** Explain how selective breeding leads to artificial selection.

**A2:** Over time, the population resulting from generations of selective breeding can change significantly, and we can then say that artificial selection has occurred.

**Q3:** Name as many organisms as you can whose current characteristics are due to artificial selection.

**A3:** Examples include any domesticated animal, including common pets and agriculturally important animals. All agricultural plants are also products of artificial selection.

## Figure 11.4

**Q1:** What is natural selection?

**A1:** Natural selection is the process by which individuals with genetic characteristics that are advantageous for a particular environment survive and reproduce at a higher rate than do individuals that have other, less useful characteristics.

**Q2:** If heavy rains caused an abundance of small, tender seeds and fewer large seeds, what do you predict would happen to the average beak size of the finches?

**A2:** Average beak size would become smaller, more like the beak of the bird on the left.

**Q3:** Compare and contrast artificial selection and natural selection. Name two ways in which they are similar. How are they different?

**A3:** They are similar in that both occur in populations, not individuals; that there must be a change or changes in the population; and that they occur over time—usually many, many generations. They are different in that artificial selection results from selective breeding performed by human beings, while natural selection occurs by the breeding of individuals in a population that survive in a particular environment.

## Figure 11.6

**Q1:** What is the general definition of a fossil?

**A1:** Fossils are the mineralized remains of formerly living organisms or the impressions of formerly living organisms.

**Q2:** How are the ancestors of modern whales different from their present form?

**A2:** They became larger, they lost their back limbs, and their front limbs became proportionally smaller.

**Q3:** What is meant by the term "intermediate fossil" when referring to the fossil record?

**A3:** Intermediate forms or fossils are fossils of species with some similarities to the known extinct ancestral group and some similarities to the descendant or currently living species. They can be thought of as "missing links" in evolution.

## Figure 11.9

**Q1:** Why do water-dwelling animals have thicker bones than land-dwelling animals?

**A1:** Thick bone is an adaptation to living in water. Thick bones are heavier and help water-dwelling animals control their buoyancy (ability to float).

**Q2:** Why does this thick-bone adaptation suggest a water-dwelling lifestyle?

**A2:** Fossil mammals with thick bones are presumed to have been water dwellers because almost all currently living water-dwelling mammals have thick bones.

**Q3:** How did this adaptation likely increase survival or reproduction in *Indohyus*?

**A3:** Thick bones probably enabled *Indohyus* to forage on the bottoms of lakes or ponds more efficiently than could species with lighter bones that had to work harder to stay submerged. A feeding advantage could have enabled *Indohyus* to eat more, live longer, and have a higher reproductive rate (producing more babies that survived) than those with lighter bones.

## Figure 11.11

**Q1:** What is meant by the term "common ancestor"? Give an example.

**A1:** A common ancestor is the species from which at least two currently living species both descended, ancestral species that two or more new species arose from through a change in the traits of a population over time.

**Q2:** Why are homologous structures among organisms evidence for evolution?

**A2:** Homologous structures are parts of an organism that have changed in size or specific form over time but are easily determined to be the same structure in the ancestral species from which the organism evolved. Species with homologous traits are all related by originally coming from an ancestor with those specific structures.

**Q3:** Aside from skeletal structural similarities, what other commonalities among organisms might be considered homologous?

**A3:** Any traits that are shared by related organisms and also shared in an ancestor could be homologous traits. Some examples include mammary glands, egg laying, structures to extract oxygen from air or water, and the use of DNA as the genetic material.

## Figure 11.12

**Q1:** Why are vestigial structures among organisms evidence for evolution? Give an example of another vestigial structure.

**A1:** Vestigial structures are evidence for evolution because they are shared among related species that all have a common ancestor. Goose bumps in humans are one example. In our furry ancestors, goose bumps fluffed the fur, thereby increasing its insulating effects and helping the animals to keep warm.

**Q2:** Are vestigial structures also homologous structures? Explain.

**A2:** Yes. Vestigial traits are shared in organisms that have a common ancestor. For example, all organisms that descended from a furry ancestor have goose bumps when they are chilled or cold.

**Q3:** Why do vestigial structures still exist if they are no longer useful?

**A3:** Only traits that harm an organism's ability to survive and reproduce disappear from the fossil record, because organisms having these traits die and do not reproduce. Traits that are merely useless and not harmful will persist in the organisms that have them. They may diminish because these structures are no longer needed and do not give organisms a selective advantage, but the organisms survive and reproduce just as well with or without them.

## Figure 11.13

**Q1:** If a sequence from another species showed a 96 percent sequence similarity to humans, would that species be more or less closely related to humans than chimpanzees are?

**A1:** The hypothetical species would be less similar to humans than chimpanzees are (96% similar versus 98.4% in chimps).

**Q2:** Should similarities in the DNA sequences of genes be considered evolutionary homology? Explain.

**A2:** All living organisms use DNA as their genetic material, suggesting that the first true ancestral cell used DNA as its genetic material. DNA sequence similarity is a homologous trait because all descendant cells use DNA; exactly the same four nucleotide molecules, in different orders, make up the DNA in all cells on Earth.

**Q3:** How is the increased similarity in the DNA sequences of genes between more closely related organisms—and the decreased similarity between less closely related organisms—evidence for evolution? Use the example in this figure to support your answer.

**A3:** DNA sequence similarity is the gold standard for determining species relatedness because it is the genetic material in all cells on Earth—the best homologous trait that exists. The changes in DNA sequences in populations over time create the changes in traits that drive evolution. We can map the changes that occur in populations or species by looking at sequence similarity and re-creating a family tree. All other evidence for evolution supports the theory that all organisms derived from an ancestral cell that used DNA as its genetic material. The more related a species is to another species, the more similar the DNA sequences are. Humans and chimps are primates, mammals, and vertebrates, and they are more similar in DNA sequence than are humans and mice, which are only both vertebrates and mammals, but not both primates. Of the examples in this figure, chickens are the least similar to humans because although they are birds, which are vertebrates, they are not mammals or primates. A nonvertebrate animal like a jellyfish or worm would be even less similar to humans than the species named here, but it would likely still show some similarity.

## Figure 11.14

**Q1:** Why should we expect to find *N. fosteri* fossils all over the world, given that it first evolved in Pangaea?

**A1:** We expect to find *N. fosteri* fossils all over the world because these organisms existed before the breakup of the mass continent Pangaea. During the breakup, these organisms traveled with their continent to the current locations.

**Q2:** Can we use biogeographic evidence to support evolution without using fossil evidence? Give examples.

**A2:** Yes. We can use the current locations of living organisms that are related to support evolution by biogeography. For example, members of the primate family are found on almost all the continents on Earth, suggesting that their common ancestor lived at the time of Pangaea.

**Q3:** How might we use DNA sequence similarities together with biogeography as evidence for evolution?

**A3:** We can couple DNA sequencing with the locations of fossil or living organisms to support evolution by common descent. We can perform DNA sequence analysis on all the primates on Earth, coupled with their current locations, to reenact the history of primate evolution.

## Figure 11.15

**Q1:** Why are the similarities among organisms during early development evidence for evolution? Give an example.

**A1:** Similarities between organisms during early development suggest that they have a common ancestor whose early development occurred in the same or similar manner. All vertebrates go through similar stages of development in the early embryo. Many invertebrate organisms also share the same steps in embryonic development.

**Q2:** Are the similar structures among vertebrate species during embryogenesis homologous structures? Explain.

**A2:** Yes. The similar structures are homologous traits because they are shared with a common ancestor.

**Q3:** Why do embryonic structures still exist at points during embryogenesis if they are not used after birth?

**A3:** These structures can be considered vestigial traits, since they are now useless to the organism in which they still exist embryonically. Remember, vestigial traits still exist because they do not harm the organisms' ability to survive and reproduce.

# CHAPTER 12

## END-OF-CHAPTER ANSWERS

1. genetic drift, establish a new distant population

2. b

3. b

4. c

5. b

6. b

7. An individual that survives well but is unattractive to potential mates or is unable to compete for access to mates will not reproduce and will not pass on genes.

8. Gene flow is the most likely mechanism because mutations are random, natural selection would have caused populations in different environments to diverge, and genetic drift is most relevant for small populations.

9. Population bottlenecks cause individuals in the resulting population to be more genetically similar to each other. In this case, the two individuals are so similar that each individual does not distinguish another devil's cells as different from its own cells.

10. b

11. b

## ANSWERS TO FIGURE QUESTIONS

### Figure 12.2

**Q1:** What is natural selection selecting for here?

**A1:** Methicillin-resistant *S. aureus* (MRSA).

**Q2:** Why do bacteria that are not genetically resistant to antibiotics die out when exposed to antibiotics?

**A2:** After entering a bacterium, an antibiotic generally blocks or poisons one or more processes of the bacterium's life cycle so that it cannot survive or reproduce. Bacteria that have a mechanism to pump out the poison generally survive the poison and live to reproduce; they are termed "resistant."

**Q3:** Why is the antibiotic represented by a kitchen strainer in this figure?

**A3:** The antibiotic is depicted as a kitchen strainer because antibiotics act like strainers: they can "catch" or kill most bacteria in a population, but there will always be at least one bacterium that can survive the antibiotic assault and slip through the strainer.

### Figure 12.4

**Q1:** What is the difference between MRSA and VRSA?

**A1:** Both of these bacteria are members of the species *Staphylococcus aureus*. The species name is the "SA" part of both names. The "MR" in MRSA stands for "methicillin-resistant," and the "VR" in VRSA stands for "vancomycin-resistant." The only difference between these two populations is that MRSA survives in the presence of the antibiotic methicillin but can be killed by vancomycin, while VRSA survives in the presence of both methicillin and vancomycin.

**Q2:** Why is there a clear zone (the "zone of inhibition") around the paper disk in the top dish but not in the bottom dish?

**A2:** The clear zone represents the area where the antibiotic has seeped into the medium from the antibiotic-soaked paper disk and killed off the bacteria. The bacteria cannot grow here, so all you see is the growth medium in the dish, with no bacteria growing on it. The rest of the dish is covered by bacteria and appears opaque.

**Q3:** Why is the lack of a clear zone around the paper disk in the bottom dish so alarming?

**A3:** The lack of a clear zone in the bottom dish suggests that the antibiotic of last resort, vancomycin, cannot kill the bacteria and they grow just as well in the antibiotic area as in the areas away from the vancomycin-soaked paper disk. If the last-resort antibiotic doesn't kill these bacteria, there is no current antibiotic

that will. VRSA is a deadly bacterial infection against which we have no good defense.

### Figure 12.5

**Q1:** What would the white-fur-pigment allele frequency be if three of the homozygous black allele mice (having two black alleles) were heterozygous (having one white and one black allele) instead?

**A1:** 16/30 = 53%.

**Q2:** What would the white-fur-pigment allele frequency be if all of the white mice died and were therefore removed from the population? Would the black-fur-pigment allele frequency be affected? If so, how?

**A2:** The white-fur-pigment allele frequency would be 3/20 = 15%. Yes, the black-fur-pigment allele frequency would be affected; there would then be 17 black alleles out of a total of only 20 alleles: 17/20 = 85%.

**Q3:** What would the white-fur-pigment allele frequency be if all of the gray mice died and were therefore removed from the population?

**A3:** 10/10 = 100%.

### Figure 12.6

**Q1:** Why does the population of *S. aureus* bacteria *not* pose a life-or-death health threat outright?

**A1:** These are the bacteria that normally live on our skin and do not harm us unless there is a major skin disturbance like a burn or a large scrape that is not cleaned and kept protected.

**Q2:** Why do vancomycin-resistant bacteria have a higher frequency in the population after treatment with vancomycin?

**A2:** All of the bacteria that do not contain the resistance allele are killed by the vancomycin and therefore no longer exist. The only bacteria left are the vancomycin-resistant bacteria (VRSA).

**Q3:** If this figure used the mouse example of allele frequency from Figure 12.5, and the white mice increased in numbers like the vancomycin-resistant bacteria here did, what would happen to the frequencies of the white-fur-pigment and black-fur-pigment alleles?

**A3:** The white-fur-pigment allele would increase in frequency, and the black-fur-pigment allele would decrease in frequency.

### Figure 12.7

**Q1:** If one extreme phenotype makes up most of a population after directional selection, what happened to the individuals with the other phenotypes?

**A1:** They were killed and eaten by predators.

**Q2:** What do you think would happen to the phenotypes of the peppered moth if the tree bark was significantly darkened again by disease or pollution?

**A2:** Since the moths that survive are more similar to the color of the bark and thus are protected from birds, which cannot as easily see them to kill and eat them, the phenotype of the population of peppered moths would become darker like the trees.

**Q3:** What do you think would happen to the phenotypes of the peppered moth if the tree bark became a medium color, neither light nor dark? (You will need to read the next paragraph to answer this question.)

**A3:** Stabilizing selection would likely occur, and only medium-colored moths would not be killed and eaten by birds.

## Figure 12.8

**Q1:** Think of another example of stabilizing selection in human biology. Has modern technology or medicine changed its impact on the resulting phenotypes?

**A1:** Stabilizing selection probably affected many human traits before modern technology and medicine played a major role in survival and quality of life. Examples include adult height and weight, which would be affected by many hormone levels and overall metabolism.

**Q2:** How do you think a graph of birth weight versus survival for a developing country with little health care would compare to the graph shown here?

**A2:** This graph would be even sharper, with less survival at either end. Evolution would be more stabilizing than in the example shown here.

**Q3:** How do you think a graph of birth weight versus survival for an affluent city in the United States today would compare to the graph shown here?

**A3:** This graph would be much wider, with more survival at both ends. Evolution would be much less stabilizing than in the example shown here.

## Figure 12.9

**Q1:** Almost all birds starved during the dry season depicted here. What type of selection would have been present if only the intermediate-beaked birds had survived (instead of the small- and large-beaked birds)?

**A1:** Stabilizing selection.

**Q2:** Describe a scenario in which African seed crackers would experience directional selection for either smaller- or larger-beaked birds. What kind of environmental conditions might bring about such a situation?

**A2:** If only small seeds were produced in a particular year, then only small-beaked birds would survive and the population would evolve toward smaller beaks. Similarly, if only large seeds were produced, then only large-beaked birds would survive and the population would evolve toward larger beaks. An environmental condition that might bring about smaller seeds would be a situation in which the faster germination time of smaller seeds was an advantage—for example, a very brief growing period. A condition that might favor larger seeds would be a situation in which the greater resources contained in the seed would allow it to survive longer—for example, an extended drought.

**Q3:** Of the three patterns of natural selection presented in this discussion, which one always results in two different phenotypes in the following generations?

**A3:** Disruptive selection.

## Figure 12.10

**Q1:** How is convergent evolution different from evolution by common descent?

**A1:** Convergent evolution is essentially the opposite of evolution by common descent. Convergent evolution begins with two distantly related organisms that, over many generations, end up with similar phenotypes because they have adapted to similar environments. Evolution by common descent begins with an original common ancestor and, over many generations, may split into many different populations that are phenotypically different.

**Q2:** What is the main difference between a homologous trait (see Figure 11.11) and an analogous trait?

**A2:** A homologous trait is shared between organisms because a common ancestor had that trait; an analogous trait performs a similar function in different organisms but is not shared by a common ancestor. Analogous traits form through convergent evolution.

**Q3:** Why are convergent traits considered evidence for evolution (see Chapter 11)?

**A3:** Convergent traits occur through changes in allele frequencies over time—essentially the definition of evolution. Convergent evolution results in organisms that are better adapted to their environment—again, evolution.

## Figure 12.13

**Q1:** If a goose with genotype *AA* had migrated instead of the goose with genotype *aa*, would the scenario described here still be considered gene flow? Why or why not?

**A1:** No, this is technically not gene flow. Although alleles are being exchanged, they are the same as the existing alleles in the population and will not change allele frequencies over time and many generations.

**Q2:** If a goose with genotype *Aa* had migrated instead of the goose with genotype *aa*, would the scenario still be considered gene flow? Why or why not?

**A2:** Yes, this is gene flow. Although the effect is not as extreme as with the *aa* genotype, the *Aa* genotype introduces a new allele into an existing population, creating offspring that can be *Aa*, and thereby changing allele frequencies over time and many generations.

**Q3:** If the goose with genotype *aa* had migrated to population 2 as shown but had failed to mate with any of the *AA* individuals, would the scenario still be considered gene flow? Why or why not?

**A3:** No, this is not gene flow. Just adding an individual with different alleles to a population does not count as gene flow. There must be an exchange of alleles between the newcomer and an existing individual.

**Figure 12.14**

**Q1:** Why do you think a genetic bottleneck is more likely to occur in a small population than in a large population?

**A1:** In a large population, it is less likely that a chance event can kill off almost all of the individuals, leaving only a few behind that randomly represent only one of multiple phenotypes. In a small population, a tsunami, hurricane, volcanic eruption, or other natural disaster could easily kill off all but a few individuals. All subsequent offspring would arise from these few individuals, whatever phenotype they might have, regardless of which phenotypes in the original population were best adapted.

**Q2:** Genetic drift is often described as a "chance event." Give other examples of chance events that could cause a genetic bottleneck.

**A2:** Examples that could cause a genetic bottleneck include deadly viruses, famine, drought, immigration of many predators, habitat loss, tsunami, or other natural disaster.

**Q3:** Which resulting population has more genetic diversity?

**A3:** The population on the left.

# CHAPTER 13

## END-OF-CHAPTER ANSWERS

1. b

2. a

3. c

4. c

5. genetic divergence, allopatric

6. Postzygotic barriers, an infertile hybrid

7. d

8. b, c, d

9. (a) 5, (b) 4, (c) 1, (d) 2, (e) 3

10. b

11. b

12. Any large geographic barrier such as a river, canyon, or mountain range might bring about allopatric speciation. In addition to the barrier, genetic divergence between the separated populations is necessary for speciation.

13. The ten species must have differed significantly in traits that were unobservable (or at least were not observed by the scientists) and that meant the fish were unable to mate and produce viable offspring.

## ANSWERS TO FIGURE QUESTIONS

### Figure 13.4

**Q1:** How do we know that these rattlesnakes are members of the same species?

**A1:** The caption specifies that they successfully mated and the resulting offspring survived and reproduced.

**Q2:** How would you design an experiment to determine whether two populations are distinct species according to the biological species concept?

**A2:** Your experiment would require mixing individuals from the two populations under conditions conducive to sexual reproduction. If the individuals did mate, the offspring would then need to be raised to maturity and be set up to also mate and produce live offspring.

**Q3:** For which types of populations does the biological species concept *not* work as a way of determining how they are related?

**A3:** Because it requires sexual reproduction, the biological species concept cannot be applied to populations that reproduce asexually (such as bacteria).

### Figure 13.5

**Q1:** List three kinds of information that scientists use to distinguish between species.

**A1:** Biogeographic information, DNA sequence similarity, and physical characteristics.

**Q2:** What differences can you observe between the individuals in the photos? Why are these differences not enough to confirm that they are from two different species?

**A2:** There are several striking differences in coloration (for example, along the side of the body) between the frogs in the two photos. These differences could simply be within-species variation, much like hair and eye color differences between humans of the same species.

**Q3:** How is genetic divergence among populations determined?

**A3:** Genetic divergence between two populations is determined by examination of the DNA sequences of many individuals in the two populations. A lot of similarity between DNA sequences suggests little genetic divergence, while the existence of many changes between the gene sequences suggests much greater genetic divergence.

### Figure 13.6

**Q1:** What is the definition of gene flow? How was gene flow blocked between these species?

**A1:** Gene flow is defined as the passing of alleles between different populations of the same species. Squirrels from the populations on either side of the river could not mate because of the geographic barrier, and thus gene flow was blocked.

**Q2:** Name as many types of geographic barriers as you can. Which do you think would be the best at blocking gene flow?

**A2:** Examples of geographic barriers include but are not limited to rivers, lakes, oceans, glaciers, mountains, canyons, brick walls, freeways, fences. The larger the barrier, the better it blocks two individuals from finding each other and mating.

**Q3:** Are geographic barriers universal for all species? If not, name a geographic barrier that might block gene flow for one species but not another.

**A3:** No, geographic barriers are not universal. A river, for example, might block gene flow between two lizard populations but not two bird populations.

## Figure 13.7

**Q1:** What factors must be present for allopatric speciation to occur?

**A1:** A geographic barrier.

**Q2:** If a geographic barrier is removed and the two reunited populations intermingle and breed, what attributes must the offspring have in order for the two populations, according to the biological species concept, to be considered still the same species?

**A2:** The offspring must be viable (alive) and fertile (be able to reproduce).

**Q3:** If the two populations in question 2 are determined to still be the same species, did allopatric speciation occur?

**A3:** If they are still the same species, no speciation occurred.

## Figure 13.11

**Q1:** Describe how coevolution, as with the hummingbird bill and hummingbird-pollinated flowers, is different from the kind of evolution described in Chapters 11 and 12.

**A1:** In coevolution, a species evolves directly to interact better with another species. Coevolution can be both species evolving to interact better with each other, or it can be just one of the species adapting to the other species.

**Q2:** Is coevolution the same thing as convergent evolution, described in Chapter 12? Why or why not?

**A2:** No. In convergent evolution, two genetically different species look more alike over time because they are adapting to similar environments. In coevolution, two different species adapt to each other's adaptations over time.

**Q3:** Do you think one species' adapting over time to feed specifically and extremely successfully on another species is an example of coevolution? Why or why not?

**A3:** Yes. Coevolution does not have to be reciprocal.

## Figure 13.12

**Q1:** What is the main difference between allopatric and sympatric speciation?

**A1:** Allopatric speciation requires a geographic barrier, and sympatric speciation cannot include a geographic barrier.

**Q2:** Name two events that must happen for both allopatric speciation and sympatric speciation to occur.

**A2:** The two populations must be reproductively isolated, and genetic change must occur.

**Q3:** Do you think all of the 500 species in Lake Victoria arose through sympatric speciation? Why or why not?

**A3:** Yes, because there are no geographic barriers to separate the populations in the lake. There would have to be a human-made wall or fence for allopatric speciation to occur.

## Figure 13.13

**Q1:** What does "prezygotic" mean?

**A1:** "Prezygotic" means "before a zygote" or "before fertilization of an egg by a sperm"—in other words, no fusion of egg and sperm.

**Q2:** How is the booby's ritual dance a prezygotic reproductive barrier?

**A2:** This dance happens before mating. If the dance is not correct, no mating happens. No mating means no fusion of egg and sperm.

**Q3:** What are some other prezygotic reproductive barriers besides a mating dance?

**A3:** Examples include all geographic barriers, inability of egg and sperm to fuse for genetic reasons (gamete incompatibility and isolation), inability to mate because the genitalia are physically incompatible (mechanical incompatibility and isolation), and ecological isolation in which two species breed in different portions of their habitat, at different seasons, or at different times of day.

## Figure 13.14

**Q1:** Which species is/are sympatric with *Echinometra lucunter*?

**A1:** *E. viridis*, because it is found on the same side of the Panama landmass as *E. lucunter*.

**Q2:** Which species is/are allopatric with *E. lucunter*?

**A2:** *E. vanbrunti*, because it is found on the opposite side of the Panama landmass as *E. lucunter*.

**Q3:** If two individuals have incompatible gametes, what will be the result of a mating event between them?

**A3:** No zygote will form. Egg and sperm will not fuse.

# CHAPTER 14

## END-OF-CHAPTER ANSWERS

1.  c

2.  b

3.  a

4.  d

5.  clade: 2, node: 3, lineage: 5, evolutionary tree: 4, shared derived trait: 1

6.  prokaryotes, Eukarya, Plantae, Animalia

7.  d

8.  e

9.  (a) Eukarya, Animalia; (b) Eukarya, Fungi; (c) Eukarya, Plantae; (d) Eukarya, Plantae; (e) Eukarya, Animalia

10. (a) kingdom, phylum, class, order, family, genus, species; (b) domain; (c) answers will vary

11. (a) 3, (b) 1, (c) 4, (d) 2, (e) 5

12. (See figure below)

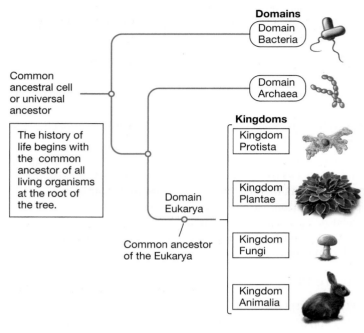

13. (a) Bacteria, Archaea, Protista; (b) Bacteria, Archaea;
(c) Bacteria, Archaea; (d) Animalia; (e) Plantae, Bacteria

14. **Xu:** We hypothesize that *Archaeopteryx* and *Xiaotingia* are dinosaurs, closely related to the deinonychosaurs.
**Godefroit:** We hypothesize that *Archaeopteryx* and *Xiaotingia* are early birds, and *Aurornis* is an earlier bird.

## ANSWERS TO FIGURE QUESTIONS

### Figure 14.2

**Q1:** Why is there a shared line from the universal ancestor for Archaea and Eukarya?

**A1:** Archaea and Eukarya share a common ancestor with each other more recently than either one does with Bacteria.

**Q2:** Where would birds be found within this figure? What about humans?

**A2:** Birds and humans would both be found as branches of Eukarya.

**Q3:** To which domain would you expect a disease-causing organism to belong? What if the organism was multicellular?

**A3:** Disease-causing organisms could be found within Bacteria, but also within Eukarya (kingdom Fungi). Multicellular organisms would be within Eukarya only.

### Figure 14.3

**Q1:** During what geologic period did life on Earth begin?

**A1:** The Precambrian.

**Q2:** How long ago did species begin to move from water to land? What period was this?

**A2:** About 480 mya, in the Ordovician.

**Q3:** In what period would *Archaeopteryx* have been alive?

**A3:** The Jurassic.

### Figure 14.5

**Q1:** In what ways were theropods the same as modern birds? Give at least two similarities.

**A1:** They ran on two legs and had hollow, thin-walled bones.

**Q2:** In what ways did theropods differ from modern birds? Give at least two differences.

**A2:** They were more variable in size and in skin covering.

**Q3:** Birds are often referred to as "living dinosaurs." Is this accurate? Why or why not?

**A3:** Birds are direct descendants of dinosaurs, so they could be argued to *be* dinosaurs.

### Figure 14.6

**Q1:** In the traditional tree, identify the node showing the common ancestor for early birds and dinosaurs.

**A1:** The common ancestor came after the split from the theropods.

**Q2:** What do both the traditional tree and Xu's tree suggest about troodontids and dromaeosaurids?

**A2:** Both trees suggest that these two groups were closely related.

**Q3:** In both trees, identify the node for the common ancestor of *Archaeopteryx* and other birds. In what way are the nodes different in the two trees?

**A3:** The traditional tree shows *Archaeopteryx* on the bird side of the split between birds and dinosaurs; the common-ancestor node is the point where the birds split into two groups. Xu's tree shows *Archaeopteryx* on the dinosaur side of the split; the common-ancestor node is the theropods.

### Figure 14.9

**Q1:** What group of organisms shares the most recent common ancestor with plants?

**A1:** Green algae.

**Q2:** Are fungi more closely related to plants or to animals? Does the answer surprise you? Why or why not?

**A2:** Fungi are more closely related to animals.

**Q3:** If you were to create an evolutionary tree in which amoebas were included within the kingdom of organisms to which they were the most closely related (rather than with protists, where they are currently placed), where would you put them?

**A3:** Either with Animalia or with Fungi.

### Figure 14.10

**Q1:** Within which category are individuals most closely related to one another?

**A1:** Species.

**Q2:** Within which category are individual species most distantly related?

**A2:** Kingdom.

**Q3:** Are individual species more closely related within the same order or within the same family?

**A3:** Family.

## Figure 14.14

**Q1:** Is *Xiaotingia* an earlier or later bird than *Archaeopteryx* in this tree?

**A1:** *Xiaotingia* is later than *Archaeopteryx*.

**Q2:** If a future study, based on more fossils or new measurements, placed *Archaeopteryx* back with dinosaurs, would this suggest that birds are not related to dinosaurs? Why or why not?

**A2:** No. What is in question is not *whether* birds are related to dinosaurs, but when exactly they split off from related dinosaur species and which species is the first example of that split.

**Q3:** If you were to create an evolutionary tree of modern birds, where would you expect to place the roadrunner (judging by its appearance in this figure) as compared to a house sparrow or pigeon?

**A3:** The roadrunner looks more like a dinosaur, so it might be argued that it belongs closer to the base of the tree.

## Figure 14.15

**Q1:** What extinction event occurred about 200 mya? What animal groups were most affected by this event?

**A1:** Triassic; reptiles.

**Q2:** Which of the mass extinctions appears to have removed the most animal groups? How long ago did this extinction occur?

**A2:** Permian; about 250 mya.

**Q3:** The best studied of the mass extinctions is the Cretaceous extinction. Why do you think it has been better studied than the other extinctions?

**A3:** Because it is more recent, more fossils may be available from this extinction than from earlier ones.

# CHAPTER 15

## END-OF-CHAPTER ANSWERS

1. d

2. e

3. c

4. b

5. Archaea, Bacteria, Prokaryotes, eukaryotes

6. b

7. e

8. (a) 1, (b) 5, (e) 3, (d) 2, (c) 4

9. a

10. Each individual eukaryote weighs much more than an individual prokaryote.

11. Prokaryotes are smaller and have simpler cellular structures, so they take less time to divide into two. Because they replicate so much more quickly, it is hard for our immune systems to respond to them in a timely way.

## ANSWERS TO FIGURE QUESTIONS

### Figure 15.3

**Q1:** If an individual prokaryote divides every 20 minutes, how many individuals will there be after an hour?

**A1:** 8.

**Q2:** If the generation time is 20 minutes, how much time will have gone by when the final generation shown has doubled?

**A2:** 20 minutes.

**Q3:** Many bacteria are able to reproduce more quickly in warmer conditions. What does this suggest to you about the importance of refrigerating foods?

**A3:** Refrigerating food slows generation time, so dangerous bacteria like *E. coli* have less of a chance to increase to threatening levels.

### Figure 15.4

**Q1:** In which of the three ways did the navel microbiome participants contribute?

**A1:** They were experimental subjects.

**Q2:** Which of the advantages listed above do you think the navel microbiome citizen scientists received?

**A2:** They may have felt a sense of contribution and purpose.

**Q3:** Would you be willing to contribute to the navel microbiome project? Why or why not?

**A3:** Various answers possible; some might express privacy concerns.

### Figure 15.5

**Q1:** The figure (and thus the study) demonstrates that Archaea and Eukarya are more closely related to each other than to Bacteria. How is that illustrated?

**A1:** The first split seen is Bacteria, apart from the other groups.

**Q2:** Where in the figure would you place the first life found on Earth?

**A2:** At the base of the tree.

**Q3:** Where in the figure is the earliest split between Archaea and Eukarya? When did that occur?

**A3:** The earliest split occurred where the cluster identified as Eukarya begins. This would have been about 2.7 million years ago.

### Figure 15.7

**Q1:** Which of these shapes do you think *Streptococcus* would take?

**A1:** Sphere, or coccus.

**Q2:** From the micrographs here, does it appear that all prokaryotes have a flagellum?

**A2:** No.

**Q3:** Which one of these shapes is most clearly capable of self-motility? Why?

**A3:** The comma, or vibrio, because of the flagellum.

## Figure 15.8

**Q1:** Which shape in Figure 15.7 corresponds to the archaeans from deep-sea thermal vents?

**A1:** Rod, or bacillus.

**Q2:** Why are many archaeans referred to as "extremophiles"?

**A2:** They are found in extreme environments (at extremes of heat, acidity, and salt level).

**Q3:** Is there anywhere you think archaeans could *not* survive? Justify your answer.

**A3:** Various answers are possible.

## Figure 15.9

**Q1:** How do individual bacteria know that they have a "quorum"?

**A1:** They detect increasing concentrations of signaling molecules.

**Q2:** There is a well-known biofilm found in your mouth. What is it?

**A2:** Dental plaque.

**Q3:** Under what conditions might bacteria want to coordinate (via quorum sensing) to increase their reproductive rate?

**A3:** If environmental conditions are very good, then it makes sense for the population of bacteria to grow as quickly as possible to take advantage of those conditions.

## Figure 15.10

**Q1:** What source of energy would you expect a cave-dwelling prokaryote to use?

**A1:** It could not rely on light, so it would most likely use chemical energy.

**Q2:** In which of these categories would you place the bacteria responsible for nitrogen fixing? Why?

**A2:** Chemoautotroph, because they use nitrogen from the atmosphere and carbon from carbon dioxide.

**Q3:** In which of these categories do decomposers belong? Explain your reasoning.

**A3:** Chemoheterotroph, because they receive both carbon and energy from dead or dying organisms.

## Figure 15.11

**Q1:** From the prokaryotic structures shown in Figure 15.7, what shape would you assign to drawing number 8 in van Leeuwenhoek's illustration?

**A1:** The comma, or vibrio, because of the flagellum.

**Q2:** Which of the large prokaryote drawings has the coccus shape?

**A2:** 24.

**Q3:** Do you think all of these "animalcules" drawn by van Leeuwenhoek are prokaryotes? Why or why not?

**A3:** No. Some have such an elaborate structure that they are most likely eukaryotes.

# CHAPTER 16

## END-OF-CHAPTER ANSWERS

1. a

2. d

3. c

4. d

5. Protista, aquatic, autotrophs

6. (a) 3, (b) 5, (c) 4, (d) 1, (e) 2

7. e

8. The kingdom Plantae contains only autotrophs. The kingdom Fungi contains only heterotrophs.

9. For each case, answers to the first part of the question will vary; here's an example for part (b): Lichens are a symbiotic relationship between a fungus and intracellular bacteria. The photosynthetic bacteria produce food for the fungus, and the fungus provides protection and anchoring for the bacteria. Answers to the second and third parts of the question are as follows:

   (a) The partners of mycorrhizae are plants (domain Eukarya, kingdom Plantae) and fungi (domain Eukarya, kingdom Fungi). The relationship is a mutualism.

   (b) The partners of a lichen are fungi (domain Eukarya, kingdom Fungi) and either algae (domain Eukarya, kingdom Protista) or cyanobacteria (domain Bacteria). The relationship is a mutualism.

   (c) The partners in this relationship are hermit crabs and the (nonliving) shells of various species of snails (both domain Eukarya, kingdom Animalia). The relationship is a commensalism.

   (d) The partners in this relationship include the malaria protozoan, its insect host, and its vertebrate host (all domain Eukarya; kingdom Protista for the malaria protozoan, kingdom Animalia for the two host types). It is a parasitic relationship.

10. Gymnosperms and angiosperms produce pollen, which is able to travel via wind or a pollinator to other individuals of the same species, thereby enabling sexual reproduction in a nonaquatic environment.

11. Possible answers include athlete's foot, fungal pneumonia, and yeast infections.

## ANSWERS TO FIGURE QUESTIONS

### Figure 16.3

**Q1:** Are ciliates more closely related to euglenoids or to diatoms? To euglenoids or to forams?

**A1:** Ciliates are more closely related to both diatoms and forams than they are to euglenoids.

**Q2:** Are all the algae groups (red, green, and brown) equally related?

**A2:** No. Red and green algae are more closely related to each other than either is to brown algae.

**Q3:** Which protist group do you think is most closely related to plants? Justify your answer.

**A3:** Green algae, because they share a common ancestor with plants.

### Figure 16.5

**Q1:** What evolutionary innovation separates all land plants from their aquatic ancestors?

**A1:** They are terrestrial.

**Q2:** How do ferns differ from bryophytes? Do they share this difference with other plant groups? (You will need to read ahead to answer this question.)

**A2:** Ferns have lignin and vascular systems, while bryophytes do not. Gymnosperms and angiosperms also have these adaptations.

**Q3:** What group(s) might a plant with seeds belong to? What about a plant with flowers?

**A3:** Seeds are found in gymnosperms or angiosperms; flowers are found only in angiosperms.

### Figure 16.6

**Q1:** In what ways are terrestrial plants and their aquatic ancestors the same? Give at least two similarities.

**A1:** Both photosynthesize and must absorb nutrients.

**Q2:** In what ways do terrestrial plants and their aquatic ancestors differ? Give at least two differences.

**A2:** Unlike their aquatic ancestors, terrestrial plants must protect themselves from dehydration and absorb nutrients from the soil.

**Q3:** Would you predict that aquatic plants (which have secondarily evolved to live in water) would be more like plants in a rainforest or more like desert plants? Explain your reasoning.

**A3:** They should be more like rainforest plants, because water is not as limiting a resource in the rainforest as in the desert.

### Figure 16.8

**Q1:** What feature(s) of the ginseng plant tell you that it is not a bryophyte?

**A1:** It has roots, and also it is able to grow higher than a bryophyte.

**Q2:** What feature(s) of the ginseng plant tell you that it is not a fern or gymnosperm?

**A2:** It has flowers, which only angiosperms have.

**Q3:** Because of the CITES classification of ginseng, you are not allowed to sell plants younger than 5 years even if they grew on your own land. Do you agree with that law? Why or why not?

**A3:** Some students will argue they should be able to do what they want on their own land; others will agree it is best to maintain the species.

### Figure 16.9

**Q1:** What group of fungi most resembles the mushrooms you buy in a grocery store?

**A1:** Basidiomycetes.

**Q2:** Are sac fungi more closely related to molds or to club fungi?

**A2:** Sac fungi are more closely related to club fungi, with which they share a common ancestor, than to molds.

**Q3:** How do we know that fungi are eukaryotes rather than prokaryotes?

**A3:** The simplest reason is that fungi are multicellular, but they also are composed of eukaryotic cells (with organelles, larger size, etc.).

### Figure 16.11

**Q1:** Why is it important that the fruiting body is aboveground?

**A1:** Because belowground, spores would not be able to travel by wind.

**Q2:** What part of a fungus is the mushroom that you can buy in the grocery store?

**A2:** A fruiting body.

**Q3:** Write a sentence in your own words that uses the terms "mycelium," "fruiting body," and "spore" correctly.

**A3:** One possible answer: The mycelium is the main body of a mushroom, while the fruiting body develops to produce and release spores into the environment, for reproduction.

# CHAPTER 17

## END-OF-CHAPTER ANSWERS

1.  a

2.  a

3.  d

4.  Mitochondrial DNA, Nuclear DNA

5.  (a) N, (b) R, (c) B, (d) B, (e) B

6.  c

7.  Monotremes lay eggs rather than developing young internally via a placenta. Marsupials have a simple placenta and so give birth to relatively undeveloped young, which then develop externally in a pouch. Eutherians have a well-developed placenta and so can support their young internally until they are more fully developed

8.  (a) 4, (b) 1, (c) 5, (d) 2, (e) 3

9. b

10. Shallow water, oxygen-poor water, areas with seasonal flooding, and nutrient-poor water but nutrient-rich land.

11. Possible answers include (a) shark; (b) hagfish; (c) bird; (d) frog.

12. It suggests that Neanderthals evolved and left Africa before modern humans evolved in Africa, and that modern humans intermingled with Neanderthal populations only after leaving Africa.

## ANSWERS TO FIGURE QUESTIONS

### Figure 17.2

**Q1:** Are mollusks more closely related to flatworms or to annelids? Explain your reasoning.

**A1:** Mollusks share a more recent common ancestor with annelids than with flatworms, so they are more closely related to annelids.

**Q2:** If you found an animal with no symmetry, to what group would it belong? Give an example of an animal that is radially symmetrical, and one that is bilaterally symmetrical.

**A2:** Animals with no symmetry are sponges. Animals with radial symmetry include cnidarians and echinoderms (secondarily). Animals with bilateral symmetry include flatworms, mollusks, annelids, nematodes, arthropods, and chordates.

**Q3:** How do we know that a sponge is an animal, and not a plant or an alga?

**A3:** Sponges are heterotrophic, not autotrophic like plants and algae.

### Figure 17.3

**Q1:** Do amphibians have amniotic eggs?

**A1:** No.

**Q2:** What group of animals has jaws but not a bony skeleton?

**A2:** Sharks and rays.

**Q3:** When people talk about animals, they are sometimes referring only to mammals. How would you explain to them their error?

**A3:** Mammals are a subset of a much larger group; animals include invertebrates, birds, and lizards, among other things.

### Figure 17.4

**Q1:** Is a sea star radially or bilaterally symmetrical?

**A1:** Radially symmetrical.

**Q2:** What advantage might a bilaterally symmetrical animal have over one that is radially symmetrical, and vice versa?

**A2:** Animals with bilateral symmetry move more efficiently. Those with radial symmetry have 360-degree access to their environment.

**Q3:** What kind of symmetry do you (a human) have?

**A3:** Bilateral.

### Figure 17.5

**Q1:** List all of the lobster's thoracic appendages.

**A1:** Walking legs, crushing claw, cutting claw.

**Q2:** Which of the lobster's appendages is most important for sensing the environment?

**A2:** The head appendage, which includes eyes and antennae.

**Q3:** What body segments do you (a human) have?

**A3:** Head, thorax, and abdomen.

### Figure 17.6

**Q1:** The Virginia opossum, or possum, is the only North American marsupial. Describe how its young are born and develop.

**A1:** They are born relatively undeveloped and crawl into the mother's pouch, where they nurse and grow.

**Q2:** If you were a monotreme, would you still produce milk and nurse your young?

**A2:** Yes; all mammals do.

**Q3:** What kind of mammal is a cow? How about a human?

**A3:** Cows and humans are eutherians.

### Figure 17.7

**Q1:** According to this evolutionary tree, which primate group is most closely related to humans?

**A1:** Chimpanzees.

**Q2:** According to this evolutionary tree, which primate group is most distantly related to humans?

**A2:** Lemurs, along with others in this branch.

**Q3:** What characteristics are common to all the primates, including humans? (You will need to read the next paragraph to answer this question.)

**A3:** The use of tools, a capacity for symbolic language, and the performance of deliberate acts of deception.

### Figure 17.8

**Q1:** Through natural selection, deleterious traits will tend to disappear from a population over time. Which traits might have been deleterious for ground-dwelling early hominins?

**A1:** Opposable toes; walking on four limbs.

**Q2:** Through natural selection, advantageous traits will tend to persist in a population over time. Which traits might have been advantageous for ground-dwelling early hominins?

**A2:** Upright posture; arched feet.

**Q3:** Adaptations to upright walking also mean that human females have more difficulty giving birth than do females of other species. What adaptation would you predict has had the greatest impact on this result?

**A3:** Reorganization of the pelvic structure.

## Figure 17.9

**Q1:** What other reason besides continuing to use trees might explain why early hominins had partially opposable big toes?

**A1:** This trait may have taken a long, long time to lose. A major change in structure cannot occur within a few generations. It was likely a gradual change over thousands of generations.

**Q2:** In what way does the pattern of footprints in this figure suggest that the print makers were walking upright?

**A2:** No hand or knuckle prints accompany the footprints.

**Q3:** Why do you think we no longer have partially opposable big toes?

**A3:** The opposable big toes would have made walking upright or running more difficult. If we no longer returned to the trees, there was no selective advantage to having them. In fact, individuals with fully opposable toes were at a disadvantage.

## Figure 17.10

**Q1:** Where would the Neanderthal species branch be on this tree?

**A1:** The *Homo neanderthalensis* lineage would branch either from the common ancestor of *H. erectus* and *H. sapiens*, or from *H. erectus* along with *H. sapiens*. The exact hereditary line for these two closely related species is still unclear.

**Q2:** How would the Neanderthal skull differ from the *Homo erectus* skull?

**A2:** The skull would be larger, more like the *H. sapiens* skull size or larger.

**Q3:** How would the Neanderthal skull differ from the *Homo sapiens* skull?

**A3:** The skull would be longer, the forehead would slope more, and there would be no chin.

## Figure 17.11

**Q1:** Why does mitochondrial DNA come only from your mother?

**A1:** Mitochondria are found in the egg but not the sperm. Eggs are from your mom, and sperm are from your dad.

**Q2:** If a Neanderthal-human hybrid was born to a human mother and a Neanderthal father, could you tell by mitochondrial-DNA sequencing?

**A2:** No. All the mitochondrial DNA would be from *Homo sapiens*.

**Q3:** If a Neanderthal-human hybrid was born to a human father and a Neanderthal mother, could you tell by mitochondrial-DNA sequencing?

**A3:** Yes. In this case, all the mitochondrial DNA would be from *Homo neanderthalensis*.

## Figure 17.12

**Q1:** If a human-Neanderthal hybrid was born to a human mother and a Neanderthal father, could you tell by whole-genome DNA sequencing that it was a hybrid?

**A1:** Yes. The offspring would have nuclear DNA from both the egg and the sperm.

**Q2:** If a human-Neanderthal hybrid was born to a Neanderthal mother and a human father, could you tell by whole-genome DNA sequencing that it was a hybrid?

**A2:** Yes. The offspring would have nuclear DNA from both the egg and the sperm.

**Q3:** Under what circumstances are scientists able to do whole-genome sequencing, and when are they restricted to mitochondrial-DNA sequencing?

**A3:** Whole-genome sequencing requires well-preserved cells or tissues with fully intact DNA. Mitochondrial DNA can be isolated from cells and tissues that aren't so well preserved and from damaged DNA.

## Figure 17.13

**Q1:** Are you surprised by the interpretations of the hominins in this picture?' Why or why not?

**A1:** This question asks for an opinion. One possible answer: *Homo sapiens* looks more primitive than expected, and the other species look strikingly like us.

**Q2:** Describe the main differences that distinguish the hominin species.

**A2:** Height, musculature, size of skull, slope of forehead, amount of hair and its location.

**Q3:** From what you've learned about these species, do you think these representations are accurate? How can you find more information about each species to help you answer this question?

**A3:** The first part of the question asks for an opinion. One possible answer for the second part: I can do much more extensive research about all of our family members online and at museums of natural history.

## Figure 17.14

**Q1:** What evidence suggests that Neanderthals never lived in Africa?

**A1:** Modern humans of African descent have neither mitochondrial nor nuclear Neanderthal DNA sequences in their genomes.

**Q2:** How does the hypothesized origin of modern humans (*Homo sapiens*) differ from the hypothesized origin of Neanderthals (*Homo neanderthalensis*)?

**A2:** Modern humans evolved from archaic humans in Africa and spread to the rest of the world. Neanderthals are thought to have evolved from archaic humans living in the Middle East.

**Q3:** What species of hominins other than the Neanderthals may have commingled with modern humans?

**A3:** *Homo erectus*.

## Figure 17.15

**Q1:** Describe the difference you observe between the modern human skull's chin and the Neanderthal skull's lower jaw.

**A1:** The human chin is higher and juts out further.

**Q2:** What other differences do you observe between the two skulls?

**A2:** The Neanderthal has a more pronounced eyebrow ridge and more sloping forehead.

**Q3:** Why would you expect a hybrid of Neanderthals and modern humans to have intermediate features?

**A3:** In hybrids, half of the DNA would be from *Homo sapiens* and half from *Homo neanderthalensis*. Hybrids have features of both parents. Some might have intermediate features, while others could have mixed features (such as a sloping forehead but a prominent chin).

# CHAPTER 18

## END-OF-CHAPTER ANSWERS

1.  e
2.  b
3.  d
4.  (a) 1, (b) 3, (c) 4, (d) 2, (e) 5
5.  Weather, climate, Climate change, global warming
6.  d
7.  Carbon sink, because more carbon is absorbed in the years between the fires than is produced by the fires.
8.  (a) 4, (b) 1, (c) 2, (d) 5, (e) 3
9.  c
10. Answers will depend on student location.
11. Answers will depend on student location.

## ANSWERS TO FIGURE QUESTIONS

### Figure 18.3

**Q1:** List as many biotic and abiotic factors in this photograph as you can.

**A1:** Biotic: all living things, such as plants, as well as all animals, including microscopic bacteria and algae. Abiotic: all nonliving things, such as rocks and water.

**Q2:** Is the forest part of the biotic or abiotic environment? Explain.

**A2:** It can be considered both. The plants and microorganisms are biotic, while the dirt and minerals are abiotic.

**Q3:** Is the river part of the biotic or abiotic environment? Explain.

**A3:** It can be considered both. The algae, plants, and aquatic organisms like fishes living in the river are biotic, while the water and rocks of the river are abiotic.

### Figure 18.5

**Q1:** Name two ways in which climate change affects the frequency and severity of floods.

**A1:** Changes in rainfall patterns and melting ice can cause rivers and lakes to overflow.

**Q2:** How has climate change caused a rise in sea level?

**A2:** Melting of glaciers and polar ice has caused sea levels to rise.

**Q3:** Give an example of an environmental effect of climate change in your state or region.

**A3:** Answers may include drought or flooding.

### Figure 18.6

**Q1:** Why is it colder at the poles than at the equator?

**A1:** The sun's rays are spread wider at the poles and strike Earth less directly than at the equator. Less direct sunlight = cooler.

**Q2:** Why is it warmer at the equator than at the poles?

**A2:** All of the sun's energy is directed at the equator, and the rays strike Earth at a direct angle there. The result is more intense heat at the equator.

**Q3:** During part of the year the Northern Hemisphere is tilted at a more direct angle to the sun than the Southern Hemisphere is, and for the other part of the year the opposite is true. How does this tilt explain temperature differences in summer and winter?

**A3:** When a hemisphere is tilted toward the sun, the sun's rays strike it more directly, causing the warmer temperatures of summer. The opposite is true in the winter.

### Figure 18.7

**Q1:** How much of the incoming solar energy is reflected back to outer space?

**A1:** About a third.

**Q2:** What kind of energy is reemitted to the atmosphere after being absorbed by Earth's surface?

**A2:** Infrared radiation.

**Q3:** How are greenhouse gases like a blanket on your bed at night?

**A3:** Greenhouse gases absorb the heat around Earth and hold it near the surface just as a blanket absorbs body heat, prevents it from escaping, and holds it near your body.

### Figure 18.8

**Q1:** What measurements do the green circles represent?

**A1:** The green circles indicate $CO_2$ levels measured from bubbles of air trapped in ice that formed many hundreds of years ago.

**Q2:** What measurements do the red circles represent?

**A2:** The red circles are direct measurements of $CO_2$ levels at the Mauna Loa Observatory in Hawaii.

**Q3:** For approximately how many years has the Mauna Loa Observatory been recording $CO_2$ levels?

**A3:** Directly for about 60 years.

### Figure 18.9

**Q1:** In what years were global temperatures the lowest?

**A1:** The years around 1910.

**Q2:** In what years were global temperatures the highest?

**A2:** The years around the late 1990s through the present.

**Q3:** What trend is apparent in this graph of actual global temperatures?

**A3:** Average global temperatures are rising.

## Figure 18.11

**Q1:** Where will fire most seriously affect the Amazon rainforest?

**A1:** Fire will most likely affect areas labeled "Dry and/or logged forest" (shown in orange).

**Q2:** Where will fire be the least damaging to the Amazon rainforest?

**A2:** Fire will least affect the old-growth forest (shown in green).

**Q3:** This map does not include an increase in pasturelands for grazing animals. Do you think more or less pastureland will be needed in 2030? Explain.

**A3:** As populations increase, the need for more pastureland to raise cattle for human consumption will increase as well.

## Figure 18.12

**Q1:** What is transpiration?

**A1:** Transpiration is the process of plants absorbing water through their roots and releasing this water through their leaves to the atmosphere.

**Q2:** Why is transpiration important to the water cycle?

**A2:** Transpiration returns water from the soil back to the atmosphere to form clouds and eventually precipitation.

**Q3:** If there are fewer plants and therefore less transpiration in a given area, what will happen to the humidity or cloud cover in this area?

**A3:** The humidity and cloud cover will decrease where there is a substantial decrease in transpiration.

## Figure 18.13

**Q1:** How do the patterns of rainfall in the Northern and Southern Hemispheres compare?

**A1:** The same patterns emerge as you move away from the equator either northward or southward.

**Q2:** How do the patterns in the kinds of environments shown in the Northern and Southern Hemispheres compare?

**A2:** As you move away from the equator, the major biomes are equivalent distances from the equator either north or south.

**Q3:** What happens at the equator to make this region so wet?

**A3:** The density of plants is very high, and therefore the amount of transpiration is very high, resulting in cloud cover and high precipitation. High precipitation results in high plant growth and high transpiration rates.

## Figure 18.14

**Q1:** What are three ways that carbon is released into the atmosphere?

**A1:** Respiration from animals; the burning of organic matter, including fossil fuels and wood; and the decomposition of dead organic material.

**Q2:** Are all of the pathways you listed for question 1 affected by human activity?

**A2:** Almost everything on our planet is affected by human activity in some way. Of the three pathways of carbon release, the one most affected by humans is the burning of organic materials for energy.

**Q3:** What are two biotic reservoirs of carbon?

**A3:** Plants and animals.

## Figure 18.15

**Q1:** How does a carbon source contribute to global warming?

**A1:** A carbon source is a reservoir of carbon that releases more than it absorbs, thereby dumping $CO_2$ into the environment, increasing greenhouse gases, and causing global warming.

**Q2:** How does a carbon sink protect against global warming?

**A2:** A carbon sink absorbs more carbon than it releases, thereby removing $CO_2$ from the environment, decreasing greenhouse gases, and blocking global warming.

**Q3:** How can trees act as both a source and a sink?

**A3:** Trees act as a carbon sink when photosynthesizing and absorbing $CO_2$ and as a carbon source when they are burned for fuel or in a wildfire.

# CHAPTER 19

## END-OF-CHAPTER ANSWERS

1. d
2. b
3. c
4. d
5. a
6. Density-dependent, density-independent
7. b
8. c
9. (a) 1, (b) 4, (c) 3, (d) 5, (e) 2
10. (See figure below)

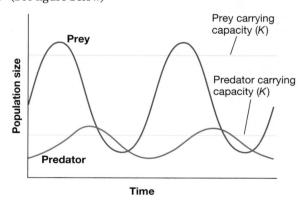

11. The food dispenser increased the carrying capacity significantly.

12. b

# ANSWERS TO FIGURE QUESTIONS

## Figure 19.1

**Q1:** What parts of the United States are within the range of the mosquito that carries Zika?

**A1:** Roughly, the southern United States and the eastern coastline.

**Q2:** What areas are *not* within the mosquito's range? Why do you think that is?

**A2:** The northern and northwestern states, probably because it is too cold for mosquitoes to overwinter there.

**Q3:** Find your own state, and where you are in the state. Are you at risk of contracting Zika?

**A3:** Answers will depend on student location. If you travel to areas with Zika-infected mosquitoes, your risk increases. If local Zika infections have not yet been reported in your state, you are at less risk than, for example, citizens of Florida.

## Figure 19.3

**Q1:** What is the main way by which someone is infected with the Zika virus?

**A1:** By being bitten by a mosquito that is infected.

**Q2:** Judging by the poster and your knowledge of mosquito behavior, what can you do to decrease your risk of being infected with the Zika virus?

**A2:** Cover up when you're outside, don't go out at dusk, wear insect repellent, and use protection when having sex with someone who has been exposed.

**Q3:** Besides the transmission methods shown on the poster, what are some other ways you could become infected with the Zika virus?

**A3:** From kissing an infected individual, or possibly from sharing a razor or toothbrush with someone. (Other answers are possible.)

## Figure 19.5

**Q1:** Which form of population growth displays a J-shaped curve?

**A1:** Exponential growth.

**Q2:** Which form of population growth displays an S-shaped curve?

**A2:** Logistic growth.

**Q3:** Describe a situation in which a population initially shows exponential growth and later shows logistic growth.

**A3:** If a population enters a new area—an island, for example— it will initially grow exponentially as it expands into the environment. Later, as it reaches the carrying capacity of the new area, it will begin to display logistic growth.

## Figure 19.6

**Q1:** According to this graph, approximately when did exponential growth begin?

**A1:** Around the year 1800 CE.

**Q2:** What milestone corresponds to the transition from logistic to exponential population growth?

**A2:** The onset of the industrial revolution.

**Q3:** What is the UN's projected carrying capacity of Earth, and when will we reach it?

**A3:** The UN projects Earth's carrying capacity to be 9–10 billion people, which we will reach in about 2050 CE.

## Figure 19.7

**Q1:** List all the stages of the mosquito life cycle.

**A1:** Egg, larva, pupa, adult.

**Q2:** Which life cycle stage is vulnerable to the GM treatment?

**A2:** The larvae, which do not survive to become pupae.

**Q3:** Why are only male GM mosquitoes released into the wild, rather than both males and females?

**A3:** Males do not feed on blood, so they are not able to transmit Zika virus, whereas females do feed and thus can transmit Zika.

## Figure 19.8

**Q1:** What factors may be limiting growth and reproduction in the plantain's crowded conditions?

**A1:** Possible factors include availability of nutrients, water, sunlight, and room for root and shoot growth.

**Q2:** Why are overcrowded conditions considered density-dependent population changes?

**A2:** Because the more organisms there are in the environment or the more densely packed they are, the more competition there is for resources and the less each individual will likely receive.

**Q3:** Relate this example of overcrowded conditions to the human population growth shown in Figure 19.6. How do you think the situations are similar? How are they different?

**A3:** Increasing human populations may lead to an increase in competition for jobs, housing, and food. So far, humans have been able to increase their carrying capacity through technology, but that may no longer be possible.

## Figure 19.9

**Q1:** In what year did the reindeer's numbers begin to rise exponentially?

**A1:** About 1933.

**Q2:** In what years was the reindeer's population growth logistic?

**A2:** Between about 1911 and 1932.

**Q3:** How do you predict the graph (population size) would change if someone had begun bringing in supplemental food for the reindeer in 1940? Draw a sketch of your prediction.

**A3:** Supplemental food would have caused the population decline to end, and the population would have either remained constant or again increased, depending on the amount of supplemental food provided.

## Figure 19.10

**Q1:** In what year did the bald eagle population rise to more than 2,000 breeding pairs?

**A1:** About 1986 or 1987.

**Q2:** Give some examples of possible density-dependent limits on bald eagle populations.

**A2:** Examples include numbers of prey available, adequate habitat for nesting and hunting, and availability of water.

**Q3:** Is the population growth of bald eagles more like logistic or exponential growth? Explain why you think so.

**A3:** The population growth of bald eagles is more like logistic growth in that it is more of an S-shaped curve—at least between 1960 and 2005, where it seemed to level off for a few years. On the other hand, between 2005 and 2010 the growth appears to be exponential.

## Figure 19.11

**Q1:** During which years did the hare likely have the greatest food supply?

**A1:** 1865 and 1888.

**Q2:** Besides the number of hare, what other factors might contribute to the number of lynx?

**A2:** Other possible factors include the quality of habitat for building dens and raising young, the availability of freshwater, and the competition with other lynx and other predators of the hare.

**Q3:** Can you draw an average carrying-capacity line on these graphs? Why or why not?

**A3:** It would be difficult to draw a line representing a carrying capacity for both of these animals because of the way they go through cycles of "boom and bust" in population size. They do not show logistic growth, where the leveling off of the S-shaped curve provides an obvious carrying capacity.

# CHAPTER 20

## END-OF-CHAPTER ANSWERS

1.  d
2.  b
3.  c
4.  c
5.  secondary consumer, primary consumer, producer
6.  e
7.  mutualism: 2, commensalism: 3, predation: 1, competition: 4
8.  (a) 4, (b) 1, (c) 2, (d) 3, (e) 5
9.  More diverse communities have low relative species abundance and high species diversity.
10. c
11. Relative species abundance will decrease, and species richness will increase.
12. (a) S, (b) S, (c) P, (d) P
13. (a) humans and killer whales, (b) phytoplankton, (c) baleen whale and sperm whale, (d) other herbivorous plankton

## ANSWERS TO FIGURE QUESTIONS

### Figure 20.2

**Q1:** List another species that is part of this community.

**A1:** One possible answer: the wolf.

**Q2:** Of which community could this aspen woodland be a smaller part?

**A2:** A larger deciduous forest community.

**Q3:** Which other small communities could be found within this larger community?

**A3:** Many other communities could exist, including soil communities (soil-dwelling insects, other invertebrates, and microbes), communities of plants and animals residing in the undergrowth of the forest, and canopy communities of animals that live in the treetops, among others.

### Figure 20.3

**Q1:** If relative species abundance increased in the first community, how would the figure look different from now?

**A1:** There would be even more trees of the white-trunked species, and fewer of the other species.

**Q2:** If species richness decreased in the second community, how would the figure look different from now?

**A2:** There would be fewer different species of trees.

**Q3:** How do relative species abundance and species richness define the species diversity of a forest community?

**A3:** High species diversity relies on having each species in reasonable abundance, and not on one species taking up the majority of space in an area (relative abundance), and on there being many different species in a given area (richness).

### Figure 20.4

**Q1:** How many species were left in 1966 in the community where sea stars were *not* removed?

**A1:** About 17 or 18.

**Q2:** How many species were left in 1966 in the community where sea stars *were* removed?

**A2:** Only 2 or 3.

**Q3:** How do your answers to questions 1 and 2 demonstrate the importance of a keystone species for the maintenance of diversity in a community?

**A3:** The species diversity plummeted as a result of the loss of this sea star. Without this species, the entire mussel community changed from many species to only a few.

## Figure 20.5

**Q1:** What species eats the coyote?

**A1:** The gray wolf.

**Q2:** What species does the coyote eat?

**A2:** The vole, the snowshoe hare, berries, the pronghorn, and the elk.

**Q3:** What do you think would happen to a community that lost its coyotes?

**A3:** The snowshoe hare and the vole would overeat the grasses and other plants, and once those food sources were destroyed, their unavailability would cause the snowshoe hare and vole populations to crash.

## Figure 20.6

**Q1:** Where do producers acquire the energy they need to perform their function in the food chain?

**A1:** From the sun. In combination with $CO_2$ and water, they produce glucose through photosynthesis.

**Q2:** Why are producers necessary for life on Earth?

**A2:** Without producers, there would be no influx of energy into Earth's biosphere—no energy source for consumers to acquire.

**Q3:** Given the lower abundance of producers in desert regions, compared to tropical rainforests, what would you predict about the abundance of consumers in the two environments?

**A3:** Consumer species would be less abundant in deserts than in rainforests.

## Figure 20.7

**Q1:** When did wolves begin to be seen again in Yellowstone?

**A1:** In the mid- to late 1980s.

**Q2:** What was the highest number of wolves observed in Yellowstone? In what year did that occur?

**A2:** About 174 individuals, in 2003.

**Q3:** Why did it take a few years after the reintroduction of wolves for the aspen population to increase, as well as the beavers and bears?

**A3:** Each individual parent plant can produce only a certain number of offspring at a time, and then those offspring have to reach maturity before they themselves can reproduce.

## Figure 20.8

**Q1:** What might happen to an anemone without a resident clownfish?

**A1:** An anemone unaccompanied by clownfish, and therefore unprotected from anemone-eating fish, could be grazed extensively and be unable to support its growth needs without the nutrients in the clownfish excrement.

**Q2:** What would happen to a clownfish that did not produce a mucous coating?

**A2:** It would be stung by the anemone's tentacles, so the mutualism would fall apart.

**Q3:** In the example of mutualism given in the text, what would happen if ticks were no longer able to feed on bison?

**A3:** There would be nothing for the birds to eat off of the bison, and no reason for the bison to allow the birds on them. The mutualism would end.

## Figure 20.9

**Q1:** How do barnacles benefit from living on a whale?

**A1:** The whale provides a home and a constant stream of water passing over the barnacles to bring them the tiny particles of food that they filter from the sea.

**Q2:** Do you think a whale could avoid being colonized by barnacles? Why or why not?

**A2:** It is probably unlikely that a whale could avoid being covered by barnacles, although since the barnacles do not help or hurt the whale, there would be no reason for a whale to try to avoid them.

**Q3:** Explain why detritivores are considered commensal to the organisms they consume. (You will need to read ahead to answer this question.)

**A3:** Since the organism is already dead, the detritivore is not harming it by eating it. The detritivore, though, benefits from the association.

## Figure 20.10

**Q1:** What kind of predator is the cheetah in the figure?

**A1:** Carnivore.

**Q2:** What kind of predator is the louse in the figure?

**A2:** Parasite.

**Q3:** What kind of predator are the elk that graze on aspen tree saplings?

**A3:** Herbivore.

## Figure 20.11

**Q1:** How do predators know that brightly colored prey are usually toxic?

**A1:** Most predators have experienced through the trial and error of tasting prey and becoming sick that brightly colored prey are toxic. Predators must learn by trying one to avoid others later. If the organism were so toxic that the predator died from eating one, then bright displays would not help prey avoid predation.

**Q2:** Do you think mimicry works if the toxic species is in low abundance? Why or why not?

**A2:** Mimicry is similar to warning coloration in that its benefit is usually accomplished through trial and error. If there are many more nontoxic mimics than real toxic individuals, predators that successfully eat the mimics will not learn to avoid them. Only if it is more likely that a predator will encounter an actual toxic and bad-tasting prey will it learn to avoid anything that looks similar to the prey it encountered.

**Q3:** Why is camouflage considered an adaptive response to predation?

**A3:** Camouflage enables prey species to blend in with their surroundings so that they're difficult for predators to detect. Random variation in coloration of a population enables those that are better hidden from predators to survive and reproduce, while those that do not match the surroundings well are easily seen and eaten. Over time, only well-camouflaged prey survive to reproduce, passing the camouflage coloration on to their offspring.

## Figure 20.12

**Q1:** What percentage of pigeons are caught when they are alone and not in a flock?

**A1:** Very close to 80%.

**Q2:** For wood pigeons, what is the minimum number of individuals that provides protection from goshawks?

**A2:** 11 individuals.

**Q3:** Why do you think a group of musk oxen versus a lone musk ox would be safer from a pack of wolves?

**A3:** The group of musk oxen could form a circle in which they faced out and use their large horns to impale the wolves, keeping their sides and rear ends protected from attack. A lone ox would be completely unprotected from a pack of wolves working together.

## Figure 20.13

**Q1:** Which species appears to be the superior competitor?

**A1:** *Aphytis lingnanensis.*

**Q2:** Why is this example considered exploitative competition?

**A2:** These wasp species feed on the same foods in the same place but never physically come in contact with each other. They are in direct competition but do not physically interact.

**Q3:** What would you predict if these species had undergone competitive exclusion? (You will need to read the next paragraph to answer this question.)

**A3:** The first species would now be exploiting another resource—for example, apples rather than citrus.

## Figure 20.15

**Q1:** What species represents the first colonizers of the sand dunes?

**A1:** Dune-building grasses.

**Q2:** What species is the intermediate species, and how does it become the dominant species?

**A2:** The pine is the intermediate species. It likely becomes dominant by outcompeting the grasses for water, nutrients, and sunlight. Once established, the pines' shade further inhibits the growth of grasses.

**Q3:** What species is the mature, climax community species, and how does it become the dominant species?

**A3:** The black oak is the mature, climax community species. It likely becomes the dominant species because it outcompetes the pines for water, nutrients, and sunlight.

## Figure 20.16

**Q1:** What other types of disturbances could you imagine destroying a forest?

**A1:** Other types of disturbances could include clear-cutting or heavy logging, abandonment of agricultural fields, flooding, clearing by tornado or hurricanes, and the creation of empty lots in a newly built subdivision, among others.

**Q2:** How is secondary succession different from primary succession?

**A2:** Secondary succession starts with some existing producers that survived a disturbance, whereas primary succession starts from nothing—pure rock or sand. A disturbance that would result in primary succession would be a lava flow that covered an entire area; succession would have to start from pure lava rock and no existing producers.

**Q3:** What is a climax community?

**A3:** A climax community is the final step in succession, in which the species in the community remain and are not replaced by any other species as time goes by.

# CHAPTER 21

## END-OF-CHAPTER ANSWERS

1. a

2. c

3. d

4. c

5. biome: 3, ecosystem: 2, net primary productivity: 1, ecological community: 4

6. chaparral, grassland, estuaries

7. b

8. (a) 4, (b) 2, (c) 1, (d) 5, (e) 3

9. An ecosystem is composed of multiple ecological communities, and it also includes physical environmental factors (communities do not).

10. (a) 100. (b) 100,000. (c) The energy is expended in metabolism, and since predation is not 100% efficient (that is, predators don't eat every last bit of the calories in their prey, leaving, for example, the bones), some energy is lost there too.

11. The water cycle is more similar to the movement of nutrients, because water cycles through the biotic and abiotic components of an ecosystem, whereas energy flows in only one direction through ecosystems because of the steady loss of heat.

12. c

## ANSWERS TO FIGURE QUESTIONS

### Figure 21.3

**Q1:** What percentage of the original 10,000 Kilocalories is available to a shark that might eat the tuna in this figure?
**A1:** 1%.
**Q2:** What trophic level and term would describe a predator of tuna?
**A2:** A predator of tuna would be at the fifth trophic level and a quaternary consumer.
**Q3:** Give an example of a primary consumer in a terrestrial environment.
**A3:** Possible answers include deer, a grasshopper, a seed-eating bird, a mouse.

### Figure 21.4

**Q1:** Which organisms are the producers in this ecosystem?
**A1:** Phytoplankton.
**Q2:** How do nutrients flow from the abiotic to the biotic components of the ecosystem?
**A2:** Via photosynthesis and uptake from the surrounding water.
**Q3:** How do nutrients flow from the biotic to the abiotic components of the ecosystem?
**A3:** Via decomposers.

### Figure 21.5

**Q1:** How is a decomposer different from a more typical consumer?
**A1:** Decomposers feed off of only dead organic matter (dead plants or animals). All other types of consumers feed off of live plants and animals.
**Q2:** What is the difference between how carbon is brought into the biotic portion of the ecosystem, and how other nutrients, such as phosphorus, are brought in?
**A2:** Carbon is pulled from the air during photosynthesis, whereas other nutrients are pulled in from the soil.
**Q3:** Describe all the points at which heat is lost in this figure.
**A3:** During the survival and reproduction of producers, consumers, and decomposers.

### Figure 21.9

**Q1:** Which terrestrial biome has the lowest NPP? Which aquatic biome?
**A1:** The lowest terrestrial NPP appears to occur in the desert and tundra biomes in the terrestrial map, and in the marine biome of the open ocean in the aquatic map.
**Q2:** Where are the most productive terrestrial biomes located?
**A2:** They appear to be mainly near the equator.
**Q3:** Give a possible reason for your answer to question 2.
**A3:** More sunlight reaches Earth at the equator, so larger populations of producers can be supported.

### Figure 21.10

**Q1:** In what years were chlorophyll levels the highest?
**A1:** 1999 and 2008.
**Q2:** In what years were the temperature changes from the average the greatest?
**A2:** 2000 (low), 2003–4 (high), and 2008 (low).
**Q3:** How do you predict this graph will look 10 years from now?
**A3:** If the world's oceans are warming, then the chlorophyll levels may continue to decrease.

# CHAPTER 22

## END-OF-CHAPTER ANSWERS

1. d

2. a

3. c

4. d

5. a

6. spermatogenesis: 2, oogenesis: 3, sexual reproduction: 4, asexual reproduction: 1

7. a

8. asexual reproduction

9. homeostatic, set point, negative, thermoregulation

10. (a) 4, (b) 1, (c) 3, (d) 5, (e) 2

11. d

12. Yes, there is a difference, and there are two possible causes. First, intercourse could happen too soon after surgery, when an egg is past the point of ligation or (more commonly) sperm are still in the vas deferens. Second, the tubes can grow back together, as happens in about one out of 2,000 surgeries for men (0.05% failure) and one out of 200 surgeries for women (0.5% failure).

13. Because newborns have a higher surface area–to–volume (SA-to-V) ratio, they can chill or overheat more quickly than older and larger people do. They also need to be fed more often, and they become dehydrated more quickly because of their high SA-to-V ratio.

## ANSWERS TO FIGURE QUESTIONS

### Figure 22.2

**Q1:** In which of the countries shown in the graph did women want the most children, on average? The least?
**A1:** Most: Finland. Least: Austria.

**Q2:** Which country has the largest gap between the number of children women would like to have and the number they actually have?

**A2:** Sweden.

**Q3:** Which country has the highest average number of children? The lowest?

**A3:** Highest: Poland. Lowest: Italy.

## Figure 22.3

**Q1:** Which tissue type is the primary component of skin?

**A1:** Epithelial.

**Q2:** Which tissue type is the primary component of bone?

**A2:** Connective.

**Q3:** Which tissue types does the hand contain?

**A3:** All four tissues are found in the hand.

## Figure 22.4

**Q1:** Which organ system defends the body from infectious diseases such as the common cold or flu?

**A1:** The immune system.

**Q2:** Which organ systems transport oxygen to cells?

**A2:** The respiratory and circulatory systems.

**Q3:** Which organ systems regulate the activities of the other organ systems?

**A3:** The nervous and endocrine systems.

## Figure 22.5

**Q1:** Why is it important to maintain a stable body temperature?

**A1:** All of the chemical processes that occur in living things have an optimal temperature range and will be less efficient (or unable to occur) if body temperature deviates too much from that range.

**Q2:** Which organ system is involved as a temperature sensor? (See Figure 22.4 for an overview of organ systems.)

**A2:** The nervous system.

**Q3:** Give another example of a homeostatic pathway in humans.

**A3:** Possible answers include systems for control of internal water content and systems for control of internal pH levels.

## Figure 22.7

**Q1:** Identify one way in which oogenesis and spermatogenesis are the same.

**A1:** One possible answer: They each produce haploid gametes through meiosis.

**Q2:** How many eggs are produced from each precursor cell? How many sperm?

**A2:** Eggs: one. Sperm: four.

**Q3:** How much time elapses between the appearance of a precursor cell and the formation of an egg? How does this process differ for a sperm? (You will need to read ahead to answer this question.)

**A3:** Whereas each egg takes approximately a month to develop, new sperm are produced daily.

## Figure 22.8

**Q1:** Which hormones important for the menstrual cycle are produced in the pituitary gland?

**A1:** Follicle stimulating hormone (FSH) and luteinizing hormone (LH).

**Q2:** Which hormone is involved in producing the uterine lining?

**A2:** Progesterone.

**Q3:** How is the egg follicle involved in producing hormones?

**A3:** The egg is carried within a follicle, which produces estradiol. After the egg is released from the follicle, the remaining corpus luteum produces progesterone.

## Figure 22.9

**Q1:** If an egg is released but no sperm enter the oviduct, what is likely to occur?

**A1:** The unfertilized egg will travel into the uterus and then out of the body, and the menstrual cycle will continue.

**Q2:** If sperm enter the oviduct but no egg is present, what is likely to occur?

**A2:** The sperm will eventually die, and no pregnancy will occur.

**Q3:** Sperm can live for up to 5 days inside a woman's body. Furthermore, eggs may be released at any point in the menstrual cycle, although midcycle ovulation is the norm. If you are trying *not* to become pregnant, when is it safe to have unprotected intercourse?

**A3:** Essentially never, since sperm can live for so long and eggs may be released at any point (and, in fact, may be released by the act of intercourse).

## Figure 22.10

**Q1:** Place these terms in the correct order of development: embryo, fetus, infant, zygote.

**A1:** Zygote, embryo, fetus, infant.

**Q2:** In what trimester is the fetus most likely to survive outside its mother's body?

**A2:** The third trimester.

**Q3:** The first trimester is the most sensitive time for exposure to mutagens. Why might that be?

**A3:** This is when most organ systems develop, so any mutations could have profound effects on the viability of the developing fetus.

## Figure 22.11

**Q1:** What is the role of estradiol in childbirth?

**A1:** It makes uterine muscles more sensitive to oxytocin.

**Q2:** Explain how the involvement of hormones in childbirth is an example of a positive feedback loop.

**A2:** Contractions of the uterus are caused by oxytocin, and increasing contractions, in turn, cause more oxytocin to be produced.

**Q3:** If a woman has been pregnant for more than 40 weeks, her doctor might give her an injection of oxytocin to precipitate labor. How would that bring about labor?

**A3:** Because of the positive feedback loop with oxytocin and uterine contractions, the initial injection of oxytocin should cause contractions—which would then kick off the feedback loop, with the woman's body now releasing more oxytocin.

# CHAPTER 23

## END-OF-CHAPTER ANSWERS

1. c

2. a

3. b

4. d

5. Adipose, dermis, epithelial

6. skeletal muscle: 3, smooth muscle: 4, cardiac muscle: 2, voluntary muscle: 1, involuntary muscle: 5

7. c

8. muscle, myosin, actin, Z, sarcomere

9. (a) 4, (b) 1, (c) 3, (d) 5, (e) 2

10. b

11. d; swimming is not weight-bearing and does not directly build muscle (as in weight lifting), so it is the least effective for increasing bone strength

12. "Cavity" = mouth; "32 protruding bones" = teeth; "meat tentacle" = tongue; "vat of acid" = stomach; "absorb its essence and transform it in[to] energy" = action of the small intestine (mainly)

## ANSWERS TO FIGURE QUESTIONS

### Figure 23.2

**Q1:** Which vitamins described in the figure are important for healthy bones?

**A1:** Vitamins A, $B_{12}$, C, and D.

**Q2:** Which vitamins are you more likely to overaccumulate?

**A2:** The fat-soluble vitamins: A, D, E, and K.

**Q3:** In your own diet, are there any vitamins that you may not be eating enough of?

**A3:** Possible answers include vitamin A if no yellow vegetables are eaten, and $B_{12}$ if no meat is consumed.

### Figure 23.3

**Q1:** List, in order, the structures of the digestive system that a piece of swallowed food would pass through, beginning with the mouth.

**A1:** Mouth, esophagus, stomach, small intestine, colon, anus.

**Q2:** What part of the digestive system hosts bacteria that produce vitamins?

**A2:** The large intestine.

**Q3:** What is the shared function of the liver and gallbladder?

**A3:** They produce, store, and release bile to digest fats.

### Figure 23.4

**Q1:** Why is a larger surface area important for absorption?

**A1:** With a larger surface area, a higher proportion of nutrients traveling through the small intestine will be absorbed into the body, since they are more likely to come into contact with the cells lining the small intestine.

**Q2:** In what way are the epithelial cells lining the villi modified to increase absorption?

**A2:** The microvilli of the cells increase surface area even further.

**Q3:** Explain the role of the capillaries within each villus.

**A3:** The capillaries are important for moving nutrients from the villi and out to the rest of the body.

### Figure 23.5

**Q1:** What organs of the digestive system are involved in processing vitamin D?

**A1:** The liver and kidneys.

**Q2:** Do you get the majority of the vitamin D you need from the sun, from natural foods, or from dietary supplements?

**A2:** Answers will vary.

**Q3:** It is true that you can increase your vitamin D levels by visiting a tanning booth (as described later in the chapter). Why is increasing vitamin D in this way considered a bad idea?

**A3:** There is a much higher risk of skin cancer from tanning booths compared to natural sunlight.

### Figure 23.6

**Q1:** The collarbone is part of which skeleton: axial or appendicular?

**A1:** Appendicular.

**Q2:** Which parts of the skeleton are made of cartilage?

**A2:** Joints, such as elbows, knees, and shoulders, as well as parts of most ribs.

**Q3:** Which skeleton—axial or appendicular—protects the central nervous system (the brain and spinal cord)?

**A3:** Appendicular.

### Figure 23.9

**Q1:** What is the function of the synovial sacs?

**A1:** They produce lubricating fluid, to allow the knee joint to move more easily.

**Q2:** Compare and contrast ligaments and tendons.

**A2:** Both connect a bone to something else. Ligaments connect bone to another bone; tendons connect bone to muscle.

**Q3:** Knee injuries are some of the most common sports injuries. Why do you think that is?

**A3:** Knees are critical for running, and many sports involve running. In addition, so many components of the knee must function together that damage to one affects functioning of the entire knee.

## Figure 23.10

**Q1:** Which types of muscles can you voluntarily contract?

**A1:** Skeletal muscles.

**Q2:** Do the muscles in the heart contract voluntarily or involuntarily?

**A2:** Involuntarily.

**Q3:** You do not have to think about breathing (otherwise, sleeping would be dangerous!), but you *can* increase or decrease your rate of breathing. Are the muscles involved in breathing, then, voluntary or involuntary?

**A3:** Involuntary, but we are able to override them for conscious control of breathing.

## Figure 23.11

**Q1:** How is skeletal muscle attached to bones in the skeleton?

**A1:** Via tendons.

**Q2:** Do skeletal muscles contract voluntarily or involuntarily?

**A2:** For the most part, skeletal muscles are involved in voluntary contractions. However, they can sometimes contract involuntarily (for example, when your leg twitches during sleep, or when your hand pulls away from a hot kettle).

**Q3:** Is a muscle fiber one cell or multiple cells?

**A3:** A muscle fiber is considered one cell, but it was created by the fusion of several cells.

## Figure 23.12

**Q1:** List muscle structures from smallest to largest, beginning with sarcomeres.

**A1:** Sarcomere, myofibril, muscle fiber, muscle fiber bundle, muscle.

**Q2:** What are the components of the sarcomere?

**A2:** Myosin, actin, Z disc.

**Q3:** Across animal species, the microscopic structure of muscles is the same. Why, then, are there differences in strength among animals?

**A3:** Animals are of different sizes and shapes, so their muscles are too. If a mouse were as large as an elephant, it would likely be as strong.

# CHAPTER 24

## END-OF-CHAPTER ANSWERS

1. d
2. c
3. a
4. b
5. d
6. (a) C, (b) P, (c) C, (d) P
7. SA node: 3, left ventricle: 2, right atrium: 1, pulmonary circuit: 4
8. b
9. nephron, reabsorb, secrete, ureter, urethra
10. (a) 4, (b) 2, (c) 1, (d) 5, (e) 3
11. Veins are larger, to transport large quantities of blood. Capillaries are smaller, to facilitate transfer of oxygen and nutrients into surrounding cells.
12. b
13. The net transport of $CO_2$ will be from the lung air space to the alveolar capillaries. Gas exchange occurs through diffusion from an area of higher concentration to an area of lower concentration.
14. (a) The decrease in blood vessel flexibility that comes with age makes it easier for an increase in blood pressure or heart rate to damage the cardiovascular system, thereby reducing the body's ability to respond to environmental changes effectively. (b) Short term: increased heart rate. Long term: decreased resting heart rate (because the heart is stronger from exercise). (c) Anxiety can increase blood pressure, and some people are afraid of the doctor's office.

## ANSWERS TO FIGURE QUESTIONS

### Figure 24.3

**Q1:** Where in the blood is the majority of carbon dioxide carried?

**A1:** In the plasma.

**Q2:** Where in the blood is the majority of oxygen carried?

**A2:** In the red blood cells, attached to hemoglobin molecules.

**Q3:** What would happen if your red blood cells carried a mutation that made the hemoglobin less effective at binding to oxygen (as in sickle-cell disease)?

**A3:** Less oxygen would be available to the body, which could have many negative effects; for example, it would be hard to exercise.

### Figure 24.4

**Q1:** Why are arteries more muscular than veins?

**A1:** They need to sustain higher pressures to move blood throughout the body.

**Q2:** What structural feature(s) of capillaries enable easier diffusion into and out of surrounding tissues?

**A2:** They are very narrow and have thin, porous walls.

**Q3:** Why do you think capillaries are not typically transplanted?

**A3:** The small size of capillaries makes them harder to transplant than larger veins and arteries.

## Figure 24.5

**Q1:** Beginning with the left atrium, list the locations of a drop of blood as it moves through the circulatory system.

**A1:** Left atrium, left ventricle, arteries, capillaries, veins, right atrium, right ventricle, lungs, left atrium.

**Q2:** Why is the left ventricle larger than the right ventricle, and why are its walls thicker and more muscular?

**A2:** It must generate higher pressures to drive blood through the long systemic circuit. The right ventricle pushes blood through the much shorter pulmonary circuit.

**Q3:** Some people have an artificial pacemaker implanted in their heart. What is its function?

**A3:** If someone's natural pacemaker (SA node) is not working, then the heart will not receive the necessary signal for it to contract in unison. An artificial pacemaker ensures that the heart receives the signal it needs.

## Figure 24.6

**Q1:** When is air richer in oxygen—as it enters the body or as it exits the body?

**A1:** As it enters.

**Q2:** The figure shows air entering through the nose. Where else can air enter the respiratory system?

**A2:** Through the mouth.

**Q3:** Explain how the body creates a change in air pressure during breathing.

**A3:** During inhalation, the diaphragm pulls down and rib muscles push out to create a larger volume and thus decreased air pressure in the lungs, bringing air into the lungs from outside the body. During exhalation, the diaphragm moves up and rib muscles pull in, decreasing volume and increasing pressure so that air moves out.

## Figure 24.7

**Q1:** Beginning with the nose, list the locations of a molecule of oxygen as it moves through the respiratory system.

**A1:** Nose, nasal cavity, pharynx, larynx, trachea, bronchi, lung.

**Q2:** From the lung, where would the oxygen molecule move?

**A2:** From alveoli in the lungs into veins in the circulatory system and then into cells through diffusion.

**Q3:** How might an upper respiratory infection like the common cold affect your respiratory system and thus your breathing?

**A3:** Tissues would become inflamed and mucus production would increase, making breathing more difficult.

## Figure 24.8

**Q1:** Why is a large surface area important for gas exchange?

**A1:** Gases are exchanged through diffusion, and the rate of diffusion is directly related to the surface area available for it to occur.

**Q2:** Does carbon dioxide move into or out of the alveoli? Into or out of capillaries at the surface of the alveoli?

**A2:** Into alveoli; out of capillaries.

**Q3:** When a person has pneumonia, the alveoli may fill with fluids. Why would this be a problem?

**A3:** Gas exchange—oxygen and carbon dioxide—would be more difficult.

## Figure 24.10

**Q1:** What is released from the kidney, and where does it go after release?

**A1:** Urine is released, and it enters the bladder and is then excreted from the body.

**Q2:** What is the difference between reabsorption and secretion?

**A2:** Reabsorption is the process of bringing substances back into the body; secretion is removing substances from the body (in urine).

**Q3:** Alcohol suppresses the kidney's ability to reabsorb water. What common consequence of drinking alcohol is related to this fact, and how might you alleviate this problem?

**A3:** Decreased reabsorption can lead to dehydration, a major cause of hangovers. Increased intake of (nonalcoholic) fluids can alleviate this problem.

## Figure 24.12

**Q1:** To which part of the nervous system—CNS or PNS—do sensory neurons in your eye belong?

**A1:** PNS; all sensory neurons are part of the peripheral nervous system.

**Q2:** To which part of the nervous system—CNS or PNS—does the spinal cord belong?

**A2:** CNS.

**Q3:** In a *reflex arc*, sensory input is processed in the spinal cord and an immediate signal is sent to activate motor output, without first routing the input signal to the brain. The "knee jerk" is one example of a reflex arc. Give another example of this kind of reflexive reaction to sensory input.

**A3:** Examples include squinting in bright light and/or flinching at a loud sound or rapidly moving object.

## Figure 24.13

**Q1:** How do neurons look different from other cells you've learned about in this book?

**A1:** They have long extensions—dendrites and axons.

**Q2:** What is the function of these differences?

**A2:** Dendrites and axons are used to receive and send information rapidly from cell to cell.

**Q3:** Some people are born without the capacity to feel pain. Although it might initially sound nice not to feel pain, it is actually quite dangerous. Describe a situation in which it would be dangerous not to feel pain.

**A3:** One possible answer: If you were going to have a bath or shower but couldn't tell that the water was too hot, you could be badly burned.

# CHAPTER 25

## END-OF-CHAPTER ANSWERS

1. c

2. b

3. d

4. a

5. b

6. adaptive immune response: 4, endocrine system: 1, innate immune response: 2, hypothalamus: 3

7. (a) A, (b) A, (c) C, (d) A, (e) C

8. primary, secondary, active, Passive, active

9. (a) 4, (b) 5, (c) 3, (d) 1, (e) 2

10. Higher levels of cortisol could be hard on the body over the longer term (since they increase both liver activity and heart rate). Decreased immune function could make a person more vulnerable to pathogens such as cold and flu viruses.

11. c

12. If you "turn off" one of the body's immune response mechanisms, it may not be able to fight a pathogen/infection as well. In addition, some people are allergic to one or more of these medications.

## ANSWERS TO FIGURE QUESTIONS

### Figure 25.3

**Q1:** What organ coordinates the endocrine system?
**A1:** The hypothalamus.
**Q2:** How does an endocrine gland differ from an exocrine gland?
**A2:** Endocrine glands are ductless; exocrine glands have ducts.
**Q3:** How do male and female endocrine systems differ?
**A3:** Different sex organs contribute to the endocrine system: in males the testes, in females the ovaries.

### Figure 25.4

**Q1:** How do hormones travel to target cells?
**A1:** Through body fluids, especially the blood.
**Q2:** Distinguish between an endocrine cell and a target cell.
**A2:** Endocrine cells produce and secrete hormones; target cells change in response to the presence of hormones.
**Q3:** Why is a hormone called a signaling molecule?
**A3:** It signals that something has changed in the organism or that its environment requires a cellular response.

### Figure 25.5

**Q1:** Describe the two ways that a hormone outside a cell can exert its effect on a cell.

**A1:** The hormone may cross the plasma membrane, or it may bind to a receptor embedded in the membrane.
**Q2:** Within the cell, how does a hormone bring about a change in cell activity?
**A2:** By changing gene expression, metabolism, cytoskeletal organization, or membrane transport.
**Q3:** It takes very little of the hormone cortisol to have large effects throughout the body. Explain why.
**A3:** The binding of cortisol to specific cells (with the appropriate receptors) activates many proteins within those cells.

### Figure 25.6

**Q1:** Describe an event (other than the one illustrated in the figure) that might cause the release of adrenaline.
**A1:** One possible answer: Taking an exam.
**Q2:** What organs does adrenaline affect?
**A2:** The liver and heart (also the immune system).
**Q3:** What do you think would happen if your adrenal glands were constantly releasing adrenaline?
**A3:** The resulting breakdown of glycogen to glucose, increase in heart rate, and suppression of the immune system could be damaging over a long time period (hence the health problems associated with constant stress).

### Figure 25.7

**Q1:** What is the main physical barrier that animals use to keep out pathogens?
**A1:** Skin.
**Q2:** Give an example of a chemical defense within the digestive system.
**A2:** Low pH in the stomach.
**Q3:** Explain why rubbing your eyes and nose during flu and cold season is not recommended.
**A3:** Viruses from your hands can move into your eyes and nose, where it is easier for them to gain access and infect you.

### Figure 25.8

**Q1:** Place these terms in order from most to least inclusive: neutrophil, white blood cell, phagocyte, innate immune system.
**A1:** Innate immune system, white blood cell, phagocyte, neutrophil.
**Q2:** Compare and contrast macrophages and neutrophils.
**A2:** Both are phagocytes, a type of white blood cell, and both engulf pathogens. Macrophages are larger than neutrophils, and neutrophils also differ from macrophages by using chemicals to attack.
**Q3:** Why would it be a problem if your innate immune system identified the insulin-producing cells in your pancreas as "nonself"?
**A3:** If your immune system thought that your insulin-producing cells were pathogens, it would attack and destroy them. (This is, in fact, what occurs in type 1 diabetes.)

**Figure 25.9**

**Q1:** What is the role of white blood cells in inflammation?

**A1:** They destroy pathogens and engulf cellular debris.

**Q2:** What would happen if histamines were not produced during inflammation?

**A2:** Macrophages could not leave the bloodstream to attack invading pathogens at the site of cellular damage.

**Q3:** Why is inflammation called a "nonspecific" immune response?

**A3:** It responds in the same way to any invading pathogen (and, in fact, to any cellular damage).

**Figure 25.10**

**Q1:** Why is blood clotting an important immune response?

**A1:** It seals a potential point of entry for pathogens, and also reduces blood loss from the wound.

**Q2:** How are the inflammatory response and blood clotting similar?

**A2:** Both are components of the innate immune system, both are rapid responses to a wound, and both seal off the site of damage.

**Q3:** Some people have a genetic disorder in which their blood cannot clot. Why would this be a problem?

**A3:** If your blood can't clot, even the smallest wound can cause enormous loss of blood.

**Figure 25.11**

**Q1:** Why are B and T cells so named?

**A1:** B cells mature in the <u>b</u>one marrow; T cells mature in the <u>t</u>hymus.

**Q2:** In what way is this immune system "adaptive"?

**A2:** It adapts to specific invading pathogens, rather than having a generalized response to all pathogens as in the innate immune system.

**Q3:** Why is the adaptive immune response considered the third layer of the immune system?

**A3:** It responds more slowly to pathogens than do either the external defenses or the innate immune system. It also evolved later and is found only in vertebrates.

# CHAPTER 26

## END-OF-CHAPTER ANSWERS

1.  a

2.  b

3.  c

4.  a

5.  d

6.  modular growth: 3, indeterminate growth: 2, primary growth: 1, secondary growth: 4

7.  c

8.  e

9.  Annual, perennial, annual, perennial, monocot, fibrous root

10. (a) 3, (b) 1, (c) 4, (d) 2

11. Color will not be as important at night, and smell should be more important, to help the bat find the plant. Since bats are generally larger than birds, bat-pollinated flowers need to be sturdier and have larger nectar rewards too.

12. Ethylene is clearly released as a gas into the air, because it doesn't require physical contact to act. This would explain why it works more effectively in an enclosed space like a bag, which would prevent it from dispersing through the air as quickly.

13. Onions are modified leaves, as the structure of the whole onion plant makes clear. Tomatoes are a fruit, as suggested by their plentiful seeds and the fact that they're sweet when ripe.

## ANSWERS TO FIGURE QUESTIONS

**Figure 26.3**

**Q1:** What is the function of the vascular tissue?

**A1:** It moves water and nutrients through the plant.

**Q2:** A plant organ is green if the cells within it contain chloroplasts. In the figure, which plant organ does not contain chloroplasts, and why do you think that is?

**A2:** The roots, because they are not exposed to sunlight and so cannot photosynthesize.

**Q3:** Which tissue type has chloroplast-containing cells? Why?

**A3:** Ground tissue, because this is where photosynthesis occurs.

**Figure 26.4**

**Q1:** How do root hairs increase the amount of water and nutrients that a plant can absorb?

**A1:** Root hairs increase the plant surface area exposed to the soil, enabling more water and nutrients to be absorbed.

**Q2:** How do roots make a plant more stable?

**A2:** Plants without broad and deep roots are more easily knocked over by wind or rain.

**Q3:** Given question 2, which do you predict would be more stable in harsh weather conditions: annuals or perennials?

**A3:** Perennials should be more stable because their roots are longer (since they grow over multiple years).

**Figure 26.5**

**Q1:** Which part of the shoot system—the bud, stem, or leaf—produces flowers?

**A1:** Buds produce flowers (or leaves).

**Q2:** Find the stoma in the figure. How is its location on the leaf important for its function?

**A2:** Stomata take in carbon dioxide and release water and oxygen, so they need to be on the surface of the leaf.

**Q3:** What nutrients does phloem move from the leaves to other parts of the plant?

**A3:** Sugars created from photosynthesis.

## Figure 26.7

**Q1:** Give an example of how plants use chemicals to defend themselves.

**A1:** One possible answer: Plants store poisons in their leaves so that herbivores won't eat them.

**Q2:** Of the plant hormones introduced in this figure, which promote growth?

**A2:** Auxins, gibberellins, and cytokinins.

**Q3:** What is the first line of plant defense? Is this a chemical defense?

**A3:** A tough outer surface and organs modified for defense (for example, thorns); these are mainly physical defenses, not chemical.

## Figure 26.10

**Q1:** Are eggs and sperm haploid or diploid? Are they part of the sporophyte or gametophyte generation?

**A1:** Haploid; gametophyte.

**Q2:** Why is the plant life cycle, but not the animal life cycle, referred to as an "alternation of generations"?

**A2:** Because plants have a complete haploid organism, the gametophyte, while in animals, only the gametes are haploid.

**Q3:** How does asexual reproduction differ from the plant life cycle diagrammed here?

**A3:** In asexual reproduction, no haploid gamete is produced, so the entire gametophyte portion of the life cycle (bottom half of the figure) does not occur.

## Figure 26.11

**Q1:** What kinds of seeds would you expect to be contained within a sugary fruit: seeds spread by water or by animals? Why?

**A1:** Seeds spread by animals, since animals must be attracted to the seeds/fruit as food.

**Q2:** What kinds of seeds would you expect to weigh less: those spread by air or those artificially selected by humans? Why?

**A2:** Those spread by air, because lighter seeds will travel farther, and humans want larger seeds or fruit because they are more edible.

**Q3:** How does the height of a tree affect its ability to spread its seeds widely?

**A3:** Seeds from taller trees should be able to disperse farther because they are higher up and better able to be lifted and carried long distances by wind.

## Figure 26.12

**Q1:** Many crops, including apples, peppers, and tomatoes, depend on insect pollinators. What would happen if no pollinator visited a plant?

**A1:** If a plant is not pollinated, the fruit won't develop.

**Q2:** Although they look very different, flower petals are actually modified leaves. Do flowers still perform the main function of leaves? Explain your reasoning.

**A2:** Most flowers aren't green, suggesting that they don't have chloroplasts and therefore cannot produce food via photosynthesis, the main function of leaves.

**Q3:** Some flowers look like an insect. How might this resemblance attract pollinators?

**A3:** Male insects in search of a mate might be attracted to the flower, thinking it is a female of their own species. Alternatively, insect predators might be attracted to the flower, thinking it is food.

## Figure 26.13

**Q1:** What traits are scientists selecting for in each new generation of Kernza?

**A1:** Larger seeds, better flavor, and deeper roots.

**Q2:** What would happen if the Kernza plants were allowed to freely pollinate, rather than being selectively pollinated by hand?

**A2:** Pollination would be random, so the desirable traits could be lost in the next generation of plants.

**Q3:** After choosing two plants to breed, researchers put the plant heads together under a bag as shown in the photo. Why do they do this before the plants begin to pollinate?

**A3:** To ensure no inadvertent pollination occurs by wind or by an insect pollinator.

# Glossary

**A**

**ABA** See **abscisic acid**.

**abiotic** Nonliving. Compare *biotic*.

**abscisic acid (ABA)** A plant hormone that mediates adaptive responses to drought, cold, heat, and other stresses.

**absorption** The uptake of mineral ions and small molecules by cells lining the cavity of the digestive tract.

**acid** A chemical compound that loses hydrogen ions (H$^+$) in aqueous surroundings. Compare *base* (definition 1).

**actin filament** One of the two kinds of protein filaments, made of the protein actin, that is found in sarcomeres. The sliding of *myosin filaments* against actin filaments enables sarcomeres to contract.

**active immunity** Immunity to a particular pathogen that is conferred by antibodies made by the body itself. Compare *passive immunity*.

**active site** The location within an enzyme where substrates are bound.

**active transport** The movement of a substance in response to an input of energy. Compare *passive transport*.

**adaptation** 1. An evolutionary process by which a population becomes better matched to its environment over time. 2. See **adaptive trait**.

**adaptive immune system** The immune system's third line of defense against pathogens, which mounts responses against specific invaders via antibodies (antibody-mediated immunity) and phagocytes and other specialized cells (cell-mediated immunity). Compare *external defenses* and *innate immune system*.

**adaptive radiation** The expansion of a group of organisms to take on new ecological roles and to form new species and higher taxonomic groups.

**adaptive trait** Also called an *adaptation*. A feature that gives an individual improved function in a competitive environment.

**adenine (A)** One of the four nucleotides that make up DNA. The other three are thymine (T), guanine (G), and cytosine (C).

**ADP** Adenosine diphosphate.

**adrenal glands** The paired glands, located atop the kidneys, that release the hormones cortisol, adrenaline (epinephrine), and noradrenaline (norepinephrine), which launch a number of rapid physiological responses, including boosting blood glucose levels.

**aerobic** Requiring oxygen. Compare *anaerobic*.

**algae** One of two main groups of protists, whose members are photosynthetic and may or may not be motile. Compare *protozoans*.

**alleles** Different versions of a given gene.

**allopatric speciation** The formation of new species from geographically isolated populations. Compare *sympatric speciation*.

**alternation of generations** The life cycle of flowering plants, in which haploid stages (gametophytes) alternate with diploid stages (sporophytes).

**alveoli (sing. alveolus)** Small clusters of minute sacs resembling a bunch of grapes where gases are exchanged in the lungs.

**amino acid** Any of a class of small molecules that are the building blocks of proteins.

**amniocentesis** A prenatal genetic screening technique in which amniotic fluid is extracted from the pregnancy sac that surrounds a fetus by means of a needle that is inserted through the abdomen into the uterus. Compare *chorionic villus sampling*.

**anabolism** Metabolic pathways that create complex biomolecules from smaller organic compounds. Compare *catabolism*.

**anaerobic** Not requiring oxygen. Compare *aerobic*.

**analogous trait** A feature that is shared across species because of convergent evolution, not because of modification by descent from a recent common ancestor. Compare *homologous trait*.

**analytical study** An observational study that looks for patterns in the information collected and addresses how or why those patterns came to exist. Compare *descriptive study*.

**anatomy** The study of the structures that make up a complex multicellular body. Compare *physiology*.

**anchorage dependence** The phenomenon in which cells stop dividing when detached from their surroundings.

**androgens** Any of the hormones produced in the testes that stimulate cells to develop characteristics of maleness, such as beard growth and the production of sperm. Compare *estrogens* and *progestogens*.

**angiogenesis** The formation of new blood vessels.

**angiosperms** Flowering plants. One of two main groups of seed-bearing plants, characterized by seeds enclosed in an ovary and by flowers. Compare *gymnosperms*.

**Animalia** The animals. One of the six kingdoms of life, in the domain Eukarya, encompassing all animals, including humans, birds, and dinosaurs.

**annual** A plant that completes its life cycle in one year. Compare *biennial* and *perennial*.

**antibody** Any of various Y-shaped proteins that recognize invaders by their *antigens* and attack them in the antibody-mediated response of the adaptive immune system.

**antibody-mediated immunity** A response mounted by the adaptive immune system in which specific invaders are attacked by antibodies. Compare *cell-mediated immunity*.

**anticodon** A unique sequence of three nitrogenous bases at one end of a tRNA molecule that binds to the corresponding *codon* on an mRNA molecule.

**antigen** Any of various nonself markers found on pathogens that mark them for destruction by *antibodies*.

**anus** The muscle-lined opening through which solid waste is eliminated from the body.

**appendicular skeleton** The part of the skeleton that has to do with motion. It is made up of the arms, legs, and pelvis. Compare *axial skeleton*.

**applied research** Research in which scientific knowledge is applied to human issues and often commercial applications. Compare *basic research*.

**Archaea** One of the three domains of life (compare *Bacteria* and *Eukarya*) and also one of the six kingdoms of life. The domain and kingdom Archaea consists of single-celled organisms best known for living in extremely harsh environments.

**artery** Any of the large vessels that transport blood away from the heart. Compare *capillary* and *vein*.

**artificial selection** The process by which individuals display specific traits through selective breeding. Compare *natural selection*.

**ascomycetes** Sac fungi, one of three main groups of fungi. Compare *basidiomycetes* and *zygomycetes*.

**asexual reproduction** The process by which clones, offspring that are genetically identical to the parent, are generated. Compare *sexual reproduction*.

**atom** The smallest unit of an element that retains the element's distinctive properties. Atoms are the building blocks of all matter.

**atomic mass number** The sum of the number of protons and the number of neutrons in an atom's nucleus. Compare *atomic number*.

**atomic number** The number of protons in an atom's nucleus. Compare *atomic mass number*.

**ATP** Adenosine triphosphate, a small, energy-rich organic molecule that is used to store energy and to move it from one part of a cell to another. Every living cell uses ATP.

**atrioventricular (AV) node** The part of the heart that sends a signal from the atria to the ventricles telling the ventricles to contract. Compare *sinoatrial node*.

**atrium (pl. atria)** Either of the two upper chambers of the heart. Compare *ventricle*.

**autosome** Any chromosome that is not one of the *sex chromosomes*.

**autotroph** A metabolic producer, an organism that makes food on its own. *Chemoautotrophs* acquire their energy from inorganic chemicals in their environment. *Photoautotrophs* absorb the energy of sunlight and take in carbon dioxide to conduct photosynthesis. Compare *heterotroph*.

**auxin** Any of a group of plant hormones that are necessary for cell division and promote root formation. Compare *cytokinin* and *gibberellin*.

**AV node** See **atrioventricular node**.

**axial skeleton** The part of the skeleton that supports and protects the long axis of the body. It includes the skull, the ribs, and a long, bony spinal column. Compare *appendicular skeleton*.

**axon** The part of a neuron that sends information to other cells. Compare *dendrite*.

**B**

**B cell** A type of lymphocyte that matures in the bone marrow. B cells are involved in antibody-mediated immunity. Compare *T cell*.

**Bacteria** One of the three domains of life (compare *Archaea* and *Eukarya*) and also one of the six kingdoms of life. The domain and kingdom Bacteria includes familiar disease-causing bacteria.

**base** 1. A chemical compound that accepts hydrogen ions ($H^+$) in aqueous surroundings. Compare *acid*. 2. The nitrogen-containing component of nucleotides. The four nitrogenous bases of DNA are adenine, cytosine, guanine, and thymine. In RNA, thymine is replaced by uracil.

**base pair** Also called *nucleotide pair*. Two nucleotides that form one rung of the DNA ladder.

**base-pairing rules** The rules that govern the pairing of nucleotides in DNA. Adenine (A) on one strand can pair only with thymine (T) on the other strand, and cytosine (C) can pair only with guanine (G).

**basic research** Research that is intended to expand the fundamental knowledge base of science. Compare *applied research*.

**basidiomycetes** Club fungi, one of three main groups of fungi. Compare *ascomycetes* and *zygomycetes*.

**behavioral trait** A characteristic of an individual's behavior, such as shyness or extroversion. Compare *biochemical trait* and *physical trait*.

**benign** Referring to a tumor, confined to one site.

**bias** A prejudice or opinion for or against something.

**biennial** A plant that takes two years to complete its life cycle, growing and maturing in the first year and reproducing in the second year. Compare *annual* and *perennial*.

**bilateral symmetry** An animal body plan in which the body can be divided by just one plane passing vertically from the top to the bottom of the animal into two halves that mirror each other. Compare *radial symmetry*.

**bile** A fluid, produced by the liver, that helps digest fats by creating a coating that enables the fat globules to interact with water molecules to partially dissolve them.

**binary fission** A type of cell division in which a cell simply divides into two equal halves, resulting in daughter cells that are genetically identical to each other and to the parent cell.

**biochemical trait** A characteristic due to specific chemical processes of an individual, such as the level of a particular enzyme. Compare *behavioral trait* and *physical trait*.

**biodiversity** The variety of all the world's life-forms, as well as their interactions with each other and the ecosystems they inhabit.

**biogeography** The geographic locations where organisms or the fossils of a particular species are found.

**biological hierarchy** A way to visualize the breadth and scope of life, from the smallest structures to the broadest interactions between living and nonliving systems that we can comprehend.

**biological species concept** The idea that a species is defined as a group of natural populations that can interbreed to produce fertile offspring and cannot breed with other such groups.

**biome** A large region of the world defined by shared physical characteristics (especially climate) and a distinctive community of organisms.

**biomolecule** Also called *macromolecule*. A large organic molecule that is critical for living cells. Four major classes of biomolecules are proteins, carbohydrates, nucleic acids, and lipids.

**biosphere** All the world's living organisms and the physical spaces where they live.

**biotic** Living. Compare *abiotic*.

**bipedal** Walking upright on two legs.

**blood** The cells, cell fragments, and plasma that circulate through the heart and blood vessels.

**blood pressure** The force of blood pushing from the heart through blood vessels.

**blood vessel** A vessel (for example, vein, capillary, artery) that transports blood throughout the body.

**boreal forest** A terrestrial biome that is found in northern or high-latitude regions, has cold, dry winters and mild summers, and is dominated by coniferous trees.

**brain** The part of the central nervous system that controls and coordinates nerve signals throughout the body and has a large capacity for processing diverse types of sensory information.

**breathing** The process of taking air into the lungs (inhaling) and expelling air from them (exhaling).

**bronchioles** A series of ever-smaller tubes that branch from the bronchi and open into the alveoli.

**bronchus (pl. bronchi)** Either of two small tubes that the trachea branches into in the chest and that lead to the lungs.

**bryophytes** A group of nonflowering plants that includes liverworts and mosses.

## C

**Calvin cycle** See **light-independent reactions**.

**Cambrian explosion** The burst of evolutionary activity, occurring about 540 million years ago, that resulted in a dramatic increase in the diversity of life. Most of the major living animal groups first appear in the fossil record during this time.

**camouflage** Any type of coloration or appearance that makes an organism hard to find or hard to catch. Compare *mimicry* and *warning coloration*.

**cancer cell** Also called *malignant cell*. A tumor cell gains anchorage independence and starts invading other tissues.

**capillary** Any of the tiny blood vessels that exchange materials by diffusion with nearby cells. Compare *artery* and *vein*.

**capsule** An additional protective layer that surrounds the cell wall in prokaryotes.

**carbohydrate** Any of a major class of biomolecules, including sugars and starches, built of repeating units of carbon, hydrogen, and oxygen.

**carbon cycle** The movement of carbon within biotic communities, between living organisms and their physical surroundings, and within the abiotic world.

**carbon dioxide ($CO_2$)** The most abundant and consequential of the greenhouse gases.

**carbon fixation** See **light-independent reactions**.

**carbon sink** A natural or artificial reservoir that absorbs more carbon than it releases. Compare *carbon source*.

**carbon source** A natural or artificial reservoir that releases more carbon than it absorbs. Compare *carbon sink*.

**cardiac muscle** The specialized muscle that helps produce the coordinated contractions known as heartbeats. It has a banded appearance and branched muscle fibers, and its contractions are entirely involuntary. Compare *skeletal muscle* and *smooth muscle*.

**cardiovascular system** A closed circulatory system consisting of a muscular heart, a complex network of blood vessels that collectively form a closed loop, and blood that circulates through the heart and blood vessels.

**carnivore** An animal (or, rarely, plant) that eats other animals. Compare *herbivore* and *omnivore*.

**carrying capacity** The maximum population size that can be sustained in a given environment.

**cartilage** A dense tissue of the skeleton that combines strength with flexibility. It is found almost everywhere that two bones meet and prevents them from grinding together.

**catabolism** Metabolic pathways that release chemical energy in the process of breaking down complex biomolecules. Compare *anabolism*.

**causation** A statistical relation indicating that a change in one aspect of the natural world causes a change in another aspect. Compare *correlation*.

**cell** The smallest and most basic unit of life—a microscopic, self-contained unit enclosed by a water-repelling membrane.

**cell cycle** The sequence of events that make up the life of a typical eukaryotic cell, from the moment of its origin to the time it divides to produce two daughter cells.

**cell division** The final stage of the cell cycle. Cell division includes the transfer of DNA from the parent cell to the daughter cells.

**cell-mediated immunity** A response mounted by the adaptive immune system in which specific invaders are attacked by phagocytes and other specialized cells. Compare *antibody-mediated immunity*.

**cell theory** One of the unifying principles of biology; a theory stating that every living organism is composed of one or more cells, and that all cells living today came from a preexisting cell.

**cellular respiration** The reciprocal process to *photosynthesis*, in which sugars are broken down into energy usable by the cell.

**cellulose** A chemical substance that provides structural strength in plant cells. Compare *lignin*.

**central nervous system (CNS)** One of two main parts of the nervous system, consisting of the brain and spinal cord. Compare *peripheral nervous system*.

**centromere** The central region of a chromosome that attaches sister chromatids together.

**cervix** The lower portion of the uterus, which narrows and connects the uterus to the vagina.

**chaparral** A terrestrial shrubland biome characterized by cool, rainy winters and hot, dry summers, and dominated by drought-resistant plants.

**chemical bond** A force that holds two atoms together.

**chemical compound** A molecule that contains atoms from two or more different elements.

**chemical reaction** The process of breaking existing chemical bonds and creating new ones.

**chemoautotroph** See **autotroph**.

**chemoheterotroph** See **heterotroph**.

**chlorophyll** A green pigment that is specialized for absorbing light energy.

**chloroplast** An organelle of plant cells and some protist cells that captures energy from sunlight and uses it to manufacture food molecules via photosynthesis.

**chordates** A large phylum that encompasses all animals with backbones, such as fishes, birds, and mammals.

**chorionic villus sampling (CVS)** A prenatal genetic screening technique in which cells are extracted by gentle suction from the villi (a cluster of cells that attaches the pregnancy sac to the wall of the uterus). Ultrasound is used to guide the narrow, flexible suction tube through a woman's vagina and into her uterus. Compare *amniocentesis*.

**chromatin fiber** A chromosome in relaxed form, made up of DNA and nucleosomes.

**chromosomal abnormality** Any change in the chromosome number or structure, compared to what is typical for a species.

**chromosome** A DNA double helix wrapped around spools of proteins.

**chromosome theory of inheritance** The theory that genes are located on chromosomes, and that these chromosomes are the basis for all inheritance.

**circulatory system** The organ system that moves oxygen from the respiratory system to the heart, which then pumps oxygen-rich blood to the rest of the body.

**citizen science** Research that is assisted by members of the public, who participate by collecting and sometimes analyzing data in cooperation with professional scientists.

**citric acid cycle** See **Krebs cycle**.

**clade** A branch of an evolutionary tree, consisting of an ancestor and all its descendants.

**class** The unit of classification in the Linnaean hierarchy above order and below phylum.

**climate** The prevailing *weather* of a specific place over relatively long periods of time (30 years or more).

**climate change** A large-scale and long-term alteration in Earth's climate, including such phenomena as *global warming*, change in rainfall patterns, and increased frequency of violent storms.

**climax community** A mature community whose species composition remains stable over long periods of time.

**CNS** See **central nervous system**.

**codominance** An interaction between two alleles of a gene that causes a heterozygote to display a phenotype that clearly displays the effects of both alleles. Compare *incomplete dominance*.

**codon** A unique sequence of three mRNA bases that either specifies a particular amino acid during translation or signals the ribosomes where to start or stop translation. Compare *anticodon*.

**coevolution** The tandem evolution of two species that results because interaction between the two so strongly influences their survival.

**collagen** A tough but pliable protein that, in addition to being the main component of cartilage, is found in a great variety of tissues.

**colon** Also called *large intestine*. The final portion of the digestive system, where water and nutrients are absorbed before the remaining solid waste is expelled.

**commensalism** A species interaction in one species benefits at no cost to the other species. Compare *competition*, *mutualism*, and *predation*.

**common ancestor** An organism from which many species have evolved.

**common descent** The sharing of a common ancestor by two or more different species.

**community** The populations of different species that live and interact with one another in a particular place.

**compact bone** One of the two major types of bone tissue. It forms the hard, white outer region of bones. Compare *spongy bone*.

**competition** A species interaction in which both species may be harmed. Compare *commensalism*, *mutualism*, and *predation*.

**competitive exclusion principle** The idea that different species that use the same resource can coexist only if one of the species adapts to using other resources.

**complementary base-pairing** The relationship between two nucleic acid strands in which each purine on one strand hydrogen-bonds with a specific pyrimidine on the opposite strand. A pairs with T or U, and C pairs with G.

**complex trait** A genetic trait whose pattern of inheritance cannot be predicted by Mendel's laws of inheritance.

**condensation** The transition from gas to liquid. Compare *evaporation*.

**conjugation** The physical process of transferring genetic material through direct contact.

**connective tissue** A tissue that binds and supports tissues and organs.

**consumer** An organism that obtains energy by eating all or parts of other organisms or their remains. Compare *producer*.

**contraceptive** Any means of preventing pregnancy.

**control group** The group of subjects in an experiment that is maintained under a standard set of conditions with no change in the independent variable. Compare *treatment group*.

**controlled experiment** An experiment that measures the value of a dependent variable for two groups of subjects that are comparable in all respects except that one group (the treatment group) is exposed to a change in the independent variable and the other group (the control group) is not.

**convection cell** A large and consistent atmospheric circulation pattern in which warm, moist air rises and cool, dry air sinks. Earth has four giant convection cells.

**convergent evolution** Evolution that results in organisms that have different genetics but appear very much alike.

**corpus luteum** The cells of the ruptured follicle that remain behind in the ovary after ovulation to produce the hormone progesterone.

**correlation** A statistical relation indicating that two or more aspects of the natural world behave in an interrelated manner: if one shows a particular value, we can predict a particular value for the other aspect. Compare *causation*.

**cotyledon** A food-storing organ that is part of the tiny, embryonic seedling that lies inside a plant seed.

**covalent bond** The sharing of electrons between two atoms. Compare *hydrogen bond* and *ionic bond*.

**credentials** Evidence of qualifications and competence to be recognized as an authority on a subject. Such evidence would include education and accomplishments.

**CRISPR** Clustered regularly interspaced short palindromic repeats. An RNA sequence that guides molecular machinery to a DNA sequence through complementary base-pairing.

**cross** See **genetic cross**.

**crossing-over** The physical exchange of chromosomal segments between homologous chromosomes. Compare *genetic recombination*.

**CVS** See **chorionic villus sampling**.

**cyclical fluctuation** A relatively predictable pattern of change in population size that occurs when at least one of two species is strongly influenced by the other. Compare *irregular fluctuation*.

**cytokine** Any of a group of signaling proteins that immune system cells use to communicate when an invader is present.

**cytokinesis** Division of the cytoplasm, the second step of mitotic division, resulting in two self-contained daughter cells. Compare *mitosis*.

**cytokinin** Any of a group of plant hormones that are necessary for cell division and promote shoot formation. Compare *auxin* and *gibberellin*.

**cytosine (C)** One of the four nucleotides that make up DNA. The other three are adenine (A), thymine (T), and guanine (G).

**cytoskeleton** The network of protein cylinders and filaments that forms the framework of a cell.

## D

**data (sing. datum)** Information collected in a scientific study.

**decomposer** A scavenger that dissolves the dead bodies of other organisms to consume them. Compare *detritivore*.

**deletion** A mutation in which a base is deleted from the DNA sequence of a gene. Compare *insertion* and *substitution*.

**dendrite** The part of a neuron that receives information from other cells. Compare *axon*.

**density-dependent population change** A change in population size that occurs when birth and death rates change as the population density changes. Compare *density-independent population change*.

**density-independent population change** A change in population size that occurs when populations are held in check by factors that are not related to the density of the population. Compare *density-dependent population change*.

**deoxyribonucleic acid** See **DNA**.

**dependent variable** Any variable that responds, or could potentially respond, to changes in an *independent variable*.

**dermal tissue** One of three types of plant tissue. Forming the outermost layer of the plant, dermal tissue protects the plant from the outside environment and controls the flow of materials into and out of the plant. Compare *ground tissue* and *vascular tissue*.

**descriptive study** An observational study that reports information about what is found in nature. Compare *analytical study*.

**desert** A terrestrial biome characterized by a lack of moisture and temperature fluctuation from very hot during the day to very cold at night, and dominated by succulent plants such as cacti.

**determinate growth** The general growth pattern of animals, in which they grow early in their lives and then level off. Compare *indeterminate growth*.

**detritivore** A scavenger that mechanically breaks apart the dead bodies of other organisms to consume them. Compare *decomposer*.

**diastole** The second pressure, or bottom number, in a blood pressure reading—the pressure at the point the heart relaxes after a contraction. Compare *systole*.

**dicot** Any plant that has two cotyledons in each of its seeds. Dicots are characterized by parallel veins in the leaves, scattered vascular bundles, fibrous roots, and floral parts in multiples of three. Examples include beans, squash, oak trees, and roses. Compare *monocot*.

**diffusion** The movement of a substance from a region of higher concentration to a region of lower concentration.

**digestion** The chemical breakdown of food.

**digestive system** The organ system that breaks down food in the mouth, stomach, and small intestine and absorbs nutrients in the small and large intestine.

**dihybrid cross** A controlled mating experiment involving organisms that are heterozygous for two traits.

**diploid** Possessing a double set of genetic information, represented by $2n$. Somatic cells are diploid. Compare *haploid*.

**directional selection** The most common pattern of natural selection, in which individuals at one extreme of an inherited phenotypic trait have an advantage over other individuals in the population. Compare *disruptive selection* and *stabilizing selection*.

**disruptive selection** The least common pattern of natural selection, in which individuals with either extreme of an inherited trait have an advantage over individuals with an intermediate phenotype. Compare *directional selection* and *stabilizing selection*.

**DNA** Deoxyribonucleic acid, the genetic code of life, consisting of two parallel strands of nucleotides twisted into a double helix. DNA is the genetic material that transfers information from parents to offspring.

**DNA polymerase** The enzyme that builds new strands of DNA in DNA replication.

**DNA replication** The duplication of a DNA molecule.

**DNA sequence similarity** The degree to which the sequences of two different DNA molecules are the same—a measure of how closely related two DNA molecules are to each other.

**domain** The highest hierarchical level in the organization of life, describing the most basic and ancient divisions among living organisms. The three domains of life are Bacteria, Archaea, and Eukarya.

**dominant allele** An allele that prevents a second allele from affecting the phenotype when the two alleles are paired together. Compare *recessive allele*.

**dominant genetic disorder** A genetic disorder that is inherited as a dominant trait on an autosome. Compare *recessive genetic disorder*.

**double helix** The spiral formed by two complementary strands of nucleotides that is the backbone of DNA.

**down-regulation** The slowing down of gene expression. Compare *up-regulation*.

## E

**Earth equivalent** The number of planet Earths needed to provide the resources we use and absorb the wastes we produce.

**ecological community** An association of species that live in the same area.

**ecological footprint** The area of biologically productive land and water that an individual or a population requires to produce the resources it consumes and to absorb the waste it produces.

**ecological isolation** The condition in which closely related species in the same territory are reproductively isolated by slight differences in habitat. Compare *geographic isolation* and *reproductive isolation*.

**ecological niche** The set of conditions and resources that a population needs in order to survive and reproduce in its habitat.

**ecology** The scientific study of interactions between organisms and their environment, where the environment of an organism includes both biotic factors (other living organisms) and abiotic (nonliving) factors.

**ecosystem** A particular physical environment and all the communities in it.

**ecosystem process** Any of four processes—nutrient cycling, energy flow, water cycling, and succession—that link the biotic and abiotic worlds in an ecosystem.

**egg** Also called *ovum*. The female gamete. Compare *sperm*.

**electron** A negatively charged particle found outside the nucleus of an atom. Compare *neutron* and *proton*.

**electron transport chain** An elaborate chain of chemical events in which electrons and protons ($H^+$) are handed over to other molecules that ultimately generates ATP and NADPH.

**element** A pure substance that has distinctive physical and chemical properties, and that cannot be broken down into other substances by ordinary chemical methods.

**elimination** The removal from the body of solid waste, consisting mostly of indigestible material and bacteria that inhabit the digestive tract.

**embryo** The earliest stage of development of an individual after fertilization, up to 2 months of age in humans. Compare *fetus*.

**embryonic development** The process by which an embryo develops. Common patterns of embryonic development across species provide evidence of evolution.

**endocrine gland** A gland that releases hormones into body fluids, such as the bloodstream, for transport to target cells throughout the body.

**endocrine system** The organ system, consisting of a number of glands and secretory tissues that produce and secrete hormones, that works closely with the nervous system to regulate all other organ systems.

**endocytosis** The process by which materials are transported into a cell via vesicles. Compare *exocytosis*.

**endoplasmic reticulum (ER)** An extensive and interconnected network of sacs made of a single membrane that is continuous with the outer membrane of the nuclear envelope. See also **rough ER** and **smooth ER**.

**endoskeleton** An internal skeleton. Compare *exoskeleton*.

**energy** The capacity of any object to do work, which is the capacity to bring about a change in a defined system.

**energy capture** The trapping and storing of solar energy by the producers at the base of an ecosystem's energy pyramid.

**energy carrier** A molecule that can store and deliver usable energy.

**energy pyramid** A pyramid-shaped representation of the amount of energy available to organisms in a food chain.

**enzyme** Any of a class of small molecules that speed up chemical reactions.

**epistasis** A form of inheritance in which the phenotypic effect of the alleles of one gene depends on the presence of alleles for another, independently inherited gene.

**epithelial tissue** A tissue that covers organs and lines body cavities.

**ER** See **endoplasmic reticulum**.

**esophagus** The food tube that connects the pharynx to the stomach.

**essential amino acid** Any of the eight amino acids that can be obtained only from food.

**estradiol** The primary estrogen. Compare *progesterone* and *testosterone*.

**estrogens** Any of the hormones produced in the ovaries that play a role in determining female characteristics such as wide hips, a voice that is pitched higher than that of males, and the development of breast tissues. Compare *progestogens* and *androgens*.

**estuary** An aquatic biome characterized by shallow depth and high productivity. It is the tidal ecosystem where a river flows into the ocean and consequently has a constant ebb and flow of fresh and salt water. It is dominated by grasses and sedges.

**ethylene** A plant hormone that stimulates the ripening of some fruits, including apples, bananas, avocadoes, and tomatoes.

**Eukarya** One of the three domains of life, including all the living organisms that do not fit into the domains *Archaea* or *Bacteria*, from amoebas to plants to fungi to animals.

**eukaryote** An organism that belongs to the Eukarya. Animals, plants, fungi, and protists are all eukaryotes. Compare *prokaryote*.

**eutherians** One of three main groups of mammals, whose members have a placenta and produce offspring that are born in a well-developed state. Compare *marsupials* and *monotremes*.

**evaporation** The transition from liquid to gas. Compare *condensation*.

**evolution** A change in the overall inherited characteristics of a group of organisms over multiple generations.

**evolutionary tree** A model of evolutionary relationships among groups of organisms that is based on similarities and differences in their DNA, physical features, biochemical characteristics, or some combination of these. It maps the relationships between ancestral groups and their descendants, and it clusters the most closely related groups on neighboring branches.

**exocytosis** The process by which materials are exported out of a cell via vesicles. Compare *endocytosis*.

**exon** A stretch of DNA that carries instructions for building a protein. Compare *intron*.

**exoskeleton** An external skeleton. Compare *endoskeleton*.

**experiment** A repeatable manipulation of one or more aspects of the natural world.

**experimental group** See **treatment group**.

**exploitative competition** Competition between species in which the two species indirectly compete for shared resources, such as food. Compare *interference competition*.

**exponential growth** A pattern of population growth in which the population increases by a constant proportion over a constant time interval, such as one year. Exponential growth occurs when there are no constraints on resources and is represented by a J-shaped curve. Compare *logistic growth*.

**external defenses** The immune system's first line of defense against pathogens, consisting of physical and chemical barriers that reduce the likelihood of harmful organisms or viruses gaining access to internal tissues. Compare *adaptive immune system* and *innate immune system*.

**F**

**F₁ generation** The first generation of offspring in a series of genetic crosses. Compare *F₂ generation* and *P generation*.

**F₂ generation** The second generation of offspring in a series of genetic crosses. Compare *F₁ generation* and *P generation*.

**facilitated diffusion** Diffusion that requires transport proteins. Compare *simple diffusion*.

**fact** A direct and repeatable observation of any aspect of the natural world. Compare *theory*.

**fallopian tube** See **oviduct**.

**falsifiable** Able to be refuted.

**family** The unit of classification in the Linnaean hierarchy above genus and below order.

**fat-soluble vitamin** A vitamin that cannot dissolve in water and therefore tends to accumulate in body fat because it cannot be so easily excreted in urine. Compare *water-soluble vitamin*.

**feces** The solid waste produced by digestion.

**feedback loop** The steps of a process that either decrease (*negative feedback*) or increase (*positive feedback*) the output of that process.

**fermentation** A metabolic pathway by which most anaerobic organisms extract energy from organic molecules. It begins with glycolysis and is followed by a special set of reactions whose only role is to help perpetuate glycolysis. Fermentation enables organisms to generate ATP anaerobically.

**ferns** A group of vascular, nonflowering plants that reproduce via spores.

**fertilization** The fusion of two gametes, resulting in a zygote.

**fetus** The second stage of development of an individual, from 2 months to birth in humans. Compare *embryo*.

**filtration** The blood-cleansing work of the kidneys.

**flagellum (pl. flagella)** A long, whiplike structure that assists bacteria in locomotion.

**flower** A structure in angiosperms that enhances sexual reproduction by bringing male gametes (sperm cells) to the female gametes (egg cells) in highly efficient ways, by attracting animal pollinators through scent, shape, and color.

**follicle-stimulating hormone (FSH)** A hormone, produced by the pituitary gland, that launches each menstrual cycle by stimulating the growth of ovarian follicles.

**food chain** A single direct line of who eats whom among species in a community. Compare *food web*.

**food web** The way in which various and often overlapping *food chains* of a community are connected.

**fossil** The mineralized remains or impression of a formerly living organism.

**founder effect** A form of genetic drift that occurs when a small group of individuals establishes a new population isolated from its original, larger population.

**freshwater biome** An aquatic biome whose character is defined by the terrestrial biomes that it borders or through which its water flows. Lakes, rivers, and wetlands are all part of the freshwater biome.

**fruiting body** In fungi, the structure resulting from mating that releases offspring as sexual spores.

**FSH** See **follicle-stimulating hormone**.

**Fungi** The fungi. One of the six kingdoms of life, in the domain Eukarya, distinguished by their modes of reproduction. Fungi are absorptive heterotrophs.

## G

**$G_0$ phase** A nondividing state of the cell.

**$G_1$ phase** "Gap 1," the first phase in the life of a newborn cell.

**$G_2$ phase** "Gap 2," the phase of the cell cycle between the S phase and cell division.

**gallbladder** An organ of the digestive system that stores bile made by the liver and dispenses the bile into the small intestine as needed.

**gamete** A sex cell. Male gametes are sperm; female gametes are eggs. Compare *somatic cell*.

**gametophyte** A haploid, multicellular structure produced by mitotic division of a plant's spore. Cells in the gametophyte differentiate into egg or sperm. Compare *Sporophyte*.

**gene** The basic unit of information, consisting of a stretch of DNA, that codes for a distinct genetic characteristic.

**gene expression** The process by which genes are transcribed into RNA and then translated to make proteins.

**gene flow** The exchange of alleles between populations.

**gene regulation** The changing of the genes that are expressed in response to internal signals or external cues that allows organisms to adapt to their surroundings by producing different proteins as needed.

**gene therapy** A genetic engineering technique for correcting defective genes responsible for disease development.

**genetic bottleneck** A form of genetic drift that occurs when a drop in the size of a population causes a loss of genetic variation.

**genetic carrier** An individual who has only one copy of a recessive allele for a particular disease and therefore can pass on the disorder allele but does not have the disease.

**genetic code** The information specified by each of the 64 possible codons.

**genetic cross** A controlled mating experiment performed to examine how a particular trait may be inherited.

**genetic disorder** A disease caused by an inherited mutation, passed down from a parent to a child.

**genetic divergence** The presence of differences in the DNA sequences of genes.

**genetic drift** A change in allele frequencies produced by random differences in survival and reproduction among the individuals in a population.

**genetic engineering** The permanent introduction of one or more genes into a cell, tissue, or organism.

**genetic modification (GM)** Altering the genes of an organism for a specific purpose.

**genetic recombination** The exchange of DNA between homologous chromosomes brought about by *crossing-over*, contributing to variation in gametes.

**genetic trait** Any inherited characteristic of an organism that can be observed or detected in some manner.

**genome** The complete set of genes of an organism.

**genotype** The allelic makeup of a specific individual with respect to a specific genetic trait. Compare *phenotype*.

**genus (pl. genera)** The unit of classification in the Linnaean hierarchy above species and below family.

**geographic isolation** The condition in which populations are separated by physical barriers. Compare *ecological isolation* and *reproductive isolation*.

**gibberellin** Any of a group of plant hormones that bring about stem growth through both cell elongation and cell division. Compare *auxin* and *cytokinin*.

**global warming** A significant increase in the average surface temperature of Earth over decades or more. Compare *climate change*.

**glycolysis** The first of three stages of cellular respiration. During glycolysis, sugars (mainly glucose) are split to make the three-carbon compound pyruvate. For each glucose molecule that is split, two molecules of ATP and two molecules of NADH are released. Compare *Krebs cycle* and *oxidative phosphorylation*.

**GM** See **genetic modification**.

**Golgi apparatus** A collection of flattened membranes that packages and directs proteins and lipids produced by the ER to their final destinations either inside or outside the cell.

**grassland** A terrestrial biome characterized by low moisture levels (but not as low as in deserts), and dominated by grasses and herbaceous plants.

**greenhouse effect** The process by which greenhouse gases let in sunlight and trap heat.

**greenhouse gas** A gas in Earth's atmosphere that absorbs heat that radiates away from Earth's surface. Examples include carbon dioxide ($CO_2$), water vapor ($H_2O$), methane ($CH_4$), and nitrous oxide ($N_2O$).

**ground tissue** One of three types of plant tissue. Forming the intermediate layer of the plant, ground tissue makes up the bulk of the plant body and performs a wide range of functions, including support, wound repair, and photosynthesis. Compare *dermal tissue* and *vascular tissue*.

**guanine (G)** One of the four nucleotides that make up DNA. The other three are adenine (A), thymine (T), and cytosine (C).

**gymnosperms** One of two main groups of seed-bearing plants, characterized by naked seeds. Compare *angiosperms*.

**H**

**halophile** A prokaryote, usually archaean, that can live in extremely salty environments.

**haploid** Possessing a single set of genetic information, represented by *2n*. Gametes are haploid. Compare *diploid*.

**heart** A muscular organ the size of a fist that works as the body's circulatory pump.

**heart rate** The number of times a heart beats per minute.

**herbivore** An animal that eats plants. Compare *carnivore* and *omnivore*.

**herd immunity** Protection against disease that is brought about by vaccination of a critical portion of a population.

**hermaphrodite** An individual that produces both functional testes and functional ovaries and is therefore both male and female.

**heterotroph** A metabolic consumer, an organism that obtains energy by taking it from other sources. *Chemoheterotrophs* consume organic molecules as a source of energy and carbon. *Photoheterotrophs* absorb the energy of sunlight but require an organic source of carbon. Compare *autotroph*.

**heterozygous** Carrying two different alleles for a given phenotype (*Bb*). Compare *homozygous*.

**histone protein** One of a class of specific proteins that, together with DNA, form nucleosomes.

**homeostasis** The process of maintaining constant internal conditions.

**homeostatic pathway** The sequence of steps that reestablishes homeostasis if there is any departure from the genetically determined normal state of a particular internal characteristic.

**hominids** The ape family, which includes humans and chimpanzees. All hominids are capable of tool use, symbolic language, and deliberate acts of deception. Compare *hominins*.

**hominins** The "human" branch of the *hominids*, including modern humans and extinct relatives such as Neanderthals.

**homologous pair** A pair of chromosomes consisting of one chromosome received from the father and one from the mother.

**homologous trait** A feature that is similar across species because of common descent. Homologous traits may begin to look different from one another over time. Compare *analogous trait*.

**homozygous** Carrying two copies of the same allele (such as *BB* or *bb*) for a particular gene. Compare *heterozygous*.

**horizontal gene transfer** The transfer of genes on plasmids from one bacterium to another.

**hormone** A signaling molecule that coordinates internal activities necessary for growth and reproduction.

**host** An organism in which a *parasite* lives.

**human microbiome** The complete collection of microbes that live in and on our cells and bodies.

**hydrogen bond** The weak electrical attraction between a hydrogen atom with a partial positive charge and a neighboring atom with a partial negative charge. Compare *covalent bond* and *ionic bond*.

**hydrologic cycle** The movement of water as it circulates from the land to the sky and back again.

**hydrophilic** Literally, "water-loving." Soluble in water. Compare *hydrophobic*.

**hydrophobic** Literally, "water-fearing." Excluded from water. Compare *hydrophilic*.

**hypertonic** Describing a fluid that has a solute concentration higher than that of the cell it surrounds. Compare *hypotonic* and *isotonic*.

**hyphae (sing. hypha)** The fine, branching threads of fungi that absorb nutrients from the environment.

**hypothalamus (pl. hypothalami)** A small organ at the base of the vertebrate brain that coordinates the endocrine system and integrates it with the nervous system.

**hypothesis (pl. hypotheses)** An informed, logical, and plausible explanation for observations of the natural world.

**hypotonic** Describing a fluid that has a solute concentration lower than that of the cell it surrounds. Compare *hypertonic* and *isotonic*.

**I**

**immune memory** The capacity of the adaptive immune system to remember a first encounter with a specific pathogen and to mobilize a speedy and targeted response to future infection by the same strain.

**immune system** The organ system that defends the body from invaders such as viruses, bacteria, and fungi.

**in vitro fertilization (IVF)** Fertilization of an egg by a sperm in a petri dish, followed by implantation of one or more embryos into a woman's uterus.

**incomplete dominance** An interaction between two alleles of a gene in which neither one can exert its full effect, causing a heterozygote to display an intermediate phenotype. Compare *codominance*.

**independent assortment** The random distribution of the homologous chromosomes into daughter cells during meiosis I.

**independent variable** The variable that is manipulated by the researcher in a scientific experiment. Compare *dependent variable*.

**indeterminate growth** The general growth pattern of plants, in which they grow throughout their lives. Compare *determinate growth*.

**induced fit** The way an enzyme changes shape when molecules bind to its active site.

**inflammation** The immediate and coordinated sequence of events mounted by cytokines in the innate immune system in response to tissue damage from a pathogen invasion or wound.

**ingestion** The taking in of food, the first stage in the processing of food by the digestive system.

**innate immune system** The immune system's second line of defense against pathogens, consisting of cells and proteins that recognize the presence of an invader and mount an internal defense to kill, disable, or isolate it. Compare *adaptive immune system* and *external defenses*.

**insertion** A mutation in which a base is inserted into the DNA sequence of a gene. Compare *deletion* and *substitution*.

**integumentary system** The largest organ system in the human body, covering and protecting the surface of the body.

**interference competition** Competition between species in which one organism directly excludes another from the use of a resource. Compare *exploitative competition*.

**intermediate fossil** A fossil that displays physical characteristics in between those of two known fossils in a family tree.

**interphase** The longest stage of the cell cycle. Most cells spend 90 percent or more of their life span in interphase.

**intron** A stretch of DNA that does not code for anything. Compare *exon*.

**invariant trait** A trait that is the same in all individuals of a species. Compare *variable trait*.

**ion** An atom that has lost or gained electrons and therefore is either negatively or positively charged.

**ionic bond** The chemical attraction between a negatively charged ion and a positively charged ion. Compare *covalent bond* and *hydrogen bond*.

**irregular fluctuation** An unpredictable pattern of change in population size. Compare *cyclical fluctuation*.

**isotonic** Describing a fluid that has a solute concentration equal to that of the cell it surrounds. Compare *hypertonic* and *hypotonic*.

**isotopes** Two or more forms of an element that have the same number of protons but different numbers of neutrons.

**IVF** See **in vitro fertilization**.

## J

**J-shaped growth curve** The type of graphical curve that represents exponential growth. Compare *S-shaped growth curve*.

**joint** A junction in the skeletal system that lets the skeleton move in specific ways.

## K

**karyotype** A depiction showing all the chromosomes of a particular individual or species arranged in homologous pairs.

**keystone species** A species that has a disproportionately large effect on a community, relative to the species' abundance.

**kidneys** The paired organs that maintain water and solute homeostasis.

**kingdom** The second-highest hierarchical level in the organization of life; the unit of classification in the Linnaean hierarchy above phylum and below domain. The six kingdoms of life are Bacteria, Archaea, Protista, Plantae, Fungi, and Animalia.

**Krebs cycle** Also called *citric acid cycle*. The second of three stages of cellular respiration. In this sequence of enzyme-driven reactions, the pyruvate made in glycolysis is broken down, releasing $CO_2$ and producing large amounts of energy carriers, including ATP, NADH, and $FADH_2$. Compare *glycolysis* and *oxidative phosphorylation*.

## L

**large intestine** See **colon**.

**larynx** Also called *voice box*. The breathing and sound-producing structure that forms the entryway to the trachea.

**law of independent assortment** The law, proposed by Gregor Mendel, stating (in modern terms) that when gametes form, the two alleles of any given gene segregate during meiosis independently of any two alleles of other genes. Compare *law of segregation*.

**law of segregation** The law, proposed by Gregor Mendel, stating (in modern terms) that the two alleles of a gene are separated during meiosis and end up in different gametes. Compare *law of independent assortment*.

**leaf** A structure in plants that produces the majority of the plant's food through photosynthesis.

**LH** See **luteinizing hormone**.

**lichen** A mutualistic association between a photosynthetic microbe (usually a green alga or cyanobacterium) and a fungus.

**ligament** A specialized, flexible band of tissue that joins bone to bone. Compare *tendon*.

**light-independent reactions** Also called the *Calvin cycle* or *carbon fixation*. The second of two principal stages of photosynthesis, in which a series of enzyme-catalyzed chemical reactions converts carbon dioxide ($CO_2$) into sugar, using energy delivered by ATP and electrons and hydrogen ions donated by NADPH. Compare *light reactions*.

**light reactions** The first of two principal stages of photosynthesis, in which chlorophyll molecules absorb energy from sunlight and use that energy for the splitting of water, which in turn produces oxygen gas ($O_2$) as a by-product that is released into the atmosphere. Compare *light-independent reactions*.

**lignin** A strengthening substance that links together *cellulose* fibers in plant cells to create a rigid network.

**lineage** A single line of descent.

**Linnaean hierarchy** A system of biological classification devised by the Swedish naturalist Carolus Linnaeus in the eighteenth century.

**lipid** Any of a major class of biomolecules built of fatty acids and insoluble in water.

**liposome** A sphere formed by a phospholipid bilayer.

**liver** A large organ of the digestive system that produces bile, stores glycogen, and detoxifies dangerous chemicals in the body.

**locus (pl. loci)** The physical location of a gene on a chromosome.

**logistic growth** A pattern of population growth in which the population grows nearly exponentially at first but then stabilizes at the maximum population size that can be supported indefinitely by the environment. Logistic growth is represented by an S-shaped curve. Compare *exponential growth*.

**lower respiratory system** The part of the respiratory system made up of the trachea, bronchi, and lungs. Compare *upper respiratory system*.

**lungs** The paired organs in which gases (oxygen and carbon dioxide) are exchanged.

**luteinizing hormone (LH)** A hormone, produced by the pituitary gland, that triggers ovulation.

**lymphatic system** The network of ducts, lymph nodes, and associated organs that are the primary sites for action by the adaptive immune system.

**lymphocyte** A type of white blood cell that confers specific immunity as part of the adaptive immune system. Immature lymphocytes differentiate into B cells and T cells.

**lysosome** An organelle in animal cells that acts as a garbage or recycling center. Compare *vacuole*.

## M

**macromolecule** See **biomolecule**.

**macronutrient** Any of nine nutrients—carbon, oxygen, hydrogen, nitrogen, phosphorus, potassium, calcium, sulfur, and magnesium—that plants need in relatively large amounts. Compare *micronutrient*.

**malignant cell** See **cancer cell**.

**mammals** A large class of animals that have body hair, sweat glands, and milk produced by mammary glands.

**marine biome** An aquatic biome characterized by salt water and encompassing both the coastal regions of all continents and the open ocean.

**marrow** A tissue found in the cavities of hollow bones that, depending on the type of bone, stores fat or produces blood cells.

**marsupials** One of three main groups of mammals, whose members have a simple placenta and produce offspring that complete development in their mother's pouch. Compare *eutherians* and *monotremes*.

**mass extinction** A period of time during which a great number of Earth's species goes extinct. The fossil record shows that there have been five mass extinctions in the history of Earth.

**matter** Anything that has mass and occupies a volume of space.

**meiosis** A specialized type of cell division that kicks off sexual reproduction. It occurs in two stages: meiosis I and meiosis II, each involving one round of nuclear division followed by cytokinesis. Compare *mitosis*.

**meiosis I** The first stage of meiosis, in which the chromosome set is reduced by the separation of each homologous pair into two different daughter cells. Each homologous chromosome lines up with its partner and then separates to the two ends of the cells. Compare *meiosis II*.

**meiosis II** The second stage of meiosis, in which sister chromatids are separated into two new daughter cells. Compare *meiosis I*.

**Mendelian trait** A trait that is controlled by a single gene and unaffected by environmental conditions.

**menstrual cycle** The process in human females by which individual eggs mature and are released in a hormone-driven sequence of events approximately monthly.

**messenger RNA (mRNA)** A type of RNA that is complementary to a DNA template strand. Compare *ribosomal RNA* and *transfer RNA*.

**meta-analysis** Work that combines results from different studies.

**metabolic heat** The heat released as a by-product of chemical reactions within a cell, typically during cellular respiration.

**metabolic pathway** Any of various chains of linked events that produce key biological molecules in a cell, including important chemical building blocks like amino acids and nucleotides.

**metabolism** All the chemical reactions that occur inside living cells, including those that release and store energy.

**metastasis** The spread of a disease from one organ to another.

**methanogen** An anaerobic archaean that feeds on hydrogen and produces methane gas as a by-product of its metabolism.

**microbe** A microscopic, single-celled organism.

**micronutrient** Any of a variety of nutrients (including iron, zinc, and copper) that plants need in relatively small amounts. Compare *macronutrient*.

**mimicry** Coloration of a nonpoisonous animal that resembles the coloration of a toxic species. Compare *camouflage* and *warning coloration*.

**mineral** Any of various small, inorganic nutrients needed by the human body for critical biological function, but only in small amounts. Compare *vitamin*.

**mitochondrial-DNA inheritance** The passing down of DNA from the mitochondria in an egg cell to a new generation. Mitochondrial DNA passes virtually unchanged from mother to child, so it can be tracked from one generation, or one species, to another. Sequencing of mitochondrial DNA can determine how related an individual is to its female ancestors on its mother's side. Compare *nuclear-DNA inheritance*.

**mitochondrion (pl. mitochondria)** An organelle that is a tiny power plant fueling cellular activities. Mitochondria are the main source of energy in eukaryotic cells.

**mitosis** Division of the nucleus, the first step of mitotic division. Mitosis is divided into four main phases: prophase, metaphase, anaphase, and telophase. Compare *cytokinesis* and *meiosis*.

**mitotic division** A type of cell division that generates two genetically identical daughter cells from a single parent cell in eukaryotes. It consists of two steps: mitosis and cytokinesis.

**modular growth** The way plants grow by repeatedly adding the same module of bud-stem-leaf unit aboveground or new lateral roots belowground.

**molecule** An association of atoms held together by chemical bonds.

**monocot** Any plant that has one cotyledon in each of its seeds. Monocots are characterized by netlike veins in the leaves, vascular bundles arranged in rings, taproots, and floral parts in multiples of four or five. Examples include all the grasses, members of the lily family, palm trees, and banana plants. Compare *dicot*.

**monomer** A small molecule that is the repeating unit of a *polymer*. For example, amino acids are the monomers that make up protein polymers.

**monotremes** One of three main groups of mammals, whose members lack a placenta and lay eggs. Compare *eutherians* and *marsupials*.

**morphology** An organism's physical characteristics.

**most recent common ancestor** The most immediate ancestor that two lineages share.

**motor output** A particular action of the body.

**mRNA** See **messenger RNA**.

**muscle fiber** A long, narrow cell that can span the length of an entire muscle because it is made up of several muscle cells that fused together during development.

**muscle tissue** A tissue that generates force by contracting.

**muscular system** The organ system that, working closely with the skeletal system, produces the force that moves structures within the body.

**mutation** A random change to the sequence of bases in an organism's DNA.

**mutualism** A species interaction in which both species benefit. Compare *commensalism*, *competition*, and *predation*.

**mycelium (pl. mycelia)** The entire bundle of hyphae that composes the main body of a fungus.

**mycorrhizal fungi** Fungi that form mutualistic associations with the root systems of plants, that help the plants absorb more water and nutrients from the soil.

**myofibril** Any of the cylindrical structures packed inside of muscle fibers containing proteins that contract by bracing against each other.

**myosin filament** One of the two kinds of protein filaments, made of the protein myosin, that is found in sarcomeres. The sliding of myosin filaments against *actin filaments* enables sarcomeres to contract.

### N

**natural selection** The process by which individuals with advantageous genetic characteristics for a particular environment survive and reproduce at a higher rate than do individuals with other, less useful characteristics. Compare *artificial selection*.

**negative feedback** The steps of a process that decrease its output. Compare *positive feedback*.

**nephron** The basic functional unit of the kidney.

**nervous system** The organ system that directs the rapid contractions of muscles and processes information received by the senses, such as touch, sound, and sight.

**nervous tissue** A tissue that communicates and processes information.

**net primary productivity (NPP)** The energy acquired through photosynthesis that is available for the growth and reproduction of producers within an ecosystem. It is the amount of energy captured by photosynthetic organisms, minus the amount they expend on cellular respiration and other maintenance processes.

**neuron** A specialized cell of the nervous system that transmits signals from one part of the body to another in a fraction of a second.

**neutron** An electrically neutral particle found in the nucleus of an atom. Compare *electron* and *proton*.

**nitrogen fixation** The process, carried out by bacteria, of taking nitrogen gas from the air and converting it to ammonia, making it available for plants.

**node** The point on an evolutionary tree indicating the moment in time when an ancestral group split, or diverged, into two separate lineages. The node represents the most recent common ancestor of the two lineages in question.

**noncoding DNA** DNA that does not code for any kind of functional RNA.

**nonspecific response** A response mounted by the immune system against pathogens that is indiscriminate as to the invaders it repels. External defenses and the innate immune system are both nonspecific responses. Compare *specific response*.

**notochord** In chordates, a flexible yet rigid rod along the center of the body that is critical for development.

**NPP** See **net primary productivity**.

**nuclear-DNA inheritance** The passing down of DNA from the nucleus in an egg or sperm cell to a new generation. Sequencing of nuclear DNA can determine how related an individual is to all of its ancestors, both male and female. Compare *mitochondrial-DNA inheritance*.

**nuclear envelope** The boundary of a cell's nucleus, consisting of two concentric phospholipid bilayers.

**nuclear pore** Any of many small openings in the nuclear envelope that allow chemical messages to enter and exit the nucleus.

**nucleic acid** Any of a major class of biomolecules, including DNA and RNA, built of chains of nucleotides.

**nucleosome** DNA wrapped around histone proteins that, in multiples, form the beads-on-a-string complex in a chromatin fiber.

**nucleotide** The basic repeating subunit of DNA, composed of the sugar deoxyribose, a phosphate group, and one of four bases: adenine (A), cytosine (C), guanine (G), or thymine (T).

**nucleotide pair** See **base pair**.

**nucleus (pl. nuclei)** 1. The dense core of an atom, which contains protons and neutrons. 2. The control center of the eukaryotic cell, containing all of the cell's DNA and occupying up to 10 percent of the space inside the cell.

**nutrient** A chemical element that is required by a living organism.

**nutrient cycling** The process by which decomposers break down dead organisms or waste products, release the chemical elements locked in the biological material, and return them to the environment.

## o

**observation** A description, measurement, or record of any object or phenomenon.

**omnivore** An animal that eats both animals and plants. Compare *carnivore* and *herbivore*.

**oogenesis** The series of cell divisions in human females that results in an egg. Compare *spermatogenesis*.

**opposable** Able to be placed opposite other digits of the hand or foot. For example, opposable thumbs can be placed opposite each of the other four fingers.

**oral cavity** The mouth.

**order** The unit of classification in the Linnaean hierarchy above family and below class.

**organ** A collection of different types of tissues that form a functional unit with a distinctive shape and location in the body.

**organ system** A network of organs that work in a closely coordinated manner to perform a distinct set of functions in the body.

**organelle** Any of the membrane-enclosed subcellular compartments found in eukaryotic cells.

**organic molecule** A molecule that includes at least one carbon-hydrogen bond.

**organism** An individual living thing composed of interdependent parts.

**origin of replication** A DNA sequence where DNA replication is initiated.

**osmosis** A form of simple diffusion in which water moves in and out of cells (and compartments inside cells).

**osteocyte** A specialized bone cell that surrounds itself with a hard, nonliving mineral matrix composed largely of calcium and phosphate.

**ovary** Either of a pair of female reproductive organs that produce eggs and estrogens in vertebrates. Compare *testis*.

**oviduct** Also called *fallopian tube*. The tube through which an egg travels from the ovary to the uterus.

**ovule** The egg-bearing structure in plants.

**ovum (pl. ova)** See **egg**.

**oxidative phosphorylation** The third of three stages of cellular respiration. During this process, the chemical energy of NADH and $FADH_2$ is converted into the chemical energy of ATP, while electrons and hydrogen atoms removed from NADH and $FADH_2$ are handed over to molecular $O_2$, creating water ($H_2O$). A large amount of ATP is generated. Compare *glycolysis* and *Krebs cycle*.

**oxytocin** A hormone—secreted by the fetus and, later in the birth process, by the mother's pituitary gland—that stimulates the uterine muscles and causes the placenta to secrete prostaglandins, which reinforce contractions.

**P**

**P generation** The first set of parents in a series of genetic crosses. Compare *F₁ generation* and *F₂ generation*.

**pacemaker** See **sinoatrial node**.

**pancreas** A gland that produces insulin and secretes fluids that aid in the digestion of food.

**parasite** An organism that lives in or on another species and harms it by stealing resources. For example, some parasites suck blood or live off the food in our intestines. Compare *host*.

**passive immunity** Immunity to a particular pathogen that is conferred by antibodies not made by the body, but received from an outside source. Compare *active immunity*.

**passive transport** The movement of a substance without the addition of energy. Compare *active transport*.

**pathogen** An infectious agent.

**PCR** See **polymerase chain reaction**.

**pedigree** A chart similar to a family tree that shows genetic relationships among family members over two or more generations of a family's medical history.

**peer-reviewed publication** The publishing of original research only after it has passed the scrutiny of experts who have no direct involvement in the research under review, or a scientific journal that follows this standard.

**penis** The male reproductive organ that introduces sperm into a female or hermaphrodite sexual partner. The penis is also involved in urination in mammals.

**perennial** A plant that lives three or more years. Compare *annual* and *biennial*.

**peripheral nervous system (PNS)** One of two main parts of the nervous system, consisting of the sensory nerves plus all the nerves that are not part of the *central nervous system*.

**PGD** See **preimplantation genetic diagnosis**.

**pH scale** A logarithmic scale that indicates the concentration of hydrogen ions. The scale goes from 0 to 14, with 0 representing an extremely high concentration of free H⁺ ions and 14 representing the lowest concentration.

**phagocyte** A type of white blood cell that functions as part of the innate immune system to mark and destroy foreign invaders by engulfing and digesting them.

**phagocytosis** Literally, "cellular eating." A large-scale version of endocytosis in which particles considerably larger than biomolecules are ingested. Compare *pinocytosis*.

**pharynx** Also called *throat*. An area where the back of the mouth and the two nasal cavities join together into a single passageway.

**phenotype** The physical, biochemical, or behavioral expression of a particular version of a trait. Compare *genotype*.

**phloem** One of two types of vascular tissue in plants. Phloem transports sugars from the leaves, where they are produced, to living cells in every part of the plant. Compare *xylem*.

**phospholipid** An organic molecule with a hydrophilic head and a hydrophobic tail.

**phospholipid bilayer** A double layer of phospholipids in which the heads face out and the tails face in. Plasma membranes are phospholipid bilayers.

**photoautotroph** See **autotroph**.

**photoheterotroph** See **heterotroph**.

**photoperiodism** The ability of plants to sense the duration of light and dark in a 24-hour cycle.

**photosynthesis** The process by which organisms capture energy from the sun and use it to create sugars from carbon dioxide and water, thereby transforming light energy into chemical energy stored in the covalent bonds of sugar molecules. Compare *cellular respiration*.

**phylum (pl. phyla)** The unit of classification in the Linnaean hierarchy above class and below kingdom.

**physical trait** An anatomical or physiological characteristic of an individual, such as the shape of an animal's head. Compare *behavioral trait* and *biochemical trait*.

**physiology** The science that focuses on the functions of anatomical structures. Compare *anatomy*.

**pili (sing. pilus)** Short, hairlike projections that cover the surface of many bacteria.

**pinocytosis** Literally, "cellular drinking." A large-scale version of endocytosis in which fluids are ingested. Compare *phagocytosis*.

**Plantae** The plants. One of the six kingdoms of life, in the domain Eukarya, encompassing all plants, which are multicellular photosynthetic autotrophs.

**plasma** The fluid portion of the blood.

**plasma membrane** A barrier consisting of a phospholipid bilayer that separates a cell from its external environment.

**platelet** A type of sticky cell fragment found in circulating blood that helps form blood clots.

**pleiotropy** The pattern of inheritance in which a single gene influences multiple different traits. Compare *polygenic trait*.

**PNS** See **peripheral nervous system**.

**point mutation** A mutation in which only a single base is altered.

**polar molecule** A molecule whose electrical charge is shared unevenly, with some regions being electrically negative and others electrically positive.

**pollen** In plants, a microscopic structure containing sperm cells that can be lofted into the air in massive quantities.

**pollinator** An animal that transports pollen from one plant to another.

**polygenic trait** A genetic trait that is governed by the action of more than one gene. Compare *pleiotropy*.

**polymer** A long strand of repeating units of small molecules called *monomers*. For example, proteins are polymers made up of amino acid monomers.

**polymerase chain reaction** (**PCR**) A technique for replicating DNA that can produce millions of copies of a DNA sequence in just a few hours from a small initial amount of DNA.

**polyploidy** The condition in which an individual's somatic cells have more than two sets of chromosomes.

**population** A group of individuals of the same species living and interacting in a shared environment.

**population density** The number of individuals per unit of area.

**population doubling time** The time it takes a population to double in size, as a measure of how fast a population is growing.

**population ecology** The study of the number of organisms in a particular place.

**population size** The total number of individuals in a population.

**positive feedback** The steps of a process that increase its output. Compare *negative feedback*.

**postzygotic barrier** A barrier that prevents a zygote from developing into a healthy and fertile individual—that is, a reproductive barrier that acts after a zygote exists. Compare *prezygotic barrier*.

**predation** A species interaction in which one species benefits and the other species is harmed. Compare *commensalism*, *competition*, and *mutualism*.

**predator** A consumer that eats either plants or animals. Compare *prey*.

**preimplantation genetic diagnosis** (**PGD**) The removal of one or two cells from an embryo developing in a petri dish, usually 3 days after fertilization, followed by testing for genetic disorders. One or more embryos that are free of disorders are then implanted into a woman's uterus.

**prey** An animal that is eaten by a *predator*.

**prezygotic barrier** A barrier that prevents a male gamete and a female gamete from fusing to form a zygote—that is, a reproductive barrier that acts before a zygote exists. Compare *postzygotic barrier*.

**primary consumer** An organism that eats producers. Compare *secondary consumer*, *tertiary consumer*, and *quaternary consumer*.

**primary growth** In plants, an increase in length by the division of cells at the tips of shoots, stems, and roots. Compare *secondary growth*.

**primary immune response** The slow response mounted by the adaptive immune system against an invading pathogen the very first time a person is exposed to that pathogen. Compare *secondary immune response*.

**primary literature** Scientific literature in which research is first published. Compare *secondary literature*.

**primary oocyte** An immature egg cell.

**primary succession** Succession that occurs in a newly created habitat, usually from bare rock or sand. Compare *secondary succession*.

**primates** The order of mammals to which humans belong. All primates have flexible shoulder and elbow joints, five functional fingers and toes, opposable thumbs, flat nails, and brains that are large in relation to the body.

**primer** A short stretch of RNA or DNA that is complementary-base-paired to a DNA template to provide a 3'-hydroxyl group for polymerase to initiate DNA replication or the polymerase chain reaction (PCR), respectively.

**process of science** See **scientific method**.

**producer** An organism at the bottom of a food chain that uses energy from the sun to produce its own food. Compare *consumer*.

**product** A substance that results from a chemical reaction. Compare *reactant*.

**progesterone** The primary progestogen. Compare *estradiol* and *testosterone*.

**progestogens** Any of the hormones produced in the ovaries that have a number of functions in the female body, including thickening the lining of the uterus and increasing the blood supply to it to create a suitable environment for a developing fetus. Compare *estrogens* and *androgens*.

**prokaryote** An organism that belongs to either the Bacteria or the Archaea. Compare *eukaryote*.

**promoter** A segment of DNA near the beginning of a gene that RNA polymerase recognizes and binds to begin transcription. Compare *terminator*.

**protein** Any of a major class of biomolecules built of amino acids.

**Protista** The protists. One of the six kingdoms of life, in the domain Eukarya; a diverse group that is composed of mainly single-celled, microscopic organisms grouped together simply because they are not plants, animals, or fungi, and that includes amoebas and algae.

**proton** A positively charged particle found in the nucleus of an atom. Compare *electron* and *neutron*.

**protozoans** One of two main groups of protists, whose members are nonphotosynthetic and motile. Compare *algae*.

**pseudoscience** Scientific-sounding statements, beliefs, or practices that are not actually based on the scientific method.

**pulmonary circuit** The circuit in the heart, consisting of the two chambers on the right side, that receives blood low in oxygen and pumps it to the lungs for gas exchange. Compare *systemic circuit*.

**Punnett square** A grid-like diagram showing all possible ways that two alleles can be brought together through fertilization.

## Q

**quaternary consumer** An organism that eats *tertiary consumers*. Compare *primary consumer* and *secondary consumer*.

**quorum sensing** A system of cell-to-cell communication used by prokaryotes that enables them to sense and respond to other bacteria in the area in accordance with the density of the population.

## R

**radial symmetry** An animal body plan in which the body can be sliced symmetrically along any number of vertical planes that pass through the center of the animal. Compare *bilateral symmetry*.

**reabsorption** The removal of valuable solutes such as sodium, chloride, and sugars from the fluid filtered by the kidneys so that they don't exit the body in the urine.

**reactant** A substance that undergoes change in a chemical reaction. Compare *product*.

**receptor-mediated endocytosis** A form of specific endocytosis in which receptor proteins embedded in the membrane recognize specific surface characteristics of substances that will be incorporated into the cell.

**receptor protein** A site where a molecule from another cell can bind.

**recessive allele** An allele that has no effect on the phenotype when paired with a *dominant allele*.

**recessive genetic disorder** A genetic disorder that is inherited as a recessive trait on an autosome. Compare *dominant genetic disorder*.

**red blood cell** A cell in the blood that greatly increases the oxygen-carrying capacity of blood.

**relative species abundance** How common one species is when compared to another.

**reproduction** The making of a new individual like oneself.

**reproductive barrier** A barrier that prevents two species from interbreeding, making them reproductively isolated.

**reproductive isolation** The condition in which barriers prevent populations from interbreeding. Compare *ecological isolation* and *geographic isolation*.

**reproductive system** The organ system that generates gametes and may also support fertilization and prenatal development.

**respiratory system** The organ system that brings in oxygen and expels carbon dioxide to support cellular respiration.

**ribonucleic acid** See **RNA**.

**ribosomal RNA (rRNA)** A type of RNA that is an important component of ribosomes. Compare *messenger RNA* and *transfer RNA*.

**ribosome** The site of protein synthesis (translation) in the cytoplasm. Ribosomes are embedded in the rough endoplasmic reticulum.

**RNA** Ribonucleic acid, a single-stranded nucleic acid transcribed from DNA and consisting of the ribonucleotides adenine, guanine, cytosine, and uracil.

**RNA polymerase** An enzyme that recognizes and binds a gene's promoter sequence and then separates the two strands of DNA during transcription.

**RNA splicing** Processing of mRNA in which the introns are snipped out of a pre-mRNA and the remaining pieces of mRNA—the exons—are joined to generate the mature mRNA.

**root** A structure in plants that absorbs water and mineral nutrients, anchors the plant, and stores food.

**root system** One of two plant organ systems. It anchors the plant, absorbs water and nutrients from the soil, transports food and water, and may store food. Compare *shoot system*.

**rough ER** A part of the endoplasmic reticulum, having a knobby appearance because of embedded ribosomes, where proteins are assembled. Compare *smooth ER*.

**rRNA** See **ribosomal RNA**.

**rubisco** The enzyme that catalyzes the first step in the light-independent reactions of photosynthesis, fixing a carbon molecule from $CO_2$.

## S

**S phase** The "synthesis" phase of the cell cycle, in which preparations for cell division begin. A critical event during this phase is the replication of all the cell's DNA molecules.

**S-shaped growth curve** The type of graphical curve that represents logistic growth. Compare *J-shaped growth curve*.

**SA node** See **sinoatrial node**.

**saliva** A fluid secreted into the oral cavity to aid in the digestion of food.

**sarcomere** Any of the contractile units of the muscular system that make up each myofibril.

**scavenger** An animal that eats dead or dying animals and plants. Scavengers are categorized as either *decomposers* or *detritivores*.

**science** A body of knowledge about the natural world, and an evidence-based process for acquiring that knowledge.

**scientific claim** A statement about how the world works that can be tested using the scientific method.

**scientific literacy** An understanding of the basics of science and the scientific process.

**scientific method** Also called *process of science*. The practices that produce scientific knowledge.

**scientific name** The unique two-word Latin name, consisting of the genus and species names, that is assigned to a species in the Linnaean hierarchy.

**secondary consumer** An organism that eats *primary consumers*. Compare *tertiary consumer* and *quaternary consumer*.

**secondary growth** In plants, an increase in thickness in either stems or roots. Compare *primary growth*.

**secondary immune response** The rapid response mounted by the adaptive immune system against an invading pathogen the second and subsequent times a person is exposed to that pathogen. Compare *primary immune response*.

**secondary literature** Scientific literature that summarizes and synthesizes an area of research. Compare *primary literature*.

**secondary succession** Succession that occurs after a disturbance in a community. Compare *primary succession*.

**secretion** The active transport by the kidneys of excess quantities of substances such as potassium and hydrogen ions, and some toxins, from the blood into the liquid passing through the kidney and out of the body in the urine.

**seed** In plants, the embryo and a supply of stored food, all encased in a protective covering.

**selective breeding** The process by which humans allow only individuals with certain inherited characteristics to mate.

**selective permeability** The quality of plasma membranes by which some substances are allowed to cross the membrane at all times, others are excluded at all times, and still others can pass through the membrane when they are aided by transport proteins.

**semiconservative replication** The mode of replication by which DNA is duplicated, where one "old" strand (the template strand) is retained (conserved) in each new double helix.

**sense** To perceive the world through a sensory system such as sight, touch, or smell.

**sensory input** Signals that are received, transmitted, and processed by the central nervous system.

**sensory organ** An organ of the body, such as the eyes or ears, that receives sensory input.

**serotype** See **viral strain**.

**set point** The genetically determined normal state of any physical or chemical characteristic of the body's internal environment.

**sex chromosome** One of the two chromosomes (X and Y) that determine gender. Compare *autosome*.

**sex-linked** Found solely on the X or Y chromosome. See also **X-linked** and **Y-linked**.

**sexual dimorphism** A distinct difference in appearance between the males and females of a species.

**sexual reproduction** The process by which genetic information from two individuals is combined to produce offspring. It has two steps: cell division through meiosis, followed by fertilization. Compare *asexual reproduction*.

**sexual selection** Natural selection in which a trait increases an individual's chance of mating even if it decreases the individual's chance of survival.

**shared derived trait** A unique feature common to all members of a group that originated in the group's most recent common ancestor and then were passed down in the group.

**shoot system** One of two plant organ systems, consisting of stems and leaves. Stems provide the plant with structural support, transport food and water, and hold leaves up to intercept light so that they can perform photosynthesis. Compare *root system*.

**simple diffusion** Diffusion in which substances such as the small, uncharged molecules of water, oxygen, or carbon dioxide, slip between the large molecules in the phospholipid bilayer without much hindrance. Compare *facilitated diffusion*.

**sinoatrial (SA) node** Also called *pacemaker*. The part of the heart that sends a signal telling the atria to contract. Compare *atrioventricular node*.

**sister chromatids** The two identical DNA molecules produced by the replication of a chromosome.

**skeletal muscle** The specialized muscle that is associated with the skeleton. It has a banded appearance, and its contractions are mainly voluntary. Compare *cardiac muscle* and *smooth muscle*.

**skeletal system** The organ system, consisting of bones, cartilage, and ligaments, that provides an internal framework to support the body of vertebrates.

**small intestine** The highly coiled tube, specialized for absorption, into which food moves from the stomach during digestion.

**smooth ER** A part of the endoplasmic reticulum, having a smooth appearance, where lipids and hormones are manufactured. Compare *rough ER*.

**smooth muscle** The specialized muscle found in the walls of the digestive system and blood vessels. It has no visible bands, and its contractions are entirely involuntary. Compare *cardiac muscle* and *skeletal muscle*.

**soluble** Able to mix completely with water.

**solute** A dissolved substance, such as sugar in water. Compare *solvent*.

**solution** Any combination of a solute and a solvent.

**solvent** The fluid, such as water, into which a substance has dissolved. Compare *solute*.

**somatic cell** A non–sex cell. Compare *gamete*.

**speciation** The process by which one species splits to form two species or more.

**species** 1. Members of a group that can mate with one another to produce fertile offspring. 2. The smallest unit of classification in the Linnaean hierarchy.

**species interaction** Any of four ecological interactions—mutualism, commensalism, predation, and competition—that occur between different species.

**species richness** The total number of different species that live in an ecological community.

**specific response** A response mounted by the adaptive immune system against a specific strain of pathogen. Compare *nonspecific response*.

**sperm** The male gamete. Compare *egg*.

**spermatogenesis** The series of cell divisions in human males that results in sperm. Compare *oogenesis*.

**spinal cord** A thick central nerve cord that is continuous with the brain, acting as a filter between the brain and the sensory neurons.

**spongy bone** One of the two major types of bone tissue. Honeycombed with numerous tiny cavities, it lies inside the *compact bone*. Spongy bone is most abundant at the knobby ends of our long bones.

**spore** 1. In fungi, a reproductive structure that can survive for long periods of time in a dormant state and will sprout under favorable conditions to produce the body of the organism. 2. In plants, a haploid cell produced by meiosis that divides mitotically to produce the gametophyte.

**sporophyte** A diploid, multicellular individual that arises from the zygote in plants. Compare *gametophyte*.

**sporulation** The formation of thick-walled dormant structures called spores.

**stabilizing selection** The pattern of natural selection in which individuals with intermediate values of an inherited phenotypic trait have an advantage over other individuals in the population. Compare *directional selection* and *disruptive selection*.

**start codon** The codon AUG; the point on an mRNA strand at which the ribosomes begin translation. Compare *stop codon*.

**statistics** A branch of mathematics that estimates the reliability of data.

**stem** A structure in plants that provides structural support, transports food and water, and holds leaves up to intercept light.

**stoma (pl. stomata)** An air pore in the leaf of a plant that controls gas exchange.

**stomach** The organ of the digestive system, located between the esophagus and intestines, in which most digestion occurs, through mechanical and chemical means.

**stop codon** The codon UAA, UAG, or UGA; the point on an mRNA strand at which the ribosomes end translation. Compare *start codon*.

**substitution** A mutation in which one base is substituted for another in the DNA sequence of a gene. Compare *deletion* and *insertion*.

**substrate** A molecule that will react to form a new product.

**succession** The process by which the species in a community change over time.

**sustainable** Able to be continued indefinitely without causing serious damage to the environment.

**swallowing reflex** The reaction of the digestive system when food comes into contact with the pharynx, in which the epiglottis seals off the entry into the trachea and food is pushed into the esophagus.

**sympatric speciation** The formation of new species in the absence of geographic isolation. Compare *allopatric speciation*.

**synovial sac** The cavity inside a joint that is filled with a lubricating fluid that reduces friction between two bony surfaces.

**systemic circuit** The circuit in the heart, consisting of the two chambers on the left side, that receives oxygenated blood from the lungs and pumps it to the body. Compare *pulmonary circuit*.

**systole** The first pressure, or top number, in a blood pressure reading—the pressure at the point the heart contracts to push blood out. Compare *diastole*.

## T

**T cell** A type of lymphocyte that matures in the thymus. T cells are involved in cell-mediated immunity. Compare *B cell*.

**temperate forest** A terrestrial biome characterized by snowy winters and moist, warm summers and dominated by trees and shrubs adapted to relatively rich soil.

**template strand** The strand of DNA that is used as a template to make a new strand of DNA.

**tendon** A specialized, flexible band of tissue, rich in collagen, that joins muscle to bone. Compare *ligament*.

**terminator** A segment of DNA that, when reached by RNA polymerase, stops transcription. Compare *promoter*.

**tertiary consumer** An organism that eats *secondary consumers*. Compare *primary consumer* and *quaternary consumer*.

**testis (pl. testes)** Either of a pair of male reproductive organs that produce sperm and androgens in vertebrates. Compare *ovary*.

**testosterone** The primary androgen. Compare *estradiol* and *progesterone*.

**theory** A hypothesis, or a group of related hypotheses, that has received substantial confirmation through diverse lines of investigation by independent researchers. Compare *fact*.

**thermophile** A prokaryote, usually archaean, that can live in extremely hot environments, such as geysers, hot springs, and hydrothermal vents.

**throat** See **pharynx**.

**thymine (T)** One of the four nucleotides that make up DNA. The other three are adenine (A), guanine (G), and cytosine (C).

**tissue** A group of cells that function in an integrated manner to perform a unique set of tasks in the body.

**trachea** Also called *windpipe*. The structure that connects the pharynx to the bronchi.

**transcription** The synthesis of RNA based on a DNA template. Compare *translation*.

**transfer RNA (tRNA)** A type of RNA that facilitates translation by delivering specific amino acids to the ribosomes as codons are read off of an mRNA. Compare *messenger RNA* and *ribosomal RNA*.

**translation** The process by which ribosomes convert the information in mRNA into proteins. Compare *transcription*.

**transpiration** The process of plants absorbing water through their roots and releasing this water through their leaves into the atmosphere.

**transport protein** A protein that acts like a gate, channel, or pump that allows molecules to move into and out of a cell.

**treatment group** Also called *experimental group*. The group of subjects in an experiment that is maintained under the same standard set of conditions as the control group but is subjected to manipulation of the independent variable. Compare *control group*.

**trimester** Any of the three defined stages of human pregnancy. Each trimester is about 3 months long.

**tRNA** See **transfer RNA**.

**trophic level** Each level of the energy pyramid, corresponding to a step in a food chain.

**tropical forest** A terrestrial biome characterized by warm temperatures, about 12 hours of daylight, and seasonally heavy or year-round rains, and containing a rich diversity of organisms.

**tumor** A cell mass that results from runaway cell division.

**tundra** A terrestrial biome containing permafrost that is found at the poles and on mountaintops. Tundra is dominated by low-growing flowering plants and a lack of trees.

**U**

**up-regulation** The speeding up of gene expression. Compare *down-regulation*.

**upper respiratory system** The part of the respiratory system that includes airways in the nose, mouth, and throat. Compare *lower respiratory system*.

**urinary system** The organ system that removes excess fluid from the body, along with waste products, toxins, and other water-soluble substances that are not needed.

**urine** The waste-carrying watery solution that is expelled from our bodies.

**uterus** The female reproductive organ in which a fertilized egg implants and develops until birth.

**V**

**vacuole** An organelle in plant cells that acts as a garbage or recycling center and that stores water. Compare *lysosome*.

**vagina** The female reproductive organ that connects the uterus to the external genitalia.

**variable** A characteristic of any object or individual organism that can change.

**variable trait** A trait that is different in different individuals of a species. Compare *invariant trait*.

**vascular system** In plants, a network of tissues that is made up of tubelike structures specialized for transporting fluids.

**vascular tissue** One of three types of plant tissue. Forming the innermost layer of the plant, vascular tissue (consisting of phloem and xylem) contains stacks of long cells forming continuous tubes that run throughout the plant body and transport materials throughout the plant. Compare *dermal tissue* and *ground tissue*.

**vein** Any of the large vessels that carry blood back to the heart. Compare *artery* and *capillary*.

**ventricle** Either of the two lower chambers of the heart. Compare *atrium*.

**vertebrae (sing. vertebra)** The strong, hollow sections of the backbone, or vertebral column.

**vertebrates** Chordates that possess a backbone.

**vesicle** A sac, formed by the bulging inward or outward of a section of the plasma membrane, that moves molecules from place to place inside a cell but also may transport substances into and out of the cell.

**vestigial trait** A feature that is inherited from a common ancestor but no longer used. Vestigial traits may appear as reduced or degenerated parts whose function is hard to discern.

**villus (pl. villi)** Any of the large number of fingerlike projections in the small intestine that are specialized for nutrient absorption.

**viral strain** Also called *serotype*. Any of the variant forms of a particular type of virus.

**virus** A small, infectious agent that can replicate only inside a living cell.

**vitamin** Any of various small, organic nutrients needed by the human body, but only in tiny amounts. Compare *mineral*.

**voice box** See **larynx**.

## W

**warning coloration** Bright coloring of an animal that alerts a potential predator to dangerous chemicals in the animal's tissues. Compare *camouflage* and *mimicry*.

**water-soluble vitamin** A vitamin that can dissolve in water and therefore tends not to accumulate in body tissues because it can be easily excreted in urine. Compare *fat-soluble vitamin*.

**weather** Short-term atmospheric conditions, such as today's temperature, precipitation, wind, humidity, and cloud cover. Compare *climate*.

**windpipe** See **trachea**.

## X

**X-linked** Found solely on the X chromosome. Compare *Y-linked*.

**xylem** One of two types of vascular tissue in plants. Xylem transports water and minerals, absorbed from the soil, upward from the roots and outward from the central stem to the leaves. Compare *phloem*.

## Y

**Y-linked** Found solely on the Y chromosome. Compare *X-linked*.

**yeasts** Single-celled fungi that belong in the group zygomycetes and are important in the rising of bread, the brewing of beer, and the fermenting of wine.

## Z

**Z disc** A structure, found at each end of a sarcomere, that contains a large protein that provides an anchor point for actin filaments.

**zygomycetes** One of three main groups of fungi, containing many species of molds. Compare *ascomycetes* and *basidiomycetes*.

**zygote** The single cell that results from fertilization.

# Credits

## Photos and Text

### Front Matter

**Page vii** top: Jonathan Mays, Wildlife Biologist, Maine Department of Inland Fisheries and Wildlife/USFWS; bottom: Alexandra Grablewski/DigitalVision/Getty Images; **p. viii** top: Peter Zurek/Shutterstock; center: (puzzle) George Diebold/Getty Images; (cell) Mopic/Shutterstock; bottom: Michael Melford/National Geographic/Getty Images; **p. ix** top: Tsuneo Yamashita/Getty Images; bottom: (background) Iaroslav Neliubov/Shutterstock; (dogs) Eric Isselee/Shutterstock; **p. x** top: (background) bitterfly/Thinkstock/Getty Images Plus; (toy) Gaby Kooijman/Thinkstock/Getty Images Plus; center: Mike Kemp/Rubberball/Getty Images; bottom: (background) shironosov/iStock/Getty Images Plus; (tobacco) joannawnuk/Shutterstock; **p. xi** top: Andrey Nekrasov/Alamy; center: (background) muss/Shutterstock; (IV bag) Thinkstock/Getty Images; (VSRA) Science Source; bottom: (lizard) MattiaATH/Shutterstock; (diver) CyberEak/Shutterstock; **p. xii** top: Lou Linwei/Alamy; center: Elke Van de Velde/Photodisc/Getty Images Plus; bottom: JimmyWrangles/iStock/Getty Images Plus; **p. xiii** top: (karyotype) Leonard Lessin/Science Source/Getty Images; (background) thomaslenne/iStock/Getty Images Plus; (table) Piotr Adamowicz/iStock/Getty Images Plus (scientist) LattaPictures/iStock/Getty Images Plus; (left skull) Leemage/Universal Images Group/Getty Images; (right skull) Sabena Jane Blackbird/Alamy Stock Photo; bottom: (background) Franz Lanting/Offset/Shutterstock; (smoke) Shutterstock; (charcoal edges) Anan Kaewkhammul/Shutterstock; (pre-burned edges) Sylwia Brataniec/Shutterstock; **p. xiv** top: Paul Starosta/Corbis Documentary/Getty Images; center: Holly Kuchera/Shutterstock; bottom: SECCHI Disk; (water) TFoxFoto/Shutterstock; **p. xv** top: SCIEPRO/Science Photo Library/Getty Images; center: (sky) Arcady/Shutterstock; (girl) Alex Mares-Manton/Getty Images; bottom: Vince Michaels/Getty Images; **p. xvi** top: Henny Boogert; bottom: Adam Gault/OJO Images/Getty Images; **p. xvii**: (Houtman) Will Prouty; (Scudellari): Sarah Der Photography; (Malone) courtesy Cindy Malone.

### Chapter 1

**Pages 2–3:** Jonathan Mays, Wildlife Biologist, Maine Department of Inland Fisheries and Wildlife/USFWS; **p. 4:** Gerrit Vyn; **p. 6** (left to right): Judith Collins/Alamy, Scott Camazine/Science Source, AS Food studio/Shutterstock, Lev Kropotov/Shutterstock, Rosa Jay/Shutterstock; **p. 7** top: Photo by Dr. Kimberli Miller, courtesy USGS National Wildlife Health Center; bottom: Stephen Alvarez/National Geographic Creative; **p. 8** left: Photo courtesy Ryan von Linden. Photo used with permission from New York State Department of Environmental Conservation. All rights reserved; right top: Kevin Wenner/Pennsylvania Game Commission; right bottom: David S. Blehert, U.S. Geological Survey—National Wildlife Health Center; **p. 9** top: Image Courtesy of The Advertising Archives; bottom: photograph by Steve Grodsky; **p. 10** (all) Amy Smotherman Burgess/Knoxville News Sentinel/ZUMAPRESS.com; **p. 13** top left: Photo used with permission from New York State Department of Environmental Conservation; center left: Greg Turner/Pennsylvania Game Commission; bottom left: Photo courtesy Ryan von Linden. Photo used with permission from New York State Department of Environmental Conservation. All rights reserved; top right: Deborah Springer; center right: Design Pics Inc/Alamy; bottom right: Dr. Mary Hausbeck; **p. 15** top (left to right): Valeriy Vladimirovich Kirsanov/Shutterstock, Valeriy Vladimirovich Kirsanov/Shutterstock, Rosa Jay/Shutterstock, Vitalii Hulai/Shutterstock; bottom right: nico99/Shutterstock; bottom left: ASA-Carlos Asanuma/Getty Images; **p. 18:** Lindsey Heffernan, PA Game Commission/Patty Stevens, USGS.

### Chapter 2

**Pages 20–21:** Alexandra Grablewski/Digital Vision/Getty Images; **p. 22:** Dr. Anna K. Eaton; **p. 23** top left: Gohil JR, Shah BA, Parekh AN. Diphtheria WebmedCentral INFECTIOUS DISEASES 2011;2(12): WMC002594 doi: 10.9754/journal.wmc.2011.002594. CC BY 3.0; center: Science Source/Colorization by Mary Madsen; top right: South West News Service; bottom left: AP Photo; bottom right: CDC; **p. 24:** Rachel Torres/Alamy Stock Photo; **p. 25** left: Dr. Anna K. Eaton; right: Ryan Collerd/The New York Times/Redux Pictures; **p. 28:** Graph (Fig. 2.6): "The Real Cause of Increasing Autism Prevalence?" by J. Emory Parker. Reprinted by permission of the author; **p. 31:** Republished with permission of Elsevier Science and Technology Journals, from "Ileal-Lymphoid-Nodular Hyperplasia, Non-specific Colitis, and Pervasive Developmental Disorder in Children." Wakefield, AJ, et al. *Lancet*, v351(9103) 1998; permission conveyed through Copyright Clearance

Center, Inc. **p. 34:** Mara008/Shutterstock; **p. 35:** Dr. Anna K. Eaton.

### Chapter 3

**Pages 38–39:** Peter Zurek/Shutterstock; **p. 40:** David McNew/Reuters/Newscom; **p. 41:** Time & Life Pictures/Getty Images; **p. 43** top: Scripps Institution of Oceanography, UC San Diego; bottom: Courtesy of Henderson James Cleaves III; **p. 45:** Scripps Institution of Oceanography, UC San Diego; **p. 46:** Photolukacs/Shutterstock; **p. 47:** Kerstin Langenberger/imagebroker/Alamy; **p. 48** top: NASA, ESA, and D. Jewitt (UCLA); bottom: NASA Ames Research Center/SETI/Peter Jenniskens; **p. 49** top: Helen Sessions/Alamy; center: Shcherbinator/Shutterstock; bottom: Helen Sessions/Alamy; **p. 52** top: Africa Studio/Shutterstock; center: Now Foods; bottom: Kateryna Kon/Shutterstock; **p. 53** top left: Deep OV/Shutterstock; top center: designelements/Shutterstock; center right: Photo Melon/Shutterstock; center left: iofoto/Shutterstock; top right: Andrey Starostin/Shutterstock; center: Abramova Elena/Shutterstock; center bottom right: tryton2011/Shutterstock; bottom right: Jacqui Hurst/Corbis Documentary via Getty Images.

### Chapter 4

**Pages 58–59** (puzzle): George Diebold/Getty Images, (cell) Mopic/Shutterstock; **p. 60** top left: A. Barry Dowsett/Science Source; top right: Biophoto Associates/Science Source/Colorization by Mary Martin; bottom left: EMBL/Johanna Höög; bottom right: CNRI/Science Source; **p. 61:** Jessica Rinaldi/Reuters/Newscom; **p. 62:** Photos courtesy J. Craig Venter Institute. From Gibson, DG et al., "Creation of a Bacterial Cell Controlled by a Chemically Synthesized Genome." *Science.* 2010 Jul 2; 329(5987):52–6. doi: 10.1126/science.1190719; **p. 63:** Courtesy of Neal Devaraj; **p. 65:** Photo Researchers, Inc./Composition by: Eric Cohen/Science Source; **p. 66:** Roger Harris/Science Source; **p. 67** top left: Don W. Fawcett/Science Source; top right: SPL/Science Source; bottom left and right: Dennis Kunkel Microscopy/Science Photo Library/Science Source; **p. 69** top: Hybrid Medical Animation/Science Source; center and bottom: Biophoto Associates/Science Source; **p. 70** top: Dennis Kunkel Microscopy/Science Photo Library/Science Source; top left; center left and bottom: Russell Kightley; bottom left: Credit: Biophoto Associates/Science Source; Colorization by: Mary Martin; **p. 71** top: Biophoto

of Dr. Xu Xing. Reprinted by permission from Macmillan Publishers Ltd: *Nature* 475:465–470 Xing, X. et al. "An Archaeopteryx-like Theropod from China and the Origin of Avialae," 2011; **p. 258** top left: Roland Birke/Getty Images; top center: Rainer Fuhrmann/Shutterstock; top right: Dr. Peter Siver/Visuals Unlimited; center left: Robert and Jean Pollock/Science Source; center: Chris Mattison/FLPA/Science Source; bottom left: Dennis Flaherty/Science Source; bottom right: Schmitz Olaf/Getty Images; **p. 259** top left: Andrew Syred/Science Source; top center: Fletcher & Baylis/Science Source; top right: Ingeborg Knol/imageBROKER/Corbis; center left: Borut Furlan/Getty Images; center: Stuart Wilson/Science Source; center right: Andrew J. Martinez/Science Source; bottom left: Andrew J. Martinez/Science Source; bottom center: Marius Dobilas/Shutterstock; bottom right: Martin Shields/Alamy; **p. 260** 1st image in each row: Hans Kim/Shutterstock; row 1 (left to right): Science Source (Colorization by: Eric Cohen), BSIP/Science Source, Georges Antoni/Hemis/Alamy, Gary Meszaros/Science Source, Picture Partners/Science Source; row 2 (left to right): Gary Meszaros/Science Source, Irochka_T/iStock/Getty Images Plus, Anthony Mercieca/Science Source, James DeBoer/Shutterstock; row 3 (left to right): Valeriy Vladimirovich Kirsanov/Shutterstock, FLPA/Alamy, James Hager/Robert Harding/Alamy; row 4 (left to right): P. Wegner/Arco Images/Alamy, Bernd Rohrschneider/FLPA/Science Source; row 5 (left to right): Image Source/Alamy Stock Photo, Eric Isselee/Shutterstock; row 6 Sabena Jane Blackbird/Alamy Stock Photo; **p. 261**: The Royal Belgian Institute of Natural Sciences, Brussels; **p. 262**: upper left and right: The Royal Belgian Institute of Natural Sciences, Brussels; top: Emily Willoughby/Stocktrek Images/Alamy; center top: Masato Hattori; center: Natural History Museum, London; center bottom: Masato Hattori; bottom: Walt Anderson/Visuals Unlimited/GettyImages.

## Chapter 15

**Pages 268–69:** Elke Van de Velde/Photodisc/Getty Images Plus; **p. 270:** image courtesy Allen Nutman, University of Wollongong. A.P. Nutman et al. "Rapid Emergence of Life Shown by Discovery of 3,700-Million-Year-Old Microbial Structures," *Nature*. Published online August 31, 2016. doi: 10.1038/nature19355; **p. 271:** photo by Amanda Ward/Dunn Lab; **p. 272** top left: Gillian Conquest, Extreme Citizen Science (ExCiteS) research group, University College London, top right: Photograph courtesy of Gardenista; bottom: Jim Davis/The Boston Globe via Getty Images; **p. 273** all: Dunn Lab; **p. 274** top left: Eye of Science/Science Source; center left: David McCarthy/Science Source; bottom left: Michael Abbey/Science Source; top right: Scimat/Science Source; bottom right: Eye of Science/Science Source; **p. 275** top left: David Henderson/caia image/Alamy Stock Photo; top left inset: Eye of Science/Science Source; top right: Jose Arcos Aguilar/Shutterstock; top right inset: image courtesy Luis R. Comolli. Comolli, L. R., Baker, B., Downing, K.H., Siegerist, C. E., and Banfield, J. (2008). "Three-Dimensional Analysis of the Structure and Ecology of a Novel, Ultra-Small Archaeon." *ISME J*, 3(2), 159–167; bottom: IFE, URI-IAO, UW, Lost City Science

Party; NOAA/OAR/OER; The Lost City 2005 Expedition; bottom inset: MichaelTaylor3d/Shutterstock; **p. 277:** Photo courtesy of Lauren Nichols; **p. 278** top left: Dr. Ron Dengler/Visuals Unlimited/Getty Images; top right: Dennis Kunkel Microscopy/Science Source; bottom left: Eye of Science/Science Source; bottom right: CDC/James Archer, Illustrators: Alissa Eckert and Jennifer Oosthuizen; **p. 279:** Universal History Archive/REX/Shutterstock; **p. 280** background: Michael Heim/EyeEm/Getty Images; top right; top left and bottom: Dennis Kunkel Microscopy/Science Source; **p. 281:** photo by Lauren Nichols, Your Wild Life/Dunn Lab.

## Chapter 16

**Pages 284–85:** JimmyWrangles/iStock/Getty Images Plus; **p. 286** left: photo courtesy of Kumar Paudel; right: Jacob Phelps; **p. 287** top: Wim van Egmond/Science Photo Library via Science Source; bottom: Plant-Success; **p. 290:** Leon Neal/AFP/Getty Images; **p. 291** left: divedog/Shutterstock; right: Casther/Shutterstock; **p. 292:** Lance Cheung/USDA; **p. 293:** Carroll & Carroll/AgStock/Design Pics Inc/Alamy Stock Photo, inset: Shawn Poynter/The New York Times/Redux; **p. 295** top left: Stefano Rellandini/Reuters/Newscom; right: Laszlo Podor/Moment Open/Getty images; bottom left: Stefano Rellandini/Reuters/Newscom; **p. 296** left: Le Do/Shutterstock; center: Suzifoo/iStock/Getty Images Plus; right: Egor Rodynchenko/Shutterstock; bottom (left to right): Amazing snapshot/Shutterstock, xpixel/Shutterstock, kzww/Shutterstock, tarapong srichaiyos/Shutterstock, Scott Sinklier/Alamy Stock Photo, Vitaly Korovin/Shutterstock; **p. 297** top: Glass and Nature/Shutterstock; bottom: Frank Kohn/USFWS.

## Chapter 17

**Pages 300–301:** (background) thomaslenne/iStock/Getty Images Plus; (table) Piotr Adamowicz/iStock/Getty Images Plus; (karyotype) Leonard Lessin/Science Source/Getty Images; (scientist) LattaPictures/iStock/Getty Images Plus; (left skull) Leemage/Universal Images Group/Getty Images; (right skull) Sabena Jane Blackbird/Alamy Stock Photo; **p. 302:** Dr. Mirjana Roksandic; **p. 306:** Edward Westmacott/Shutterstock; **p. 307** left: Steve Bloom Images/Alamy; center: John W Banagan/Getty Images; right: Dave Watts/Nature Picture Library/Alamy Stock Photo; **p. 311:** John Reader/Science Source; **p. 313:** Frank Vinken; **p. 318** top: Sabena Jane Blackbird/Alamy Stock Photo; bottom: Silvana Condemi; **p. 319:** Mark Thiessen/National Geographic Creative.

## Chapter 18

**Pages 322–23:** (background) Franz Lanting/Offset/Shutterstock; (smoke) Shutterstock; (charcoal edges) Anan Kaewkhammul/Shutterstock; (pre-burned edges) Sylwia Brataniec/Shutterstock; **p. 324** top and bottom: Jennifer Balch; **p. 325** left: NASA; right: Fotos593/Shutterstock; **p. 326** top: Paul Lefebvre, Woods Hole Research Center; bottom: Dr. Michael T. Coe; **p. 327** top left: David Greedy/Getty Images; top right: John SommersII/Reuters/Newscom; center: Greenpeace/Ges.oek.Forschung; bottom left:

James Steinberg/Science Source; bottom right: AP Photo/Nick Ut; **p. 330** top: Hervé Collart/Sygma via Getty Images; bottom: Ricky Rogers/Reuters/Newscom; **p. 331:** (leaves) Valentyn Volkov/Shutterstock, (tree section 1) My Life Graphic/Shutterstock, (tree section 2) andersphoto/Shutterstock; **p. 337:** Andrew Orlemann/Shutterstock.

## Chapter 19

**Pages 340–41:** Paul Starosta/Corbis Documentary/Getty Images; **p. 343** top: Estefan Radovicz/Agencia o Dia/AGENCIA ESTADO/Xinhua/Alamy Live News; bottom: photo courtesy of Dr. Matthew DeGennaro; **p. 344:** CDC; **p. 345:** CDC/Cynthia Goldsmith; **p. 346:** 7th Son Studio/Shutterstock; **p. 349** left: Dyakova Yulia/Shutterstock; right: Mark Newman/Tom Stack & Associates; **p. 350:** Tom Brakefield/Getty Images; **p. 351** top: Alan G. Nelson/Dembinsky Photo Associates; bottom: photo courtesy of Dr. Margareth Capurro; **p. 352:** Angel Valentin/The New York Times/Redux; **p. 353:** (mosquito) Antagain/E+/Getty Images Plus, (animal silhouettes) nikiteev_konstantin/Shutterstock; **p. 355:** Graph: "Oxitec Approach Suppresses Mosquitos in Piracicaba, Brazil." Appeared in "No GM Mosquitoes Didn't Start the Zika Outbreak," Discovermagazine.com, January 31, 2016. Reprinted by permission of Oxitec Ltd.

## Chapter 20

**Pages 356–57:** Holly Kuchera/Shutterstock; **p. 358** top: photo by Dr. William Ripple; bottom: Pat & Chuck Blackley/Alamy; **p. 359:** Dr. William Ripple; **p. 361:** Don Johnston_IH/Alamy Stock Photo, inset: Kevin Ebi/Alamy Stock Photo; **p. 362:** Ksanawo/Shutterstock; **p. 363** left: Martin Harvey/Corbis via Getty Images; right: Luc Novovitch/Alamy Stock Photo; **p. 364:** Daniel Stahler/NPS; **p. 365** top: Mark Conlin/Oxford Scientific/Getty Images; bottom: David Fleetham/Visuals Unlimited/Getty Images; **p. 366:** (wolves) rachisan alexandra/Shutterstock, (deer) KatarinaF/Shutterstock, (trees) Vertyr/Shutterstock; **p. 367:** PhotoStock-Israel/Alamy; **p. 368** top: Joe McDonald/Visuals Unlimited; center left: Nancy Nehring/Getty Images; center right: CHAINFOTO24/Shutterstock; bottom: Gerry Bishop/Visuals Unlimited; **p. 369:** Norbert Rosing/National Geographic Creative; **p. 371:** Anne-Marie Kalus; **p. 372:** Adam Burton/Alamy; **p. 373** left: Stan Osolinski/Dembinsky Photo Associates; center: Howard Garrett/Dembinsky Photo Associates; right: Walt Anderson/Visuals Unlimited.

## Chapter 21

**Pages 376–77:** (water) TFoxFoto/Shutterstock; SECCHI Disk; **p. 378:** Gary Corbett/Alamy; **p. 379** top: Jacques Descloitres, MODIS Rapid Response Team, NASA/GSFC, inset: D.P. Wilson/FLPA/Science Source; bottom left: Tania Wong; bottom right: Dr. Boris Worm; **p. 382:** NASA image by Jesse Allen & Robert Simmon; **p. 383:** Chris Linder/Visuals Unlimited; **p. 384** top left: Bernd Zoller/imagebroker/Alamy; top right: John E Marriott/Alamy; center: Ken Lucas/Visuals Unlimited; bottom left: Pixtal Images/Media Bakery; bottom right: Design Pics Inc/Photolibrary/Getty Images;

p. 385 top left: Michael P. Gadomski/Earth Sciences/Animals Animals; top right: Mark De Fraeye/SPL/Science Source; center: Willard Clay/Dembinsky Photo Associates; bottom left: Mark Goodreau/Alamy; bottom right: Jim Zipp/National Audubon Society Collection/Science Source; **p. 387:** Abigail McQuatters-Gollop.

## Chapter 22

**Pages 392–93:** SCIEPRO/Science Photo Library/Getty Images; **p. 394:** Spies 2015; **p. 399** left: Reinhard Dirscherl/agefotostock; center: Matteo photos/Shutterstock; right: cbpix/Alamy; **p. 400** left: Clouds Hill Imaging Ltd/Getty Images; right: Dennis Kunkel Microscopy/Science Source; **p. 402:** Dr. Tina Kold Jensen; **p. 403:** Lukiyanova Natalia/frenta/Shutterstock; **p. 404:** Dr. Niels Jorgensen; **p. 406** left: MedicalRF.com/Corbis; center: Joo Lee/Corbis Documentary via Getty Images; right: Anatomical Travelogue/Science Source.

## Chapter 23

**Pages 412–13:** (sky) Arcady/Shutterstock; Alex Mares-Manton/Getty Images; **p. 414:** BSIP/Science Source; **p. 415** top left: Gtranquillity/Shutterstock; top center: Juliya Shangarey/Shutterstock; top right: Media Bakery; bottom left: Nik Merkulov/Shutterstock; bottom center: D7INAMI7S/Shutterstock; bottom right: Shutterstock; **p. 421:** James L. Amos/Science Source; **p. 424:** Philippe Autier; **p. 426** top (left to right): jeehyun/Shutterstock, Maks Narodenko/Shutterstock, Petr Malyshev/Shutterstock, Nattika/Shutterstock; bottom (left to right): Levent Konuk/Shutterstock, Binh Thanh Bui/Shutterstock, Viktor1/Shutterstock, (pill bottle) HeinzTeh/Shutterstock; **p. 427:** Chloe Johnson/Alamy.

## Chapter 24

**Pages 430–31:** Vince Michaels/Getty Images; **p. 432** top and bottom right: Shawn Rocco/Duke Medicine; **p. 433:** Laura Niklason; **p. 434** top and center: Dennis Kunkel/Science Source; bottom: Richard Kessel & Dr. Randy Kardon/Visuals Unlimited; **pp. 439, 440, 442** (both) The Ott Lab/Massachusetts General Hospital, Boston MA; **p. 444:** Alfred Pasieka/Science Source; **p. 445:** Christine E. Schmidt, Ph.D.; **p. 448:** Scott Lewis Photography, LLC.

## Chapter 25

**Pages 452–53:** Henny Boogert; **p. 454** top: Matthijs Kox, PhD; **p. 454** center: Henny Boogert; bottom: Enahm at Innerfire; **p. 456:** Marcel Rekers; **p. 461** left: Don W. Fawcett/Science Source; right: MedImage/Science Source; **p. 462:** Volker Steger/Science Source; **p. 466:** Henny Boogert.

## Chapter 26

**Pages 470–71:** Adam Gault/OJO Images/Getty Images; **pp. 472, 473:** courtesy of The Land Institute; **p. 478** left: Dragomir Radovanovic/Shutterstock; right: Photograph by Kathleen Bauer, GoodStuffNW.com; **p. 479** top: courtesy of The Land Institute; bottom: University of Minnesota, College of Food, Agricultural and Natural Resources Science, Department

of Agronomy and Plant Genetics; **p. 482** (left to right): Vasilius/Shutterstock, bergamont/Shutterstock, natu/Shutterstock, Madlen/Shutterstock; **p. 484** left: Stephen Ausmus/USDA ARS; center: Anthony Mercieca/Science Source; right: courtesy of The Land Institute; **p. 485:** courtesy of The Land Institute.

## Infographics

### Chapter 1

**Page 15,** "Bug Zappers": Data from Justin G. Boyles, Paul M. Cryan, Gary F. McCracken, and Thomas H. Kunz, "Economic Importance of Bats in Agriculture," *Science*, April 1, 2001, www.sciencemag.org/content/332/6025/41.

### Chapter 2

**Page 34,** "Safety in Numbers": Data from Caroline L. Trotter and Martin C. J. Maiden, "Meningococcal Vaccines and Herd Immunity: Lessons Learned from Serogroup C Conjugate Vaccination Programs," Expert Review of Vaccines, July 2009, www.ncbi.nlm.nih.gov/pubmed/19538112.

### Chapter 3

**Page 54,** "What's It All Made Of?": Earth's Atmosphere data from Wikipedia, en.wikipedia.org/wiki/Abundance_of_the_chemical_elements. The Universe data from WebElements, www.webelements.com/periodicity/abundance_universe/. The Human Body data from Wikipedia, en.wikipedia.org/wiki/Chemical_makeup_of_the_human_body. Earth's Crust data from Wikipedia, en.wikipedia.org/wiki/Abundance_of_elements_in_Earth%27s_crust.

### Chapter 4

**Page 73,** "Sizing Up Life": Data from Genetic Science Learning Center at the University of Utah, "Cell Size and Scale," learn.genetics.utah.edu/content/cells/scale/.

### Chapter 5

**Page 92,** "Making Way for Renewables": Data from REN21, "Renewables 2016 Global Status Report," www.ren21.net/status-of-renewables/global-status-report/.

### Chapter 6

**Page 112,** "Cancer's Big 10": Data from Centers for Disease Control and Prevention, "United States Cancer Statistics: 2013 Top Ten Cancers," nccd.cdc.gov/uscs/toptencancers.aspx.

### Chapter 7

**Page 130,** "Does Bigger Mean Better?": Data from National Center for Biotechnology Information, U.S. National Library of Medicine, "Genome Information by Organism," www.ncbi.nlm.nih.gov/genome/browse/.

### Chapter 8

**Page 143,** "Genetic Diseases Affecting Americans": Data from Cold Spring Harbor Laboratory,

DNA Learning Center, "Your Genes, Your Health," www.ygyh.org.

### Chapter 9

**Page 168,** "The Meteoric Rise of CRISPR": PubMed search results data from National Center for Biotechnology Information, U.S. National Library of Medicine, www.ncbi.nlm.nih.gov/pubmed/?term=crispr. Timeline data from Broad Institute, "CRISPR Timeline," www.broadinstitute.org/what-broad/areas-focus/project-spotlight/crispr-timeline, and from CRISPR Update, "CRISPR Timeline," www.crisprupdate.com/crispr-timeline/.

### Chapter 10

**Page 185,** "The Deadly Price of a Pandemic": Data from U.S. Department of Health and Human Services, "Pandemic Flu History," www.flu.gov/pandemic/history/index.html, and from World Health Organization, "Global Influenza Programme," www.who.int/influenza/en/.

### Chapter 11

**Page 206,** "Watching Evolution Happen": Data from Oliver Tenaillon, Alejandra Rodríguez-Verdugo, Rebecca L. Gaut, Pamela McDonald, Albert F. Bennett, Anthony D. Long, and Brandon S. Gaut, "The Molecular Diversity of Adaptive Convergence," Science, January 27, 2012, www.ncbi.nlm.nih.gov/pubmed/22282810.

### Chapter 12

**Page 227,** "Race against Resistance": Antibacterial drugs data from Infectious Diseases Society of America, "Bad Bugs, No Drugs: As Antibiotic Discovery Stagnates... A Public Health Crisis Brews," www.idsociety.org/uploadedfiles/idsa/policy_and_advocacy/current_topics_and_issues/antimicrobial_resistance/10x20/images/bad%20bugs%20no%20drugs.pdf, and from Dalia Deak, Kevin Outterson, John H. Powers, and Aaron S. Kesselheim, Annals of Internal Medicine, September 6, 2016, "Progress in the Fight Against Multidrug-Resistant Bacteria? A Review of U.S. Food and Drug Administration–Approved Antibiotics, 2010–2015," annals.org/aim/article/2526197/progress-fight-against-multidrug-resistant-bacteria-review-u-s-food/. Timeline data from Centers for Disease Control and Prevention, "Antibiotic Resistance Threats in the United States, 2013," www.cdc.gov/drugresistance/threat-report-2013/pdf/ar-threats-2013-508.pdf.

### Chapter 13

**Page 244,** "On the Diversity of Species": Data from A.D. Chapman, "Numbers of Living Species in Australia and the World, 2nd Edition, A Report for the Australian Biological Resources Study, September 2009," www.environment.gov.au/node/13875.

### Chapter 14

**Page 264,** "The Sixth Extinction": Data from International Union for Conservation of Nature and Natural Resources, The IUCN Red List of Threatened Species, Summary Statistics, www.iucnredlist.org/about/summary-statistics.

## Chapter 15

**Page 280,** "The Bugs in Your Belly Button": Data from Robert R. Dunn, Julie Horvath-Roth, Sarah P. Council, Holly Menninger, and Matthew Fitzpatrick, Your Wild Life, November 25, 2013, "The Belly Buttons Will be Revealed, Slowly," yourwildlife.org/2013/11 /the-belly-buttons-will-be-revealed-slowly/.

## Chapter 16

**Page 296,** "Food Banks": Data from Food and Agriculture Organization of the United Nations, "Biodiversity for Food Security and Nutrition: 30 Years of the Commission on Genetic Resources for Food and Agriculture," www.fao .org/resources/infographics/infographics -details/en/c/174199/, and "Commission on Genetic Resources for Food and Agriculture: Plant Genetic Resources," www.fao.org/nr/cgrfa /cthemes/plants/en/.

## Chapter 17

**Page 309,** "Hereditary Heirlooms": Data from National Geographic, "Genes Are Us. And Them," ngm.nationalgeographic.com/2013/07 /125-explore/shared-genes, and from Stefan Lovgren, *National Geographic News*, August 31, 2005, "Chimps, Humans 96 Percent the Same, Gene Study Finds," news.nationalgeographic .com/news/2005/08/0831_050831_chimp _genes.html.

## Chapter 18

**Page 331,** "Forest Devastation": Deforestation and forest degradation data from G. Kissinger, M. Herold, and V. De Sy, "Drivers of Deforestation and Forest Degradation: A Synthesis Report for REDD+ Policymakers," Lexeme Consulting, Vancouver, Canada, August 2012, www.gov.uk/government/uploads/system /uploads/attachment_data/file/65505/6316 -drivers-deforestation-report.pdf. Top 5 country data from Food and Agriculture Organization of the United Nations, "Change in Extent of Forest and Other Wooded Land 1990–2005," www.fao .org/forestry/32033/en/.

## Chapter 19

**Page 353,** "World's Deadliest Animals": Data collected from multiple sources at Bill Gates, GatesNotes, April 25, 2014, "The Deadliest Animal in the World," www.gatesnotes.com /Health/Most-Lethal-Animal-Mosquito-Week/.

## Chapter 20

**Page 366,** "Cause and Effect": Data from William J. Ripple and Robert L. Beschta, "Restoring Yellowstone's Aspen with Wolves," *Biological Conservation* 138 (2007), www.cof .orst.edu/leopold/papers/Restoring%20 Yellowstone%20aspen%20with%20wolves.pdf.

## Chapter 21

**Page 388,** "Productive Plants": Data from Oak Ridge National Laboratory Distributed Active Archive Center, "NPP and the Global Carbon Cycle," daac.ornl.gov/NPP/html_docs /npp_ccycle.html.

## Chapter 22

**Page 403,** "Preventing Pregnancy": Data from James Trussell, "Contraceptive Failure in the United States," *Contraception*, May 2011; 83(5): 397–404, www.ncbi.nlm.nih.gov/pmc/articles /PMC3638209/.

## Chapter 23

**Page 426,** "Nutritional Needs": Multivitamin data from One A Day® Men's Pro Edge and Women's Pro Edge vitamins. RDA data from Jennifer J. Otten, Jennifer Pitzi Hellwig, and Linda D. Meyers (editors), Dietary Reference Intakes: The Essential Guide to Nutrient Requirements, The National Academies Press, Washington, DC, 2006, www.nap.edu /catalog/11537.html; www.nal.usda.gov/fnic /DRI/Essential_Guide/DRIEssentialGuide NutReq.pdf. Food data from National Institutes of Health, Office of Dietary Supplements, "Dietary Supplements Fact Sheets," ods.od.nih .gov/factsheets/list-all/.

## Chapter 24

**Page 447,** "Have a Heart": Waiting list registration data from United Network for Organ Sharing, "Data Slides for Spring Regional Meetings: 2013 Data," www.unos.org/docs/DataSlides _Spring_2014.pdf. Completed transplants data from U.S. Department of Health and Human Services, Organ Procurement and Transplantation Network, optn.transplant.hrsa.gov.

## Chapter 25

**Page 465,** "Driven by Hormones": Melatonin data from B. Selmaoui and Y. Touitou, "Reproducibility of the Circadian Rhythms of Serum Cortisol and Melatonin in Healthy Subjects: A Study of Three Different 24-h Cycles over Six Weeks," *Life Sci.*, 73 (2003): 3339–3349. Epinephrine data from University of Mississippi Medical Center, "Physical Exercise—Epinephrine," www.umc.edu/Education /Schools/Medicine/Basic_Science/Physiology _and_Biophysics/Core_Facilities(Physiology) /Physical_Exercise_-_Epinephrine.aspx. Cortisol data from J. H. Wilmore, D. L. Costill, and W. L. Kenney, Physiology of Sport and Exercise, Human Kinetics, Champaign, IL, 2007.

## Chapter 26

**Page 482,** "GM Crops Take Root": Data from Clive James, "Global Status of Commercialized Biotech/GM Crops: 2012," ISAAA Brief No. 44, ISAAA, Ithaca, NY, 2012, www.isaaa.org /resources/publications/briefs/44/download /isaaa-brief-44-2012.pdf.

# Index

Page numbers set in *italic* type refer to materials in figures or tables.

dietary fiber, 414
diffusion, *64*, 65, *440*, 444
diffusion MRI, 444
digestion, 416, 417, 418
digestive system, *397*, 416–20, *459*
dihybrid crosses, *123*
dinosaurs
    appearance in Triassic, 253
    *Archaeopteryx*, 250, 254–56, 261, 262
    deinonychosaurs, *254*, *255*, 256
    dinosaur-bird evolutionary tree,
        253–56, 261, *262*
    *Eosinopteryx brevipenna*, *254*, 261, *262*
    feathered dinosaurs, 250, 253–54, 256,
        261, *262*
    *Protoceratops*, *196*
    raptors, *196*, 256, 261
    theropods, 253–54, *255*, 256
    *Tyrannosaurus rex*, *254*
    *Xiaotingia zhengi*, *254*, 255–56, 262
diphtheria, 22, *23*
diploid cells
    diploid chromosome set, 105–6, *107*,
        240
    diploid primary oocytes, 399, *400*
    diploid sporophytes, 480, *481*
    diploid zygotes, 399, *400*, 480, *481*
directional selection, 216, *217*
disaccharides, structure, *53*
disease, defined, 127
disruptive selection, 216, 218, *219*
DNA (deoxyribonucleic acid). *See also*
        genetic code; viral DNA
    about, 156
    base-pairing rules, 157–58
    cell cycle and, *99*, 100
    as characteristic of life, 6, 51, 60, 201
    compact DNA in chromosomes, 102,
        160, *161*
    complementary base-pairing, 157, *158*,
        164, *167*
    crossing-over, *109*
    double helix, 60, 157, *158*, *161*
    in eukaryotic chromosomes, 101
    evolution and DNA sequences, 201–2,
        *203*
    exons, 178, *179*
    introns, 178, 179
    mitochondrial DNA, 312–13, *314*, 319
    nuclear and mitochondrial DNA
        inheritance, 312, *313*, *314*
    nucleotide sequence and phenotypes,
        158, 159
    origins of replication, 162, 163

plasmids, 221
polymerase chain reaction (PCR),
        163–64
    primers, 162, 163–64
    promoters, 177–78, 179
    repair, 61, *160*, *164*
    replication, 100, 102, 162–64
    replication errors, 165–66
    semiconservative replication, 162
    structure, 51, *52*, 60, 157–58
    synthetic DNA, 60
    template strands, 157, 162–63, *164*,
        178–79
    terminator sequence, 178, 179
    transcription, *174*, 176–79
DNA polymerase, 162, *163*, 164–65
DNA repair proteins, *164*, 165
DNA replication, 100, 102
DNA sequence similarities, 202, *203*, 235
DNA sequencing machines, 164
Doberman pinschers, *125*
dogs and genetics. *See also* Portuguese
        water dogs
    body shape gene, 126
    body size gene (*IGF₁*), 126
    border collies, 129, 131
    boxers, 124, *125*
    Chihuahuas, *117*, 120, 124, *126*, 193
    coat color gene, 120, 127–28
    cocker spaniel, 126
    Doberman pinschers, *125*
    dog genome, 124, *125*, *130*
    expression gene for melanin
        deposition, 128
    genetic diseases, 124, *125*
    genetic diversity, 119, 124, 193, *194*
    golden retrievers, *125*
    Great Danes, *117*, 120, 124, *126*, 193,
        *194*
    gum color gene, 126
    Labrador retrievers, *124*, 127, *128*
    pedigree analysis, 119, 126
    Pekingese, 119
    Pembroke Welsh corgis, *125*
    percentages of genes shared with
        humans, *309*
    poodles, *120*, *125*
    relationship to gray wolf, 193, *194*
    Saint Bernard, 119
    Scottish terriers, 125
    selective breeding, 193
    species, 119, 193
dolphins, 192, *193*, 195, 205, *207*, 218,
        *220*

domains, 251–52
dominant alleles, 119–20, 146
dormancy, 478, 480
*Dorudon*, 196, *197*
double helix, 60, 157, *158*
Doudna, Jennifer, 159–60
down-regulation, 183
Down syndrome, 98, 139, *140*, *143*, 148
*DprA* gene, 223
Duchenne muscular dystrophy, *143*, 145
Dunn, Rob, 270, 271, 273–76, 275,
        277–78, 280, 281
Dworkin, Jason, 44
dysbiosis, 281


**E**

eagles, *350*, 367
Earth equivalents, 332, *333*
Earth's age, 250
Earth's atmosphere, gasses in, *54*
Earth's early atmosphere, 40–43, 49, 252,
        278
Earth's human population, 347
Eaton, Anna, 22–23, 25, 26, 28, 31, 35
Eaton, Caroline, 25, 31, 35
Ebola virus, 66, 186
echidna, 307
ecological communities. *See also*
        biological communities; food
        chains; food webs; species
        interactions; succession
    climax communities, 372, *373*
    in ecosystems, 378
    energy movement through, *362*
    overview, 359–60, 378
    relative species abundance, 359, 360
    species richness, 359, 360
    succession in, 371–72, *373*, 382
ecological footprint, 332
ecological isolation, 238–39, *242*
ecological niche, 369
ecology, defined, 325
ecosystem processes
    carbon cycle, 336–37
    defined, 382
    energy flow, 362, *363*, 380, *382*
    nutrient cycling, *259*, 279, 380–82,
        389
    succession, 371–72, *373*, 382
    water cycle (hydrologic cycle), 47,
        332–35
ecosystems
    abiotic factors, 325, 336–37, 378,
        380–82

# H

hair follicles, *416*
halophiles, 275
haploid chromosome set, 105–6, *107, 109,* 240
heart, 433, 436, *437, 447*
heart rate, 436
*Heliobacterium, 278*
Hellberg, Michael, 238–39
hemagglutinin, 176–78, *179,* 181, 184, 186
hemoglobin, *165, 434*
hemophilia, *143,* 144, 150
herbivores, 72, 257, 365, *384–85,* 476
herd immunity, 32, *33,* 34, 35
hermaphrodites, *399*
Herrel, Anthony, 233, 234, 236
heterotrophs, 278
heterozygotes, 126
heterozygous genotype, defined, 120, 140, *141,* 145
Hicks, Alan, 4–7, 8–9, 12, 16
hindbrain, 444
hippocampus, 444
hippopotamus, 198, 199, 200, 202
histone proteins, 160, *161*
HIV cocktails, 460
HIV (human immunodeficiency virus), 459
Hof, Wim ("Iceman")
  autonomic nervous system modulation, 454, 458–59, 466
  cold exposure, *453,* 454, 455, 466
  cortisol levels in blood, 456, 457
  immune system modulation, 454–55, 456, 457, 463
  response to endotoxin injection, 457, 458
  training of volunteers, 454, 459, 460–61, 466
homeostasis
  body temperature, 398
  as characteristic of life, 6, 66
  defined, 6, 396
  feedback loops, 398–99, 404, 407
  homeostatic pathways, 396, 398–99
  requirement for organ systems, 396, 399
  set point, 396, *398*
homeostatic pathways, 396, 398–99
hominids (apes), *260,* 307, 309
hominins, 307–8, 311–12, 315, 317
*Homo erectus, 312,* 315–16, 317
*Homo* genus. *See also* Neanderthals
  anatomically modern *Homo sapiens, 312,* 317, 318

archaic *Homo sapiens,* 316–17
biological classification, 303, 311
*Homo erectus, 312,* 315–16, 317
*Homo habilis, 312,* 315–16
origins, 315
out-of-Africa hypothesis, 316, 317
*Homo habilis, 312,* 315–16
homologous pairs of chromosomes, 106–8, *109, 110, 119,* 122, 124
homologous traits, 200–201
homozygous genotype, defined, 120, 140, *141*
honeybees, *309, 484*
horizontal gene transfer, 221–22, *223,* 277
hormones. *See also* endocrine system
  abscisic acid (ABA), *477*
  adrenaline (epinephrine), 456, 457–58, 463, *465,* 466
  androgens, 404
  auxins, *477*
  cortisol, 456, 457, 458, *465*
  cytokinins, *477*
  estradiol, *401,* 404, 407, *408*
  estrogens, 98, 101, 105
  ethylene, *477*
  feedback loops, 399, 404, 407
  in flowering plants, 476, *477*
  follicle-stimulating hormone (FSH), *401*
  functions, 455–56
  gibberellins, *477*
  insulin, 51, 99, 398
  luteinizing hormones (LH), *401*
  menstrual cycle, 401, 404
  noradrenaline (norepinephrine), 456, 458
  ovaries and, *401,* 404
  oxytocin, 407, *408*
  progesterone, *401,* 404
  progestogens, 404
  at puberty, 399, 402, 404
  release into circulatory system, 455–56
  signal amplification in cells, 456, *457*
  sleep and wake cycles, *465*
  testes and, 402, 404
  testosterone, 404
hosts
  defined, 365
  humans and bacteria, 275, 276, 277
  mycorrhizal fungi and, 287
  parasites and, 365, *367*
  pathogenic bacteria and, 276, 365
  sea fans and algae, 239–40
  of viruses, 66

human body
  birth weight in humans, 216, *218*
  brain volume, 311
  elements in, *54*
  number of cells, 60
  organ systems, 396, *397,* 432
human egg cell size, *73,* 402
human evolution. *See also* evolution; *Homo* genus; Neanderthals
  anatomically modern *Homo sapiens, 312,* 317, 318–19
  archaic *Homo sapiens,* 316–17
  *Ardipithecus ramidus,* 311
  *Australopithecus afarensis,* 311, *312*
  *Australopithecus africanus, 312,* 315
  belief in, 195
  bipedalism and upright posture, 308, *310,* 311
  brain volume, 311
  characteristics of modern humans, 316
  chimpanzees, DNA sequence similarity to humans, 202, *203, 309*
  differences between humans and other apes, 307, 309, *310*
  DNA sequence similarities to other species, 202, *203*
  hominin branch of the ape family, 307–8, 311–12, 315, 317
  *Homo erectus, 312,* 315–16, 317
  *Homo* genus, biological classification, 303, 311
  *Homo* genus, origins, 315
  *Homo habilis, 312,* 315–16
  Linnaean hierarchy of human classification, *260*
  Neanderthal–modern human interbreeding, 302, 303, 312, 314, 318–19
  opposable thumbs, 307
  opposable toes, 309, *310,* 311
  out-of-Africa hypothesis, 316, 317
  percentages of genes shared by humans and other species, *309*
  *Sahelanthropus tchadensis,* 311
  skulls of hominin species, *312*
human immunodeficiency virus (HIV), 459
human microbiome, 270, 271, 280, 281. *See also* belly button swabs
hummingbirds, 240, 360, *484*
Hunt, Patricia, 98–99, 100, 104–5, 106, 108, 111
Huntington disease, *144,* 147, 225

Portuguese water dogs (PWD)
  black-fur allele (*B*), 120
  breed characteristics, 118, 131
  brown-fur allele (*b*), 120
  Chou, 131
  genetic research, 119–20, 126, 131
  Georgie, 118, 124, 129
  Georgie Project, 119, 120, 131
  Mopsa, 118, 119, 131
  pedigree analysis, 119, 126
positive feedback, 398, 407
postzygotic barriers, 241–42
potassium, 279, *382*, 441, 479
Prada, Carlos, 237–39, 242
prairie dogs, 360
Precambrian period, *252*, 257, *261*
predation, 365, 369
predators. *See also specific animals*
  carnivores as, 365, *367*
  defined, 365
  density-dependent population changes
    and, 349
  effect on diversity, *361*
  evolution and, 216, *217*, 257, 305
  herbivores as, 365
  moth color and, 216, *217*
  parasites as, 365, *367*
  top predators, 307, *367*, 368, 378,
    *381*
pregnancy, trimesters, *406*, 407
preimplantation genetic diagnosis (PGD),
    148
premature infants, 148, 414, 415, 416, 420
prenatal genetic screening, *147*, 148
prezygotic barriers, 241–42
primary consumers, 362, *380*
primary growth, 477
primary immune response, 463–64
primary literature, 26, *27*, 424
primary oocytes, 399, *400*, *401*, 402
primary succession, 372
primates, *260*, 307, *308*, *310*
primers, 162, 163–64
probability and genetics
  overview, 124
    Punnett squares, 122–23, 124, 142,
      *146*, *147*
process of science. *See* scientific method
producers. *See also* plankton
  aquatic producers, *336*
  cellular respiration, *82*
  defined, 362
  energy capture, 380, 383, 386
  in food chains, 362, *363*

net primary productivity (NPP), 386,
    *387*, *388*, 480
  plants as basis of terrestrial food webs,
    *258*, 291, 362, *363*
  primary producers, 379–80, *388*
products, defined, 41
progesterone, *401*, 404
progestogens, 404
prokaryotes. *See also* archaea; bacteria
  asexual reproduction, 100, 277
  cell division by binary fission, 100, 277
  circular DNA loop, 100, 271
  comparison to eukaryotes, 68, *69*, 271
  energy sources and carbon sources,
    277–79
  evolutionary history, 252
  in extreme environments, 271, *275*
  horizontal gene transfer, 221–22, 223,
    277
  overview, 271
  prokaryotic cells, characteristics, 68, *69*
  quorum sensing, 276
  rapid population growth, 271
promoters, 177–78, 179
prophase I in meiosis, *107*, *109*
prophase II in meiosis, *107*, *109*
prophase in mitosis, 102
prostaglandins, 407, *408*
Protalix Biotherapeutics, 186
proteins
  amino acids in, 42, 48, 51, *52*
  digestion, 417, 418
  functions, 51
  macronutrients, 414
  structure, *52*
  synthesis during translation, *174*, 179–83
  in vaccines, 174–75
protists (Protista). *See also* algae
  asexual reproduction, 289
  decomposers, *258*, 279, 288
  kingdom Protista, *256*, 257, *258*, *259*,
    288–89
  mixotrophs, 289
  pathogenic protists, 289
  protozoans, 288
*Protoceratops*, *196*
protocetids, 205
proton pumps, *86*
protons, 40, *41*
protozoans, 288
*Pseudogymnoascus destructans*
    (*Geomyces destructans*), 7, 9,
    11, 12, *13*. *See also* white-nose
    syndrome

pseudopodia, 288
pseudoscience, 28, *29*, *30*, 31, 35
pulmonary circuit, 436, *437*
Punnett squares, 122–23, 124, 142, *146*,
    *147*
pyruvate, *89*, 90
pythons, *202*

**Q**

Quail Botanical Gardens, 289–90
quaternary consumers, 362
Quaternary period, *253*, *263*
quorum sensing, 276

**R**

rabies vaccine, 186
raccoons, 365
radial symmetry, *303*, 305
rainforest (Amazon). *See* Amazon
    rainforest
randomization, 11
Ranga Rao, A. (and his widow), 196, 198
raptors, *196*, 256, 261
reabsorption, 441
reactants, defined, 41
receptor-mediated endocytosis, *67*, 68
receptor proteins, 66–67
recessive alleles, 120, 142, 145–46
red blood cells
  in isotonic solution, 66
  oxygen transport, 434, *440*
  production, *421*, 456
  sickle-cell disease, *144*
  size, *73*
red-green color blindness, 145
red-spotted newt, genome size, *130*
reindeer, *349*
relative species abundance, 359, 360
renewable energy, 81, 92
reproduction as characteristic of life, 6, 66
reproductive barriers, 241–42, *243*
reproductive system
  anatomy in humans, *405*
  defined, 395
  flowering plants, 475–76
  functions, *397*
  homeostasis, 396, 399
  variation in animals, 399
reproductively isolated populations, 234,
    238, 241
reptiles
  evolution and classification, 252–53,
    257, 260, 305
  number of known species, *244*

upper respiratory system, 438, *439*

up-regulation, 183

uracil (U), 178

Urey, Harold, 40, 41, 42–43, 49. *See also* Miller experiments

urinary system, *397*, 433, 439–42. *See also* kidneys

urine, 440, 441–42

U.S. Food and Drug Administration (FDA), 108, 111, 176, 186, 227, 352, *403*

uterus
  anatomy, *405*
  labor and childbirth, 407, *408*
  menstrual cycle events, *401*, 404
  pregnancy, 148, *406*, 407, *408*

**V**

vaccines
  as cause of autism, arguments against, 25, 28, 30–31
  as cause of autism, scientific claims, 23, 26–27
  Centers for Disease Control and Prevention (CDC) recommendations, 22, 31, *32*, 34
  diphtheria, 22, *23*
  evaluation for effectiveness, safety, and side effects, 31, *32*
  flu vaccine from tobacco, 174–76, *177*, 179–81, 184, 186
  flu vaccine recommendations, 22
  hepatitis B, 31
  and herd immunity, 32, *33*, 34, 35
  history, 22
  how vaccines work, 22, 174
  immune system response, 22, 174
  measles, mumps, and rubella (MMR) vaccine, 27–28, 30–31
  meningitis, 34
  pertussis (whooping cough), 22, *23*
  production in chicken eggs, 174–75, 176, 186
  production methods, 176
  proteins in, 174–75
  rabies, 186
  RotaTeq vaccine, 26
  rotavirus, 22, *23*, 26, 186
  side effects, 31
  smallpox, 22, *23*
  tetanus, *23*
  thimerosal in, 30
  vaccination, defined, 22

*Vaccines and Your Child* (Offit), 25

vacuoles, *69*, *71*, 72

vagina, 396, *405*, 407, *408*

vancomycin, 212

vancomycin-resistant *Staphylococcus aureus*. *See* VRSA

van Leeuwenhoek, Antonie, 272, 279

variable traits, 118

variables, 9–10

vascular system in plants, 292

vascular tissues, 272, 474, *475*

vas deferens, *405*

vectors, for diseases, 343, 345

veins, 435

*Velociraptor*, 196

Venter, J. Craig, 60, 61–62, 72, 74

ventricles, 436, *437*

vertebrae, 305

vertebrates
  defined, 304
  evolution, *252*, *257*, 305

vesicles, 67–68, *70*, 72

vestigial traits, 201, *202*

Vézina, Louis-Phillippe, 175–76, *177*, 184, 186

Vibram, 24

viceroy butterfly, *368*, 369

villi (singular "villus"), 148, *418*

viral DNA. *See also* DNA
  active DNA in pig genome, 161–62
  in human genome, 161
  porcine endogenous retroviruses (PERVs), 157, 158–59, 160, 164–65, 166, *167*
  protein coat, 66

viral strains, 66

viruses
  characteristics, 61, 66
  envelope, 66
  evolution, 66
  living or nonliving, *6*, 61, 66
  protein coat, 66
  sizes of viruses, *73*
  synthetic genome, 61

vitamin D
  absorption in small intestine, 418
  blood levels in infants, 414, 415, 416, 420
  bone health and, 414, 415, 420, 421, 422
  calcium absorption and, 414, 416, *419*, 421, 427
  correlation with disease risk, 416, 423, 424–25, 427
  deficiency, 414, 415–16

fortified foods, *415*, 418–19
inflammation effect on, 425
modification in liver and kidneys, 419
muscle health and, 423
recommended daily allowance (RDA), *426*, 427
sources, *415*, 419, *426*
supplements, 415–16, 423, 424–25, 427
synthesis in skin, 414, 416, 419
ultraviolet (UV) rays and synthesis, 414, *419*, 424

vitamins. *See also* vitamin D
  absorption in small intestine, 418–19
  B vitamins, *415*
  folic acid, *415*, *426*
  fortified foods, *415*, 418–19
  multivitamins, *426*
  nutritional needs and sources, *415*
  overview, 414
  recommended daily allowance (RDA), *426*
  vitamin A, 24, *415*, *426*
  vitamin $B_{12}$, *415*
  vitamin C, 127, *415*, *426*
  vitamin E, *415*
  vitamin K, *415*

volcanoes, 47–48

vom Saal, Frederick, 111

VRSA (vancomycin-resistant *Staphylococcus aureus*)
  clonal cluster 5 (CC5) strains of *S. aureus*, 222–23
  discovery, 212–13
  *DprA* gene, 223
  evolution, 213–15, 218–19, 226
  foot infections in diabetic patients, 212, 214, 219, 225
  genes of VRSA strains, 223, 226
  horizontal gene transfer from *Enterococcus*, 221–22, 223, 226
  natural selection and antibiotic resistance, 215–16, 223, 226
  treatment of infections, 225–26
  vancomycin-resistance test, *213*

**W**

Wakefield, Andrew, 27, 28, 31

Wallace, Alfred Russel, 214, 215

Wanner, Michael, 174, 175, 184

warning coloration, 368–69

wasps, exploitative competition, *370*

water, properties and structure, 44–47

water, use in agriculture, 472

*Raised from the Ground*

Also by José Saramago in English translation

# Raised *from* *the* Ground

## José Saramago

*Translated from the Portuguese by Margaret Jull Costa*

HOUGHTON MIFFLIN HARCOURT

BOSTON ◆ NEW YORK

2012

First U.S. edition

For information about permission to reproduce selections from this book,
write to Permissions, Houghton Mifflin Harcourt Publishing Company,
215 Park Avenue South, New York, New York 10003.

www.hmhbooks.com

First published with the title *Levantado do Chão* in 1980
by Editorial Caminho, SA, Lisbon

*Library of Congress Cataloging-in-Publication Data*
Saramago, José. [Levantado do chão. English] Raised from the ground /
José Saramago ; translated from the Portuguese by Margaret Jull Costa. — 1st U.S. ed.
p. cm.
"First published with the title Levantado do chão in 1980 by Editorial Caminho,
SA, Lisbon" — T.p. verso.
ISBN 978-0-15-101325-8 (hardback)
I. Title. PQ9281.A6614813 2012
869.3'42 — dc23      2012017326

Book design by Melissa Lotfy

Printed in the United States of America
DOC 10 9 8 7 6 5 4 3 2 1

This publication was assisted by a grant from the Direcção-Geral do Livro
e das Bibliotecas / Portugal

To the memory of Germano Vidigal
and José Adelino dos Santos,
both of whom were murdered

I ask the political economists and the moralists if they have ever calculated the number of individuals who must be condemned to misery, overwork, demoralization, degradation, rank ignorance, overwhelming misfortune and utter penury in order to produce one rich man.

— ALMEIDA GARRETT

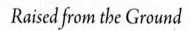

*Raised from the Ground*

HERE, IT'S MOSTLY countryside, land. Whatever else may be lacking, land has never been in short supply, indeed its sheer abundance can only be explained by some tireless miracle, because the land clearly predates man, and despite its long, long existence, it has still not expired. That's probably because it's constantly changing: at certain times of the year, the land is green, at others, yellow or brown or black. And in certain places it is red, the color of clay or spilled blood. This, however, depends on what has been planted or what has not yet been planted, or what has sprung up unaided and died simply because it reached its natural end. This is not the case with wheat, which still has some life left in it when it is cut. Nor with the cork oak, which, despite its solemn air, is full of life and cries out when its skin is ripped from it.

There is no shortage of color in this landscape, but it isn't simply a matter of color. There are days as harsh as they are cold, and others when you can scarcely breathe for the heat: the world is never content, the day it is will be the day it dies. The world does not lack for smells either, not even here, which is, of course, part of the world and well provided with land. Were some insignificant creature to die in the undergrowth, it would smell of death and putrefaction. Not that anyone would notice if there were no wind, even if they were to pass close by. The bones would be either washed clean by the rain or baked dry by the sun, or not even that if the creature were very small, because the worms and the grave-digger beetles would have come and buried it.

This, relatively speaking, is a fair-sized piece of land, and while it begins as undulating hills and a little stream-water, because the water that falls from the skies is just as likely to be feast as famine, farther on it flattens out as smooth as the palm of your hand, although many a hand, by life's decree, tends, with time, to close around the handle of a hoe, sickle or scythe. The land. And like the palm of a hand, it is crisscrossed by lines and paths, its royal or, later, national roads, or those owned by the gentlemen at the town hall, three such roads lie before us now, because three is a poetical, magical, spiritual number, but all the other paths arise from repeated comings and goings, from trails formed by bare or ill-shod feet walking over clods of earth or through undergrowth, stubble or wild flowers, between wall and wasteland. So much land. A man could spend his whole life wandering about here and never find himself, especially if he was born lost. And he won't mind dying when his time comes. He is no rabbit or genet to lie and rot in the sun, but if hunger, cold or heat were to lay him low in some secluded spot, or one of those illnesses that don't even give you time to think, still less cry out for help, sooner or later he would be found.

Many have died of war and other plagues, both here and in other parts, and yet the people we see are still alive: some perceive this as an unfathomable mystery, but the real reasons lie in the land, in this vast estate, this latifundio, that rolls from high hills down to the plain below, as far as the eye can see. And if not this land, then some other piece of land, it really doesn't matter as long as we've sorted out what's mine and thine: everything was recorded in the census at the proper time, with boundaries to the north and

south and to the east and west, as if this were how it had been or-
dained since the world began, when everything was simply land,
with only a few large beasts and the occasional human being, all
of them frightened. It was around that time, and later too, that the
future shape of this present land was decided, and by very crooked
means indeed, a shape carved out by those who owned the largest and sharpest knives and according to size of knife and quality
of blade. For example, those of a king or a duke, or of a duke who
then became his royal highness, a bishop or the master of an order,
a legitimate son or the delicious fruit of bastardy or concubinage, a
stain washed clean and made honorable, or the godfather of a mistress's daughter, and then there's that other high officer of the court
with half a kingdom in his grasp, and sometimes it was more a case
of, this, dear friends, is my land, take it and populate it to serve
me and your offspring, and keep it safe from infidels and other
such embarrassments. A magnificent book-of-hours-cum-sacred-
accounts-ledger presented at both palace and monastery, prayed to
in earthly mansions or in watchtowers, each coin an Our Father,
ten coins a Hail Mary, one hundred a Hail Holy Queen, Mary is
King. Deep coffers, bottomless silos, granaries the size of ships,
vats and casks, coffers, my lady, and all measured in cubits, rods
and bushels, in quarts, pottles and tuns, each piece of land according to its use.

Thus flowed the rivers and the four seasons of the year, on
those one can rely, even when they vary. The vast patience of time
and the equally vast patience of money, which, with the exception
of man, is the most constant of all measurements, although, like
the seasons, it varies. We know, however, that men were bought
and sold. Each century had its money, each kingdom its man to
buy and sell for maravedis, or for gold and silver marks, reals, dou-
bloons, cruzados, sovereigns or florins from abroad. Fickle, various
metal, as airy as the bouquet of a flower or of wine: money rises,
that's why it has wings, not in order to fall. Money's rightful place

is in a kind of heaven, a lofty place where the saints change their names when they have to, but not the latifundio.

A mother with full breasts, fit for large, greedy mouths, a womb, the land shared out between the largest and the large, or, more likely, joining large with larger, through purchase or perhaps through some alliance, or through sly theft, pure crime, the legacy of my grandparents and my good father, God rest their souls. It took centuries to get this far, who can doubt that it will always remain the same?

But who are these other people, small and disparate, who came with the land, although their names do not appear in the deeds, dead souls perhaps, or are they still alive? God's wisdom, beloved children, is infinite: there is the land and those who will work it, go forth and multiply. Go forth and multiply me, says the latifundio. But there is another way to speak of all this.

T HE RAIN CAUGHT UP with them toward the end of the afternoon, when the sun was barely a half-span above the low hills, to the right, however, the witches were already combing their hair, for this is their favorite weather. The man reined in the donkey and, to relieve the animal's load on the slight incline, used his foot to shove a stone under one wheel of the cart. This rain is most unseasonable, whatever can have got into the ruler of the celestial waters. That's why there's so much dust on the roads as well as the occasional dried cow pat or lump of horse dung, which no one has bothered to pick up, this being too far removed from any inhabited place. No young lad, basket over his arm, has ventured this far out in search of some natural manure, tentatively picking up the crumbling sphere, which is sometimes cracked like a ripe fruit. In the rain, the hot, pale earth became spattered with sudden dark stars, falling dully onto the soft dust, and then a torrent flooded everything. The woman, however, still had time to lift the child down from the cart, out of the concave nest formed by the striped mattress squeezed in between two large chests. She held him to her breast, covered his face with the loose end of her shawl, and said, Good, he's still asleep. This was her first concern, the second was,

Everything's going to get drenched. The man was looking up at the high clouds, wrinkling his nose, and then, in his male wisdom, he declared, It'll pass, it's only a shower, but just in case, he unrolled one of the blankets and draped it over the furniture, Why did it have to rain today of all days, damn it.

A flurry of wind sent the now sparse drops flying. When the man gave the donkey a slap on the back, it shook its ears vigorously and tugged at the shafts and the man helped by pushing against the wheel. They set off again up the slope. The woman followed behind, her child in her arms, and, pleased to see him sleeping so soundly, she peered down at him, murmuring, There's a good boy. The ground to either side of the cart track was thick with undergrowth, in which a few lost, choked holm oaks stood, trunk-deep, abandoned or perhaps born there. The wheels of the cart gouged and squelched a path through the sodden earth, and now and then gave a sudden violent jolt whenever a stone raised a shoulder above the surface. The furniture creaked beneath the blanket. The man, walking beside the donkey, his right hand resting on the reins, was silent. And so they reached the top of the hill.

A great mass of dense, towering clouds was heading toward them from the south over the straw-colored plain. The path plunged downward, barely distinguishable between the crumbling ditches planed almost flat by the winds sweeping in across the empty expanse. At the bottom, the path would join a wide road, a rather ambitious word to use in a place so ill served by roads. To the left, almost hugging the low horizon, a small settlement turned its white walls to face the west. As we said before, the plain was vast and smooth, interrupted only by a few holm oaks, alone or in pairs, and little else. From that modest vantage point, it was not difficult to believe that the world had no known end. And seen from there, in the yellowish light and beneath the great leaden sheet of the clouds, the settlement, their destination, seemed unreachable. São Cristóvão, said the man. And the woman, who had never traveled so far south, said, Monte Lavre is bigger, which, while apparently a merely comparative statement, hinted perhaps at homesickness.

They were halfway down the hill when the rain returned, at first in the form of a few plump drops threatening a downpour, so much for it being a passing shower. Then the wind swept across the plain, pushing everything before it like a broom, scooping up straw and dust, and the rain advanced from the horizon, a grayish curtain that soon obscured the distant landscape. It was a steady rain, of the sort that looks set in for many hours, one that arrives and is reluctant to leave, and when, finally, the earth can't cope with all that water, it's hard to know then if it's the sky or the earth doing the drenching. The man again said, Damn it, the kind of thing people say if they have learned no grander expressions. Shelter is far away, and with no coats to put on, they have no alternative but to receive on their backs whatever rain may fall. From there to the village, given the speed at which this weary and somewhat reluctant donkey is traveling, it will be at least another hour's journey, and by then it will be dark. The blanket, which barely protects the furniture, is soaked and dripping, the water falling in drops from the white threads, what hope for the clothes in the chests, the few migratory possessions of this family who, for reasons of their own, are making this cross-country trek. The woman looks up at the sky, an ancient, country way of reading the great blank page above our head, this time in order to see if the sky is clearing, which it isn't, looking, rather, as if it were heavy with dark ink, the weather won't change this evening. The cart travels onward, it's a boat plunging into the deluge, it'll go under at any moment, that seems to be why the man is driving the donkey forward, but it's only so that they can reach that holm oak, which will shelter them from the worst of the storm. Man, cart and donkey have arrived, and the woman is nearly there, sliding about in the mud, she can't run, she would wake her child, that's how the world is, we never notice other people's problems, not even when the people involved are as close as mother and son.

Underneath the oak tree, the man was gesticulating impatiently, he obviously doesn't know what it's like to carry a child in his arms, he'd be better employed checking that the ropes on the cart haven't slackened, because traveling at that speed, the knots are sure to have slipped or the furniture shifted, and the last thing we need now is for the little furniture we have to fall and break. Under the tree, the rain is lighter, but large drops still fall from the leaves, this is no dense orange tree, standing beneath these enormous, wide-spread arms is like standing beneath a porch full of holes, indeed, it's hard to know where to stand, but just then the child began to cry, prompting the mother to perform a more urgent task, unbuttoning her blouse and giving him her breast, almost empty of milk now, barely enough to stave off hunger. His crying stopped at once, and mother and child were at peace, wrapped about by the steady murmur of the rain, while the father walked around the cart, untying and retying knots, bracing his knee on the side of the cart to pull the ropes tight, while the donkey, abstracted, shook his ears hard and gazed out at the puddles and the flooded path. Then the man said, We were so near, and then all this rain, these were words spoken in mild anger, uttered almost unthinkingly and hopelessly, as if to say, the rain won't stop just because I'm angry, well, that's the narrator speaking, which we can quite do without. We would be better off watching the father, who asks at last, How's the child, and goes over and peers under the shawl, he is her husband after all, but so quickly and modestly does his wife cover herself up that he can't be sure now whether it was his son he wanted to see or her bare breast. He just had time to make out, in the tepid darkness, in the scented warmth of crumpled clothes, his son's intensely blue eyes watching him from that private interior, with that strange pale light that usually stared out at him from the cradle, transparent and stern, an exile among the dark brown eyes of the family he was born into.

The heavy clouds had thinned a little, the first torrent of rain had slowed. The man stepped out onto the path, looked up questioningly at the sky, turning to the four cardinal points, and said to his wife, We'd better go, we can't stay here until it's dark. And his wife said, Let's go then. She withdrew her nipple from the baby's mouth, the child sucked air for a moment, seemed about to cry, then stopped, rubbed his face against the now withdrawn breast and, sighing, fell asleep. He's a quiet child, good-humored, and a friend to his mother.

They were walking along together now, wrapped about by the rain, so wet that not even a cozy barn would tempt them, they'll stop only when they reach their new home. The night was coming on fast. In the west, there was only a last faint glow that grew gradually red, then was gone, and the earth was a dumb, black well, full of echoes, how large the world seems at nightfall. The squeaking of the wheels seemed louder, the stuttering breath of the donkey as unexpected as a secret suddenly spoken out loud, and the whisper of their wet clothes was like a continuous murmured conversation between friends, with no awkward silences. For leagues around, not a light was to be seen. The woman crossed herself, then made the sign of the cross over her son's face. At this hour of night, it's best to defend the body and protect the soul, because ghosts begin to appear on the roads, either passing in a whirlwind or sitting down on a rock to await the traveler, of whom they will ask three questions to which there is no answer, who are you, where do you come from and where are you going. The man walking alongside the cart would like to sing, but he can't, all his energy is going into pretending that the night doesn't frighten him. Not much farther, he said when they reached the road, we just keep going straight now and this is a better road too.

Ahead, far away, a flash lit up the clouds, no one could have guessed they were so low. Then a pause and, finally, the low rumble of thunder. That's all we need. The woman said, Holy Saint Barbara save us, but the thunder, if it wasn't a remnant of some distant storm, seemed to be taking a different route, either that or Saint Barbara had shooed it away to places of lesser faith. They were on the road now, they could tell because it was wider, although any other differences could only be found with great patience and by the light of day, they had come through mud and potholes, and through mud and potholes they continued, and now it was so dark that they couldn't see where they were putting their feet. The donkey advanced by instinct, walking alongside the ditch. The man and the woman skidded along behind. Now and then, if the road curved, the man ran blindly ahead to see if he could catch a glimpse of São Cristóvão. And when they saw, amid the darkness, the first white walls, the rain suddenly stopped, so abruptly that they barely noticed. One moment it was raining, the next it wasn't. It was as if a great roof had stretched out over the road.

It's hardly surprising that the woman should ask, Where's our house, a perfectly understandable question in someone who needs to take care of her child and, if possible, put the furniture in its proper place before laying her weary body down in bed. And the man answers, On the other side. All doors are closed, only a few faint chinks of light betray the presence of the other inhabitants. In a yard somewhere, a dog barks. There's always a dog barking when someone walks past, and the other dogs, caught unawares, pick up the first sentinel's word and fulfill their canine duty. A gate was opened, then closed. And now that the rain had stopped and the house was near, husband and wife were more aware of the cold wind that came running along the street, before plunging down the narrow alleys, where it shook the branches that reached out over the low roofs. Thanks to the wind, the night grew brighter. The great cloud was moving off, and here and there you could see patches of clear sky. It's not raining now, said the woman to her child, who was sleeping and, of the four, was the only one not to know the good news.

They came to a square in which a few trees were exchanging brief whispers. The man stopped the cart and said to the woman, Wait here, and walked under the trees toward a brightly lit doorway. It was a bar, a taberna, and inside three men were sitting on a bench while another was standing at the bar, drinking, holding his glass between thumb and forefinger as if posing for a photograph. And behind the bar, a thin, shriveled old man turned his eyes to the door, through which the man with the cart entered, saying, Good evening, gentlemen, the greeting of a new arrival wishing to gain the friendship of everyone in the room, either out of fraternal feeling or for more selfish commercial reasons, I've come to live here in São Cristóvão, my name's Domingos Mau-Tempo* and I'm a shoemaker. One of the men sitting on the bench joked, Well, you certainly brought the bad weather with you, and the man who was drinking and had just emptied his glass smacked his lips and added, Let's hope his soles are better than the weather he brings, and the others, of course, laughed. These were not intended as rude or unwelcoming words, but it's nighttime in São Cristóvão, all the doors are shut, and if a stranger arrives bearing a name like Mau-Tempo, only a fool could resist making a joke of it, especially after that heavy downpour. Domingos Mau-Tempo responded with a reluctant smile, but that's to be expected. Then the old man opened a drawer and produced a large key, Here's the key, I was beginning to think you weren't coming, and everyone stares at Domingos Mau-Tempo, taking the measure of this new neighbor, every village needs a shoemaker and São Cristóvão is no exception. Domingos Mau-Tempo offered an explanation, It's a long way from Monte Lavre, and it rained while we were on the road,

---

* Mau-Tempo means bad weather.

not that there's any need for him to account for himself, but he wants to be friendly, and then he says, Let me buy you all a drink, which is an excellent way of touching the pockets of men's hearts. The men who were seated stand up and watch the ceremony of their glasses being refilled, and then, unhurriedly, each man again picks up his glass with a slow, careful gesture, this is wine, after all, not cheap brandy to be drunk down in one gulp. Won't you have a drink yourself, sir, says Domingos Mau-Tempo, and the old man, who knows the ways of the big city, answers, Here's wishing good health to my new tenant. And while the men are engaged in these niceties, the woman comes to the door, although she doesn't actually come in, the taberna is reserved for men only, and she says quietly, as is her wont, Domingos, the child is restless, and what with the furniture and everything being so wet, we need to get unloaded.

She is quite right, but Domingos Mau-Tempo disliked being summoned by his wife like that, what will the other men think, and as they cross the square, he scolds her, If you do that again, I'll be very angry. The woman did not respond, too busy trying to quiet the baby. The cart went slowly on, jolting over the bumps. The donkey had stiffened up with the cold. They went down a side street where the houses alternated with vegetable gardens, and they stopped outside a low hovel. Is this it, asked the woman, and her husband replied, Yes.

Domingos Mau-Tempo opened the door with the large key. In order to enter, he had to lower his head, for this is no palace with high doors. There were no windows. To the left was the fireplace, with the hearth at floor level. Domingos Mau-Tempo made a small, flickering torch from a sheaf of straw and held it up so that his wife could see their new home. There was a bundle of firewood by the chimney breast. Enough for their immediate needs. In a matter of minutes, the woman had laid the child down in one corner to sleep, gathered together some logs and some kindling, and the fire had sprung into life, like a flower on the whitewashed wall. The house was once again inhabited.

Domingos Mau-Tempo led the donkey and the cart in through the gate to the yard, and started unloading the furniture and carrying it into the house, where he set it down willy-nilly, until his wife could come and help him. The mattress was wet on one side. The water had got into the clothes chest, and one leg of the kitchen table was broken. But on the fire was a saucepan of cabbage leaves and rice, and the baby had suckled again and fallen asleep on the dry side of the mattress. Domingos Mau-Tempo went out into the yard to do his business. And standing in the middle of the room, Sara da Conceição, Domingos's wife and João's mother, stood quite still, staring into the flames like someone waiting for a garbled message to be repeated. She felt a slight movement in her belly. And another. But when her husband came back in, she said nothing. They had other things to think about.

DOMINGOS MAU-TEMPO will not make old bones. One day, when he has given his wife five children, although not for that most mundane of reasons, he will put a rope around the branch of a tree, in a desolate place almost within sight of Monte Lavre, and hang himself. Before he does this, however, he will carry his house on his back to other places, run away from his family three times, but fail to make his peace with them on that third occasion because his hour will have come. His father-in-law Laureano Carranca had predicted just such an unfortunate end when he was forced to give in to Sara's stubbornness, for, so besotted was she with Domingos Mau-Tempo that she swore that if she could not marry him, she would marry no one. Laureano Carranca would roar furiously, He's a ne'er-do-well and a drunkard and will come to no good. And so the family war raged on until Sara da Conceição fell pregnant, a conclusive and usually highly effective argument when persuasion and pleading have failed. One morning, Sara da Conceição left the house, in May it was, and walked across the fields to the place where she had arranged to meet Domingos Mau-Tempo. They were there for half an hour at most, lying amid the tall wheat, and when Domingos returned to his lasts and Sara to her parents' house, he went off whistling with satisfaction, while she was left shivering despite the hot sun beating down on her. And when she crossed the stream by the ford, she had to crouch down beneath some willows and wash away the blood flowing from between her legs.

João was made, or to use a more biblical term, conceived, on that same day, which, it would seem, is most unusual, because, in the haste and confusion of the moment, semen does not necessarily do its job the first time, only later. And it's true that there was considerable consternation, not to say suspicion, regarding João's blue eyes, for no one else in the family had such eyes nor, as far as they could recall, had any relative, close or distant, we, however, know that such thoughts were grossly unfair to a woman who, after much soul-searching, had deviated from the straight virginal path and lain down in a wheatfield with that one man alone and, by her own choice, opened her legs to him. It had not been the choice, almost five hundred years before, of another young woman, who, standing alone at the fountain filling her jug with water, was approached by one of the foreigners who had arrived with Lamberto Horques Alemão, the governor of Monte Lavre appointed by Dom João the First, by a man whose speech she couldn't understand, and who, ignoring the poor girl's cries and pleas, carried her off into the bracken where, purely for his own enjoyment, he raped her. He was a handsome fellow with pale skin and blue eyes, whose only fault was the fire in his blood, but she, naturally enough, could not bring herself to love him, and when her time came, she gave birth alone. Thus, for four centuries those blue German eyes appeared and disappeared, like the comets that vanish and return when we least expect them or simply because no one has bothered to record their appearances and thus discover a pattern.

This is the family's first move. They came from Monte Lavre to São Cristóvão on a strangely rainy summer's day. They traversed the whole district from north to south, what on earth can have made Domingos Mau-Tempo decide to move so far away, well, he's a bungler and a good-for-nothing, and things were getting difficult for him in Monte Lavre because of drink and certain shady deals, and so he said to his father-in-law, Lend me your cart and your donkey, will you, I'm going to live in São Cristóvão, By all means go, and let's hope you acquire a little common sense, for your own good and for the sake of your wife and son, but be sure to bring that donkey and cart back promptly, because I need them. They took the shortest route, following cart tracks, or highways when they could, but mostly heading across country, skirting the hills. They lunched in the shade of a tree, and Domingos Mau-Tempo gulped down a whole bottle of wine that he soon sweated out again in the heat of the day. They saw Montemor in the distance, to the left, and continued south. It rained on them when they were just one hour from São Cristóvão, a deluge that presaged no good at all, but today it is sunny, and Sara da Conceição, sitting in the garden, is sewing a skirt, while her son, still rather unsteady on his legs, is feeling his way along the wall of the house. Domingos Mau-Tempo has gone to Monte Lavre to return the donkey and cart to his father-in-law and tell him that they're living in an excellent house, that customers are already beating a path to his door and that he won't lack for work. He will return on foot the following day, as long as he doesn't get drunk, because apart from his drinking, he isn't a bad man, and God willing he'll sort himself out, after all, there have been worse men than him and they've turned out all right in the end, and if there's any justice in the world, what with one small child and another on the way, he'll shape up to be a respectable father too, and as for me, well, I'll do what I can to give us all a good life.

João has reached the end of the wall, where the picket fence begins. He grips it hard, his arms being stronger than his legs, and peers out. His horizon is quite limited, a strip of muddy road with puddles that reflect the sky, and a ginger cat sprawled on the doorstep opposite, sunning its belly. Somewhere a cock crows. A woman can be heard shouting out, Maria, and another, almost childish voice answers, Yes, Senhora. And then the silence of the great heat settles again, the mud will soon harden and return to the dust it was. João lets go of the fence, that's quite enough looking at the landscape for the moment, executes a difficult half-turn and commences the long journey back to his mother. Sara da Conceição sees him, puts her sewing down on her lap and holds out her arms to her son, Come here, little one, come here. Her arms are like two protective hedges. Between them and João lies a confusing, uncertain world with no beginning and no end. The sun sketches a hesitant shadow on the ground, a tremulously advancing hour. Like the hand of a clock on the great expanse of the latifundio.

When Lamberto Horques Alemão stepped out onto the terrace of his castle, his gaze could not encompass all that lay before him. He was the lord of the village and its lands, ten leagues long and three leagues wide, and he had the right to exact a tribute, and although he had been charged to go forth and multiply, he had not ordered the rape of the girl at the fountain, it happened and that was that. He himself, with his virtuous wife and his children, will scatter his seed where he pleases, depending on how the mood takes him, This land cannot remain as uninhabited as it is now, for you can count on the fingers of one hand the number of settlements in the whole estate, while the uncultivated areas are as many as the hairs on your head, Yes, sir, but these women are the swarthy cursed remnants of the Moors, and these silent men can be vengeful, besides, our king did not call on you to go forth and multiply like a Solomon, but to cultivate the land and rule over it so that people will come here and stay, That is what I am doing and will do, and whatever else I deem appropriate, for this land is mine and everything on it, although there are sure to be people who will try to hamper my efforts and cause trouble, there always will be, You are quite right, sir, you obviously gleaned such knowledge from the cold lands of your birth, where people know far more than we natives of these remote western lands, Since we are in agreement, let us discuss what tributes should be imposed on these lands I am to govern. Thus, a minor episode in the history of the latifundio.

T HIS SO-CALLED SHOEMAKER is really nothing but a cobbler. He soles and heels and dawdles over his work when he
isn't in the mood, often abandoning last, awl and knife to go to the
taberna, he argues with impatient customers and, for all these reasons, beats his wife. Not just because he is obliged to sole and heel,
but also because he can find no peace in himself, he's a restless man
who has no sooner sat down than he wants to get up again, who
as soon as he has arrived in one place is already thinking about
another. He's a child of the wind, a wanderer, this bad-weather
Domingos, who returns from the taberna and enters the house,
bumping into the walls, glancing sourly at his son, and for no reason at all lashes out at his wife, wretched woman, let that be a lesson to you. And then he leaves again, goes back to the wine and
his carousing mates, put this one on the slate, will you, landlord,
of course, sir, but there's quite a lot on the slate already, so what, I
always pay my debts, don't I, I've never owed anyone a penny. And
more than once, Sara da Conceição, having left her child with the
neighbor, went out into the night to search for her husband, using the shawl and the darkness to conceal her tears, going from
taberna to taberna, of which there weren't many in São Cristóvão,

but enough, peering in from outside, and if her husband was there, she would stand waiting in the shadows, like another shadow. And sometimes she would find him lost on the road, abandoned by his friends, with no idea where his house was, and then the world would suddenly brighten, because Domingos Mau-Tempo, grateful to have been found in that frightening desert, among hordes of ghosts, would put an arm about his wife's shoulders and allow himself to be led like the child he doubtless still was.

And one day, because he had more work than he could cope with, Domingos Mau-Tempo took on an assistant, thus giving himself more time with his fellow drinkers, but then, on another ill-fated day, he got it into his head that his wife, poor, innocent Sara da Conceição, was deceiving him in his absence, and that was the end of São Cristóvão, which the guiltless assistant had to flee at knifepoint, and Sara, pregnant, quite legitimately, for a second time, underwent her own painful via dolorosa, and the cart was loaded up again, another trek to Monte Lavre, more toing and froing, We're fine, and your daughter and grandson are happy, with another on the way, but I've found a better job in Torre da Gadanha, my father lives there and will be able to help us out. And once more they set off north, except that this time the landlord was waiting for them on the way out of São Cristóvão, Just a moment, Mau-Tempo, you owe me for the rent and the wine that you drank, and if you don't pay up, me and my two sons here will make you, so pay me what you owe or die.

It was a short journey, which was just as well, because almost as soon as Sara da Conceição set foot in the house, she gave birth to her second son, who, for some forgotten reason, was named Anselmo. He was fortunate from the cradle on because his paternal grandfather was a carpenter by trade and very pleased to have his grandson born so close to home, almost next door. His grandfather worked as a carpenter and had no boss and no apprentice, no wife either, and he lived among lengths of timber and planks, permanently perfumed by sawdust, and used a vocabulary particular to laths, planes, battens, mallets and adzes. He was a serious man of few words and not given to drinking, which is why he disapproved of his son, who was hardly a credit to his name. Given Domingos Mau-Tempo's restless nature, however, his father had little time in which to enjoy being a grandfather, just long enough to teach his oldest grandson that this is a claw hammer, this is a plane and this a chisel. But Domingos Mau-Tempo could bear neither what his father said nor what he didn't say, and like a bird hurling itself against the bars of its cage, what prison is this in my soul, damn it, off he went again, this time to Landeira, in the extreme west of the district. Preferring this time not to approach his father-in-law, who would find such wanderings and uncertainties odd, he had, at some expense, hired a cart and a mule, intending to keep quiet about his plans and tell his father-in-law later. We never seem to settle anywhere, we go from one place to another like the wandering Jew, and it's not easy with two small children, Be quiet, woman, I know what I'm doing, there are good people in Landeira and plenty of work, besides, I'm a craftsman, not like your father and brothers tied to their hoes, I learned a trade and have a skill, That's not what I'm saying, you were a shoemaker when I married you and that's fine, but I just want some peace and to stop all this moving around. Sara da Conceição said nothing about the beatings, nor would it have been appropriate, because Domingos Mau-Tempo was traveling toward Landeira as if to the promised land and carrying on his shoulders his eldest son, holding on to his tender little ankles, which were a bit grubby, of course, but what does that matter. He barely felt the weight, because years of sewing leather had given him muscles and tendons of iron. With the mule trotting along behind, with a sun as warm as a cozy blanket, Sara da Conceição was even allowed to ride in the cart. But when they reached the new house, they found that their furniture was once again badly damaged, At this rate, Domingos, we'll end up with no furniture at all.

It was in Landeira that João, who already had his real godparents in Monte Lavre, found a new and more illustrious godfather. He was Father Agamedes, who, because he lived with a woman he called his niece, provided João with a borrowed godmother too. The child did not lack for blessings, being as protected in heaven as he had been on earth up until then. Especially when Domingos Mau-Tempo, encouraged by Father Agamedes, took on the duties of sacristan, helping at mass and at funerals, because thanks to this, the priest befriended him and adopted João. Domingos Mau-Tempo's sole aim in being received into the bosom of the church was to find a respectable reason for avoiding work and a respite from his persistent vagabond restlessness. But God rewarded him as soon as he saw him at his altar, clumsily performing the ritualistic gestures he was taught, for Father Agamedes also liked his drink, and thus celebrant and acolyte came together over that other sacrifice. Father Agamedes owned a grocery store, not far from the church, where he worked whenever his priestly duties allowed, and when they didn't, his niece would come down to the square and, from behind the counter, rule over the family's earthly business. Domingos Mau-Tempo would drop in and drink a glass of wine, then another and another, alone, unless the priest was there, and then they drank together. God, meanwhile, was up above with the angels.

But all heavens have their Lucifers and all paradises their temptations. Domingos Mau-Tempo began to look on his neighbor's companion with covetous eyes, and she, as niece, took offense and mentioned it to her uncle, and that was enough to create bad feeling between those two servants of the holy mother church, one permanent and the other temporary. Father Agamedes could not speak frankly for fear of giving credence to the evil thoughts of those parishioners who had their doubts about that niece-uncle relationship, and so, to drive away the threat to his own honor, he focused on the married status of the offending party. Deprived of his easy access to wine and weary of plying his trade here, there and everywhere, Domingos Mau-Tempo declared his intention at home of avenging himself on the priest. He did not say exactly what he was avenging himself for, and Sara da Conceição did not ask. She continued to suffer in silence.

The church had few parishioners and not all of them regular. It provided no remedy for their ills, nor was it obliged to, since, as far as one could see, it didn't increase them either. That was not the problem. The lack of apostolic action was not conducive to increased devotion, not so much because Father Agamedes lived with his so-called niece and ran a grocery store, because only those who are not of the people are ignorant of such basic needs, but because he mangled the words of the prayer book, and dispatched newborns, newlyweds and the dead with the same cold-blooded indifference with which he slaughtered and ate his pig and with equally scant attention to the letter or spirit of the holy writ. Ordinary people can be strangely sensitive. Domingos Mau-Tempo knew how to ensure that the church would be full. He let it be known that the next mass would be something special, that Father Agamedes had told him that in future he was going to take particular pains over the holy precepts, and would make use of sublime pauses and even vibrato, you'd be a fool to miss it, so don't come complaining to me afterward if you do. Father Agamedes was amazed when he saw the church packed with people. It wasn't the church's name day and the drought had not been so bad as to require celestial intervention, but he said nothing. If the flock came to the pen of its own free will, so much the better for the shepherd when it came to rendering accounts to his master. In short, so as not to appear ungrateful, he outdid himself and, all unknowing, confirmed Domingos Mau-Tempo's prediction. However, the shoemaker raised up to the position of sacristan, and already planning another escape, had his revenge prepared. When it came to the point in the mass where he had to ring the sanctus bell, he calmly raised the bell and shook it. It was as if he had waved a chicken feather in the air. At first, the faithful thought that they must all have gone deaf, others, out of habit, bowed their heads, while others watched distrustfully as Domingos Mau-Tempo, in dramatic silence, his face a mask of innocence, continued to shake the bell. The priest looked puzzled, the faithful muttered to each other, the younger members laughed. It was shameful, what with all the saints, not to mention all-seeing God, looking down on them. Father Agamedes could contain himself no longer, and he stopped the communion service there and then, grabbed the bell and felt inside it. There was no clapper. And yet no thunderbolt fell to punish such impiety. Terrible in his holy fury, Father Agamedes slapped Domingos Mau-Tempo hard about the face, right there in that sacred place, it scarcely seemed possible. But Domingos Mau-Tempo responded in kind, as though this were all part of the mass. And it was not long before the priest's vestments and the sacristan's surplice were caught up in a furious maelstrom, one on top, the other underneath, rolling sacrilegiously about, bruising their ribs on the altar steps, beneath the round-eyed gaze of the monstrance. The congregation rushed to separate the two warring powers, and some took advantage of that tangle of arms and legs to slake an ancient thirst for revenge on either one side or the other. The old ladies had gathered in one corner, praying to all the hosts of heaven, and, finally summoning up physical force and spiritual courage, advanced on the altar in order to save their priest, however unworthy. It was, in short, a triumph of faith.

The next day, Domingos Mau-Tempo left the village, followed by a noisy cortege of boys, who accompanied him and his family as far as the barren outskirts. Sara da Conceição bowed her head in shame. João looked about him with his stern blue eyes. The other boy was sleeping.

THEN THE REPUBLIC arrived. The men earned twelve or thirteen vinténs, and the women, as usual, less than half. Both ate the same black bread, the same cabbage leaves, the same stalks. The republic rushed in from Lisbon, traveled from village to village by telegraph, if there was one, advertised itself in the press for those who knew how to read, or passed from mouth to mouth, which was always by far the easiest way. The king had been toppled, and according to the church, that particular kingdom was no longer of its world, the latifundio got the message and did nothing, and the price of a liter of olive oil rose to more than two thousand réis, ten times a man's daily wage.

Long live the republic. So how much is the new daily rate, boss, Let's see, I pay whatever the others pay, talk to the overseer, So, overseer, how much is the daily rate, You'll earn an extra vintém, That's not enough to live on, Well, if you don't want the job, there are plenty more who do, Dear God, a man could die of hunger along with his children, what can I give my children to eat, Put them to work, And if there is no work, Then don't have so many children, Wife, send the boys off to collect firewood and the girls for straw, and come to bed, Do with me as you wish, I am my master's slave, and there, it's done, I'm pregnant, with child, in the family way, I'm going to have a baby, you're going to be a father, I've missed a period, That's all right, eight can starve as easily as seven.

And because far from there being any visible differences, only similarities, between the latifundio under the monarchy and the latifundio under the republic, and because the wages they earned could buy so little that they only served to increase their hunger, some innocent workers got together and went to the district administrator to demand better living conditions. The person with the best handwriting wrote out their request, remarking on the new joy felt among the Portuguese people and the new hopes that had sprung up with the coming of the republic, we wish you good health and send fraternal greetings, sir, and await your reply. Once the supplicants had been dismissed, Lamberto Horques sat down in his Hanseatic chair, meditated deeply on what would be best for the farms, himself and the people he governed, and having perused the maps on which the various parcels of land were marked, he placed his finger on the one most densely populated and summoned the captain of the guard. The captain had formerly belonged to the civil police force and now cut a martial figure in his new uniform, but he had a short memory and had, therefore, forgotten the days when he had worn the blue-and-white ribbon on his left sleeve.* Thanks to the captain's zeal and vigilance, Lamberto learned that the workers were agitating for change and protesting about the forced loans and other such impositions, they were complaining, too, about the poor-quality food, which, after paying the various taxes and tributes, was all they could afford, these complaints were all there in the letter of petition, albeit expressed in measured tones, but perhaps those tones only disguised other, worse intentions. An ill wind of insurrection was blowing

---

* Blue and white were the colors of the old monarchical flag.

through the latifundio, the snarling of a cornered, starving wolf that could cause great damage if it should turn into an army of teeth. It was necessary, therefore, to set an example, to teach them a lesson. Once the interview was over and he had received his orders, Lieutenant Contente clicked his heels and ordered the bugle to be sounded on the parade ground. There the republican national guard lined up, sabers at their side and reins tight, harnesses, mustaches and manes gleaming, and when Lamberto appeared at the window of his room, the guards saluted him as if they were waving goodbye, thus uniting in one gesture both affection and discipline. Then he withdrew to his chamber and summoned his wife, with whom he took his pleasure.

See how the guards go flying through the countryside. They trot, they gallop, the sun beats down on their armor, the saddle cloths swirl about the horses' legs, O cavalry, O Roland, Oliveros and Fierabras,* happy the country that gave birth to such sons. The chosen village is within sight, and Lieutenant Contente orders the squadron to prepare to charge, and when the bugle sounds, the troops advance in lyrical, warlike fashion, sabers unsheathed, the whole nation comes to the balcony to observe the spectacle, and when the peasants emerge from their houses, from barns and cattle sheds, they are mown down by the charging horses and struck from behind by the blades of the soldiers' swords, until Fierabras, frisky as an ox stung by a gadfly, grips his saber in his hand and cuts, scythes, slices, pierces, blind with rage, although quite why he doesn't know. The peasants lay moaning on the ground, and when finally carried back into their huts, they did not rest but tended their wounds as best they could, with lavish use of water, salt and

---

* In French and Italian chivalric literature, Fierabras was a Saracen giant who sacked Rome and stole two containers of the fluid in which Christ was said to have been embalmed. This fluid was reputed to be able to heal any wound. Fierabras was, in turn, defeated by Oliveros, who gave the containers to Charlemagne so that he could return them to Rome. Roland became, in legend, Charlemagne's chief paladin.

cobwebs. We'd be better off dead, said one. Our time has not yet come, said another.

The national guard, belovèd child of the republic, is leaving, the horses are still trembling, and flecks of foam still fill the air, and now they move on to the second phase of the battle plan, which is to ride into the hills and gullies and hunt down the workers who are inciting the others to rebellion and strikes, leaving the work in the fields undone and the animals untended, and thus thirty-three of them were taken captive, along with the main instigators, who ended up in military prisons. The guards led them off like a train of mules, their backs clothed in lashes, blows and mocking remarks, you bastards, mind you don't trip over your cuckold's horns, long live the republican guard, long live the republic. The farm workers were all individually bound and then tied as well to a single rope, like galley slaves, can you believe it, as if these were tales from barbarous times, from the days of Lamberto Horques Alemão, from the fifteenth century, at most.

And who is going to take the leaders of the mutiny to Lisbon? Eighteen soldiers from the seventeenth infantry, led by their lieutenant, also called Contente, set off secretly on the night train, thirty-eight eyes keeping watch over five farm laborers accused of sedition and incitement to strike. They will be handed over to the government, our solicitous correspondent informs us, this government is a regular almshouse, always eager to receive such deliveries. And it's May again, gentlemen, the month of Mary. There goes the train, there it goes, whistling away, there go the five farm laborers, to rot in Limoeiro prison. In these barbarous times the trains travel slowly, they stop for no apparent reason in the middle of nowhere, perhaps at some halt perfect for an ambush and sudden death, and the locked carriage in which the malefactors are traveling has its curtains closed, if there are curtains in the days of Lamberto Horques, if such extravagances are commonplace in third-class carriages, and the seventeenth infantry have their rifles cocked, perhaps even their bayonets fixed, who goes there, getting off the train ten at a time whenever it stops, to prevent any attacks or attempts to free the prisoners. The poor soldiers are under orders not to sleep, and they stare nervously at the hard, grimy faces of those five criminals, so like you. And when I get out of the army, my friend, who knows, perhaps another soldier will arrest me and carry me off to Lisbon on the night train, in the dark, We know our place now, but tomorrow, who can say, They lend you a rifle, but they never say anything about turning it on the estate workers, All that training, all that take aim and fire, is actually turned against yourself, the barrel of your weapon is staring at your own deceived heart, you have no idea what you're doing, and one day they'll give the order to fire, and you'll shoot yourself, Shut your mouth, you seditious bastards, you'll learn your lesson, who knows how many years you'll spend inside, Yes, Lisbon is a big city, the biggest in the world they say, as well as home to the republic, which should, by rights, set us free, We're perfectly within the law.

There are now two groups of workers face to face, a mere ten paces apart. Those from the north are saying, We're perfectly within the law, we were hired and we want to work. Those from the south say, You've agreed to work for less money, you come here to do us harm, go back where you came from, you rats,* you black-legs. Those from the north say, Where we come from there is no work, it's all stones and scrub, we're from the Beira, so don't insult us by calling us rats. Those from the south say, But you are rats, you come here to gnaw at our bread. Those from the north say, We're hungry. Those from the south say, So are we, but we refuse to accept this poverty, if you agree to work for such a low wage, we'll be left with nothing. Those from the north say, That's your fault, you shouldn't be so proud, accept what the boss offers you,

---

* *Ratinhos* in Portuguese; these were temporary workers from northern and central Portugal who went to seek work in the Alentejo.

better something than nothing, and then there'll be work for everyone, because there aren't many of you and we've come to help. Those from the south say, That's just a trick, they want to trick us all, we don't have to accept that wage, why not join forces with us and then the boss will have to pay everyone a better wage. Those from the north say, Each man knows his own heart and God knows them all, we don't want to make alliances, we've traveled a long way, we can't stay here and make war on the boss, we want to work. Those from the south say, Well, you're not going to work here. Those from the north say, Yes, we are. Those from the south say, This land is ours. Those from the north say, But you don't want to work it. Those from the south say, Not for this wage, no. Those from the north say, The wage is fine with us. The overseer says, All right, you've had your chat, now stand aside and let these men get to work. Those from the south say, Don't do it. The overseer says, Get working, if you don't do as I say, I'll call the guards. Those from the south say, There'll be blood spilled before the guards arrive. The overseer says, If the guards do come, still more blood will be spilled, so don't say I didn't warn you. Those from the south say, Brothers, listen to what we're saying, for pity's sake, join us. Those from the north say, Like we said, we want to work.

Then the first man from the north walked over to the wheat with his sickle, and the first man from the south grabbed his arm, and they grappled clumsily, awkwardly, roughly, brutishly, hunger against hunger, poverty against poverty, how dearly we buy our daily bread. The guards arrived and broke up the fight, attacking one side only, driving back with their sabers those from the south and corralling them as if they were animals. The sergeant says, Shall I arrest the lot of them. The overseer says, It's not worth it, leave the bastards there for a while to cool off. The sergeant says, But one of the other men has a wound to the head, he was attacked, and the law is the law. The overseer says, It's not worth it, Sergeant, why worry over spilling a mere animal's blood, it doesn't matter whether they're from the north or the south, they're worth about as much as the boss's piss. The sergeant says, Speaking of the boss, I need some firewood. The overseer says, We'll send you a cartload. The sergeant says, And a few roof tiles too. The overseer says, Well, we can't have you without a roof over your head. The sergeant says, Life is very expensive. The overseer says, I'll send you some sausages.

The men from the north are in the field now. The blond ears of wheat fall onto the dark earth, how lovely, it smells like a long-unwashed body, then, in the distance, a passing tilbury stops. The overseer says, It's the boss. The sergeant says, Give him my thanks, and let me know if you need any more help. The overseer says, Keep an eye on those rascals. The sergeant says, Don't worry, I know how to handle them. Some of the men from the south say, Let's set fire to the wheatfield. Others say, That would be a terrible shame. They all say, That lot don't know what shame is.

THEY HAD BEEN to Landeira, and to Santana do Mato, in and out of the parish, to Tarrafeiro and Afeiteira, and in the midst of all this traveling their third child was born, a daughter this time, Maria da Conceição, and a fourth, a boy named Domingos, like his father. May God give him better fortune, because there was nothing good to be said about his progenitor, who, caught between the wine and the cheap brandy, the mallet and the shoe stud, was going from bad to worse. And as for the furniture, the least said the better, for it continued to be bumped over hills and ditches as it was transported from house to cart and from cart to house and from village to village. A new shoemaker's arrived, his name's Mau-Tempo, let's go and see what this master craftsman is like, mind you, he drinks wine all year round the way you drink water in August, he's certainly a master at that. While living in Canha with her husband and children, Sara da Conceição suffered from tertian fever for two years, which, for those unfamiliar with the disease, is the sort of fever that comes and goes every four days. That is why, when his mother was ill in bed, João Mau-Tempo, he of the blue eyes, which were not repeated in his siblings, used to go to the well, and once, as he plunged the jug in, he lost his footing, proof that

32

no one watches over the innocent, and fell into the water, which was very deep for a little seven-year-old. He was carried home by the woman who saved him, and his father beat him while his mother lay in bed trembling with fever, shaking so hard that even the brass balls on the bed shook, Don't hit the boy, Domingos, but she might as well have been talking to a brick wall.

Then came the day when Sara da Conceição called her husband and he did not answer. That was the first time Domingos Mau-Tempo spurned his family and went wandering off. Then, Sara da Conceição, who had kept silent for so long about her life, asked a literate neighbor to write a letter for her, and it was as if she were pouring her whole soul into it, because such behavior certainly wasn't what had made her fall in love with her husband. Dearest Father, for the love of God, please come and fetch me with your donkeys and your cart and take me back home with you where I belong and I beg you please to forgive me all the trouble and grief I've caused you as well as all you've had to put up with and believe me when I say how I regret not following the advice you gave me over and over not to make this unfortunate marriage to a man who has brought me only sorrow because I've suffered so much poverty disappointments beatings I was well advised but ill fated, this final phrase was drawn from her neighbor's literary treasure trove, marrying the classical and the modern with admirable boldness.

What would any father worthy of the name do, regardless of previous scandals? What did Laureano Carranca do? He sent his gloomy, ill-tempered son, Joaquim, to Canha to fetch his sister and however many grandchildren there might be. Not because he loved them dearly, they were, after all, the children of that drunkard cobbler, no, he didn't love those chips off the old block, and besides, he had other grandchildren he preferred. And so, that poor woman and her children, abandoned by husband and father, arrived in Monte Lavre with the ruins of their furniture piled high on the cart, and some were given house-room by parents or grandparents, out of a somewhat tetchy sense of pity, while others were deposited in a hayloft until a home could be found for them. And when they had to find shelter, mats on the floor served them as beds, and for food the older children went begging, as Our Lord once did, for it is a sin to steal. Sara da Conceição worked hard, of course, because she wasn't just there to bring children into the world, and her parents helped her out a little, her mother rather more generously, as is only natural, well, she was her mother. And thus they scraped along. A few weeks later, though, Domingos Mau-Tempo reappeared, prowling around Monte Lavre, trailing after his wife and children and finally ambushing them, contrite and repentant, to use his words, doubtless learned while he was sacristan. Laureano Carranca flew into a great rage, saying that he never wanted to see his daughter again if, heaven forbid, she went back to that useless, drunken scoundrel of a son-in-law. A much-chastened Domingos Mau-Tempo went to talk to him and assured him that he had changed his ways, and that this absence had shown him, blind as he was, how much he loved his wife and his dear children, I swear this to you, sir, on bended knee if necessary. Having somewhat assuaged their anger with all his tears, he and his family set off for a nearby hamlet, Cortiçadas de Monte Lavre, almost within sight of the paternal home. Having lost all the equipment that had allowed him, as he preferred, to work for himself, Domingos Mau-Tempo was forced to take employment with Master Gramicho, while Sara da Conceição labored away stitching uppers to soles, to help out her husband and keep her children fed and clothed. And the fates? Domingos Mau-Tempo once again began to slide into sadness, like a monster in exile, for that is the worst of all sadnesses, as you can see from the tale of Beauty and the Beast, and it wasn't long before he said to his wife, It's time to move on, I don't feel comfortable here, wait for a few days with the children while I go and look for work elsewhere. Sara da Conceição, not believing that her husband would come back, waited for two months, what else could she do, and was once more the abandoned widow, then up he popped again, happy as a lark, full of sweet words, Sara, I've found work and a really nice house in Ciborro. And so they left for Ciborro, and things went quite well for them, because the people there were pleasant and paid their bills promptly. There was no shortage of work, and the shoemaker seemed to have lost his taste for the taberna, not entirely, that would be asking too much, but enough to make him seem a respectable man. And this happened at an opportune moment because, meanwhile, a primary school had been set up there, and João Mau-Tempo, who was the right age, went there to learn to read and write and count.

And the fates? For some reason werewolves are drawn to crossroads, the poor wretches, not that I claim to understand such mysteries, dear reader, it's as if they were under an evil spell, but on a particular day of the week, they leave their houses and at the first crossroads they come to, they take off their clothes, throw themselves on the ground, roll around in the dust, and are transformed into whatever animal has left its trail there, You mean any trail, or only the trail left by a mammal, Any trail, sir, once, a man was transformed into a cartwheel, and he went spinning and spinning along, it was terrible, but it's more common for them to be changed into animals, as was the case with a man, whose name I can't now recall, who lived with his wife in Monte do Curral da Légua, near Pedra Grande, and his fate was to go out every Tuesday night, but he knew what would happen, and so he warned his wife never to open the door when he was outside, no matter what noises she heard, because he uttered cries and howls that would freeze the blood of any Christian, no one could sleep a wink, but one night, his wife screwed up her courage, because women are very curious and always want to know everything, and resolved to open the door. And what did she see, oh dear God, she saw before her a huge pig, like a rampant boar, with a head this size, this big, and it hurled itself at her like a lion ready to devour her, but luckily she managed to slam the door shut, although not before the pig had bitten off a piece of her skirt, and imagine her horror when her husband returned home at dawn with that same piece of cloth still in his mouth, but at least it gave him an opportunity to explain that whenever he went out on Tuesday nights, he was changed into an animal, and that night he had been a pig, and he could have done her real harm, so next time she must on no account open the door, because he couldn't answer for his actions, How dreadful, Anyway, his wife went to speak to her in-laws, who were most upset to learn that their son had become a werewolf, because there weren't any others in the family, and so they went to a holy woman who recited the prayers of exorcism appropriate to such cases, and she told them that the next time he was changed into a werewolf, they must burn his hat, and then it would never happen again, and this proved to be a sovereign remedy, because they burned his hat and he was cured, Do you think burning his hat cured him because the sickness was in his head, I have no idea, the woman never said, but let me tell you of another, similar case, a man and his wife lived on a farm near Ciborro, why these things only happen between couples, I don't know, where they raised chickens and other livestock, and every night, because it happened every night, her husband would get out of bed, go into the garden and start clucking, can you imagine, and when his wife peered around the door, she saw that he had been turned into a huge chicken, What, the same size as that pig, You may laugh, but just hear me out, this couple had a daughter, and when their daughter was about to get married, they killed a lot of chickens for the wedding feast, because that was what they had most of, but that night, the wife didn't hear her husband get out of bed or hear him clucking, and you'll never guess what happened, the man went to the place where the chickens had been killed, picked up a knife, knelt down by a bowl, and stuck the knife in his own throat, and there he stayed until his wife woke to find the bed empty, went in search of her husband and found him dead in a great pool of blood, you see, like I said, it's the fates.

Domingos Mau-Tempo went back to his old ways, wine, idleness, beatings, fights and insults. Mama, is Papa cursed, Don't say such things about your father. These are words often spoken in such circumstances, and neither those intended as an accusation nor those intended to absolve should be taken seriously. Poverty was casting a dark shadow over the faces of these people, and the children who were old enough to do so went begging. However, there are still some kind, conscientious people, such as the owners of the house in which the Mau-Tempo family lived, who often gave them food, but children can be cruel, and although when bread was being baked in the owners' house they always reserved a bread roll for João Mau-Tempo, the boys of the family, who went to the same school and were all friends, used to play a practical joke on João Mau-Tempo, tethering him with a rope to the trough with the bread roll before him and refusing to let him go until he had eaten it. And people say there's a God.

Then, what had to happen, happened. Domingos Mau-Tempo reached the last of his misfortunes. One afternoon, he was sitting on his bench polishing the heel of a shoe when he suddenly put everything down, untied his apron, went into the house, made up a bundle of clothes, took some bread out of the bread bin, put everything in a knapsack and left. His wife was working, along with her two youngest children, João was at school, and the other one was idling about somewhere. This was the last time Domingos Mau-Tempo left home. He will still appear to say a few words and to hear others, but his story is over. He will spend the next two years as a wanderer.

N ATURE DISPLAYS REMARKABLE callousness when creating her various creatures. Apart from those who die or are born crippled, some do manage to escape and thus guarantee the results of nature's engeneration, to coin an ambivalent and therefore equivocal noun that combines generation and engendering, with just the right cozy margin of imprecision that surrounds the many mutations of what one says, does and is. Nature does not itself parcel out the land, but uses the system to its advantage. And if after harvest time the granaries of the thousand anthills of the fields are not all equally full, the profits and losses feed into the great accounting department of the planet and no ant is left without its statistical quota of food. In the settling of accounts it matters little that millions of ants have died from being flooded out, dug up or urinated on: those who lived ate, and those who died left the others behind. Nature doesn't count its dead, it counts the living, and when there are too many of those, it organizes a new slaughter. It's all very easy, very clear and very fair, and as far as the memory of ants and elephants can recall, no one in the animal kingdom has as yet complained.

Fortunately, man is the king of the beasts. He can therefore do his accounts with pen and paper or by other, subtler means, murmured comments, hints, glances and nods. Such mimicry and onomatopoeia come together, in cruder form, in the songs and dances of struggle, seduction and enticement that certain animals use to obtain their goals. This may help in understanding Laureano Carranca, that rigid man of principle, think only of his inflexibility, his chill disapproval of his daughter's marriage, and the game of emotional weights and measures that he practiced daily, now that he has his grandson João at home with him, an act of reluctant charity, and another, much more favored grandson called José Nabiça. Let us explain why, although it won't really contribute much to our understanding of the story, only enough for us to know each other better, as the gospels urge us to do. José Nabiça was the child born to one of Sara da Conceição's sisters and a man whose anonymity consisted in everyone pretending not to know who he was, when in fact his identity was public knowledge. In such cases, there is often a general complicity, based on everyone knowing the truth but feeling curious as to how the protagonists will behave, and what's wrong with that, given how few distractions life provides. Such love children are often abandoned, sometimes by both mother and father, and consigned to the foundling hospital or left out on the road to be devoured either by the wolves or the Brothers of Mercy. Fortunately for José Nabiça, however, despite the taint of his birth, he was blessed with a father who had a little money and with grandparents who had an eye on a future inheritance, a remote possibility but of some substance nevertheless, enough to be a promise of wealth for the Carranca family. They treated João Mau-Tempo as if he wasn't of the same blood at all, and so he, as the son of a cobbler-turned-vagrant, would inherit neither money nor land. The other grandson, though he was the son of a sin unpurged by marriage, was treated like a prince by his grandfather, who remained deaf to what people said and blind to the evidence of his daughter's besmirched honor, and all because he had hopes of a legacy that never materialized. Proof perhaps that divine justice does exist.

João Mau-Tempo had more than a year of schooling, and that was the end of his education. His grandfather eyed that skinny little body, pondered for the nth time those blue eyes that were immediately lowered in fright, and decreed, You're to help your uncle in the fields, so behave yourself, because if you don't, you'll feel the weight of my hand. By work in the fields he meant clearing land and digging, a kind of brute labor quite unsuitable for a child, but it was as well for him to find out now what his place in the world would be when he grew up. Joaquim Carranca was himself a brute, and would leave João out all night in the fields, on guard in the cabin or on the threshing floor, when such duties were completely beyond the strength of a child. Worse still, during the night, out of pure malice, he would go and see if the boy was sleeping and then throw a sack of wheat on top of him and make the boy cry, and as if that were not enough, or, indeed, too much, he would prod him with a metal-tipped stick, and the more his nephew screamed and wept, the more the heartless wretch would laugh. These things really happened, which is why they're hard to believe when set down as fiction. In the meantime, Sara da Conceição gave birth to another daughter, who died eight days later.

There were rumors in Monte Lavre that a war was being waged in Europe, a place that few people in the village knew much about. They had their own wars to wage, and not small ones either, working all day, when there was work, and feeling sick with hunger all day, whether there was work or not. Not quite so many people died though, and generally speaking, any corpses entered the grave in one piece. However, as previously announced, the time had come for one of them to die.

When Sara da Conceição heard that her husband had been seen in Cortiçadas, she gathered together the children who lived with her and, putting little trust in her father's ability to protect

João, she picked him up en route and sought shelter in the house of some relatives, the Picanços, who were millers in a place called Ponte Cava, about half a league away, the place taking its name from the bridge that crossed the river there. The bridge in question, however, was now nothing but a crumbling arch and some large boulders on the riverbed, but João Mau-Tempo and the other children would bathe naked there, and when João lay on his back staring up at the sky, all he could see was sky and water. It was there in Ponte Cava that the family chose to hide, fearful of the threats emanating from Cortiçadas via the mouths of well-known tattletales. Domingos Mau-Tempo might never have come to Monte Lavre if the messenger, on his return journey, had not told him that his family had fled in terror. One day, he slung a saddlebag over his shoulder and, blinded by fate, set off along cart tracks and across plains, and when he reached the mill, he stood outside, demanding satisfaction and the return of his family. José Picanço came out to speak to him while, in the depths of the house, his wife kept guard over the refugees. Domingos Mau-Tempo says, Good morning, Picanço, And José Picanço says, Good morning, Mau-Tempo, what do you want. And Domingos Mau-Tempo says, I've come for my family, who, it seems, have run away from me, and someone told me that they're living in your house. And José Picanço says, Whoever told you that was quite right, they are living in my house. And Domingos Mau-Tempo says, Then send them out to me, because my wandering days are done. And José Picanço says, Who are you trying to fool, Mau-Tempo, you certainly can't fool me, I know you too well. And Domingos Mau-Tempo says, They're my family, not yours. And José Picanço says, Well, they're certainly in far better hands here, anyway, no one is coming out, because no one wants to go with you. And Domingos Mau-Tempo says, I'm the father and the husband. And José Picanço says, Get out of here, I saw how you treated your honest, hard-working wife when we were neighbors, and your poor children, and the misery you put them through, in fact, if it hadn't been for me and a few others, they would have died of hunger, and there would be no need for you to be here now, because they would all be dead. And Domingos Mau-Tempo says, Yes, but I'm still the father and the husband. And José Picanço says, Like I said before, get out of here and go where no one can hear or see or speak to you, because you're a hopeless case, a lost cause.

It's a beautiful day. A sunny morning after rain, because we're in autumn now, you see. Domingos Mau-Tempo draws a line on the ground with his stick, an apparent challenge, a sign that he is ready to fight, at least that is how Picanço interprets it, and so he, too, picks up a stick. These are not his problems, but often a man cannot choose, he simply happens to find himself in the right place at the right time. At his back, behind the door, are four frightened children and a woman who, if she could, would defend them with her own body, but the forces are so unequal, which is why Picanço draws his own line on the ground. He needn't have bothered. Domingos Mau-Tempo says nothing, makes no other gesture, he is still absorbing what has been said to him, but if he is truly to absorb it, he cannot stay there. He turns and goes back the way he came, taking the path that follows the river past Monte Lavre. Someone sees him and stops, but he doesn't respond. He might perhaps murmur, Wretched bloody place, but he says it with great sadness, with the grief of having been born, because he has no particular reason to hate this place, or perhaps all places are wretched and all are cursed, condemned and condemning. He goes down a grassy slope, crosses a fast-flowing stream via three steppingstones and climbs up the bank. There is a hill opposite Monte Lavre, each man has his mount of olives and his reason for going there. Domingos Mau-Tempo lies down in the sparse shade and looks up at the sky without knowing that he's looking at it. His eyes are dark, as deep as mines. He isn't thinking, unless thought is this slow parade of images, back and forth, and the occasional indecipherable word dropping like a stone that suddenly rolls for no reason down a hillside. He sits up and leans on his elbows, Monte Lavre is there before him like a nativity scene, at its highest point, above the tower, a very tall man is hammering at the sole of a shoe, raising his hammer and bringing it crashing down. Fancy seeing such things, and he's not even drunk. He is merely sleeping and dreaming. Now it's a cart passing by, piled with furniture and with Sara da Conceição perched precariously on top, and he is the one who's going to have to be the mule, fancy hauling all that weight, Father Agamedes, and around his neck is a bell without a clapper, he shakes it hard to make it ring, it must ring, but it's made of cork, oh, to hell with mass. And coming toward him is cousin Picanço, who removes the bell and replaces it with a millstone, you're a hopeless case, a lost cause.

He felt as if he had spent the whole afternoon daydreaming like this and yet it took only a few minutes. The sun has barely moved, the shadows haven't changed. Monte Lavre has neither grown nor shrunk. Domingos Mau-Tempo got up, ran his right hand over his beard and, when he did so, a piece of straw got caught in his fingers. He rubbed it between his fingertips, broke it in two and threw it away. Then he put his hand into his bag, produced a length of rope and walked in among the olive trees, out of sight now of Monte Lavre. He walked, looking about him as he went, like a landowner sizing up the harvest, he calculated heights and resistance, and finally decided where he would die. He slung the rope over a branch, secured it well, then climbed onto the branch, put the noose about his neck and jumped. No hanged man ever died so quickly.

JOÃO MAU-TEMPO IS now the man of the house, the old-est son. The firstborn with no firstborn's legacy, the owner of nothing at all, he casts a very brief shadow. He clomps around in the clogs his mother bought for him, but they're so heavy that they fall off his feet, and so he invents some rough-and-ready suspend-ers, which he loops under the soles of the clogs and through the holes he has made in his trouser bottoms. He cuts a grotesque fig-ure, with his mattock, much larger than him, over his shoulder, as he rises from his thin mattress at dawn, in the cold, oily light of the lamp, so confused, so heavy with sleep, so clumsy in his ges-tures, that he probably leaves his bed with the mattock already on his shoulder and his clogs on his feet, a small, primitive machine capable of only one movement, raising the mattock and letting it fall, heaven knows where he gets the strength. Sara da Conceição said, Son, they've given me work for you so that you can earn a little money, because life is hard and we have no one to help us. And João Mau-Tempo, who already knows about life, asked, Shall I go and dig, Mama. If she could, Sara da Conceição would have said, No, my son, you're only ten years old, digging is no work for a child, but what is she to do when there are so few ways of earning a living on the latifundio and when his dead father's trade proved so ill fated. It is still pitch-black when João Mau-Tempo gets up, but luckily for him, his path to the farm of Pedra Grande passes through Ponte Cava, a fortunate place for him despite all, the place where they, poor things, were saved from the wrath of Domingos Mau-Tempo, indeed, a doubly fortunate place because, even though his father killed himself in that cruel fashion, and despite his many sins, if that shoemaker is not at God's right hand, then there is no such thing as mercy. Domingos Mau-Tempo was a sad, unfortunate wretch, so let not good souls condemn him. His son, then, is setting off in the dim light of a still distant sun when Picanço's wife comes out to meet him and says, So, João, where are you off to. The blue-eyed lad answers, I'm off to Pedra Grande to clear the fields. And Picanço's wife says, You're far too small to use a mattock and the weeds are far too tall. One can see at once that this is a conversation between poor people, between a grown woman and a man still growing, and they speak of these lowly and insubstantial matters because, as you have seen, they are rough-and-ready types, with no education to enlighten them, or if they have, any light once shed is rapidly burning out. João Mau-Tempo knows what answer he will give, no one taught it to him, but any other reply would be out of time and place, That may be so, but I have to help my poor mother, well, you know what our life is like, and my brother Anselmo is going out begging for alms so that he can bring me something to eat in the fields, because my mother hasn't even enough money to buy food. Picanço's wife says, You mean you're going off to work without anything for your lunch, you poor lad. The poor lad answers, Yes, Senhora, I am.

This would be an appropriate moment for a Greek chorus to declare its horror and to create a suitably dramatic atmosphere for large, generous gestures. The best charity is that which one poor person gives to another, for, that way, at least it's between equals. Picanço was working in the mill and his wife called to him, Come

here, husband. He came, and she said, Just look at João here. They
had the same conversation over again, and it was decided there and
then that on the days when he worked at Pedra Grande, he should
stay in their house, and Picanço's wife, like the good woman she
was, filled his lunch basket with food. She, too, is seated at God's
right hand, doubtless in earnest conversation with Domingos
Mau-Tempo, as they try together to understand why misfortune
so outweighs reward.

João Mau-Tempo earned two tostões, which would have been
the wage of a grown man four years earlier, but which was now a
pittance, given how expensive life had become. He benefited from
the good graces of the foreman, a distant relative, who pretended
not to notice the boy battling with the roots of the weeds, far too
tough for a small child. He spent the whole day, hours on end, half
hidden in the undergrowth, slicing away with the mattock at those
recalcitrant roots, why, Lord, do you make even children suffer so.
Foreman, what's that boy doing here, you're not going to get much
work out of him, commented Lamberto one day as he was passing. And the foreman answered, We took him on out of kindness,
sir, his father was that wretch Domingos Mau-Tempo. I see, said
Lamberto, and went into the stables to visit his horses, of which
he was very fond. It was warm in there and smelled of straw, This
one is called Sultão, this one Delicado, this one Tributo and this
one Camarinha, and this as yet unnamed colt will be called Bom-
Tempo, Fair-Weather.

When the land had been cleared, João returned to his mother's
house. But he was in luck, because just two weeks later, he had
found work again, on an estate belonging to another man, Norberto by name, and under the orders of a foreman called Gregório
Lameirão. This Lameirão fellow was an utter brute. For him, the
temporary workers were a mutinous rabble who would only respond to the stick and the whip. Norberto saw none of this, and

yet he was said to be an excellent person getting on in years, a white-haired gentleman with a distinguished bearing and a large family, who were refined folk, albeit of the country kind, and who went sea bathing in Figueira in the summer. They owned property in Lisbon, and the younger members of the family were gradually moving away from Monte Lavre. The world lay before them like a vast landscape, although they knew this only by hearsay, of course, and the time was approaching when they would take their feet out of the mud and go in search of the paved streets of civilization. Norberto did not oppose them, and this new trend in his descendants and their collaterals even gave him a certain modest contentment. Thanks to cork trees and wheat, acorns and grubbing pigs, the latifundio rewarded the family with large surpluses, which were quickly converted into money, as long, of course, as the day laborers played their part, they and all the others. That is what the foremen were for, like rustic copies of Lieutenant Contente, with no right to a horse or a saber but invested with just as much authority. With a slender cane under his arm, which he used as a horsewhip, Gregório Lameirão would walk along the line of workers, keeping an eagle eye out for the slightest sign of slacking or sheer exhaustion. Fortunately he was a man who stuck to the rules and used his own sons as examples. They all suffered there, the younger ones, that is, because hardly a day passed without one of them getting a sound beating, or two or three if their father was in his angry vein. When Gregório Lameirão set out from his house or barracks, he left his heart hanging behind the door and thus walked with a lighter step, his only desire being to deserve the boss's confidence in him and to earn the larger wage and better food that were his due as foreman and scourge of his troops. He was also an arrant coward. Once, the father of one of his unfortunate victims met him on the road and made it quite clear that if he unjustly punished his boy one more time, he would see, if he could still see, his own brains spattering the door of his house. The threat worked in that case, but this only meant that he increased the number of punishments he meted out to the others.

In Norberto's household, the ladies had all the refinements of the female sex, they drank tea, knitted, and were godmothers to the daughters of the maids closest to them. Fashion magazines lay on the sofas in the living room, ah, Paris, a city the family was determined to visit once there was an end to this stupid war, which, quite apart from other inconveniences of a greater and lesser degree, was delaying their plans. It is not in our power, of course, to do anything about such problems. And when old Norberto listened to his foreman giving his mumbled report on how the work on the land was going, a report whose sole object was to make himself look good, Norberto would grow as impatient as if he were listening to communiqués from the front. His imperial tendencies, and perhaps some trace memory of the birthplace of Lamberto Horques, who might well have been his ancestor, meant that he was a natural Germanophile. And one day, in a playful spirit, he said as much to Gregório Lameirão, who simply stared at him, eyes wide, not understanding what he had heard, because he was stupid and illiterate. Just in case, he humbled himself still more and was even more rigorous with his workers. His oldest sons now refused to work for him and went in search of employment on other estates, which offered more humane foremen and more security, although more security meant only that they would not die quite so soon.

These were good times for discipline. Sara da Conceição, who, understandably enough, could not forget the bad example set by her husband nor the worm of guilt that gnawed away inside her for the unfortunate manner of his death, was always saying, João, if you don't toe the line, I'll give you a sound beating, we've got a living to make. That is what his mother told him, a sentiment reinforced by Lameirão, who used to say, According to your mother, all she wants from you are your bones to make a chair with and your skin to make a drum. When two such authorities were so clearly of one mind, what could João do but believe them. But one day, worn down by beatings and overwork, he braved the threat of being flayed and boned, and spoke frankly to his astonished mother. Poor Sara da Conceição, who knew so little of the world. Amid screams and sighs, she said, That wretched man, I never said any such thing, a mother doesn't give birth to a child in order to be the death of him, oh, how the rich despise the poor, that monster doesn't even love his own children. But we ourselves have said as much before.

João Mau-Tempo is not the stuff of heroes. He's a skinny little ten-year-old runt, a scrap of a boy who still regards trees as shelters for birds' nests rather than as producers of cork, acorns or olives. It's unfair to make him get up when it's still dark and have him walk, half asleep and on an empty stomach, the short or long distance to wherever his place of work happens to be, and then slave away all day until sunset, only to return home, again in the dark, mortally tired, if something so like death can be called tiredness. But this child, a word we use only for convenience's sake, because this is not how the latifundio categorizes its population, people are either alive or dead, and all one can do with the dead is bury them, you certainly can't make them work, anyway, this child is just one among thousands, all the same, all suffering, all ignorant of what evil they committed to deserve such a punishment. On his father's side, he comes from tradesmen's stock, his father a shoemaker, his grandfather a carpenter, but see how destinies are forged, there is no bradawl here, no plane, nothing but dry earth, killing heat, deathly cold, the great droughts of summer, the bone-deep chill of winter, the hard morning frost, lace, Dona Clemência calls it, cracked, bloody, purple chilblains, and if that swollen hand rubs against a tree trunk or a stone, the soft skin opens, and who can say what misery and pain lies beneath. Is there no other life than this drudgery, an animal living on the earth alongside other animals, the domestic and the wild, the useful and the harmful, and he himself, along with his human brothers, is treated as either harmful or useful, depending on the needs of the latifundio, now I want you, now I don't.

And sometimes there is no work, first the youngest are dismissed, then the women, and finally the men. Caravans of people set off along the roads in search of a miserable wage somewhere else. At such times, there's not a foreman or an overseer to be found, far less a landowner, they're all shut up in their houses, or far away in the capital or some other hiding place. The earth is either a dry crust or pure mud, it doesn't matter. The poorest boil up some weeds and live on those, their eyes burn, their stomachs bloat, and this is followed by long, painful bouts of diarrhea, the sense that the body is letting go, detaching itself, becoming fetid, an unbearable weight. You feel like dying, and some do die.

As we said before, there is war in Europe. And war in Africa too. But these things are like shouts from a hilltop, you know you shouted, and sometimes it might be the last thing you do, but down below, that shout grows fainter and fainter until it vanishes into nothing. Monte Lavre hears about these wars from the newspapers, but they are only for those who can read. When those who can't read see prices going up or basic foodstuffs running short, they ask why, It's the war, say those in the know. War ate a great deal and war grew fat and rich. War is a monster who empties men's pockets, coin by coin, before devouring the men themselves, so that nothing is lost and all is changed, which is the primary law of nature, as one learns later on. And when war has eaten its fill, when it is sated to the point of vomiting, it continues its skillful pickpocketing, always taking from the same people, the same pockets. It's a habit acquired in peacetime.

In some places, people put on mourning clothes because a relative had died in the war. The government sent condolences, deepest sympathies, and spoke about the nation. The usual mentions were made of Afonso Anriques and Nuno Álvares Pereira,* about how we Portuguese were the ones who discovered the sea route to India, and how Frenchwomen have a weakness for our soldiers, but nothing was said about African women apart from what we know already, the czar was deposed, the neighboring powers are concerned about the situation in Russia, there's a big offensive on the western front, aviation is the weapon of the future, but the infantry still reigns supreme in battle, you can't do anything without the artillery, dominion of the seas is indispensable, revolution in Russia, Bolshevism. Adalberto read his newspaper, looked anxiously out at the foggy weather, shared the indignation expressed in the newspaper and said out loud, It will pass.

It isn't all roses for one side or the other, although, as we have explained before, the distribution of thorns is made according to the old familiar rules of disproportion and gives the lie to the dictum, which may be true in the world of navigation, The larger the ship, the bigger the storm. On land, it's different. The Mau-Tempo family have only a tiny, flat-bottomed boat, and it's only by chance and because of the demands of the story that they haven't all drowned. However, their small craft was giving every sign of breaking up on the next reef or the next time the store cupboard was empty, when, unexpectedly, Sara's brother, Joaquim Carranca, lost his wife. He wasn't of a mind to remarry, nor did he have a list of potential brides, plus he had three children to bring up and a very bad temper, but hunger joined forces with a desire to eat, and this prompted brother and sister to unite lives and children. It balanced out nicely, the brother provided a new father, the sister a new mother, but it was all kept in the family, so let's see how things turned out. It wasn't any worse than what could have happened,

---

* Afonso Anriques, or Henriques, was the first king of Portugal, nicknamed "the Conqueror." Nuno Álvares Pereira (1360–1431) was a Portuguese military leader who played a crucial role in assuring Portugal's independence from Castile.

and possibly better. The Mau-Tempo children stopped begging from door to door, and Joaquim Carranca had someone to wash his clothes, which is something every man needs, and, in addition, someone to look after his children. And since it is not the custom for brothers to beat their sisters, or at least not as often as it is for husbands to beat their wives, this was the beginning of better days for Sara da Conceição. Some may consider this to be very little. They, we would say, clearly don't know much about life.

EVERY DAY HAS ITS story, a single minute would take years to describe, as would the smallest gesture, the careful peeling away of each word, each syllable, each sound, not to mention thoughts, which are things of great substance, thinking about what you think or thought or are thinking, and about what kind of thought it is exactly that thinks about another thought, it's neverending. It would be best to say that for João Mau-Tempo, these years will provide his professional education, in the traditional country sense that a workingman has to know how to do everything, from scything to harvesting cork, from clearing ditches to sowing seeds, and he needs a good strong back for carrying loads and for digging. This knowledge is transmitted across the generations with no examinations and no discussions, and it has always been the same, this is a hoe, this is a scythe, and this is a drop of sweat. It is also the thick white saliva you get in your mouth on furnace-hot days, it's the sun beating down on your head, and your knees going weak with hunger. Between the ages of ten and twenty you have to learn all this very fast, or no one will employ you.

Joaquim Carranca remarked one day to his sister how good it would be to find a boss who would take them all on, and she agreed, a habit born of years as a submissive married woman, but in this case what flickered before her was the hope of spending a whole year safe from unemployment, that would be her one modest but sure ambition, for they could hardly aspire to anything more. At this time, three brothers inherited Monte de Berra Portas following the death of the old owner, their father, who had sowed his seed in the womb of a very canny mistress, who, while appearing to submit to the patriarch's terrible whims and to his thunderous rants and rages, had gradually tamed him, like a lamb, so much so that he agreed, at the last, to disinherit his closest relatives in favor of his three natural sons. Pedro, Paulo and Saul took turns presiding over the estate, each taking a different season, and when Pedro was giving the orders, the other two obeyed, a system that could have worked well if each brother hadn't chosen to spy on his other siblings, with Saul declaring that when he wasn't in charge, the household went to rack and ruin, with Paulo stating that he was the only really capable administrator, and with all three becoming embroiled in domestic alliances and plots, as often happens in families. The story of this triumvirate would, alone, be enough to make an opera. And then there was the mother, who screamed that she had been plundered by her own sons, or to speak more plainly, robbed, after all she had done for them, putting up with being the servant of that old pig and now finding herself the slave of her own children, who kept her short of money and a virtual prisoner in the house. At night, when the countryside drew the silence up about it like a blanket, the better to hide itself away in the great secrets of the dark, you would hear what sounded like a sow having its throat slit and the loud stamping of feet, it was the war between mother and sons.

Joaquim Carranca found employment with these bosses, and João Mau-Tempo worked as a day laborer. All in all, they earned a pittance, enough, just about, for them not to be constantly hungry, but there was at least the advantage that they could all be to-

gether and have access to a vegetable patch where they could break
their backs toiling away on high days and holidays. Joaquim Carranca's wage at this time consisted of lodging, firewood, sixty kilos
of maize flour, three liters of olive oil, five liters of cowpeas, one
hundred escudos and, at the end of the year, a modest handout.
As for the younger members of the family, they earned forty kilos
of maize flour, a liter and a half of olive oil, three liters of cowpeas
and fifty escudos. And so it went on, month after month. They
would take their sacks and bags to the granary, their jug to the cellar, where the foreman would measure out their rations of food
and oil, and the administrator would pay their wages, and that was
all they had to keep body and soul together and to recoup the energy expended every day. Of course, not all of them did recover,
and they accepted this, time would inevitably take its toll, the skull
beneath the skin becoming ever more evident, but then we are all
born in order to die. Joaquim Carranca died, without having had
a single day's illness, after coming back from working in his vegetable patch on one of those Sundays when it's easy to believe in
the existence of God, even without the aid of Father Agamedes, it
was just a shame that the mattock was so heavy that he had to sit
down on a log at the front door, feeling unusually tired, and when
Sara da Conceição came out to tell her brother that supper was
ready, he had lost all appetite. There he was, eyes wide, his hands
open on his lap, more peaceful than he could ever have dreamed of
being when alive, and he wasn't a bad man really, despite his sudden rages, despite his cruelty to his oldest nephew, what's done is
done. Death is like a great strickle that passes over the measuring
jug of life, discarding any excess, although it is often hard to make
out what exactly its criteria are, as in the case of Joaquim Carranca,
who was still needed by his family.

Life, or whoever rules over life, with either a sure or an indifferent hand, expects us to acquire both our professional and our sentimental education at the same time. This conjunction is clearly a

mistake, doubtless made necessary by the brevity of life, which is not long enough for things to be done in a more leisurely, timely manner, which means that one neither acquires enough nor feels enough. Since the world was not going to change its ways, João Mau-Tempo, as he acquired his working skills, also went courting in the local villages and dancing wherever the sound of an accordion was to be heard, and he was a good dancer too, and, who would have thought it, much sought after by the girls. As we know, he had inherited his blue eyes from that ancestor of four hundred years before, the same one who, not far from here, lying on the forebears of this same bracken, raped a young girl who had gone to the well for water, watched by birds whose plumage remains unchanged, and who gazed down on the pair struggling amid the greenery, a scene with which those creatures of the air had been familiar since the world began. And his blue eyes troubled the hearts of the young girls, which would melt when those eyes grew suddenly dark, though he himself was unaware of any ancient amorous rage rising up in him, such is the hidden force of past actions. Ah, youth. The fact is that João Mau-Tempo may have flirted a lot but he rarely went further than that. When he had had a few drinks, he might touch a girl rather more boldly or give her a clumsy kiss devoid of all the knowledge that the century was gradually accumulating for future general use.

In the eclogues of old, the shepherds played their lutes and the shepherdesses wove their garlands of flowers, but in this modern version, João Mau-Tempo, during a ten-week contract that took him off to Salvaterra to cut cork, ate a whole string of garlic in the hope of preserving himself from the mosquitoes, as a result, you could smell him ten paces away. He was learning the cork trade in the hope that he might one day earn the eighteen escudos paid to master cork cutters, and fortunately enough, he was far from his would-be girlfriends, who, while they might have been pretty tolerant of most smells, would perhaps have drawn the line at garlic. Happiness, as we know, depends on such small details.

And now João Mau-Tempo has received his call-up papers. He is full of daydreams, he imagines himself far from Monte Lavre, in Lisbon perhaps, having completed his military service, only a fool would miss the chance to find a job on the trams or on the police force or with the national guard, he has a smattering of education, he just has to push himself forward, he wouldn't be the first. Call-up day is a day of celebration, with fireworks and wine, the young lads who finally deserve to be called men are all there in their freshly washed clothes, and when they're lined up stark naked, they make macho jokes to disguise their embarrassment and stand at attention, red-faced, to answer the doctor's questions. Then the draft board meets and makes its selections. A few men were chosen, and of the four who weren't, only one went away downhearted. That was João Mau-Tempo, who watched his dream of wearing a uniform vanish into the realm of the impossible, his dream of standing on the platform of a tram, ringing the bell, or becoming a policeman and policing the streets of the capital, or, as a guard, guarding, ah, but on whose behalf, the very fields in which he labored now, and he found this possibility so troubling that it helped him get over his disappointment. One cannot think of everything all at the same time.

So what is João Mau-Tempo to do? He has just turned twenty, he has been let off military service, he hasn't filled out much since the days when he, tiny as a dwarf, battled with the weeds in Pedra Grande and ate the maize porridge that Picanço's wife used to make for him out of familial charity. In Salvaterra, he buys his first cape and struts about in it like a tomcat with its tail in the air. It's very full and reaches down to his heels, but the village doesn't expect people to be dressed in the height of fashion, he has reached heights enough simply by owning a new item of clothing, regardless of what it's like. When João Mau-Tempo plunges his mattock into the earth, he thinks about that cape, about the dances he goes to, about the girlfriends in his life, some more serious than others, and he forgets the pain of living here, bound to this place, so far from Lisbon, if he ever really had aspired to living there, if that wasn't all just a youthful dream, for what else is youth for but to dream.

A time of great storms is approaching, some will arrive with their natural boom and bluster, others more quietly, without a shot being fired, coming from far-off Braga, but we will hear more of these later on, when there is nothing to be done about them. However, although one should deal with each event in its proper order, and although, as we feel we should point out so as not to keep offending against the rules of storytelling, we have, in fact, already anticipated the death of Joaquim Carranca, which actually happened a few years later, let us nonetheless talk about the storm that remained fixed in people's memories for reasons of grief and loss. It was summer, ladies and gentlemen, when one doesn't really expect such things, though occasional solemn rolls of thunder boomed across the stubble, catapum, one moment distant and almost sleepy, the next flickering right above our heads and pounding the earth, whatever would we do without Saint Barbara's help. Now, the Mau-Tempo family may seem to have been singled out for grim happenings, but only someone of little understanding could possibly believe that. After all, so far only one member of the family has died, and if we're talking hunger and poverty, then any other family could serve as an example, for hunger and poverty are hardly in short supply. Besides, the uncle in question was not even a blood relative. Augusto Pintéu was married to one of Sara da Conceição's sisters, and although he was a farm laborer, he chose, in his spare time, to work as a carter. He, naturally, had his appointment with death, but how oddly things turn out, for this simple, mild-mannered, soft-spoken man met a very dramatic end, with much celestial and terrestrial brouhaha, like a character in a tragedy. This serene man did not leave life as serenely as Joaquim Carranca. And such contradictions provide much food for thought.

As we said, Augusto Pintéu also worked as a carter, traveling between Vendas Novas and Monte Lavre to be exact. The former had a train station, to which, with his pair of mules and his cart, he would take cork, coal and wood and bring back groceries, seeds and whatever else was needed, not many men enjoyed such a good life. On that day, which, being a summer's day, should have been long and bright, the sky suddenly filled with black clouds and there was an almighty thunderclap. The heavens opened and unleashed all God's store of water. Augusto Pintéu wasn't particularly worried, because these summer storms come and go, and so he continued his work of loading and unloading, fearing nothing worse than arriving home soaked to the skin. When he left Vendas Novas, night had already closed in, lit by lightning so bright that there seemed to be some celebration going on up above, some holy procession. The mules knew the route blindfold and could find and recognize it even when it was flooded, as the lower parts already were. With two thick sacks on his head to protect him, Augusto Pintéu consoled himself with the thought that, in such weather, there was, at least, little danger of being ambushed by thieves, as had happened in the past. In a storm like this, highwaymen would all be safe in their lairs, roasting their stolen slices of pork and drinking coarse wine, because they rarely stole anything else. It's three leagues from Vendas Novas to Monte Lavre, but Augusto Pintéu would not travel the last league, nor would his mules. By the time they reached the stream, the darkness had grown as black as pitch, and the waters roared and thundered loudly enough to frighten anyone. This was usually the place where, in good weather, you could ford the stream, with the water up to your knees, but for those on foot there was a broad wooden plank that went from shore to shore, past a giant ash tree that had been born there and grown up in the days before the course of the river had changed. In the midst of the water, the ash tree rustled furiously, defending with its thick roots its vital patch of earth, threatened now by the speed and force of the current. Augusto Pintéu had crossed there with his cart and his mules many times. He would not cross it again. Right at the beginning of the ford, the bed of the stream dropped away to form a deep, deep chasm, which was called, because everything has to have a name, Pego da Carriça, Wren's Pool. Augusto Pintéu put his trust in the Holy Virgin and in his mules and thus managed to reach the middle, where the water lapped against the bottom of his cart. At that point, fearing the current seething about them and fearing that he would be swept away with no hope of salvation, he tried to drive the mules upstream. They resisted as best they could, but being subject to the whip and the bit, they finally submitted. At one point, the right-hand mule lost its footing, one wheel slipped off the edge and into the chasm, and amid screams and rumbles of thunder, Augusto Pintéu and his mules, along with the cart, the groceries and the other merchandise, were all drowned, plunged forever into the thick blackness of the waters, into mortal silence. They touched bottom and there they remained, with Augusto Pintéu still held fast to the reins, and the mules to the cart, because down below the waters were absolutely calm, as if they had been like that since the world began. The following day, accompanied by the widow's screams and the orphaned children's tears, they were pulled out, thanks to some lengths of rope and the efforts of some very brave men, while a crowd, come from far and wide, gathered on the banks of the river. It had stopped raining by then. That was a summer of great afflictions. So great were the storms that men working in the cork forests fell from the trees and, as they fell, cut themselves on their axes. This is a life more filled with tribulations than one can say.

At the time, the Mau-Tempo family lived in Monte de Berra

Portas with their uncle and brother, Joaquim Carranca. The next year, when Portugal had been following the Braga road* for some six months, João Mau-Tempo, along with his siblings Anselmo and Maria da Conceição, went to work in the winter pastures for a different boss, in a place called, for some reason, Pendão das Mulheres, Ladies' Pennant. It was four long leagues away, on foot and on bad roads, from Monte de Berra Portas that is, whereas from Monte Lavre it was only a league and a half. There were quite a few girls in the party, which explained why the boys were so pleased, up there all week with those young women and only going home every other Saturday. The workers were mostly youngsters. The place was a hotbed of flirtations and dalliances, and quite a few got burned. At the time, João Mau-Tempo had a girlfriend elsewhere, but he didn't care, and pretended that he was a free agent, and his skill as a dancer made him a most attractive prospect.

What with work and romance, the weeks flew past, and then a girl from Monte Lavre joined them, a girl he knew well from having danced and sung with her countless times. But they had never been in love. Half serious, half joking, they addressed each other as Friend João and Friend Faustina, for that was her name. There would seem to be nothing more to say about them. However, this turned out not to be the case. Whether it was because of the freedom they enjoyed or because the time had come to tie that particular knot, João fell in love with Faustina, and Faustina with João. In matters of love, it can as easily bloom in diamond rings in shop windows as grow wild among the castor-oil plants, only the language differs. This love began to put down roots, and João Mau-Tempo forgot all about his other girlfriend, but since this new love was serious, they agreed to say nothing for the moment to Faustina's family, because although João Mau-Tempo himself had done

---

* On May 28, 1926, General Gomes da Costa led an uprising in Braga, which was the prelude to the so-called National Dictatorship, which, in turn, paved the way for Salazar's dictatorship.

nothing to be ashamed of, he had inherited his father's bad name, and these things stick, for as the saying goes, He who is born of a cat will run after mice. The secret nonetheless reached the ears of Faustina's family, and they made her life a misery. He can't be any good, he looks shifty with those strange blue eyes, and then there was that father of his, a loose-living, drunken fellow who only ever did one good thing, which was to hang himself. That is how some village evenings are spent, beneath the starry sky, while the male genet pursues the female genet and copulates with her amid the bracken. The lives of human beings are far more complicated, for we are, after all, human.

It was January and very cold, the sky was one solid sheet of cloud, the laborers were walking along the road toward Monte Lavre for their fortnightly rest, and as befitted a courting couple, João was talking to Faustina, and she, fearing the domestic storm awaiting her, was telling him her problems. Then suddenly they were assailed by the angry shouts and violent gestures of one of Faustina's sisters, who, given their mother's advanced age, had taken over as family spokesperson, and it was her treacherous ambush that so startled the couple. Natividade, for that was her name, said, Have you no shame, Faustina, you stubborn creature, it seems that no amount of good advice and beatings has any effect, Lord knows what will become of you. She said other things too, but Faustina did not leave João's side. Natividade stood in front of them, intending to block their path and their destiny, if it is in a sister's power to do such a thing, and it was then, so to speak, that João Mau-Tempo took hold of the world and felt its weight, because from then on, it would be a matter of world and man, house, children, the shared life. He placed one hand on Faustina's shoulder, for she would be his world, and said, trembling at his own daring, We can't go on like this, we either finish right here and now so that you don't suffer anymore, or you come and live with me in my mother's house, until I can get us a house of our own, and from now on I will do all I can to protect you. As we said earlier, the sky was one solid sheet of cloud, and it stayed like that, thus providing natural proof that the heavens care nothing about us, if they did, the clouds would have opened in glory. Because Faustina, a brave, trusting lass, the color of whose eyes and the expression of whose face we haven't described yet, said in a firm, loud voice, João, where you go, I go, if you will promise to love me and care for me always. And Natividade said, You ungrateful wretch, and with that, she turned abruptly and shot off home like an arrow to announce this latest catastrophe. The two lovers were left alone, evening was coming on, and João Mau-Tempo took her hands in his and said, I will care for you for as long as we live, in sickness and in health, but now let us go our separate ways, and when we reach the village, we'll set a time for our escape.

João Mau-Tempo's brother Anselmo and his sister Maria da Conceição were with him and had witnessed some of what had happened. He went over to them and said in a firm voice, Go to the village and tell our mother that I'm bringing my girl home with me, that I count on having her permission to do so, and that I'll explain everything later. Anselmo said, Think carefully before you act, don't get into something you can't get out of. And Maria da Conceição said, I hate to think what our mother and our uncle will say. And João Mau-Tempo said, I'm a grown man now, I've been turned down for the army, and if my future is to take a new direction, then why wait, better sooner than later. And Anselmo said, One day, Uncle Joaquim Carranca could get an idea in his head and simply go off, you know what he's like, and you're needed at home. And Maria da Conceição said, You might be doing the wrong thing. But João Mau-Tempo said, Be patient, these things happen. When they left him, Maria da Conceição had a tear in her eye.

During this time of weekly comings and goings between Pendão das Mulheres and Monte de Berra Portas, the Mau-Tempo children had lodgings in the house of Aunt Cipriana, who was the woman we saw weeping by the river after the waters of Pego da Carriça had swept her husband away. She is dressed in mourning and will remain so until she dies, many years later, lost from our sight. Her nephew's bold move, however, gave her a taste for acting as go-between, an honest one, of course, not a procuress, and as a protector of star-crossed lovers, and she never regretted this or suffered public censure for her actions. But that is another story. When João Mau-Tempo arrived, he said to his aunt, Aunt, will you please let Faustina come and meet me here until we can leave for my mother's house in Monte de Berra Portas. And Cipriana answered, Think about what you're doing, João, I don't want any problems, and I don't want to besmirch your late uncle's memory either. And João replied, Don't worry, we'll only be here until it gets dark.

This was what João agreed with Faustina afterward, when he went to meet her, and she had deliberately dawdled, well, that's only normal when you're in love, but he can't dissuade her from seeing her mother before they run away together, even if she doesn't tell her where she's going. João Mau-Tempo, not wanting to start his new life with a fortnight's growth of beard, decided to visit the barber's, where he got himself done up like a bridegroom, that is, with a clean-shaven face. Whenever such usually thickly bearded faces are shaven, they look somehow innocent, defenseless, their very fragility touches the heart. When he returned to Aunt Cipriana's house, Faustina was there waiting for him, still tearful from her sister's angry words, her father's terrible rage and her mother's grief. She had crept away unnoticed, but since her family would doubtless be scouring Monte Lavre to find out where the couple had gone, João and Faustina decided they had better make their escape as soon as possible. Cipriana said, It's going to be a very tiring journey, and we're in for a wet, dark night, take this umbrella and some bread and sausage to eat on the way, now that you've played this very unfunny joke on everyone, be sure to behave yourselves in future, that was what Cipriana said, but in her heart she was blessing them, vicariously enjoying this youthful transgression, ah, to be young again.

It was two and a half leagues from there to Monte de Berra Portas, the night had closed in completely, and rain was threatening. Walking two and a half leagues along paths that are all shadows and alarming shapes and noises, your thoughts inevitably turn to stories about werewolves, what's more, because there is no other way, they have to cross the plank bridge at Pego da Carriça. Let's say a prayer for my uncle, he was a good man and did not deserve such a sad death. The ash tree rustled gently, the water flowed like dark, whispering silk, and to think that in this very place, who would believe it. João Mau-Tempo was holding Faustina's hand, his calloused fingers trembled, he guided her beneath the trees and through the dense undergrowth and the wet grass, and suddenly, quite how they didn't know, perhaps it was due to exhaustion after so many weeks of work, perhaps to an unbearable shaking, they found themselves lying down. Faustina soon lost her maidenhood, and when they had finished, João remembered the bread and sausage, and it was as man and wife that they shared the food.

As WE HAVE SEEN, Lamberto, regardless of whether he's German or Portuguese, is not a man to work his vast estate with his own hands. When he inherited it or bought it from the friars or, since justice is blind, stole it, he found, clinging to the estate like a tree trunk to its roots, a few creatures with arms and legs who were created for precisely such a fate, by producing children and bringing them up to be useful. Even so, whether out of pragmatism, custom, etiquette or pure self-interested prudence, Adalberto has no direct contact with those who will work his land. And that is a good thing. Just as the king in his day, or the president of the republic in his, did not and does not go about bandying words and gestures with the common people in an overly familiar manner, it would seem quite wrong on a large estate, where the owner has more power than either president or king, were Floriberto to be too forward. However, this intentional reserve did allow for certain deliberate exceptions, intended as a more refined way of bending wills and attracting perfect vassals, namely, the subservient creatures who, receiving as they do both caresses and beatings, enjoy the former and respect the latter. This matter of relations between employer and employee is a very subtle thing which cannot be determined or explained in a few words, you have to be there and eavesdrop like a fly on the wall. Add to this, brute force, ignorance, presumption and hypocrisy, a taste for suffering, a large dollop of envy, guile and a taste for intrigue, and you have a perfect training in diplomacy, for anyone who cares to learn. However, a few empirical rules, tried and tested over the centuries, will help us understand such cases better.

As well as land, the first thing Lamberto needs is a foreman, the foreman being the whip that keeps order in the pack of dogs. He is a dog chosen from among the others to bite his fellow dogs. He needs to be a dog because he knows all a dog's wiles and defenses. You wouldn't go looking for a foreman among the children of Norberto, Alberto or Humberto. A foreman is, first and foremost, a servant, who receives privileges and payments in proportion to the amount of work he can get out of the pack. He is, nonetheless, a servant. He is placed among the first and the last, a kind of human mule, an aberration, a Judas, who betrays his fellows in exchange for more power and a slightly larger chunk of bread.

The biggest and most decisive weapon is ignorance. At his birthday supper, Sigisberto said, It's just as well that they know nothing, that they can't read or write or count or think, that they assume and accept that, as Father Agamedes will explain, the world cannot be changed, that this is the only possible world, exactly as it is, that they will find paradise only after death, and that work alone brings dignity and money, but they mustn't go thinking that I earn more than they do, the land, after all, is mine and when the time comes to pay taxes and contributions, I don't go to them asking for a loan, it's always been like that and always will be, if I didn't give them work, who would, it's them and me, I'm the land and they are the work, what's good for me is good for them, that is how God wanted it, as Father Agamedes will explain in simple terms, we don't want to make them even more confused than they are already, and if Father Agamedes doesn't do the trick, then we'll ask the guards to ride around the villages on their horses, just as a reminder that they exist, a message they're sure to understand. But tell me, Mama, do the guards beat the estate owners as well, You're clearly not quite right in the head, my boy, the national guard was created and is maintained in order to beat the people, But how is that possible, Mama, do you mean that the guard was made simply in order to beat the people, but what do the people do, They don't have anyone who can, in turn, beat the estate owner when he sends out the guards to beat them, Well, I think the people should ask the guards to beat the estate owner, If you want my advice, Maria, the child is slightly mad, don't let him go around saying such things, we have our work cut out as it is, keeping the guards in check.

The people were made to be hungry and dirty. People who wash regularly are people who don't work, well, maybe it's different in the cities, I don't deny that, but here on the estate they're hired to work away from home for three or four weeks, sometimes months, if that's what Alberto wants, and during that time it's a point of honor and of manhood with them to wash neither face nor hands and to remain unshaven. If they did wash and shave, if such a hypothesis were not so laughably improbable, they would be the butt of jokes from bosses and fellow workers alike. That's the great thing about this day and age, the sufferers glory in their suffering, the slaves in their servitude. This beast of the earth must remain a beast who never rubs the sleep from his eyes from morning to night, indeed the dirt on his hands, face, armpits, groin, feet, arsehole must be for him the glorious aura surrounding work on the latifundio, man must be lower than the beasts of the field, for they, at least, lick themselves clean, man, however, must degrade himself so that he respects neither himself nor his fellows.

More than that. The workers boast of the beatings they get when working the land. Each beating is a medal to be bragged about at the inn, between drinks, I got beaten *x* number of times when I was working for Berto and Humberto. That's a good worker for you, one who, when he gets whipped, will show off his raw welts, and if they're bleeding, all the better, these are the same sort of boasts that the urban rabble make, taking as proof of their virility the number of cankers and sores acquired from their labors in a hired bed. Ah, you people preserved in the grease or honey of ignorance, you have never lacked for exploiters. So, work, work yourself to death, yes, die if necessary, that way you'll be remembered by the foreman and the boss, but woe betide you if you get a reputation for being an idler, no one will ever love you then. You can go and stand at the doors of inns with your companions in misfortune, and they, too, will despise you, and the foreman, or the boss, if he deigns to notice, will eye you with disgust and you'll be given no work, just to teach you a lesson. The others have already learned their lesson, they go off every day to slave away on the latifundio, and when you get home, if the hovel you live in can be called a home, how are you going to explain that you have no work, that the other men have but you haven't. Mend your ways while there's still time, and swear that you've taken twenty beatings, crucify yourself, hold out your arm to be bled, open your veins and say, This is my blood, drink it, this is my body, eat it, this is my life, take it, along with the church's blessing, the salute to the flag, the march past, the handing over of credentials, the awarding of a university diploma, thy will be done on earth as it is in heaven.

Ah, but life is a game too, a playful exercise, playing is a very serious, grave, even philosophical act, for children it's part of growing up, for adults it's a link with their childhood, advantageous for some. Whole libraries of books have been written on the subject, all of them solid, weighty tomes, only a fool could fail to be convinced. The mistake lies in thinking that such profundity can be found only in books, when in fact a quick glance, a moment's attention, is all it takes to see how the cat plays with the mouse, and how the latter is eaten by the former. The question, the only one that matters, is knowing who exploits the initial innocence of the game, this game that was never innocent, for example, when the foreman says to the workers, Let's run, and see who gets there last, And the innocents, blind to the obvious deceit, run, trot, gallop, stagger from Monte Lavre to Vale de Cães, merely for the glory of arriving first or for the smug satisfaction of not being last. Because the last man, well, someone always has to be last, will have to put up with the jeers and mockery of the winners, who are already panting and breathless, they haven't even started work yet but the poor fools waste their breath on this explosion of scorn. Poor João Mau-Tempo won the booby prize, not that anyone knows what that is, a prize that marks you out as idle or not being fast enough on your feet, that says you're not a man but a mere nothing. Portugal is a country of men, there's certainly no lack of them, only the one who comes last in the race is not a man, get away, you lazy brute, you don't even deserve the bread you eat.

But the games have not ended. The last to arrive, if he has any self-respect, will offer to carry the first load, well, it's some compensation. The pile of wood that will eventually become charcoal is being prepared, and having placed a sack on your back to dull the pain to come, you say, Give me that big trunk, I'll carry that. The foreman is watching, you have to prove to your colleagues that you're as good a man as they, and besides, you can't afford to be without work next week, you have children to think of, and then, groaning with effort, two men lift the trunk, they're not your children but it's as if they were, and they place the trunk on your shoulders, you kneel down like a camel, if you've ever seen one, and when you feel the weight, your knees sag, but you grit your teeth, brace your back and gradually draw yourself up, it's a huge trunk, like the leg of a giant, it feels like a hundred-year-old cork oak on your shoulders, you take the first step, and how far away that pile of wood is, your colleagues are watching, and the foreman says, Let's see if you can do it, if you can you're a brave man. That's what it's about, being brave, bearing the weight of that trunk and the pain in your creaking spine and in your heart, just so as to look good in the foreman's eyes, who will say to Adalberto, He's a brave fellow that Mau-Tempo, although it could be any other name, you should have seen the piece of wood they gave him to carry, sir, it really was a sight to see, oh yes, he's a real man all right. Possibly, but so far you've taken only three steps. What you really want to do now is put the load down on the ground, at least that's what your tormented body is asking. Your soul, your spirit, if you have the right to one, tells you that you can't, that you would rather die than be humiliated in your own village and dubbed a weakling, anything but that. People have been going on for two thousand years or more about how Christ carried the cross to Golgotha with help from the Cyrenian, but no one has a word to say about this crucified man who dined last night on very little and has had almost nothing to eat today, and he's still only halfway there, his vision grows blurred, it's a real torment, ladies and gentlemen, with everyone watching and shouting, He can't do it, he can't do it, and although you have ceased to be yourself, at least you haven't yet been reduced to being an animal, a great advantage, because an animal would have fallen to its knees, crushed by the load, but you haven't, you're a man, the dupe at the universal gaming table, why not place a bet, your wage may not pay you enough to feed you, but life is this merry game, He's nearly there, you hear someone say, and you feel as if you were not of this world, carrying a load like that, have pity on me, help me, comrades, if we all carried it together it would be so much easier, but that's not possible, it's a matter of honor, you would never again speak to the man who helped you, that is how deceived you are. You deposit the trunk in precisely the right place, a huge achievement, and your comrades all cheer, you're no longer the last in the race, and the foreman says gravely, Well done, man. Your legs are shaking, you're as exhausted as an overladen mule, you have difficulty breathing, you have a stitch, dear God, it's not a stitch, you fool, what you have is a strain, a pulled muscle, you don't even know the words, you poor creature.

Work and more work. Now they travel far from Monte Lavre, some take their families with them, to work as charcoal burners in the area around Infantado, those men without wives bed down in this big hut, and those who brought their wives set up house in another, using mats or cotton curtains or improvised panels to separate the couples, with the children, if they have them, sleeping with their parents. The midges bite furiously, but it's worse during the day, when the mosquitoes come in clouds, so many you can barely see, and they fall upon us, whining, like a rain of ground glass, our grandmothers, who knew so much about life, were quite right when they said, I'll never see my grandchildren again, they'll die far from home. They know, these are not things one forgets, that the children's little bodies will become a running sore, a torment to them, little lepers who will lie down among rags at night, their stomachs crying out for food, it's never enough, they're growing up without any consolation from their parents, who very slowly touch each other, move and sigh, as if this were something they had to do in order to keep their senses more or less placated, while beside them another couple echoes that touching, moving and sighing, either because they want to or by suggestion, and all the children in the great hut lie listening, eyes open, experiencing their own gestures and disappointments.

From the tops of these hills, on a clear day, you can see Lisbon, who would have thought it was so close, we imagined that we lived at the end of the world, the mistaken ideas of those who know nothing and have had no one to teach them. The serpent of temptation slithered up the branch from which João Mau-Tempo can see Lisbon and promised him all the marvels and riches of the capital in exchange for the very modest price of a ferry ticket, well, not that modest for someone with nothing, but, in for a penny, in for a pound, he'd be a fool to refuse. We will disembark in Cais do Sodré and declare, wide-eyed, So this is Lisbon, the big city, and the sea, look at the sea, all that water, and then we walk through an archway into Rua Augusta, so many people, so much traffic, and we're not used to walking on pavements, we keep slipping and sliding in our hobnail boots, and we cling to each other in our fear of the trams, and you two fall over, which makes the Lisbonites laugh, What bumpkins, they cry, And look, there's Avenida da Liberdade, and what's that thing sticking up in the middle, that's Restauradores, oh, really, and I think to myself, Well, frankly, I'm none the wiser, but ignorance is always the hardest and most embarrassing thing to own up to, anyway, let's walk up Avenida da Liberdade and find our sister, who's working as a maid, this is the street, she's at number ninety-six, isn't that what you said, after all, you're the one who can read, No, there must be some mistake, it goes from ninety-five to ninety-seven, there is no ninety-six, but he who seeks always finds, here it is, they laughed at us because we didn't know that ninety-six was on the other side, the people in Lisbon laugh a lot. Here's the building where our sister works, it's really tall, the owner and resident of the first-floor apartment is Senhor Alberto, our sometime boss, everything belongs to the same family, Well, look who's here, Maria da Conceição will say, what a surprise, and how plump she's got, there's nothing like being a maid. We'll all go out together afterward, because the lady of the house is very generous and gives her time off, although it will be discounted from her next bit of leave, because normally she gets an afternoon off once a fortnight, between lunch and supper. We'll visit some cousins who live in the area, in streets and back streets, and there'll be the same joyful greeting, Well, look who's here, and we'll arrange to go to the show tonight, but first, you mustn't miss the zoo, the monkeys are so funny, and that's a lion over there, and look at the elephant, if you came across a monster like that in the countryside, you'd die of fright, and the show is called *The Clam*, starring Beatriz Costa and Vasco Santana, the man almost had me crying with laughter. We'll sleep here in the kitchen and in the corridor, don't worry, we're used to all sorts, the nights are different in Lisbon, it's the silence, the silence isn't the same, So, did you sleep well, and no one dares to say No, we spent all night tossing and turning, but now let's have a cup of coffee and then go for a stroll around the city, but this isn't a city, it's the size of a county, and in Alcântara we'll meet a group of men working on the railway line, and they say, Morning, bumpkins, and that's it, our brother-in-law takes umbrage and goes over to them, What did you say, a few blows are exchanged, then we flee in shame, and the men shout, Look at the one in the jacket, Look at that bumpkin run, but we're not bumpkins, and even if we were, that would still be no reason to scorn us. We will cross the river again, Look at the sea, and a gentleman traveling with us in the boat says very politely, Actually, this is the river, the sea starts over there, and he points, and then we realize that you can't see land in that direction, how is that possible. When we disembark in Montijo, we'll still have a few kilometers to walk, eight to be precise, until we reach our work camp, we spent an awful lot of money but it was worth it, and when we get back to Monte Lavre, we'll have a lot to tell, because life has its good points too.

Sometimes when people get married, there's already a baby on the way. The priest blesses the couple and the blessing falls on three, not two, as you can see by the sometimes quite prominent bump beneath the woman's skirt. But even when that isn't the case, whether the bride is a virgin or not, it would be very unusual for a year to pass without a child being born. And if God so wills it, it'll be one child out, another one in, for as soon as the woman gives birth, she falls pregnant again. They're real brutes these people, ignorant, worse than animals, because animals aren't in heat all the time, they follow the laws of nature. But these men arrive home from work or from the inn, get into bed, their blood inflamed by the smell of their wife or by the wine they've drunk or by the sexual appetite that comes with tiredness, and they get on top of her, they don't know any other way, they huff and puff, they're not exactly subtle, and leave their sap to soak in the mucous membrane inside the woman's incomprehensibly intricate innards. This is a good thing, better than going with other women, but the family is growing, more and more children are born, because they don't take precautions, Mama, I'm hungry, the proof that God does not exist lies in the fact that he did not make men sheep so that they could eat the grass in the fields, or pigs so that they could eat acorns. But even if they did eat acorns and grass, they couldn't do so in peace, because there's always a warden or the guards around, with eye and rifle cocked, and if the warden, in the name of Norberto's lands, doesn't shoot you in the leg or kill you right off, the guards, who will do the same if they're ordered to, or even if they're not, can choose the more benign options of prison, a fine and a beating. But this, ladies and gentlemen, is a bowl of cherries, you pull out one and three or four come out together, there are even estates that have their own private prison and their own penal code. Justice is done every day on the latifundio, what would become of us if the authorities weren't here.

The family grows, though many children die of diarrhea, dissolving in their own shit, poor little angels, snuffed out like candles, with arms and legs more like twigs than anything else, their bellies distended, until the moment comes and they open their eyes for the last time to see the light of day, unless they die in the dark, in the silence of the hovel, and when the mother wakes and finds her child dead, she starts to scream, always the same scream, these women whose children have died aren't capable of inventing anything, they're speechless. As for the fathers, they say nothing and, the following night, go to the taberna looking as if they're ready to kill someone or something. They come back drunk, having killed nothing and no one.

The men go far away to work, wherever they can earn some

money. At bottom they're all itinerants, they go here and there, and come home weeks or months later to make another child. Meanwhile, as they labor on the cork plantations, watched by the overseers, each drop of sweat is a drop of spilled blood, and the wretches suffer all day and sometimes all night as well, counting the number of hours worked on the fingers of three hands, except when they have to resort to a fourth hand, like the four-legged beasts they are, to count the rest, their clothes don't dry on their backs for a whole two weeks. To rest, if such a word can be used in the circumstances, they lie down on beds of heather with some straw on top of them, and, dirty and bruised, they moan all night, it's quite wrong, how can they believe in Father Agamedes when they see him coming back from his Sunday lunch at Floriberto's house. Judging from the loud belch that echoes around the estate, it was a very good lunch indeed.

This is the power of the heavens. It is, besides, an oft-repeated story. The men are in the hut, exhausted, still clothed, some are sleeping, others can't sleep at all, and through the gaps in the cane walls there enters a never-before-seen light, the morning is still far off, so it's not the morning light, one of the men goes outside and stands frozen with fear, because the whole sky is a shower of stars, falling like lanterns, and the earth is lit more brightly than by any moonlight. Everyone comes out to look, some are really terrified, and the stars fall silently, the world is going to end, or perhaps begin at last. One man, with a reputation as a sage, says, When the stars are restless, so is the earth. They are standing close together, looking up, their heads right back, and they receive on their grubby faces the luminous dust from the falling stars, an incomparable rain that leaves the earth with a different and much greater thirst. And a rather dim laborer who passed through there the following day swore on his mother's life that those celestial signs were announcing that in a ruined shepherd's hut, three leagues from there, a child had been born of another mother, probably not a virgin, a child who couldn't be said to be Jesus Christ only because he had been baptized with another name. No one believed him, and that general skepticism aided Father Agamedes, who, on the following Sunday in a church unusually full and abuzz with excitement, mocked the fools who believe that Jesus will return to the earth just like that, I, your priest, am here to tell you what he would say, I have my holy orders and instructions and am mandated by the Holy Roman Catholic Apostolic Church, do you hear, because if you can't, I'll open another ear on the top of your head.

He was quite right, that wise man who predicted that if the stars are restless, then the earth will be too, the Abyssinians were the first to confirm this, immediately followed by the Spaniards, and later by half the world. Here, the earth is moving according to the old customs. Saturday comes and brings with it the market, but so poorly stocked that it's hard to know how one will fill next week's lunch sacks, it makes you shudder to think of it. A woman went to the grocer and said, Can I owe you for this week's groceries, we've had a terrible week because of the bad weather. Or she would say the same thing in different words, but starting in the same way, Can I owe you for this week's groceries, there was no work this week and my husband hasn't earned a thing. Or perhaps, staring shamefaced at the counter, like someone with not a penny more to her name, Sir, my husband will earn more come the summer, then he'll sort things out with you and pay what he owes. And the grocer, thumping his account book with his fist, would reply, Don't come to me with that old story, I've heard it before, the summer comes and goes and the dog will still be barking, because debts are like dogs, a funny idea, I wonder who first thought of it, this is a people who come up with these sharp, urgent images, they imagine the account book of the grocer or the baker, the large numbers written in pencil, this much and this much, yes, it's like a small, soft puppy that can grow into a beast with wolf's teeth, last year's still unpaid debt, Pay up or I'll cancel your credit, But my children are hungry, and some are ill, my husband has no work, where can we find money, Too bad, you get nothing without paying for it. Everywhere the dogs are barking, we can hear them at the doors, they pursue those who can't pay, bite their shins, bite their souls, and the grocer comes out into the street and says loudly enough for everyone to hear, Tell your husband, and we know the rest. Some people peer out of their doors to see who is being shamed, a poor person's malice, today it's me, tomorrow it could be you, you can't really blame them.

When a man complains, it's because something must be hurting him. We are complaining about this nameless cruelty, and it's a pity it has no name, What will become of us today, this is all the money we have, and we're weeks behind in paying, the grocer won't give us any more credit, and every time I go there, he threatens to cut it off completely, not a penny more, Go and try again, the husband says, but that's just for the sake of saying something, he doesn't really have a stone for a heart, No, not on my own, I can't face going through that door again, only if you come with me, Then we'll both go, but men are not much good at these things, their job is to earn the money and the wife's to make it stretch, besides, women are used to it, they protest, swear, bargain, cry, are capable even of falling to the floor, give the poor woman a drink of water, she's fainted, but a man goes in there and he's shaking, because he should be earning and he isn't, because he should be keeping his family in food and he isn't, How can I do what I said I'd do when I married, Father Agamedes, tell me that. We reach the shop, and other customers are there, some are leaving, some are going in, not all are there simply to buy, and we keep getting pushed to the back of the queue, standing here in this corner next to a sack of beans, let's hope he doesn't think we're going to steal it. Finally, the other customers have all gone, and we make our move, I, the man, step forward, my hands shaking, Senhor José, you were kind enough to give me some credit, but I can't pay it all back today,

I've had a dreadful week, but believe me, as soon as my earnings increase I'll pay it back, then I won't owe you a penny. Needless to say, these are not new words, they were spoken on the previous page, spoken on every page of the book that is the latifundio, how could one expect the answer to be any different, No, I won't give you any more credit, but before the grocer said these words, his hand greedily snatched up the money I had put on the counter to placate him. And I said, with all the calm I could muster, and God knows that wasn't much, Senhor José, don't do this to me, how am I going to feed my children, have pity on me. And he said, I don't want to know, I won't give you any more credit, you already owe me a lot. And I said, Senhor José, please, at least give me something for the money you've taken from me, just so that I can give my children something to eat, until I can sort something out. And he said, I can't give you any more, this money won't pay even a quarter of what you owe me. He thumps the counter, defying me, and I make as if to hit him, perhaps with the strickle, or else to stick a knife in him, this penknife, or yes, this curved blade, this Moorish dagger, What are you doing, man, think of our children, take no notice of him, Senhor José, don't take it the wrong way, such is the despair of the poor. I'm bundled toward the door, Let me go, woman, I'll kill the bastard, but my thoughts are thinking, I won't kill him, I don't know how to kill, and from inside the shop he says, If I give credit to everyone and none of them ever pays me back, how will I live. We are all in the right, who, then, is my enemy.

It's because of these and other, similar deficiencies that we invent stories about hidden treasure, or search out ones that have been invented already, proof of a very ancient need, it's nothing new. There are always warnings that must be attended to, one false move and the gold turns into a fish and the silver into smoke, or a man goes blind, it's happened before. Some say that one cannot trust dreams, but if, on three consecutive nights, I dream of a treasure and tell no one about it nor about the place I saw in my dream, it's certain that I'll find it. But if I speak about it, I won't, because treasures have their fate too, they can't just be distributed as man wishes. There's that old story about a girl who dreamed three times that on the branch of a particular tree she would find fourteen coins and beneath the tree's roots a clay pot full of gold pieces. One should always believe these things even when they're invented. The girl told her dream to her grandparents with whom she lived, and they went together to the tree. There on the branch were the fourteen coins, so half the dream had come true, but they didn't want to dig down into the roots because it was a lovely tree, and with its roots exposed it would die, well, the heart has its reasons. Anyway, the news spread, no one knows how, and when she and her grandparents went back, having thought better of their scruples, they found the tree had been dug up and in the hole was a clay pot split in two, and nothing else. Either the gold had disappeared by magic or someone, less scrupulous or with a harder heart, had taken the treasure and made off with it. Anything is possible.

A still clearer case is that of the two stone chests buried by the Moors, one containing gold and the other containing the plague. It is said that, fearful of opening the wrong chest, no one had had the courage to look for them. But if that's true, how is it that the plague has spread throughout the world.

J OÃO MAU-TEMPO AND FAUSTINA are married, a peaceful conclusion to the romantic episode which, on a rainy, overcast night in January, with no moon and no nightingales, in a tangle of half-unfastened clothes, satisfied the desires of both parties. They have three children. The oldest is a boy called António, who is the very image of his father, although he is of a stronger build and lacks his father's blue eyes, which have not as yet reappeared, where can they have gone to. The other two are girls, as gentle and discreet as their mother was and continues to be. António Mau-Tempo is already working, he helps out keeping pigs, for he isn't old enough nor his arms strong enough to do any heavier work. The foreman doesn't treat him well, but that's the custom in this place and this time, so don't let's get steamed up over nothing. As is also traditional, António Mau-Tempo's lunch sack is light as a feather, a banquet consisting of half a mackerel and a hunk of maize bread. As soon as he leaves the house, the mackerel vanishes, because some hungers simply cannot wait, and his is a very old hunger. The bread is all he has left for the rest of the day, just a mouthful now and then, as he nibbles away at the crust, taking scrupulous care not to let a single crumb fall into the grass, where the ants, their noses in the air like dogs, are desperate to fill their stores with any leavings and leftovers. The foreman, in his role as foreman, would stand on a patch of bare ground and shout, Run over there, boy, and see to those animals on the other side, and António Mau-Tempo, like a small broom, would run around the herd of pigs as if he were a sheepdog. The foreman, now that someone else was doing all the work, passed the time picking ripe pine cones, which he would first roast and peel and then extract the kernels, which he would carefully toast and put away in his haversack, all the while enjoying the rustic peace of the lovely trees. The fire would glow red, the resin-scented pine cones would open in the heat of the fire, and if António Mau-Tempo, mouth watering, found a pine cone that had by chance fallen within sight of his yearning eyes, he quickly hid it, so that it didn't immediately go to increase the other man's wealth, as happened on a few dramatic occasions. Childhood has its just revenge. One day, near some wheatfields, when the foreman was engaged in roasting pine cones, he said to António Mau-Tempo, as he often did, Keep an eye on those pigs over there and make sure they don't get into the wheat. A really cutting wind was blowing that day, and, dressed as he always was in the skimpiest of clothes, António Mau-Tempo decided to give the pigs a holiday, while he took shelter behind a machuco, What's a machuco, A machuco is a young chaparro, everyone knows that, And what's a chaparro, A chaparro is a young cork oak, of course, So a machuco is a cork oak, Isn't that obvious, Ah, As I was saying, António Mau-Tempo sat down behind a machuco, wrapped in the sack that served as his coat in all weathers, come rain or ice, a guano sack was all he had, may God suit the cold to the covering, anyway, there was, for once, general contentment, the pigs in the wheatfield, the foreman roasting pine cones and António Mau-Tempo in his shelter, gnawing away at his crust of bread. And to think that some people still have nothing but bad things to say about the latifundio. Now the trouble was that the foreman had a dog, a clever creature who, suspicious of what António Mau-Tempo was up to behind the tree, started barking furiously, It's true what they say, that a dog is man's best friend, It was no friend of António Mau-Tempo's, however. The foreman leapt up in alarm and when he found the boy, he cried, So you're asleep, are you, and threw a stick at him, which, had it hit its mark, would have been the end of António Mau-Tempo. No boy worth his salt would have given him a second chance, so António Mau-Tempo grabbed hold of the stick himself and hurled it into the middle of the wheatfield, there, go and find it if you can, and then he legged it. The pigs' fun did not last very long either. Isn't it always the way.

Such episodes are all part and parcel of the pastoral life and of a happy childhood. You just have to see for yourself how easy it is to live happily on the latifundio. The pure air, for example, I'll give a prize to anyone who can find better. And the birds, singing away above our heads when we stop to pick a little flower and study the behavior of the ants or this slow, black stag beetle afraid of nothing, impassively crossing the path on his long legs, but who dies beneath our boot, if we so choose, it depends on our mood, at other times, we might be more disposed to consider all life sacred and then even the centipedes escape with their lives. When the foreman comes to complain, António Mau-Tempo's father is there to defend him, Don't hit the boy, I know exactly what goes on, you sit there toasting pine kernels, talking to whoever happens by, and he has to play sheepdog, running from one side to the other, the boy isn't a beetle for you to crush. The foreman went off and found another assistant, and António Mau-Tempo went to keep pigs for a new boss, until he grew stronger.

Man has many jobs to do. We've mentioned some already, and now we add others for the purpose of general enlightenment, because townspeople think, in their ignorance, that it's all a matter of sowing and harvesting, well, they're much mistaken unless they learn all the other verbs involved and realize just what they mean,

harvesting, carrying sheaves, scything, threshing either by machine
or by hand, flailing the barley, covering the hayrick, baling up straw
or hay, shucking the maize, spreading manure, sowing seeds, digging, clearing land, cutting up the maize stalks and digging them
in, shoeing, pruning, ringing, leveling, digging ditches and trenches,
hoeing, making terraces, grafting vines, taping up the graft, spraying with copper sulfate, carrying the grapes, working in the cellars,
laboring in the vegetable plots, preparing the ground, beating the
olive trees, working the oil presses, cutting cork, shearing sheep,
cleaning wells, hacking undergrowth, chopping firewood, staking, covering with straw, earthing up, plugging, bagging and whatever else needs doing, all those lovely terms enriching our lexicon,
blessed be the workers, and if we were to start explaining how each
task is performed and in which season, and the tools and the implements needed, and whether it was men's work or women's and
why, we would never end.

Anyway, a man is hard at work, in this case he happens to be a
man, or rather, he is at home after work, when a hunting hound
comes in through the door, his name isn't Ranter or Ringwood,
he has two legs and a man's name, but he's a vicious beast all the
same, and he says, I've got a piece of paper here for you to sign,
you're to go to Évora on Sunday to a rally in support of the Spanish nationalists, it's an anticommunist rally, transport's free, you'll
be taken there in a truck, all expenses paid by the bosses or the
government, it comes to the same thing. The man feels like saying no, but can't find the will to say it, he sits there chewing, pretending he hasn't heard, but there's no point, the other man repeats
what he's said, but in a different, somewhat threatening tone, and
João Mau-Tempo looks at his wife, who is there as well, and Faustina looks at her husband, who wishes he weren't there, and at the
hound grasping the piece of paper, waiting for a reply, what shall
I say to him, what do I care about such things, I don't know anything about communism, well, that's not quite true, last week I

found some papers wedged under a stone, with one corner sticking out, as if they were trying to attract my attention, and I dropped behind and picked them up, no one saw me, but what's this hound doing here, baring his teeth, perhaps someone told him, perhaps he came here to see if I would dare to say that I don't want to go to Évora, that I won't sign, the worst thing is that afterward, because everyone knows this dog, his name's Requinta, he'll go and tell on me, there's sure to be someone with a grudge against me, but if I come up with an excuse, tell him I've got a pain in the gut or that I have to mend the rabbit hutch, he won't believe me, and they might arrest me, All right, Requinta, I'll sign.

João Mau-Tempo signed where others had signed before him, or put their mark because they didn't know how to write, which was most of them. And when Requinta left to continue collecting signatures, his nose in the air, sniffing the wind, the impudent creature, João Mau-Tempo felt a great thirst and drank straight from the jug, drowning in water the sudden fire that was merely a wave of unexplained embarrassment, other men would have drunk wine. Faustina had heard something of the conversation and hadn't liked what she heard, but she preferred to console her husband, Well, at least it will mean a trip to Évora, it will be a distraction, and it's free too, with transport there and back, it's a shame you can't take António, he'd love it. This wasn't all that Faustina said, she continued to murmur something or other without really thinking what she was saying, and João Mau-Tempo knew that her words were like gestures that bring no hope of salvation, but which the patient receives gratefully like a soft hand on his brow, or rather a rough hand, given that we're in the country, but all the same. All the same, they shouldn't force a man to go, because that's what they're doing, I'd rather pretend to be ill. Faustina said, It's not so dreadful, treat it like an outing, I'm sure the government knows what it's doing. João Mau-Tempo said, Yes, you're right. Anyone overhearing this conversation might declare that these people are a lost cause, but he or she has no idea what it's like here, the people live miles from anywhere, they either get no news at all or don't understand it when they do, and only they know what a struggle it is simply to survive.

The day came, and at the appointed time, the men gathered in the street, and while they waited, some went into the taberna and drank as much wine as their pockets could afford, each drinker sticking out his lips to catch the surface bubbles bursting under his nose, ah, wine, blessed be the man who invented you. The more refined and better-informed among them were expecting great things of Évora and kept their appetites for later, but they soon learned their lesson, because they were dropped at the door of the bullring and picked up from there at the end of the rally. Forewarned is forearmed, a bird in the hand is worth two in the bush, that's what people say, some live their whole lives according to such wisdom, and it does them no harm. This time the drinkers were right and were pleasantly merry by the time the trucks arrived, with their bellies singing hosannas and uttering the holy belch of wine, and enjoying the aftertaste that lingers in the mouth, the taste of paradise.

It's quite a journey. On the bends, even when not taken at speed, the truck leans to one side and the men have to cling to each other so as not to be thrown out, they totter about, the wind catches their hats and they have to hang on to them so they don't fly away, Go more slowly, driver, we don't want a man overboard. One of the wittier men said this, well, that's what gives a little spice to life, if not, life would be very dull indeed. They stopped in Foros to take on more people, and then it was plain sailing, they glimpsed Montemor, but there was no time to visit, and Santa Sofia and São Matias, I've never actually been there myself but I have family there, a cousin of my sister-in-law, he's a barber and has done really well for himself, it would be a different story, of course, if men's beards stopped growing, it would be the same for prostitutes if men's cocks stopped growing too. The man who says this knows what he's talking about, well, once in a while never hurts, I haven't been to a whorehouse since I did my national service, this time, though, I'm going to fill my boots. Men's talk. Humanity has done its best to improve communications, even the estate has trucks at its disposal, Évora lies before them, and the hound Requinta, because he came too, barks, When we get out, follow me, and those fateful words cast a pall over the various appetites for wine and women, for that imagined long, restless night in bed with some woman, but dreams are never to be trusted.

The bullring is packed. Hordes of farm laborers have been herded in there, sometimes by a landowner, smiling and chatty, and there's always some lackey toadying up to him, shaming those who came for the sensible reason that they feared being left jobless. On the whole, though, they do their best to appear happy. That's the kindness of the crowd, not wishing to disappoint the person who expects us to be contented, and although it's true that this doesn't look much like a party, it's no one's funeral either, so tell me what face to wear, should I cheer or boo, cry or laugh, tell me. They're sitting on the benches in the stands, others fill the arena, it would be better if there were some bulls there, and they still have no idea what's going to happen or what exactly a rally is. Where has Requinta got to, Requinta, when does the party begin. Friends and acquaintances wave to each other, the more timid among them change places in search of some braver souls, Come over here, and then Requinta says, Keep together and pay attention, this is serious business, we came here to find out who is on the side of good and who on the side of evil, that would be useful, wouldn't it, to have Requinta lead us by the hand toward a knowledge of good and evil, who would have thought it could be so easy, Father Agamedes, all you have to do is stop thinking and plunk your bum down on a bench, Where do we go to take a piss, Requinta, such talk is the first sign of a lack of respect, and Requinta frowns and pretends not to hear, but now the rally is about to start, Ladies and gentlemen, that's funny, so in the bullring in Évora I'm a gentleman, am I, I don't remember being a gentleman anywhere else, not even by my own choice, what's he saying, Viva Portugal, I can't hear him, We are gathered here today, united by the same patriotic ideal, in order to say to our government that we are pledged to continue the great Lusitanian adventure and that we promise to follow in the footsteps of those ancestors who gave the world whole new worlds and spread both faith and empire, and when the trumpet sounds, we will come together, as one man, around Salazar, the genius who has dedicated his life, here there are shouts of salazar salazar salazar, the genius who has dedicated his life to the service of the country, against the barbarous threat from Moscow, against those wretched communists who threaten our families and who would kill your parents, rape your wives and daughters, who would send your sons to labor camps in Siberia and destroy the holy mother church, for they are atheists, godless men with no morals and no shame, down with communism, death to all traitors, the bullring bawls out the slogan, some still have no idea what they're doing here, others have begun to understand and are saddened, some are convinced, or deceived, a worker makes a speech, then another speaker, he's from the Portuguese legion, he stretches out one arm and bawls, Who gives the orders, who gives us life, well, that's a good question, the boss gives the orders, and as for life, what's that. But the obedient bullring gives the expected response, and no sooner has the legionnaire stopped speaking than another man is there, mouth open, they certainly talk a lot, these people, something about Spain, about how the nationalists are fighting the reds, and how the lands of Castile and Andalusia are defending the sacred, eternal values of western civilization, that it's every man's duty to help our fellow believers, and that the remedy for communism is to be found in a return to the Christian morality whose living symbol is Salazar, goodness me, we have a living symbol, we must not be soft on our enemies, words words words, and then he goes on to talk about the good people of the region, expressing their gratitude to that immortal statesman and great Portuguese citizen who has devoted his whole life to serving his country, may God preserve him, and I will tell the president what I saw today in this historic city of Évora, and promise him that each of the thousands of hearts was beating in unison with that of the fatherland, that each heart is the fatherland, that deathless, sublime and most beautiful of all fatherlands, because we are blessed with a government that places the interests of the nation above the interests of any one social class, because men pass and the nation remains, death to communism, or is it down with communism, who cares, among so many people who's going to notice, we must remember that life in the Alentejo, contrary to what many may think, is not propitious to the development of subversive ideas, because the workers are the true partners of the landowners, sharing the profits and losses, ha ha, ha, Where do I go to take a piss, Requinta, that's just a joke, no one here would dare say such a thing at a moment of such gravity, when the nation, which never has to take a piss, is being evoked by that well-dressed gentleman on the platform, who is opening wide his arms as if he wanted to embrace us all, and since he can't do that, the men on the platform embrace each other, the commander of the legion, the major from Setúbal, the members of parliament, the man from their national union, the captain of cavalry regiment five, a man from the en-i-double-u-double-u, if you don't know what that means, just ask, the national institute of work and welfare, and all the others who have traveled from Lisbon, they look like rooks perched on top of a holm oak, but that's where you're wrong, we are all rooks, lined up on the benches, flapping our wings, cawing away, and now it's time for the music, it's the national anthem, everyone stands up, some because they know it's the thing to do, others out of pure imitation, Requinta reviews his men, Come on, sing, I wish I could, who knows the national anthem, if it was some popular song we all knew, that would be another matter, oh, are we leaving, no, it's not time to leave yet, if only we could fly, spread our wings and fly far from here, over the fields, watching from on high the trucks driving back, how sad, it was all so sad, and we shouted as if we had been paid to do it, I don't know what's worse, it's not right, it was like a carnival farce. So you didn't enjoy yourself, João, Not a bit, Faustina, we went like sheep and we came back like sheep. By the time they're in the truck again, evening is falling, an aid to melancholy, someone tries singing and two men join in, but when sadness weighs heavy, even that sad voice falls silent, and then they hear only the sound of the engine, and they sit in silence, being thrown about, a badly tied load, a loose load, this was no work for men, João Mau-Tempo. The truck drops the men off outside Monte Lavre, like a flock of dark birds who scatter, not knowing quite where to go, some go to the taberna to slake their thirst and their bitterness, others mumble to themselves, the saddest go back to their houses, We're just like dolls to be traipsed back and forth, who's going to pay us for today, I had work to do in the vegetable garden, it's that wretch Requinta's fault, I'll find some way to get my own back, words and promises born only of the pain underneath, but they can give full expression to little of that pain, it's too vague, it may not hurt but it cripples. That's why Faustina asks, Are you ill. João Mau-Tempo says no, he isn't, and if he says nothing more, it's because he doesn't know how to explain how he feels. Lying in bed, they talk a little more, So you didn't enjoy yourself, Not a bit, and by way of pouring out his heart and confessing his feelings, João Mau-Tempo rests his head on Faustina's shoulder and falls asleep.

The gentlemen of the estate go up the hill so that the sun will warm them alone, at least they do in João Mau-Tempo's rough-and-ready dream, because the gentlemen have no faces and the hill has no name, but that's how it is when João Mau-Tempo wakes up, and when he falls asleep again, a procession of gentlefolk are walking along and he goes ahead of them, digging up weeds with his mattock, clearing the way for that gay company of men, he pulls up the gorse with his bare hands, his hands are bleeding, and the gentlemen of the estate are laughing and talking, they are generous and patient when he falls behind in his weeding, they wait, they don't mistreat him or summon the guards, they simply wait, and while they wait, they picnic, and João Mau-Tempo dredges up the strength from somewhere and lays in with his mattock, breaking the earth and slicing through the roots, he's a man now, and above him, on the side of the hill, he sees trucks passing, bearing a sign that says Surplus Goods from Portugal, they're heading for Spain, don't give the reds an inch, as for those others, the saints, the pure ones, those who save me, João Mau-Tempo by name, from falling into hell, down with them, death to them, and now a man on horseback is coming after me, and the horse is the only thing in the dream to have a name, it's called Bom-Tempo, well, horses have a long life, Wake up, João, it's time to go to work, says his wife, and yet it's still pitch-black outside.

OTHERS, THOUGH, HAD already got up, not in the sense of someone who, sighing, drags himself from the dubious comfort of a mattress, if he has one, but in that other, peculiar sense of waking in the middle of the day to discover that, only a minute before, it was still black night, for man's true time and the changes to which he is subject are not ruled by the rising of the sun or the setting of the moon, objects that are merely part of the celestial and terrestrial landscape. It is true that there is a time for everything, and this particular event was fated to happen during harvest time. Sometimes, a physical impatience, not to say exasperation, is required for souls finally to move, and when we say soul, we mean that thing with no real name, which is perhaps merely the body, the whole body. One day, if we don't give up, we will all know what these things are and how far they are from the words that attempt to explain them, and how far those words are from the things themselves. But this looks far more complicated when you try to write it down.

This machine looks complicated too, and yet it is so simple. It's a thresher, never better named because that is precisely what it does, it removes the grain from the ear of wheat, separating stalks and husks from grain. From the outside, it looks like a large wooden box on metal wheels, connected by a chain to an engine that trembles, roars, rumbles and, if you'll forgive the word, pongs. It was originally painted egg yolk yellow, but the color has faded beneath the dust and the harsh sun, and now it looks more like another feature of the landscape, alongside others, like the piles of straw, in this sun it's even hard to tell them apart, nothing is still or quiet, the engine is throbbing, the thresher is vomiting out straw and grain, the slack chain is vibrating, and the air shimmers as if it were the reflection of the sun in a mirror held in the sky by the small, unsteady hands of angels with nothing better to do. A few shapes can be seen in the midst of this mist. They have been working all day, and yesterday and the day before, and before that, ever since the threshing began, there are five of them, one older man and four younger men, whose seventeen or eighteen years are not enough to cope with such strenuous work. They sleep on the threshing floor, in between the bales, but it's already dark by the time the engine falls silent, and the sun is still far off when this beast fed on cans of sticky, black liquid first groans into life and then proceeds to batter their ears with noise all the blessed day. It's the machine that sets the pace of work, the thresher cannot chew on nothing, as becomes immediately obvious when the foreman emerges from his hiding place and bawls at them to keep it fed. The inside of the machine's mouth is a volcano, a giant gullet, and the older of the five men tends to be in charge of feeding the monster. The others are responsible for helping the piles of straw to grow higher and higher, they spin like mad things in that fog of chopped-up straw, they haul the rough, dry wheat, the stiff stalks, the bearded ears, the dust, where is the tender springtime green of the fields when the earth really does seem like paradise. The heat is unbearable. The older man steps down and one of the younger men takes his place, and the machine is like a bottomless pit. All it needs is for a man to fall in. The bread would then take on the correct blood-red color, rather than its usual innocent white or neutral brown.

The foreman comes over and says, Go and work down at the chaff end. The chaff is that weightless monster, that straw-cum-dust that blocks your nostrils, that creeps in through every gap in your clothing and sticks to your skin like a layer of mud, it itches like crazy and gives you the very devil of a thirst. The water they drink from the clay jug soon grows lukewarm and slimy, as if you were drinking directly from a swamp full of worms and bloodsuckers, which is what we call leeches around here. The lad goes down to the chaff end and receives it full in his face like a punch, and his body begins slowly to protest, it doesn't have the strength to do more than that, but then, and only those who have experienced this themselves will know what I mean, the despair feeds on the body's exhaustion, grows steadily stronger, and that strength feeds back into the body, and finally, with that redoubled energy, the lad, whose name is Manuel Espada, and who will reappear later in this story, steps away from the chaff, calls to his colleagues and says, I'm off, this isn't work, it's slow death. The older man is once more standing on the thresher, What about the straw bales, but he's left with his words hanging in the air and his arms by his sides, because the four lads leave together, brushing off their clothes, they're like clay figures ready for the kiln, grayish brown, their faces striped with sweat, they look just like clowns, except that they're not funny at all. The older man jumps down from the thresher and turns off the engine. The silence is like a blow to the ears. The foreman comes running over, panting, What's going on, and Manuel Espada says, I'm leaving, and the others say, We're leaving too, the threshing floor is stunned, So you don't want to work. Anyone looking around can see the air trembling, it's only heat haze but it feels as if the whole estate were trembling, and yet it's just these four young men, who are free to leave, having no wife and no children to feed, for as João Mau-Tempo says to Faustina, That's the reason why I agreed to go to Évora. His wife answers, Don't think about that now, get up, it's time.

Manuel Espada and his friends go to the overseer, squint-eyed Anacleto, to ask for the money they're owed for the days they've worked, and to tell him that they're leaving because they can't take any more. Anacleto fixes his wandering eye on the four young rascals, ah, if only he could give them a good whipping, You're not getting any money, and be warned, I'm going to put you down as strikers. The insurrectionists are too young and innocent and ignorant to know what this word means. They walk back to Monte Lavre, which is a long way, taking the most direct route they can along old paths, feeling neither happy nor regretful, that's how it is, a man cannot spend his whole life obeying orders, and these four men, if we can call them that, stroll along talking and saying the kind of things lads of their age always say, one of them even throws a stone at a hoopoe that fluffs up its wings as it crosses their path, the only thing they regret is leaving behind those women from the north who worked alongside them on the threshing floor, there being a great shortage of labor that season.

Anyone traveling by foot has all the time in the world, but when speed is of the essence, and especially when one is athirst for justice, when evil deeds and evildoers threaten to put the latifundio at risk, it's understandable that Anacleto should go by cart to Montemor, he is furious and trembling, his face tinged with the holy blush that marks the faces of all those who struggle passionately for the preservation of the world, yes, it's understandable that he should rush to Montemor where these matters can be dealt with properly and that he should inform the guards that four men from Monte Lavre have declared themselves to be on strike, What will become of me, what shall I tell the boss when he wants to know how the threshing is going, now that I've lost these men. Lieutenant Contente said, Don't worry, we'll take care of it, and Anacleto returned to the threshing floor with his mind at rest, and as he was driving back, in less of a hurry now, enjoying the warm glow of one who has performed a pleasant duty, a car laden with men passed him, and someone inside waved, it was the district administrator, and with him, shouting, Goodbye, Anacleto, were the lieutenant and a whole patrol, bearing down on the enemy in a panzer sherman tank bristling with weapons of all calibers, from the standard-issue pistol to the recoilless rifle, and off they go, with the nation watching them, they offer their breasts to bullets, sound their horn, and it's like a bugle giving the order to charge, while somewhere on the estate, walking, as we have said, along old paths, those four hardened criminals have stopped for a moment to see who can pee highest and farthest.

At the entrance to Monte Lavre, the dogs bark at the would-be tank, it would seem unreal without that detail, and since it's a steep road, the patrol gets out and advances in formation, with the administrator at the front this time and his back protected. Their first call, carried out with the efficiency of someone on maneuvers, in the knowledge that they are only firing blanks, leads them to the local parish councilor, who is, so to speak, dumbstruck when he sees the lieutenant and the administrator coming into his shop, while outside, the patrol scans the surrounding area with suspicious eyes. On the other side of the street, some boys have gathered, and in places invisible or unidentifiable, mothers call for their children, as they did at the time of the massacre of the innocents. Let them call, much good may it do them, and let's go to the shop, where the parish councilor has recovered his voice and is now all politeness and flourishes, unctuously addressing both the administrator and the lieutenant as sir, he stops short of calling the soldiers sir, because that would sound odd, and the administrator takes from his pocket Anacleto's statement, on which he had noted the names of the criminals, Can you tell me where Manuel Espada, Augusto Patracão, Felisberto Lampas and José Palminha live, and not contented with his role as informer, the parish councilor summons his wife to keep watch over the counter and the cash drawer, and then the company, enlarged by one, sets off into the labyrinths of Monte Lavre, with one eye peeled for ambushes, just like the Spanish civil guard, may God preserve them. Monte Lavre is a desert under the blazing heat of the sun, even the boys have lost interest, it's like an oven, all doors are shut, but some are open just a crack, cracks being the resort of those who do not wish to show themselves, and when the guards march past, they are followed by the eyes of women and by those of the occasional inquisitive old man with nothing else to do. Imagine if now we were to launch into a detailed explanation of the expression in those eyes, we'd never get to the end of the story, and yet all those things, the seemingly unimportant and the seemingly important, form part of the same narrative, and might be as good a way as any to explain the latifundio.

Some things are innately funny, for example, the armed forces and the civil authority coming to arrest four dangerous agitators and finding none of them. The strikers are still a long way off. You wouldn't be able to see them from the highest point in Monte Lavre, even from the tower, if it is the tower, which it is, from which Lamberto Horques witnessed the charge of his cavalry in that fifteenth century we mentioned earlier. In the midst of that tangled landscape, not even the sun would help them spot the four tiny ruffians, who are probably lying down in the shade, perhaps dozing, waiting for the relative cool of evening. Not everyone finds their exploits so amusing, their mothers, for example, who have been informed by the lieutenant and the administrator that their sons are to present themselves in Montemor the next morning, if not, the guards will come to Monte Lavre and drag them to Montemor kicking and screaming, as they rather extravagantly put it. The tank sets off down the road, throwing up dust all around, but before it does, the administrator goes to present his respects to

the largest landowner resident there, whether Lamberto or Dagoberto it doesn't matter, who receives them all, apart from the soldiers, who are dispatched to the cellar, but Lieutenant Contente
and that bestower of respects, the administrator, are ushered into
a cool reception room on the first floor, how delightful it is here
in the dark, your wife and daughters are well, I hope, and yourself, have another glass of liqueur, and on the way out, the lieutenant stands at attention and gives the most perfect salute, the administrator is trying to speak man to man, but the latifundio is so
very large, and Alberto holds out one strong hand and says, Don't
let them get away with it, and the administrator, who bears the
singular name of Goncelho, says, I can't understand them, when
there's no work, they complain there's no work, and when there
is, they're not up to it. He's not exactly eloquent, but that's how it
came out, among neighbors one can speak freely on the latifundio,
and Norberto smiles sympathetically, The poor devils don't know
what they want, Ungrateful wretches, says the administrator, and
the lieutenant salutes again, he doesn't know what else to do, well,
his knowledge lies elsewhere, especially in military matters, but he
lacks opportunities to apply it.

The condemned men arrived at sunset. No sooner had they arrived than their mothers cried, What have you done, and they replied, We haven't done anything, we left because we couldn't face
working with that machine anymore. There seems nothing wrong
with that, but if you did do wrong, then what's done is done, tomorrow you must go to Montemor, don't worry, they won't arrest
you, said their parents. And so the night passed in stifling heat, the
lads would have been sleeping on the threshing floor now, and perhaps some woman from the north would have come out for a pee
and lingered there, breathing in the night air and perhaps hoping
that the world might take a turn for the better, Shall I go or will
you, until one of the lads decides to chance it, his heart beating fast
and his groin tense, well, he is only seventeen, what do you expect,

and the woman doesn't move away, she stands there, perhaps the world really is going to take a turn for the better, and this space between the bales seems tailor-made for the purpose, big enough for two bodies lying one on top of the other, it's not the first time, the boy doesn't know who the woman is, and the woman doesn't know who the boy is, it's better like that, come morning, there'll be no need for embarrassment if there was none at night, it's a game played fairly, with each player giving his or her all, and the slight giddiness they feel when they slip in between the bales, the sweet smell, and then the flailing limbs, the trembling body, but that way we'll get no sleep, and tomorrow I have to go to Montemor.

The four travel in a small cart pulled by José Palminha's parents' most precious possession, a rather rickety-looking mule, who nevertheless trots tirelessly on, they are silent, their hearts filled with dread, they cross the bridge and go up the hill beyond, and now they're in Foros, with one house here, another one there, that's what these far-flung hamlets are like, and then on the left-hand side is Pedra Grande, and shortly afterward, rising above the horizon, in the already hot morning air, stands the castle of Montemor, what remains of the city walls, it makes you sad. A man of seventeen starts speculating about the future, what will become of me, denounced as a striker by Anacleto, and the only thing my three friends are guilty of is keeping me company, our only other unforgivable fault being that we lacked the strength to keep up with the killing pace set by a thresher that was threshing me as it threshed the wheat, in I go through the machine's mouth and out it spits my bare bones, turning me into straw, dust, chaff, I'm being forced to buy the wheat at a price not of my choosing. Augusto Patracão, who is a great whistler, does so to calm his nerves, but his stomach hurts, he's no hero and doesn't even know what a hero is, and José Palminha keeps his mind occupied driving the mule, a task he performs to perfection, as if the mule were a high-stepping steed. Felisberto Lampas may be called Felisberto, but that's just a coincidence, and he sits sulkily, legs dangling, his back turned on his destiny, as he will do for the rest of his life. Then suddenly Montemor is upon them.

They leave the cart under a plane tree, and the mule with its nosebag on, what more can life have to offer a mule, and the four of them go up to the barracks, where a corporal tells them brusquely that they're to be at the town hall at one o'clock. The four lads kick their heels in Montemor for the rest of the morning without even the possibility, given their youth, of waiting inside the local taberna, it's impossible to describe the hours that precede any interrogation, so much happens in them, all the fear and dread inside each person's head, ill-disguised anxiety etched on every face, and the knot in the throat that neither wine nor water can dissolve. Manuel Espada says, It's all my fault you're here, but the others shrug, what difference does it make, and Felisberto Lampas answers, We just have to put on a brave front and show no weakness.

For these callow youths, things turned out well. At one o'clock, they were waiting in the corridor of the town hall listening to administrator Goncelho's voice booming around the building, Are the men from Monte Lavre here. Manuel Espada answered as he should, after all, he was the leader of the rebellion, Yes, sir, we're here, and they stood in a line, waiting to see what would happen next. The administrator played his part as the representative of the authorities, and Lieutenant Contente stood by him, You young rascals, do you have no shame, you're going to be sent across the sea to Africa, that will teach you to respect authority, Manuel Espada, come here, and the interrogation began, Who taught you to be strikers, who taught you, because you've obviously had good teachers, and Manuel Espada answered, with all the force of his innocence, No one taught us, we don't know anyone, we know nothing about strikes either, it was the machine, it kept eating and eating and the piles of straw were getting higher and higher. And the administrator said, I know your sort, that's what they taught you to say, and who is going to speak on your behalf, the administrator was preparing the ground because, when it was known in Montemor that some lads from Monte Lavre had been accused of being strikers, a few people of good sense had already spoken to him and to Lieutenant Contente, There's no point taking these things too seriously, they didn't mean any harm, what do they know about strikes. Nevertheless, all four were questioned, and once this was over, the administrator made a speech, in which, of course, he stated the obvious, Be more sensible in the future, learn to respect the people who give you work, we'll let it go this time, but don't let me see you here again or you'll end up in prison, so be careful, and if anyone comes along wanting to give you things to read or to engage in subversive conversations, tell the guard and they'll deal with it, and be grateful to the people who spoke up for you and don't let them down, you can go now, say goodbye to Lieutenant Contente here, he is your friend, as am I, for I only want what's best for you, don't forget that.

That's what this part of the country is like. The king said to Lamberto Horques, Cultivate and populate it, watch over my interests without forgetting your own, I give you this counsel because it suits me too, and if we follow this advice to the letter, we will all live in peace. And to his pastured sheep, Father Agamedes said, Your kingdom is not of this world, I suffered so that you might enter heaven, the more tears you shed in this vale of tears, the closer to the Lord you will be when you cast off the world, which is nothing but perdition, the devil and the flesh, and you can be sure that I'll be keeping my eye on you, for you are greatly deceived if you think that the Lord Our God has left you free to do both good and evil, everything will be placed in the balance come the day of judgment, better to pay in this world than be in debt in the next. These are excellent doctrines and are probably the reason why the four from Monte Lavre had to accept that the wages they had earned but not been paid, nine escudos a day, for three and a quarter days during the week in which they committed their crime, would go to the old folks' home, although Felisberto Lampas did mutter on the journey home, They'll probably spend our money on beer. We must forgive the young, who so often think ill of their elders. Far from being spent on beer, those one hundred and seventeen escudos given into the hand of the administrator meant that the old people enjoyed better food, a positive orgy, you can't imagine, all these years later they still talk of that feast, and one very ancient resident was heard to say, Now I can die.

They're strange creatures, men, and boys are perhaps stranger, for they are quite a different species. We have said enough about Felisberto Lampas, who is in a bad mood, and for whom the matter of the stolen wages is just a pretext. However, they all returned to Monte Lavre feeling sad, as if something more valuable had been stolen from them, perhaps their sense of pride, which they hadn't lost, of course, but there had been something offensive about the whole situation, they had been treated with scorn, stood in line to hear the administrator's sermon, while the lieutenant watched from the sidelines, memorizing their faces and features. They were even angry at the people who had interceded on their behalf, and whose pleas probably wouldn't have helped at all if the incident hadn't taken place two days before a bomb attempt on Salazar's life, from which he escaped unharmed.

That Sunday, the four went to the square, but could find no one to take them on. The same thing happened on the following Sunday and the Sunday after that. The estate has a long memory and good communications, it misses nothing and passes on the word, it will forgive only when it chooses to, but it will never forget. When they finally did find work, they each went their separate ways. Manuel Espada had to go and tend pigs, and during his time as a pigherd, he met António Mau-Tempo, who, later on, when the time comes, will become his brother-in-law.

S ARA DA CONCEIÇÃO IS not well. She has taken to dreaming about her husband, barely a night passes when she doesn't see him lying on the ground in the olive grove with the purple mark of the rope on his neck, she can't let his body go to the grave like that, and then she starts washing his neck with wine, because if she can make the mark disappear, she will have her husband back again, alive, which is the last thing she would want when awake, but that, inexplicably, is how it is in the dream. This woman, who traveled around so much when young, lives a very quiet, stable life now, but then she always did really, she helps out in the house of her son João Mau-Tempo and her daughter-in-law Faustina, she takes care of her granddaughters, Gracinda and Amélia, tends to the chickens, darns and redarns the clothes, patches up trouser seats, a skill learned during her short time as a stitcher of uppers to soles, and she has a strange habit that no one can understand, which is to go out walking at night when all her family is sleeping. True, she doesn't go very far, fear won't let her, the end of the street is quite far enough. The neighbors say she's slightly mad, perhaps she is, because if all the old mothers came out into the street at night so that their sons and daughters-in-law or their daughters and sons-in-law could take their pleasures in peace, it would be worthy of being recorded in the very brief history of small human gestures, imagine seeing lots of old ladies wandering about in the shadows or in the moonlight or sitting on the ground next to the low walls or on the steps outside the church, waiting silently, what would they talk about, remembering their own past pleasures, what it had been like or what it had not been like, how long those pleasures had lasted, until one of them says, We can go back now, and they all get up, See you tomorrow, and return to their houses, quietly lifting the latch, and the young couple are perhaps sleeping, quite innocent of any conjugal activities, which can't happen every night. But Sara da Conceição prefers to err on the side of caution, finding it difficult only when the weather is bad, and then she stands under a porch in the garden, but thanks to Faustina, who understood her, that's women for you, they would call her in, a sign that the night would be as pure as the cold stars, unless it was on one of those starry nights when João Mau-Tempo sought his legitimate wife beneath the sheets.

Perhaps Sara da Conceição, with all that coming and going, is merely fleeing the dreams that await her, but one thing is sure, at dawn, she will once more find herself in the olive grove, the day after the death, which was when they found the body, as she knows in her dream, and with a bottle of wine and a rag she rubs and rubs, and the head sways from side to side, and when it turns in her direction, her husband fixes her with his cold eyes, and when it turns away, the corpse has no face, which is even worse. Sara da Conceição wakes up in a cold sweat, hears her son snoring, her grandson tossing and turning, but not her granddaughters or her daughter-in-law, they're women after all, and therefore silent, and she moves closer to the two girls, with whom she sleeps, who can say what fate awaits them, let's hope a better fate than that of the woman who dreams such dreams.

One night, Sara da Conceição went out and did not come back. They found her in the morning, outside the village, quite lost and talking about her husband as if he were still alive. So sad. Her daughter, Maria da Conceição, who was working as a maid in Lisbon, asked her employers to help, and they did, and yet still people speak ill of the rich. Sara da Conceição traveled from Monte Lavre and, for the first time, took a taxi from the boat in Terreiro do Paço, south and southeast, to the insane asylum in Rilhafoles, where she lived until she died like a wick burning out for lack of oil. Sometimes, but not often, well, we all have our own lives to lead, Maria da Conceição went to visit her mother, and they would sit looking at each other, what else could they do. When, some years later, João Mau-Tempo was brought to Lisbon for reasons we will learn in due course, Sara da Conceição had died, surrounded by the laughter of the nurses, because the poor fool kept humbly asking for a bottle of wine, imagine that, for some task she had to finish before it was too late. Isn't that sad, ladies and gentlemen.

I N THE INVENTORY OF WARS, the latifundio plays its part, although not a large one. Those Europes, where another war has
just begun, play a far greater part, and from what one can ascertain, which is not very much in a land of such ignorance, so removed from the rest of the world, Spain is in such a state of ruin it
would break your heart. But any war is a war too many, that would
surely be the view of those who died in a war they never wanted.

When Lamberto Horques took charge of the lands in Monte
Lavre and environs, the soil was still fresh with the blood of Castil-
ians, although as to freshness, that is merely a rather bloodthirsty
image when set beside the far more ancient blood spilled by Lusitanians and Romans, or by the confusing tumult of Alanis, Van-
dals and Swabians, if they got this far, as the Visigoths certainly
did, followed later by the infernal, swarthy caravan of Moors, and
then the Burgundians arrived to spill their blood and that of others, and a few crusaders, not all of them heroes like Osberno,* and
then more Arabs, how much death these lands have seen, and the

---

* Osberno was a crusader who took part in the siege of Lisbon in 1147, when the city was
taken from the Moors by King Afonso I, and who received decisive help from crusaders
from northern Europe. Osberno left a written record of the siege.

only reason we haven't mentioned Portuguese blood is because all the blood spilled was Portuguese, or came to be, once enough time had passed for it to be naturalized, which is why we haven't mentioned the French or the English, for they truly are foreigners.

Things did not change after Lamberto Horques took over. The frontier is an open door, you can almost step across the Caia river, and the plain seems to have been deliberately and lovingly made smooth by warrior angels so that combatants can face each other with no obstacles to get in the way of arrows or, later on, all the many different kinds of bullets. The vocabulary of the armory is very beautiful, from the helmet to the cuirass, from the halberd to the harquebus, from the bombard to the ballista, and if the knowledge that such an arsenal walked, trod and fought in these lands sends a tremor of fear through you, you would tremble again if you saw the efficacy of such inventions. Anyway, blood was made to flow, whether from this wound in the throat or from that belly slit open, and would make an excellent ink in which to write such secret enigmas as whether those people were resigned to their deaths and aware of why they were dying. The bodies are carried away or buried where they fell, the latifundio is swept clean, and the land is left ready for the next battle. That is why the relevant trades have to be learned thoroughly and practiced assiduously, without a thought for expense, as when the Conde de Vimioso wrote this detailed letter to his majesty, Sir, the men of the cavalry should be armed with a carbine and two pistols per soldier, the carbines will take musket bullets or smaller and the barrel will be no more than three spans in length, which will be quite sufficient, because if they had to be reinforced, as such bullets require, thus making the barrel longer, the carbines would no longer be manageable, they will also need a metal charger for their powder flask, the pistols, too, will be of good quality, with a two-span barrel, and come with saddle holsters and two chains to hang them from, it would be useful if I could have some spare pistols and carbines so that we can make more of them, and a good quantity of iron should be sent to Vila Viçosa to be distributed to the riflemen, some of the iron can stay in Montemor and in Évora, those are my requirements for the cavalry, however, I leave it up to your majesty to decide what is most convenient.

His majesty, because of financial difficulties, did not always prove to be a prompt and generous paymaster, In Montemor we have been working on the fortifications with the two thousand cruzados that your majesty was kind enough to send and with the further two thousand donated by the people, and since the agreement was that your majesty would give six and the people another six, the town council has written to say that your majesty needs to give a further two thousand which they will then match, I told them that they should try to come up with that amount, and I, meanwhile, would ask your majesty to send your two thousand so that the people can then make their contribution. These are bureaucratic negotiations with distrust on both sides and a lot of buck-passing, but there is no haggling over blood, no one says, Why doesn't your majesty give a liter of your own blood, red or blue, it doesn't matter, because within half an hour of being spilled on the ground, it will be the same color as the earth. People don't dare go that far, because even if the blood of the whole royal household, including that of all the heirs to the throne and any bastard children the king or queen may have had, were poured into the same vat, it would still not be enough for the necessities of war. Let the people give their blood and their money, and his majesty will give the same amount of money that the people paid him earlier in taxes and tributes.

There are, of course, always calamities. All this talk of cavalry, crusades and fortifications, as well as the blood that binds them all together, belongs to the seventeenth century, a long, long time ago, but things have never improved, that's how, during the war of the oranges, we lost Olivença and never got it back, and thus, embarrassingly, without a shot being fired, Manuel Godoy, meeting with no resistance, marched in, and to our shame and his gallantry, he sent a fruit-laden branch from an orange tree to his lover, Queen María Luisa,* all that was lacking was for us to lie back and serve as their bed and mattress. Infinite misfortune, inconsolable grief, both of which lasted from the nineteenth century to the day before yesterday, there's something about oranges, they have a bad effect on both personal and collective destinies, if not, why would Alberto order any windfall oranges to be buried and say to the overseer, Bury the oranges, and if anyone picks them up and eats them, they'll be dismissed as of Saturday, and some men were dismissed because, in secret, they did eat the oranges, that forbidden fruit, while they were still good, rather than leaving them to rot beneath the earth, buried alive, poor things, what did we or the oranges do wrong. But there is a reason for everything, let us take a closer look at the situation, because, toward the end of the war that has just begun in Europe, a certain Hitler, Germany's very own Horques Alemão, will send children of twelve or thirteen to form the last battalions of the defeat, wearing uniforms so big they fall from their shoulders and hang about their ankles, carrying recoil rifles that their shoulders are too frail to withstand, and that's precisely what the owners of the large estates complain about, that there are no longer any children of six or seven who can tend the pigs or the turkeys, what will happen if they can't earn their daily bread, they say to the brutalized parents who have already given their blood and their money and still haven't caught on, or are just

* In 1800, Bonaparte and his ally, the Spanish prime minister Manuel de Godoy, issued an ultimatum to Portugal demanding that it enter into an alliance with France in the war against Britain and cede to France most of its national territory. Portugal refused, and in April 1801, French troops arrived in Portugal. On May 20, they were joined by Spanish troops under the command of Godoy. In a disastrous battle for Portugal, Godoy took the Portuguese town of Olivença and, following his victory, picked some oranges and sent them to his mistress, the queen of Spain. The conflict thus became known as the War of the Oranges.

beginning to feel the stirrings of mistrust, as, in another century, they distrusted the king's scornful rebuffs.

Wars are the least of it. A man can get used to anything, and between one war and another, he has time to make a few children and hand them over to the latifundio, without a spear thrust or a rifle shot cutting short the dream that the boy might be lucky enough to be made a foreman or an overseer or a trusted servant, or might choose to go and live in the city, which provides at least for a cleaner death. The worst of all things are the plagues and the famines that occur most years, and which are the ruin of the people, leaving the fields empty, whole villages closed down, you can travel for leagues without seeing a soul, although now and again you might spot ragged, wretched bands walking paths that the devil would walk only if carried on the shoulders of men. Some fall by the wayside, it's an itinerary of corpses, and when the plague relents and the famine eases and the living are counted, you don't have to count very high, because there are so few left.

These are all evils, and great evils at that. One might say, to use the language of Father Agamedes, that they are the three horsemen of the apocalypse, of whom there were once four, and if you start to count, on your fingers if you know no better system, the first is war, the second is plague and the third is famine, and there's always the fourth, the wild beasts of the earth. The last is the most commonly seen and has three faces, the face of the latifundio, the face of the guards who defend property in general and the latifundio in particular, and then there's the third face. He's a serpent with three heads and but one desire. He who gives orders is not necessarily the best fitted to do so, and the best fitted to give orders does not necessarily look the part. But we should perhaps speak more clearly. This horse can be seen in all the cities, towns, villages and hamlets, and he trots along with his leaden eyes and his legs that resemble human hands and feet, but are not human. What human being would say to Manuel Espada, years later, when on military service in the Azores, and forgive us if we jump forward a little in the story, When I get out of here, I'm going to join the police for the vigilance and defense of the state,* and Manuel Espada asked, What's that, and the other man answered, It's the political police, and it's just great, say there's someone you don't like, you simply arrest him, haul him off to the civil authorities and, if you like, shoot him in the head before you get there and say he tried to resist. This horse kicks down doors, eats at the same table on the latifundio as Father Agamedes and plays cards with the guards, while the colt called Bom-Tempo kicks in the prisoner's head. You can find these horses in cities, towns, villages, everywhere, they neigh, rub noses, exchange secrets and allegations, invent persuasive tortures and tortuous persuasions, which is what first made us realize that they did not belong to the equine race, Father Agamedes is a fool to believe that the horses he read about in the bible were real horses, a fundamental error revealed to Manuel Espada in the Azores by his promising fellow recruit. The roots of the tree of knowledge are not fussy about where they grow and are not put off by distance.

Father Agamedes bawls from the pulpit, There are certain men sneaking around who are intent on undermining your common sense, and yet in Spain, by the grace of God and the Virgin Mary, they were crushed, vade retro Satanas et abrenuncio, you must flee from them as if from plague, famine and war, for they are the worst misfortune that could befall our holy land, like the plague of locusts in Egypt, and that is why I will never tire of saying to you that you must heed and obey those who know more about life and the world, look upon the guards as your guardian angels, don't resent them, because sometimes even a father is obliged to beat the child he so loves and cares for, and we all know that sooner

* The PVDE was created in 1933 by Salazar himself in order to prevent, repress and punish crimes of a social or political nature. In 1945, this body was dissolved and replaced by the PIDE (Polícia Internacional e de Defesa do Estado), whose task was to investigate, detain and arrest anyone suspected of plotting against the State.

or later the child will say, It was for my own good, the only blows
that were wasted were those that struck the ground, that, my children, is how it is with the guards, not to mention the other authorities, both civil and military, the mayor, the administrator, the
regimental commander, the civil governor, the commander of the
legion and all those other gentlemen in positions of power, beginning with those who give you work, yes, what would become of
you if there were no one to give you work, how would you feed
your families, tell me that, answer me, all right, I know that the
congregation does not normally speak during mass, but it is your
own conscience you must answer to, and for all these reasons, I
urge, demand and order you to pay no heed to those red devils
who want only your unhappiness, because that was not why God
created the earth, he created it that it might be rocked in the lov-
ing arms of the Virgin Mary, and if you believe that someone is
trying to lead you astray with seductive words, then go straight to
the guards' barracks, for then you will be carrying out God's work,
but if you lack the courage, if you are afraid of reprisals, I will hear
you in the confessional and do with your confession what my soul
and my conscience deem to be the right thing, and now let us say a
pater noster for the salvation of our country, a pater noster for the
conversion of Russia and a pater noster for those who govern our
nation, who have so sacrificed themselves and who so love us, our
father, who art in heaven, blessed be thy name.

Father Agamedes is quite right. There are men roaming the latifundio, they can be found hiding away in groups of three or four,
in solitary places or abandoned houses, where they keep watch, or
in the shelter of a valley, some from here, some from elsewhere,
and they hold long conversations. They take turns to speak, and
the others listen, anyone seeing them from a distance would say,
They're itinerant workers, gypsies, apostles, and when they have
finished talking, they scatter, taking out-of-the-way paths and carrying with them papers and decisions. This is what is called or-

ganization, and Father Agamedes is purple with rage, with righteous anger, May they be damned, may their souls fall into the very depths of hell, they are a harmful infection that seeks to destroy you, only yesterday I was talking to the president of the council, and he said to me, That fatal disease is already afflicting our village, Father Agamedes, we must do something to counteract the pernicious doctrines being spread among our families by these enemies of faith and civilization, O ingrates, do you not realize that the peace and order we enjoy in our country is the envy of other nations, and you come to me saying that you are willing to lose all that, you're just spoiled, that's your problem.

João Mau-Tempo has never been a man for going to mass, but now that he lives in Monte Lavre, he goes to church now and then both to please his wife and out of necessity. He hears Father Agamedes's fiery sermons and compares them in his head with what he has picked up from the papers handed to him in secret, he makes his own judgment as a simple man, and while he believes some of the things written on those papers, he doesn't believe a word the priest says. It seems that Father Agamedes himself finds it hard to believe, with all that ranting and raving and foaming at the mouth, which does not look good on one of God's ministers. When mass is over, João Mau-Tempo goes out into the square along with the rest of the congregation, and there he finds Faustina, who had been sitting with the other women, and he walks part of the way home with her before going to join some friends to have a drink, just one, though the others laugh at him, You drink like a little boy, Mau-Tempo, but he merely smiles, a smile that says everything, so much so that the others say nothing more, it's as if the body of a hanged man had suddenly dropped down from one of the beams in the inn. Then one of his friends says, Did Father Agamedes give a good sermon today, a question that has no answer, because he is one of the few men in Monte Lavre who never goes to mass, he only asks in order to provoke, João Mau-Tempo smiles again and says, Oh, it was the usual thing, then says nothing more, be-cause he's nearly forty now and never drinks so much that he loses control of his tongue. It was that same friend who gave him the papers, and they look at each other, and Sigismundo, for that is his friend's name, winks and raises his glass of wine to him, Good health.

I T WAS WHILE ANTÓNIO Mau-Tempo was employed tending pigs that he met Manuel Espada, who had been forced to
take such unskilled work because he could find nothing else once
he and his companions had become dubbed locally, and for two
leagues around, as strikers. Like everyone else in Monte Lavre, António Mau-Tempo knew what had happened, and in his still childish imaginings, he found some similarities with his own rebellion
against the pine-nut-roasting, stick-wielding foreman, although
he never confessed as much, especially given that Manuel Espada
was six years older than him, long enough to separate a mere child
from a lad and a mere lad from a man. The foreman of these pigs
didn't work any harder than the other one, but he, at least, had age
as an excuse, and the lads he employed didn't mind taking orders
from him, after all, someone has to be in charge, him in charge
of us and us in charge of the pigs. The working day of the swineherd is very long, even in winter, the hours pass so slowly they positively dawdle, like a shadow moving from here to there, and pigs
are creatures of little imagination, their snouts always pressed to
the ground, and if they do wander off, they mean no mischief, and
a well-aimed stone or a sharp thump on the back with a stick will

bring them, ears twitching, back to the rest of the herd. The pig soon forgets such incidents, having a poor memory and being little prone to bearing grudges.

There was, then, more than enough time to talk, while the foreman dozed under the holm oak or tended the animals farther off. Manuel Espada spoke of his adventures as a striker, although he never exaggerated, that wasn't in his character, and he shed a little light on the kind of thing that can happen on the threshing floor at night with the female workers, especially the ones from the north who have no men with them. The two became friends, and António Mau-Tempo greatly admired the older lad's serenity, a quality he lacked, for, as we will see later on, he was always itching to be up and off. He had inherited the vagabond tendencies of his grandfather Domingos Mau-Tempo, with the great and praiseworthy difference that he had a naturally sunny temperament, which didn't mean, however, that he was always laughing and joking. Nevertheless, he had the same tastes and anxieties of any lad his age, and took on the ancient and never resolved question of what separates boys from sparrows, he always spoke his mind and was, on occasions, impetuous, and those qualities will make him a touch impatient and something of a wanderer. He'll enjoy dances, as his father did in his youth, but will care little for large gatherings. He will be a great teller of tales about things he has either seen or invented, experienced or imagined, and he will possess the supreme art of being able to blur the frontiers between the two. But he will always work hard at acquiring all the rural skills. We're not reading this future in the palm of his hand, these are simply the elementary facts of a life that contained many other things, including some that appeared not to be promised to his generation.

António Mau-Tempo did not spend long with the pigs. He left Manuel Espada there and went off to learn skills that the latter, being older, already knew, and at thirteen, he was working with grown men, burning undergrowth, digging ditches, building dams, tasks requiring good strong arms. By the time he was fifteen, he had learned to cut cork, a precious skill at which he became a master, as, to be frank, he did with everything he turned his hand to. When he was still very young, he left his mother and father and traveled to places where his grandfather had left his mark and a few bad memories. But he was so very different from his grandfather that it never occurred to anyone that they could possibly belong to the same family, despite having the same surname. He was very drawn to the sea, he discovered the banks of the river Sado, and walked its whole length, which is no small journey, just to earn a bit more money than the pittance being offered in Monte Lavre. And one day, much later, as we will describe in due course, he will go to France to exchange a few years of life for a little hard currency.

The latifundio has its pauses, the days are indifferent or so it seems, what day is it today, for example. It's true that people die and are born just as they did in more remarkable times, hunger still doesn't always take account of the needs of the stomach, and the heavy workload hasn't grown much lighter. The biggest changes happen outside, there are more roads and more cars on them, more radios and more time to listen to them, understanding them is another skill entirely, more beers and more fizzy drinks, but when a man lies down at night, in his own bed or on the straw in a field, the pain in his body is just the same, and yet he should consider himself lucky to be employed. There's nothing much to say about the women, their fate as beasts of burden and bearers of children remains the same.

However, when one looks at this apparently lifeless swamp, only someone born blind or choosing not to see could fail to notice the watery tremor rising suddenly from the depths to the surface, the result of accumulated tensions in the mud, caught up in a chemical process of making, unmaking and remaking, until the liberated gas explodes. But to notice this, you have to look hard and not say as you pass by, There's no point hanging around here, let's go. If we were to go away for a while, distracted by different landscapes and picturesque events, we would notice, on our return, how, contrary to appearances, everything is finally changing. That is what will happen when we leave António Mau-Tempo to his life and return to the thread of the story we began, though all this is merely hearsay, including the story about José Gato and the misfortune that befell him and his companions, as António Mau-Tempo can witness and testify.

This isn't one of those tedious tales about the Brazilian bandit Lampião, nor of others nearer to home, such as João Brandão or José do Telhado, who were bad people or, who knows, just wrongheaded. I don't mean by this that there had never been any shady characters on the latifundio, no bandits who would leave a traveler dead and stripped of everything he had, regardless of how little that was, but the only one I knew of was José Gato, he and his companions, or should I say gang, whose names, if I remember rightly, were Parrilhas, Venta Rachada, Ludgero, Castelo and others whose names I've forgotten, well, one can't remember everything. I'm not even sure they were bandits. Itinerant workers, yes, that's what they were. If they wanted to work, they would work as hard as anyone else, they weren't criminals, but one day, it was as if they suddenly got the wind up their tails or something, and they put down their hoes or their axes, went to the overseer or foreman to receive what was owed to them, because no one ever dared to withhold their pay, and then they vanished. At first, they went their separate ways, each silent, solitary man for himself, and only later did they get together and form a gang. When I met them, José Gato was already the head of the gang, and I don't think anyone would have tried to take his place. They stole mainly pigs, of which, it must be said, there was no shortage. They stole in order to eat and also to sell, of course, because a man cannot live only by what he eats. At the time, they had a boat anchored in the Sado river, and that was their slaughterhouse. They slaughtered the animals and placed the meat in the salting trough for times of need. And speaking of the salting trough, they once ran out of salt and were discussing what to do and what not to do, and José Gato, who was a man of very few words indeed, told Parrilhas to go to the saltworks. Normally, José Gato only had to say, Do this, and like the word of God, it was done, but for some reason Parrilhas refused to go, a decision he lived to regret. José Gato snatched off Parrilhas's hat, threw it in the air, picked up his rifle and blasted the hat to pieces with two shots, then he said to Parrilhas in the quietest of voices, Go and get the salt, and Parrilhas saddled up the donkey and went to get the salt. That was the kind of man José Gato was.

For anyone living in one of the work camps nearby and who was brave enough, José Gato was the main supplier of pork meat. One day, Venta Rachada turned up, in secret, of course, at the place where I was working, to ask if anyone wanted to buy some meat. I did, as did two of my companions, and we arranged to meet Venta Rachada in a place called Silha dos Pinheiros. We went there, each carrying our coarse linen bag and a little money, and, just in case, those of us who had some money put by left it back at the camp, we didn't want to go looking for wool and come back shorn. I had fifty mil réis on me, and the others had more or less the same amount. It was pitch-black outside, and the place where we were to meet Venta Rachada was enough to give anyone the creeps, in fact, he was hiding there, waiting, and he played a trick on us by leaping out, pointing his rifle at us and saying, I could rob you of all you have, we all laughed, of course, and I, my heart thumping, managed to say, It would hardly be worth the bother, and then it was Venta Rachada's turn to laugh and say, Don't worry, I won't hurt you, follow me.

At the time, José Gato was based in the Loureiro hills, near Palma, you probably know the place. It was full of cane apple trees as big as a house and no one ever went there. An abandoned farm laborers' hut served as their slaughterhouse. They all lived there and moved on only when they noticed any suspicious activity, strangers prowling around, or heard rumors that the guards were closing in. We walked and walked and when we got within sight of the hut, we saw two men on guard, rifles at the ready. Parrilhas gave his name, we went in and found José Gato and the other men playing the mouth organ and dancing the fandango, now I don't know much about such things, but I thought they danced pretty well, and besides, everyone has a right to enjoy themselves now and then. Looped over one of the beams above the fire were some wires from which hung a large stewpot containing pigs' innards. José Gato said, So these are our buyers, are they. Venta Rachada said, They are, and the only ones, too. José Gato said, Don't worry, boys, before we do business, join us for a bite to eat, these were welcome words indeed, because the smell was already beginning to make my mouth water. They had wine, they had everything. To sharpen our appetite, we had some slices of ham and a few glasses of wine, José Gato played the mouth organ and kept an eye on the pot, he was wearing chaps made of donkey skin, with big buttons on them, as was the fashion, the rascal looked just like any other farmer. In one corner of the hut there were various rifles, the gang's arsenal, one was a five-shot rifle and belonged to Marcelino, but more of that later. We were happily engaged in eating and drinking when suddenly we heard a bell ringing, *ting-a-ling*, and I must confess that I shuddered, this could all end very badly indeed. José Gato noticed my unease and said, Don't worry, they're friends, they've come to buy meat. It was Manuel da Revolta, so called because he owned a shop in Monte da Revolta, and I could tell you a few stories about him too, but another time. Anyway, Manuel da Revolta arrived, loaded six pigs onto his cart and carried them off, the next day, of course, he would be doing the rounds of the work camps, selling them, pretending he had slaughtered them himself, even the guards would buy meat from him, and I still don't know to this day whether the guards were suspicious or whether it simply suited their purposes to say nothing. Then a fishmonger we all knew arrived, he kept us all supplied with fish and tobacco and a few other things that José Gato needed. He loaded one pig onto his bicycle, but left the head behind. Then someone else arrived, without a bell this time, he simply whistled and those on guard responded, that was the arrangement, just in case. He took away two pigs, one slung on either side of his mule, again with no head, the pigs that is, because obviously the mule needed his head to see where he was putting his feet. In the end, there were only two pigs left, lying on a couple of old sacks. A few rashers of bacon were fried and added to the stew, along with the seasonings, onions and so on, and then down it went into our stomachs, and boy, was it good that stew, washed down with a fair bit of wine. Then José Gato said, addressing me, António Mau-Tempo, Right, to business, how much money did you bring with you, and I said, I've brought fifty escudos, that's all I've got. Said José Gato, It's not a lot, but you won't leave empty-handed, and he sliced a pig in two, a piece weighing four and a half or five arrobas, Open your bag, but first he made sure to take the money and slip it into his pocket. It was the same with the others, to all of whom he said, Not a word to anyone, if you tell a soul, you'll live to regret it, and so we left, laden down with meat, and his warnings and threats stood us in good stead, because it turned out later that the pigs had been stolen from the very estate we were working on. The overseer bombarded us with questions, but all three of us kept our word. I dug a hole in the ground, lined it with cork, put in the meat and covered it with a cloth, having first sliced it up and salted it. It kept really well, too, and we had meat for a good long time.

That's just one story. Had it been João Brandão, I'm not sure how it would have turned out, but the man I dealt with was José Gato, with someone else it might have been different. Later, the gang moved to Vale de Reis, you city folk just can't imagine how wild it is around there, grottoes and caves and evil-looking swamps, no one else would go anywhere near, not even the guards, they didn't dare. The gang set up camp there, and they had a warning system in Monte da Revolta, whenever the guards appeared, Manuel da Revolta's mother would stick a pole up the chimney with a rag tied on top, and that was the sign. One of the gang always kept an eye on that chimney, and as soon as he saw that old rag, he would warn the others and they would all vanish, disappear without trace. The guards never caught any of them. Those of us who knew the signal, when we were out in the fields working, we'd say, Something's up.

Let me tell you now about Marcelino. He was the overseer in Vale de Reis and owned a famous rifle that the boss had bought him so that he could shoot any member of José Gato's gang he caught stealing. But before I tell you about that, I want to tell you another story about a rifle. Once, when Marcelino was out riding, José Gato ambushed him and, with his gun pointing straight at him, said mockingly, which was very much his style, Just open your arms nice and wide and I'll take the rifle, and Marcelino had no alternative but to do as asked, however much it galled him. José Gato was a small man, but he had a very big heart. Then it was the turn of the five-shot rifle, you know how it is, you start telling one story and other stories get in the way. Marcelino was riding along a path, no one bothered to clear the paths then, they were too busy cutting cork and slicing it up into small pieces, so the undergrowth was really thick. Marcelino was riding proudly along with his five-shot rifle loaded with five cartridges, thinking, If anyone tries to attack me now, that'll be their goose well and truly cooked, but José Gato was hiding behind a slender holm oak, aiming straight at him, Give me that rifle, I need it, and off he went. Later, the boss said to Marcelino, I'll buy you a carbine, I don't want you being made to look a fool, and Marcelino replied tartly, I don't want a carbine, from now on, it'll be just me and my stick, that's the best way to keep watch.

Marcelino had no luck at all with rifles. He even lost the one he owned himself and kept at home. The swineherd's dogs started barking, they could smell that something was up, and the swineherd went to Marcelino and said, The dogs are barking, there's someone trying to steal the pigs. Marcelino immediately picked up his rifle and his cartridge box and stood there guarding the pigs. Now and then he fired a shot, and José Gato's men, hiding in the bushes, knew that these shots were intended for them and responded, although without wasting much ammunition. And where was José Gato all this time, why, up on the roof, onto which he had climbed unnoticed and where he remained all night, crouched like a lizard so that no one would spot him, he was nothing if not bold. Come the morning, at daybreak or shortly afterward, just as it was beginning to grow light, and when any shots from the other side had long since ceased, Marcelino said, They must have run away, I'll just go home and have my breakfast, I'll be back in a jiffy. And the swineherd, whose own appetite was stirred by those words, thought, Yes, I'll go and have a bite to eat as well, why not. With his enemies gone, José Gato jumped down from the roof, ah, I forgot to mention that Marcelino had left his rifle inside the swineherd's hut, anyway, José Gato jumped down from the roof, took the rifle and the swineherd's new boots and a blanket, perhaps they were short of those as well, and meanwhile, his companions, there were five of them at the time, grabbed a pig each and carried them off into the undergrowth. Sows are like us, they have a joint just here, and if you cut it, they can't move, and that's what happened with these, only about a hundred and fifty yards from the pen, if that. And with someone keeping watch all the time. The boars noticed the sows were missing, but went look-

ing for them far away, down the road, and none of them thought
of looking closer to home. That night, José Gato went to fetch the
sows, and so Marcelino's third rifle was lost.

There's another, more important story. Marcelino was standing guard, without his rifle this time, for they had all been stolen,
and José Gato decided to set about stealing the broad beans, which
had all been harvested and were lying on the threshing floor. It was
close to the gang's current hideout which we found out was there
only when we were felling trees in the area, by which time they had
moved on. They had dug a deep ditch and carved out caves along
the walls. There were some high hills all overgrown with willows,
and they had cut a path through them, rather the way mongoose
do, and created alcoves furnished with comfortable beds made out
of reeds and twigs. Anyway, José Gato went out nightly to steal
some of the beans, and Marcelino realized that someone had been
taking them because some had been crushed underfoot and you
could see the empty shells underneath. Marcelino said to himself,
The bastards, they're after my beans, and so what did he decide
to do, I'm going to confront them, he said, and so he tethered his
horse out of sight, took a large sack with him, because in summer
you don't need a blanket, and a big stick. Shortly afterward, he
heard rustling, it was José Gato tossing three or four bundles of
beans in a cloth to shell them, but everything was so dry that the
beans crunched underfoot, and then, at the agreed hour, a member of the gang came to help him carry away the beans, about a
hundred liters of them. They were probably going to sell them to
Manuel da Revolta in exchange for bread and other essentials, I'm
not sure. José Gato was completely absorbed in his work, and Marcelino, barefoot, was drawing closer and closer, his own description of it was very funny, I was barefoot, you see, gradually edging
nearer, and I got within about six or seven meters of the guy, another three or four meters and I could have hit him with my stick,
but he was too sharp and he heard me, and just when I thought I'd

deal him a blow with my stick, in two hops he was gone, now you see him, now you don't, and I was pretty quick off the mark myself, but there he was pointing his rifle at me. José Gato said, or so Marcelino said, You're lucky, you were kind to a friend of mine once, that was at a time when the guards were doing their worst and Marcelino had given shelter and food to one of the gang, You're lucky, otherwise, I would have shot you dead. But Marcelino was a brave man too in his own way, Hang on, this calls for a smoke, and he pulled out his tobacco pouch, rolled himself a cigarette, stuck it in his mouth, lit it, then said, Right, I'm off now.

Later, the gang were all arrested. It started in Piçarras, in an out-of-the-way place between Munhola and Landeira. There was a showdown with the guards, shots were fired, it was like a war. The guards caught them, but every one of them was given a job by local farmers, Venta Rachada became a watchman on a vineyard in Zambujal, and others the same. I would love to have heard one of those conversations between guards and farmers, We've arrested a man, Oh good, I'll have him, I don't know who was the more brazen of the two. José Gato was only arrested some time later, in Vendas Novas. He was living with a woman who sold vegetables there and he always went about in disguise, which is why the guards never caught him. Some say she gave him away, but I don't know. He was taken prisoner at his lover's house, in the cellar, when he was sleeping, in fact, he had said once, If they don't catch me while I'm sleeping, they won't catch me at all. Rumor had it that he was taken to Lisbon, and just as the others were given jobs by farmers, it was said that he had been sent to the colonies as a member of the PVDE. I don't know if he would ever have agreed to that, I find it hard to believe, or perhaps they killed him and that was the story they made up, it wouldn't be the first time.

José Gato had many good qualities. He never stole from the poor, his intention being to steal only from the rich, as people say José do Telhado used to. Once, Parrilhas came across a woman who had gone shopping for her family, and he robbed her, the wicked devil. Unfortunately for him, José Gato found the poor woman sobbing. He asked what was wrong and realized from what she said that Parrilhas had been her attacker. He gave the woman enough money for three loads of shopping and Parrilhas got the worst beating of his life. Quite right, too.

José Gato was a man with no illusions, small in stature but brave, as you'll see from something that happened in Monte da Revolta. At the time, it was a very international place, you got people there from all over, suffice it to say that a man from the Algarve who was working on clearing the land managed to build a little cabin for himself, and there were others like him, with no house and no home, or if they had one, they kept quiet about it. A man there tried to provoke an argument between Manuel da Revolta and José Gato, telling Manuel da Revolta that José Gato had boasted about how he was going to sleep with Manuel's wife. But Manuel da Revolta, who trusted José Gato, said to him straight out, So-and-so told me this. José Gato said, The bastard, let's go and see him, and so they did, and when they got there, he said, This is what you told Manuel, and I'd like to hear you say the same to my face. The other man answered, Look, I was a bit drunk at the time, but you never said anything of the kind, and that's the honest truth. José Gato said very calmly, Walk a hundred paces ahead of me, that way he knew he had no chance of killing the man, then he fired two or three shots at his back, so that a couple of pellets just stuck in his flesh while the others ricocheted off, then he gave him a couple of lashes with a whip as he lay on the ground, Behave like a man from now on, and don't go playing any more childish pranks on people. It always seemed to me that José Gato got involved in a life of crime only because he couldn't earn enough to eat.

He was in this area when I was a little boy. He was the foreman in charge of clearing the area between Monte Lavre and Coruche. The road was built entirely by itinerant laborers, lots of people worked like that, putting in three or four weeks until they had earned enough cash and then others came to take their place. José Gato arrived and clearly knew what he was doing, so he was made foreman, although he kept away from the low-lying valleys. I was herding pigs at the time, before I got to know Manuel Espada, so I saw it all. It came to be known that he'd had a few run-ins with the guards, and then the guards learned, or someone told them, that he was in the area, and they hunted him down and caught him. They didn't quite have the measure of him though. He was at the head of the patrol, looking all meek and mild, and the guards were following behind him, looking smug, then suddenly he bent down, grabbed a handful of earth and threw it in the eyes of one of the guards, and was gone. Until his final arrest, they never saw him again. José Gato was a true wanderer. And I reckon he was always a very solitary man.

THE WORLD WITH ALL its weight, this globe with no beginning and no end, made up of seas and lands, crisscrossed by rivers, streams and brooks carrying the clear water that comes and goes and is always the same, whether suspended in the clouds or hidden in the springs beneath the great subterranean plates, this world that looks like a great lump of rock rolling around the heavens or, as it will appear to astronauts one day and as we can already imagine, like a spinning top, this world, seen from Monte Lavre, is a very delicate thing, a small watch that can take only so much winding and not a turn more, that starts to tremble and twitch if a large finger approaches the balance wheel and seems about to touch, however lightly, the hairspring, as nervous as a heart. A watch is solid and rustproof inside its polished case, shockproof up to a point, even waterproof for those who have the exquisite taste to go swimming with it, it is guaranteed for a certain number of years, possibly many years if fashion does not laugh at what we bought only yesterday, for that is how the factory maintains its outflow of watches and its inflow of dividends. But if you remove its shell, if the wind, sun and rain begin to spin and beat inside it, among the jewels and the gears, you can safely bet that the happy days are over. Seen from Monte Lavre, the world is an open clock, with its innards exposed to the sun, waiting for its hour to come.

Having been sown at the right time, the wheat sprouted, grew and is now ripe. We pluck an ear from the edge of the field and rub it between the palms of our hands, an ancient gesture. The warm, dry husk crumbles and we hold cupped in our hand the eighteen or twenty grains from that ear and we say, It's time to harvest. These are the magic words that will set in motion both machines and men, this is the moment when, to abandon the image of the watch, the snake of the earth sheds its skin and is left defenseless. If we want things to change, we must grab the snake before it disappears. From high up in Monte Lavre, the owners of the latifundio gaze out at the great yellow waves whispering beneath the gentle breeze, and say to their overseers, It's time to harvest, or, if informed of this in their Lisbon homes, indolently say the same thing, or, more succinctly, So be it, but having said these words, they are trusting that the world will give another turn, that the latifundio will respect the regularity of its customs and its seasons, and they are relying, in a way, on the urgency with which the earth accomplishes these tasks. The war has just ended, a time of universal fraternal love is about to begin. They say that soon the ration books will be unnecessary, those little bits of colored paper that give you the right to eat, if, of course, you have the money to pay with and always assuming there is something for that money to buy. These people aren't much bothered really. They have eaten little and badly all their lives, they have known only scarcity, and the hunger marches practiced here are as old as tales of the evil eye. However, everything has its moment. As anyone can see, this wheat is ripe and so are the men.

There are two slogans, not to accept the daily rate of twenty-five escudos and not to work for less than thirty-three escudos a day, from morning to night, because that's how it must be, fruits do not all ripen at the same time. If the wheatfields could speak, they would say in astonishment, What's going on, aren't they going to harvest us, someone isn't doing his job. Pure imagination. The wheatfields are ripe and waiting, it's getting late. Either the men come now or, when the season is over, the stems will break, the ears crumble, and all the grain will fall to the ground to feed the birds and a few insects, until, so that not everything is lost, they let the livestock into the fields, where they will live as if in the land of Cockaigne. That is pure imagination too. One side will have to give in, there is no record of the wheat ever being left to fall to the ground like that, or if it did, it was the exception that proved the rule. The latifundio orders foremen and overseers to stand firm, the language is warlike, No going back, the imperial guard will die rather than surrender, oh, if only they would die, but there are faint echoes here of bugle calls, or are they merely a nostalgia for battles lost. The guards are beginning to emerge from their cocoons, the corporals and sergeants appear at the windows of their barracks to sniff the air, some are oiling their rifles and giving their horses double rations from the emergency reserves. In the towns, men stand shoulder to shoulder, muttering. The overseers come to talk to them again, So, have you reached a decision, and they reply, We have, and we won't work for less. In the distance, on this hot evening, a warm wind blows as if it came from the earth itself, and the hills continue to hold tight to the roots of those dry stalks. Hidden in the forest of the wheatfield, the partridges are listening hard. No sound of men passing, no roaring engine, no tremulous shaking of the ears of wheat as the sickle or the whirlwind of the harvester approach. What a strange world this is.

Saturday comes. The overseers have been to speak to the owners, They're very determined, they said, and the owners of the latifundio, Norberto, Alberto, Dagoberto, replied in unison, each from his particular place in the landscape, Let them learn their lesson. In their houses, the men have just had supper, the little or nothing they dine on every day, the women are looking at them in

silence, and some ask, What now, while some men shrug glumly
and others say, They're sure to come to their senses tomorrow, and
there are those who have decided to accept what they are being offered, the same pay as last year. It's true that from all sides comes
news that many men are refusing to work for such a pittance, but
what is a man to do if he has a wife and children, the little urchins
who are all eyes and who stand, chin resting on the edge of the table, using one saliva-moistened fingertip to hunt breadcrumbs as
if they were ants. Some of the luckier men, although they might
not seem so to those who know little of such things, have found
employment with a smallholder, a man who cannot risk losing his
harvest and who has already agreed to pay them thirty-three escudos. The night will be a long one, as if it were winter already.
Above the rooftops is the usual wasteful sprawl of stars, if only
we could eat them, but they're too far away, the ostentatious serenity of a heaven to which Father Agamedes keeps returning, he
has no other topic, stating that, up above, all our hardships in this
vale of tears will end and we will all stand equal before the Lord.
Empty stomachs protest, grumbling away at nothing, proof of that
inequality. Your wife beside you isn't asleep, but you don't feel like
rolling over on top of her. Perhaps tomorrow the bosses will come
to an agreement, perhaps we'll find a pot full of gold coins buried
at the back of the fireplace, perhaps the chicken will start laying
golden eggs, or even silver would do, perhaps the poor will wake
up rich and the rich poor. But we do not find such delights even in
dreams.

Dearly beloved children, says Father Agamedes at mass, because it's Sunday already, Dearly beloved children, and he pretends
not to notice how sparse the congregation is and how ancient most
of its members are, nothing but old ladies and altar boys, Dearly
beloved children, and it's only natural that the old ladies should be
thinking vaguely that they long ago ceased to be children, but what
can one do, the world belongs to men, Dearly beloved children, be

very careful, the winds of revolution are blowing across our happy lands, and once more I say to you, pay them no heed, but why bother writing down the rest, we know Father Agamedes's sermon by heart. The mass ends, the priest disrobes, it's Sunday, that holiest of days, and lunch, blessings be upon it, will be served in the cool of Clariberto's dining room, although Clariberto goes to mass only when he really wants to, which is rare, and the ladies are equally lazy, but Father Agamedes doesn't take it to heart, if they should be overcome by devotion or overwhelmed by fears of the beyond, they have a chapel in the garden, with newly varnished saints, including a Saint Sebastian generously sprinkled with arrows, may God forgive me, but the saint does seem to be enjoying it rather more than virtue should allow, and Father Agamedes enters through the same door that the overseer Pompeu has just left, carrying in his ear the consoling message, Not a penny more, there's nothing quite like a man with authority, be it on earth or in heaven.

A few men are hanging around outside, and although the labor market normally starts later on, some of them go to the overseer and ask, So what has the boss decided, and he replies, Not a penny more, well, why waste a nice turn of phrase or spoil it with redundant variations, and the men say, But some farmers are already paying thirty-three escudos, and Pompeu says, That's their business, if they want to bankrupt themselves, good luck to them. This is when João Mau-Tempo opens his mouth, and the words come out as naturally as water flowing from a good spring, The wheat won't get harvested then, because we're not working for less. The overseer did not reply, because his lunch was waiting for him and he wasn't in the mood for such unsettling conversations. And the sun beat down hard, glinting like a guard's saber.

Those who could eat ate, and those who couldn't starved. The labor market has begun now, all the rural workers from Monte Lavre are there, even those who have already been hired, but only the ones who are being paid thirty-three escudos, anyone who accepted the old rate is sitting at home, chewing on his own shame, getting annoyed with his children who can't keep still and giving them a clip on the ear for no reason, and the wife, who is always the voice of justice in any punishment, protests, We're the ones who bore them, besides, you shouldn't hit an innocent child, but the men in the square are innocent too, they're not asking for the moon, just thirty-three escudos for a day's work, it's hardly an outrageous amount, by which they mean that the boss isn't going to lose out. This isn't what Pompeu and the other overseers say, but perhaps he speaks more brusquely because of his Roman name, What you're asking for is outrageous, you'll be the ruin of agriculture. Various voices cry, Some farmers are already paying that, and the chorus of overseers replies, That's their choice, but we're not paying it. And so the haggling continues, retort and counter-retort, who will tire of it first, it's hardly a dialogue worthy of setting down, but there is nothing else.

The sea beats on the shore, well, that's one way of describing it, but not everyone would know what we meant, because there are many around here who have never been to the sea, the sea beats on the shore and if it meets a sandcastle in its path or a rickety fence, it will flatten both, if not at the first attempt, then at the second, and the sandcastle will have been razed to the ground and the fence reduced to a few planks being washed back and forth by the waves. It would be simpler to say that many men accepted the twenty-five escudos, and only a few dug in their heels and refused. And now that they are alone in the square, asking each other if it was worth it, and Sigismundo Canastro, who is one of those men, says, We mustn't get discouraged, this isn't happening only in Monte Lavre, if we win, then everyone will benefit. What makes him think this, when there are just twenty men unemployed. If only there were more of us, says João Mau-Tempo gloomily. And these twenty men seem about to go their separate ways, with nowhere to head but

home, which is not a good place to be today. Sigismundo Canastro tells them his idea, Tomorrow, let's go together to the fields and
ask our comrades not to work, tell them that everywhere people
are fighting for their thirty-three escudos, we in Monte Lavre can't
be seen to weaken, we're as brave as they are, and if the whole district refused to work, the bosses would have to give in. Someone in
the group asks, What's happening in those other places then, and
someone answers, either Sigismundo Canastro or Manuel Espada
or someone else, it doesn't matter, It's the same in Beja, in San-
tarém, in Portalegre, in Setúbal, this isn't just one man's idea, either
we all work together or we're lost. João Mau-Tempo, who is one of
the older men present and therefore has a greater responsibility,
stares into the distance as if he were gazing inside himself, judging
his own strength, and then he says, We should do as Sigismundo
says. From where they are standing, they can see the guards' barracks. Corporal Tacabo appeared at the door to enjoy the cool of
the evening, and it was doubtless purely by chance that the first bat
also appeared at the same moment, cutting smoothly through the
air. It's a strange animal, almost blind, like a rat with wings, and it
flies as fast as lightning and never bumps into anything or anyone.

A scorching June morning. Twenty-two men left Monte Lavre,
separately, so as not to attract the guards' attention, and met up
on the riverbank, just beyond Ponte Cava, among the reeds. They
discussed whether they should set off together and decided that,
since there were so few of them, it would be best not to break up
the group. They would have to walk farther and more quickly, but
if things went well, they would soon find others to join them. They
drew up an itinerary, first Pedra Grande, then Pendão das Mulheres, followed by Casalinho, Carriça, Monte da Fogueira and Cabeço do Desgarro. They would see how they felt after that, assuming there was sufficient time and enough people to send to other
places. They crossed at the ford, where the water formed a sort of
natural harbor, and they were like a band of boys, wearing very se-

rious smiles, or playful recruits with few weapons, taking off their shoes and putting them on again, with someone saying, as a joke of course, that he'd rather spend the day swimming. It's three kilometers to Pedra Grande, along a bad road, then another four to Pendão das Mulheres, three to Casalinho, and beyond that, it's best not to count, otherwise people might give up before they take the first step. Off they go then, the apostles, they could certainly do with a miracle of the fishes, preferably grilled over hot coals, with a drizzle of olive oil and a pinch of salt, right here underneath this holm oak, if duty were not calling to us so softly that it's hard to know whether it's coming from inside us or from outside, if it's pushing us from behind or is there up ahead, opening its arms to us like Christ, how amazing, it's the first comrade to leave the fields of his own free will, without waiting for someone to give him a reason, and now they are twenty-three, a veritable multitude. Pedra Grande comes into sight, and the fields lie before us, they've nearly cleared them already, as if they were working out their rage, who is this talking to them, it's Sigismundo Canastro, who knows more than the others, Comrades, don't be deceived, we workers must remain united, we don't want to be exploited, what we are asking for wouldn't even pay for a filling in one of the boss's teeth. Manuel Espada steps forward, We cannot be shown to be weaker than our comrades in other towns, who are also demanding a fairer wage. Then a Carlos, a Manuel, an Afonso, a Damião, a Custódio, a Diogo and a Filipe speak, all saying the same thing, repeating the words they have just heard, repeating them because they have not yet had time to invent their own, and now it's João Mau-Tempo's turn, My only regret is that my son António isn't here, but I hope that wherever he is, he will be saying the same things his father is saying, let us join together to demand a decent wage, because it's high time we spoke out about the value of the work we do, it can't always be the bosses who decide what they should pay us. Appetite comes with eating, and the ability to talk comes with talking.

The foremen arrive, gesticulating, they look like scarecrows frightening off sparrows, Get out of here, if these people want to work, let them, you're nothing but troublemakers, you lot, you deserve a good thrashing. But the workers have stopped, they have set down the sheaves, men and women are coming toward them, dark with dust, too baked dry with heat even to sweat. Work has stopped, the two groups join together, Tell the boss that if he wants us here tomorrow, all he has to do is pay us thirty-three escudos a day. Christ's age when he died, says one joker who knows about religious matters. There may have been no multiplying of the fishes, but there was a multiplying of men. They split into two groups and divided up the itinerary, with some going to Pendão das Mulheres and others to Casalinho, and they will meet back here on this hill to divide up again.

In heaven, the angels are leaning on the windowsills or over that long balcony with the silver balustrade that runs right around the horizon, you can see it perfectly on a clear day, and they are pointing and calling mischievously to each other, well, it's their age, and one angel higher up the scale runs off to summon a few saints formerly linked with agriculture and livestock, so that they can see what's going on in the latifundio, such upheavals, dark knots of people walking along the roads, where there are roads, or along the almost invisible tracks across the fields, taking shortcuts, in single file, around the edges of the wheatfields, like a string of black ants. The angels haven't enjoyed themselves so much in ages, the saints are giving gentle lectures about plants and animals, although their memory isn't what it used to be, but still they expound on how to grow wheat and bake bread, and how you can eat every bit of a pig, and how if you want to know about your own body, just cut open a pig, because they're just the same as us. This statement is both daring and heretical, it brings into question the whole of the Creator's thinking, had he run out of ideas when it came to creating man and so simply copied the pig, well, if enough people say so, it must be true.

The saints live so high up and so far away, and have so completely forgotten the world in which they lived, that they can find no explanation for the trail of humans walking from Casalinho to Carriça, from Monte da Fogueira to Cabeço do Desgarro, and now, while some head off in that direction, others are going farther afield, to Herdade das Mantas, to Monte da Areia, all of which are places where the Lord never trod, and even if he had, what would he or we have gained. They're heretics, Father Agamedes will bawl each day, and he's bawling these words out now from the window of his house, because the pilgrims are beginning to arrive in Monte Lavre, can this be the new Jerusalem, it's like the morning procession on Ascension Day, and the corporal has just run across the road, heading who knows where, someone must have summoned him, The boss wants to speak to you, and he pulls on his beret and tightens his belt, that's military discipline for you, because the guards fall just short of being an army, and it is precisely that shortfall that makes them feel hard done by, he enters the perfumed cool of the cellar where Humberto is waiting, Right, you know what's been happening, and Corporal Tacabo does know, it's his duty to know, that's what he's paid for, Yes, sir, the strikers have been visiting the workers on the estates and now they're back, So what are we going to do, I've asked for orders from Montemor, we're going to find out who's behind the mutiny, Don't worry, I have a list of names here, twenty-two of them, they were seen at Ponte Cava before they set off, and while he's saying this, Corporal Tacabo has poured himself a drink, Norberto paced back and forth, bringing his heels down hard on the flagstones, They're troublemakers, idlers, that's what they are, they don't want to work, if the right side had won the war, they wouldn't dare to so much as wag a finger, they'd be quiet as mice, happy to be working for whatever we were prepared to pay them, this is what Alberto says, and the confused corporal doesn't know what to say, he doesn't like the Germans and wants nothing to do with the Russians but he has a soft spot for the English, and when he thinks about it, he's not quite sure who it was who won the war, but he takes the list of names, it will look good on his service record, twenty-two proven strikers is no small thing, even though the angels find it all terribly amusing, they're young, you can't really blame them, one day they will learn the harsh realities of life, if they start having children, always supposing there are girl angels, as is only right and proper, and then they'll have to feed them, and if heaven becomes a latifundio, then they'll see.

But the ants won. In the fading evening light, the men gathered in the square and the overseers came, grim-faced and silent, but defeated, Tomorrow you can work for thirty-three escudos, that was all they said and then they withdrew, humiliated, thinking vengeful thoughts. That night, joy was unconfined in the tabernas, João Mau-Tempo, most unusually, dared to drink a second glass of wine, the shopkeepers are hoping to get some of their debts repaid and are considering raising their prices, at the mention of money the children cannot even think of what they would want to buy, and since the body is sensitive to the contentments of the soul, the men moved closer to the women, and the women closer to the men, and they were all so happy that if heaven understood anything about human lives, you would have heard hosannas and the clamor of trumpets, and the moon was its usual bright, lovely June self.

And now it's morning again. Each day's work is worth an extra eight escudos, which is less than a ten-tostão increase per hour or almost nothing per minute, so little that there isn't a coin small enough to represent it, and each time the sickle cuts into the wheat, each time a left hand grasps the stems and a right hand deals a final, decisive blow with the blade at ground level, only someone versed in higher mathematics could say how much that gesture is worth, how many zeros you would have to add to the right of the decimal point, in what thousandths we could measure out the sweat, the tendon in the wrist, the muscle in the arm, the strained back, the eyes fogged with fatigue, the broiling noonday heat. So much suffering for so little reward. And yet there are still some who sing, although not for long, because they soon hear the news that yesterday, in Montemor, the guards rounded up agricultural workers in the area and put them in a bullring, penned in like cattle. Those with long memories remembered what had happened in Badajoz,* the carnage that took place there, again in the bullring, it doesn't seem possible, they machine-gunned the whole lot of them, but it won't be like that here, we're not that cruel. Dark presentiments fill the countryside, the line of reapers advances hesitantly, unrhythmically, and the furious foremen take out their anger on the workers, anyone would think it was their money, Now that you're earning more, I've suddenly got a fieldful of malingerers. The line grows livelier, they don't want to seem to be in the boss's debt, they move more quickly, but then their imaginations turn back to the bullring in Montemor full of our people, from all over the latifundio, and fear so dries the mouth that some call to the water carrier to let them drink, Who knows what will happen to us. The guards know, as they walk over the clods of earth, a few at each end of the line, rifles at the ready and fingers on the trigger, If anyone makes a run for it, shoot in the air first, then aim at their legs, and if you have to fire a third time, make sure you don't have to shoot again. The reapers straighten up when they hear the names, Custódio Calção, Sigismundo Canastro, Manuel Espada, Damião Canelas, João Mau-Tempo. These are the local mutineers, the others are being rounded up right now, or they already have

---

* In August 1936, during the Spanish Civil War, between 1,300 and 4,000 Republicans — civilian and military — were rounded up and killed by Nationalist troops.

been or soon will be, if they thought they wouldn't have to pay the price for their insubordination, they were roundly deceived, they clearly didn't know the latifundio. Those left behind lower head and arms, bow their whole trunk with heart and lungs, their back struggling to keep them upright, and the sickle again slices through the wheat, cutting what, why, the dry stalks of course, what else. And beside the workers, the foreman growled like a wolf, You're lucky you weren't all taken away, that's what you deserve, if it was up to me, I'd teach you a lesson you wouldn't forget.

The five conspirators are flanked by the guards, who taunt them, So you thought you could lead a strike and get off scot-free, did you, well you've got another think coming. None of the five men replies, they hold their heads high, but have pangs in their stomachs that are not hunger pangs, and they're strangely unsteady on their feet, that's what fear does, it takes you over and it makes no difference if you speak or keep silent, but it will pass, a man is a man, whereas, even today, we can't be quite sure whether a cat is an animal or a human. João Mau-Tempo makes as if to say something to Sigismundo Canastro, but we never find out what it is because, as one man, one commander, with one will, the guards say, If you open your gob, we'll hit you so hard you'll leave teeth marks in the road, and so no one else dares say a word, and they arrive in Monte Lavre in silence, go up the ramp to the guards' post, because they had been arrested by then, all twenty-two of them, so someone had obviously betrayed us. They put them in an enclosure in the yard at the back, piled them in with nowhere to sit but the ground, although what does that matter, they're used to it, weeds can survive the hardest of frosts, they have skin as thick as donkey hide, which is just as well, because that way they get fewer infections, if it were us, frail city dwellers, we wouldn't stand a chance. The door is open, but in front of it, under a porch, stand three guards, rifles at the ready, one of them doesn't seem too happy in his sentry box, he averts his gaze, the barrel of his rifle pointing at the ground, and he doesn't have his finger on the trigger, He looks quite sad, who would have thought it. The prisoners only think this, they don't speak, they're under strict orders, but Sigismundo Canastro does manage to murmur, Courage, comrades, and Manuel Espada says, If we're questioned, the answer is always the same, we simply want to earn a just wage, and João Mau-Tempo says, Don't worry, they're not going to execute us or send us to Africa.

From the street comes a sound like that of waves breaking on a deserted beach. It's their relatives and neighbors come to ask for news, to plead for the men's impossible release, and then the voice of Corporal Tacabo is heard, a roar, Get back all of you or I'll order my men to charge, but this is purely a tactical threat, how are they going to charge if they have no horses, and one can hardly imagine the guards advancing with fixed bayonets to pierce the bellies of children or women, some of whom aren't bad-looking as it happens, and old ladies who can barely stand and who are about ready for the grave anyway. But the crowd draws back and waits, and all you can hear is the soft weeping of the women, who don't want to cause a scandal for fear that it might redound on their husbands, sons, brothers, fathers, but they are suffering too, what will become of us if he goes to prison.

Then, as evening comes on, a truck arrives from Montemor with a large company of guards, they're strangers here, we're used to the local ones, but so what, it's not as if we're going to forgive them, how can they have sprung from the same suffering womb only to turn on ordinary people who have never done them any harm. The truck reaches the fork in the road, and one branch leads off to Montinho, where João Mau-Tempo once lived, as did his late mother Sara da Conceição and his brothers and sisters, some of whom live here and others over there, but none in Monte Lavre, but this is the story of those who stayed, not those who left, and before we forget, the other road is the one the owners of the lati-fundio usually drive along in their cars, now the truck turns and comes bumping down toward them, belching out smoke and kicking up dust from the parched road, and the women and children, the older people too, find themselves pushed out of the way by the truck's swaying carcass, but when it stops, right by the wall that surrounds the guards' barracks, they cling to the sides in desperation, a foolish move, because the guards inside use the butts of their rifles to strike the people's dark, dirty fingers, they don't wash, Father Agamedes, it's true Dona Clemência, they're impossible, worse than animals, and Sergeant Armamento from Montemor shouts, If anyone comes too near, we'll shoot, so we can see at once who is in charge. The rabble falls silent, retreats to the middle of the road, between the barracks and the school, O schools, sow your seeds,* and it is then that the prisoners are called out, with the patrol forming up in two lines from the door of the barracks to the truck and inside it, too, like a hedge, or like a net into which the fish, or men, were drawn, for when men or fish are caught, there are few differences between them. All twenty-two came out, and each time one appeared on the threshold, there came from the crowd an irrepressible shout or cry, or, rather, shouts, because by the time the second or third man had appeared, there was an incessant clamor, Oh, my dear husband, Oh, my dear father, and the rifles were trained on the malefactors, while the local garrison kept their eyes fixed on the crowd, in case there should be a rebellion. It's true that there are hundreds of people there and that they are desperate, but there are the barrels of the rifles saying, Come any closer and you'll see what happens. The prisoners emerge from the barracks, look frantically around them, but there's no time, they are forced onward and when they reach the edge of the wall, they have to jump into the truck, it seems like a spectacle put on to terrify the people, and meanwhile the light is fading, and in the gloom they can't make out individual faces, barely has the first man

---

* The opening lines of the Republican hymn to public education.

emerged than they are all in the truck and the truck is setting off, it swerves wildly as if to scythe through the crowd, someone falls, but fortunately suffers only a few scratches, downhill it's easy, the men sitting in the back of the truck are thrown around like sacks, and the guards hang on to the sides, forgetting all about keeping their rifles trained on the crowd, and only Sergeant Armamento, with his back to the cab, legs straddled, faces the crowd running after the truck, the poor things are getting left behind, they gain on it slightly at the bottom, when the truck has to slow down to turn left, but then they can do nothing more, for the truck accelerates in the direction of Montemor, and the poor, panting people wave and shout, but both cries and gestures are lost as the vehicle moves away, they can't hear us now, the faster runners among them try to keep up, but what's the point, the truck disappears around the first bend, we'll see it later on going over the bridge, there it is, there it is, what kind of justice is this and what kind of country, why is our portion of suffering so much greater, they might as well kill the whole lot of us, thus sealing our fate once and for all.

Each man is immersed in his own thoughts. From what they heard while they were waiting to leave the barracks, Sigismundo Canastro, João Mau-Tempo and Manuel Espada know that they have been named as the main leaders of the strike. Of the three, Sigismundo Canastro is the calmest. Sitting on the floor along with all the other men, he began by resting his head on his folded arms, which were, in turn, resting on his knees, you get the picture. He wants to be able to think more clearly, but suddenly it occurred to him that his companions might think, from his posture, that he was discouraged, and he didn't want that, so he unfolded his arms and sat up straight, as if to say, here I am. Manuel Espada is remembering and comparing. He recalls how, eight years ago, he made the same journey in a smaller truck with his youthful companions, only Augusto Patracão is with him this time, Palminha had come to his senses and made other plans, and Felisberto Lampas became an itinerant worker and hasn't been seen since. Manuel Espada says to himself that there's really no comparison, this time things are serious, then they were just a bunch of boys, this time they're grown men, the level of responsibility, as no one would deny, is quite different. These three, for we cannot speak for every man there, are caught up in a never-ending stream of thoughts, a mixture of determination, fear and bravery, a trembling in hands and legs, no one's immune from that, João Mau-Tempo is lost in a kind of dream, it's almost dark now, and if his eyes fill with tears, so be it, no man is made of stone, his comrades mustn't see this though, he doesn't want them to lose courage too. Once past Foros, there is only open countryside, soon the moon will rise, well, it's June and the moon rises early, and ahead lie some large rocks, what giants could have rolled them there, a good place for an ambush, imagine if José Gato was there along with his fellow gang members, Venta Rachada, Parrilhas, Ludgero and Castelo, suddenly leaping out from behind the log they've rolled across the road, after all, they've had plenty of practice, and shouting, Stop, and the truck braking sharply and skidding on the tarmac, bloody hell, I hope the tires don't burst, and then, One move and you're dead, each bandit with his rifle at the ready, and they're not joking either, you can tell from their faces, there's the five-shot rifle that José Gato stole from Marcelino, Sergeant Armamento does make a move, well, it's what his superiors would expect of him, but he falls from on high with a hole right through his heart, and José Gato puts a second cartridge in the chamber and says, The prisoners can get out, meanwhile, the guards are standing with their hands in the air like in a Wild West film, and Venta Rachada and Castelo start collecting the rifles and the cartridge belts, behind the rocks they've tethered two of the mules they use to carry sides of pork, a little more dead weight won't bother them. João Mau-Tempo ponders whether to go straight back to Monte Lavre or to stay there in hiding until things quiet down a little, but he would have to send a message to his family to reassure them that everything has turned out for the best.

Everyone jumps out, Quick, quick, says a resuscitated Sergeant Armamento, with no hole through his heart. They're at the gate of the barracks in Montemor, and there's no sign of José Gato. The guards line up, they're not so tense now they're back on home ground, and there's no danger of riots or armed attacks, and as you'll have guessed, well, it wasn't that hard, José Gato's bold intervention was all in João Mau-Tempo's imagination. The rocks are still there at the side of the road, where they've been for centuries and centuries, but no one leapt out from behind them, the truck passed by with its usual mechanical calm, dropped the men off at the barracks and left, having done its duty. The twenty-two men are bundled down a corridor and across a courtyard, where two guards are standing by a door, one of them opens the door to reveal a room packed with people, some standing, some sitting on the floor, on the straw from two bales that have been pulled apart and strewn about to serve as bedding. The floor is made of concrete, and the room is strangely cold, considering how many people are crammed inside and that this is the hottest time of the year, perhaps it's because the back wall is built onto the side of the castle. Including those who were there already, there are nearly sixty men, who would make a good gang of workers. The door clangs shut, deliberately loud, and the sound of the key turning in the lock grates on the nerves like one of those bits of broken glass that the latifundio places on top of the walls surrounding its gardens, when the sun catches them, they look quite pretty, glinting away, and beyond lie trees heavy with oranges, and not just oranges, but pears, another fine fruit, and roses twine about the arches that line the orchard paths, any worker passing through would smell the perfume, but frankly, Father Agamedes, I doubt they have soul enough to appreciate such beauty. The ceiling is very low and is lit by one lightbulb, twenty-five watts at most, we haven't yet lost our frugal habits, and in the end, there's no denying it, the heat becomes unbearable. The men recognize each other or introduce themselves, there are people from Escoural and Torre da Gadanha, they say that the men from Cabrela were taken to Vendas Novas, but that's not certain, and so what are they going to do with us now. Whatever it is, says one of the men from Escoural, they can't take those thirty-three escudos away from us, now we just have to wait.

They wait. The hours pass. Now and then the door opens, more men are bundled in, the dungeon is beginning to be too small for so many people. Most have had nothing to eat since morning, and there's no sign that the guards have any intention of feeding their prisoners. Some lie down on the straw, the more trusting or those with the strongest nerves fall asleep. They hear the town hall clock strike midnight, nothing more will happen today, it's too late, they'd better get some sleep, their empty stomachs are protesting but not too much, and as the men are about to abandon themselves to slumber, made drowsy by the smell and the heat from all those bodies, the door is flung open and Corporal Tacabo and six guards appear, the corporal is clutching a piece of paper and the guards their rifles as if they had emerged fully armed from their mothers' wombs, and the corporal bawls, João Mau-Tempo from Monte Lavre, Agostinho Direito from Safira, Carolino Dias from Torre da Gadanha, João Catarino from Santiago do Escoural. The four men, four shadows, stand up and go out through the door. Their companions feel as if their hearts were in their mouths, what will happen to the poor things. Then comes the voice of a man who can no longer keep the secret, Apparently they killed a man here yesterday.

This time, they do not cross the courtyard. They continue along by the wall, between the guards, before being pushed toward a door. The light from the lamp there is much brighter, the prisoners screw up their eyes against the aggressive brightness, the first aggression of the night. The guards left, leaving only the corporal, who went over and put the piece of paper down on a desk behind which were seated two men, one in uniform, Lieutenant Contente, and the other in plain clothes. João Mau-Tempo, Agostinho Direito, Carolino Dias and João Catarino were ordered to stand next to each other in a line. Lift your snouts up high so we can see if you resemble your whores of mothers, said the man wearing civilian clothes. João Mau-Tempo couldn't resist retorting, My mother is dead, to which the man responded, Do you want your face smashed in, you may speak only when I tell you to, it won't be long before you lose your taste for talking, but that's precisely when you'll have to talk. Then Lieutenant Contente began to give orders, Stand up straight, you're not at home in your nice soft bed now, the usual military talk, and pay attention to the policeman here. The other man stood up, reviewed the ragged troop, staring at them hard, damn the man, it's as if he were trying to look right inside me, fixing me with a lingering, intimidating look, What's your name, and the man questioned answered, João Catarino, and you, Carolino Dias, and you, Agostinho Direito, and you, the one with the dead mother, what's your name, João Mau-Tempo. The PIDE agent smiled broadly, That's a fine name and very appropriate for the situation. Then he strode over to the desk, took his pistol out of its holster, slammed it down and turned angrily on the poor men, I want you to know that you won't get out of here alive unless you vomit up everything you know about this strike, about the organization, the people who gave you orders, the propaganda they've fed you, everything, I want it all out in the open, and woe betide you if you don't talk. Lieutenant Contente picked up four school exercise books that were in a pile at one end of the desk, You are each going to be locked in a room with one of these exercise books and a pencil, and you're to write down everything you know, names, dates, meeting places and houses, how and when any leaflets and so on were delivered, do you understand, and you won't be let out until it's all there in black and white. The PIDE agent returned to the desk, put his pistol back in the holster, having completed his show of force, and said, It's enough to drive a man crazy, you see before you an exhausted man, unable to sleep because of this wretched strike, so be sensible and write down everything you know and hide nothing, because if I find out later that you have left anything out, all the worse for you. João Catarino says, I can barely write, Agostinho Direito says, I can only write my name, João Mau-Tempo says, I can hardly write at all, Carolino Dias says, Nor can I. You know enough for our purposes, says the agent, we chose you because you know how to read and write, if you don't like it, tough, you shouldn't have learned, now you're going to regret not having stayed as stupid as you were born. The agent laughed at his own joke, the corporal laughed as did the private, and Lieutenant Contente, of course, laughed contentedly. The lieutenant gives an order to the corporal, the corporal tells the private, the private opens the door, and the four rascals leave, outside are the other troops, it's a public event, and like someone putting pigs in a pigsty, they march the four men down the corridor, opening doors and shoving them in, each with his own exercise book, Dias, Direito, Catarino and Mau-Tempo, they're just scum, Father Agamedes, if you'll forgive the expression.

In the barracks a great silence falls, full of noises as silences always are. The men locked up in the dungeon moan and sigh, unable to sleep, as is usual with weary bodies, and even when they do sleep, there's that ache from the day when they were working at the charcoal pit and tried to carry a great heavy log, if it was now, they'd tell them to piss off, I wonder what's happening to our comrades, I can't hear anything, only the footsteps of the sentries outside, and the clock chiming, I wish that bloody owl would shut up, it gives you gloomy thoughts. Locked in their rooms, the four make the same gestures, they look around them, there's the table and the pencil, it felt like a game, like being back at school and hav-

ing to do a dictation, except that there was no teacher to read and
mark the lesson, their conscience would have to be their teacher,
deciding what they would write in their slow, crooked hand, and
each of them, at some point, wrote his name on the first line of
the first page, right in the margin, as if they wanted to make sure
they had enough paper to write down all they were going to write,
my name is Agostinho Direito, my name is João Mau-Tempo, my
name is João Catarino, my name is Carolino Dias, and then they
sat staring at the page, all those lines to fill, and then on and on until the final page, it was like a wheatfield, but for some reason this
pencil-cum-sickle won't cut, won't move forward, it gets stuck on
this root, this stone, what on earth am I supposed to write, they're
waiting for me to tell them all I know, here on these crooked lines,
or do they only look crooked because I'm so tired, João Catarino is
the first to push the exercise book to one side, he wrote his name,
he will write nothing more, his name will stay there so that people
will know that the owner of that name wrote nothing more than
his name, not a word more, and then, at different times, each of the
other men pushed the exercise book to one side with a large, dark
hand, some closed the book, others left it open so that the name
was the first thing that would be seen when they came for them,
and nothing more.

At the first crack, which is a very picturesque and rural way of
speaking that came into being perhaps along with the unboarded
roof, especially the thatched variety, in which cracks and holes appear with wear and tear and no thanks to the skills of the thatcher,
and it is through those cracks and holes that the dawn light enters,
although the light could have entered earlier from a star which, on
its journey, was caught there by the eyes of some sleepless person.
The idea of the exercise books was probably a ruse on the part of
the PIDE agent and the lieutenant to be able to get a decent night's
sleep while the criminals made their confessions, or a subtle way of
dispensing with a scribe and getting the work done for free. We'll

never know the truth until it is confirmed, or not, in this account of prison and interrogation. At the first crack, we have to go back to that phrase because the sentence was left unfinished and the meaning lost, when the doors opened and the dapper PIDE agent, as dapper and fresh as if he really had slept at home and in a good bed, went from room to room, his anger growing, because each exercise book told him only what he knew already, that this villain is called João Catarino, that this turd is called Agostinho Direito, that this piece of shit is called Carolino Dias, and that this son-of-a-bitch, yes, son-of-a-bitch, is called João Mau-Tempo. They must have planned it together, the bastards, Come here, there's to be no more joking now, I want to know who organized the strike, who your contacts are, or the same thing will happen to you as happened to that other man. They don't know who that other man is, they don't know anything, they shake their heads, determined, weary, brave, hungry heads, oh dear, my eyes are filling with tears. And Lieutenant Contente, who was also there, says, You'll end up being sent to Lisbon, you'd be better off confessing here on your home territory, among people who know you. But for some reason the agent softened, Send them back inside, we'll decide what to do with them later. The four were almost dragged down the corridor into the courtyard, look up there, my friend, at the sky, it's bright even though the sun's not out, and then were plunged, stumbling over the bodies on the floor, into the darkness of the dungeon where their comrades were still being kept. Those who were asleep had to wake up, or else, grumbling, turn over, but all finally settled down again, because the four men, before they, too, lay down and slept, as was their perfect right, all said, hand on heart, that they had told them nothing, not a single word. That sleep did not last long, for these are people accustomed to sleeping little and rolling up their blanket when the sun is still hidden among the mountains in Spain, and besides, there is the nagging, cruel anxiety that slips in between the folds of the unconscious mind, shakes and distends them, breaking the chrysalis, and on top of that is the hollow ache in the stomach, which has not been fed for who knows how many hours, you wouldn't even treat an animal like this.

It's midmorning when the door opens again, and Corporal Tacabo says, João Mau-Tempo, you have a visitor, and João Mau-Tempo, who was talking to Manuel Espada and Sigismundo Canastro about what fate might await them, jumps to his feet in surprise and sees the astonishment on his companions' faces too, it's only natural, everyone knows that in situations like this there are no visitors, such kindness is unheard of, and there are even those who wonder if their comrade really did say nothing, which is why João Mau-Tempo leaves, flanked by two silent, serious groups of men, and why he drags his feet as if he were carrying the guilt of the world on his shoulders. He is like a spinning wheel, going round and round, with the sky above full of sunlight, who can possibly have come to visit me, it must be Faustina and the kids, no, it can't be, the lieutenant wouldn't give permission, and there's no way that the PIDE agent, that foul-mouthed dog, would allow it.

The corridor seems far shorter, it was behind that door that he spent the night gazing at a school exercise book, a particularly hard lesson, my name is João Mau-Tempo, and now, while the guard is knocking at the next door and waiting for the order to enter, it must be Faustina, or else they're just saying that to get my hopes up, when in fact they're going to question me again, perhaps beat me, what did that policeman mean when he said that if we didn't talk, the same thing would happen to us as happened to the other man, what other man. Thoughts move quickly, which is why João Mau-Tempo had time to think all this while he was waiting, but when the door opened, his brain emptied of ideas, as if his head were filled with the blackness of night, and then he felt a great sense of relief, because standing between the agent and the lieutenant was Father Agamedes, they wouldn't beat me up in front of a priest, but what's he doing here.

This is how it will be in heaven, with me in the middle as befits the spiritual obligation that has been mine ever since I have known myself and you have known me, with you, Lieutenant, at my right hand as protector of the law and those who make the law, and you, Senhor Agent, on my left hand as the man who does the dirty work, about which I would really rather not know. The door to this house of discipline opens, and what do I see, O my poor eyes, better to have been born blind than to see this, tell me you're deceiving me, can this be João Mau-Tempo from Monte Lavre, the home of my somewhat troublesome flock, you must be mad, according to the lieutenant and the policeman, or the policeman and the lieutenant, you have refused to tell them all that you know, well, it would be best if you did, for your own sake and that of your family, they are not to blame for the mistakes and follies of their father, you should be ashamed of yourself, João Mau-Tempo, a grown man, a respectable man caught up in such foolishness, this so-called insurrection, how often have I told you and the other men at the church, Beloved brethren, the road you are taking will lead you only to perdition and to hell, where there shall be wailing and gnashing of teeth, I have told you that so often, I've grown weary of repeating it, but what good did it do, João Mau-Tempo, it's not that I don't care about the others, I don't know them, but the policeman and the lieutenant told me that of the men from Monte Lavre, you were the one they asked to write in that exercise book, but you wrote nothing, you refused to help, as if you were mocking them, mocking these poor, patient gentlemen, who spent a sleepless night, because they have families too, you know, sitting at home waiting for them, and because of you, they had to say to them, I won't be home until late or I have to work tonight, don't wait up for me, have your supper and go to bed, I won't be home until morning, or not even then, because it's almost lunchtime now, and the lieutenant and the policeman are both still here, I just can't believe it, João Mau-Tempo, you clearly have no consideration for the authorities at all, if you did, you wouldn't behave like this, what would it cost you to tell them who organized the strike and who distributed the leaflets, where they come from and how many there are, what would it cost you, you wretched man, what could be simpler than to give them the names, the policeman here and the lieutenant would do the rest, you could then go home to your family, what could be nicer, a man in the bosom of his family, tell me, although, obviously, as a priest, I can't reveal the secrets of the confessional, but was it So-and-so and Whatsisname, was it, tell me, a nod will do if you prefer not to speak, only we four will ever know, was it them or wasn't it, that's what I've heard, but I can't be sure and I'm not saying it was them, I'm simply asking, really, João Mau-Tempo, I find your attitude most disappointing, aren't you ashamed to make your family suffer like this, speak, man.

Speak, man, there's no one else here, just me, Father Agamedes, the lieutenant, the policeman and you, there are no other witnesses, why can't you tell us what you know, which probably isn't much, but each man does what he can, you can't do more than that, Look, Father Agamedes, I don't know anything, I can't repent of something I didn't do, I would give anything to be back with my wife and my daughters, but I can't give you what you're asking me, I can't say anything because I don't know anything, and even if I did, I'm not sure I would tell you, Now you've shown your true colors, you bastard, shouts the policeman, Stop, says Father Agamedes, as I never tire of saying, they're nothing but poor brutes, I said as much the other day when I was at Dona Clemência's house, he probably really doesn't know anything and was just led astray by the others, He's down as one of the strike leaders, says Lieutenant Contente, Right, says the policeman, send him back inside.

João Mau-Tempo leaves, and as he walks down the corridor for what seems like the nth time, he sees, coming out of another

door, flanked by a large escort of guards, So-and-so and Whatsisname, their eyes meet in recognition, they've been badly beaten,
poor things, and João Mau-Tempo, as he walks across the courtyard, feels his eyes fill with tears, not because he's dazzled by the
sun, he's used to that now, but out of an absurd feeling of contentment and relief to know that the two men have already been arrested, and that he wasn't the one who betrayed them, no, it wasn't
me, what a relief that they've been arrested, but what am I saying,
and he weeps twice over, once out of contentment and once out of
sorrow, and both times he weeps to have seen them here, they've
obviously been beating them up, as sure as my name is João Mau-
Tempo, that policeman was spot-on when he said I have the right
name for the times we're living through.

He went back into the dungeon and told the others what had
happened. They saw that he had tears in his eyes and asked if he
had been beaten. He said no, he hadn't, but continued to weep, his
heart filled with sorrow, any contentment he had felt quite gone, replaced by a feeling of mortal sadness. The men from Monte Lavre
gathered around him, those of the same age, that is, because the
younger ones moved away, embarrassed to be near a man whose
hair was already white but who was crying like a child, is that what
awaits us as well, they thought. These are scruples it would be best
to accept without further analysis or discussion.

It was after midday when the situation took a turn for the better. They were led out into the courtyard, where they were reunited with their families, who had come from far and wide, those
who could, and who were only now allowed into that anteroom of
authority, having been kept waiting outside the barracks, penned
in by the guards, where they redoubled their sighing and sobbing,
but when Corporal Tacabo turned up to give the order to let them
in, they were filled with hope, and there were Faustina and her two
daughters, Gracinda and Amélia, who had walked the four leagues
from Monte Lavre, what a wearisome life they lead, along with

others, mostly women, There they are, and the guards finally relaxed their security measures, ah, what hungry kisses might then
have been heard throughout the glade,* what do you mean glade,
the poor creatures embraced and wept, it was like the resurrection
of the dead, and as to kisses, that is not something in which they
have much practice, but Manuel Espada, who had no family there,
stood looking at Gracinda, who had her arms about her father, but
she was already taller than him and so could look at Manuel over
her father's shoulder, of course, they had met before, and this was
hardly love at first sight, but afterward she said, Hello, Manuel,
and he replied, Hello, Gracinda, and that was that, and anyone
who thinks more is required is quite wrong.

The families were still engaged in this festival of embraces when
Lieutenant Contente and the PIDE agent came out into the courtyard, and the speech they gave emerged from their two mouths
simultaneously, it was impossible to know which of them was im-
itating the other, or if there was some mechanism at work, connected to Lisbon perhaps by electric cables, that made them speak
like that, like two phonographs, Lads, be careful from now on, this
time we're letting you walk free, but be warned, if you get involved
in any such terrorist activities again, you will pay twice over, so
don't be so foolish as to be taken in by false doctrines, doctrines
spread by the enemies of our nation, if you come across pamphlets
on the roads or in the streets of a village, don't read them, or if you
do read them, burn them immediately afterward, don't give them
to anyone else or repeat what you read, because that is a crime, and
then both you and your innocent families will suffer, if you have a
problem to resolve, don't go on strike, go to the authorities, who
are there to inform and help, that way you will be given whatever
is fair and lawful, with no need for fuss or upsets, that's why we're
here, and now go and work in peace, and may God go with you,

---

* A reference to canto IX, verse 83, of *The Lusiads* by Luís Vaz de Camões.

but before you leave, you have to pay the cost of gas for the truck that brought you from Monte Lavre to Montemor, you're the ones who did wrong and you're the ones who have to pay, the State can't be expected to do that.

They scraped together the necessary money, having rummaged around in bags and pockets and handkerchiefs, there's the money, Lieutenant Contente, at least we won't be in debt to the State, because we know how hard up it is, it's just a shame the trip wasn't longer, because we already know the Monte Lavre road. These words were not, in fact, spoken, the narrator took the liberty of adding them, but the following words were spoken, by the PIDE agent, alone this time, Now that you've settled your bill, go back to your homes and may God go with you, and be sure to thank the priest here, who has shown what a friend he is to you all. At these words, Father Agamedes raises his arms, as if he were standing before the altar, and people don't know quite what to do, some go over and thank him, others pretend to have neither heard nor seen him and gaze off into space or talk to their wives and families, and Manuel Espada, who, by some strange coincidence, is standing right next to Gracinda Mau-Tempo, mutters, as if the words were biting into his heart, I feel quite ashamed, and just when he thought things could get no worse, Father Agamedes, smiling broadly, says, And now for some good news, there is transport for everyone outside in the street, provided by your employers, with no charge either, you're to be driven home in your employers' cars and carts, and to think that some people still speak ill of them. And off Father Agamedes goes, his black, wax-spattered cassock fluttering in the breeze, carrying along in his blessed wake his wretched flock frantically chewing on the tiny quantity of food brought from home, and Manuel Espada, who, by some strange coincidence, is still standing right next to Gracinda Mau-Tempo, said, And they expect us to be grateful to them, it's just despicable. Gracinda Mau-Tempo did not reply, and Manuel Espada returned to his theme, Well, they're not taking me, I'm walking. Then the anxious girl did move and said, part shyly, part boldly, It's an awfully long way, but immediately corrected herself, unsure who to praise and who to censure, whether those who accepted the offer of a lift or this rebel, It's up to you, of course, Manuel Espada replied that he knew it was a long way, took three steps, then turned back, Would you be my girl, and she responded with a look, which was all that was needed, and when Manuel Espada had already turned the first corner, that was when Gracinda Mau-Tempo said Yes in her heart.

During the days that followed, Father Agamedes stocked up his already well-stocked larder with the gratitude of his parishioners, It's not very much, I'm afraid, but it comes from the heart, this is for all you did for us, a pint of beans, a little bag of maize, a laying hen, a bottle of olive oil, three drops of blood.

O LÉ. ON THE ORDERS of the president of the bullring, the constable enters the arena, inspects the locks on the corrals, counts the number of halters, decides that there are enough, takes a turn about the arena to get a good view of the whole thing, the tiered benches, the boxes, the bandstand, the seats in the shade and in the sun, sniffs the odor of fresh dung on the air and says, They can come in now. The doors are opened and the bulls enter, these are the bulls that will be fought today according to the rules and precepts of the art, taunted with a cape, stuck with darts, beaten with sticks and finally crowned with the hilt of the sword, whose point and blade pierce my heart, olé. They are brought in by the guards, they come from near and far, from places we have already mentioned, but not, as chance would have it, from Monte Lavre, and gradually the ring fills up, not the benches, the very idea, no, the audience is composed entirely of guards, who stand around, in the shade where possible, their rifles at the ready, well, they don't feel like men without them. The ring starts filling up with dark cattle, captured from leagues around in heroic combat, with the guards on the attack, at the charge, there they are bearing down on those beastly strikers, those lions of the sickle, those men of sorrows, These are the captives from the battle, and at your feet, lord, we lay the flags and cannon seized from the enemy, see how red they are, but not as red as they were at the beginning of the war because, meanwhile, we have heaped dust and spit upon them, you can hang them in the museum or in the regimental chapel where the recruits kneel, waiting to have revealed to them the mystical fate of being a guard, but perhaps it would be preferable, lord, to burn them, because the sight of them offends the feelings you taught us to have, and we want no other feelings. The constable, with the benign authorization of the president of the bullring, had ordered the arena to be scattered with straw, so that the men, because they are men not lions and have neglected to bring their sickles with them, can sit or lie down, grouped more or less according to their village of origin, such gregarious instincts are hard to give up, but there are a few others, too, who go from group to group, offering a word here, a hand on the shoulder there, a glance or a discreet gesture, so that everything, as far as possible, is safe and clear, and now it's just a matter of waiting.

The guards are keeping watch from their viewing platform, and one of them says to the other, with a hearty, military laugh, It's like the monkey house at the zoo, all we need are some nuts to throw to them, that would be funny, watching the monkeys scrabbling for food. This implies that some of the guards have traveled, that they have visited a zoo, practiced the rules of summary observation and of expeditious classification, and if they say that the men of sorrow herded into the bullring in Montemor are monkeys, who are we to contradict them, especially when they are pointing their riffles in our direction, we say riffle to provide a sort of half rhyme with pistol, although piffle would be funnier, and there's plenty of that about. The men talk to pass the time or to prevent it from passing, it's a way of putting your hand on your heart and saying, Don't go forward, don't move, if you take another step, you'll crush me, what did I ever do to you. It's also like bending down, placing one hand on the earth and saying, Stop turning, I want to see the sun for a while longer. While all this is going on, this heaping of words one upon another, just to see if they come out differently, no one has noticed that the constable has entered the ring in search of a man, just one, who is not even a lion with a sickle and who has not even come very far, and that man, if he were given an exercise book in which to write down all he knows, as the four from Monte Lavre, Escoural, Safira and Torre da Gadanha will do the following day, that man would write on the first line or on every line, so that there could be no doubt and so that there could be no change of heart from one page to the next, as I say, if he were to write his name, he would write Germano Santos Vidigal.

They have found him. Two guards lead him away, and whichever way we turn, that is all we see, they lead him out of the ring, to the exit door from sector six, where two more guards join them, and now it seems deliberate, it's uphill all the way, as if we were watching a film about the life of Christ, up there is Calvary, and these are the centurions in their stiff boots and warriorlike sweat, their spears cocked, it's suffocatingly hot. Halt. A few men are coming down the road, and Corporal Tacabo, fearing that they might be José Gato and his gang, says, Keep walking, this man is under arrest. The passersby stay as far away as they can, pressed against the wall, they're in no danger, it's almost as if they were grateful for that order and for the information, and the cortege has only a hundred meters or so to go now. Up above, we can see her over the wall, a woman is hanging a sheet out on the line, it would be funny if that woman was called Verónica, but she isn't, her name is Cesaltina and she's not much of a one for churches. She sees the man pass by under guard, follows him with her eyes, she doesn't recognize him, but she has a presentiment and presses her face to the damp sheet as if it were a shroud, and says to her son, who insists on playing outside in the sun, Let's go indoors.

The guards cross the road that leads up to the castle, where it widens out to form a square, only a few more steps to go and so little profit in them, but if you think that is what the prisoner is thinking, you're quite wrong, we can't know what his thoughts are or will be, but now it's up to us to start thinking. If we were to stay outside, if we followed that woman, Cesaltina, and sat down, for example, to play with her son, well, who doesn't like children, but then we wouldn't find out what is about to happen, and we can't have that. Two sentries are at the door, the guards are on a war footing, raise up once again the grandeur of Portugal,* you get a good view of the countryside from here, the chapel of Our Lady of the Visitation, who is as miraculous as they come, but we don't want any pilgrimages here, and a few gardens, but in this cramped space there's no room to see more. Let's go indoors, says Cesaltina to her son, Let us go indoors too, through here, past the sentries, they can't see us, that's our privilege, let's cross the courtyard, no, don't go in there, that's a kind of dungeon, a kind of wholesale warehouse for criminals, tomorrow the men from Monte Lavre and elsewhere will come here, minor cases, this is the way, but don't take that corridor, it's around this corner, another ten paces or so, mind you don't trip over that bench, here it is, we need go no further, we've arrived, it's just a matter of opening the door.

We have missed the preliminaries. We lingered to look at the landscape, to play with the little boy who so loves to play in the sun, however often his parents call him indoors, and to ask questions of Cesaltina, whose husband is not involved in these troubles, he works for the council and is called Ourique, but all these things were merely excuses, delaying tactics, ways of averting our eyes, but now, in between these four whitewashed walls, on this tiled floor, notice the broken corners, how some tiles have been worn smooth, how many feet have passed this way, and look how interesting this trail of ants is, traveling along the grooves between tiles

---

* A line from the Portuguese national anthem.

as if they were valleys, while up above, projected against the white
sky of the ceiling and the sun of the lamp, tall towers are moving,
they are men, as the ants well know, having, for generations, experienced the weight of their feet and the long, hot spout of water
that falls from a kind of pendulous external intestine, ants all over
the world have been drowned or crushed by these, but it seems
they will escape this fate now, for the men are occupied with other
things. The hearing apparatus and musical education of ants do
not allow them to understand what men say or sing, so they cannot catch every detail of the interrogation. But that doesn't matter,
in the morning, in this same barracks, albeit in a less secret place,
the men from Monte Lavre, Torre da Gadanha, Safira and Escoural will be questioned too, and then we'll hear everything, along
with the insults, son-of-a-bitch, bastard, son-of-a-bitch, piece of
shit, son-of-a-bitch, faggot, all of which is very trivial, and we won't
be offended by such trifles, it's like a scurrilous form of the tittle-
tattle exchanged by gossips, she said this, then she said, who cares,
in a couple of days they'll have made up, but not in this case.

Let's take this ant, or, rather, let's not, because that would involve picking it up, let us merely consider it, because it is one of
the larger ones and because it raises its head like a dog, it's walking along very close to the wall, together with its fellow ants, it will
have time to complete its long journey ten times over between the
ants' nest and whatever it is that it finds so interesting, curious or
perhaps merely nourishing in this secret room, before this episode
doomed to end in death is over. One of the men has fallen to the
ground, he's on the same level as the ants now, we don't know if he
can see them, but they see him, and he will fall so often that, in the
end, they will know by heart his face, the color of his hair and eyes,
the shape of his ear, the dark arc of his eyebrow, the faint shadow
at the corner of his mouth, and later, back in the ants' nest, they
will weave long stories for the enlightenment of future generations,
because it is useful for the young to know what happens out there

in the world. The man fell and the others dragged him to his feet again, shouting at him, asking two different questions at the same time, how could he possibly answer them even if he wanted to, which is not the case, because the man who fell and was dragged to his feet will die without saying a word. Only moans will issue from his mouth, and in the silence of his soul only deep sighs, and even when his teeth are broken and he has to spit them out, which will prompt the other two men to hit him again for soiling State property, even then the sound will be of spitting and nothing more, that unconscious reflex of the lips, and then the dribble of saliva thickened with blood that falls to the floor, thus stimulating the taste buds of the ants, who telegraph from one to the other news of this singularly red manna fallen from such a white heaven.

The man fell again. It's the same one, said the ants, the same ear shape, the same arc of eyebrow, the same shadow at the corner of the mouth, there's no mistaking him, why is it that it is always the same man who falls, why doesn't he defend himself, fight back. This is the reasoning of the ant and of ant civilization, they do not know that Germano Santos Vidigal is not fighting with those two thugs Escarro and Escarrilho,* but with his own body, with the searing pain between his legs, or his testicles, to use the language of a physiology manual, or his bollocks, to use the more easily acquired and cruder language of the street, fragile balls, balloons full of some imponderable ether which raise us men up to ecstasy, that carry us between heaven and earth, but not these pathetic objects anxiously protected by hands that suddenly release them when the heel of a boot thuds brutally into the small of the back. The ants are surprised, but only fleetingly. After all, they have their own duties, their own timetables to keep, it is quite enough that they raise their heads like dogs and fix their feeble vision on the fallen man to check that he is the same one and not some new variant in the

---

* Literally, Big Spit and Little Spit.

story. The larger ant walked along the remaining stretch of wall, slipped under the door, and some time will pass before it reappears to find everything changed, well, that's just a manner of speaking, there are still three men there, but the two who do not fall never stop moving, it must be some kind of game, there's no other explanation, let's hope Cesaltina's son never plays this game, they are engaged in hurling the other man against the wall, they grab him by the shoulders and propel him willy-nilly in the direction of the wall, so that sometimes he hits his back, sometimes his head, or else his poor bruised face smashes into the whitewash and leaves on it a trace of blood, not a lot, just whatever spurts forth from his mouth and right eyebrow. And if they leave him there, he, not his blood, slides down the wall and he ends up kneeling on the ground beside the little trail of ants, who are startled by the sudden fall from on high of that great mass, which doesn't, in the end, even graze them. And when he stays there for some time, one ant attaches itself to his clothing, wanting to take a closer look, the fool, it will be the first ant to die, because the next blow falls on precisely that spot, the ant doesn't feel the second blow, but the man does, and his stomach, not he, gives a lurch, and again he collapses, retching, from that violent kick to the stomach, followed by another to his private parts, which is an expression too widespread to cause offense.

One of the men leaves the room to rest from his exertions. His name is Escarrilho, he has a mother and father and is married with children, which isn't saying much, because the one who stayed behind to guard the prisoner, Escarro, also has a mother and father, and is married with children, the men are distinguishable only by their features, although only just, and by their names, one is Escarro and the other Escarrilho, they are not related and yet they belong to the same family. He walks down the corridor and, in his weariness, stumbles over the bench, These guys who won't talk will be the death of me, but screw the bugger, I'll get something out of him or my name's not Escarrilho. He takes a long, long drink of water, he's burning up with fever, then a kind of nervous fit comes upon him, and, energies replenished, he irrupts into the room again like a typhoon, and launches himself at Germano Santos Vidigal like a dog, he is a dog called Escarrilho, and it's as if Escarro were urging him on, Go on, bite him, and perhaps he really does bite him, later on they'll find teeth marks here and there, but whether they're from a man or a dog is hard to tell, for sometimes, as everyone knows, men are born with dogs' teeth. Poor dogs, trained to bite those they should respect and to bite parts of the body they should never bite, here, for example, the place that marks me out as a man, no more than they should bite a man's arm or jaw, or this other place, the heart, our inner eye, or the head, where our real eyes are. But I was told as a child that this restless piece of machinery is what makes me a man, and although I didn't really believe them, I'm fond of it, and it isn't something that a dog should bite.

The large ant is on its fifth journey, and still the game continues. This time it was Escarro's turn to go out for a rest, he went into the courtyard to smoke a restoring cigarette, then visited Lieutenant Contente in his office to ask about the progress of the field operations, the great maneuvers, and the lieutenant told him they were making a general sweep of strikers in the area, deploying all their manpower, it was good that they finally sent us reinforcements, he said, enough to arrest as many men again as we've got penned up in the bullring. And has that guy Germano Vidigal talked yet, asks Lieutenant Contente discreetly, because it really has nothing to do with him and Escarro is under no obligation to answer, but he does, Not yet, he's a tough nut to crack, and the lieutenant, solicitously, helpfully, adds, You'll have to tighten the screws still more. This mini-Torquemada of Montemor makes a good adjutant, offering them a roof over their heads and protection, and also throwing in a little free advice, but as he lights a cigarette, he hears Escarro's ill-tempered response, We know what we're doing, he snarls, then leaves, slamming the door and muttering, Imbecile, and, feeling perhaps put out by this exchange, he went into the room where the ants were and removed from the drawer a deadly weapon, a steel-tipped cat-o'-nine-tails, he looped the handle about his wrist to get a better grip, and as Germano, that foolish man of sorrows, tried to crawl away from his attacker, Escarro unleashed the whistling whip upon his shoulders, moving slowly down his back, centimeter by centimeter, as if he were threshing green rye, as far as the kidneys, where he lingered, blind even though his eyes were open, for there is no more dangerous form of blindness, rhythmically thrashing the man now lying on the floor, beating him methodically so as not to tire himself too much, because tiredness is the real killer, but gradually he began to lose all self-control and became a kind of manic whipping machine, a drunken automaton, until Escarrilho placed one hand on his arm, Don't get carried away, man, you'll kill the guy. Ants know about death, because they're used to seeing their own dead and to making instant diagnoses, sometimes, on their travels, as they're dragging along a grain of wheat, they stumble upon some small, shriveled, almost indecipherable thing, but they don't hesitate, despite being encumbered by their load, they thoroughly investigate the object with their antennae, but their Morse code is quite explicit, This is a dead ant, and you only have to glance away for a moment, and when you look again, the corpse has gone, that's what ants are like, they don't leave behind those who fall in the line of duty, and for all these reasons, the large ant, which was on its seventh trip back and forth and happened to be passing, raises its head and studies the great cloud before its eyes, but then makes a special effort, adjusts its visual mechanism and thinks, How pale this man is, he doesn't look the same at all, his face is all swollen, his lips are cut, and his eyes, poor eyes, you can't see them for the bruises, he's so different from when he first arrived, but I know him by his smell, because smell is the keenest of the ants' senses. The ant is still thinking all this when the face is removed from view because the other two men turn the man over and lay him on his back, they throw water on his face, a whole jug of cool water, pumped up from the deep, dark well, little did that water suspect the fate awaiting it, coming as it did from the depths of the earth, after who knows how many years traveling underground, having known other places, the stony steps of a spring, the harsh brilliance of sand, the soft warmth of mud, the putrid stagnation of the swamp, and the fire of the sun that slowly erased it from the earth, vanished, gone, until it reappears in a passing cloud long, long afterward and suddenly falls to earth, falling helplessly from above, the earth seems beautiful to the water, and if the water could choose the places where it fell, if it could, there would be far less thirst or far less surfeit, yes, long, long afterward it fell to earth and went traveling, gradually evolving into pure, crystal-clear water, until it found a course to follow, a secret stream, this dark, echoing well, this surface perforated by a suction pump, and suddenly it's trapped inside a transparent trap, a jug, is its fate perhaps to slake someone's thirst, no, it's being poured from on high onto a face, an abrupt fall, abruptly broken as it runs slowly over lips, eyes, nose and chin, over gaunt cheeks, over a forehead drenched in sweat, another kind of water, and thus it comes to know this man's as yet still-living mask. But the water drips onto the floor, spattering everything around and the tiles turn red, not to mention the ants who were drowned, apart from this larger one tirelessly making its eighth journey.

Escarro and Escarrilho grab Germano Santos Vidigal under the arms, lift him bodily, he hates to be a bother, and sit him on a chair. Escarro is still holding the cat-o'-nine-tails, the handle is still looped over his wrist, the fury that had gripped him has passed, but he still yells, Bastard, and spits in the face of the man who sits slumped in the chair like an empty jacket. Germano Santos Vidigal opens his eyes, and, incredible though it may seem, what he sees is the trail of ants, perhaps because there are so many of them in the place where his gaze happens to fall, it's hardly surprising, human blood is a delicacy for ants, when you think about it, they live on nothing else, and three drops of blood have fallen there, Father Agamedes, and three drops of blood make a well, a lake, an ocean. He opened his eyes, if you can use the word open to describe the narrow slits through which light barely penetrates, and what light does enter is too much, piercing his pupils with pain, which he is aware of only because it is a new pain, a knife sticking into flesh already pierced by another one hundred revolving knives, and then with a moan he stammered out a few words that Escarro and Escarrilho both hastened to hear, regretting now having beaten him so badly that they may have rendered him incapable of speech, but what Germano Santos Vidigal wants, poor man, still subject to his bodily needs, is to relieve his bladder, which for some reason is suddenly sending out an urgent signal, and will, if not heeded, empty itself right here and now. Escarro and Escarrilho don't want to get the floor any dirtier than it already is, and they also cherish the hope that they have finally broken this stubborn man's resistance and that this request is the first sign, one of them goes to the door to check that no one is in the corridor, nods, then goes back inside, and together the two men help Germano Santos Vidigal to walk the five meters that separate them from the latrine, they lean him up against the urinal and leave the poor man to unbutton his fly with clumsy fingers, feeling for and extracting his tortured penis, his cock, not daring to touch his swollen testicles, his torn scrotum, and then he concentrates, calls on all his muscles to help him, asking them first to contract and then to relax so that the sphincters soften and relieve the terrible tension, he tries once, twice, three times, and out it spurts, blood, mingled perhaps with urine, although it's impossible to tell from that one red stream, as if every vein in his body had burst and found an outlet there. He tries to hold it back, but the stream continues to pour forth as strongly as ever. It's his life pouring out of him, and it's still dribbling out when he finally puts his cock away, lacking the strength now to rebutton his fly. Escarro and Escarrilho lead him, feet dragging, back to the room of the ants and sit him down again on the chair, and Escarrilho asks, in a voice full of hope, So now will you talk, he has the idea that having been allowed to go to the toilet, the prisoner has a duty to speak, after all, one good turn deserves another, but Germano Santos Vidigal's arms drop to his sides, his head slumps onto his chest, and the light goes out inside his brain. The larger of the ants disappears under the door, having completed its tenth journey.

When it returns from the ants' nest, it will find the room full of men. Escarro and Escarrilho are there, along with Lieutenant Contente, Sergeant Armamento, Corporal Tacabo, two nameless privates and three specially chosen prisoners who state that the policemen left the room for a minute, no more than that, to deal with some urgent matter, and when they returned, found the prisoner had hanged himself on a piece of wire, just as we see him now, with one end tied around that nail there, and the other wound twice around Germano Santos Vidigal's neck, yes, his name's Germano Santos Vidigal, it's important to know that for the death certificate, the official doctor must be called, yes, as you can see, he's kneeling, yes, kneeling, but there's nothing odd about that, if someone wants to hang himself, even if it's only from a bedstead, it's all a matter of will, does anyone have any questions, Not me, say the lieutenant, the sergeant and the corporal, and the two privates and the three prisoners, who thanks to this stroke of luck will probably be set free today. There is great indignation among the ants, who witnessed everything, at different times, but meanwhile they have joined forces and pieced together what they saw, they know the whole truth, even the larger of the ants, who was the last to see the man's face close up, like a vast landscape, and it's a well-known fact that landscapes die because they are killed, not because they commit suicide.

The body has been removed. Escarro and Escarrilho put away the tools of their trade, the stick, the cat-o'-nine-tails, they rub their knuckles, inspect the tips and heels of their shoes, in case some thread of clothing or some bloodstain should reveal to the sharp eyes of Sherlock Holmes the weakness of their alibi and the conflicting times, but there's no danger of that, Sherlock Holmes is dead and buried, as dead as Germano Santos Vidigal, buried as deep as Germano soon will be, and the years will pass and these cases will remain swathed in silence until the ants acquire the gift of speech and tell the truth, the whole truth and nothing but the truth. Meanwhile, if we hurry, we'll still be in time to catch up with Dr. Romano, he's over there, head bowed, small black bag over his left arm, which is why we can ask him to raise his right hand, Do you swear to tell the truth, the whole truth and nothing but the truth, that's how it is with doctors, they're used to such solemn acts, Speak up, Dr. Romano, doctor of medicine, you who have sworn the Hippocratic oath with its various modern revisions to form and sense, speak up, Dr. Romano, here beneath the bright sun, is it really true that this man hanged himself. The doctor raises his right hand, looks at us with candid, innocent eyes, he's a much-respected man in the town, a regular churchgoer and punctilious in carrying out his social duties, and having shown us what a pure soul he is, he says, If someone has a wire wound twice around his own neck, with the other end tied to a nail above his head, and if the wire is pulled taut enough, even by only the partial weight of the body, then there is no doubt that, technically speaking, the man has hanged himself, and having said this, he lowered his hand and went about his business, Not so fast, Dr. Romano, doctor of medicine, it's not time for supper yet, if you still have any appetite after what you've just seen, I envy you your strong stomach, tell me, didn't you see the man's body, didn't you see the welts, the bruises, the battered genitals, the blood, No, I didn't, they told me the prisoner had hanged himself and he had, there was nothing else to see, You're a liar, Dr. Romano, medical practitioner, how and why and when did you acquire the ugly habit of lying, No, I'm not a liar, it's just that I can't tell the truth, Why, Because I'm afraid, Go in peace, Dr. Pilate, sleep in peace with your conscience, and give her a good screwing, because she deserves both you and the screwing, Goodbye, Senhor Author, Goodbye, Senhor Doctor, but take my advice, keep well away from ants, especially those that raise their heads like dogs, they're very observant creatures, you can't imagine, you will be watched from now on by all ants, don't worry, they won't harm you, but you never know, one day your conscience might make a cuckold of you, and that would be your salvation.

The street we are on is Rua da Parreira, or the street of the vine trellis, presumably because in days gone by, it was shaded by a trellis of fine grapes, and since the council couldn't come up with the name of a saint or a politician or a benefactor or a martyr to bestow on the street, it will for the time being continue to be called Rua da Parreira. What shall we do now, given that the men from Monte Lavre, Escoural, Safira and Torre da Gadanha only arrive tomorrow, given that the bullring is closed and no one can get in, what shall we do, let's go to the cemetery, perhaps Germano Santos Vidigal has arrived there already, the dead, when they choose to, can move very fast, and it's not that far and it's cooler now, you go down this street, turn right, as if we were going to Évora, it's easy enough, then left, you can't go wrong, there are the white walls and the cypresses, the same as everywhere else. The mortuary is here, but it's locked, they lock everything and they've taken away the key, we can't go in, Good afternoon, Senhor Ourique, no rest for the wicked, eh, That's true, but what's a man to do, people may not die every day, but you still have to straighten their beds and sweep the paths, Yes, I saw your wife Cesaltina and your son earlier on, he's a lovely child, That's true, True is a good word, Senhor Ourique, That's true, Tell me, is it true that the body in the mortuary died of a beating or simply because its former owner decided to hang himself, It's true that my son is a lovely boy, always wanting to be out playing in the sun, it's true that the body in there is that of a hanged man, it's true that given the state he was in, he wouldn't have had the strength to hang himself, it's true that his private parts were battered and bruised, it's true that his body was caked in blood, it's true that even after death the swellings didn't go down, the size of partridge eggs, they were, and it's true that I would have died of far less, and I'm used to death, Thank you, Senhor Ourique, you're a gravedigger and a serious man, perhaps because you're so fond of your son, but tell me, whose skull is that you're holding in your hand, does it belong to the king's son, That I don't know, I wasn't working here then, Goodbye, Senhor Ourique, it's time to close the gates, give my regards to Cesaltina and my love to your boy who so likes to play in the sun.

We have said our farewells, from down here you can see the castle, who could recount all its stories, those from the past and those to come, it would be quite wrong to think that just because wars are no longer fought outside castles, such military actions, however petty, however inglorious, are a thing of the past, as the Marquis de Marialva put it, Have you noticed, your majesty, how poorly Manuel Ruiz Adibe, governor of Montemor, runs the barracks there, because quite apart from his general incompetence, if the workers give him enough money, he excuses them from having to help build the fortifications, which is why so little work has been done, as anyone can see, and so I am asking your majesty if I might suggest someone more suited to the post, notably the lieutenant general of artillery, Manuel da Rocha Pereira, who possesses all the necessary qualities, efficiency, energy, zeal, as well as a desire to occupy said post, so if your majesty would be so kind as to write the

requisite letter of appointment, giving him the title of field mar-
shal, then Manuel Ruiz Adibe can still enjoy his salary, as do the
other cavalry captains whom your majesty has retired, he's not so
needy nor does he have so many responsibilities that he need live
uncomfortably, even if his salary isn't always paid promptly. Devil
take Adibe, who took such poor care of your majesty's service and
such good care of his own, the times have changed, now there are
zealous functionaries willing to kill a man in the Montemor barracks, then go outside to smoke a cigarette, wave goodbye to the
sentry courageously gazing out at the horizon to make sure no
Spaniards are approaching, and set off down the road with a firm
step, chatting serenely and totting up their day's work, so many
punches, so many kicks, so many blows with a stick, and they feel
proud of themselves, neither of them is called Adibe, their names
are Escarro and Escarrilho, they're like twins, they pause outside
the cinema, where the film being shown tomorrow, on Sunday, is
advertised, the summer season is getting off to a good start with
an interesting comedy called The Magnificent Dope. Bring your
wives, they'll enjoy it, poor ladies, when things calm down a bit,
it's sure to be worth seeing, but if you want a really good film, don't
miss the Thursday showing, with Estrellita Castro, the goddess of
song and dance, starring alongside Antonio Vico, Ricardo Merino
and Rafaela Satorrés in that marvelous musical Mariquilla Terre-
moto, olé.

THESE MEN ESCAPED from among the dead and the wounded. We will not name them one by one, it's enough to know that some went to Lisbon to languish in prisons and dungeons, and others returned to the threshing machine, being paid the new wage for as long as the harvest lasted. Father Agamedes issues a paternal admonishment to these madmen, reminds them directly or indirectly how much they owe him and how they, therefore, have still more of an obligation to fulfill their Christian duties, for did not the Holy Mother clearly demonstrate her power and influence by touching the bolts on the prison doors and making them fall away and by prying open the bars on the windows, hallelujah. He makes these grand statements to a church almost empty apart from old ladies, because the other parishioners are still brooding over how much that gratitude has cost them and are not consoled. In Monte Lavre, they know little of the arrests, it's all very vague, however often Sigismundo Canastro tells them how many there were, and only tomorrow will it become known how many deaths there were, as worker talks to worker, but the weariness of the living seems to hang heavier than a death about which they can do nothing, My father is ill and I don't know what to do with him, these are private concerns particular to each household, not to mention that the harvest is coming to an end, and then what will happen. It will be no different from other years, but now Norberto, Alberto and Dagoberto are saying, through the mouths of the overseers, that this rabble will regret ever going on strike, and the extra money they earned will cost them dearly. Adalberto has already sent written instructions from Lisbon to the effect that, once the harvest and the threshing are over, he will keep on only the swineherds and shepherds and the watchman, because he doesn't want his land trampled by strikers and idlers, later we'll see, it depends on the olives, how are the olives doing, by the way. The overseer will reply, but this is the kind of correspondence no one bothers to keep, you receive the letter, do what it tells you to do or send an answer to the question asked, and then it's, Now where did I put that letter, it would be amusing to base a whole history on such letters, it would be another way of doing it, our problem is that we think only the big things are important, and so we talk about them, but then when we want to know how things really were, who was there and what they said, we're in trouble.

Her name is Gracinda Mau-Tempo and she is seventeen. She will marry Manuel Espada, but not just yet. She's young, she can't simply get married from one day to the next, with no trousseau, they will have to be patient. Quite apart from these obvious social obligations, they have nowhere to live, It would mean having to move somewhere else, You don't want to be like your brother, always having to live so far away, I know it's not the same thing, because you're a girl, but it's bad enough never seeing one child, ah me, that boy of mine. These are Faustina's words, and João Mau-Tempo nods, he always feels a pang in his heart whenever they talk about his son, the little devil, who was only eighteen when it became clear that he had inherited his late grandfather's wanderlust. Gracinda Mau-Tempo will tell Manuel Espada the substance of these conversations later, and he will say, I want to marry you and I don't mind waiting, and he says this gravely, as is his custom on all occasions, a manner that makes him seem older than his years, and there was already quite an age difference, as Faustina had pointed out to Gracinda when Gracinda told her that Manuel Espada had asked her to be his girl, But he's much older than you, What's that got to do with it, Gracinda had replied, offended, and quite right, too, because that wasn't what mattered, what mattered was that she had liked Manuel Espada ever since that June day in Montemor, what did age have to do with anything, although Manuel Espada, when he spoke to her, had also pointed this out, I'm seven years older than you, and she, smiling, not sure quite what she was saying, had replied, The husband should be older than the wife, and then she had blushed because she realized that she had said yes without actually saying yes, as Manuel Espada realized, and he passed on to the next question, So that's a yes, is it, and she said, Yes, and from then on they spent time with each other as the rules of courtship demanded, at the front door of her house, because it was too soon for him to be allowed into the house, but where Manuel Espada did not follow the rules was in speaking to her parents right away, rather than waiting until both he and she were sure of their feelings and of their ill-kept secret. It was then that João Mau-Tempo and Faustina explained, and this was hardly news, that marriage was an economic impossibility just then, and that they would have to wait, I'll wait as long as I have to, said Manuel Espada, and then he left, determined to work and save, although he also had to help his own parents, with whom he still lived. These are the problems of ordinary life, which change little or so little in two generations that one hardly notices, and Gracinda Mau-Tempo knows that in future she will have to agree, by negotiation with her mother, how much of her wage she can put aside for her trousseau, as is her duty.

We have spoken a great deal about men and a little about women, but only in passing, as fleeting shadows or occasionally essential interlocutors, as a female chorus, albeit usually silent because weighed down either by some burden or by the weight in their bellies, or else, for various reasons, in the role of mater dolorosa, a dead son or a prodigal son, or a daughter dishonored, there's never any shortage of them. We will continue to talk about men, but also more and more about women, and not because of this particular courtship and future marriage, because we have already witnessed the respective courtships and marriages of Sara da Conceição and Faustina, Gracinda's grandmother now long gone and her mother happily still alive, and we said little about them, there are other reasons, as yet somewhat vague, and that's because the times are changing. Declaring their feelings at the door of a prison, or, rather, in a barracks and a place of death, which comes to the same thing, goes against all the traditions and conventions, and at a time of such suffering too, doubtless compensated by the joys of an as yet timorous freedom, fancy a young man saying to a young woman, Will you be my girl, ah, it's all very different from when I was their age.

Gracinda was born two years before her sister Amélia, who, because she had filled out early, looked, to the ill informed, about the same age. There was little physical resemblance, perhaps because the family blood was so mixed and so prone to produce singularities. We have only to think of that ancestor who came from the cold north and raped a girl at the fountain, a crime that went unpunished by his lord and master, Lamberto Horques, who was more concerned with origins of another kind and with horses. However, so as to confirm how small and modest a world this is, here we have Manuel Espada asking Gracinda Mau-Tempo to marry him next to that very same fountain, next to a field of bracken, which will not this time be trampled and broken until the body of the rape victim gives in, defeated. If only we could tie up all the loose ends, the world would be a stronger and better place. And if the fountain could speak, for example, which it would be perfectly justified in doing, given that it's been a constant source of pure, bubbling water for over five hundred years, or longer if it was a Moorish fountain, anyway, if it could speak, we think it would say, This girl has been here before, an understandable mistake, over time even fountains get confused, not to mention the vast difference in how Manuel Espada behaves toward Gracinda, merely taking her hand and saying, So that's a yes, is it, and then the two of them walking back, leaving the bracken for another occasion.

These three children know a lot about many different things. There are only four years between António Mau-Tempo, the oldest, and Amélia Mau-Tempo, the youngest. Once, they were just three bundles of ill-nourished, ill-dressed flesh and bone, as they continue to be today as adolescents, if that word isn't too refined for these lands and these latifundios. They were carried on the backs of father and mother or in baskets on their parents' heads, when they could still not walk or their little legs got tired quickly, or on their father's shoulders or in their mother's arms, or on their own two feet, they traveled more, given their age, than the wandering Jew. They battled with mosquitoes in the ricefields, poor, defenseless innocents who didn't even know to brush from their faces the squadrons of flying lancers that whined with pure, intense pleasure. However, since mosquitoes have very short lives and since none of the children died, it is of them that we speak, not of some others who died of malaria, so if there were any winners in the war, they were those who practiced passive resistance. It doesn't often happen, but in this case it did.

Look at these children, it doesn't matter which one, the oldest boy, or the middle child, or the youngest, lying in a box in the shade of the holm oak while her mother, let's say the child is a girl, works nearby, not so near that she can see her, and like all children, especially when they can't yet talk, she gets a pain in her belly, or not even that, just the usual outpouring of poo, at least she hasn't got dysentery this time, and by the time Faustina comes back, it's lunchtime, and Gracinda is covered in excrement and flies like the dung heap she has, alas, become. By the time her mother has washed, and washed not just Gracinda's little body, which is smeared all over, but also the rags covering her and which she hopes will dry draped over this pile of firewood, lunchtime has passed and so has her appetite. At this point, we don't know who to take care of first, Gracinda, who, though clean and fresh, is all alone, or Faustina, who returns to work, gnawing on a bit of dry bread. Let's stay here, beneath this holm oak, fanning the child's face with this branch as she tries to sleep, because the flies are back again, but also to save the parents any grief, because you never know, a cortege of kings and knights might pass by, and the barren queen's nursemaid might spot this little angel and carry her off to the palace, and how awful it would be if, later, she didn't recognize her real parents, because in the palace she wears only velvets and brocades and plays the lute in her room in a tower, with its view of the latifundio. Later on, Sara da Conceição used to tell such stories to her grandchildren, and Gracinda wouldn't believe us if we told her what danger she would have been in if we weren't here, sitting on this stone, fanning her with this branch.

But children, if they get the chance, grow up. Until they are of an age to work, they are left in the care of their grandmother or their mother, if there's no work for the mother, or with their mother and father, if there's no work for the father either, and if, when they're older, there are no children and all are workers, if there's no work for fathers, mothers, children or grandmothers, there you have it, ladies and gentlemen, the ideal Portuguese family gathered around the same hunger, depending on the season. If it's acorn time, then the father goes to gather them, as long as Norberto, Adalberto or Sigisberto doesn't send the guards to patrol at night, which is why, as soon as it came into being, the dear republic set up the national republican guard. That's all a very long story. But nature is prodigal, a generous teat that spills forth its milk in every ditch. Let's go gathering thistles, dockweed, watercress, what better diet could there be. Dockweed is just the same as spinach, it looks the same, although it tastes quite different, but once cooked, fried with a little of the onions we have left, it's enough to make your mouth water. And as for thistles. Strip those thistles, add a few grains of rice, and you have a banquet, please, Father Agamedes, help yourself, he who ate the meat can gnaw the bones. Every Christian, and even a non-Christian, needs his three meals a day, breakfast, lunch and supper, or whatever you choose to call them, what matters is having a full plate or bowl, or, if it's only bread and scrape, then it should be rather more than just a nice smell. It's a rule as golden as any other noble rule, a human right for both parents and children, which means that I don't have to eat only once in order for them to eat three times, although those three meals serve more to keep hunger at bay than to fill the stomach. People talk and talk, but they don't know what real need is, it means going to the bread bin knowing that the last crust of bread was eaten yesterday, and yet still opening the lid, just in case there's been another miracle of the roses,* which would, in any case, be quite impossible, because neither you nor I can remember putting roses in the bread bin, to do that we would have had to pick them, and have you ever seen roses growing on a cork oak, if only they did, hunger, as you see, can bring on delirium. Today is Wednesday, Gracinda, take your sister Amélia and go up to the big house, hold her hand, Gracinda, António won't go this time. Encouraging children to beg, that's the kind of education the parents give their children, I don't know why my tongue doesn't form a knot in my mouth or fall to the floor and leap about like a lizard's tail, that would teach me to be more care-

---

* Queen Isabel of Aragon, the wife of the Portuguese king Dom Dinis, was devoted to helping the poor. When her husband upbraided her for giving money to beggars, the queen drew back her cloak to reveal not money but a magnificent bunch of roses. The king, seeing this miracle, allowed her to continue her good works.

ful what I say and not speak about hunger on a full stomach, because it's not polite.

Wednesday and Saturday are the days when Our Lord God comes down to earth consubstantiated into bacon and beans. If Father Agamedes were here, he would cry heresy, call for the holy inquisition, and all because we said that the Lord was a bean and a slice of bacon, but the trouble with Father Agamedes is that he has little imagination, he has grown used to seeing God in a wafer and was never able to think of him in any other way, except, of course, as the Father with the full beard and dark eyes and the Son with the short beard and pale eyes, was there perhaps some incident involving a fountain and bracken at some point in the sacred story, do you think. Dona Clemência knows more about such transfigurations, having been the wife and fount of virtue from Lamberto down to the last Berto, because on Wednesdays and Saturdays she presides over how much food should be given to whom, advising on and checking the thickness of the slice of bacon, the piece with the least meat, of course, because if it's pure fat, all the better, so much more nourishing, she also levels off the measure of beans with the strickle, purely in the interests of fairness and charity, you understand, we don't want the children to quarrel, You've got more than me, I've got less than you. It's a lovely ceremony, it quite makes one's heart melt with saintly compassion, not a dry eye in the house, or a dry nose, well, it's winter now, especially outside, where the children of Monte Lavre are leaning against the wall, waiting to receive alms, how they suffer, barefoot, in pain, see how the girls lift first one foot and then the other to escape the icy ground, they would lift both at once if they ever grew the wings it's said they will have once they're dead, if they have the sense to die early, and see how they keep tugging at their dresses, not out of injured modesty, because the boys are too young to notice such things, but because they're terribly cold. They form a queue, each holding a small tin, all of them snotty and snuffling, waiting for the window above finally to open and for the basket to descend on a rope from the skies, very slowly, magnanimity is never in a hurry, oh no, haste is plebeian and greedy, just don't eat the beans as they are, because they're raw. The first child in the queue places his tin in the basket, and then the basket ascends, off you go and don't be long, the wind cuts along the wall like a barbed razor, who can possibly bear it, well, they all do in the name of what is to come, and then the maid sticks her head out of the window, and down comes the basket with the can full or half full, just to show any smarty-pants or novices that the size of the tin has no influence over the donor of this cathedral of beneficence. Anyone seeing this would think he had seen just about everything. But that's not true. No one leaves until the last one has received his ration and the basket has been taken in until Saturday. They have to wait until Dona Clemência comes to the window, warmly wrapped up, to make her gesture of farewell and blessing, while the dear little children chorus their thanks in various ways, apart from those who merely move their lips, Oh, Father Agamedes, it does my soul so much good, and if someone were to assert that Dona Clemência was nothing but a hypocrite, they would be much mistaken, because only she can know how different her soul feels on Wednesdays and Saturdays, in comparison with other days. And now let us recognize and praise Dona Clemência's Christian act of mortification, for although she has both the time and the money to hand out bacon and beans every day of the week, as well as the permanent, assured comfort of her immortal soul, she doesn't do it, and that, dear readers, is her personal penance. Besides, Dona Clemência, these children mustn't be allowed to acquire bad habits, imagine what demanding creatures they would grow into.

When Gracinda Mau-Tempo grew up, she did not go to school. Nor did Amélia. Nor had António. A long time before, when their father was a child, the propagandists for the republic urged the people, Send your children to school, they were like apostles sporting goatees, mustaches and trilby hats and proclaiming the good news, the light of education, a crusade they called it, with the signal difference that it wasn't a matter of driving the Turk out of Jerusalem and from the tomb of our Lord, it wasn't a question of absent bones, but of present lives, the children who would later set off with their bag of books slung over one shoulder with a piece of twine, and inside it, the primer issued to them by the same republic that ordered the national guard to charge if these same children's progenitors demanded higher wages. That is how João Mau-Tempo learned to read and write, enough to have misspelled his name in that exercise book in Montemor as João Mau-Tenpo, although, unsure which was the correct version, he sometimes wrote João Mautempo, which is better but still not quite right, Mau-Tempo, of course, being clear evidence of grammatical presumption. The world progresses, but within certain limits. In Monte Lavre it didn't advance enough for him to be able to send his own three children to school, and now how will Gracinda Mau-Tempo write to her fiancé when he's far away, a good question, and how will António Mau-Tempo send news if the poor thing never learned how to write and has apparently joined forces with a gang of ne'er-do-wells, I just hope he's leading a respectable life, says Faustina to her husband, You always set such a good example.

João Mau-Tempo nods, but in his heart of hearts he's not sure. It wounds him not to have his son by his side, and to see only women around him. Faustina is so different from what she was as a young woman, and she was never pretty, and his daughters, whose freshness and youth still survive despite a life of hard labor in the fields, it's just a shame that Amélia has such dreadful teeth. But João Mau-Tempo isn't so sure about having set a good example. He has spent his whole life simply earning his daily bread, and some days he doesn't even manage that, and this thought immediately forms a kind of knot inside his head, that a man should come into a world he never asked to be born into, only to experience a greater than normal degree of cold and hunger as a child, if there is such a thing as normal, and grow up to find that same hunger redoubled as a punishment for having a body capable of withstanding such hardship, to be mistreated by bosses and overseers, by guards both local and national, to reach the age of forty and finally speak your mind, only to be herded like cattle to the market or the slaughterhouse, to be further humiliated in prison, and to find that even freedom is a slap in the face, a crust of bread flung to the ground to see if you'll pick it up. That's what we do when a piece of bread falls to the ground, we pick it up, blow on it as if to restore its spirit, then kiss it, but we won't eat it there and then, no, I'll divide it into four, two large pieces and two small, here you are Amélia, here you are Gracinda, this is for you and this is for me, and if anyone asks who the two larger pieces are for, he is lower than the animals, because I'm sure even an animal would know.

The parents cannot do everything. They bring their children into the world, do for them the little they know how to do and hope for the best, believing that if they're very careful, or even when they're not, for fathers often deceive themselves and think they have been attentive when they haven't, no son of theirs will become a vagabond, no daughter of theirs will be dishonored, no drop of their blood poisoned. When António Mau-Tempo spends time in Monte Lavre, João Mau-Tempo forgets that he is his father and older than him and starts dogging his footsteps, as if he wanted to find out the truth behind those absences, as far away as Coruche, Sado, Samora Correia, Infantado and even the far side of the Tejo river, and the true stories he hears from his son's mouth both confirm and confuse the legend of José Gato, well, legend is perhaps an exaggeration, because José Gato is nothing but an inglorious braggart, he allowed us to be driven from Monte Lavre to prison, the stories are important more because they involve António Mau-Tempo, who either was there himself or heard about it later, than because they are picturesque facts that contribute to the history of minor rural crimes. And João Mau-Tempo sometimes has a thought that he cannot really put into words, but which, from the glimpse we've had of it, seems to say that if we're talking about good examples, perhaps that of José Gato is not so very bad, even if he is a thief and doesn't turn up when he's needed. One day, António Mau-Tempo will say, In my life I've had a teacher and an explainer, but now I've gone back to the beginning to learn everything over again. If you need an explanation, let's say that his father was the teacher, José Gato the explainer and that what António Mau-Tempo is learning now he will not be learning alone.

This Mau-Tempo family learn their lessons well. By the time Gracinda Mau-Tempo marries, she will know how to read. This formed part of her engagement, a reading primer by João de Deus,* with the words in black and gray so that you could distinguish the syllables, but it's not natural that such refinements should take root in memories born to remember other things, she just has to continue hesitantly reading and pausing between the words, waiting for her brain to light up her understanding, It's not *acega*, Gracinda, it's *acelga*.† Manuel Espada is now allowed into the house, if it wasn't for the primer he would still be lingering on the threshold, but it seemed wrong that they should sit outside learning to read where other people could see them, and besides, their relationship is clearly a serious one, Manuel Espada's a good lad, Faustina would say, and João Mau-Tempo watched his future son-in-law and saw him walking from Montemor to Monte Lavre, scorning cars and carts so as to stay true to his beliefs and not be in debt to the very people who had refused him his daily bread. That, too, was a lesson, and João Mau-Tempo took it as such, although

* João de Deus (1830–1896) was the greatest poet of his generation, but he turned his attention to education. His *Cartilha maternal,* a reading primer published in 1876, was used in schools for more than fifty years.
† Chard.

Sigismundo Canastro had said, What Manuel Espada did was good, but that doesn't mean we acted wrongly either, he gained nothing by walking, and we lost nothing by traveling back in the cart, one has to act according to one's conscience. And Sigismundo Canastro, who had a mischievous, albeit rather toothless smile, added, And of course he's still a young man, whereas our legs are getting old and heavy. That may well be, but even if there were thirty-three other reasons why Gracinda's parents should welcome Manuel Espada's courtship of their daughter, the very first, if João Mau-Tempo were ever to confess as much to himself, would be those twenty kilometers, Manuel Espada's out-and-out rejection of help, his affirmation of himself as a man during the almost four hours it took him to walk, with the sun beating down and his boots pounding the tarmacadam road, it was as if he were carrying a large flag that would not submit to being carried in the cars and carts owned by the latifundio. In this way, and as has always happened since the world began, the old learn from the young.

MAY IS THE MONTH of flowers. Let the poet go on his way in search of the daisies he has heard of, and if he doesn't come up with an ode or a sonnet, he'll produce a quatrain, which is much more to the common taste. The sun hasn't reached the crazy temperatures it does in July and August, there is even a cool breeze, and wherever you look, from this high vantage point, which would once have served as a lookout post, everything is green fields, no spectacle can more easily soften souls, only someone very hard of heart would not feel a tremor of joy. Over there, the thick growth of bushes resembles a garden lacking both irrigation and a gardener, these are plants that have had to learn by themselves how best to adapt to nature, to the brute stone that resists their roots, and perhaps for that very reason, because of the stubborn energy expended in these places that men avoid, here where the struggle is between vegetable and mineral, the scents are so penetrating, and when the sun blazes down upon the hillside, all the perfumes open and might lull us to sleep forever, we might perhaps die with our face to the earth, while the ants, raising their heads like dogs, advance, protected by gas masks, for this is their home as well.

These are easy poems to write. The odd thing is that there are no men to be seen. The fields grow green and lush, the undergrowth is steeped in peace and perfume, but a second look tells us that the wheat has lost its first tender freshness, there are tiny dabs of yellow in that vast space, barely noticeable, and the men, where are the men in this happy landscape, perhaps they are not, in fact, the serfs of this glebe, tethered to a stake like goats so that they can eat only what is within reach. There are long periods of idleness while the wheat grows, man has sown the seed in the earth, and if the year is favorable, then lie down and sleep, and call me when it's harvest time. It's hard to understand that this May of flowers is actually a sullen month, we don't mean the weather, which is lovely and seems set fair, but these faces and eyes, this mouth, this frown, There's no work, they say, and if nature sings, good luck to her, we're not in the mood for singing.

Let's go for a walk in the country, up into the hills, on the way the sun glints on this one stone, and we, who are suckers for happiness, say, It's gold, as if all that glittered were gold. We see no men working and immediately declare, What an easy life, the wheat is growing and the workers are resting. However, the truth is rather different. The winter passes, as we have described, in grand banquets and feasts of thistles, dockweed and watercress, with a little fried onion, a few grains of rice and a crust of bread, taking the food from our own mouths so that our children don't go hungry, we shouldn't really need to repeat this, you'll think we're boasting about the sacrifices we make, the very idea, it was the same for our parents and for their parents, and for the parents of those parents, in the days of Senhor Lamberto and before, as far back as anyone can remember, the winter passed, and while some died of starvation, there are plenty of other ways of describing the cause of death, names that are far less offensive to modesty and decency. It's mid-January, men are needed to prune the trees, whether for Dagoberto or Norberto, it doesn't matter, we start to earn a little money, but there's not enough work for everyone, Make a choice, don't get into arguments, and then, once the trees have been pruned, there's the wood on the ground, and the charcoal burners arrive to buy from here and there, and then it's time for them to perform their fiery art, and while we savor the vocabulary of charcoal-burning, staking, earthing up, plugging and firing, the words are doing what they say, it's nothing to do with us, we just know the words, but we didn't know them before, we had to learn them fast, out of necessity, and if everything's ready, let's bag the charcoal up and carry it away, and that's that until next year, my name's Peres, I own twenty-five charcoal kilns in the Lisbon area, as well as several others in the environs, and you can tell your mistress that my charcoal is good stuff, oak, so it burns nice and slow, which is why, of course, it's more expensive. We're burning up in this dryness, this dust, this smoke, there's water to drink over there, I put the jug to my lips, lean back my head, the water gurgles down, a shame it's not cooler, it dribbles from the corners of my mouth and traces rivers of pale skin among the banks of coal dust. We must all have experienced such things and others, because life, despite being short, has room for these and many more, but there are some who lived but briefly, and their whole lives were consumed in this one task.

The charcoal burners and sellers have gone, and now it's May, the month of flowers, may those who write verses try eating them. There are sheep to be sheared, who knows how to do that, I do, I do, cry a few, and the others return to the good life, so called, to weeks of the bad life, going in and out of their houses, until the wheatfields are ready to be harvested, earlier here, later there, yes, we need you now or we might need you later, the goat is tethered to the stake and has no more to eat. It hasn't for some time. So what's the daily rate, ask the workers in the labor market, and the overseers stroll along the unarmed battalions, the sickle has been left at home and we don't use hammers in our trade, and as they stroll along or pause, drumming with their fingers on their waist-coat pockets, they say, We pay whatever the others pay. This is a very old conversation, that's what they said in the days of the monarchy, and the republic changed nothing, these are not things that can be changed by replacing a king with a president, the trouble lies elsewhere, in other monarchies, Lamberto gave birth to Dagoberto, Dagoberto gave birth to Alberto, Alberto gave birth to Floriberto, and then came Norberto, Berto, and Sigisberto, and Adalberto and Angilberto, Gilberto, Ansberto, Contraberto, it's no surprise that they all have such similar names, they simply mean the latifundio and its owners, names don't count for much, which is why the overseer mentions no names but simply says "the others," and no one will ask who those others are, only city folk would make that mistake.

And so when someone asks, How much are we going to be paid, the overseer will say only Whatever the others pay, thus closing the circular conversation of I asked and you didn't answer with a non-response, You'll find out when you go to work. The man says more or less the same thing to his wife, I'll go to work and see what happens, and she thinks, or says out loud, and perhaps she shouldn't say anything, because such things hurt, Well, at least you've got work, and on Monday, the workers are out in the fields doing their duty, and they say to each other, How much do you reckon it will be, and they don't know, What about them over there, I've asked, but they don't know either, and so we arrive at Saturday, and the foreman comes and says, The wages are this much, and they have worked the whole week not knowing how much their work was worth, and at night their wives will ask, Do you know yet, and the husbands will reply irritably, impatiently, No, I don't, stop asking me, and she will say, I'm not asking for myself, the baker wanted to know when we could pay off our debt, such wretched conversations, which continue, That's not much, Well, when the others pay more, so will I. Pure lies, we all know that, but they are lies agreed upon between Ansberto and Angilberto, between Floriberto and Norberto, between Berto and Latifundio, which is another way of saying everything and everybody.

E VERY YEAR, ON CERTAIN dates, the nation summons its
sons. That's a somewhat exaggerated way of putting it, a skillful imitation of some of the proclamations used in time of national
need, or by the person speaking, when necessary, on the nation's
behalf, for overt and covert reasons, to show that we are all one
big happy family of brothers, with no distinction made between
Abel and Cain. The nation summons its sons, can you hear the
voice of the nation calling, calling, and you, who up until now were
worth nothing, not even the bread you need to satisfy your hunger,
nor the medicine for any illness you might have, nor the knowledge to end your ignorance, you, the son of this great mother who
has been waiting for you ever since you were born, you see your
name on a piece of paper at the door of the town hall, not that you
can read it, but someone who can indicates the line where a black
worm coils and uncoils, that's you, you discover that the worm is
you and your name, written by the clerk at the local recruitment
office, and an officer who doesn't know you and is interested in you
for only this one purpose writes his name under yours, an even
more tangled and confusing worm, you can't make out what the officer's name is, and from now on, there's no running away, the na-

tion is staring at you hard, hypnotizing you, to flee would be to offend against the memory of our grandfathers and the Discoveries.
Your name is António Mau-Tempo, and since you came into this
world, I have been waiting for you, my son, for I am, you see, a devoted mother, and you must forgive me if, during all these years, I
haven't paid you much attention, but there are so many of you, and
I can't possibly keep my eye on everyone, I've been preparing my
officers who will be in charge of you, one can't live without officers,
how else would you learn to march, one two left right, right turn,
halt, or to use a gun, careful when you load the breech, country
boy, make sure you don't get your finger caught, and yet they tell
me you can't read, I'm astonished, didn't I set up primary schools
in all the strategic places, not secondary schools, of course, because
you wouldn't need them for the kind of life you lead, and yet you
come and tell me that you can't read or write or do arithmetic, well,
you're putting me to a lot of trouble, António Mau-Tempo, you're
going to have to learn while you're in the army, I don't want illiterate sons bearing my standard, and if, later on, you forget what I'm
ordering you to learn, never mind, that won't be my fault, you're
the one who's stupid, a bumpkin and a yokel, truth be told, my
army is full of bumpkins, but it's not for long, and once your military service is over, you can go back to your usual job, although if
you want another, equally difficult job, that can be arranged too.

If the nations were telling the truth, this is the speech we would
hear, give or take a comma or two, but then we would have to suffer the disappointment of ceasing to believe in the sweet fairy tales
of yesterday and today, which are sometimes clothed in armor and
gauntlets, sometimes in epaulettes and jambeaux, for example, the
story about the little soldier in the trenches who missed his real
mother, the heavenly one having already died, and who would gaze
for hours at the portrait of she who brought him into the world,
until a stray bullet, or an extremely well-aimed one from a skilled
marksman on the other side, shattered the portrait, and the young

soldier, mad with grief, clambered over the parapet and ran toward the enemy trenches, brandishing his rifle, he didn't get very far, though, he was mown down in a hail of bullets, that's what they say in war stories, said hail of bullets coming from a German soldier, who also had a portrait of his dear old mother in his pocket, we add this information so as to round out these stories about mothers and nations and about who dies or kills for such stories.

António Mau-Tempo left his work unfinished and headed for Monte Lavre, getting off the train in Vendas Novas, where he viewed from outside the barracks he would have to enter in three days' time, before setting off to walk the three leagues home, and since it was fine weather, he walked at a steady but relaxed pace, leaving behind him on his left the firing range, an ill-fated place, and, like certain men, punished with sterile upheavals, he finally loses sight of it, or, to be more exact, when he can no longer see it, he puts it out of his mind, and feels upset just to think that he is about to lose his freedom for a year and a half. He thinks of José Gato, and wonders if he ever did his military service, and feels a great weight lifting off his heart, as if destiny were opening a door to the empty road before him and saying, Leave it all behind, why be stuck in a barracks, trapped between four walls, only to go back afterward to cutting cork, digging, scything, don't be a fool, look at José Gato, now that's what I call living, no one dares lay a finger on him, and besides, he has his gang, he's the boss, what he says goes, and though you're not the boss now, you could learn, you're young, it wouldn't be a bad beginning. Temptations, we all have the temptations available to us according to our class and background, as well as those we have learned. His plan may seem rash in a lad who comes from an honest family, apart from the stain left by the life and death of his grandfather Domingos Mau-Tempo, but you can't spend your whole existence thinking about that, let he who has never dreamed of these and worse things throw the first stone, especially since, at this point, António Mau-Tempo doesn't yet know the whole of José Gato's story, which is still to come, and all he can think of is the delicious smell of pork bought clandestinely with his honestly earned money.

With fifteen kilometers to walk, a man has plenty of time to think, to weigh up his life, yesterday he was just a kid and soon he'll be a recruit, but the young man walking determinedly along the road is the best cutter of cork of his nine fellow trainees, perhaps he'll meet one of them in the army. The weather has warmed up, his bag doesn't weigh that much, but it jogs against him and keeps sliding from his shoulder, I'll stop here for a rest, a few meters off the road, not too far, but out of sight, on the grass because the earth is damp, I'll lay my head on my bag and sleep, I've plenty of time to get to Monte Lavre. An old lady sits down next to me, bad luck for me and good luck for her, I don't know what she wants from me, what power she has, perhaps she's a witch, she takes my hand, opens my closed fingers, and says, According to your hand, António Mau-Tempo, you will never marry or have children, you will make five long journeys to distant countries and will ruin your health, you will never own any land of your own apart from the plot that will be your grave, you're no different from other men, and that plot will be yours only until you're nothing but dust and a few bones the same as everyone else's, which will end up somewhere or other, my predictions don't go that far, but as long as you're alive, you will do no wrong, even if others tell you otherwise, but you must get up now, it's time. But António Mau-Tempo, who knew he was dreaming, pretended he hadn't heard this order and continued sleeping, a bad move because he never knew that a weeping princess had been sitting beside him, and that she had held his hand, so hard and calloused though he was still young, so young, and then, having waited a long time, the princess left, trailing the satin of her gown over the gorse and the rockroses, which

196 • *José Saramago*

is why, when António Mau-Tempo finally woke up, the shrubs and bushes were covered in white flowers such as he had never seen before.

These apparently impossible but entirely true incidents often occur on the latifundio. However, the reason António Mau-Tempo was deep in thought from there to Monte Lavre was that he had found two drops of water in the palm of his hand and couldn't work out where they had come from, especially since they refused to mingle into one, but rolled around like pearls, such prodigies are also common in the latifundio, and only the presumptuous would doubt them. António Mau-Tempo would, we believe, still have those drops of water if, when he arrived home and embraced his mother, they had not slipped from his hand and flown out of the door, fluttering white wings, What birds were those, António, I don't know, Ma.

Some people sleep very heavily, some lightly, some, when they fall asleep, detach themselves from the world, some have to sleep in a particular position in order to dream. We would say that Joana Canastra belongs to the latter category. If she's left to sleep peacefully, which is the case when she's ill, and if she's not in too much pain, she lies there just as she did in the cradle, or so someone who knew her then would say, resting her dark, weary cheek on her open palm and immersed in a long, deep sleep. But if she has things to do, things that have to be done at a particular time, then fifteen minutes before the designated hour, she abruptly opens her eyes, as if in obedience to an internal clock, and says, Get up, Sigismundo. Now, if this story were being told by the person who lived it, you would see that already dastardly changes have been made, some involuntary, some premeditated and in accordance with certain rules, because what Joana Canastra really said was, Get up, Sismundo, and one can see how little such minor errors matter when both parties know what they're talking about, the proof being that Sigismundo Canastro, who has his own doubts about how his name should be spoken or spelled, throws off the blanket, jumps out of bed in his long johns, walks over to open the shutters and peers out. It's still black night, and only a very sharp eye, which Sigismundo no longer has, or millennia of experience, which he has in abundance, could distinguish the imponderable change taking place in the east, perhaps it is the fact, and who can comprehend such natural mysteries, that the stars are shining more brightly, when you would expect quite the opposite to be the case. It's a cold night, which is hardly surprising, November is a good month for cold, but the sky is clear and will remain so, for November is also a good month for clear skies. Joana Canastra gets up, lights the fire, puts the blackened coffeepot on to heat up the coffee, the name that continues to be given to this blend of barley and chicory or ground toasted lupine seeds, for even they are not always sure what they are drinking, then goes over to the bread bin to fetch half a loaf and three fried sardines, leaving little if anything behind, and places them on the table, saying, Coffee's ready, come and eat. These may seem trivial words, the poor talk of people with little imagination, who have never learned to enlarge life's small actions with superlatives, compare, for example, the words of farewell spoken by Romeo and Juliet on the balcony of the room in which she has just become a woman, and the words spoken by the blue-eyed German to the no less maidenly, albeit plebeian girl who became a woman against her will after being raped amid the bracken, and, of course, the words she said to him. If these dialogues were being held on the elevated level demanded by the circumstances, we would know that, although this is hardly the first time Sigismundo Canastro has left the house, there's more to this departure than meets the eye, which is why we're telling you about it. Sigismundo ate half a sardine and a crust of bread, with no plate and no fork, slicing into the sardine and cutting chunks off the loaf with the keen blade of his penknife, and once this pap was safely in his stomach, he topped it off with the comforting warmth of that ersatz coffee, there are those who swear blind that the existence and harmonious coexistence of coffee and fried sardines is sufficient proof that God exists, but these are theological matters that have nothing to do with early-morning journeys. Sigismundo then put his hat on his head, laced up his boots, pulled on a worn sheepskin coat and said, See you later, and if anyone asks for me, you don't know where I've gone. There was no need for him to give this advice, it's always the same, besides, Joana Canastra would have little to say, because, although she knows what her husband is going to do, and she wouldn't tell anyone that, even if they killed her, but since she doesn't know where he's going, she couldn't tell them even if they did kill her. Sigismundo will be out all day and won't return until after dark, more because of the path taken and the distance covered than because of the actual time it takes, although one never knows. The woman says, Goodbye, Sismundo, she insists on calling him this, and we shouldn't laugh or even smile, after all, what's in a name, and once he'd gone out through the gate, she went and sat down on a cork stool by the fire and stayed there until the sun came up, her hands clasped, but there's no evidence that she was praying.

Faustina Mau-Tempo, at the other end of Monte Lavre, is not used to this, it's the first time, which is why, although she knew her husband wouldn't have to leave the house until sunrise, she couldn't sleep all night, alarmed that the usually restless João Mau-Tempo should be sleeping so peacefully, like a man afraid of nothing, though he should be afraid. This is the body's way of soothing the troubled soul. It's daybreak, not daylight, when João Mau-Tempo wakes up, and the memory of what he is about to do suddenly enters his eyes, so much so that he immediately closes them and feels a pang in his stomach, not of fear but of quiet respect, the kind one feels in a church or a cemetery or when a child is born. He's alone in the room, he can hear the sounds of the house and those outside, the cold trilling of a lone bird, the voices of his daughters and the crackle of burning wood. He gets up, he is, as we have said before, a small, wiry man with ancient, luminous blue eyes, and at forty-two his hair is thinning and what hair he has is turning white, but before standing up, he pauses to accommodate his body to the sharp pain in his side that always resuscitates after he's been lying down all night, when it should be quite the opposite if his body has rested properly. He dresses and goes over to the fire in the kitchen, as if still wanting to savor the warmth of bed, you would never think he was a man accustomed to bitter weather. He says, Good morning, and his daughters come and kiss his hand, it's good to see the family all together, all are currently unemployed, although they have plenty of things to do to fill the day, be it darning clothes or, in Gracinda's case, working on her trousseau as best she can, though the marriage won't be until next year, and that afternoon, she'll go with her sister to wash clothes in the stream, a whole load of laundry from the big house, well, twenty escudos is better than nothing. Faustina, who is going deaf, didn't hear her husband, but she felt him, perhaps the seismic tremor of the earth as he approached or the movement through the air that only his body makes, each body is different, but these two have been together for twenty years, probably only a blind man would make a mistake, and she has no problems with her eyesight, it's her hearing that's going, although it seems to her, and this is her usual excuse, that people nowadays gabble when they speak, as if they were doing so on purpose. This may sound like the sort of thing only the very old complain about, but these are simply people tired before their time. João Mau-Tempo is stoking up for the day, he drinks his coffee, which is as disgusting as Sigismundo Canastro's coffee, eats some bread made from various flours, just which part of the wheat do they use in the flour, he wonders, and devours a raw egg, making a hole in each end, one of life's great pleasures, when he can get it. The tightness in his stomach has gone, and now that the sun has risen, he's suddenly in a great hurry, See you later, he says, and if anyone asks for me, you don't know where I've gone, this is no pre-planned formula, they are merely the words that come naturally, and there's no need to search for other reasons. Neither Gracinda nor Amélia know where their father is going, they ask their mother when he's left, but she makes the most of her deafness and pretends not to hear. We shouldn't blame the girls, they're young and curious, certainly not irresponsible, an imputation that would doubtless offend Gracinda, who knows all about the exploits of Manuel Espada and his friends when they were only lads, and he was Monte Lavre's first known striker.

The meeting is in Terra Fria. Places are given names doubtless for some comprehensible reason, but to find out why this place was called Terra Fria, Cold Land, on a latifundio that is as hot in summer as it is cold in winter, you would have to go right back to the origins, and those, as lazy people say, are lost in the mists of time. Before they get to Terra Fria, Sigismundo Canastro and João Mau-Tempo will meet at Atalaia hill, not on the very top, of course, they wouldn't want to make themselves too visible, although in this particular area and on this occasion, the latifundio is not exactly as busy as the main square in Évora. They will meet in the dense woods at the foot of the hill. Sigismundo Canastro knows the place well, João Mau-Tempo less well, but all roads lead to Rome. And they will travel on to Terra Fria together, along paths that God never walked and along which the devil would walk only if forced to.

There is no one on the circular balcony of the sky, which is the angels' usual viewing platform above the horizon whenever there is any significant activity on the latifundio. This is the great and fatal mistake made by the heavenly hosts, they measure everything against the crusade. They ignore small patrols, bold sorties, like these tiny dots, the volunteers for this mission, two men here, another farther off, another up ahead and another as yet far away and lagging behind, but all converging, even when they seem not to be, on a place that has no name in heaven, but which down here on earth is called Terra Fria. Perhaps above, in the peaceful empyrean, they think these humans are merely going to work, though there's none to be had, as even heaven must know thanks to the occasional messages sent by Father Agamedes, and it's true that the meeting is work-related. This is a different kind of work, however, and such a great responsibility that João Mau-Tempo will ask Sigismundo Canastro when he meets him and they have taken their first few steps together, or when he has finally managed to overcome his shyness, Do you think they'll accept me, and Sigismundo Canastro will answer, with the confidence of someone older and more experienced, You've been accepted already, you wouldn't be coming with me today if there was any doubt.

One man has come on his bike. He will hide it in the bushes, in some easily identifiable place, just in case he gets disoriented afterward and can't find it. No need, of course, to worry about number plates, if he was in a car, the guards might stop him out of sheer pigheadedness or because they felt a sudden twinge of suspicion, Where are you going, where have you come from, show me your license, and that wouldn't be good, this man happens to be called Silva, but he's also Manuel Dias da Costa, Silva to those he's going to meet in Terra Fria, Manuel Dias da Costa to the guards, with a different name in the registry office and known by a different name again to Father Agamedes, who baptized him far from here. There are those who say that without a name we wouldn't know who we are, which seems a perceptive and philosophical view to take, but this man Silva or Manuel Dias da Costa pedaling along a muddy cart track, for he's now left the road where the guards occasionally appear or else don't appear for days on end, but you never know, your guess is as good as ours, this cyclist is utterly at peace with his soul, quite untouched by these subtle questions of identity. Although that's not quite true, he is actually far more certain of who he is than of the documents that name him. And since he is a thoughtful fellow, he thinks how odd it is that the guards put more faith in a piece of stamped paper, worn thin from being unfolded and refolded, than in what they can actually see, a man and his bicycle, All right, on your way, but as the man puts his foot on the pedal and presses down, he thinks that it would be best not to come this way again in the near future, this is his first time here, and he's been lucky, no one has ordered him to stop.

Some come by train, getting off at São Torcato, on the Setil line, or at Vendas Novas, or even Montemor, if the meeting is being held in Terra da Torre, and at the nearer stations if they're meeting in Terra Fria. It's just a hop and a jump for anyone coming from São Geraldo, but anyone leaving São Geraldo on similar business today will have gone farther afield, and this is not just chance, but doubtless in accordance with very sensible rules. It's midmorning now and there's no bicycle to be seen, the trains are far away somewhere, you can hear them whistling, and a red kite is hovering over Terra Fria, a lovely sight to see, but even lovelier is first seeing it and then hearing its cry, the thin, piping call that no one can quite put into words, but when we hear it, we immediately want to say what it sounded like and can't, there's no shortage of singing birds, but that cry of the red kite is different, so wild it almost sends a shiver down your spine, it wouldn't surprise me to learn that if you heard it often enough, you would sprout wings yourself, well, stranger things have happened. Hovering high up, the red kite drops its head a little, the smallest of movements, because it doesn't need to be that tiny bit closer to see, we're the ones troubled by myopia and astigmatism, a word that should be used with caution on the latifundio, in case the angels mistake it for stigmatism and rush to the balcony expecting to see Francis of Assisi and finding instead a red kite calling and five men approaching Terra Fria, some nearer, some farther off. Only the red kite sees them from on high, but it's never been a telltale.

The first to arrive were Sigismundo Canastro and João Mau-Tempo, who have made a special effort to be early because one of them is new. While they waited, sitting in the sun so as not to get too cold, Sigismundo Canastro said, If you take off your hat, always place it on the ground crown uppermost, Why, asked João Mau-Tempo, and Sigismundo Canastro replied, So as not to reveal your name, we shouldn't know each other's names, But I know yours, Yes, but don't say it, the other comrades will do the same, it's just in case anyone should be arrested, if we don't know each other's names, we're safe. They talked of other things too, just for talking's sake, but João Mau-Tempo was still thinking about how careful they had to be, and when the man with the bicycle arrived, he realized at once that here was someone whose real name he would never know, perhaps because of the great respect with which he was treated by Sigismundo Canastro, who nevertheless addressed him as *tu*, but then perhaps that was the most respectful thing he could do. This is our new comrade, said Sigismundo Canastro, and the man with the bicycle held out his hand, it wasn't the large, coarse hand of an agricultural worker, but strong and with a firm grip, Comrade, the word is not a new one, that's what one's work colleagues are, but it's like saying *tu*, it's the same and, at the same time, so utterly different that João Mau-Tempo's knees buckle and his throat tightens, which is odd in a man past forty who has seen a great deal of life. The three men chat together while they wait for the others to arrive, We'll wait half an hour, and if they don't come, we'll start anyway, and at some point João Mau-Tempo takes off his hat and, before putting it down on the ground, crown uppermost as Sigismundo Canastro had recommended, he quickly looked inside it and saw his name written on the band, in the hatter's fine lettering, as was the custom in the provinces at the time, whereas city folk were already favoring anonymity. The man with the bicycle, as we know him, although João Mau-Tempo assumes that he has come all the way on foot, the man with the bicycle is wearing a beret, which might or might not have his name in it, and if it did, what would it be, after all, you can buy berets at markets and from cheap tailors who don't take such pride in their craft and have no tools for doing poker work or gilding, and who don't care whether their client loses a beret or finds it.

The other two men arrived within a few minutes of each other. They had all met on other occasions, apart from João Mau-Tempo, who was there as the prime exhibit, if you like, and at whom the others stared long and hard in order to memorize his face, which was easy enough, you certainly wouldn't forget those blue eyes. The man with the bicycle asked gravely and simply for better punctuality in future, although he recognized that it was hard to calculate precisely how long it would take to cover such long distances. I myself arrived after these two comrades, and I should have been here first. Then money was handed over, only a few coins, and each man received small bundles of pamphlets, and if names had been permitted, or if the red kite had overheard and repeated them, or if the hats had sneaked a furtive look at the names on each other's respective hatbands, we would have heard, These are for you, Sigismundo Canastro, these are for you, Francisco Petinga, these are for you, João dos Santos, none for you this time, João Mau-Tempo, you just help Sigismundo Canastro, and now tell me what's been happening. The person he addressed was Francisco Petinga, who said, The bosses have found a new way of paying us less, when they have to take us on by order of the workers' association,* they dismiss us all on the Saturday, every single one of us, and say, On Monday, go to the workers' association and tell them I said I want the same workers back, that's the boss speaking, and the result is that we waste all of Monday going to the workers' association, and the boss only has to start paying us on Tuesday, what are we supposed to do about that. Then João dos Santos said, Where

* *Casa do povo* in Portuguese. The organization was set up in 1933 to protect the rights and welfare of agricultural workers.

I live, the workers' associations are in cahoots with the bosses, if they weren't, they wouldn't act the way they do, they send us off to work, we go where we're sent, but the bosses won't accept us, and so back we go to the workers' association, but they won't accept us either and tell us to leave, and that's the way things stand now, the bosses won't accept our labor, and the workers' association either has no power to force them to or is simply having fun at our expense, what are we supposed to do about that. Sigismundo Canastro said, The workers who do get jobs are earning sixteen escudos for working from dawn to dusk, while many can't get any work at all, but we're all of us starving, because sixteen escudos doesn't buy you anything, the bosses are just playing with us, they have work for us to do, but they're allowing the estates to go to rack and ruin and doing nothing about it, we should occupy the land, and if we die, we die, I know you say that would be suicide, but what's happening now is suicide too, I bet there's not a man here can boast of having eaten anything you might call supper, it's not just a matter of feeling downhearted, we must do something. The other men nodded their agreement, they could feel their stomachs gnawing, it was past midday, and it occurred to them that perhaps they could eat the bit of bread and scrape they had brought with them, but at the same time they felt ashamed to have so little, though they were all equally familiar with such dearth. The man with the bicycle is wearing clothes so threadbare that it's as plain as day he has no lunch concealed in his pockets, and what we know and the others don't is that the ants could walk up and down his bicycle all they liked, but they wouldn't find a single crumb, anyway, the man with the bicycle turned to João Mau-Tempo and asked, And what about you, do you have anything to add, this unexpected question startled the novice, I don't know, I have nothing to say, and he said no more, but the other men sat silently looking at him, and he couldn't let the situation continue like that, five grave-faced men sitting under an oak tree, and so, for lack of anything else to say, he added, When there is work, we wear ourselves out working day and night, and still we starve, I keep a few bits of land they give us to cultivate, and I work until late into the night, but now there's no paid work to be had, and what I want to know is why are things like this and will it be like this until we die, there can be no justice as long as some have everything and others nothing, and all I want to say really is that you can count on me, comrades, that's all.

Each man gave his arguments, they are sitting so still that from a distance they look like statues, and now they are waiting to hear what the man with the bicycle will say and what he's already saying. As before, he speaks first to the men as a group, then to Francisco Petinga, then to João dos Santos, more briefly to Sigismundo Canastro and then at length to João Mau-Tempo, as if he were putting together stones to make a pavement or a bridge, a bridge more like, because over it will pass years, footsteps, heavy loads, and below it lies an abyss. From here, it's like watching a dumb show, we see only gestures, and there are few enough of those, everything depends on the word and the stress laid upon it, and on the gaze too, but from here, we cannot even make out João Mau-Tempo's intensely blue eyes. We don't have the keen vision of the red kite, which is still circling around, hovering over the oak tree, sometimes dropping down whenever the air current slackens, and then with a slow, languid beat of its wings rising up again in order to take in the near and the far, this and that, the excesses of the latifundio and just the right measure of patience.

The meeting has ended. The first to leave is the man with the bicycle, and then, in a single expansive movement, like a sun exploding, the other men head off to their respective destinations, at first keeping within sight of each other, as they would know if they were to turn around and look, which they don't, that's another of the rules, and then they are hidden, they don't hide, but are hidden by a dip or vanish into the distance behind a hill, or simply into the distance and the intense cold, which they are aware of now, and which makes them screw up their eyes, you have to look where you're putting your feet too, you can't just amble along willy-nilly. The red kite utters a loud cry, which echoes throughout the celestial vault, then it moves northward, while the startled angels rush to the window, bumping into each other, only to find no one there.

MEN GROW, AND WOMEN grow, everything in them grows, both the body and the area occupied by their needs, the stomach grows commensurate to our hunger, the sex grows commensurate to our desire, and Gracinda Mau-Tempo's breasts are two billowing waves, but that's just the usual lyrical tosh, the stuff of love songs, because the strength of her arms and the strength of his arms, we are referring here, by the way, to Manuel Espada, for three years have passed and there has been no inconstancy of feelings, but, rather, great steadfastness, anyway, the strength of their arms, male and female, is, by turns, required and rejected by the latifundio, after all, there is not such a big difference between men and women, apart from the wages they are paid. Mother, I want to get married, said Gracinda Mau-Tempo, here's my trousseau, it's not much to look at, but it will have to do if Manuel Espada and I are ever to lie down together in a bed that is his and mine, and in which we can be husband and wife, and for him to enter me and for me to be in him, as if we had always been together, because I don't know much about what happened before I was born, but my blood remembers a girl who, at the fountain in Amieiro, was violated by a man who had blue eyes like our father, and I know, al-

though quite how I don't know, that out of my womb will come a
son or a daughter with the same eyes.

If Gracinda Mau-Tempo really had said these words, there
would have been a revolution on the latifundio, but it is our duty
to understand what her actual words meant, mean or will mean,
because we know how hard it is to express the little we do say
each day, sometimes because we don't know which word best fits
which meaning, or which of the two words we know is the more
exact, often because no word seems right, and then we hope that
a gesture will explain, a glance confirm and a mere sound confess.
Mother, said Gracinda Mau-Tempo, the little I have is enough for
us to make a home, or perhaps she said, Mother, Manuel Espada
says that it's time we married, or perhaps she said neither of those
things, but gave the great cry of a solitary red kite, Mother, if I
don't marry now, I'm going to lie down in the bracken by the fountain and wait for Manuel Espada to come and enter my body, and
then I will lift up my dress and wash myself in the stream, and
my blood will flow off to some unknown place, but at least I will
know who I am. And perhaps it wasn't like that either, perhaps one
night Faustina said to João Mau-Tempo, possibly interrupting his
thoughts about leaving some pamphlets in the hollow of a particular tree, She should get married now, she has her little trousseau
ready, and João Mau-Tempo would have replied, It'll have to be
a modest affair, I'd like it to be a really special occasion, but that's
not possible, and António won't be able to help now that he's doing
his national service, tell Gracinda to sort out the paperwork and
we'll do what we can. As ever, it's still the parents who have the last
word.

They have a house, one that suits their pocket, and since their
pocket is small, the house is small too, and rented of course, just
in case you were thinking that Gracinda Mau-Tempo and Manuel Espada were about to announce proudly, This is our house,
no, they would rather hide the fact and say, I live over there some-

where, as if they were playing hide-and-seek or hunt the thimble, except, of course, those are games played at school or in the city, simply so that no one will know exactly where they live, in this house which is just walls and a door, with one room up and one down, a rickety ladder that wobbles when you climb it and no fire in the grate when we're out. We're going to live on the side of this hill in Monte Lavre, in this little yard, there's not enough space to swing a hoe if we wanted to grow some cabbage, after all it does get the sun all day, although I don't know that it's worth the trouble, we're hardly going to get fat on cabbage. We'll sleep downstairs, in the kitchen, except it won't be a kitchen when we're sleeping in it, just as it won't be a bedroom when we're up and about, what should we call it then, a kitchen when we're cooking, a sewing room when Gracinda Mau-Tempo is doing the darning, and a waiting room when I'm sitting looking at the hills opposite, with my hands in my lap, this may seem as if they're just playing with words, but it's simply their mutual excitement, each tumbling over the other in their eagerness to speak.

If we start to get too far ahead of ourselves, we'll soon be talking about children and the problems they bring. Today is a holiday, Manuel Espada is going to marry Gracinda Mau-Tempo, there hasn't been a marriage like this in Monte Lavre for a long time, that is, with such an age difference between bride and groom, he's twenty-seven and she's twenty, but they make a handsome couple, he's the taller of the two, which is as it should be, although she's not short either, she doesn't take after her father in that respect. I can see them now, she's wearing a pink, calf-length dress with a high neck and long sleeves buttoned at the cuff, if it's hot, she's not aware of it, as far as she's concerned it might as well be winter, and he's wearing a dark jacket, more like a three-quarter-length coat than the jacket of a suit, a pair of rather tight trousers and shoes that no amount of polishing will bring a shine to, a white shirt and a tie bearing a pattern of branches as indecipherable as the tops of trees no one has bothered to prune, but let there be no misunderstanding, the trees are just a simile, nothing more, because the tie is new and will probably never be worn again, unless it's at another wedding, should we be invited. It's not a big wedding party, but there are plenty of friends and acquaintances, and children attracted by the prospect of sweets, and old ladies at the door talking about heaven knows what, one never knows what old ladies talk about, whether they are uttering blessings or reproaches, poor things, what is the point of their lives.

The ceremony takes place after the mass, as is the custom, and people look a bit cheerier than usual because, luckily, there's plenty of work around at the moment, plus it's a nice day. Doesn't the bride look lovely, the boys don't dare make many jokes about marriage, because, after all, Manuel Espada is older, nearly thirty, a different generation from us, that's a bit of an exaggeration, of course, since he's only twenty-seven, but it's an interesting situation, even the married men refrain from teasing him, the bridegroom is hardly a boy, and he always looks so serious, he was the same when he was a child, you can never tell what he might be thinking, just like his mother, who died last year. They're quite wrong, though, it's true that Manuel Espada has a grave face or countenance, as people used to say, but even if he wanted to, he wouldn't be able to explain quite what he is feeling, it's like water singing as it rushes over the rocks up there in Ponte Cava, which is a bleak place and a bit frightening at night, but then, come the dawn, you see there was no reason to be afraid, it was just the water singing among the rocks.

Great injustices are committed because of how people look, that was the case with Manuel Espada's mother, a woman who seemed to be made of granite, but who melted sweetly at night in bed, which is perhaps why Manuel Espada's father is slowly weeping, some say, It must be tears of joy, and only he knows that it isn't. Let's see, how many people are here, twenty, and each one of them would make a story, you can't imagine, years and years of living is a lot of time, and a lot of things can happen in that time, if we were all to write our life story, think how big that library would be, we would have to store the books on the moon, and when we wanted to find out who So-and-so is or was, we would travel through space to discover not the moon, but life. It makes us feel, at the very least, like turning back and recounting in detail the life and love of Tomás Espada and Flor Martinha, if we weren't driven on by events and by the new life and love of their son and Gracinda Mau-Tempo, who have now entered the church, surrounded by a throng of excited children, take no notice, boys will be boys, while the older people, who are familiar with rituals and sermons, enter, looking composed and slightly constrained, wearing old clothes from a time when they were slimmer. Just this coming into church and being here, these faces, feature by feature, each line and wrinkle, would merit chapters as vast as the latifundio that laps around Monte Lavre like a sea.

Father Agamedes is at the altar, and I don't know what exactly has got into him today, what fair wind greeted him when he got up, perhaps it was the Holy Spirit, not that Father Agamedes is one to boast of his closeness to the third person of the Holy Trinity, he himself doubts the simplicity of these theological formulae, but for whatever reason, this old devil of a priest is in a good mood, he's very composed, but his eyes are shining, and that can't be because he's looking forward to satisfying his greedy appetite, there will hardly be an abundance of food at the wedding feast. Perhaps it's simply the pleasure of blessing this marriage, Father Agamedes is a very human priest, as we have seen throughout this story, and even if, for the moment, he chooses not to think about the latifundio's variable need for workers, he must be pleased that this man should join flesh with this woman and make children who will then grow up and who are sure to bring some benefit to the church by being born, marrying and dying, as the other people here present have and will. This is a flock that brings him little wool, but it's better than nothing, out of these crumbs comes a sponge cake, Have another slice, Father, and drink this glass of port, and then another slice, I couldn't eat another thing, Senhora Dona Clemência, Go on, make a sacrifice, Father Agamedes, after all, that's what he does every day, the sacrifice of the holy mass, come closer now and I will make you man and wife.

There is some confusion among the witnesses, none of whom can remember which side they should be standing on, and Father Agamedes says the necessary words, folds and unfolds his stole, steals a suspicious glance at the sacristan, who arrives late, but what are you thinking, he's not Domingos Mau-Tempo, that was years ago, and this isn't the same priest, people don't live forever. Nothing happened, the light didn't change, the church didn't fill up with thrones and seraphim, and a turtledove cooing in the garden continues to coo, preoccupied perhaps with other weddings, and Gracinda Mau-Tempo can now look at Manuel Espada and say, This is my husband, and Manuel Espada can look at Gracinda Mau-Tempo and say, This is my wife, which, as it happens, will only be true from this moment on, because the bracken at the fountain has never received these two bodies, though that once seemed a distinct possibility.

The bride and bridegroom are just crossing the tiny nave when the door of the church opens and in comes António Mau-Tempo in his army uniform, he's late for his own sister's wedding, a matter of delayed trains, missed connections, which left him furiously counting the kilometers between him and home, but finally, after António Mau-Tempo had uttered oaths capable of melting the bronze bearings on a train and alternately run and strode along the verge of the road, the driver of a passing fish truck succumbed to the magic spell of his uniform and asked, Where are you going, To Monte Lavre for my sister's wedding, and dropped him off at the bottom of the hill, saying, Congratulate the happy couple for me, and António Mau-Tempo bounded up that hill like a mountain goat, walked straight past the big house and the guards' barracks without so much as a glance, bastards, and then it suddenly occurred to him that perhaps the wedding was over, but no, there are still people outside, only a few more meters, up the steps in two strides, and there's my sister and there's my brother-in-law, I'm glad you could make it, brother, Oh, I'd have made it if I had to set fire to the whole regiment. Out in the street now, the main topic of conversation isn't the wedding but António Mau-Tempo, who was given leave to come to his sister's wedding, and since he then has to embrace everyone, father and mother, relatives and friends, the wedding cortege is slightly disrupted, patience, not that Gracinda Mau-Tempo is jealous, she has her magnificent husband, Manuel Espada, by her side, she stands arm in arm with him the way couples at the very smartest weddings stand, and she's blushing furiously, Lord in heaven, why can you not see these things, these men and women who, having invented a god, forgot to give him eyes, or perhaps did so on purpose, because no god is worthy of his creator, and should not, therefore, see him.

The disruption was short-lived. Manuel Espada and Gracinda Mau-Tempo are once more the king and queen of the party, António Mau-Tempo having now joined his childhood friends, with whom he always needs to reinforce and refresh the bonds of friendship after his long absences in such places as Salvaterra, Sado and Lezírias, farther north toward Leiria, and now, during his national service. The wedding feast is being held in someone else's house, lent for the day. There is wine, lamb and bread stew, with more bread than lamb, bride cakes, two bottles of fortified wine and a few tasty pigs' ears, this is no banquet but the wedding of poor people, so poor that João Mau-Tempo would clutch his head if we were cruel enough to mention the expense and the quadrupled debt at the grocer's and the haberdasher's, the all too familiar dogs that will soon be snapping at the debtor's heels, but which for the time being remain treacherously silent, Is there anything else you need, after all, it's not every day your daughter gets married.

Until Father Agamedes joins them, no one can eat, wretched priest, he's obviously not as hungry as I am, the smell of that stew is making my stomach rumble, I don't know how I've lasted this long really, I deliberately didn't eat supper last night so that I'd have more appetite today. One doesn't own up to such feelings, of course, admitting that one didn't have supper so as to be able to eat more at other people's expense, but we're all familiar enough with such human frailties, and therefore with our own, to be able to forgive them in others. Especially now that Father Agamedes has finally arrived and goes over to say a few words to Tomás Espada and the Mau-Tempos, words that Faustina doesn't quite understand, although she nods vigorously and adopts an expression of filial unction, not that she's a hypocrite, poor woman, it's just that the timbre of Father Agamedes's voice makes her ears buzz, otherwise she would be able to understand him perfectly. Father Agamedes is very fatherly with the bride and groom, he gestures with his right hand, blessing people on either side, and they forget about their hunger for a moment, but now it comes roaring back, at last we're going to start. In came the platters and tureens, all borrowed, well, two of them weren't, and as for Gracinda Mau-Tempo's own meager collection of crockery, her mother was very firm, You're not taking that to the wedding, we'll sort something out, don't you worry, you can't start married life with a load of broken crockery, it might bring bad luck. Finally they ate, at first greedily, then more slowly, because everyone knew that there wouldn't be much more to eat, and common sense dictated, therefore, that they make the stew and the pigs' ears last, at least there was plenty of wine, which was something.

After a while, Father Agamedes got to his feet, made a gesture calling for silence, just a gesture, nothing more, he was a tall, extremely thin man, indeed it was a matter of great perplexity among his parishioners as to just where Father Agamedes put the considerable quantity of food that he ate, as was evident at the weddings and christenings he presided over, he got to his feet, looked at the people seated around him, wrinkled his sensitive nose at the sight of the dirty, disorderly table, oh, they're so ill bred, Senhora Dona Clemência, but then felt himself filled with charity, doubtless Christian charity, and said, My dear children, I address myself to you and especially to the newlyweds on this happy day on which I have had the great good fortune to unite in holy matrimony Gracinda Mau-Tempo and Manuel Espada, she, the daughter of João Mau-Tempo and Faustina Gonçalves and he, the son of Tomás Espada and the late Flor Martinha. You have made the vows of faithfulness and mutual support that the holy mother church requires of all those who come to her in order to sanctify the joining together of man and wife until death do them part. Father Agamedes was wrong to mention death at this point, because Tomás Espada closed his eyes to hold back his tears but failed, tears are like water oozing out from a painful crack in a wall, everyone, very wisely, pretended not to notice, and Father Agamedes proceeded regardless, This land of ours may be small, but fortunately we share a great friendship, there are no dissensions or disputes such as I have seen in other places, and although it's true that the people here are not great frequenters of our beloved mother church, who is always waiting patiently for her children to come to her, it is also true that almost no one omits to attend the sacraments, and those who don't attend are lost sheep whom I, alas, have long given up all hope of saving, may God forgive me, for a minister should never lose hope of leading his entire flock into the arms of God. One of those stray sheep was present, as was his wife, who compared very favorably with her husband in the stray-sheep stakes, namely Sigismundo Canastro and Joana Canastra, both of whom were beaming as if Father Agamedes's words were bouquets of roses, far be it from me to boast, but I have proven to be a constant and caring shepherd, for example, three years ago, at the time of the strike, as I hope you will all remember, some of you here today were among those I freed from prison, as you yourselves can attest, and were it not for Monte Lavre's good standing with the Lord, all twenty-two of you could have been taken to the bullring, as happened to other men in lands less blessed by Our Lord and the Virgin Mother, although I know, of course, that I, poor repentant sinner than I am, cannot take the credit for such things.

At this point, João Mau-Tempo turned red and, needing to look at someone, he looked at Sigismundo Canastro, whose grave and now unsmiling eyes were fixed on the priest, and then António Mau-Tempo spoke up, This is my sister's wedding, Father Agamedes, it's no time to speak of strikes or who should take credit for what, and his voice was so serene that he didn't seem the least angry, although he was, and everyone else kept silent, waiting to see what might happen next, but the priest merely proposed a toast to the health of the newlyweds and sat down. That was not a good idea, Father Agamedes, Norberto said afterward, what possessed you to say such things, it's like mentioning rope in the house of a hanged man, You're quite right, said Father Agamedes, I don't know what came over me, I just wanted to show them that if it weren't for us, the church and the latifundio, the two persons of the Holy Trinity, of which the third is the State, that purest of doves, if it were not for us, how would they keep body and soul together, and, come election time, who would they give their votes to, but I confess I was wrong, mea culpa, mea maxima culpa, that's why I didn't stay there much longer, I gave my pastoral duties as an excuse and left, I was, admittedly, slightly tipsy, though I didn't drink much of that rough fortified wine of theirs, far too acidic for my stomach, not like the excellent wine from your cellar, Senhor Lamberto.

Then António Mau-Tempo, as spokesman, said, Right, now that Father Agamedes has gone and we're among family, we can say what we like, according to our inclinations and our hearts' choosing, so Manuel Espada will talk to Gracinda, his wife and my sister, while my other sister, Amélia, doubtless has her eye on someone too, though she might not be free to speak to him, and if he's not here, then she can think about him, and we will all understand, because sometimes that's the most we can do, and my parents will think back over their lives and over ours and what they were like when they were young, and they will forgive us our mistakes, and the rest of you will think about yourselves and your nearest and dearest, some of whom are already dead, I know, but if you call them, they will come back, that's all the dead are waiting for, indeed, I can already feel the presence of Flor Martinha, someone must have summoned her here, but since I'm the one speaking, I will keep the floor, and don't be surprised at my fine way of speaking, you don't only learn about fighting in the army, if you really want to, you can learn how to read and write and do arithmetic, and that way you can begin to understand the world and a little about life, which isn't simply a matter of being born, working and dying, sometimes we have to rebel, and that's what I want to talk to you about.

Any conversations going on around him stopped, Gracinda Mau-Tempo and Manuel Espada ceased gazing at each other, although they continued to hold hands, Flor Martinha said her farewells, Goodbye, Tomás, the guests put their elbows on the table, they have no manners these people, and if someone sticks a finger in his mouth to extract from some cavity in his teeth a bit of gristle from the lamb, don't be angry, we live in a land where food cannot be wasted, and António Mau-Tempo, in his cotton uniform, is talking about just that, about food. It's true that there's a lot of hunger hereabouts, sometimes we're obliged to eat weeds, and our stomachs are as swollen and tight as drums, and perhaps that's why the commander of the regiment believes that if a donkey is hungry enough it will eat thistles, and since we are donkeys, because we hear nothing else on the parade ground, well, actually we hear far worse than that, we do eat thistles, but I can tell you that I would rather eat thistles than the food they serve at the barracks, which is fit only for pigs, although even they might turn up their snouts at it.

António Mau-Tempo paused, took a sip of wine to clear his throat, wiped his mouth with the back of his hand, after all, what more natural napkin is there, and resumed his speech, They believe that because we are starving at home, we should accept anything, but that's where they're wrong, because our hunger is a clean hunger, and the thistles we have to strip, we strip with our own hands, which even when they're dirty are still clean, no one has cleaner hands than us, that's the first thing we learn when we enter the barracks, it's not part of the weapons drill, but you sense it, and a man can choose between outright hunger and the shame of eating what they give us, they came to Monte Lavre to summon me to serve the nation, or so they said, but I don't know what that means, the nation is my mother and my father, they said, well, I, like everyone else, know my real mother and father, who took the food from their own mouths so that we could eat, in that case, the nation should also take the food from its own mouth so that I can eat, and if I have to eat thistles, then the nation should eat them too, if not, that means that some are the sons of the nation and others are the sons of whores.

Some of the women were shocked, some of the men frowned, but António Mau-Tempo, who has something of the vagabond about him despite his uniform, will be forgiven anything for having put Father Agamedes firmly in his place, and besides, he says these other words that taste to his listeners like the excellent wine from Senhor Lamberto's cellar, although that's purely a hypothesis, because our lips have never actually touched the stuff, Anyway, in the barracks we decided to hold a hunger strike, we wouldn't eat a single crumb of what they put before us, just like pigs who refuse to eat from the trough in which there's more rubbish in the swill than even a pig will eat, we don't mind eating two quarts of earth a year, the earth is as clean as us, but not that food, and I, António Mau-Tempo, speaking to you now, was the one who had the idea, and I'm proud of that, you don't know how different you feel until you've done these things, I talked to my comrades and they agreed that the situation could only be worse if they were actually spitting on us, and when the day came, the cookhouse bell rang and we went and sat down as if we were going to eat, but the food arrived and it stayed there on the plates uneaten, the sergeants bawled and yelled, but no one picked up his spoon, it was the revolution of the pigs, and then the officer on duty turned up, made a speech like the one Father Agamedes made, but we pretended we didn't understand a word of it, as if he were talking Latin, first he tried to win us over with sweet words, but then he lost his rag and started screaming at us, ordering us to form up on the parade ground, an order we did understand, because what we wanted more than anything was to get out of that cookhouse, so out we went, whispering words of encouragement to each other, Don't give up, courage, my friend, stick to your guns, we're all in this together, and there we stood for half an hour, and that, we assumed, was the punishment until we saw them setting up three machine guns trained on us, all in accordance with the regulations, with gunners and their assistants, and boxes of ammunition, and then the officer said that if we didn't go and eat, he would give the order to fire, that was the voice of the nation speaking, it was as if my mother had said to me, either eat your food or I'll slit your throat, none of us believed he would do it, but then they started loading the machine guns, and from that point on we had no idea what was going to happen, I can tell you I felt a shiver go down my spine, what if they really did shoot and there was a bloodbath over a bowl of soup, was it worth it, not that we were weakening, but in situations like that, you can't help such thoughts running through your mind, and then, from

within the ranks, we never did find out who it was, even the comrades standing nearest never said, we heard a very calm voice say, as
if it were someone politely inquiring after our health, Comrades,
stand your ground, and then another voice at the other end of the
line said, Go on, shoot us, and then, even now it brings a lump to
my throat, every single soldier in the ranks repeated those defiant
words, Go on, shoot us, I don't think they would have fired on us,
but if they had, I know that we would all have stood our ground,
and that was our real victory, rather than getting them to improve
the food, it's odd how sometimes you start out fighting for one
thing and end by winning something else, and that second thing
was the best of the two. António Mau-Tempo paused again, and
then, much wiser than his years, he added, But to win that second
thing, you have to start fighting for the first.

The women are weeping and the men's eyes are filling with
tears, this is the best wedding you could possibly imagine, Monte
Lavre has never seen the like, and then Manuel Espada stood up
and went to embrace António Mau-Tempo, thinking how different this army is from the one he served in, and he remembers his
national service in the Azores and hearing his fellow soldier issu-
ing that vague threat, When I get out of here, I'm going to join the
police for the vigilance and defense of the state, it's great, say there's
someone you don't like, well, you simply arrest him, haul him off to
the civil authorities, and if you like, shoot him in the head before
you get there and say he tried to resist.

Now Sigismundo Canastro, tall and thin, has got to his feet, he
toasts the newlyweds, and when everyone has downed some of the
fortified wine, he announces he's going to tell a story which, while
not quite the same as António Mau-Tempo's, is nevertheless similar, because with stories and anecdotes you can always find some
similarity, however unlikely, Many years ago, and at this point he
pauses, just to make sure everyone is listening, and they are, their
eyes fixed on him, some are rather sleepy, it's true, but can still

manage to keep awake, and then he goes on, Many years ago, I was
out hunting, oh, no, not another hunting story, all lies and exaggeration, but Sigismundo Canastro isn't joking and doesn't respond to
this interruption, he merely looks around him as if pitying such a
lack of seriousness, and whether it was that look or mere curiosity
to find out how big a lie this will be, silence falls, and João Mau-
Tempo, who knows Sigismundo Canastro very well, is sure there
will be more to this story than meets the eye, the problem will
be understanding it, At the time, I didn't have a rifle of my own,
I used to borrow one from whoever I could, and I was a pretty
good hunter too, just ask the people who knew me then, and I had
a little dog I was training up, a real gem with a really keen nose,
and one day I went out with some friends, there were quite a few
of us, each of us with our dog, and we had already walked a long
way and were somewhere over near Guarita do Godeal when a
partridge suddenly flew up, as fast as you like, I put my rifle to
my eye, and the bird fell just as I was about to pull the trigger, I
certainly didn't hit it, fortunately, though, for my good name as a
hunter, there was no one else around, but Constante, my dog, ran
to where the partridge had fallen, thinking perhaps that the bird
was wounded, lost amid the gorse, because the undergrowth was
really thick, and there were some large rocks blocking your view,
but anyway, the dog disappeared, and I called and called, Constante, Constante, and I whistled and whistled, but no response, it
would be even more embarrassing having to return home without
the dog, besides, I was really fond of him, he was one of those dogs
who could almost speak. His audience was hanging on his every
word now, listening and digesting, it doesn't take much to make a
man happy and a woman content, and even if the story turned out
to be pure hokum, it was a good story well told, as Sigismundo
Canastro went on to show, Two years later, I happened to be in
those parts again, and I came across a vast area of land which they
had begun to clear but then, for some reason, abandoned, and I

remembered what had happened with Constante, and I plunged
in among the rocks and the undergrowth, it was the devil of a job,
but something was leading me on, as if someone were saying, don't
give up, Sigismundo Canastro, and suddenly what did I see but the
skeleton of my dog standing there, guarding the skeleton of the
partridge, and they had been like that for two years, both equally
determined. I can see it now, my dog Constante, his nose pointing
forward, his front leg poised and lifted, and no wind could knock
him over and no rain dissolve his bones.

Sigismundo Canastro said no more and sat down. No one else
spoke a word, no one laughed, not even the younger people, who
belong to a less credulous generation, and then António Mau-
Tempo said, They're still there, the dog and the partridge, I dreamt
about them once, what more proof could you want, and having
said that, everyone cried out together, They're still there, they're
still there, and then they believed the story and burst out laughing. And after they had laughed, they carried on talking, and they
talked all afternoon, about this and that, come on, have another
drink, at this same hour, the parade ground at the barracks will be
deserted, while the empty sockets of Constante the dog stare at the
empty sockets of the partridge, both equally determined. When
night fell, they said their goodbyes, some accompanied Gracinda
Mau-Tempo and Manuel Espada to the door of their house, tomorrow there's work to be done, we're lucky to have it, Don't be
long, Gracinda, Just coming, Manuel. In the next yard a dog barks,
surprised to have new neighbors.

J OSÉ CALMEDO IS JUST one guard among many. You wouldn't
notice him on parade, he's no more striking than any of his colleagues, and when he's not on parade, but on patrol or otherwise
on duty, he's a quiet, easygoing fellow who always seems to have
his mind on other things. One day, quite unexpectedly, perhaps
even to himself, he will hand in his application for discharge to the
commander of the Monte Lavre barracks, who will then begin the
necessary procedures, and he will take his wife and two children
far from there, learn to live the life of a civilian and spend his remaining years forgetting that he was once a guard. He is, however,
a man with a history, which we do not, alas, have time to go into
here, except to mention his family name, a story that is both brief
and charming, and illustrative of the beauty of names and their
unusual origins, for it is the fault of our feeble memory and our
lack of curiosity that makes us ignore or forget, for example, that
the name Sousa* means wild dove, isn't that lovely, and is not just
the very ordinary name set down in the register of births, which

---

* Sousa was Saramago's father's name, Saramago being a family nickname, meaning wild
radish, accidentally incorporated into Saramago's name when his birth was recorded in the
register of births.

immediately clips its wings, that's the trouble with writing things down. But best of all are the names born out of the distortion of other names or of words that never had any intention of becoming names, for example, Pantaleão became Espanta Leões, pity the poor family who has to go through life with a duty to drive away lions in city and country. But we were talking about the guard José Calmedo and the brief and charming history of his name, born, or so the story goes, from the unintentional bravado of an ancestor of his, who, unaware of the very real danger he was in, was not as frightened as he ought to have been, and responded to the person asking about his lack of fear, *Qual medo*, what fear, and people were so amazed by the spontaneous effrontery of that question that this unintentional hero and his descendants, including this guard and his children, were known ever after as Calmedo, although later, another version was born, because Calmedo means very hot, windless weather, which is what it's like as he leaves the barracks now, carrying his secret orders.

It's three kilometers there and three kilometers back on foot, but that's what a guard's life is like, although not, of course, for the mounted guard, anyway, there's José Calmedo heading down the hill in Monte Lavre to the valley, he skirts around a village toward the west, then heads north along the road, with ricefields to his left, it's a beautiful July morning and hot, as we said before, but it will be even hotter later. There's a little stream down below, much thirst and little water, his boots strike the surface of the road, and he feels very much a man as he strides it out, while his head is full of stray clouds, words that once had meaning but have lost it, well, we were walking along the road, but now we've gone down the bank to the right, into the cool shade beneath the viaduct, and are now sitting beneath the whispering branches of the poplars, the place is deserted, who would have thought it, the empty pool, the ruined water wheel and, beyond it, the brick kiln with the broken roof, it seems that the latifundio corrodes everything that gets in its way. José Calmedo rests his rifle on his shoulder, takes off his cap and uses his handkerchief to wipe the sweat from his brow, where the dark and the light skin show the effect of the sun or the lack of it, it's almost as if the top part of his head belonged not to him but to his cap, although these, of course, are pure imaginings.

It's not much farther, he's going to Cabeço do Desgarro and should arrive there in time for lunch. He will return with João Mau-Tempo, on the pretext of some insignificant matter that has nothing whatever to do with João Mau-Tempo, it doesn't need to be a complicated story, the simpler the better, the more credible. He can see the hut among the trees and the men, who stand around the fire, removing the pot before it boils over or burns, it won't take long, he just has to go over to him and say, Come with me to the barracks, but José Calmedo doesn't take the few steps that would place him where he could be seen by everyone, should they look. He hides behind some high bushes and stays there, allowing time for João Mau-Tempo to finish his sparse lunch, while in the sky occasional clouds continue to pass, so few that they don't cast a shadow. José Calmedo is sitting on the ground smoking a cigarette, he has propped his rifle against the trunk of a tree. He has a good life, this guard, with few duties, simply watching the days pass by, only occasionally are there a few more serious cases, although there will be more, otherwise, the months come and go, the latifundio is calm and peaceful, the barracks and his beat are calm and peaceful, apart from report writing and patrols, court proceedings and the kind of complaints bickering neighbors always have. Life goes by, and before you know it you've reached retirement age. These are the thoughts of a peace-loving man, you would never think he had a rifle and a cartridge belt at his side and was wearing seven-league boots, above his head a bird is singing, its name doesn't matter, it hops from branch to branch, you can see its silhouette from here, just the fan of its tail and a wing. If we looked down at the ground, we would see the crawling fraternity of insects, the ant that raises its head like a dog, the other that always keeps it lowered, the tiny spider, wherever does it put its food, but we mustn't let ourselves be distracted, we have to go and arrest a man, we're simply letting him finish his lunch, well, just because we're guards doesn't mean we don't have a heart, you know.

There are no great feasts on the latifundio. José Calmedo peers between the bushes, everyone has finished eating. He gets up, sighing perhaps at the effort made or about to be made, puts his rifle over his shoulder with measured gestures, not because these gestures are important, but because they are crutches, things a man can hold on to in order not to get lost amid the meaninglessness of his actions, and then he heads off down the hill toward the men. They see him coming from a distance, their hearts perhaps beat a little faster, the laws of the latifundio are strict, whether they're to do with who owns the acorns or where you can collect firewood, or far worse misdemeanors. José Calmedo approaches, then stops and summons the foreman, he doesn't want everyone to hear, men may not be girls, but they have their modesty, Tell João Mau-Tempo I want to have a word with him.

João Mau-Tempo's heart beats as fast as that of a little bird. Not that he feels himself guilty of any heinous crime, of the kind that merits rather more than just a fine or a beating. He senses that he is the man the guard has come for, that from the moment the foreman says, João Mau-Tempo, go and talk to the guard, it will be like removing a layer of cork, you hear the creak and know that the efforts of both man and tree are working as one, all that's lacking is the man's grunt, Uh, and the scream of the bark as it comes away, craaack, So, Senhor José Calmedo, what can I do for you, asks João Mau-Tempo with the apparent calm of someone congratulating the guard on his appearance, fortunately our hearts are hidden, otherwise all men would be condemned sooner or later, either for their innocence or their crimes, because the heart is an impulsive, impatient thing, incapable of restraint. The person who made hearts clearly didn't know what he was doing, but fortunately one can learn to be sly, otherwise how could José Calmedo say, without anyone having told him to, Oh, it's nothing important, we just want to clear up a case involving two guys who stole a couple of sheaves of wheat, the owner swears it was them, but they say you're a witness to the fact that it wasn't, I don't really understand the situation myself, to be honest. It's always the same, however well intentioned, a man tends to get in a tangle when he shouldn't and whatever he says becomes like the devil's cape, which, being short, both covers and uncovers, but even when João Mau-Tempo, who is, in this case, completely innocent, even when he says, But what have I got to do with it, why should I get involved, the guard responds with the old argument, You have nothing to worry about, just come along with me, say your piece and leave.

So be it. João Mau-Tempo is about to go off and pick up his few tools and what's left of his lunch, but José Calmedo, still carried along on the wave of his invention, says, Don't bother doing that, you'll be back soon, it won't take long. And having fulfilled his quota of lies, he moves off, with an uneasy João Mau-Tempo following behind, clacking along in the clogs he wore when working. From there to Monte Lavre, José Calmedo's face is a picture of rage, as befits a guard who has made an arrest and is escorting his prisoner, but that wasn't the reason, rather, the sadness of having won such a pathetic victory, is this what two men were born for. And João Mau-Tempo, deep in his own thoughts and anxieties, was trying to convince himself that some sheaves of wheat really had been stolen and that his testimony really could save two innocent men.

João Mau-Tempo enters the same barracks where he had been held prisoner some four years before. Everything looks the same, as if time hadn't passed. José Calmedo goes to tell the corporal that the arrested man is here, that there have been no hitches, mission completed, but please, keep the medals for another occasion, just leave me alone to get on with my life and my cloud-thoughts, one day, I will present a sheet of paper bearing an official stamp and addressed to the Commander General of the National Republican Guard, Sir, meanwhile, Corporal Tacabo orders João Mau-Tempo to come in and says, Sit down, Senhor Mau-Tempo, such politeness is not so very odd, guards don't always behave like cruel executioners, Do you know why you've been summoned. João Mau-Tempo is about to say that if it's about some sheaves of wheat, he has nothing to tell, but he doesn't have time to open his mouth, and it's just as well, because if he had, José Calmedo would have been found to be a liar, fortunately, Corporal Tacabo went on, well, it's best to get this over with quickly, What were you doing in Vendas Novas, You must be mistaken, I wasn't doing anything, Well, I have an order from the Vendas Novas barracks to arrest you as a communist.

Here is an example of simple, straightforward dialogue, with no harmonies and no arpeggios, unaccompanied and unadorned by thoughts and subtleties, it's as if they weren't dealing with serious matters, but were saying, So, how have you been, Very well, thank you, and yourself, A friend of yours in Vendas Novas sends you his regards, Do give him mine when you next see him. A bell has just rung inside João Mau-Tempo's head, there is a great clanging sound like that of castle doors being slammed shut, no one enters here. But the owner of the castle is trembling, his hands and voice are trembling, Defend yourself, my soul, but this lasted only a second, time enough to feign horror, surprise, offended, outraged innocence, How can you say that, sir, I haven't been involved in anything like that for four years, not since I was arrested and taken to Montemor, there must be some mistake, and Corporal Tacabo says, Well, I certainly hope so, if you know nothing about it, the authorities will release you immediately. Perhaps it will be all right, perhaps it was a false alarm, perhaps no one is drowning, perhaps the fire will die down without him burning his hands, Could you send for my wife so that I can speak to her, Corporal. A perfectly natural request, but Monte Lavre, you see, is such an unimportant place, just a tiny village in the latifundio, that here the corporal is the commander, and he responds as firmly as the commander in Lisbon who gives him his orders, No, your wife can't speak to you, nor can anyone else, regardless of what you may say, you have been judged to be a dangerous element, a soldier will go and fetch anything you need from your house.

João Mau-Tempo, a dangerous element. He was taken to the room that served as a cell, again it was José Calmedo who took him, there didn't seem to be anyone else on duty, and João Mau-Tempo, before he was locked up, says, So you tricked me, and José Calmedo doesn't respond at first, he feels offended, he was only doing his job and this is his reward, but he can't remain silent as if he really had committed some crime, and so he replies, I didn't want to worry you, José Calmedo really shouldn't be wearing the uniform of the national guard, which is why he will take it off one day and go and live in a place where no one knows he was a guard, and that is all we will ever learn about his life.

Faustina Mau-Tempo and her two daughters come and go outside the barracks. They are in tears and anxious, they don't know what their husband and father stands accused of, only that he will be taken to Vendas Novas, and as ill luck would have it, as the saying goes, the three of them, for one reason or another, are absent when the jeep from Vendas Novas arrives, complete with a patrol armed with rifles and bayonets, to fetch the criminal. When mother and daughters return, they will learn that João Mau-Tempo is no longer there, and they are left standing out in the street, at the door of the barracks, to which entry is barred, He's not here now, that's all we know, go home and you'll be kept informed, they say these words to the poor women, but they are pure mockery, just as the guards who came from Vendas Novas to get João Mau-Tempo mocked him when they said, Hop in, we're off on a little trip. The guard would never normally invite him to go anywhere, with transport paid for by the nation, which pays for all these things and out of our pockets too, and João Mau-Tempo would love to travel, to leave the latifundio and see other lands, but now that he has been dubbed a dangerous element, no thought is given to the inconvenience caused to the guards, who enjoy their rest, nor to the price of gasoline, nor to the depreciation in the value of the vehicle, and they immediately provide a jeep and a patrol complete with rifles and bayonets to go to Monte Lavre to find the malefactor and bring him safely to Vendas Novas, Hop in, we're off on a little trip, if that isn't mockery, I don't know what is.

The journey is brief and silent, the guards' fount of jokes, always the same, soon runs dry, and João Mau-Tempo, after much thought, says to himself that he might as well be hanged for a sheep as for a lamb, and that no one will get any compromising information out of him, if I were to talk, it would be better if all the mirrors in the world were shattered and all the eyes of those who come to see me were closed, so that I would never have to see my own face again. This road has many memories, it was here that Augusto Pintéu died crossing the stream with his mule cart, and over there, behind that hill, was where I first lay with Faustina, it was winter and the grass was wet, I wouldn't do it now, but that's youth for you. And he can taste in his mouth the bread and chorizo they ate afterward, their first meal as man and wife according to the laws of nature. João Mau-Tempo puts his hand to his eyes as if they were burning, all right, they're tears, and a guard says, Don't cry, man, and another adds, His sort only cry when they're caught, but that isn't true, I'm not crying, retorts João Mau-Tempo, and he's right, even though his eyes are full of tears, it's not his fault that the guards lack any understanding of their fellow man.

João Mau-Tempo is in the barracks at Vendas Novas now, the journey was all a dream, and this PIDE agent, there's no mistaking him, once you've seen one, you've seen them all, and João Mau-

Tempo has more than enough experience of them, this agent says, while the barracks commander is picking at his teeth, Yep, this is the gentleman who'll be coming on a little trip to Lisbon with me, what is it with these people, they all talk about going on trips, let's go on a little trip, they say, and sometimes these are trips from which you don't return, that's what you hear anyway, but meanwhile, the agent turns to a guard and gives an order, the commander is here to obey, he's a stooge, a toady. Take this man to the recreation room so he can rest until tomorrow, and João Mau-Tempo feels someone grab his arm roughly and take him out the back, into a garden, the guards love gardens, perhaps their many sins will be forgiven them because of their love of flowers, which means that not everything is lost in their hardened souls, a moment of beauty and grace redeems the worst of crimes in the eyes of the supreme judge, like this crime of taking João Mau-Tempo from Monte Lavre and throwing him into a temporary dungeon and into other, more permanent ones, not to mention what will happen later. For now, it's a provincial cell, and over there is a truckle bed with a mat and a bundle of foul blankets, and here's a jug of water, he's so thirsty, he raises it to his lips and finds that the water is warm, but he drinks only after the guard has left, and now I can cry, don't laugh at me, I'm forty-four, but forty-four is nothing, you're still a lad, in the prime of life, don't say that on the latifundio and to my face, when I feel so tired and when there's this pain in my side that never leaves me and these lines and wrinkles that the mirror can still, for the time being, show me, if this is the prime of life, then allow me to weep.

We will pass over the night during which João Mau-Tempo did not sleep but merely paced up and down, not wanting to rest his body on the bed. Day dawns, he is weary and anxious, what will become of me, and when nine o'clock struck, the door opens and the guard says, Come out where I can see you, that's how he speaks, he hasn't been taught any other way, and the PIDE agent says, It's time to catch the train and set off on our little trip. And they leave, accompanied to the door by the commander of the barracks, who is very scrupulous and polite in such matters, See you, then, he says, and although João Mau-Tempo may be innocent, he is not so innocent as to think that this farewell is intended for him, but on the way to the station, he says, Sir, I swear I'm innocent. If the train wasn't about to leave, we could sit here and debate what it means to be innocent, and whether João Mau-Tempo truly believes in that oath and how he can believe in what appears to be a perjury, and we would discover, if we had time and intelligence enough, the difference between being innocently blameless and blamelessly innocent, although such subtleties are lost on João Mau-Tempo's companion, who responds angrily, Stop your bellyaching, they'll straighten you out in Lisbon.

Let us pass over the journey too, since it does not appear in the history of railways in Portugal. Such is the body's sovereign power over us that João Mau-Tempo even dozed a little, lulled to sleep by the swaying carriage and the clatter of the wheels over the rails, clackety-clack, but each time, he started awake, terrified to discover that he wasn't dreaming. Then there was the boat to Terreiro do Paço, what if I threw myself into the water, these are black thoughts, I want to die and not heroically either, what is unusual about João Mau-Tempo is that he has never seen a film and therefore doesn't know how easy and much applauded is that leap from the side of the boat, the impeccable dive and the swim American-style that carries the fugitive to the mysterious chartered yacht that waits at a distance, along with the veiled countess who, in order to be there, has broken the sacred bonds of family and the rules of her aristocratic heritage. But João Mau-Tempo will only learn later that he is the son of the king and sole heir to the throne, three cheers for King João Mau-Tempo, king of Portugal, the boat moors at the pontoon, and the man who was asleep wakes up, and by the time he does so, there are two men standing over him, Is he the only one, they ask, and the man who came with João Mau-Tempo answers, Yes, he's the only one this time.

Let us also pass over without much comment the journey through the city, the trams, the many cars, the passersby, the statue of Dom José on his horse,* now which one is the horse's right leg, João Mau-Tempo recognizes the various places, how could one forget such a big square and the arches, bigger than those in Giraldo square in Évora, but then suddenly everything is new to him, these steep, narrow streets, and just when he is finding the journey long, it becomes all too short, this half-door opening obliquely, the fly has been caught in the spider's web, we need no better or more original image.

And now there are stairs to climb. João Mau-Tempo is still flanked by the two men, well, you can't be too careful, high security, he is, after all, a dangerous element. Above and below, it's like a termites' nest, a hive of buzzing drones and ringing telephones, but as they go up, first floor, second floor, across wide landings, the noise and bustle diminish, they meet fewer people, and on the third floor there is almost complete silence, only the muted sounds of car engines and the vague murmur of the city in the heat of the afternoon. These are the attic rooms and this corridor leads to a long, low chamber where the ceiling is almost at head height, and some other men are sitting on long benches, and I am going to sit down next to them, I, João Mau-Tempo, native and inhabitant of Monte Lavre, forty-four years old, the son of Domingos Mau-Tempo, shoemaker, and Sara da Conceição, madwoman, and I have been dubbed a dangerous element, as Corporal Tacabo at the local barracks was kind enough to inform me. The other men sitting there look at João Mau-Tempo, but no one says a word. This is the house of patience, and here we await our immediate destiny. The roof is

---

* In Portuguese, this is an untranslatable joke: "Qual é a pata direita do cavalo de Dom José?," "Which is the right leg of Dom José's horse?," because the horse's left leg is straight (*direita*) and the right leg (*direita* also means right) is bent.

right above our heads, it creaks in the heat, if you poured water on it, it would boil, and João Mau-Tempo hasn't eaten for more than twenty-four hours, and for him there is no heat, it's a winter's day, he shivers as if he were exposed to the December wind blowing across the latifundio, with no more protection than his own bare skin. That is exactly what it is like, for this is the bench of the naked, every man for himself, they will not help each other, you must clothe yourself in strength and determination, in the loneliness of the moors, in the high soaring flight of the red kite who finally descends to ground level to count his own and test their courage.

However, the victims must be fed, we don't want to lose them sooner than would be convenient. Half an hour passed, and another, and finally in came some kitchen servant or other, bringing each prisoner a bowl of prison soup and two deciliters of wine, a kind thought from the nation to these her stepchildren, I hope they're grateful. And as João Mau-Tempo was scraping the bowl with his spoon, he heard one policeman say to the other, they were standing by the door keeping watch over the flock and shuffling papers, That guy's being handed over to Inspector Paveia, and the other replied, Rather him than me, and João Mau-Tempo said to himself, That's me they're talking about, and, as he found out later, it would have been far better not to have known. The plates and glasses were taken away, and the waiting continued, what will become of us, it was nearly night when they got their marching orders, some were being sent here and some there, Caxias or Aljube, provisional billets, there would be further moves, all of them to worse places, as the name became a face, so the face became a target. And the voice of Dona Patrocínio, a functionary in this socially useful service, was definitely the voice of the nation, So-and-so is to go there, So-and-so somewhere else, she could not have a better name as patron of displacements, it's the same with Dona Clemência, who is now doubtless chatting with Father Agamedes, I hear that João Mau-Tempo has been arrested, Yes, Senhora, he's paid for all his sins at once, and to think I went out of my way to help him and others, He seemed such a decent fellow, They're always the worst, Senhora Dona Clemência, they're always the worst, He wasn't even a drinking man, If only he had been, then he wouldn't have been tempted into such evil actions, What evil actions, Ah, that I don't know, but if he was innocent, they wouldn't have arrested him, Perhaps we should give his wife some help, You're a saint, Senhora Dona Clemência, if it wasn't for your kind patronage, I don't know what would become of these wretches, but leave it for a while, and see if they learn to be less proud, because that's their worst defect, pride, You're quite right, Father Agamedes, and pride is a mortal sin, The worst of all sins, Senhora Dona Clemência, because it is pride that causes a man to rise up against his employer and his god.

On the way back, the truck passed through Boa-Hora to pick up some prisoners who were being tried there. All of this is carefully measured and calculated, according to the order of service, the police van must be used to capacity, it's like saying, you have to take the rough with the smooth, and given how poor the nation is, the prisoners would be the first to agree, indeed, they might even suggest it, Let's pass through Boa-Hora, and some will think, Hmm, Boa-Hora, Good-Hour, what an inappropriate name, and pick up those who are being judged by the worthy judges, and then we can all go together, it'll make for better company, it's just a shame we don't have a guitar with which to accompany our sorrows. João Mau-Tempo has never traveled so much in his life. Or, rather, as much as any other man in the latifundio, but not as much as his son António, now a soldier, but who traveled a lot in the past, driven by life's obligations and the needs of his stomach, with his knapsack on his back, with hoe and scythe, ax and adze, but the latifundio is the same everywhere, some parts have more cork oaks or holm oaks, some have more wheat or rice, some have guards or overseers or managers or foremen, it makes no difference, this, however, is quite different, a good tarmacked road, and if it were daytime, you'd be able to see more clearly. The nation really looks after its disobedient sons, as one can tell from these high, secure walls and the care the guards take over their work, they're a real plague, they're everywhere, or were they cursed at birth and this is their fate, to be wherever the suffering are, although not to minister to their misfortunes, that is why they have neither eyes nor hands, but say, Hop into the jeep, we're off on a little trip, or Move along, or Go on, we're off to the barracks, or You stole some acorns, so pay the fine and take a beating, they must have studied, otherwise they wouldn't be guards, because no one was born a guard.

Which, do you think, are the narrator's thoughts, and which are João Mau-Tempo's, both are right, and if there are any mistakes, they are shared mistakes. This bureaucracy of registers, index cards and papers is there from the day we're born, we take no notice of it, unless one day we're allowed to come here and find out in detail what actually went on, from the dotted line on which his name is written, João Mau-Tempo, forty-four years old, married, native and inhabitant of Monte Lavre, where's that, in the district of Montemor-o-Novo, well, you must be a good sort. They take João Mau-Tempo into a room along with other prisoners, sleep if you can, and if you're hungry, tough, because suppertime is long gone. The door closes, the world vanishes. Monte Lavre is a dream, and Faustina is deaf, poor thing, however, let us not say, out of some foolish superstition, that this is the hour of bats and owls, poor creatures, it's not their fault they're ugly, you perhaps are convinced that you're handsome, now who's a fool.

João Mau-Tempo will be here for twenty-four hours. He won't have much opportunity to talk, although the following day, a prisoner will come up to him and say, Listen, friend, we don't know why you're here, but for your own sake, take my advice.

THIRTY DAYS IN SOLITARY confinement is a month that doesn't fit in any normal calendar. However carefully you make your calculations, there are always too many days, it's an arithmetic invented by mad people, you start counting, one, two, three, twenty-seven, ninety-four, then find you've made a mistake, only six days have passed. No one interrogates him, they brought him from Caxias, this time during the day, so he at least knew where he was, although trying to see the world through those cracks was like trying to see it through the eye of a needle, and then he was ordered to undress, the nation does things like that, it happened to me once before, the doctors did it when I was called up, to decide whether or not I was good enough, well, I'm obviously good enough for these people, they're not going to send me away, they empty my pockets, they rummage and search and ransack, they even remove the insoles in my shoes, these clever folk know where we stash our secrets, but they find nothing, of the two handkerchiefs I brought with me, they take one, of the two packs of cigarettes, they take one, farewell, knife, these police aren't always so thorough, only now do they take my knife off me, what if

I'd tried to kill myself. They read me the rubric, While in solitary confinement, you will not be allowed any visitors nor can you write to your family, and so on and so forth, otherwise, you will be punished. But one day, much later, he was given permission to write a letter, and back came some clean clothes, washed and ironed by Faustina herself and sprinkled with a few tears, for they're a sentimental people whose fountains of tears have not as yet dried up.

On the twenty-fifth day, at three o'clock in the morning, João Mau-Tempo was, as usual, sleeping badly, and so he woke at once when the cell door opened and the guard said, Get dressed, Mau-Tempo, you're leaving. What, you're going to let me go, the imaginations of the wretched know no bounds, they always think the best or the worst, depending on their mood, that's the attraction of extremes, let's hope he's not disappointed. He's taken down to the ground floor, where there are people waiting, plus a fierce-looking hound, Here's that good-for-nothing you're taking for a walk, jokes the guard, they're clearly obsessed with this idea of walks and trips and rides, we know exactly what they mean, they're not fooling anyone, but they keep saying it, with a few minor variations, as if they didn't know what else to say. The hound goes on ahead, To show you the way to brigade headquarters, that's what the dog barks at João Mau-Tempo, and the guard from Aljube prison is such a card, just fancy, at this time of the morning and in these painful circumstances he can still manage to say, Have a good journey. Words were not presented to mankind as a gift, far from it, each word was hard won and occasionally abused, and there are some words that should only be sold at a high price, bearing in mind who is saying them and to what end, as in this case, Have a good journey, he says, when he knows full well that the journey will be far from good, animals are kinder to each other, for at least they don't speak. But here is this hound leading me through the deserted streets, at least it's a lovely night, although all I can see of it is this corridor of sky between the buildings, and to the left the cathedral and to the right another, smaller church, Santo António, and farther on the Madalena, neither small nor large, it's a street of churches, I am under the protection of the heavenly host, and perhaps that's why the hound speaks rather gently, Don't tell anyone I told you, but things aren't looking good, apparently a comrade of yours gave them your name, you'd best tell them everything you know, that's the only way to get back to your family, you won't gain anything by being stubborn. This street is called São Nicolau and the one over there São Francisco, and if I left some saint or other behind me along the way, you can have him, Sorry, I don't know what you're talking about, officer, I haven't done anything wrong, I've been working ever since I was born, I don't know anything about these things, I was arrested once but that was years ago, and I've had nothing to do with politics since, these are João Mau-Tempo's words, some true, some false, and he won't say anything else, that's the good thing about words, they're like a river flowing over rocks, it always does so in the same way, be careful not to stumble, the water flows so quickly it can dazzle you, watch out. The hound barks, João Mau-Tempo recognizes the place, this slope with the tram lines shining, Ah, so that's it, well, just you wait, and the soft dawn is bruised by the bad words hurled at him, you this and you that, words barely known in the latifundio. And now João Mau-Tempo feels his strength leaving him, he's been stuck in a cell for twenty-five days, scarcely moving, or only from cell to latrine and from latrine to cell, with his poor mind working overtime, tying up loose ends that immediately come undone again with more anxious thinking, not to mention the sleepless nights, and now there's this walk, which seems so long and yet it's nothing compared to the distances his legs used to cover on the latifundio, and suddenly he's afraid he won't make it, afraid he'll tell all he knows as well as what he could never possibly know, but then he hears again the prisoner in Caxias, Listen, friend, we don't know why you're here, but for your own sake, take my advice, and he remembered this just in time, he covered the final meters as if in a dream, he's through the door, going up the steps, up to the first floor again, there's no one to be seen, a terrifying silence reigns, second floor, third floor, we're here, João Mau-Tempo's fate has been waiting for him, legs crossed, that's the trouble with fates, they do nothing but wait, and we are the ones who have to do everything, for example, learn when to speak and when to keep silent.

The hound shoved João Mau-Tempo into a room and remained on guard outside. After a few minutes, the door burst open and in came a very spruce-looking gentleman, freshly shaved and smelling of cologne and brilliantine, he gestured to the other man to leave and immediately started shouting, Because of this bastard, this bloody communist, I can't go to mass today, that really is what he said, although I doubt anyone will believe me, but it's true, probably the influence of the ecclesiastical neighbors mentioned earlier while we were walking over from the Aljube prison, not to mention the Church of the Martyrs and the Square of the Two Churches, the Church of the Incarnation and that other one, now what the devil is it called, Father Agamedes would love it here, he'd be able to hear the confession of this Inspector Paveia, who is so upset about having missed mass you would think he'd have his own chaplain really, and now, to complete this edifying picture, imagine if João Mau-Tempo were to say, Oh, sir, please don't miss mass on my account, if you like, I'll go with you. We can't believe our ears, and not even João Mau-Tempo knows why he said it, but we don't have time now to examine these bold or spontaneous words, because Inspector Paveia doesn't give us time to think, Bastard, faggot, swine, I'm sorry, Father Agamedes, but that's exactly what he said, it's not my fault, and, Shut up or it's the trapeze for you, what circus arts these are João Mau-Tempo has no idea, but he sees Inspector Paveia go over to a desk, he's rather ill named really, when you think that *paveia* means a sheaf of wheat of the kind I used to clutch to my chest, and he takes a pistol out of the drawer, along with a stick and a heavy ruler, He's going to kill me, thought João Mau-Tempo, and the inspector said, See this, it's for you if you don't tell me the whole story, and be warned, you won't leave here until you've told me everything you know, stay standing, don't move, not so much as a finger, if you move, you're in for it.

Every three hours, one man leaves and another enters. The victim doesn't change his story, So what were you up to in your village, Working to earn enough money to feed my family, the first question and the first answer, the question is as predictable as the answer is true, and this man should be allowed to go free because he has told the truth, Do you mean working or do you mean distributing communist newspapers, you can't fool us, you know, But I wasn't involved in that kind of thing, sir, All right, so you weren't distributing newspapers, you were taking it up the bum, you and your friends were taking it up the bum from the man in charge so that he would teach you the Moscow doctrine, isn't that right, look, if you want to go back to Monte Lavre and see your children again, tell us the full story, don't cover up for the buddies you held meetings with, think of your family, think of your own freedom. And João Mau-Tempo is thinking about his family and his freedom, but he remembers the story about the dog and the partridge told by Sigismundo Canastro, and says nothing, Go on, tell us the story, what is it you lot say, you bastards: Those thieves in government won't give us what we want, so we're going to get rid of them, we're going to rebel against them and against Salazar's laws, isn't that what you say to each other, isn't that what you intend doing, tell me the truth, commie, don't cover up, if you tell us the whole story, you can leave for Monte Lavre tomorrow and see your children again, and João Mau-Tempo, thinking of the skeleton of the dog face to face with the partridge, says again, Sir, I've told you my story, I was arrested in nineteen forty-five, but since then I've never been involved in anything political, and if someone has told you otherwise, he's lying. They hurled him against the wall, beat him, called him every name under the sun, and this they did over and over, without letup, but the victim still did not change his story.

João Mau-Tempo will stand there like a statue for seventy-two hours. His legs will swell up, he'll feel dizzy, and every time his legs give way, he'll be beaten with the ruler and the stick, not that hard, but enough to hurt. He didn't cry, but he had tears in his eyes, his eyes swam with tears, even a stone would have taken pity on him. After a few hours, the swelling went down, but beneath his skin, his veins became as thick as fingers. His heart shifts position, it's a thudding, deafening hammer echoing inside his head, and then finally his strength deserts him, he can no longer remain on his feet, his body droops without his realizing it, and he's crouching now, he's a poor farm laborer from the latifundio, squeezing out a final turd, the turd of cowardice, Get up, you swine, but João Mau-Tempo couldn't get up, he wasn't pretending, this was another of his truths. On the last night, he heard screams and moans coming from the room next door, then Inspector Paveia came in, accompanied by a large number of policemen, and when the screams started again, growing ever shriller, Paveia walked over to him with calculated slowness and said in a voice intended to terrify, So, Mau-Tempo, now that you've been to Monte Lavre and back, you can tell us your story. From the depths of his misery, his hunched body almost pressed against the floorboards, his eyes clouded, João Mau-Tempo answered, I have no story to tell, I've said all I have to say. It's a modest sentence, it's the skeleton of the dog after two years, a sentence barely worth recording compared to what others have said, From the top of those pyramids, forty centuries look down on you, I'd rather be queen for a day than duchess for a lifetime, Love one another, but Inspector Paveia's blood is boiling, And what about the twenty-five newspapers you distributed in your village, if you deny it, I'll kill you right now. And João Mau-Tempo thought, Life or death, and said nothing. Maybe Inspector Paveia was once again late for mass, or perhaps leaving his prisoner seventy-two hours on his feet was enough for the first round, but what he said was, Take the bastard back to Aljube and let him rest there, then bring him back here again to tell his story, otherwise he goes straight to the cemetery.

Two dragons approach, grab João Mau-Tempo by the arms and drag him down the stairs, from the third floor to the ground floor, and while they're hauling him along, they say, Tell him your story, Mau-Tempo, it will be better for you and for your family, besides, if you don't, the inspector will pack you off to Tarrafal,* he knows everything, a friend of yours from Vendas Novas told him, all you have to do is confirm what he said. And João Mau-Tempo, who can barely stand, who feels his feet flopping from step to step as if they belonged to someone else, answers, If they want to kill me, let them, but I have nothing to tell. They bundled him into the police van, it was a short journey, there had been no earthquake, all the churches were still triumphantly standing, and when they reached Aljube and opened the door, Out you jump, he missed the step and fell, and again was dragged inside, his legs were slightly steadier now, but not much, and then they shoved him into a cell, which, either by chance or on purpose, was the one he had been in before. Almost fainting, he collapsed face-forward onto the rolled mattress, but although he felt as if he were in a dream, he had just enough strength to unroll and fall on top of it, and there he lay for forty-eight hours, as if dead. He is clothed and shod, a broken statue held together only by his internal wiring, a puppet from the latifundio who peers over the top of the curtain and makes faces while he dreams, his beard continues to grow, and from one corner of his mouth a trickle of saliva forges a slow path through the stubble and the sweat. During those two days, the guard will look in now and then to see if the cell's occupant is alive or dead, the

---

* A prison camp in Cape Verde, known as the Camp of Slow Death, where Salazar sent opponents of his regime.

second time he looks in, he feels relieved, because the sleeper has, at least, changed position, but the guard knows the routine, whenever these men come back from playing statues, they always sleep like this, they don't even need to eat, but now the prisoner has slept enough, he's sleeping less profoundly, Wake up, your lunch is here on the shelf, and João Mau-Tempo sat up on the mattress, uncertain as to whether he had dreamed those words or not, because although no one else is in the cell, he can smell food, he feels a great and urgent hunger, but when he makes a first attempt to stand, his legs buckle and his eyes grow dim from the sheer effort, he tries again, it's only two steps from there to the shelf, the worst thing is that he won't be able to sit down to eat, because in prison you eat standing up so as to get the food down more quickly, and João Mau-Tempo, who had been small for his age as a child and never grew much taller, has to stand on tiptoe, a torment for someone in his weakened state, and if he drops any food on the floor, he knows he'll be punished, he who gives the food gives the orders.

Five days passed, which would have had as much to tell as any of the others, but that is the trouble with stories, sometimes they have to leapfrog over time, because suddenly the narrator is in a hurry, not to finish, not yet, but to reach an important episode, a change of plan, for example, the beat that João Mau-Tempo's heart skips simply because the guard comes into his cell and says, Mau-Tempo, get ready to leave, I want those blankets returned to the stores, along with the bowl and the spoon, have this place shipshape by the time I come back. The problem with these men from the latifundio, especially when they're innocent, is that they take everything so literally, call a spade a spade, which is why João Mau-Tempo is so happy, hoping for the best, Perhaps they're going to set me free, the man's a fool, as becomes immediately evident when the policeman returns to accompany him to the quartermaster's store, where he deposits blankets, spoon and bowl, and where he receives the few personal items that have been kept there, and now,

We're taking you to the mixed cell, you're not incommunicado any-more, which means you can write to your family and ask them to send anything you need, and then he opened the door and inside was a whole world of people, of all nationalities, well, that's just a manner of speaking, meaning that there were a lot of people, some of whom were foreigners, but João Mau-Tempo is too shy and too constrained by his strong Alentejo dialect ever to be on friendly terms with them, however, as soon as the door closed, all the other Portuguese men surrounded him, wanting to know why he was there and if he had any news from outside. João Mau-Tempo has nothing to hide, he tells them everything that has happened to him, and so steadfast is he in his declaration that he hasn't been involved in any political activities since nineteen forty-five that he repeats it there and then, though there's no need, because no one has asked him.

João Mau-Tempo proved very popular, and once, coming across a fellow prisoner smoking, he asked him for a cigarette, which was rather cheeky given that he didn't know him from Adam, but other prisoners immediately offered him tobacco too, and best of all was when another man, who had overheard their conversation, came bearing an ounce of superior tobacco, a pack of cigarette pa-pers and a box of matches, Just say if you need anything, comrade, it's share and share alike here, you can imagine João Mau-Tem-po's feelings, with the first puff he grew six inches, with the sec-ond he returned to his normal height, but greatly fortified, a di-minutive figure among the other men, who smiled as they watched him smoking. And since even in the lives of prisoners there are happy events and coincidences, two days later, João Mau-Tempo was summoned to a room outside the mixed cell, where the guard, beaming as if he himself were the donor, for guards are contra-dictory creatures, said, Mau-Tempo, a gentleman from your village has brought you these clothes, four ounces of tobacco and twenty escudos. João Mau-Tempo was touched, more by the reference to

Monte Lavre than by the unexpected gift, and he asked, Who was the gentleman, and the guard answered, It doesn't matter, to us a donor is a donor, which was something João Mau-Tempo didn't know. He went back into the room clutching his treasure, and as soon as he did, he let out a shout that could have been heard all over the latifundio, Right, comrades, if anyone wants to smoke, here's tobacco for you, and another voice responded equally loudly, for these are things that need to be trumpeted abroad, That's how it is, comrades, share and share alike, we're all brothers here and we all have the same rights. Normally, one would choose quite different substances as proof of solidarity, but everyone takes what he needs or gives what he has, in this case cigarettes, little threads of tobacco rolled up inside the white cigarette paper, and now the tremulous tip of the tongue running along the edge and sealing it up, job done, humanity would be in a bad way indeed if it failed to understand such large gestures.

Some leave, others do not, new faces arrive, but they are rarely strangers, there is always someone who says, Well, fancy seeing you here, and after a few days, a policeman comes to the door of the room and says, Mau-Tempo, put your jacket on, we're going for a stroll, but you'll be right back, no need to take anything else. Perhaps he'll be back, perhaps he won't, but João Mau-Tempo is there to say that his heart dropped into his boots, and this is far truer than his statement that he hasn't been involved in political activities for four years. He repeats the journey with a hound at his side, this time a big, almost beardless lad who seems nervous, perhaps he's not yet used to this work, he keeps reaching around to feel his back pocket and doesn't utter a word, but at least João Mau-Tempo can look at the people passing by, they must know I'm a prisoner, he can see the trams, peer in at shop windows, this time it really is almost a stroll, so much so that he nearly forgets to feel afraid, then the fear comes flooding back, jumbling his thoughts, troubling his blood, and he misses the mixed cell, the shared ciga-

rettes and the conversations. He feels a sudden sympathy for statues, because for all we know the bronze and marble variety might
find it hard to stay standing, how is it that they don't get a cramp,
those men with their arms outstretched, those animals frozen in
the same posture, never giving in, never running away, though they
lack man's strength of will, despite which a man so often weak-
ens and crouches down, and no amount of kicking will make him
stand, he could die for all he cares, as long as his tongue doesn't
speak, except to repeat the same lie over and over. But the idea that
the torment is about to start again, that he is going to experience
the same pain or worse, this is what fills João Mau-Tempo's mind,
and suddenly a great darkness falls on the city, despite its being
broad daylight and hot, as it always is in August, but all the ungrateful creature can think is, what will become of me, what tor-
ture awaits me.

The door opened again, João Mau-Tempo went up the steps,
followed by the hound, and entered the room, and look who's here,
it's the man from Vendas Novas who traveled with João Mau-
Tempo as far as Terreiro do Paço, his name is Leandro Leandres,
and now he says in a scornful voice, Do you know why you've been
brought here, and João Mau-Tempo, always polite and respectful, says, No, sir, I don't, and Leandro Leandres says, You've come
to tell the rest of your story, but there's no point now describing
what happens next, it's the same old thing, the same conversation,
how many newspapers were distributed, and how local commit-
tees were formed and why did they stop meeting, and how many
members were there, and who, look, someone here gave us your
name, so it must be true, and if you don't confess, you won't get out
of here alive, it would be best for you if you talked, you know, but
João Mau-Tempo isn't at all sure about that, and even if he was, It's
been four years since I so much as touched any political papers,
and they were only ones I picked up in the streets or along the
way, apart from that I can't remember anyone actually giving them

to me, that was years ago, all I think about is my work, I swear. The conversation was always the same, the same questions and answers, the same grilling and the same lying, but this time there were no beatings and the statue that is João Mau-Tempo remained in his natural position, sitting on a chair, he looked as if he were posing for a portrait, except that his soul was jumping about inside his heart like a poor, frightened lunatic, and his pale but constant will kept saying, You mustn't talk, lie all you want but don't talk. There was another difference too, the fact that a hound of a lower category was typing out all the questions and answers, but after a few pages there was nothing more to write, because the conversation was like dredging up water with a wheel equipped with bottomless buckets, going round and round, the mule was treading in its own dung now and the sun sinking, and that was where the statements ended, and the man at the typewriter asked, Where's this guy's original statement, and Leandro Leandres, not realizing what he was saying, answered, It's over there, along with Albuquerque's statement, João Mau-Tempo had tormented himself over and over as to who had given them his name, and now he knew, it was Albuquerque, and knowing this was so painful and so sad, what must they have done to him to make him talk, or did he do so willingly, what could he have been thinking, well, it happens, and João Mau-Tempo cannot know that some years later, he will see Albuquerque, the squealer, in Monte Lavre, and remember that he was the one who once said, If they turn up here, I'll shoot them, I mean it, and yet in the end he had squealed on him, and when Albuquerque got out of prison, he became a Protestant priest, not that we have anything against religion, but why go around proclaiming the salvation of all men when he couldn't even save his few comrades, what will he have to say for himself at the hour of his death, but now all that João Mau-Tempo feels is a great sorrow and a great sense of relief that at least he has not talked, perhaps they won't beat me again or make me play the statue, I'm not sure I could take it.

João Mau-Tempo returned to Aljube, then, after a few days, he was taken from there to Caxias, and news of this finally reached Monte Lavre. Letters will come and go, everything has to be meticulously arranged between Faustina and João Mau-Tempo, because these things are no joke, everything has to be worked out to the last detail if a person is traveling a long way to be at a certain place at a certain time, even when the meeting is not a clandestine one, even when it's the police themselves who open the door and say, Come in. No, you have to take account of every eventuality, from Monte Lavre to Vendas Novas by cart, then from Vendas Novas to Barreiro by train, possibly in the same carriage that brought João Mau-Tempo and Leandro Leandres, and then by boat, it's only the second time Faustina Mau-Tempo has seen the sea, this vast estuary, and then again by train to Caxias, where the sea is suddenly much bigger, This is real sea, she says, and the woman who met her at Terreiro do Paço and who lives in the city smiles sympathetically and kindly at her friend's limited experience and says, Yes, that's the sea all right, but says nothing of her own ignorance about what the sea really is, not this modest opening of arms between two towers, but an infinite, liquid longing, a continuous sifting of glass and foam, a mineral hardness that softens and chills, the home of the great fish and of sad shipwrecks and of poems.

It is very true that while one may know some things, one cannot know everything, and Faustina Mau-Tempo's friend knows where to get off the train in Caxias but not where the prison is, however, she doesn't want to admit that she doesn't know and sets off in one direction, saying, It must be down here. It's August, and it's baking hot at this hour, which is fast approaching the hour so laboriously communicated and memorized, the hour of the visit, in the end, they had to ask a passerby, realized they had gone entirely the wrong way and turned back, already weary with all the toing and froing, and Faustina Mau-Tempo took off her tight shoes, to which her feet were unaccustomed, and was left in her stocking feet, this, however, was a big mistake, and only someone with no heart could laugh, this is the kind of humiliation that burns itself into the memory for the rest of one's life, the tarmac had half melted in the heat, and her stockings stuck fast as soon as she planted her feet on it, and the more she pulled, the more the stockings stretched, it was like a circus act, the funniest of the season, enough, enough, the clown's mother has just died, and everyone is crying, the clown isn't funny, he's frightened, and that is how we feel about Faustina Mau-Tempo, and we form a screen so that her friend can help her off with her stockings, modestly, for women who have only ever known one man are incurably shy, and now she's barefoot, and we can go home, and if any of us do smile, it's out of tenderness. But when Faustina Mau-Tempo arrives at the fort, her feet are in a terrible state, made worse by wearing shoes without stockings, they are black with tarmac and bleeding from where the skin has been rubbed raw, what a hard life the poor have.

The visitors have left, the hour has passed, and no one came to see João Mau-Tempo, his companions tease him in that stupid, joshing, manly way, She's forgotten all about you, You certainly weren't expecting to be stood up, while at the entrance, poor Faustina Mau-Tempo is demanding to be allowed in, Is this where my husband is, she asks, his name's João Mau-Tempo, and the man at the door responds jokingly, No, there's no one here by that name, and the other says mockingly, Do you mean your husband's in prison, this is their way of passing the time, for they lead very dull lives, they don't even get to beat up the prisoners, other men do that, but Faustina Mau-Tempo can't tell the difference, Yes, he is, and you're the ones who brought him, so he must be here, but her anger was like the fury of a sparrow, the rage of a chicken, the wrath of a lamb, finally, though, the man leafed through a book and said, Yes, you're right, he is here, in room six, but you can't visit him now, visiting time is over. Faustina Mau-Tempo is perfectly within her rights to burst into tears. She is a pillar crumbling, we can see the cracks appear and bits break off, and this pillar of the latifundio has painful feet too, she can cry about that now as well, and for everything else she has suffered and will suffer in her life, now is the moment to cry out your tears, pull out all the stops, Faustina Mau-Tempo, dissolve into tears, perhaps you'll manage to touch the hearts of these two iron dragons, or, if they have no heart, they might at least prefer not to be embarrassed, and since you're just a poor woman, they're not going to throw you out bodily, so weep, demand to see your husband, All right, all right, woman, I'll go and see if they'll let you in on a special dispensation, but this is an expression Faustina Mau-Tempo doesn't understand, what is a dispensation, let alone a special one, and how will they let her in on one, will it help her to see her husband. Those who travel by crooked paths also arrive, and though the visit will last no more than five minutes, that's long enough for two people who haven't seen each other for far too long, João Mau-Tempo is there, full of hope, his comrades tell him, It must be your wife, and it is, Faustina, João, and they embrace, and both shed copious tears, and he wants to know about the children and she wants to know how he is, and three minutes have gone, and are you well, and how have you been, have you had work, and Gracinda, and Amélia, and António, they're all fine, but you're a lot thinner, be sure not to get ill, five minutes, goodbye, goodbye, send them my love, ah, so much love, come back soon, I will, I know where it is now and I won't get lost, and I won't either, goodbye.

There will be other visits, different, less rushed, his daughters will come, his brother Anselmo will come, and António Mau-Tempo will come, only to leave feeling angry, no one made him angry, but that is how he feels when he leaves, he will stand some way off, a picture of rage, staring at the prison, he is not the António

Mau-Tempo we know, Manuel Espada will come, he will enter looking grave-faced and leave lit up by a serene light, and cousins and uncles will come too, some of whom live in Lisbon, but they will only be allowed in the corridors, peering in through a screen so fine that it's hard to see the people on the other side, and with a policeman patrolling up and down, listening for any complaints from the prisoners. And the months will pass, the long days and the even longer prison nights, the summer will end, the autumn will go and winter will arrive, João Mau-Tempo is still there, he is no longer summoned for interrogation, they have forgotten he exists, perhaps he'll stay there for good. Sigismundo Canastro had also been arrested, as João Mau-Tempo will find out later, when he is back in Monte Lavre and hears that Sigismundo has been freed, and then there he is, they embrace with free hearts, I didn't talk, Neither did I, It was Albuquerque. Sigismundo has had a much harder time, and yet he laughs, while João Mau-Tempo cannot help but feel a certain melancholy about the injustice done to them. There is much talk in room six, they discuss politics and other matters, some men study, others teach, there are lessons in reading, arithmetic, some do drawings, it's a people's university, about which we will say no more, because eternity would not be long enough.

Today he is to be freed. Six months have passed, it's January. Only last week, João Mau-Tempo was working on the access road along with other residents of room six, working in rain so cold it was like melted snow, and now he is sitting wondering what life holds in store for him, many men have been tried, but he has not, and some say this is a good sign, then the door opens and a guard says in his usual arrogant tone, João Mau-Tempo, and João Mau-Tempo stands at attention as the prison rules dictate, and the guard says, You're leaving prison, get your things ready and be quick about it. Those who are staying are so delighted, it's extraordinary, it's as if they themselves were being freed, and one says, The sooner they empty the prisons, the better, we're not achieving anything here, it's as logical as saying, The sooner they give me the tools, the sooner I can get down to work, and then everyone joins in, it's like a mother dressing her child, someone is putting on his shoes for him and another is pulling on his shirt or shaking out his jacket, anyone would think João Mau-Tempo was going to meet the Pope, it's amazing, they're like children, any moment now they'll all burst out crying, well, if they don't, João Mau-Tempo soon will when they ask, Now, Mau-Tempo, have you enough money to get home, and he replies, I have a little, comrades, but I'll be all right, and they start collecting money, five escudos here, ten there, and they manage to scrape together enough for the journey with a little left over, and then, when he sees that a little money can also be great love, João Mau-Tempo will no longer be able to hold back his tears, and he will say, Thank you, comrades, and goodbye, I wish you all the best, and thank you again for everything you've done for me. This party atmosphere is repeated each time someone is released, ah, the joys of prison life.

It was dark when the van dropped João Mau-Tempo at the door of Aljube prison, it seems that this devilish Black Maria knows no other roads, and when João Mau-Tempo steps out, a free man this time, the policeman says to him, Go on, get lost, he seems almost sorry to see him leave, but that's what they're like, they grow fond of the prisoners and find it hard to lose them. João Mau-Tempo runs down the road as if the devil were after him, so much so that he glances over his shoulder to see if anyone is following him, perhaps the police indulge in such amusements, pretending to set a prisoner free and then mounting a hunt for him, and however hard the poor man runs, there'll be a net waiting for him down some passageway, and he'll be caught again, shoved into the police car, with all the policemen laughing and clutching their bellies, God, it was funny, oh, I haven't laughed so much in ages, not even at the circus. They're perfectly capable of such tricks.

The street is completely deserted, it's black night, it's not raining, which is fortunate, but the wind whips in between these tall buildings like the blunt razor of a barber in a hurry, it keeps cutting through João Mau-Tempo's thin clothes, the wind is as naked as he is, or so it seems. He has stopped running, his legs feel awkward and he's out of breath, he's forgotten how to walk, he leans against a wall with his bag and a suitcase tied up with string, and although both are quite light, his arms can barely carry them, which is why he puts them down on the ground, who would think it, this same man who once carried such enormous weights couldn't swing a cat by its tail, and if it weren't so cold, he would lie down, he has too much suffering on his shoulders to remain standing, and yet he does. People pass him, there's always someone out and about, but they don't look at him, each thinking about his own life, I've got quite enough problems of my own, thank you, they have no idea that the man standing at the corner has just been let out of Caxias prison, where he has spent the past six months, and where he was beaten and made to play the statue for seventy-two hours, they wouldn't believe that such things could happen in our lovely country, such stories are doubtless greatly exaggerated. What will João Mau-Tempo do in a city he doesn't know, there is no door he can knock on, Comrades, give me shelter for the night, I've just got out of prison, that would be an unusual conversation, and he has no idea whose houses these are, he was arrested in Monte Lavre by the guard José Calmedo, and that is where he must return, not now, because it's dark, but tomorrow, with the money given him by men who needed it themselves, he knows he has comrades there, but he can hardly go to Caxias now, knock on the door of room six, assuming he would be allowed back into the prison, and, when they opened, say, Comrades, give me shelter for the night, he is clearly mad, or perhaps he fell asleep despite the cold, yes, he must have fallen asleep, for he's no longer standing up as he thought, but sitting on his suitcase, and now it occurs to him, well, it had occurred to him before but now it occurs to him again, to go and knock on the door of the house where his sister works as a maid and say, Maria da Conceição, do you think your employers would let me sleep here tonight, but he won't do this, although in other circumstances he might have, they would have told Maria da Conceição to put a mattress down in the kitchen, you can't leave a fellow Christian to sleep out in the street like a stray dog, but given that he has just come out of prison, out of that prison and for those particular reasons, even if they agreed, they might view his sister differently afterward, poor thing, she never married and has always worked for the same boss, it's as if that's what she was born for, who knows what they will have told her already, it's not hard to imagine, The ungrateful wretches, they would starve to death if it wasn't for us, your brother will pay dearly for those ideas of his, they're against us, you see, they're all against us, but we're your friends and we won't punish you for his crazy ideas, but from now on, it would be best if he didn't come to the house, so you be careful now, you've been warned.

This is the domestic litany repeated by the mistress of the house, the master of the house is more categorical and more to the point, He's never to set foot in this house again, and I'm going to tell our estates in Monte Lavre that he is to be given no more work, let him go to Moscow. It seems that João Mau-Tempo went back to sleep, he must be very tired indeed to be able to sleep in this bitter cold, he stamps his feet on the ground, and the noise echoes and re-echoes in the icy air, a policeman might come along and arrest him for disturbing the peace, then João Mau-Tempo picks up his bag and his suitcase and heads back down the road, he can barely walk, he's limping, he seems to recall that the station is off to the left, but he's afraid he might get lost, that's why he asks a passerby, who tells him, Yes, you're going in the right direction, and adds a few more details, João Mau-Tempo, holding his suitcase and his bag in numb hands, is about to carry on, but the passerby asks, Do you want some help, here we all tremble, what if this man is a thief and has decided to rob this poor farm laborer, what could be easier, even in the dark it's clear he can barely walk, No, sir, thank you, says João Mau-Tempo politely, and the man does not insist, he isn't a ruffian after all, he says only, You look like you've been in prison, and we, who know João Mau-Tempo and how sensitive he is to kind words, can hear him telling his whole story, how he was in Caxias for six months and has just arrived in Lisbon, how they dumped him there, and how he has to get back to his village, to Monte Lavre, in the parish of Montemor, yes, I'm from the Alentejo, he doesn't know if there's a boat at this hour or a train, I'm going to the station to see, no, he has nowhere to sleep, although a sister of his works as a housemaid, But I don't want to bother her, her employers might not like it, and the other man asks, he's a very inquisitive fellow, And what if there isn't a boat and a train, and João Mau-Tempo says simply, Then I'll spend the night in the station, there's sure to be a bench there, it's a shame it's so cold, but I'm used to that, thanks very much for your help, and having said this, he moves off, but the other man says, I'll come with you, let me carry that bag for you, and João Mau-Tempo hesitates, but having spent six months with humane and generous men, who looked after him, taught him, gave him tobacco and money for the journey, it seems churlish to distrust this man, so he hands him the bag, the city can be full of surprises, and off they go, down the remaining streets, as far as the big square, under the arcade and into the station, João Mau-Tempo has difficulty reading the timetable, all those tiny figures, and the man helps him, running his finger down the columns, No, there's no train, the earliest one is tomorrow morning, and when he hears this, João Mau-Tempo immediately starts looking for a place where he can curl up, but the man says, You're tired, and you're obviously hungry too, come and sleep at my house, have a bowl of soup and rest, you'll die of cold if you stay here, that's what he said, no one believes such things can happen, and yet it's true, João Mau-Tempo can only answer, Thank you very much, it's a real act of charity, Father Agamedes would cry hosanna were he here, he would praise the kindness of man to man, and he's quite right, this man carrying the bag on his back deserves to be praised, though he's not a churchgoer, not that he's said as much, but the narrator knows these things, as well as others that have nothing to do with the story, because this is a story about the latifundio, not about the city. The man is older than João Mau-Tempo, but stronger and quicker on his feet, indeed, he has to slow down to accommodate the painful pace of this man raised from the dead, and to cheer him up, he says, I live near here, in Alfama, and he turns onto Rua da Alfândega, and João Mau-Tempo is already feeling better, and they set off down damp, steep-sided alleyways, well, in this weather it's hardly surprising they're damp, a door, the narrowest of stairways, an attic room, Hi, Ermelinda, this gentleman is sleeping here tonight, he's going home tomorrow and has nowhere to stay, Ermelinda is a plump woman who opens the door to them as if she were opening her arms, Come in, and João Mau-Tempo, and sensitive readers, forgive me, and those who only appreciate large, dramatic events, but the first thing he notices is the smell of food, a bean and vegetable soup bubbling on the stove, and the man says, Make yourself at home, and then, What's your name, and João Mau-Tempo, who is already sitting down, overcome by a sudden weariness, tells him his name, Well, I'm Ricardo Reis,* and this is my wife Ermelinda, these are ordinary enough names, and that's pretty much all we know about them, that and these bowls of soup on the kitchen table, Eat up, the cold is easing now, Lisbon has turned out to be a kindly place, this window at the back looks out over the river, there are a few small lights on boats, but fewer on the farther shore, who would have thought that one day, seen from here, they would be a feast for the eyes.

---

* One of the heteronyms of the Portuguese writer Fernando Pessoa and the main character in Saramago's later novel *The Year of the Death of Ricardo Reis*.

Have another glass of wine, and perhaps that's why, after a second glass of strong wine, João Mau-Tempo is smiling so broadly, even when he tells them what happened to him in prison, by the time he's finished it's getting late, and he can barely keep his eyes open, Ricardo Reis is looking serious, and Ermelinda is drying her tears, and then they say, It's time you went to bed, you need to rest, and João Mau-Tempo doesn't notice that it's a double bed they've given him, he hears footsteps in the corridor, but they're not the guard's footsteps, not the guard, not the guard, and, free at last, he falls asleep.

THERE HAVE BEEN six months of changes, which sometimes seem too few and sometimes too many. They are barely noticeable in the landscape, apart from the usual seasonal variations, but it's frightening to see how people have aged, both the men just let out of prison and those who never left Monte Lavre, and how the children have grown, only João Mau-Tempo and Sigismundo Canastro seem unchanged in each other's eyes. Sigismundo Canastro arrived yesterday and has already said that they must meet and talk, he's as stubborn and determined as ever, you see, but that's the way he is. Some people, however, are a pleasure to look at, as is the case with Gracinda Mau-Tempo, who has grown into a beautiful young woman, marriage clearly suits her, or so say the gossips, both the kindly and the cruel, but that's as far as the latter will go, and there are other changes too, for example, Father Agamedes has gone from being tall and thin to being short and fat, and the amount of money owed at the shop has grown enormously, as is to be expected when the man of the house has been away. That's why, when the time came, João Mau-Tempo set off with his daughter Amélia for the ricefields near Elvas, and to give you an idea of the geographical sensibilities of these rustic inhabitants, it's said in

Monte Lavre that beyond Elvas lies Spanish Extremadura, heaven knows how they stumbled upon this knowledge of a larger universe, which pays no heed to frontiers or borders, and if we want to know what lies behind João Mau-Tempo's excursion to Évora, it's largely to do with the latifundio's suspicions about the ways and wiles of João Mau-Tempo, political prisoner. It's true that he was never tried, but that's the fault of the police, who are not as efficient as they ought to be. After a few months, things will get back to normal, but meanwhile, it's best if he keeps his distance, then he won't contaminate our beloved land, and as for Sigismundo Canastro, they tell him there's no work and that he'll have to find it elsewhere.

So off João Mau-Tempo went to Elvas, taking with him his daughter Amélia, the one with the bad teeth, although if she had good teeth, she would easily be a match for her sister. Let it be said now that hell is not far away. There are one hundred and fifty men and women, divided into five groups, and this torment will last sixteen weeks, it's a veritable harvest of scabies and fevers, a labor not of love but of pain, weeding and planting from before the sun rises until after it has set, and when night falls, one hundred and fifty ghosts trudge up to the place where they have their lodgings, the men to one side, the women to the other, all of them scratching at the scabies caught in the flooded fields, all suffering from the fevers picked up from the rice paddies. You need sugar, milk, rice and a few eggs to make that delicacy rice pudding, how often do I have to tell you, Maria, it should be fluffy, not this stodgy pap, you should be able to taste every grain. All through the night in the dormitories, you can hear these poor people sighing and moaning, the anxious scratching of hard, black nails on skin that is already bleeding, while others lie, teeth chattering and staring up at the roof with eyes glassy with fever. There is little difference between this and the death camps, except that fewer people actually die, doubtless due to the Christian charity and concomitant self-interest with which the bosses, almost every day, load up the trucks with all that mangy, feverish misery and transport it to the hospital in Elvas, some today, others tomorrow, an endless coming and going, the poor things set off close to death, but are saved by the miraculous medicine, which, in a matter of days, has them as good as new, with very weak, tremulous legs, it's true, but who cares about such trivia, you can go back to work, the doctors say, addressing us contemptuously as *tu*, and the truck disgorges its load of broken-backed laborers, there's work to be done, there's no time to waste, Are you better, father, asked Amélia, and he answered, Yes, daughter, what could be simpler.

There haven't been that many changes. Weeding and planting the ricefields is done exactly as it was in my grandfather's day, the creepy-crawlies in the water haven't changed their stings or their slime since the Lord God made them, and if a hidden sliver of glass cuts a finger, the blood that flows is still the same color. You would need a lot of imagination to invent any extraordinary incidents. This way of life is made up of repeated words and repeated gestures, the arc made by the sickle is precisely adjusted to the length of the arm, and the sawing of the blade through the dry stems of wheat produces the same sound, always the same sound, how is it that the ears of these men and women do not grow weary, it's the same with that hoarse-voiced bird that some say lives in the cork oak, between the bark and the trunk, and that screams whenever they tear off its skin or perhaps pluck out its feathers, and what is left is painful, goosepimpled flesh, but this idea that trees cry out and feel pain comes purely from the narrator's private imaginings. We would do far better to notice Manuel Espada perched, barefoot, at the top of this cork oak, for he is a serious, barefoot bird, hopping from branch to branch, not that he sings, he doesn't feel like singing, the real boss here is the cork ax, chop, chop, chop, making circular incisions around the larger boughs and vertical ones on the trunk, then the handle of the ax serves as a lever, go on, push, there, can't you hear the hoarse-voiced bird that lives in the cork oak, it's screaming, not that anyone feels any pity. The cylinders of bark rain down, falling on the cork already cut from the trunks, there is no poetry in this, we'd like to see someone make a sonnet out of one man losing his grip on his ax and watching it skitter down the branch, catching the bark as it falls, and impaling itself in a bare foot, coarse and grubby but so fragile, because when it comes to skin and the blade of an ax, there is little difference between the delicate, rosy foot of some cultivated maiden and the calloused hoof of a cork cutter, it takes the same time for the blood to spurt out.

But here we are talking about work and the working day, and we nearly forgot to describe the night when João Mau-Tempo arrived back in Monte Lavre, when his house was filled almost to bursting with his closest friends and their wives, those who had them, as well as a cacophony of young lads, some of them intruders, unrelated to any of those present, not that anyone cared, and António Mau-Tempo, who had finished his national service and was now working on the cork plantations, and his sisters Gracinda and Amélia, and his brother-in-law Manuel Espada, a whole crowd of people. Faustina cried all the time, out of joy and grief, she had only to recall the day on which her husband was arrested, who knows why, and taken to Vendas Novas and to Lisbon, with no idea when, if ever, he would come back. She didn't talk about the sad case of her stockings ruined by the tarmac, not a word, that would remain forever a secret between the couple, both of whom felt slightly ashamed, knowing that, even in Monte Lavre, someone might make fun of them, the poor woman with her stockings stuck fast to the tarmac road, dreadful, who wouldn't do their best to avoid such mockery. João Mau-Tempo described his misfortunes and spared no detail, so that they would know just what he had suffered at the hands of the dragons of the PIDE and the national guard. All of this would be confirmed and repeated later by Sigis-

mundo Canastro, but although he wasn't so insensitive as to treat
the matter lightly, he did tell the most alarming stories as if they
were perfectly natural, and recounted everything with such an air
of simplicity that not even the women wept for pity, and the young
boys moved away, disappointed, he might as well have been talking
about the state of the wheatfields, and perhaps, who knows, they
were, indeed, one and the same thing. Maybe that is why, one day,
Manuel Espada approached Sigismundo Canastro, with all the respect that their difference in age required, and said, Senhor Sigismundo, if I'm needed, I can help. We would be much deceived if
we were to think that this impulse came from Sigismundo Canastro's quiet way of describing his experiences, which, in tempera-
ments like Manuel Espada's, might well have provoked such an important decision, the proof of this is that Manuel Espada went on
to say, No one should treat a man the way they treated my father-
in-law, and Sigismundo Canastro answered, No one should treat
men the way we were treated, but let's talk later, these arrests and
imprisonments have muddied the waters, best to let a little time
pass to allow the mud to settle, because these things, like fishing
nets, take longer to mend than to break, and Manuel Espada responded, I'll wait as long as is necessary.

Sometimes, when you sit down to read the history of this Portuguese land, you come across such silly things they make you smile,
although in this case, outright laughter seems to be called for, and
I mean no offense, each person does what he can or as the hierarchy orders him to do, and if it was a fine and praiseworthy thing
for Dona Filipa de Vilhena* to arm her sons so that they could go
and fight for the restoration of the fatherland, what can one make
of Manuel Espada, who, with no cavalry to back him up, says sim-

---

* Filipa de Vilhena was a Portuguese noblewoman who became a symbol of patriotism
in Portugal when, in 1640, she urged her sons to fight for the restoration of the country's
independence from Spain. Almeida Garrett wrote a play about her, which further contributed to her heroic image.

ply, Here I am, he has no mother to urge him on, she, of course, is dead, only his own will. Dona Filipa did not lack for people to sing her praises and describe her heroism, there was João Pinto Ribeiro, the Count of Ericeira, Vicente Gusmão Soares, Almeida Garrett, and Vieira Portuense painted her portrait, but Manuel Espada and Sigismundo Canastro have no one to take their part, it's simply a conversation between two men, they have said what they have to say and now each goes his own way, there is no call for oratory or paintbrushes, this narrator is all they need.

Indeed, as an aid to our understanding of these events, let us take another slow walk about the latifundio, with no particular goal in mind apart from picking up a stone or a branch and giving it a proper name, seeing what animals live there and why, and since we can hear guns firing over there, although what that's about we have no idea, let us begin right here, well, what a coincidence, this is the same road that José Calmedo took when he went to arrest João Mau-Tempo, indeed, given how easy it is to find oneself back where one was before, the latifundio seems more like a minifundio, not a large estate but a very small one. True, the last time we came this way it wasn't quite as noisy, but there's the ruined water wheel and, beyond, invisible, the brick kiln, don't worry about the shooting, it's probably just target practice or something, with proper bullets, mind, none of those lead pellets fit only for a little light hunting, quite a different kettle of fish.

The firing has stopped and we can walk on quite happily now, but look, there's a man coming from the same direction as the shooting, by his looks we would say he's one of us, and he crosses the valley, that smooth expanse of dark earth, goes over a small bridge with a low handrail, it's only a tiny stream, and starts to climb up this side of the valley, through thick, thorny undergrowth marked by the faintest of trails, Why is he going over there, with no hoe and no mattock, with no ax and no pruning hook, let's sit here and rest awhile, he'll have to go back down again and then we'll know, you were saying that this is a wilderness, Well, it is, and don't go thinking that the track through the brambles will be much use to the lackey who just passed, You mean he's a lackey, He certainly is, But he's not wearing livery, No, livery's a thing of the past, from the days when the countess armed her own sons, if you know who I mean, no, nowadays lackeys dress like you and me, well, not like you, you're from the city, even we can tell them apart by the way they behave, But why do you say that the track through the brambles won't be much use to him, Because what he's looking for lies off the beaten track, and he can't turn around, he has to go straight ahead, those are his orders, using his crook to beat a path through the undergrowth, that's worse than useless, But why is he doing it, Because he's a lackey, and the more scratches he has on him when he goes back, the better, So that old rule applies here as well, does it, It does, but to go back to our conversation, I was saying what a wilderness it is here, but it wasn't always like this, believe you me, there was a time when this whole area was cultivated right down to the bottom of the valley, it's good soil, and there are springs aplenty, not to mention the stream, So how did it become like this, Let's see now, the father of the present owners, the ones who were doing the shooting, eventually took over this whole area, it was the usual thing, a few small farmers got into financial difficulties, and he, I can't remember what his name was now, Gilberto or Adalberto or Norberto, something like that, lent them money, which they couldn't pay back, well, times were hard, and he ended up owning the lot, That doesn't seem possible, It's perfectly possible, it's what's happened all the time on the latifundio, the latifundio is like one of those mules that's always biting the mule next to it, You amaze me, Oh, if I told you everything I know, we'd be here for the rest of our days, and the story would have to be passed on to our grandchildren, if you have any, but here's the lackey, let's follow him.

The sound was that of slithering feet and of some heavy object being dragged along. Once, he fell and went rolling back down to the bottom, he could have been killed, What's that he's carrying on his back, It's a barrel, the owners of both barrel and lackey use the barrel as their target, But I thought slavery had been abolished, That's what you think, How can a man submit to such a thing, Ask him, Oh, I will, excuse me, friend, what is that you're carrying on your back, It's a barrel, But it's full of holes, you couldn't use it to store water or other liquids, or are you going to fill it up with stones, It serves as a target for my masters Alberto and Angilberto, they shoot at the barrel and I go and find it to check the number of hits, and then I put it back in the same place until it's so full of holes, it's like a sieve, at which point I fetch a new one, And you submit to their orders. The world suddenly becomes unfit for conversation, with Alberto and Angilberto on the far side of the valley shouting, impatient at the lackey's delay, it's getting late, they're saying, and we've still got two boxes of bullets left, and the poor slave trots across the valley and over the bridge, the barrel is like a huge rust-red hump, and now, as he climbs the hill on the far side, he looks more like a beetle than a man, So, do you still believe slavery was abolished, It seems impossible, And what do you know about impossibilities, Oh, I'm beginning to learn, Let me tell you about another impossibility, over there, on the right bank of the stream, beyond the viaduct, are some fields that extend as far as the foot of the hills, do you see, fine, well, those same marksmen sold that land to some small farmers, and if they had been decent, honorable men, as they should have been, they would have included the stream in the sale, but no, they kept back the ten or twenty meters of land that bordered the stream, so if the farmers wanted water, they had to dig wells, what do you say to that, Again, it seems impossible, Yes, it does, doesn't it, it would be like me refusing you a drink of water when you were thirsty, and telling you, if you want water, then dig a well with your bare hands, while I empty my glass and amuse myself watching the water flow, So a dog could go and drink from the stream, but not the farmers, Ah, now you're beginning to understand, look, there's another lackey bringing a new barrel, Your masters are obviously good shots, Yes, sir, but they wanted to know who you are, and when I said I didn't know, they said that if you don't leave right now, they'll call the guards. The two walkers withdrew, the threat had its effect and the argument a certain authority, trespassing on private property, even when it isn't fenced off, would be taken very seriously indeed, especially if the guards happened to be in a bad mood, there would be no point explaining to them that they didn't know the boundaries, in fact, given that there's no right of way here, they were very lucky not to get shot, Just pretend it was a stray bullet, Alberto, those two were asking for it anyway.

But there are times when one would be perfectly justified in roaring with laughter at what happens on the latifundio, if, that is, we were in the mood for some fun, but I'm not sure it would be worthwhile, we're so used to laughter turning into tears or a howl of rage so loud it could be heard in heaven, not that there is any bloody heaven, Father Agamedes is more easily accessible, and he never hears or else pretends that he doesn't, yes, a howl that would be heard throughout the earth, although I wonder if it would be heard and if anyone would come to our aid, unless, of course, the reason they couldn't hear us was because they were shouting so loudly themselves. Let's tell one such story, and laugh if you can, especially since that is what the guards are for, not to be laughed at, heaven forfend, but to be summoned and dispatched, and although it's usually the governor or some other official who does the summoning and dispatching, the latifundio has a great deal of power and authority over them as well, as you will see in this excellent tale involving Adalberto, a shepherd, his two assistants, three dogs, six hundred sheep, a jeep and a patrol of republican guards, rather than a whole squadron, that would be excessive, shoulder arms, quick march.

These sheep have strayed. They are on land belonging to Berto and are heading for more land belonging to Berto, well, that's a generalization and not entirely true, because while the lands do belong to Adalberto, en route the sheep will pass through Norberto's lands, and as they pass, they graze, because sheep aren't like a pack of dogs you can muzzle, and even if this were practicable and the sheep allowed it, the shepherd would never do it, it wouldn't be worth the fuss, but one should perhaps add another hypothesis, in situations where the shepherd has no real excuse for traveling from one man's property to another, he could claim to have become disoriented and to have crossed the boundary accidentally, for the real skill lies in taking advantage of those vague boundaries, making any such incursion seem purely accidental and putting on an air of wounded innocence if suspicions are aroused, Oh, I am sorry, I didn't notice, I was just walking along with my flock and kept going straight, thinking I was still on my master's land, that's all, no offense intended. Those who are quicker on the uptake will be thinking, He's lying, and they're not far wrong, but something more subtle is at work here, and the first thing to ascertain would be this, in carrying out this highly irregular act, was the shepherd thinking more about his sheep's bellies than about the interests of his master Berto. And this being noted, so that all eventualities will be covered, let us return to the story, to the six hundred sheep trotting briskly along, under the protection of the shepherd, his two assistants and his dogs, and let us city dwellers withdraw into the shade, it's wonderful to see the sheep pouring down the hillside or across the plain, so peaceful, far from the insalubrious urban hubbub, from the disorderly tumult of the metropolis, Begin, O muses, begin your bucolic song, and we're in luck, because the flock is coming over here, so we'll be able to savor the episode right from the start, let's just hope the dogs don't bite us.

On that day, as chance would have it, Adalberto had set out in his car to go for a spin in the countryside to view his estate, sometimes a love of nature requires such outings, and although the car cannot plunge in among the foliage, down tracks and over fallow fields, it nonetheless gives him freedom enough to roam along these cart tracks, as long as his car has a good suspension and he keeps a light touch on the steering wheel and doesn't attempt to drive too fast. Adalberto is alone, the better to enjoy the rural solitude and the birdsong, although the car engine does somewhat disturb both the peace and the music, but it's all a matter of knowing how to combine ancient and modern, rather than clinging to past pleasures, the easy gait of the horse pulling the tilbury, and the straw hat in profile beneath the limber length of the whip, which now and then caresses the horse's rump, which is all it takes for the horse to understand what you want. These are delights rarely encountered now, a horse costs a fortune and eats even when it isn't working, the horse, needless to say, is a distinguished beast, with its somewhat feudal echoes, but times inevitably change, and not only is the car much cleaner, it impresses the populace and saves one from unnecessary familiarities, we haven't got time for that.

Today, however, Adalberto is at peace, following the gentle curves of the road, his elbow nonchalantly leaning on the open window, since Lamberto died, all this land is mine, although, in fact, not all of Lamberto's land went to him, because that would make another good story, the divisions and redivisions, the amalgamations and accretions, but we don't have time right now, we should have started earlier, now Adalberto's car appears among the trees, the sun glinting on its polished body and on the chrome, and suddenly he stops. He's probably seen us, we'd better go a little farther down the hill, just to avoid any awkward questions, because I'm a peace-loving man and a respecter of other people's property, and when we look back to see if a furious Adalberto is following hard on our heels, we see, with horror, that he is getting out of his car and staring, enraged, at the languid flock, which takes no notice of him, just as they took no notice of us, not even the dogs see him, intent as they are on sniffing out rabbits, and then, shaking his fist, he gets back in the car, turns around, jolting over the rough ground, and, as they say in novels, disappears in a cloud of dust. We, needless to say, have legged it already, something is about to happen, why did he storm off like that, after all, this is a flock of sheep, not a pride of lions, but only Adalberto knows why as he hurtles back to Monte Lavre in search of reinforcements, namely the guards, who, at this very moment, are dying of boredom at the barracks, but that's what the latifundio is like, it's either man the barricades or complete idleness, such is the fate of those who choose the military life, and the reason why their superiors put on maneuvers and exercises, otherwise, Corporal, it's all or nothing.

Adalberto arrives at the entrance to the barracks in the aforesaid cloud of dust, and although his body is heavy with age and other excesses, he steps lightly in, it's not a large space but it coped easily enough, as I'm sure you'll recall, with all those orchestrated entrances and exits during that business over the thirty-three escudos, and when he leaves, he is not alone, he's joined by Corporal Tacabo and by a private, and all three climb into Adalberto's car, Holy Mother, where are those guards off to in such a hurry, the old ladies standing at their doors do not know, but we do, they are coming here where the flock is grazing, while the shepherd rests beneath a holm oak, and his assistants, with the help of the dogs, watch the sheep, it's not a major operation but it's not without its problems either, keeping such a large flock together, without too many gaps, after all, a sheep, too, needs a little breathing space, And what next, while we wait for Adalberto, there's something I don't quite understand, why this close relationship between the latifundio and the guards, You're either very naïve or you haven't been paying attention, how can you still be asking such questions at this stage in the story, or are you just play-acting, pretending that you don't know, perhaps it's a mere rhetorical device, the effective use of repetition, be that as it may, even a child knows that the guards are here to guard the latifundio, To guard it from what, it's not going anywhere, From the risk of theft, looting and other such wickedness, because the ordinary people we've been talking about until now have bad blood, by which I mean that the wretches and their parents and grandparents and the parents of their grandparents have known nothing but hunger all their lives, how could they not covet another's wealth, And is that wrong, It's the worst sin there is, You're kidding, Of course I am, but there are plenty of people who genuinely believe that this band of rustics want to steal their land, these sacred lands that go way back, and so the guards were posted here to maintain order, to suppress the slightest murmur of discontent, And do the guards like that, Oh, they do, the guards have their reward, a uniform, boots, a rifle, the authority to use and abuse, and the gratitude of the latifundio, let me give you one example, in payment for this extraordinary military operation, Corporal Tacabo will receive a few dozen liters of olive oil and a few cartloads of firewood, and while he may receive seventy of something, the mere guard will receive less because he's lower down the ranks, but he'll nonetheless receive some thirty or forty of whatever is on offer, because the latifundio is very reliable on that score, it always repays a favor, and the national guard is pretty easy to please, just imagine what must go on in Lisbon behind closed doors, How sad, Don't start crying now, imagine coming back from a day spent clearing land and walking miles with a sack of kindling on your back, panting like a beast of burden, and the guards ambush you, rifles cocked, hands up, what have you got in that sack, and you say, I've been working in such-and-such a place, and they'll check to see if you're telling the truth, and if not, you're in trouble, Personally, I'd rather be ambushed by José Gato, for at least he, Yes, José Gato would be preferable, but even worse would be to find, farther on, a whole cartload of six or seven hundred or a thousand kilos of firewood set aside for the guards, a gift from the latifundio in payment for their good and loyal service, They sell people very cheap, Whether they sell them cheap or dear doesn't matter, the problem isn't how much or how little.

This conversation went no further, what would be the point, although the narrator is free to say what he likes, that's his privilege, but now Adalberto has arrived along with his army, he stops the car, the doors open, it's an invasion, a landing, and from high up they wave to the shepherd, but he's a lazybones, a native of these parts, seated he is and seated he remains, then, finally, he gets to his feet, making it quite clear what an effort this entails, and yells, What's the problem, and Corporal Tacabo gives the order to charge, to attack, to release the bombs, take no notice of these warlike exaggerations, what do you expect, they have so few opportunities, by now, the shepherd has understood the situation, the same thing once happened to his father, laughter bubbles up inside him, the lines around his eyes betray him, it's enough to make you split your sides, Do you have permission to be on this land, the question comes from Corporal Tacabo, who, as master of the law and the carbine, thunders, That's a fine of five escudos per sheep, let's see, six hundred sheep at five escudos each, six times five is thirty, add the zeros, why that's three thousand escudos, that's very expensive grazing, and the shepherd says, There must be some mistake, the sheep belong to the boss here and I'm on his land, What did you say, asks Corporal Tacabo foolishly, and the private with them gazes up at the skies, and Adalberto, backtracking, says, You mean this is mine, Yes, sir, I'm in charge of these sheep, and these sheep are yours, Go, beloved muses, my song is ended.

The troops returned to the barracks, the three men on the expedition said not a word, and when Adalberto arrived home, he issued orders about the olive oil, while Corporal Tacabo and the private put away their weapons, totting up how much they would earn and praying to Saint Michael the archangel for more such dangerous but profitable adventures. This is the kind of minor incident that occurs on the latifundio, but many pebbles go to make a wall and many grains make a harvest, What's that noise, It's an owl, any moment now the other owl will respond, Domingos, he's the one nearest the nest.

J UST BECAUSE Sigismundo Canastro told that story about the dog Constante and the partridge doesn't mean he has a monopoly on strange hunting tales. António Mau-Tempo has his own tales to tell, as well as those he has picked up from others, indeed, so many and so various are they that he could easily have told the aforementioned story, with Sigismundo Canastro chipping in to confirm its truth with the irrefutable proof that he had dreamed about it. To those surprised at the freedom with which people add to, subtract from and generally alter stories, we need only remind them of the vastness of the latifundio, of the way in which words are lost and found, whether mere days or centuries later, when you sit beneath a cork oak, for example, and listen in on the conversation between that tree and its neighbor, ancient, albeit somewhat confused stories, because cork oaks do get muddled as they grow older, but whose fault is that, ours perhaps, because we've never bothered to learn their language. Anyone who has ever got lost on the latifundio always ends up being able to distinguish between the landscape and the words it conceals, which is why we sometimes come across a man standing in the middle of the countryside, as if, as he was walking along, someone had suddenly grabbed his hand and said, now listen to this, he is sure to be hearing words, stories and ripostes, simply because he happened to be in the right place at the right time, when the air unleashed its story, whether it was the magnificent tale of Constante the dog or one about the proven curiosity of hares, as explained by António Mau-Tempo and backed up by all of Sigismundo Canastro's dreams, unless there's someone else here eager to tell us about his dreams.

First, find a good, flat stone, about a span high, and wide enough to cover half a sheet of newspaper. You can't do it on a windy day, mind, because the wind will blow away the little pile of pepper that, among the tangle of headlines and the tiny italic and roman type, will form the trigger of this particular rifle. Now the hare, as everyone knows, is a curious creature, What, you mean even more than the cat, Oh, there's no comparison, the cat isn't interested in what's going on in the world, he simply doesn't care, whereas if a hare sees a newspaper lying on a path, he'll immediately go over to find out the latest news, so much so that some hunters have come up with a game plan, they stand behind a hedge and, when the hare approaches to read the news, bang, they shoot him, the trouble is that the newspaper gets completely shredded by the lead shot and you have to buy yourself another one, some hunters have been seen with their cartridge belts stuffed with newspapers, it's not right, But why the pepper, Ah, yes, the pepper, that's the secret ingredient, but it's essential to choose a windless day, because if you were to leave a newspaper on the path, the wind would catch it and send it flying, and the hare wouldn't be interested, because he likes to read the news in peace, You don't say, Oh, I could say much more on the subject if you have the time, anyway, once you've laid the trap, stone, newspaper and pepper, all you have to do is wait, and if you have to wait a long time, that's because it's not a good place for hares, it can happen, but don't go complaining that you didn't kill any hares, that's entirely your fault, because when you know the area, it never fails, anyway, in a little while, up will hop the first hare, a nibble here, another nibble there, and suddenly its ears go up because it's seen the newspaper, And what does he do then, Poor creature, he never learns, he's so keen to get the latest news that he runs over to the newspaper and starts reading, he's a really happy, contented hare, he doesn't miss a line, but then he sniffs the pepper and sneezes, And what happens next, Exactly what would happen to you if you were him, he sneezes, hits his head on the stone and dies, And then, You just have to go and find him, or, if you like, go a few hours later and you'll find a whole line of hares, one after the other, they're so curious that they can't see a newspaper without wanting to read it, Is that true, Ask anyone, even a babe in arms knows about these things.

António Mau-Tempo had no rifle at the time, which was just as well. If he'd had one, he would have been just another ordinary hunter with a ready-made weapon, rather than the inventor of Saint Hubert's pepper, but this doesn't mean that he scorned the art of marksmanship, the proof is the muzzle-loading rifle he bought one day for twenty escudos from a spendthrift tenant farmer and with which he performed miracles. City dwellers are brought up to be suspicious of miracles, they always want proofs and oaths, which is quite wrong, for example, there was the time when António Mau-Tempo, by then the proud owner of said muzzle-loading rifle, found himself with plenty of gunpowder but no lead shot. We should perhaps mention that it was rabbit-hunting season, in case someone should come along and ask why he didn't use the same stone-newspaper-pepper method as he did with hares. Only someone ignorant of the art of hunting could fail to be aware that rabbits have no curiosity at all, seeing a newspaper lying on the ground or a cloud in the sky is all the same to them, except that rain falls from clouds and not from newspapers, which is why the rabbit hunter still needs a rifle, trap or stick, but in this case we're talking rifles.

There is no greater misfortune for a hunter than to be in possession of a good weapon, even if it's only a flintlock, plenty of gunpowder but no lead shot, Why didn't you buy some, No money, that was the problem, So what did you do, At first I didn't do anything, I just thought, And did you come up with an idea, Of course, that's what thinking is for, So tell me how you solved the problem, because I still don't see how you did it, Well, I had a box of tacks for my boots and I loaded them into the rifle, What, you loaded your rifle with tacks, You may not believe me, but I did, Oh, I believe you, it's just that I've never heard of such a thing, At some point, you'll have to start believing in things you've never heard of, Tell me the rest of the story, then, All right, I was heading out into the countryside when I had a thought that almost made me turn back, What was that, It occurred to me that any rabbit I hit would be reduced to a pulp, torn to shreds, inedible, So, So I started thinking again, And you came up with an idea, Of course, like I said, that's what thinking is for, anyway, I positioned myself opposite a big old tree with a really thick trunk and I waited, Did you wait long, As long as I had to, one never waits too long or too little, So you waited until the rabbit appeared, Yes, when he spotted me, he ran away from me and toward the tree, I had studied the lay of the land, you see, and as soon as he passed close by, I shot him, And he wasn't shot to pieces, No, why else did I do all that thinking, the tacks pierced his ears and nailed him to the tree, which was a holm oak, by the way, Amazing, Yes, it was, and all I had to do was give him a quick blow to the back of the head with my stick, and once I'd eaten the rabbit, I still had the tacks to mend my boots with.

Men are made in such a way that even when they're lying, they tell a kind of truth, and if, on the contrary, it's the truth they want to blurt out, it's always accompanied by a kind of lie, however unintentional. That's why if we started debating what was true and what was false in António Mau-Tempo's hunting tales, we would never reach a conclusion, we should simply be man enough to recognize that everything he described could be touched with one's fingers, be it the hare or the rabbit, the muzzle-loading rifle, of a kind that still exists, gunpowder, which is cheap, the tack with which we shoe the poverty of the ill shod, the boot, which is witness to that, the miraculous pile of pepper all the way from India, the stone of course, the newspaper that hares can read better than humans, and António Mau-Tempo, who is right here, the teller of tales, because if there was no one to tell them, there would be no tales.

I've told you one story, I've told you two, and I'll give you a third, because three is the number God made, the Father, the Son and the Holy Ghost of the ear by which the rabbit was caught in the excellent tale I'm about to tell you, You've spoiled it for us now that we know how the story ends, So what, we all die, what matters is the life we've led or will lead, not the end, All right, tell me about the rabbit, Well, I still had the same rifle, in fact, I'd got so used to it that the sight of those double-barreled ones used to make me laugh, let alone the ones with four barrels, like cannons they are, they should be banned, Why, Think how much nicer it is for a man to slowly and quietly prepare his rifle, loading the gunpowder, tamping it down, measuring out the lead shot, when you have it that is, and watching one of the animals you're hoping to hunt pass you by, saying to itself, phew, that was a close shave, and you feel full of friendly feelings for the feathered or furry creature moving off, it's all a question of believing in fate, for their hour had clearly not yet come, That's one way of looking at it, anyway, what happened next, Next, you mean before, well, on that occasion, too, I had no money to buy lead shot, You never seem to have any money, To listen to you, anyone would think you had never lacked for it yourself, Don't change the subject, my finances are my affair, carry on with the story, All right, so I had no money to buy some shot, but I had a steel ball, one of those ball-bearing things, I found it among the rubbish in a workshop, and I used the same method, but without the tree this time, the tree worked only with the tacks, What do you mean, It seemed to me that if I could somehow sharpen the ball bearing, it would be like a bullet, and wouldn't destroy the animal's flesh or skin, it was all a matter of marksmanship, and, if I do say so myself, I'm a pretty good shot, And then, Then I went into the countryside, to a place I knew, a sandy area where I'd seen a rabbit as big as a baby goat, he was obviously the father rabbit, because no one has ever seen the mother rabbit, she never leaves the burrow, which is as deep as the pool at Ponte Cava, she goes underground and no one knows where she is, Fine, but that's another story, That's where you're wrong, it's exactly the same story, but I don't have time to tell it now, So what happened next, This rabbit had given me the slip on other occasions, and had a way of scooting out of sight as soon as I raised my rifle, but that had been when my rifle was loaded with shot, Ah, so you weren't bothered about spoiling its skin, With a rabbit that big, it wouldn't matter, But you just said, Look, if you're going to keep interrupting, All right, carry on, So I waited and waited, one hour passed, then another, and finally it hopped into view, well, leapt really, because, as I said, it was the size of a small goat, and when it was airborne, I pretended to myself that it was a partridge and shot it, Did you kill it, No, it just shook its ears, gave another hop and then another, and of course I had no more ammunition left, anyway, it ran off into some bushes, gave another leap, one of those really long ones, from here to over there, say, and what did I see, What, The rabbit was caught, squirming and wriggling, as if someone were holding it up by one ear, and then I went over and saw what had happened, Don't keep me in suspense, I'm dying of curiosity, Just like the hares, Stop playing around and tell me the rest of the story, Well, it so happened that someone had been cutting back the bushes, and a few twigs the size of a finger had been left sticking up, and, can you believe it, the rabbit had got caught on a twig through the hole the ball bearing had made in its ear,

So presumably you freed him and hit him hard on the back of the neck, No, I freed him and let him go, You don't say, I do, catching him in the ear like that had nothing to do with marksmanship, it was chance, sheer luck, and the father rabbit couldn't be allowed to die by chance, It's a great story, And it's all true, just as it's true that on that same night, the rabbits came out to dance into the small hours, by the light of the full moon, Why, Because they were so pleased that the father rabbit had escaped, You saw them, did you, No, but I dreamed it.

That's how it is. The fish dies by its mouth unless, because it looks too small on the hook or will cut a sad figure in the frying pan, the man throws it back in the water, an act of compassion for its youth, perhaps, or mere self-interest on his part, hoping that it will grow larger and reappear later on, but the father rabbit, who would certainly not grow anymore, was saved partly by the honesty of António Mau-Tempo, who, although he was perfectly capable of inventing good stories, did not need to invent a better one, given that it was far harder to hit a rabbit in the ear than in the body, and in the silence of the latifundio, once the sound of firing had died away in the undergrowth, he knew that he could not have lived with the memory of the rabbit's wide, angry eye as it watched him approach the bush.

The latifundio is a whole field of twigs, and from each one hangs a squirming rabbit with a hole in one ear, not because it has been shot, but because it has been like that since birth, they stay there all their lives, scrabbling at the earth with their claws, fertilizing it with their excrement, and if there's any grass, they eat as much of it as they can, nose pressed to the ground, while all around they hear the footsteps of hunters, I could die at any moment. One day, António Mau-Tempo freed himself from the bush and crossed the frontier, he did so for five years running, going to France once a year, to northern France, to Normandy, but he was being led by the ear, caught by the bullet hole of necessity, it's true he had never married nor had children who needed bread to eat, but his father wasn't at all well, a consequence of his time in prison, they might not have killed him, but they broke him, and there was an employment crisis in Monte Lavre, whereas in France work was guaranteed and well paid, compared with the norm on the latifundio, in a month and a half he could earn fifteen or sixteen thousand escudos, a fortune. Possibly, but as soon as he arrived home in Monte Lavre, most of that disappeared in back payments, and the little that remained was set aside for the future.

And what exactly is France. France is an endless field of sugarbeet in which you work a double shift of sixteen or seventeen hours a day, that is, all the hours of daylight and quite a few of the night. France is a family of Norman French, who see three Iberian creatures come through their door, two Portuguese men and one Spaniard from Andalusia, António Mau-Tempo and Carolino da Avó from Monte Lavre and, from Fuente Palmera, Miguel Hernández,* who knows a few words of French, picked up as an emigrant worker, and with those words he explains that they have been hired to work there. France is a cheerless barn where one sleeps little and dines on a dish of potatoes, it's a land where, mysteriously, there are no Sundays and no public holidays. France is a bent and aching back, like two knives pressing in here and here, an affliction and a martyrdom, a crucifixion on a piece of land. France is to be viewed with one's eyes a few inches from a sugarbeet stem, the forests and the horizons in France are all made of sugarbeet, that's all there is. France is this scornful, mocking way of speaking and looking. France is the gendarme who comes to check our pa-

---

* Miguel Hernández, after whom this character is apparently named, was a self-taught Spanish poet who spent his childhood working as a farm laborer and goatherd. He published his first book of poetry at twenty-three and achieved considerable fame. He was active on the Republican side in the Spanish Civil War, and after the Republicans were defeated, he was arrested several times and finally sent to prison, where he died of tuberculosis.

pers, line by line, comparing and interrogating, keeping three paces away because of the smell we give off. France is an ever-watchful distrust, a tireless vigilance, it's a Norman Frenchman inspecting the work we've done and placing his foot as if he were stepping on our hands and enjoying it. France is being meanly treated as regards food and cleanliness, certainly compared with the horses on the farm, who are fat, large-footed and proud. France is a bush bristling with twigs, each with a rabbit dangling by the ear like a fish on the end of a rod, slowly suffocating, and Carolino da Avó is the least able to take it, bent double and limp as a penknife in which the spring has suddenly snapped, his blade is blunt and his point broken, he will not return next year. France is long train journeys, an immense sadness, a bundle of notes tied up with string and the stupid envy of those who stayed behind and now say of someone who left, He's rich, you know, these are the petty jealousies and selfish malice of the poor.

António Mau-Tempo and Miguel Hernández know about such things, they write to each other in the meantime, Mau-Tempo from Monte Lavre, Hernández from Fuente Palmera, they are simple letters with spelling mistakes in nearly every word, and so what Hernández reads is not quite Portuguese and what Mau-Tempo reads is not quite Spanish, but a language common to them both, the language of little learning and much feeling, and they understand each other, it's as if they were signaling to each other across the frontier, for example, opening and closing their arms, the unmistakable sign for an embrace, or placing one hand on the heart, signifying affection, or merely looking, which indicates a readiness to reveal one's thoughts, and both sign their letters with the same difficulty, the same grotesque way of holding the pen as if it were a hoe, which is why it looks as if a physical effort were needed to form each letter, Miguel Hernández, uh, António Mau-Tempo, uh. One day, Miguel Hernández will stop writing, two of António Mau-Tempo's letters will go unanswered, and however hard you try to explain these things, they still hurt, it's not exactly a great misfortune, I'm not going to lose my appetite over it, but this is merely what one says to console oneself, perhaps Miguel Hernández has died or been arrested, as happened with António Mau-Tempo's father, if only he could go to Fuente Palmera to find out. António Mau-Tempo will remember Miguel Hernández for many years to come, whenever he speaks of his time in France, he will say, My friend Miguel, and his eyes will fill with tears, he'll laugh them off and tell a story about rabbits or partridges, just to amuse people, none of it invented, you understand, until the wave of memory calms and ebbs away. Only then does he feel any nostalgia for France, for the nights spent talking in the barn, the stories told by Andalusians and those who came from the other side of the Tagus river, from Jaén and Évora, stories about José Gato and Pablo de la Carretera, and those other crazy nights when their work contract had ended, and they went whoring, stealing hasty pleasures, allez, allez, their unslaked blood protesting, and the more exhausted they were, the more they wanted. They were driven out into the street by a rapid-fire language they couldn't understand, allez, nègres, that's how it is with dark-skinned races, everyone's a black for those born in Normandy, where even the whores think they're pure-bloods.

Then one year, António Mau-Tempo decided not to return to France, partly because his health was suffering. After that, he went back to being nothing but a latifundio rabbit, caught on a twig, scratching away with his claws, the ox returns to the furrow, the stream to its familiar course, alongside Manuel Espada and the others, cutting cork, scything, pruning, hoeing, weeding, why do they not weary of such monotony, every day the same as the last, at least as regards the scant food and the desire to earn a little money for tomorrow, which hangs over these places like a threat, tomorrow, tomorrow is just another day, like yesterday, rather than being the hope of something new, if that's what life is.

France is everywhere. The Carriça estate is in France, that's not what it says on the map, but it's true, if not in Normandy, then in Provence, it really doesn't matter, António Mau-Tempo no longer has Miguel Hernández by his side, but Manuel Espada, his brother-in-law and his friend, though they are very different in character, they are scything, doing piecework, as we shall see. Gracinda Mau-Tempo is here too, pregnant at last, when it seemed that she would never have children, and the three of them are living, for as long as the harvest lasts, in an abandoned laborer's hut, which Manuel Espada has cleaned up to make comfortable for his wife. No one had lived there for five or six years, and it was a real ruin, full of snakes and lizards and all kinds of creepy-crawlies, and when it was ready, Manuel Espada, having first sprinkled the floor with water, went to fetch a bundle of rushes to lie down on, and it was so cool inside that he almost fell asleep, it was just an adobe wall with a covering of gorse and straw to serve as a roof, then suddenly a snake slithered over him, as thick as my wrist, which is not of the slenderest. He never told Gracinda Mau-Tempo, and who can say what she would have done had she known, perhaps it wouldn't have bothered her, the women in these parts are made of stern stuff, and when she arrived at the hut, she found it all neat and ready, with a truckle bed for the couple, another for António Mau-Tempo and a large sack to share as a blanket, that is how intimately people live on the latifundio. Oh, don't get all hot under the collar, Father Agamedes, where have you been, by the way, these men are not really going to sleep here, if they do lie down on the bed, they will do so simply in order not to die, and now is perhaps the moment to speak about pay and conditions, they're paid so much a day for a week, plus five hundred escudos for the rest of the field, which must all be harvested by Saturday. This may seem complicated, but it couldn't be simpler. For a whole week, Manuel Espada and António Mau-Tempo will scythe all day and all night, and you need to understand exactly what this means, when they are utterly exhausted after a whole day of work, they will go back to the hut for something to eat and then return to the field and spend all night scything, not picking poppies, and when the sun rises, they will again go back to the hut to eat something, lie down for perhaps ten minutes, snoring like bellows, then get up, work all day, eat whatever there is to eat and again work all night, we know no one is going to believe us, these can't be men, but they are, if they were animals they would have dropped down dead, only three days have passed, and the two men are like two ghosts standing alone in the moonlight in the half-harvested field, Do you think we'll make it, We have to, and meanwhile Gracinda Mau-Tempo, heavily pregnant, is weeding in the ricefield, and when she can't weed, she goes to fetch water, and when she can't fetch water, she cooks food for the men, and when she can't cook, she goes back to the weeding, her belly on a level with the water, her son will be born a frog.

The harvest is done, and in the agreed time too, Gilberto came to pay these two ghosts, but he's seen plenty of ghosts in his time, and António Mau-Tempo has now gone to work on the other side of this France, this killing field. Manuel Espada and his wife Gracinda Mau-Tempo stayed on in the hut until it was time for her to give birth. Manuel Espada took his wife home and then went back to the Carriça estate, where, fortunately, there was work. Anyone who remains unsurprised by all this needs to have the scales removed from his eyes or a hole bored in his ear, if he hasn't got one already and sees them only in the ears of others.

GRACINDA MAU-TEMPO gave birth in pain. Her mother Faustina came to help her during labor, along with old Belisária, who had long practiced as a midwife and been responsible for a fair few deaths in childbirth, of both mothers and babies, but to make up for this, she did create the finest navels in Monte Lavre, and while this may sound like a joke, it isn't, rather, it deserves to be the subject of obstetric research into just how Belisária managed to cut and suture umbilical cords in such a way that they resembled goblets straight out of the thousand and one nights, which, opportunity and audacity allowing, one could verify by comparing them with the bare bellybuttons of the Moorish dancers who, on certain mysterious nights, cast off their veils at the fountain in Amieiro. As for the pain suffered by Gracinda Mau-Tempo, it was neither more nor less than that suffered by all women since Eve's fortunate sin, fortunate, we say, because of the earlier pleasure enjoyed, a view that does not sit well with Father Agamedes, who disagrees out of duty and possibly conviction, as the upholder of the most ancient punishment in human history, meted out by Jehovah himself, You will give birth in pain, and so it has been all the days of all women, even those who didn't know

Jehovah's name. The rancor of the gods lasts much longer than that of men. Men are poor wretches, capable of terrible vengeance, but capable, too, of being moved to tears by the slightest thing, and if the time is right and the light propitious, they will fall into their enemy's arms and weep over how strange it is to be a man, a woman, a person. God, Jehovah or whoever, never forgets anything, the sinner must be punished, which is why there is this endless line of gaping vulvas, dilated, volcanic, out of which burst new men and new women, all covered in blood and mucus, all equal in their misery, but so instantly different, depending on the arms that receive them, the breath that warms them, the clothes that cover them, while the mother draws back into her body that tide of suffering, even while the last flower of blood drips from her torn flesh, and while the flabby skin on her now empty belly slowly stirs and hangs in folds, that is when youth begins to die.

Meanwhile, up above, the balconies of heaven are deserted, the angels are taking a nap, of Jehovah and what remains of his wrath there is no news that makes any human sense, and there is no record that the celestial fireworks were summoned to conceive, create and launch some new star to shine for three days and three nights above the ramshackle hut that is home to Gracinda Mau-Tempo, Manuel Espada and their first child, Maria Adelaide, for that is the name she will bear. And we are in a land that does not lack for shepherds, some who were shepherds as children and others who continued to be and will be until the day they die. The flocks are large too, we saw one of six hundred sheep, and there's no shortage of pigs either, although the pig is not really a suitable animal for nativity scenes, it lacks a sheep's elegance, thick coat, soft woolly caress, pass me my ball of yarn, will you, darling, such creatures are made to bend the knee, whereas the pig rapidly loses its sweet look of a pink, newborn bonbon and becomes instead a bulbous-nosed, malodorous lover of mud, sublime only in the meat that it gives us. As for the oxen, they are busy working, nor are there so many of them on the latifundio that they can afford to attend belated scenes of adoration, and as for the donkeys, beneath their saddle cloths there are only sores, around which buzz bluebottles excited by the smell of blood, while in Manuel Espada's house the flies swarm feverishly above Gracinda Mau-Tempo, who smells like a woman who has just given birth, Keep those flies off, will you, says old Belisária, or perhaps she doesn't, so used is she to this accompanying halo of winged, buzzing angels, who appear as soon as summer arrives and she has to go off to help some woman in labor.

Miracles do happen, though. The child is lying on the sheet, they smacked her as soon as she came into the world, not that this was necessary, because her first cry was already forming in her throat, and one day she will shout other things that now seem quite impossible, she cries, although she sheds no tears, merely screws up her eyes, making a face that would frighten any visiting Martian, but which, nonetheless, makes us sob our hearts out, and since it is a bright, warm day and the door is open, there falls onto this side of the sheet a kind of reflected light, where it comes from doesn't really matter, and Faustina Mau-Tempo, so deaf that she cannot even hear her granddaughter crying, is the first to see her eyes, which are blue, as blue as João Mau-Tempo's eyes, two drops of sky-bathed water, two round hydrangea petals, but neither of these vulgar comparisons serves, they merely reveal a lack of imagination, no comparison will serve, however hard future suitors may struggle to come up with one that does justice to these eyes, which are blue, not aquamarine or azure, not some botanical caprice or the product of some subterranean forge, but bright, intense blue, like João Mau-Tempo's eyes, we can compare them when he arrives, and then we will know what kind of blue it is. For now, though, only Faustina Mau-Tempo knows, which is why she can proclaim, She has her grandfather's eyes, and then the other two women want to see as well, Belisária, much put out at being deprived of her midwife's privileges, and Gracinda Mau-Tempo, a jealous she-wolf to her cub, but Belisária takes her time, which is why Gracinda Mau-Tempo is last, not that it matters, she will have time enough to be attached by her nipples to that sucking mouth, she will have time to lose herself in contemplating those blue, blue eyes while the milk flows from her breast, whether here beneath these badly laid roof tiles, or beneath a holm oak in the middle of the countryside, or standing up when there's nowhere to sit, or hurriedly when she can't dawdle, milk that flows, in small and large quantities, from that breast, that life, like white blood made out of the other, red variety.

Then the three kings arrived. The first was João Mau-Tempo, who came on foot when it was still light, so he needed no star to guide him, and the only reason he didn't arrive earlier was male modesty, because, were such things the norm in that time and place, he could easily have been there at the birth, after all, what's wrong with seeing your own daughter give birth, but it's simply not done, people would talk, such ideas belong in the future. He arrived early because he's currently out of work, and has been clearing a piece of land he's been given to cultivate, and when he went into the house, his wife wasn't there, but his neighbor informed him that he was the grandfather of a little girl, and he was pleased, but not as pleased as you might expect, because he would have preferred a boy, men do, in general, prefer boys, and then he left the house, walked at his usual slow, swaying pace, caught between two different pains, one here, the other there, the old pain acquired carrying logs to the charcoal pit, and the other, a dull ache, was the result of being forced to stand for hours like a statue, he looks like a sailor fresh off the boat after a long voyage, disconcerted to find that the ground he walks on doesn't move, or as if he were riding on the back of a camel, the ship of the desert, a comparison that paints exactly the right picture, for, given that João Mau-Tempo is the first of the magi to arrive, it is only proper that he should travel according to his condition and tradition, the others can choose their own mode of transport, and he brings no gifts to speak of, unless the ark of suffering that João Mau-Tempo carries in his heart could be considered a gift, fifty years of suffering, but no gold, and incense, Father Agamedes, is for the church, and as for myrrh, that's been used up on those who died along the way. It seems rather mean and in somewhat bad taste to give such a gift to a newborn, but these men from the latifundio can only choose from what they, in turn, were given, as much sweat as one could want, enough joy to fill a toothless smile and a plot of land large enough to devour their bones, because the rest of the land is needed for other crops.

João Mau-Tempo, then, arrives empty-handed, but on the way, he remembers that his first grandchild has just been born, and from a garden he plucks a single geranium flower on its knotty stem, with its acrid smell of poor households, but what a pretty sight it is to see one of the magi mounted on his camel with its gold and crimson saddle cloth, humbly bending down to pick a pelargonium, he didn't order a slave to do this, of the many who accompany and serve him, what a fine example he sets. And when João Mau-Tempo reaches the door of his daughter's house, it seems that the camel knows its duty, for it kneels down to let this lord of the latifundios dismount, while the republican guards from the local barracks present arms, although Corporal Tacabo has his doubts about whether a large, exotic beast like a camel should be allowed to travel the public highway. These are fantasies born of the harsh sun, now sinking in the sky but still beating down on the stones along the road, which are as hot as if the earth had just given birth to them. My dear daughter, and it is then that João Mau-Tempo sees that his eyes are immortal, for there they are again after a long peregrination, even he doesn't know where they started out, where they came from and how, he knows only that there are no other such eyes in Monte Lavre, in his own family or elsewhere, my daughter's children are definitely my grandchildren, while those of my son might and might not be, none of us is free of such popular malice, but who can doubt those eyes, look at me, look into my blue eyes, and now look at those of my granddaughter, who is to be called Maria Adelaide and is the image of her ancestor more than five hundred years ago, for those eyes come from that foreigner, that deflowerer of virgins. All families have their myths, some of which they do not know, as is the case with this Mau-Tempo family, who should, therefore, be very grateful to the narrator.

The second of the magi arrived when night had already fallen. He came straight from work to find no light on in the house, the fire out, and so no hope of a full stewpot, then his heart turned over and did so again when he met the same neighbor who told him, Your sister has had a little girl, your father and mother are there, because by now the whole of Monte Lavre knows that it's a baby girl and is vastly amused to know she has blue eyes, but the neighbor says nothing about this last point, she's a kind woman who believes that surprises have their time and place, why tell António Mau-Tempo, Your niece has blue eyes, she would then be denying him the pleasure of seeing this with his own brown eyes. The guards have returned to the barracks, no one is there to present arms to António Mau-Tempo, well, if you thought there would be, more fool you, but he is, nevertheless, a flesh-and-blood king walking down the street, as dirty as befits someone who has come straight from work, he hasn't had time to wash, but he doesn't forget his brotherly obligations and picks a daisy from a whitewashed can-cum-flowerpot beside a door, and so that it doesn't wilt in his fingers, he puts it between his lips to water it with his saliva, and when he finally goes into the hut, he says, For you, sister, and gives her the marguerite, what could be more natural than for a flower to change its name, it happened earlier with the geranium and the pelargonium, and will happen again one day with the carnation.*

It's just as well that António Mau-Tempo wasn't expecting to see those blue eyes. The child is sleeping peacefully, her eyes are closed, her decision, and she will open them again only for the third wise man, but he will arrive much later, in the dead of night, because he is coming from far away and on foot, he's made this same journey for the past three days or three nights, because for those who like to know the facts, Manuel Espada is now on his third night with little sleep, and he's used to that, as all these people are, but perhaps we should explain. Because Manuel Espada works far from home, he usually sleeps there, in a shepherd's hut or a cabin, it doesn't really matter, but as the time of the birth drew nearer, what did Manuel Espada do, he stopped work at sunset, reached home after midnight, where his child was still nothing but a swollen belly, lay down for an hour or so beside Gracinda Mau-Tempo, then got up halfway between night and morning and went back to work, and this is the third such night, but third time lucky, for when he arrives, he will see his wife and his newborn child, isn't that good.

Faustina, João and António Mau-Tempo killed a chicken to celebrate, and Gracinda Espada drank some of the broth, which is good for mothers who have just given birth, and meanwhile, more uncles and aunts and other relatives came and went, Gracinda needs to rest, at least today, bye, see you tomorrow, what a lovely little girl, the image of her grandfather. The church clock chimed midnight, and if no misfortune has befallen the traveler, if he has not slipped down a hill or into a ditch, if no impatient ne'er-do-

---

* The Portuguese revolution in 1974 came to be known as the Carnation Revolution because, despite its being a military coup, no shots were fired and, in the streets, people handed the soldiers red carnations, which they pinned on their uniforms or placed in the barrels of their guns.

well has broken the rule about not attacking someone as poor as himself, then it should not be long before this third wise man arrives, what gifts will he bring with him, we wonder, what cortege, perhaps he'll be mounted on an Arab steed with hooves of gold and a bridle of silver and coral, perhaps, instead of some bearded scoundrel stepping out onto the road, he will meet his fairy godmother, who will say, Your daughter has been born, and because she has blue eyes, I give you this horse so that you can see her all the sooner, before life drains the color from those eyes, but even were that fairy godmother to intervene, which is, after all, pure fantasy, these paths are difficult, and even more so at night, the horse might tire or break a leg, and then Manuel Espada would have to make the journey on foot anyway, through the great, vast, starry night full of terrors and indecipherable murmurings, but the three kings still have the magical powers they learned in Ur and Babylon, how else explain the two fireflies that go ahead of Manuel Espada, he can't get lost, he simply has to follow them as if they were the two sides of the path, how are such things possible, how can such creatures guide a man, they go up hill and down dale, they skirt ricefields and fly across plains, we can see the first houses in Monte Lavre now, and there the fireflies have alighted on top of the door frame, at head height, to light his way, glory be to man on earth, and Manuel Espada passes between them, a suitable guard of honor for someone who has just come from hours of hard labor to which he will have to return before sunrise.

Manuel Espada brings no gifts from near or far. He reaches out his hands, and each hand is a large flower, then he says, Gracinda, and can say no more, but kisses her on the cheek, just once, but what is it about that one kiss that brings a lump to our throat, and we're not even family, if we did have something to say at this juncture, we wouldn't be able to, and just when those gestures are being made and that word is being spoken, Maria Adelaide opens her eyes, as if she had been waiting for that moment, her first childish trick, and she sees a large shape and large open hands, it's her father, she doesn't yet understand what this means, as Manuel Espada well knows, so much so that he feels as if his heart were going to leap out of his chest, his hands are shaking, how can he pick up this child, his daughter, men are so useless, and then Gracinda Espada says, She looks like you, well, it's possible, although at that age, only a few hours old, you can never tell, but João Mau-Tempo is quite right when he proclaims, But she has my eyes, and António Mau-Tempo says nothing, because he is merely the uncle, and poor deaf Faustina can only guess at what is being said, and says in turn, My love, quite why she doesn't know, because, for reasons of modesty and reserve, these are not words normally used on the latifundio and in these situations.

Two hours later, and however much time he had spent there it would have seemed too short, Manuel Espada left the house, he is going to have to walk very fast to get to work before sunrise. The two waiting fireflies set off again, flying close to the ground now and shining so brightly that the ants' sentinels shouted to their fellow ants inside the nest that the sun was coming up.

THE HISTORY OF THE wheatfields is one that repeats itself with remarkable regularity, but it has its variants too. It's not that sometimes the wheat is ready to be harvested later or earlier, that depends on whether there has been too much or too little rain, or on the sun, which can transgress by sending too much heat or not enough, nor is it that the seeds were sown on a steep slope or on low ground, in clayey or in sandy soil. The men of the latifundio have long been accustomed to the perversities of nature and to their own mistakes, and are unlikely to be thrown by such slight and inevitable occurrences. And although it is true that the aforementioned variants, individually and as a whole, deserve to be dealt with at greater length, unhurriedly, with time to go back and discuss perhaps a forgotten lump of soil, without having to worry about our listeners' growing impatience, it is also true, alas, that such considerations are out of place when telling a story, even when it's a story about the latifundio. Let us accept, then, that we must keep quiet about all these subtle differences and let us add to less serious defects the far graver one of pretending that everything remains the same in the wheatfields from one year to the next, and let us merely ask why this delay, why have the harvesters, human and mechanical, not yet entered the fields, when even we ignorant city dwellers can clearly see that the moment is here and is passing us by, that the dry whisper of the wheat in the wind is like the whisper of dragonfly wings, in short, let us ask what damage is being done here and to whom.

The history of the wheatfields repeats itself, with variants. In the present case, it isn't because the men are kicking up their usual ruckus, demanding more money. Well, it's the same cry every year at every season and about every job, It's as if they don't know how to say anything else, Father Agamedes, instead of worrying about the salvation of their immortal soul, if they have one, they care only about bodily comforts, they have learned nothing from the ascetics, no, all they think about is money, they never ask if there is any or if I can afford to pay. The church is a great source of consolation in these situations, it takes a tiny sip of wine from the chalice, just another drop, please, do not remove this cup from me, and raises remorseful eyes to the heavens from which it hopes one day to receive rewards for the latifundio, when the time comes, of course, but the later the better, Tell me, Father Agamedes, what do you make of these idlers going around cheering this general,* it seems that these days one can trust nobody, I mean, a military man of all people, and he seemed so trustworthy, so well loved by the regime that made him, yet here he is, traveling around stirring up the populace, how did the government allow things to get this far. Father Agamedes has no answer to this question, his kingdom is not always of this world, and yet he has been a witness to and a personal

---

* General Humberto da Silva Delgado was initially a staunch supporter of Salazar's right-wing dictatorship and the youngest general in Portuguese history. However, he later became a defender of democratic ideals and decided to run for the Portuguese presidency in 1958. When asked what he would do about Salazar if he won the election, Delgado famously remarked, "Obviously, I'll sack him." As it happened, he won only twenty-six percent of the vote, losing to the government's preferred candidate, Américo Tomás, amid widespread allegations of vote rigging. On February 13, 1965, Delgado and his secretary were murdered after being lured into an ambush by PIDE agents.

victim of this great national terror, this hothead shouting wildly, I'll sack him, I'll sack him, and who was he referring to, why, Professor Salazar, of course, hardly the behavior of a candidate, a candidate should be polite at all times, but his behavior backfired on him in the end, and they say he's on the run, and to think what a quiet life we had until now, before all this fuss, But between you and me, Father Agamedes, because no one's listening, things could have been worse, it took a lot of skill not to let the situation get out of control, nonetheless, we must remain vigilant, and the first thing we should do is teach these idlers a lesson, which is why not a sheaf of wheat will be harvested this year, That'll teach them, Senhor Norberto, Yes, that'll teach them, Father Agamedes.

It's not known where this spirit of didacticism came from, whether from Lisbon, Évora, Beja or Portalegre, or if it was used in a jocular mode at the club in Montemor or after too much cognac, or if Leandro Leandres brought it home with him from the house of dragons, but whatever its origin, it quickly spread throughout the latifundio, passing from Norberto to Gilberto, from Berto to Lamberto, from Alberto to Angilberto, and, once it had found general acceptance, the overseers were summoned and given their orders, Any harvesting already begun must stop, and don't start work on any new fields. Some calamity must have occurred, perhaps the wheatfields have become leprous, and the latifundio has taken pity on its harvester-children and doesn't want to see them disfigured, their fingers mere stubs, their legs stumps, their noses absences, their lives are unfortunate enough as it is. This bread is poisoned, at the end of each field place scarecrows with horrible gaping skulls for heads, that should put the fear of death into even the most resolute souls, and if that doesn't scare them off, then call the guards, they'll sort them out. And the overseer says, That won't be necessary, no one is likely to go to the fields unless he's sure to get paid, and certainly not if he's going to get a bullet between the shoulder blades, but think of the loss of income. And Alberto says,

Better cut the shoe than pinch the foot, it won't ruin us to leave the wheat unharvested for a year. And the overseer says, They want more money, they say the price of food is going up and up and that they're starving. And Sigisberto says, That's nothing to do with me, we pay them what we want to pay them, food is expensive for us as well. And the overseer says, According to them, they're going to get together to talk to the boss. And Norberto says, I don't want any dogs following behind me, barking.

All over the latifundio one hears only the barking of dogs. They barked when, from the Minho to the Algarve, from the coast to the eastern border, the people rose up in the general's name, and they were barking a new bark, which, in the language of ordinary folk, translated as, If you want better pay, vote Delgado on the day, this taste for rhymes goes back a long way, well, we are, after all, a nation of poets, and they barked so much that soon they were barking at people's doors, it won't be long, Father Agamedes, before they start profaning churches, that's always the first thing they do, spitting in the face of the holy mother church, Please, Dona Clemência, don't even talk about it, not that I'm afraid of martyrdom, but Our Lord will not allow a repetition of the kind of outrage that took place in Santiago do Escoural, where, can you imagine, they turned the church into a school, I didn't see it with my own eyes, of course, it was before my time, but that's what I've been told, It's true, Father Agamedes, as true as we're sitting here now, ah, the follies of the republic, which, God willing, will not be repeated, be careful when you leave, mind the dogs don't bite you. When Father Agamedes peers around the door on the way out, he asks in a shrill, tremulous voice, Are the dogs under control, and someone replies dully, These ones are, but put like that, how do we know which dogs are loose and which not, but Father Agamedes feels certain that this information will ensure the safety of his delicate calves and steps out into the courtyard, where he finds, to his relief, that the dogs are indeed safely tethered, but when he goes out through the street door, he finds a crowd of people, not barking, all we need is for men to start barking, but if this murmuring doesn't sound like a dog growling, I'll eat my hat, and Father Agamedes doesn't see the line of ants marching along the wall of the building, raising their heads like dogs, they're quiet now, but whatever will become of us if the whole pack of them were to join forces.

As mentioned above, this year's harvest has been canceled, as a punishment for the usual impertinence of asking for more wages and for the exceptional crime of supporting Delgado, everywhere and anywhere. I really don't care, said Adalberto, I just need to be sure that the government is in agreement, Oh, it is, said Leandro Leandres, and so are we, we think it's a magnificent idea. And what about the losses, sir, what about the losses, you can count on our good will, but we'll have to be compensated, everything has its price, a perfectly justifiable remark made in some unnamed place on the latifundio, it must have been in a town, because what would the civil governor be doing in a tiny village unless he was there to attend some inauguration ceremony, but wherever it was, the remark was made, perhaps on a balcony looking out over the countryside, Don't worry, Senhor Berto, we're studying how best to assist agriculture, the nation is aware of farmers' concerns and will not forget this patriotic gesture. They almost hoist the flag, but why bother, election day has been and gone, and while Tomás* may be our brand-new president, he acts the same as the former resident, well, if the others can rhyme, why shouldn't I, I'm as important as they are, and I can make very pretty rhymes, for example, I'm always hungry year on year, Through winter and through spring, While down in hell, Death rings his bell: Your scythe awaits you here, and after this ditty sung in chorus, a great silence falls

---

* Américo de Deus Rodrigues Tomás became minister of the navy in 1944, was elected president of the Portuguese republic in 1958 and was reelected in 1965 and 1972. As president, he was a mere figurehead and widely lampooned. His one bold act was to dismiss Salazar after the latter suffered a crippling stroke in 1968.

over the latifundio, what's going on, and while we're thinking this, our eyes fixed on the ground, a shadow passes rapidly overhead, and when we look up, we see the great red kite hovering above, so the moan that emerged from my throat was actually his cry.

That night, Sigismundo Canastro went over to João Mau-Tempo's house to talk to him and António Mau-Tempo, and from there he went to Manuel Espada's house, where he spent some time. He visited another three houses, two of which were far out in the country, he spoke to people in this way and that, used these words and those, because you can't talk to everyone in the same way, if you do, your words might be misconstrued, and his message, in essence, is to meet in Montemor in two days' time to demonstrate outside the town hall, we want as many people to be there as possible, to demand work, because there's plenty of work to be done, but they're refusing to let us do it. En route they will discuss what the men of the latifundio think about the farce of handing the presidency of this wretched republic over to an out-and-out imbecile and yes man, surely one was enough, how many more will there be. These bitter words come not from excessive drinking or eating, neither of which is much practiced on the latifundio, although having said that, there's no shortage of men who bend the elbow rather too much, but that can be excused, for when a man finds himself tethered to a stake all his life, smoking and drinking are ways of escape, especially drinking, though each drink is another step toward death. This bitterness comes from the frustrated hope that they were finally going to be able to speak freely, had freedom come, but it didn't, someone once caught a glimpse of that much-vaunted freedom, but she is not one to be seen out walking the highways, she won't sit on a stone and wait to be invited in to supper or to share our bed for the rest of our life. Groups of men and some women had been out and about, cheering and shouting, and now we are left with a bitter taste in our mouths as if we, too, had been drinking, our eyes see ashes and little more, only wheatfields as yet unharvested, What are we going to do, Sigismundo Canastro, you who are older and more experienced than us, On Monday we'll go to Montemor to demand bread for our children and for the parents who have to bring them up, But that's what we always do, and to what end, We've done it in the past, we must do it now and must continue to do it until things change, It feels like a never-ending struggle, But it will end, When we're dead and buried and our bones are there for all to see, if there are any dogs around to dig them up, There'll still be enough people around when the time comes, your daughter, you know, gets prettier by the day, She has my father's eyes, these words were spoken by Gracinda Mau-Tempo, all the conversation prior to this having been with her husband Manuel Espada, and it is he who says, I'd sell my soul to the devil to see that day come, not tomorrow, but now, and Gracinda Mau-Tempo picks up her three-year-old daughter and scolds her husband, Don't say such things, Manuel, and Sigismundo Canastro, older in years and experience, smiles, The devil doesn't exist and so can't make any deals, and no amount of oaths and promises will change anything, work is the only way to get what we want, and our work now consists in going to Montemor on Monday, people will be coming from all over.

These June nights are beautiful. If there's a moon, you can see the whole world from high up in Monte Lavre, well, let's pretend you can, we're not that ignorant, we know the world is much bigger, I've been to France, António Mau-Tempo would say, and that's a long way away, but in this silence, anyone, even I, would believe it if someone said, There is no other world apart from Montemor, where we're going on Monday to ask for work. And if there is no moon,* then the world is simply this place where I put my feet and all the rest is stars, perhaps there's a latifundio up there too, which

---

* A possible reference to Luís de Sttau Monteiro's 1962 play *Felizmente há luar!* (*Fortunately There's a Moon!*), which was banned because it was deemed to be critical of the Salazar regime and, in particular, of the Delgado-Tomás presidential campaign.

is why our new president is a rear admiral who's never been to sea and who won the election game with four aces and a few more besides, because nothing trumps being an apparent pillar of society and a cheat. Had Sigismundo Canastro thought such wicked, witty thoughts, we would have stood back at the edge of the road, hat in hand, astonished at the worldly wisdom of the latifundio, but what he is really thinking is that he has spoken to everyone he needed to speak to and that he was right to speak to them today rather than leave it until tomorrow, which is why we don't know what to do with our hat, or even if we should be holding it in our hand, Sigismundo Canastro has done his duty, and that's that. However, despite the gravity of the steps to be taken, he also has a spritely, mischievous side, as we have seen before, and so, noticing that the door to the guards' barracks was locked and in darkness, he went over to the wall and peed long and pleasurably as if he were peeing on the whole lot of them. The childish tricks of an old man whose cock no longer serves him for very much except to make this lovely little stream that finds its way among the cobbles, I wish I had liters of urine in me so that I could stay here peeing all night, like the dam at Ponte Cava, perhaps we should all pee at the same time and flood the latifundio, I wonder who would be saved. It's a fine, starry night. Sigismundo Canastro buttons up his fly, the comedy is over, and sets off home, sometimes the blood still stirs, you never know.

In the days when people made pilgrimages, we used to say that all roads lead to Rome, you just had to walk and ask the way as you went, that's how sayings come into being and are then unthinkingly repeated, and it's not true, for here all the many roads and paths lead to Montemor, and although no one is speaking, only a deaf man could fail to hear the lofty speech echoing around the latifundio. Some, if they can find no better mode of transport and regardless of whether they come from near or far, are on foot, others are pedaling along on ancient bicycles that wobble and creak like mule carts, while those who can, have come by bus, and all are converging on Montemor, arriving from all the points of the compass rose, and carried there by a strong wind. The sentinels on the castle ramparts watch the Moorish host approach, the flag of the prophet folded in their bosom, O Holy Mother of God, the infidels are coming, lock up your wives and daughters, gentlemen, close the doors and raise the drawbridge, for in truth I say unto thee, today is the day of judgment. The narrator is, of course, exaggerating, doubtless the result of too much time spent immersed in medieval studies, fancy imagining armies and pennants when there is only this disparate band of rustics, probably not even a thousand of them, and yet the final gathering will be far larger. But one thing at a time, there's another two hours yet, for the moment Montemor is just a town with more people in the streets than usual, they wander about in the main square talking to each other in low voices, and those with a little money to spend buy themselves a drink. Has the party from Escoural arrived, I don't know, we're from Monte Lavre, there aren't many of them, it's true, but at least they're here, and they've brought a woman with them, because Gracinda Mau-Tempo wanted to come too, there's no stopping women nowadays, that's what the older, more old-fashioned men think, although they say nothing, imagine what they would have said if they had overheard the following conversation, Manuel, I'm going with you, and Manuel Espada, despite himself, thought she must be joking and responded, or, rather, all the Manuels in the world answered for him, This isn't women's business. What did you say, a man should be careful when he speaks, it's not just a matter of saying the words, you can end up looking ridiculous and losing all authority, fortunately Gracinda and Manuel really love each other, nevertheless, the discussion continues for the rest of the evening and even when they're lying in bed, The child can stay with my mother and then you and I can go together, we don't just share a bed, you know, and finally Manuel Espada gave in and, glad to give in, put his arm around his wife and drew her to him, the man invites and the woman consents, the little girl is sleeping and hears nothing, Sigismundo Canastro, too, is asleep in his bed, having tried and succeeded, perhaps the next time will be even better, a man can't just give up, damn it.

What went on between man and wife last night or the night before, and what they will do later, are not matters to be discussed in Montemor, or, indeed, when this day is over, for who knows how it will end. The cavalry, as usual, rides forth from the guards' barracks, while inside, Lieutenant Contente and Leandro Leandres are deep in conversation, the order to mobilize has been issued, now they must await events, although others have decided to wait elsewhere, they are the owners of the latifundio who live in Montemor, and there are quite a few of them, so we were not far off when we spoke of sentinels, for there is a stockade along the walls of the castle, with the braver of the infantes perched on the reconstructed ramparts, and a rosary of fathers and mothers, the former dressed as knights and the latter clad in suitably light colors. The more malicious commentators will say that they have taken shelter there because they are afraid of this invasion of farm laborers, a hypothesis that has a certain ring of truth about it, but let us not forget how few distractions there are here, apart from bullfights and the cinema, this time it's rather like a picnic in the country, there's plenty of shade and, if necessary, there is the safe haven of the convent of Our Lady of the Annunciation, pray for us. It is, however, true and verifiable that they left their houses out of a hitherto unknown fear, the servants remained behind on guard, because if you take on servants when they're young, they tend to be loyal, as is doubtless the case with Amélia Mau-Tempo, who also works as a maid in Montemor, these facts are at once contradictory and inevitable, but given the times we live in, one cannot really trust anyone, not because the workers of the latifundio have joined together to make their demands, it's not the first time they've asked for work, but because one can all too easily imagine those hands closing into fists, there's a lot of anger out there, a lot of conspiracies, dear aunt, a lot of conspiracies. From up here, you can see them walking down the narrow lanes and converging on the square outside the town hall. They look like ants, says an imaginative child heir, and his father corrects him, They may look like ants, but they're dogs, now there's the whole situation summed up in one brief, clear phrase, and then silence falls, we don't want to miss anything, look, there's already a squadron of guards in front of the town hall, and there's the sergeant, what's that he is holding, a machine gun, that's what Gracinda Mau-Tempo thought too, and glancing up at the castle, she saw that it was full of people, who can they be.

The square fills up. The people from Monte Lavre are standing in a group. Gracinda, the only woman, her husband Manuel Espada, her brother and father, António and João Mau-Tempo, and Sigismundo Canastro, who says, Stick together, and there are two other men called José, one is the great-grandson of the Picanços, who kept the mill at Ponte Cava, and the other is José Medronho, whom we haven't had occasion to mention until now. They are in a sea of people, the sun beats down on this sea and burns like a nettle poultice, while up in the castle, the ladies open their sunshades, anyone would think it was a party. Those rifles are loaded, you can tell by the look on the guards' faces, a man carrying a loaded weapon immediately takes on a different air, he grows hard and cold, his lips tighten, and he looks at us with real rancor. People who like horses sometimes give them the name of a person, like that colt called Bom-Tempo, but I don't know if the horses at the end of the street have names, perhaps they simply give them numbers, they do everything by numbers in the guards, call out number twenty-seven, and the horse and the man riding it both step forward, how confusing.

The shouting has begun, We want work, we want work, we

want work, that's about all they say, apart from the occasional insult, you thieves, but spoken so quietly that it's as if the person hurling the insult were ashamed, then someone else shouts, Free elections, what's the point of saying that now, but the great clamor of voices rises up and drowns out everything else, We want work, we want work, what kind of world is it that divides into those who make a profession of idleness and those who want work but can't get it. Someone gave the signal, or perhaps it was agreed that the meeting could go on for a certain number of minutes, or perhaps Leandro Leandres or Lieutenant Contente made a telephone call, or maybe the mayor peered out of the window, There they are, the dogs, but whatever the sequence of events, the mounted guards unsheathed their sabers, oh good heavens, such courage, such heroism, it sends a shiver down the spine, I had quite forgotten about the sun until it glinted on those polished blades, a positively divine light, enough to make a man tremble with patriotic fervor, well, doesn't it you.

The horses break into a trot, there's not enough space for anything bolder, and those who try to escape from beneath the hooves and the saber thrusts immediately fall to the ground. A man could perhaps stand such humiliation, but sometimes he chooses not to, or is suddenly blinded with rage, and then the sea rises, arms are raised, hands grab reins or throw stones picked up from the ground or brought with them in their bags, it's the right of those who have no other weapons, and the stones come flying from the back of the crowd, probably without hurting anyone, horse or rider, because a stone hurled at random like that, if it was, simply drops to the ground. It was a battle scene worthy of a painting on the wall of the commander's office or in the officers' mess, the horses rearing up, the imperial guard, sabers unsheathed, striking with either the flat of the sword or the edge, the rebellious workers retreating then advancing like the tide, the wretches. This was the charge of June twenty-third, fix that date in your memory, children, although other dates also adorn the history of the latifundio and are deemed glorious for the same or similar reasons. The infantry also excelled itself, especially Sergeant Armamento, a man with a blind faith and a wrong-headed view of the law, there's the first burst of machine-gun fire, and another, both of them fired into the air as a warning, and when the people in the castle hear these shots, they cheer and clap, the sweet girls of the latifundio, faces scarlet with heat and bloodthirsty thoughts, and their mothers and fathers, and the boyfriends trembling with the desire to get out there themselves, lance in hand, and finish the job just started, Kill them all. The third burst of fire is aimed low, all that target shooting is proving its worth, let the smoke clear, not bad, although it could have been better, there are three men lying on the ground, one of whom is getting up, clutching his arm, he was lucky, another is dragging himself painfully along, one leg incapacitated, and that one over there isn't moving at all, It's José Adelino dos Santos, it's José Adelino, says someone from Montemor, who knows him. José Adelino dos Santos is dead, he got a bullet in the brain and couldn't believe it at first, but shook his head as if he had been bitten by an insect, then he understood, Those bastards have killed me, and he fell helplessly backward, with no wife there to help him, his own blood formed a cushion under his head, a red cushion, if you please. The people in the castle applaud again, they sense that this time it's serious, and the cavalry charges, scattering the crowd, someone should pick up the body, but no one approaches.

The people from Monte Lavre heard the whistle of the bullets, and José Medronho is bleeding from his face, he was lucky, it was just a graze, but he'll be scarred for the rest of his life. Gracinda Mau-Tempo is weeping, clinging to her husband, she heads off with other people down the narrow streets, how terrible, they can hear the triumphant cry of the guards as they make their arrests, and suddenly Leandro Leandres appears along with other dragons from the PIDE, a half dozen of them, João Mau-Tempo saw them and turned pale, and then he did something quite mad, he stood in the path of the enemy, trembling, but not with fear, ladies and gentlemen, let us be quite clear about that, but the other man either did not see him or did not recognize him, though those eyes are not easy to forget, and when the dragons had passed him by, João Mau-Tempo could no longer hold back his tears, tears of rage and deep sadness too, when will our suffering end. José Medronho's wound is no longer bleeding, no one would think that he had been within a millimeter of having the bones in his face shattered, what would he look like now if that had happened. Sigismundo Canastro is breathing hard, but the others are fine, and Gracinda Mau-Tempo is a girl again, sobbing, I saw him, he fell to the ground, dead, that's what she's saying, but some disagree, no, they say, he was taken to the hospital, although how we don't know, whether on a stretcher or in someone's arms, they wouldn't dare just drag him there even if they wanted to, Kill them all, comes the cry from the castle, however, one must respect the formalities, a man is not dead until a doctor says he is, and even then. Dr. Cordo is here, dressed in his white coat, let us hope his soul is of the same color, and as he is about to approach the body, Leandro Leandres blocks his path and says in a voice of urgent authority, Doctor, this man is wounded, he must be taken to Lisbon at once, and you must go with him for his greater safety. Those of us who have been listening to these stories from the latifundio are amazed to see the dragon Leandro Leandres taking pity on a victim and expressing a wish to save him, Take him, Doctor, an ambulance is already on its way, a car, quickly, there's no time to lose, the sooner he leaves here the better, hearing him talk like this, so urgently, so briskly, it's hard to believe what happened to João Mau-Tempo, or what he claims happened, when he was taken prisoner eight years earlier, that's how long ago it was, they obviously couldn't have treated him so very badly, apart from that statue business, the proof being that he came from Monte Lavre to take part in this demonstration, he clearly hasn't learned his lesson, he was lucky a bullet didn't find him.

Dr. Cordo goes over to José Adelino dos Santos and says, This man is dead, there is no denying such words, after all, a doctor studies for many years, and in that time he must have learned how to tell a dead man from a live one, however, Leandro Leandres was taught from a different primer and is, in his own way, a connoisseur of the living and the dead, and based on that knowledge and on his own self-interest, he insists, This man is wounded, Doctor, and must be taken to Lisbon, even a child would understand that these words are intended as a threat, but the doctor's soul clearly is as white as the coat he's wearing, and if it's stained with blood, that's because the soul has blood in it too, and he responds, I take the wounded to Lisbon, but I do not accompany the dead, and Leandro Leandres loses his temper and propels the doctor into an empty room, I'm warning you, if you don't take him, you'll pay for it, and the doctor answers, Do what you like, but I'm not taking a dead man to Lisbon, and having said that, he left the room to deal with the genuinely wounded, of which there were many, and some of whom went straight from there to prison, in fact, more than a hundred men, whether wounded or unscathed, were arrested, and if José Adelino dos Santos did end up being transported to Lisbon, it was simply a drama put on by the PIDE, a way of pretending that they had done everything they could to save him, a form of mockery really, and along with José Adelino dos Santos, they took other men arrested on that same day, and each one suffered as João Mau-Tempo suffered and as we have described.

The party from Monte Lavre escaped the patrols scouring and encircling the town, and all returned save one, António Mau-Tempo, who told his father, I'm going to stay here in Montemor, I'll be back tomorrow, and there was no point arguing with him, he replied to all their arguments with, Don't worry, I'm not in any danger, and though he had no clear idea of what he was going to do, he felt he had to stay, and then the others set off along ancient paths into the countryside, they're going to be tired out by the time they get home, although perhaps, farther on, when they rejoin the road, someone will come along and give them a lift to Monte Lavre, where the news of what happened has already spread, and oddly enough, when they did arrive, Faustina Mau-Tempo immediately heard their knock at the door and understood everything they said as if she had the keenest hearing in the world, even though she's deaf as a post, although some might hint that she sometimes only pretends to be deaf.

That night, which was again starry but moonless, while many women were grieving in Montemor, one woman more than all the others, there was a great uproar at the guards' barracks. More than once, patrols were dispatched to search the surrounding area, they entered houses, woke people up, in an attempt to solve the mystery of the stones or pebbles that kept falling onto the roof, some tiles had been broken and some windows too, constituting damage to public property, perhaps it was revenge on the part of the angels or mere mischief-making born of sheer boredom up there on heaven's balcony, because miracles shouldn't only involve restoring sight to the blind and giving new legs to the lame, a few well-aimed stones have their place in the secrets of the world and of religion, or so thinks António Mau-Tempo, because that's why he stayed behind, in order to perform that miracle, hidden away high up on the hill, in the pitch-black shadow of the castle, hurling the stones with his strong right arm, and whenever a patrol came by, he hid away in a cave from which he would later emerge as if from the dead, and fortunately no one spotted him. At around one in the morning, his arm grown weary, he threw one last stone and felt as sad as if he were about to die. A tired and hungry man, he went around the south side of the castle and down the hill, then spent the rest of the night walking the four leagues to Monte Lavre, following the road but keeping well away from it, like some malefactor afraid of his own conscience, occasionally having to go around the edge of some of the unharvested wheatfields blocking his path, because he couldn't risk walking through them and had to remain hidden from both the latifundio guards with their hunting rifles and the uniformed national guards armed with carbines.

When he was within sight of Monte Lavre, the sky was beginning to grow light, a glow so faint that only expert eyes would notice. He forded the stream, not wanting to be seen by anyone watching from the bridge, and then he followed the course of the stream, keeping close to the willows, until he reached a point where he could climb up the bank and into the village, taking great care in case any insomniac guards should still be out and about. And when he drew near the house, he saw what awaited him, a light, a lantern, like the lantern on a small fishing boat, where the mother of this boy of thirty-one was watching and waiting for him to return home, late from playing at throwing stones. António Mau-Tempo jumped over the fence and into the yard, he was safe now, but this time Faustina Mau-Tempo, absorbed in tears and dark thoughts, did not hear him arrive, but she did notice the sound of the door latch or perhaps felt a vibration that touched her soul, My son, and they embraced as if he had returned from performing great deeds in a war, and knowing herself to be hard of hearing, she did not wait for his questions, but said, as if she were reciting a rosary, Your father got home safe and so did Gracinda and your brother-in-law, and all the others, you were the only one who had me sick with worry, and António Mau-Tempo again embraced his mother, which is the best and most easily understood of answers. From the next room, still in darkness, João Mau-Tempo asks, and not in the voice of someone who has just woken up, You're back safe then, and António Mau-Tempo answers, Yes, Pa. And since it's nearly time to eat, Faustina Mau-Tempo lights the fire and puts the coffeepot on the trivet.

THE LATIFUNDIO IS AN inland sea. It has its shoals of tiny, edible fish, its barracuda and its deadly piranha, its pelagic fish, its leviathans and its gelatinous manta rays, blind creatures that drag their bellies along in the mud and die there too, as well as other great, strangling, serpentine monsters. It's a Mediterranean sea, but it has its tides and undertows, gentle currents that take time to complete the circuit, and occasional sudden churnings that shake the surface, provoked by winds that come from outside or by unexpected inflows of water, while in the dark depths the waves slowly roll, bringing with them nourishing ooze and slime, for how much longer, one wonders. Comparing the latifundio with the sea is as useful as it is useless, but it has the advantage of being easily understood, if we disturb the water here, the water all around will move, sometimes too far away to be seen, that is why we would be wrong to call this sea a swamp, and even if it were, it would still be a great mistake to believe in mere appearances, however dead this sea might seem to be.

Every day, the men get out of their beds, and every night, they lie down in them, and by beds we mean whatever serves them as a bed, every day, they sit down before their food or their desire to have enough food, every day, they light and extinguish a lantern, there is nothing new under the rose of the sun. This is the great sea of the latifundio, with its clouds of fish-sheep and predators, and if it was ever thus, why should it change, even if we accept that some changes are inevitable, all it needs is for the guards to remain vigilant, that's why every day the armed boats put out to sea with their nets intending to catch fishermen, Where did you get that bag of acorns, or that bundle of firewood, or What are you doing here at this hour, where have you come from, where are you going, a man cannot choose to step out of his usual rut, unless he has been employed to do so and is, therefore, being watched. However, each day brings some hope with its sorrow, or is that just laziness on the part of the narrator, who doubtless once read or heard these words somewhere and liked them, because if sorrow and hope come along together, then the sorrow will never end and the hope will only ever be just that and nothing more, this is what Father Agamedes would say, for he lives off sorrow and hope, and anyone who thinks differently is either mad or foolish. It would be nearer the truth to say that each day is the day it is, plus the day just gone, and that the two together make tomorrow, even a child should know such simple things, but there are those who try to divide up the days like someone cutting slices of melon rind to give to the pigs, the smaller the pieces, the greater the illusion of eternity, that's why pigs say, O god of pigs, when will we ever eat our fill.

This sea of the latifundio is subject to undertows, pounded by storms, lashed by waves, enough sometimes to knock down a wall or simply leap over it, as we understand happened in Peniche, and now you can see how right we were to mention the sea, because Peniche is both a fishing port and a prison-fortress, but still they escaped,* and that escape will be much discussed on the latifun-

---

* A reference to the dramatic escape in January 1960 of ten leading members of the Portuguese Communist Party being held in the high-security prison in Peniche.

dio, but what sea are we talking about, this land is usually as dry as dust, that's why men ask, When will we ever slake our thirst and the thirst of our parents, not to mention the thirst waiting under this stone for any children we might have. The news arrived and was impossible to hide, and there was always someone to fill in what the newspapers didn't say, let's sit down beneath this holm oak and I'll tell you what I know. It's time for the red kites to fly still higher, they cry out over the vast earth, anyone who can understand them will have much to tell, but for the moment we must make do with our human language. That's why Dona Clemência can say to Father Agamedes, The peace we never had is over, which may seem like a contradiction and yet this lady never spoke a truer word, these are new times and they're approaching very fast, It's like a stone rolling down a hill, that is what Father Agamedes says, because he prefers to use secondhand words, a habit acquired at the altar, but let us have enough evangelical charity to try and understand him, what he means is that if we don't get out of the way of the stone, God knows what will happen, and let us forgive him this new ruse, because it's quite clear that we don't need to wait for God in order to know what will happen to someone who fails to get out of the way of a rolling stone, which gathers no moss and spares no Lambertos.

And no sooner had this conversation ended, well, that's not quite true, because there were a few anxious months when negligence joined forces with sacrilege, because it was sheer negligence to allow those prisoners to escape and sacrilege to see a ship once named the *Santa Maria* sailing the seas under the new name of *Santa Liberdade*,* Dona Clemência is, of course, praying fervently

---

* A reference to the hijacking of the Portuguese liner *Santa Maria* by the DRIL (Directorio Revolucionário Ibérico de Liberación) in 1961. The "pirates" sailed the ship out into the Atlantic and renamed it the *Santa Liberdade*, gaining the attention of the world's press and hoping to undermine the Salazar and Franco dictatorships.

and passionately for the salvation of the church and the nation, at the same time demanding punishment for the ruffians, We wouldn't be in this situation if they had better examples to follow, you can't play with other people's lives, still less with my wealth. However, this is merely what the lady of the house says while safe within her four walls, always assuming Norberto is willing to listen to her, she would have no one to talk to if it wasn't for Father Agamedes, for she barely leaves the house now, or only rarely for a trip to Lisbon to see the latest fashions, or to Figueira for the traditional family holiday by the sea, and to be honest, her mind seems to be wandering, it must be her age, talking about her wealth and some ship sailing the sea, it's certainly not sailing on the inland sea of the latifundio, she must be going soft in the head, but there you'd be quite wrong, because she inherited shares in the colonial navigation company from her father, Alberto, God rest his soul, and that's what bothers her.

This bitter cold isn't just because it's January on the latifundio. All the windows are shut, and if this were Lamberto's castle rather than Norberto's palatial mansion, we would see armed men on the ramparts, just as, not so long ago, we saw fearful, bloodthirsty people filling the ruins of Montemor, the times are changing, platoons of guards patrol the latifundio, in their boots and on a war footing, while Norberto reads the newspapers and listens to the radio and shouts at the maids, that's what men do when they get upset. What really angers him is the air of sly contentment he sees on the faces of ordinary people, as if spring had arrived early for them, they don't seem to feel the cold, at least their contentment proved short-lived, for two days later they had to change their tune, God does not sleep and they will be punished, the *Santa Maria* has risen from the deep, pray for us, and let us not think too badly of Father Agamedes, who succumbed to the sin of envy, it was a long time coming in such a holy creature, and all because he couldn't hold a solemn Te Deum Laudamus as an act of thanksgiving, but that would not have gone down well in this wretched village of Monte Lavre with its godless inhabitants.

This is a bad year for the latifundio. There goes the maiden out for a ride on her fine steed, her skirt and saddle cloth flapping, her veil fashionably loose in the wind, the picture of composure, when suddenly the beast stumbles, for these are medieval roads, sir, and she falls flat on the ground, revealing all her most private penumbras, she doesn't seem too seriously hurt, poor love, the worst thing was the way the animal reared up and kicked as it scrambled to its feet. They say that pride goeth before a fall, which is a horsey version of the more melancholy dictum, Misfortunes never come singly, why, only yesterday those prisoners escaped from Peniche, bloody communists, baby eaters, have you seen my children, neighbor, only yesterday souls and seas were all stirred up by that new tale of pirates, we should shoot the lot of them, such a lovely ship too, all dressed in white, *Santa Maria* walking on the water like her divine son, and now there's news from Africa as well, about the blacks, Well, I always said we were too lenient with them, I said as much, but no one would believe me, you have to live there to know how to deal with them, they don't like work, you see, they're shirkers, they'll always go to the bad, and now you see the result, we treated them too kindly, as if they were Christians, but all is not lost, we won't lose Africa if we send in the army and have a proper war, remember Gugunhana,* brave words from the mayor, spoken quickly and boldly, he could have been a general if he'd had the military training, but at least he spoke out. The imperial dream soon faded, best to run away from the mess we made, the black man is now a Portuguese citizen, long live the black man who comes bear-

---

* Gugunhana, or Ngungunyane, was a tribal king of a territory in Mozambique. He rebelled against the Portuguese and was defeated by General Joaquim Mouzinho de Albuquerque in 1895. He lived the rest of his life in exile, first in Lisbon and then in the Azores, where he died in 1906.

ing no weapon, but keep your eye on him nonetheless, and down with the other sort, and one day, if we happen to wake up in a good mood, we'll declare that these overseas provinces, our former colonies, are now independent states, well, what's in a name, what matters is that the shit stays the same and that those who have eaten nothing else should continue to eat shit, whites and blacks, and anyone who can spot the difference wins a prize.

It would seem, Father Agamedes, that God and the Virgin have turned their benign eyes away from Portugal, look how discontented and restless people are, the devil has clearly taken hold of the gentle hearts of the Lusitanians, perhaps we didn't pray enough, the priests told us as much, and I've done what I can, and I've always been ready with good advice, both in the pulpit and in the confessional, this is, in fact, a dialogue, in which two people take turns to speak, but when Father Agamedes returns to his house, he is thinking something quite different, something more suited to a man of this time or of that other time when souls were conquered with the sword and with fire, What they need is a sound beating, that's telling them.

One really doesn't know where to turn, now it's the fortresses in India, weep, O souls of da Gama, Albuquerque, Almeida and Noronha,* no, that's all we need, for grown men to weep, we must hold out to the last man, we will show the world what we Portuguese are worth, anyone who retreats is betraying the nation, better cut the shoe than pinch the foot, the government calls on everyone to do his duty. It's a sad Christmas in Alberto's house, not that there is any shortage of food or of the Lord's blessings, at least it was a good year for cork, so that's something, but there are black clouds with thunder in their bellies gathering over the country and over the latifundio, what will become of Portugal and of us, true, we have someone to protect us, for a start, there are the guards, to

---

* All are names of Portuguese viceroys of India in the fifteenth and sixteenth centuries.

each of whom we give a gift, to the captain, lieutenant, sergeant and corporal, poor things, it's only right, they earn so little and are always so ready to defend our property, imagine if we had to pay them out of our own pockets, it would cost us a fortune. It's just as well, now that the last vestiges of a Portuguese presence in the East are being removed, along with our soldiers and sailors, that we never really took much interest in Goa, Daman and Diu anyway, a gift, you say, what an idea, I don't mean that kind of gift, we've already mentioned the ones we gave to the captain, the lieutenant, the sergeant and the corporal, each of whom either came to fetch his own or, out of discretion and a desire to avoid prattling tongues, had it brought to him, no, this is a different kind of gift, that given by the soldiers and sailors who, on the point of death, raise themselves up on their elbow and, dying, cry out in response to the roll call, absent, an ancient practice, for when necessary even the dead can vote. The other good thing is that all this is happening a long way off, India and Africa are not exactly close, the fires are burning far from my borders, the sea, lots of sea, separates us from them, they won't come over here, and Portugal won't lack for sons to defend the latifundio from afar, don't bite the hand that feeds you, you've been warned.

Tomorrow, said Dona Clemência to her children, and her nieces and nephews, is New Year's Day, or so she had gleaned from the calendar, placing her hopes in the brand-new year and sending her best wishes to all the Portuguese people, well, that isn't quite what she said, Dona Clemência has always spoken rather differently, but she's learning, we all choose our own teachers, and while these words are still hanging in the air, news comes that there has been an attack on the barracks of the third infantry regiment in Beja, now Beja is not in India or Angola or Guinea-Bissau, it's right next door, it's on the latifundio, and the dogs are outside barking, though the coup was put down, they will speak of little else over the next weeks and months, so how was it possible for a barracks to be attacked, all it took was a little luck, that's all it ever takes, perhaps that's what was lacking the first time around, and no one noticed, that's our fate, if the horse carrying the messenger bearing orders to commence battle loses a shoe, the whole course of history is turned upside down in favor of our enemies, who will triumph, what bad luck. And in saying this we are not being disrespectful to those who left the peace and safety of their homes and set off to try and pull down the pillars of the latifundio, though Samson and everyone else might die in the attempt, and when the dust had settled and we went and looked, we found that it was Samson who had died and not the pillars, perhaps we should have sat down under this holm oak and taken turns telling each other the thoughts we had in our head and heart, because there is nothing worse than distrust, it was good that they hijacked the *Santa Maria*, and the attack in Beja was good too, but no one came to ask us latifundio dogs and ants if either the ship or the attack had anything to do with us, We really value what you're doing, though we don't know who you are, but since we are just dogs and ants, what will we say tomorrow when we all bark together and you pay as little heed to us as did the owners of this latifundio you want to surround, sink and destroy. It's time we all barked together and bit deep, captain general, and meanwhile check to see that your horse doesn't have a shoe missing or that you have only three bullets when you should have four.

THESE MEN AND WOMEN were born to work, like good to average livestock, they leave or are dragged out of their mother's womb, left to grow up one way or another, it doesn't really matter, what matters is that they should be strong and good with their hands, even if they can make only one gesture, so what if, within a few years, they become stiff and heavy, they are walking logs who, when they arrive at work, give themselves a shake and produce from their rigid bodies two arms and two legs that move back and forth, you see how kind and competent the Creator was in making such perfect instruments for digging, scything, hoeing and generally making themselves useful.

Since they were born to work, it would be a contradiction in terms for them to have too much rest. The best machine is the one most capable of continuous work, properly lubricated so that it doesn't jam up, frugally fed and, if possible, given only as much fuel as mere maintenance requires, and, in case of breakdown or old age, it must, above all, be easily replaceable, that's what those human scrap yards, cemeteries, are for, or else the machine simply sits, rusting and creaking at its front door, watching nothing at all pass by or gazing down at its own sad hands, who would have thought it would come to this. On the latifundio, generally speaking, men and women have short lives, it's astonishing that any of them ever reach old age, but when we happen to pass some apparently old man, we learn that he is only forty, and that the shrunken woman with the leathery face is not yet thirty, so living in the country doesn't exactly extend your life, that's an urban myth, as is that most sensible of sayings, Early to bed and early to rise makes a man healthy, wealthy and wise, it would be amusing to see those same urbanites standing with one hand on the handle of their hoe, staring at the horizon, waiting for the sun to come up and, utterly exhausted, longing for a dusk that never comes, because the sun is an awkward so-and-so, always in such a hurry to rise and always so reluctant to set. Just like us.

However, the days of acceptance and resignation are coming to an end. A voice is traveling the roads of the latifundio, it goes into towns and villages, it talks on the hillsides and on cork plantations, a voice that consists of two essential words and many others that serve to explain those two words, eight hours, this may not appear to mean very much, but if we say eight hours of work, then the meaning becomes clearer, there are sure to be those who protest at this scandalous idea, what is it these workers want, if they sleep eight hours and work another eight, what will they do with the remaining eight hours, it's an invitation to idleness, they clearly don't want to work, these are modern ideas, it's all the fault of the war, customs have changed out of all recognition, first they stole India from us, now they want to take Africa away, then there was that ship that sailed the seas causing an international scandal, and the general who rose up against those who gave him his stars, who can one trust, tell me that, and now there's this disastrous business of the eight-hour day, they should have stuck to the law of God, give or take an hour, twelve hours of daylight and twelve hours of night, depending on when the sun rises and sets, of course, and if that isn't God's law, then let's say it's the law of nature, which must be obeyed.

The voice roaming the latifundio may not hear these mutterings, and if it does, it ignores them, these are old-fashioned ideas from the days of Lamberto, Work keeps them busy, if they weren't working, they'd be getting drunk in bars and then going home to beat their wives, poor things. Don't go thinking these are easy paths to follow. This voice has been pounding roads and streets for a whole year now, eight hours, eight hours of work, and some don't believe it, some believe that this will happen only when the world is about to end and the latifundio wants to save its soul and be able to appear at the final judgment and say to the angels and archangels, I took pity on my serfs, they were working far too many hours, and for the love of God, I ordered them to work only eight hours a day, with a rest on Sunday, and because I did this, I expect nothing less than a place in paradise at God's right hand. That is what some skeptics think, afraid that it will be a change for the worse. But the carriers of the voice did not rest all year, they traveled the whole latifundio proclaiming those words, while the guards and the PIDE agents twitched their ears uneasily, the way donkeys do when tormented by flies. Then they unleash furious, martial patrols, all that's missing are the bugles and drums, and they would have loved that, but it didn't fit in with the battle plan, imagine if the conspirators were gathered together on some lonely hillside or deep in the woods and they heard the trumpets blaring in the distance, tantararatantan, we'd never catch anyone. The guards were given reinforcements, so were the PIDE agents, and any village without a doctor was given the medicine of twenty or thirty guards and accompanying weapons, and these guards were, of course, in permanent communication with the dragons defending the State and who don't like me at all, pity the real dragons, they're as ugly as toads and lizards, but they don't do any real harm, the proof of which is that paradise is full to bursting with fire-breathing dragons.* And given that guards tend to be astute rascals, they invented the subtle art of placing a pamphlet beneath a stone, yet visible enough for a blind man to see, the kind of pamphlet left by those commies who travel around the latifundio saying subversive things about eight-hour days and so forth, they might as well hand over the country to Moscow right now. Anyway, having laid this trap, they hide behind a hedge or in a hollow or behind an innocent tree or boulder, and when some unsuspecting man comes along, he perhaps picks up the pamphlet and puts it in his pocket or inside his hat or between skin and shirt, it's one of those white sheets covered in small black lettering, not only does he not read very well, his eyesight is poor too, anyway, he hasn't gone ten steps when the guards ambush him on the path, Halt, show us what you've got in your pockets, if this doesn't strike you as a show of great astuteness, then all we can say is that there is clearly a lot of ill feeling against the guards, who deserve only praise for their expert application of the principles of hypocrisy and petty mendacity, rammed into them at the same time as they were being taught how to use a gun and organize an ambush.

Surrounded by carbines, the finder of the pamphlet has no choice but to empty his pockets of a gypsy knife, half an ounce of tobacco, a book of cigarette papers, a piece of string, a gnawed crust of bread and ten tostões, but this doesn't satisfy the guards, who have other ambitions, Take another look, it's for your own good, you might get hurt if we were to frisk you, and then, from between skin and shirt, he produces the pamphlet, already damp

---

* Presumably a reference to Salazar's entirely cynical view of rural life as some kind of idyll or paradise (for example, there were often competitions to find the most Portuguese village in Portugal). This paradise may not have a doctor, but there will always be a dragon or two in the form of a PIDE agent.

with sweat, not that it's so very hot, but the poor man isn't made of steel, marooned as he is amid these guffawing guards, things are getting serious, Corporal Tacabo, or some private temporarily promoted to lead the patrol, knows very well what the pamphlet is, but he pretends to be surprised and examines it carefully, before saying slyly, Now you're in for it, we've caught you carrying communist propaganda, we're taking you to the barracks, it's Montemor or Lisbon for you, my lad, I certainly wouldn't want to be in your shoes. And when the man tries to explain that he has just found the pamphlet a moment ago, that he hasn't had time to read it, that he doesn't know how to read, that he happened to be passing by, saw the pamphlet and, out of natural curiosity, picked it up as anyone would, but he doesn't finish what he has to say, because he receives a blow to the chest or the back with the butt of a rifle, or else a kick, get a move on or I'll shoot you, arms are my theme and these matchless heroes.*

Talking is like eating cherries from a bowl, you take hold of one word and others immediately follow, or perhaps they're like ticks, which are equally hard to disentangle if they're attached one to the other, because words never come singly, even the word loneliness needs the person who's feeling lonely, which is just as well, I suppose. These guards are so steadfast and loyal that they go wherever the latifundio sends them, they never question, never argue, they are mere minions, you only have to consider what happened on May Day, when men and women duly celebrated the day of the worker, but when they returned to their labors the following day, the guards were waiting for them, Only those who didn't miss work yesterday can work today, those are our orders, although there was little point in saying this, since everyone had missed

---

* An ironic reference to the first line of Luís Vaz de Camões's poem *The Lusiads*, Portugal's great epic, published in 1572. It glorifies the great Portuguese navigators, who set sail to discover new worlds. It is, of course, also an echo of the opening line of *The Aeneid*.

work. What's going to happen now, the workers draw back, how are they going to resolve this, and because the guards had occupied the terrain, and the overseer was hiding behind them rather than taking his due part in the negotiations, the workers decided to go back to their houses, it was still early in the morning, you see, and enjoy another day's holiday, and the guards stayed behind to keep an eye on the ants, who were going about their business and raising their heads in surprise like dogs. But before the workers left, the sergeant, standing next to the overseer or foreman or manager or whatever, made intelligent use of his interrogation methods, Why didn't you come to work yesterday, Because it was the first of May, the day of the worker, and we're workers. It was an innocent enough reply, there they were, standing before me, Corporal, looking at me with grave eyes, thinking they could deceive me, as if I would be so easily taken in, that's what these shameless wretches do, they look at you gravely like that and you can't tell what they're really thinking, but I gave it to them straight, I know how to deal with them, I said, you'd better tell me the truth, you can't fool me, the reason you didn't come to work was political, but they said, No, sir, it wasn't political, the first of May is the day of the worker, and when they said that, I gave a little mocking laugh, And what would you know about that, and someone at the back, unfortunately I couldn't see his face, said, It's the same all over the world, and that, as you can imagine, really got my goat, This isn't the world, this is Portugal and the Alentejo, we have our own laws, and at this point, the foreman whispered a secret to me, well, it wasn't a secret exactly, it was simply what we'd already agreed I would say, and I declared, with all the authority with which I had been invested, Only those who didn't miss work yesterday can work today, and as soon as I said this, they all moved away, all together as they usually do, it's the same when they sing, and off they went back home, their hoes on their shoulders, because it was hoeing they had come to do, and I couldn't help but feel a certain respect for them really, although I'm not sure why. Words are like ticks, or like the cherries that ripen in May, and if respect is not the final word, it is at least the right one.

April is the month of a thousand words.* Meetings are held at night in the fields, the men can barely see each other's faces, but they can hear each other's voices, slightly muffled if the place isn't deemed particularly safe, or louder and clearer if they're in the middle of nowhere, but they always keep sentinels posted, in accordance with the strategic art of prevention, as if they were defending an encampment. On their side, they are waging a peaceful war. The guards don't come in pairs anymore, but in dozens or half dozens, and when the roads allow, they arrive in jeeps or trucks, or they advance in a line, like beaters, so if in the dark of night the guards are heard approaching, the workers' sentinels draw back to give the alarm, and either the guards pass right by, in which case silence is the best defense, with every man seated or standing, holding both breath and thoughts, turned suddenly to stone like ancient megaliths, or the guards head straight for them, and the order then is to scatter along the beaten tracks, fortunately the guards don't yet have dogs.

The following night, they will pick up the conversation where they left off, in that same place or somewhere else, their patience is infinite. And when they can, they meet by day as well, in smaller groups, or go to someone's house and talk by the fire while the women silently wash the dishes and the children sleep in the corner of the room. And if one man happens to be standing next to another on the threshing floor, each word spoken and heard is like a mallet striking a stake, driving it a little further in, and when, in the fields, it's time to eat, they sit on the ground with their lunch

---

* The expression in Portuguese is "em Abril aguas mil" — literally, "in April a thousand waters" — because April is traditionally a rainy month. This is Saramago's version of that familiar expression.

pails between their legs, and while the spoon rises and falls and the cool breeze chills their body, their words return to the same theme, and they say slowly, Let's demand an eight-hour day, enough of working from dawn to dusk, and the more prudent among them speak fearfully of the future, What will happen if the bosses refuse to give us work, and the women washing the dishes after supper, while the fire burns, feel ashamed of their husbands' caution and agree with the friend who knocked on the door to say, Let's demand an eight-hour day, enough of working from dawn to dusk, because that is how long the women work too, except they often do so when in pain or menstruating or heavily pregnant, or with their breasts overflowing with the milk that should have been suckled, they're lucky it hasn't dried up, so those who believe that all one has to do is raise a banner and say, Right, let's go, are much mistaken. Yes, April has to be the month of a thousand words, because even those who are certain and convinced have their moments of doubt, of soul-searching and despair, there are the guards, the dragons of the PIDE and the black shadow that spreads over the latifundio and never leaves, there is no work, and are we, with our own hands, going to shake the sleeping beast awake and say, Tomorrow we will only work for eight hours, this is not the first of May, the first of May is the least of it, no one can force me to go to work, but if I were to say, Eight hours and no more, it would be like baiting a rabid dog. And the friend says, sitting here on the cork plantation or by my side on the threshing floor or in the night so dark I can't see his face, he says, It's not just about working eight hours, we're going to demand a minimum wage of forty escudos too, if we don't want to die of exhaustion and hunger, these are fine things to ask for and to do, but difficult to achieve. It's good that there are a lot of conversations and a lot of voices, but out of the meeting comes one voice, and this isn't just a manner of speaking, it's true, some voices stand up on their own two feet, What kind of life have we led, tell me that, in the last couple of years, two of my children have died of hunger, and the one child I have left will be brought up to be a beast of burden, and I don't want to carry on being the beast of burden I am, these are words that might wound delicate ears, but there are no such ears here, although no one present at this meeting likes to look in the mirror and see himself stuck between the shafts of a cart or wearing a saddle and a yoke, That's how it's been ever since we were born.

Then another voice emerges, and on the shadow of the night falls another shadow come from who knows where, what can he be thinking of, he's not talking about the eight-hour day or about the forty-escudo minimum wage, that's what we came here to discuss, but no one has the heart to interrupt him, They have always done their best to strip us of our dignity, and everyone knows who he means by they, they are the guards, the PIDE, the latifundio and its owners Alberto or Dagoberto, the dragon and the captain, gnawing hunger and broken bones, anxiety and hernias, They have always done their best to strip us of our dignity, but it can't go on, it must stop, listen to what happened to me and to my father, now dead, it was a secret between us, but I can stay silent no longer, if my story doesn't convince you, then we are lost, there's nothing more to be done, once, many years ago, it was a dark night just like tonight, my father went with me and I with him to pick acorns, because there was nothing to eat at home, and I was already a young man and thinking about getting married, we had a bag with us, just an ordinary bag, and we went together to keep each other company, not because of the heavy load, and when the bag was nearly full, the guards appeared, I'm sure the same thing has happened to many of you here tonight, it's nothing to be ashamed of, picking up acorns from the ground isn't stealing, and even if it was, hunger is a good enough reason to steal, he who steals out of hunger will find forgiveness in heaven, I know that isn't quite how the saying goes, but it should, and if I'm a thief because I stole some acorns, then so is the owner of the acorns, who neither made the

earth nor planted the tree nor tended it, anyway, the guards arrived and said, well, there's no point repeating what they said, I
can't even remember, but they called us every name under the sun,
how have we put up all these years with being called such names,
and when my father begged them for the love of God to let us take
the acorns we had picked, they started laughing and said we could
keep the acorns, but on one condition, and do you know what that
was, they wanted us to fight each other and to let them watch, and
my father said he wasn't going to fight with his own son, and I said
the same, that I wouldn't fight with my father, but they said, in that
case, we would be taken to the barracks, where we would have to
pay a fine and possibly get a beating too, just to teach us some manners, and then my father said, all right, we would fight, but please,
comrades, don't think ill of that poor old man, now dead, and God
forgive me if, in telling you this story, I'm dragging him from his
grave, but we were starving, you see, anyway, my father pretended
to give me a shove, and I pretended he had pushed me over, we
wanted to see if we could fool them, but they said that if we didn't
fight properly, with a real intent to hurt each other, they would arrest us, I don't have words to describe what happened next, my
father became desperate, I saw it in his eyes, and he hit me and
really hurt me, but not because he had hit me that hard, and I responded in kind, and a minute later, we were rolling around on the
ground, and the guards were laughing like mad things, and, once,
I touched my father's face and it felt wet, but not with sweat, and I
was filled with rage, and I grabbed him by the shoulders and shook
him as if he were my worst enemy, and he, underneath me, kept
punching me on the chest, God, what a state to be in, and still the
guards were laughing, it was a dark night like tonight and so cold
that it ate into your bones, all around lay the countryside, and yet
the stones did not rise up, is this what men were born to do, and
when finally we stopped fighting, we were alone, the guards had
left, doubtless in sheer disgust, which was what we deserved, and

then my father started crying and I rocked him in my arms as if he were a child and swore I would never tell anyone, but I can no longer remain silent, it's not just a matter of eight hours or forty escudos, we must do something now if we are not to lose ourselves, because that isn't a life, two men fighting each other, father and son or whoever, purely to amuse the guards, it's not enough that they have weapons and we have none, we are not men if we do not now raise ourselves up from the ground, and I say this not for my own sake but for the sake of my dead father, who won't ever have another life, poor man, only the memory of me beating him and the guards laughing, as if they were drunk, if there was a God, surely he would have intervened then. When the voice stopped speaking, everyone stood up, there was no need to say anything more, each man set off to follow his own destiny, determined to be there on the first of May, determined to hold out for the eight-hour day and the wage of forty escudos, and even today, after all these years, no one knows which of them it was who fought with his own father, our eyes cannot bear the sight of too much suffering.

From hillside to woodland, these and other words did the rounds of the latifundio, although no one ever mentioned that father-son fight, because no one would believe it, and yet it was true, and meetings were arranged in Monte Lavre too, some people were afraid, but others were not, and so when the first of May arrived, everyone was ready, and those who felt afraid stuck fast to those who showed no fear, that's how it is even in time of war, said someone who had been in the war, although whether as one of the brave or of the timorous we don't know. A lot of gasoline and diesel was consumed that day, the spring air was full of fumes from the endless stream of jeeps and trucks laden with rifles and masked guards, they wear masks so as not to feel ashamed, and when they reached some town or village where there was a barracks, they would stop for a conference with the general staff, exchange orders and discuss the situation, how are things over in Setúbal, and in Baixo Alentejo and in Alto Alentejo, and in Ribatejo, which, don't forget, is also the latifundio. Armed patrols roamed the main streets and side streets, hoping to sniff out subversion, and from high vantage points they surveyed the inland sea like fish eagles, to see if they could spot the black or red flag of a pirate ship, as if anyone were going to run up such a flag, but the guards are obsessed, they can think of nothing else, and what they saw was perfectly innocuous, men strolling up and down in the squares, talking, all dressed in their skillfully darned and patched Sunday best, because the women of the latifundio are experts at patching the seats and knees of trousers, you should see them rooting around in the rag basket in search of just the right scrap of fabric, then placing it on the offending trouser leg before carefully cutting the fabric to size and sewing it on, it's a job requiring great precision, I'm sitting on the step outside my front door patching my husband's trousers, well, he can't go to work naked, it's enough that he's naked between the sheets.

Some will think this has nothing to do with the first of May and the eight-hour day and the forty escudos, but they are people who pay little attention to what goes on in the world, they think the world is this sphere rolling through space, pure astronomy, they might as well be blind, for there is nothing more closely connected to the first of May than this needle and this thread in the hand of this woman called Gracinda Mau-Tempo, who is patching these trousers so that her husband Manuel Espada can celebrate the first of May, the day of the worker. The guards pass right by the front door in a military-looking jeep, and Gracinda Mau-Tempo draws her only daughter, Maria Adelaide, closer to her, and the girl, who is seven and has the bluest eyes in the world, watches the jeep pass, these children seem singularly unimpressed by the sight of a uniform, there she is with her stern gaze, she has seen enough of life already to know who these guards are and what that uniform means.

After dark, the men return home. They will spend a restless night, like soldiers on the eve of battle, who knows who will return alive, strikes and demonstrations are one thing, they're used to that and know how bosses and guards usually respond, whereas this is more of a challenge, denying the latifundio a power that has been passed down to them from their great-great-great-grandparents, You will work for me from dawn to dusk all the days of your life, in accordance with my wishes and my needs, on the other days you can do as you please. From now on, Sigismundo Canastro won't need to get up so early, nor will João Mau-Tempo or António Mau-Tempo or Manuel Espada, nor any of the other men and women, who are still awake, thinking about what will happen tomorrow, it's a revolution, an eight-hour day on the latifundio, It's a gamble, win or lose, in Montargil they won, and we can't be seen to be less than them, in the middle of the night, they hear the guards' jeep prowling the streets of Monte Lavre, they want to frighten us, but they'll see.

These words are spoken by other mouths as well, those of Gilberto and Alberto, They'll see, and it was a great moment in the history of the latifundio, for even the owners of the land got up early to be present at the dawning of the day, if you don't look after what's yours, the devil will take it, the sun is up and not a single devil has turned up for work, overseer, foreman and manager are nervous, but the countryside is a balm to the eyes, May, glorious May, and Norberto consults his watch, half past seven, still no one, This has all the look of a strike, says a lackey, but Adalberto responds angrily, Shut up, he is furious, he knows what he intends to do, they all know, it's just a matter of waiting. And then the men begin to arrive, all together at the agreed hour, they politely say Good morning, why be bitter, and when it's eight o'clock, they start work, this is what they had decided to do, but Dagoberto bawls, Stop, and they all stop and look at him with innocent eyes, What is it, sir, such sang-froid is enough to drive a man mad. Who told you to come to work at this hour, Norberto asks, and it is Manuel Espada who speaks for the other workers, We did, on some estates they're already working an eight-hour day, and we are no less than our comrades on those other estates, and Berto strides over to him as if he were about to hit him, but he doesn't, he wouldn't go that far, On my land, the timetable is the same as it has always been, from dawn to dusk, it's up to you, you either stay and tomorrow make up for the time lost this morning, or you leave, because I don't want you here, That's telling them, Dona Clemência will say later, when her husband boasts of his deeds, and then what happened, Then, Manuel Espada, who is married to Mau-Tempo's daughter, he was the group's spokesman, said, Fine, we'll leave, and they all left, and when they were walking back up to Monte Lavre, António Mau-Tempo asked, What next, what do we do now, not because he was worried or afraid, his question was intended to help his brother-in-law, Now we do as we agreed, we gather together in the square, and if the guards turn up looking for trouble, we go home, and tomorrow we go back to work, we start scything at eight o'clock, like today, that, more or less, was what João Mau-Tempo said to another group of laborers, and Sigismundo Canastro to his group, and so they all gathered in the square, and the guards turned up, and Corporal Tacabo came over to them, So you don't want to work, then, We do, but only for eight hours, and the boss doesn't want to give us those eight hours of work, Sigismundo Canastro is speaking the honest truth, but the corporal wants to know more, So this isn't a strike, No, we want to work, but the boss sent us away, he says we can't work for just eight hours, and that clear response will cause Corporal Tacabo to say later on, I don't know what to do with them, Senhor Dagoberto, the men say they want to work, and that it's you who, but before he can finish his sentence, Dagoberto roars, They're idlers, that's what they are, either they work from dawn to dusk or they can die of hunger, there's no work for them here, as far as I know the gov-

336 • *José Saramago*

ernment has issued no edict regarding an eight-hour day, and even if they have, I'm in charge here, I own the land, and that was the end of his conversation with Corporal Tacabo, and so the day concluded, with each man going home to his house, and the women wanting to know what had happened, as we saw with Dona Clemência, and as is the right of the other women too.

The men do their calculations, they have earned no money today, and how many more days like this will there be, it depends on the place, elsewhere, the latifundio gave in after two days, in others three, in others four, and in some places they spent weeks embroiled in this tug of war, to see whose strength or patience would win out, in the end, the men didn't bother turning up for work to find out if their conditions would be accepted, they stayed in the towns and villages, on strike, and this was all that was needed for the guards to return to their old ways, beating the workers and patrolling the latifundio on a war footing, but why repeat what everyone knows. Dagoberto and Alberto, Humberto and another Berto held out in their castles, however, the sacred alliance was beginning to unravel, and from other places came news of surrender, What should we do, Oh, leave them to it, they'll pay for it in the end, Yes, I know, Father Agamedes, such vengeful thoughts are most unchristian, and I'll do penance for it later, Well, it's not quite that clear-cut, Senhor Alberto, in Deuteronomy the Lord says, Vengeance is mine, and I will repay, Father Agamedes is a real fount of knowledge, how is it that from a book as big as the bible he managed to glean that one vital passage, what further justification do we need.

Here in Monte Lavre, though, they were fortunate in that the shopkeepers were willing to extend their credit, and in other places too, but this story is of particular interest to us, because João Mau-Tempo has had to walk these streets filled with the shame of owing money he could not pay back, with his wife Faustina weeping in misery and grief, and now he is going from shop to shop to pass on the message, and when he is received rudely, he pretends to feel nothing, suffering has given him a thick skin, he is not dealing here only with his own needs, Senhora Graniza, we are engaged in a struggle to gain the right to work an eight-hour day and the bosses refuse to agree to this, which is why we're on strike, I've come to ask if you could wait another three or four weeks, and as soon as we return to work, we'll start repaying what we owe, no one will be left owing you anything, it's a very big favor we're asking you, and the owner of that shop, a tall woman with pale eyes and a dark gaze, places her hands on the counter and says, respectfully, as a younger person to an older one, Senhor João Mau-Tempo, as surely as I hope that you will one day remember me, my house stands open, and these sibylline words are characteristic of the woman, who holds long mystical and political conversations with her customers and recounts tales and instances of miraculous cures and intercessions, well, all kinds of things happen on the latifundio, not just in the cities. João Mau-Tempo left with this good news, and Maria Graniza prepared a new slate, let's hope they all repay their debts, for they owe her twice over.

The birds of dawn wake up and see no one working. The lark says, How the world has changed. But the red kite, soaring high above, cries out that the world has changed far more than the lark suspects, and not just because the men are working only eight hours now, as the ants know, for they have seen many things and have good memories, which is hardly surprising, since they're always together. What do you say to this, Father Agamedes, I really don't know what to say, Senhora Dona Clemência, apart from farewell to a world that's going from bad to worse.

JOÃO MAU-TEMPO IS in bed. Today will be the day of his death. The illnesses that poor people die from are almost always indefinable, so much so that doctors find it extremely hard to fill in the death certificate unless they drastically simplify things, generally people die of some obscure pain or in childbirth, but how to translate this into clear nosological terms, all those years of studying count for nothing. João Mau-Tempo was in the Montemor hospital for two months, although this did him little good, not that this was the fault of the care he received, some cases are beyond salvation, and in the end, he was brought back home to die, and while his death here will be much the same, it will, at least, be quieter, there'll be the smell of his own bed, the voices of people passing by in the street, the sounds made by the poultry at dusk, when the chickens go to their roosting places and the cockerel vigorously shakes its wings, who knows, he might miss these things in the next world. During the time that João Mau-Tempo languished in the hospital, he lay awake all night listening to the sighs and moans and sufferings of the ward, and fell asleep only toward dawn. He doesn't sleep much better at home, but at least he has just his own pain to worry about now, and that is something to be resolved as a matter of confidence between his body and the spirit that still sustains it, with only his family as witness, and although one day their time will come, for they will not be left unharvested, even they will not be able to understand what it means to be a man alone with his own death, knowing, without anyone having to tell him, that today is the day. These are certainties that come into the mind when one wakes very early in the morning to hear the rain falling and dripping from the eaves like the threads of water from a spring, as children we used to perch on the lintel and, leaning against the door frame, hold out our hand to catch the drips, that's what João used to do and others who are not João. Faustina sleeps on top of the chest, at her insistence, so that her husband can have the double bed to himself, and there is no danger that she will forget her duties, you can see her eyes shining in the night, catching either the gleam of the dying fire or the glow from the oil lamp, perhaps her eyes shine so brightly as a compensation for being deaf. But if she falls asleep and João Mau-Tempo's pain becomes such that he cannot bear it alone, there is a piece of string linking his right wrist to his wife's left wrist, having reached a certain age, they are not going to be separated now, he only has to give the string a tug and Faustina will wake from her lightest of sleeps, get up fully dressed, go over to the bed and in the great silence of her deafness take her husband's hand in hers and, unable to do anything more, say a few comforting words to him, not everyone can boast of being able to do so much.

It isn't Sunday today, but in this rain, with the fields waterlogged, no one can go to work. João Mau-Tempo will have all of his small family around him, apart from those who live far away and cannot come, his sister Maria da Conceição, who still works as a maid in Lisbon, still with the same employers, for such examples of loyalty do exist, give them some gold dust and, when you come back, you'll find it all still there and possibly more besides, and his brother Anselmo, who went to live up north and was never heard of again, perhaps he's dead, perhaps he's gone on ahead, like Domingos in whatever year it was he died, who remembers now and who cares. Some lives are erased more completely than others, but that's because we have so many things to think about, we end up not noticing those lives until there comes a day when we regret our neglect, I was wrong, we say, I should have paid more attention, exactly, if only we'd had those feelings earlier, but these are merely twinges of remorse that arise and, fortunately, are almost immediately forgotten. His daughter Amélia will not be there either, as we know, she has worked as a maid in a house in Montemor ever since she was a girl, she was lucky, though, to have been able to visit him in the hospital and keep him company, and she has been able to save enough money to buy false teeth, her one little luxury, alas, her smile came too late to save her. Some friends will be missing too, Tomás Espada, who long withstood the absence of his wife Flor Martinha, no one ever saw their wrists bound together by a piece of string, but then some things that are invisible nevertheless exist, perhaps the people themselves would be unable to explain how, but Sigismundo Canastro, the oldest friend of all, will come, and Joana Canastra will help as much as she can, if only to console Faustina, for they have known each other so long they do not even need to speak, but will simply exchange a look, with no tears shed, because Faustina won't be able to cry and Joana never has, these are mysteries of nature, who can say why it is that this woman can't weep and the other doesn't know how.

António Mau-Tempo, my son, will be here too, he has just got up and is still barefoot, How are you feeling, Father, and I, who know that today is the day of my death, answer, Fine, perhaps he'll believe me, he's leaning on the frame at the foot of the bed, looking at me, he obviously doesn't believe me, you can't convince someone of something if you don't believe it yourself, he's still a long way from fifty, but France really finished him off, as everything does in the end, this pain, this pang, or perhaps it isn't the pang of pain itself but some underlying ache, even I don't know. And my son-in-law Manuel Espada will come, and my daughter Gracinda, they will both be here at my bedside, beside this bed from which I will be carried out, probably by Manuel and António, because they have more strength, but the women will wash me, that's usually women's work, to wash the corpse, ah, the things women have to do, at least I won't hear them crying. And there'll be my granddaughter Maria Adelaide, who has the same blue eyes as me, well, not quite, why am I boasting, my eyes are like dull ashes compared to hers, perhaps when I was younger, when I used to go to dances and was courting Faustina, when I stole her from her parents' house, then my eyes must have been as blue as those that have just walked into the room, Your blessing, Grandfather, how are you feeling, better, I hope, and I make a gesture with my hand, that's all that remains of blessings, none of us believes in them, but it's the custom, and I answer that I'm feeling fine and turn my head toward her so as to see her better, Ah, Maria Adelaide, my granddaughter, although I don't say those words, I think them, it does me good to see her, she's wearing a scarf on her head and a little knitted jacket, her skirt is wet, the umbrella didn't protect her entirely from the rain, and suddenly I feel a terrible urge to weep, because Maria Adelaide took my hand in hers, it was as if we had exchanged eyes, what a daft idea, but a man who is about to die can have whatever ideas he likes, that's his prerogative, he's not going to have many more opportunities to have new ideas or repeat old ones, I wonder what time I will die. And now Faustina is coming over with my bowl of milk, she's going to spoon-feed it to me, I might as well stay hungry today, I would leave the world more lightly, and someone else could drink the milk, what I would really like is for my granddaughter to feed me, but I can't ask that, I can't upset Faustina on my last day, who would console her afterward, when she said, Ah, my dear husband, I didn't give him his milk to drink on the day he died, the grandmother might resent the granddaughter for the rest of her days, perhaps in a little while Maria Adelaide can give me my medicine, according to the doctor's instructions, half an hour after eating, but these are impossible desires. Maria Adelaide is leaving, she just looked in to see how I am, and I'm fine, her father and mother will be here soon, but she's gone already, she's still too young to be a witness to such spectacles, she's only seventeen and has the same blue eyes as me, or have I said that already.

When João Mau-Tempo wakes from the torpor into which he slipped after taking his medicine, which was a real boon, affording him a prolonged respite from the pain and allowing him to sink into what seemed like a natural sleep, but now the pain has returned, and he wakes up moaning, it's like a stake piercing his side, when he recovers full consciousness, he finds himself surrounded by people, there isn't room for anyone else, Faustina and Gracinda are bending over him, Amélia too, so she did come after all, it was the moan that summoned them, and Joana Canastra is standing farther back because she's not a family member, and the men keep their distance too, this is not their moment, they are by the door that opens onto the yard and are blocking the light, Sigismundo Canastro, Manuel Espada and António Mau-Tempo.

If João Mau-Tempo had any doubts, they end here, they all know that today is the day of his death, some of them must have guessed and then passed on the word, but in that case they're not going to hear me groaning, so thought João Mau-Tempo and gritted his teeth, well, that again is a manner of speaking, he can't grit the few teeth he has left, above and below, he has to grit his gums, ah, old age, old age, and yet this man is only sixty-seven, all right, he's no stripling, the years haven't passed in vain, but other men who are older than him are in far better health, yes, but they live far from the latifundio. Anyway, it isn't a matter of having or not having teeth, that isn't the point, the point is stopping the moan or groan when it's still in its infancy and allowing the pain to grow, because that is something one cannot avoid, the point is to take away its voice, to silence it, just as he did more than twenty years ago when he was a prisoner and forced to play statues and withstand the pain in his lower back when they hit him without caring where they struck, his face is drenched in sweat, his limbs tensed, well, his arms at any rate, because he can't feel his legs at all, indeed, at first he thinks perhaps he isn't properly awake, but when he realizes that he is, in fact, fully conscious, he tries to move his feet, just his feet, but they don't move either, he tries to bend his knees, but it's useless, no one has any idea what's going on beneath this sheet, this blanket, it's death, death has lain down with me and no one else has noticed, somehow you imagine that death will walk in through the door or the window, but instead it's actually here in bed with me, and how long has it been here, What time is it. This is a question that everyone asks and which always has an answer, asking what time it is distracts people from thinking about the time left or the time that has already passed, and once the question has been answered, no one thinks any more of it, it was simply the need to interrupt something or to set something else in motion again, there isn't time now to find out, the thing we have been waiting for is here. João Mau-Tempo looks vaguely around him, there are his closest relatives and friends, three men and four women, Faustina, with the string wound around her wrist, Gracinda, who saw men killed in Montemor, Amélia, submissive, but for how much longer, Joana, ever the tough nut, Sigismundo, his comrade, Manuel, grave-faced, António, my son, ah, my son, and these are the people I am about to leave, Where's my granddaughter, and Gracinda answers, her voice tearful, João Mau-Tempo really is about to die, She's gone home to fetch some clothes, someone thought it best she shouldn't be here, she's still so young, and João Mau-Tempo feels a great relief, there's no danger then, if they were all here that would be a bad sign, but now that his granddaughter is missing, he can't die, he will die only when they are all here, if they knew that, they would make sure one of them was always out of the room, what could be simpler.

João Mau-Tempo uses his elbows to drag his body into an upright position, the others rush to help him, but he alone knows that this is the one way he will be able to move his legs, he is sure he will feel better sitting up, it will relieve the tightness he suddenly feels in his chest, not that he's frightened, he knows that nothing will happen until his granddaughter returns, and then perhaps one of the others will leave the room to go and see if the rain is clearing up, it's so hot in here, Open that door, it's the door that opens onto the yard, it's still raining, only in novels do the heavens open like this on these occasions, a white light enters, and suddenly João Mau-Tempo can no longer see it, and even he doesn't know how or why.

MARIA ADELAIDE IS WORKING away from home, over toward Pegões. It's too far for her to travel back and forth, a glance at the map will tell you that it's at least thirty kilometers from Monte Lavre, and the work is killing, as anyone who has ever set foot in a vineyard with a hoe in his hand will tell you, Now get hoeing. And this isn't the kind of work you can finish in a week or so, Maria Adelaide has been here for three months now, and however blue her eyes may be, that counts for nothing. She goes home only every two or three weeks, on a Sunday, and while she's there, she rests in the way women on the latifundio have always rested, by doing some other kind of work, then it's back to the vineyard and the hoe, under the watchful eye of some neighbors who are working there too, much to the relief of her parents, well, Manuel Espada was bound to be concerned about what his only daughter might get up to, especially coming as she does from Monte Lavre, a place rife with distrust when it comes to romantic relationships, a boy can't be seen so much as talking to a girl, and if Maria, say, and Aurora turn out to be flighty creatures who chat away quite happily with boys and laugh at their jokes, you can be sure that they're nothing but flibbertigibbets and hussies. And all because a boy and a girl have been seen talking for a couple of minutes in broad daylight and in the middle of the street. Who knows what they might be hatching, mutter the old and the not so old ladies, and when the gossip reaches the maternal and paternal ears, the usual admonitory questions are asked, who was that boy, what did you say, you be careful, young lady, even if the parents have their own charming love story to tell, as is the case with Manuel Espada and Gracinda Mau-Tempo, although we did not perhaps give the story the detailed description it deserved, but that's what parents are like, they forget so quickly and customs change so slowly. Maria Adelaide is only nineteen and, up until now, has given them no cause for concern, her sole concern being the hard work she has to cope with, but what alternative is there, women weren't born to be princesses, as this story has more than demonstrated.

All days are the same and yet none resemble each other. About halfway through the afternoon, troubling news arrived at the vineyard, no one knew quite what had happened, Something about the army in Lisbon, I heard it on the radio, but if that was the case, you would expect them to know all about it, but it's a mistake to think that it would be easy to find out the facts in a forest of vines only a few short meters from hell, people don't have a radio dangling around their neck as if it were a cowbell, or stuck in their pocket like some singing, talking creature, such frivolities are not allowed, the news came from someone who chanced to be passing and mentioned to the foreman what he had heard on the radio, hence the confusion. The rhythm of work immediately slows, the rise and fall of the hoe seems but an embarrassing distraction, and Maria Adelaide is just as curious as the others, she has her nose up, like a hare that has sensed the presence of a newspaper, as her uncle António Mau-Tempo would say, what's happened, what's happened, but the foreman is no town crier, his job is to watch over and guide the workforce. Come on now, back to work, and since there is no more news, the hoes return to their labors, and those who care about such matters recall that, a month before, the troops in Caldas da Rainha came out onto the streets, although with little result. The afternoon continues and ends, and if they did hear further news, they didn't believe it any more than they had the first lot. In the latifundio, so far from the barracks in the Largo do Carmo in Lisbon,* not a shot has been heard and no one is wandering the fields shouting slogans, it's hard to understand what revolution means and what it involves, and if we were to try and explain, someone would probably comment, with the air of someone who doesn't believe a word we're saying, Ah, so that's what a revolution is.

It is true, however, that the government has been overthrown. When the workers gather together in their barracks, their civil rather than military barracks, everyone knows much more than they had imagined, at least they now have a small radio, one that runs on batteries, screeching and whistling so loudly that, from a meter or so away, you can't understand a word, but it doesn't matter, you get the gist, and then the fever spreads, they're all very excited, talking wildly, So what do we do now, these are the hesitations and anxieties of those waiting in the wings to go on stage, and although there are some who feel happy, others feel not sad exactly, rather, they don't know quite what to think, and if that strikes you as odd, imagine yourself in the latifundio with no voices and no certainties, and then think again. A few more hours of the night passed, and things became clearer, well, that's just a manner of speaking, because, put simply, they knew what had ended, but not what had begun. Then the neighbors who were keeping an eye on Maria Adelaide, the Geraldo family, husband, wife and daughter, an older girl, decided to go back to Monte Lavre the next day, you

---

* The Largo do Carmo barracks was where Marcelo Caetano (appointed prime minister after Salazar suffered a stroke) took refuge, only to find himself surrounded by revolutionary troops.

might say this was a whim of theirs were it not for the very sensible reason they give, namely, that they wanted to be at home, they might lose two or three days' work but at least they'd have a better idea of what was going on, rather than being stuck here in the back of beyond, they asked Maria Adelaide if she wanted to go with them, she had, after all, been entrusted to their care, Your father will be glad to have you back, but this was said simply to say something, because all they knew of Manuel Espada was that he was a good man and a good worker, and as for any suspicions they might have about him, these were only of the kind that arise in all small villages, where people are always guessing at what they don't know. Others had also decided to return to their villages, they would go and come straight back, so many of them, in fact, that the foreman had no choice but to let them, what else could he do. Unfortunately, in the midst of what seemed to be the best possible news, the radio suddenly lost its voice and became a catarrhal growl so low that you couldn't make out a single word, why did the stupid thing have to pick today of all days to go wrong. For the rest of the night, the workers' barracks was an island lost in the latifundio sea, surrounded by a country that did not want to go to bed, exchanging news and rumors, rumors and news, as tends to happen in these situations, until finally, having nothing more to hope for from the defunct machine, they went to their respective mats and tried their best to sleep.

Early the next morning, they set off for the nearest road, a good league from where they were working, praying to the celestial powers who rule over such things that the bus would come along with a few empty seats, and when it appeared, they saw that it had, with practice you can tell these things at once, from a quick head-count and from the driver's oddly obliging air. This is the bus that goes to Vendas Novas, and only the Geraldo family and Maria Adelaide get on, the others from Monte Lavre have decided not to go, preferring to err on the side of caution or unwilling to commit themselves, or perhaps it's that they need the money even more than their colleagues do. Those heading for other destinations remain by the side of the road, what happened to them, what fate, good or bad, awaited them, we will never know. There is little traffic, and so the journey passes quickly, and their more urgent anxieties are dissipated right there and then, for driver, conductor and passengers are all of one mind, the government has been overthrown, no more Tomás and no more Marcelo, but who's in charge now, on that point the general harmony founders, nobody quite knows, someone mentions a junta but others weren't so sure, what's a junta, what kind of a name is junta for a government, there must be some mistake. The bus drives into Vendas Novas, and given the number of people in the street, you'd think it was a public holiday, the horn has to really open its lungs to make its way down the narrow street, and when we finally reach the main square, seeing the troops there with their martial air is enough to give a person goosepimples all over, and Maria Adelaide, who is young and has the dreams appropriate to someone of her age and condition, feels as if her legs had been cut from under her as she gazes out the window at the soldiers outside the barracks, at the cannons decorated with sprigs of eucalyptus, and the Geraldo family are saying to her, Aren't you coming, it's as if she had lived her entire life with her eyes closed and has only now opened them, first she has to understand the nature of light, and these are things that take much longer to explain than to feel, the proof being that when she reaches Monte Lavre and embraces her father, she will discover that she knows everything about his life, even though those things had only ever been spoken of obliquely, Where's Pa gone, Oh, he had some business to deal with some way away, he won't be back tonight, and when he did come home, there was no point asking him what that business was, first because daughters don't interrogate their fathers, and second because when mysteries belong in the outside world, it's best to leave them there. The narrator would like to recount events as they happened, but he can't, for example, just a moment ago, Maria Adelaide was sitting glued to her seat on the bus, apparently feeling quite faint, and suddenly here she is standing in the square, having been the first to get off the bus, well, that's youth for you. And although she is with the Geraldo family, she doesn't live under their wing, she is free to cross the road and take a closer look at the soldiers and wave to them, and the soldiers see her, struggle briefly with the awkwardness of being men trained to respond with weapons and possibly answer for that response, then, having won that battle and thrown discipline to the winds, wave back, well, it isn't every day you see such a pair of blue eyes.

Meanwhile, Geraldo Senior had found some transport to take them to Monte Lavre, a normally difficult enterprise, but today, ah, if only it was always like this, everyone is our friend, it's only a small truck and a bit of a squeeze, but we can cope with a little discomfort, these people are accustomed to sleeping on a board, with a plow handle for a pillow, the driver will charge them only the price of the gas, if that, At least let us buy you a drink, All right, but only because I don't want to be rude, and no one is surprised when Maria Adelaide bursts into tears, she will weep tonight as well when she hears a voice say over the radio, Viva Portugal, either then, or perhaps it was yesterday, when they first heard the news, or when she crossed the street to take a closer look at the soldiers, or when they waved to her, or when she embraced her father, she doesn't know herself, but at that point she realizes that life has changed and says, I just wish Grandpa, but she can't finish her sentence, gripped by the despair of knowing that she cannot bring him back.

We mustn't think, though, that the whole of the latifundio is singing the praises of the revolution. Let us remember what the narrator said about this Mediterranean sea with its barracudas and other perils, as well as the occasional unctuous monkfish.

The whole Lamberto Horques dynasty is gathered together, sitting at their respective round tables, with glum, scowling faces, the less furious members speak hesitantly, cautiously, if, nevertheless, yet, however, perhaps, this is what passes for the great unanimity of the latifundio, What do you think, Father Agamedes, this is a question that would normally never lack for an appropriate answer, but the prudence of the church is infinite, and Father Agamedes, though he is God's humble servant sent to evangelize souls, knows a lot about the church and about prudence, Our kingdom is not of this world, render unto Caesar the things which are Caesar's and unto God the things that are God's, a sower went out to sow his seed, pay no attention, when confronted by such tricky questions, Father Agamedes does tend to go off on a tangent and speak in parables, to gain time until he receives his instructions from the bishop, still, you can always count on him to say something. One cannot, alas, count on Leandro Leandres, who died last year in his bed, having received the sacraments as he deserved, meanwhile, all over the country, his many successors, associates, brothers or superiors have, we learn, been arrested, those, that is, who did not flee, and in Lisbon, we hear, shots were exchanged before they surrendered, people died,* what, I wonder, will happen to them now. There is little news of the national guard, except that it is keeping a low profile and awaiting orders, Corporal Tacabo went, shamefaced, to Norberto's house to say just this, cringing as he did so, as if he were naked, and he left as he had arrived, with eyes downcast, struggling to find an appropriate face to wear as he walked through Monte Lavre, past these men who look at him and watch him from afar, not that he's afraid, a corporal in the national guard

---

* Presumably a reference to an incident that took place during the Carnation Revolution in 1974 near the PIDE headquarters in Rua António Maria Cardoso, where a few desperate PIDE agents opened fire on the troops and the crowds surrounding the building. Four people in the crowd were killed. These were the only casualties in an otherwise bloodless coup.

is never afraid, but the air of the latifundio seems suddenly to have become unbreathable, as if a storm were brewing.

And then the talk turns to the first of May, a conversation that is repeated every year, but now it's a vociferous public debate, with people recalling how only last year the celebrations had to be organized in secret, with the organizers constantly having to regroup, getting in touch with those in the know, encouraging the undecided, reassuring the fearful, and there are those who still can't believe that the first of May will be celebrated as freely as the newspapers claim, the poor distrust charity. It's not charity, declare Sigismundo Canastro and Manuel Espada, opening a newspaper from Lisbon, It says here that the first of May is to be openly celebrated as a national holiday, And what about the guards, insist those with good memories, They'll have to watch us go strolling past them, who would think such a thing would ever happen, the guards standing silently by while we shout hooray for the first of May.

And since we always have to overlay what we are allowed with what we imagine, if not, we do not deserve the bread that we eat, people started saying that we should hang bedspreads out the windows and deck everything with flowers, as we do for religious processions, any moment now they'll be sweeping the streets and whitewashing the houses, that's how easy it is to climb the steps of contentment. This, however, is also how human dramas are created, well, it's an exaggeration to call them dramas, but they are genuine quandaries, what if I have no bedspreads in my house and no garden full of carnations and roses, whose idea was that. Maria Adelaide partly shares this anxiety, but being young and optimistic, she tells her mother that they must do something, if they don't have a bedspread, then a large white tablecloth will do, draped over the door, a flag of peace in the latifundio, any civilian passing by should, out of respect, doff his hat, and any guard or soldier stand at attention and salute in homage outside the door of Manuel Espada, a good worker and a good man. And don't worry about flowers, Mother, I'll go to the spring at Amieiro and pick some of the wild flowers that cover the valleys and hills in May, and I'll bring back some orange blossoms too, that way our front door will be as finely decked out as any castle balcony, we won't be seen to be inferior to anyone, because we are the equal of everyone.

Then Maria Adelaide went down to the spring, although why she chose that particular place she herself doesn't know, after all, as she said, the hills and valleys are covered in flowers, she takes the path that leads between two hedges, and even there she had only to reach out her hand, but she doesn't, these are ancient decisions that run in the blood, she wants flowers picked in this cool place, with its abundant bracken, and farther on, in an especially sunny spot, there are daisies, the very daisies whose name changed when António Mau-Tempo picked one for his niece, Maria Adelaide, on the day she was born. She has her arms full of greenery now, a constellation of suns with yellow hearts, now she will go back up the path, she will cut some orange blossoms from the branches overhanging the wall, but she feels a sudden strange pang, I don't know quite what I feel, I'm not ill, I've never felt so well, so happy, perhaps it's the smell of the ferns clasped to my chest, I do sweet violence to them and they to me. Maria Adelaide sat down on the low wall by the spring, as if she were waiting for someone. Her lap was full of flowers, but no one came.

They're interesting, these stories of enchanted springs, with Moorish girls dancing in the moonlight and Christian girls left raped and weeping on the bracken, and all I can say is that anyone who doesn't think so has clearly lost the key to his own heart. However, only a short time after April and May, the same harsh measures returned to the latifundio, not as applied by the guards or the PIDE, for the latter has been abolished and the former live shut up in their barracks, peering at the street through closed windows, or, if they must go out, they keep close to the houses, hoping not to be seen. The harsh measures are the usual ones, it makes one feel like turning back the pages and repeating the words we said previously, The wheat was ripe but no one was harvesting it, they weren't allowed to, the fields have been abandoned, and when the men go to ask for work, they are told, There is no work, what kind of liberation is this, people are saying that the war in Africa is nearly over, and yet the war on the latifundio rages on. All that talk of change and hope, the soldiers leaving their barracks, the cannons decked with eucalyptus and scarlet carnations, call them red, madam, say red, because we can now, on the radio and the television they preach democracy and equality, and yet when I want work, there is none, tell me, what kind of revolution is this. The guards are lolling in the sun now, the way cats do when they're sharpening their claws, the same people continue to dictate the laws of the latifundio so that the same people obey them, I, Manuel Espada, I, António Mau-Tempo, I, Sigismundo Canastro, I, José Medronho of the scarred face, I, Gracinda Espada and my daughter Maria Adelaide, who wept when she heard them say Viva Portugal, I, the man and the woman of this latifundio, heir to only the tools of my trade, if they're not as spent and broken as I am, desolation has returned to the fields of the Alentejo, there will be more blood spilled.

Don't give them any work and then we'll see who's strongest, says Norberto to Clariberto, it's simply a matter of letting time pass, and the day will come when once again they'll be eating out of our hand, these are the scornful, rancorous words of someone who has just had a nasty fright and who, for some time, remained meekly closeted in his own little domestic shell, whispering with his wife and relatives about the dreadful news of revolution emanating from Lisbon, the rabble in the streets, the demonstrations about everything and nothing, the flags and banners, about how, on the very first day, the police were forced to hand over their weapons, poor things, a grave insult to a fine body of men, who had rendered them so many services and could still do so, but it's like a wave, you see, you mustn't confront it full on, because while that might look like courage, it is, in fact, rank foolishness, no, crouch down as low as you can and it will pass right over, almost without noticing you, having found no obstacle to strike, and now you're safe, out of reach of the break line, the foam and the current, these are fishermen's terms, but how often must we repeat that the latifundio is an inland sea, with its barracuda, its piranha, its giant squid, and if you have workers, dismiss them, keep only the man in charge of the pigs and the sheep, and the estate guard in case the herdsman gets uppity.

The fate of the wheatfields is clear, the crops are lying here on the ground, and it won't be long before it's time to start sowing, what will Gilberto do, let's go to his house and ask him, after all, it's a free country and we all have to give an account of ourselves, Tell your master there are some people here who want to know what he's doing, the first rains have fallen and it's time for sowing to begin, and while the maid goes off to get an answer, we stand at the door, because we haven't been asked inside, and the maid returns and declares rudely, I hope she isn't the Amélia Mau-Tempo mentioned in this story, The master says it's none of your business, the land is his, and if you come here again, he'll call the guards, and with that she slams the door in our face, they wouldn't even do that to a vagrant, because the masters are scared stiff of such wanderers, fearing that they might have a knife. There's no point asking again, Gilberto isn't sowing, Norberto isn't sowing, and if someone of another name is sowing, that's because he's still afraid that the soldiers might come and start asking questions, What's going on here, but there are other ways of swatting these flies, by smiling and pretending to oblige, yes, of course, and then doing exactly the opposite, encouraging intrigue, withdrawing money from the bank and sending it abroad, there's always someone happy to do this in exchange for a reasonable commission, or stashing it away in the car somewhere, the border guards will close their eyes, poor things, they don't want to waste their time crawling underneath the chassis or removing the mudguards, they're not young lads anymore, they're worthy public servants, they have to keep their uniforms clean, and thus five or ten or twenty million escudos or the family jewels, the family silver and gold or whatever, slip out of the country, no problem. What hopeless fools they were, these workers, who, seeing the olive trees laden with fruit, ripe and black and glossy, as if the oil were already oozing out, finally, after much thought and discussion, what's the best way to go about this, picked and sold the olives, then took the money they would have earned, charging the going rate, and gave the rest to the latifundio owner. Who gave them permission, it's a shame the estate guard didn't catch them, they should have been shot, that would teach them to meddle in other people's business, Sir, the olives were ready to be picked, if we had waited any longer, they would have all gone to waste, here's the money left over after we've taken our day's pay, it's more than the amount we set aside for ourselves, the sums are easy enough to do, But I didn't give you permission, and wouldn't have even if you'd asked, We gave ourselves permission. This was one case, a sign that the wind had changed direction, but how could we save the fruits of the earth if Adalberto cut the corn down with machines, if Angilberto let the cattle into the fields, if Ansberto set fire to the wheat, so much bread lost, so much hunger.

Standing at the top of the tower, resting his warrior's hands on the ramparts, his conquistador's hands, grown calloused from gripping his sword, Norberto looked on everything he had made and saw that it was good, and then, as if he had lost track of the number of days, he did not rest, Those devils in Lisbon may be willing to ruin the legacy our grandparents left us, but here on the latifundio we respect the sacred fatherland and the sacred faith, send in Sergeant Armamento, Things are going much better, sir, send in

Father Agamedes, You're looking very well, Father Agamedes, you look younger, That must be because I have been praying for your excellency's health and the preservation of our land, Of my land, Father Agamedes, Yes, sir, of your land, that's what the sergeant here says too, Yes, those were the orders I received from Dom João the First, and I have passed them on to generation upon generation of sergeants, but while these three have been talking in the warmth and shelter of the house, winter has arrived and bitten the workers, and just because they're used to it doesn't mean they feel it any less keenly, The bosses are the owners of the land and of those who work it, We are even less important than the dogs that live in the big house and in all the big houses, they eat every day and from a full bowl, no one would let an animal die of hunger, Well, if you don't know how to look after an animal, you shouldn't keep one, But with men and women it's different, I'm not a dog and I haven't eaten in two days, and these men who have come here to make their demands are a pack of dogs who have been barking for a long time, any day now we'll stop barking and we'll bite, just like those red ants, the ones that raise their heads like dogs, yes, we'll learn from them, see those pincers, if my skin wasn't so tough and calloused from wielding a scythe, I'd be bleeding.

This is empty talk, which relieves one's frustrations but changes nothing, for the moment, it makes no difference whether I'm unemployed or not, I mean what is the point of working, the overseer arrives with the cynical air of one who doesn't care, who knows how deep his cynicism goes, and says, There's no money this week, you'll have to be patient, maybe next week, meanwhile, in his pocket, Dona Maria the First and Dom João the Second* are singing a duet, and a week later, he says exactly the same thing, and one or two or three or four or six weeks later, there's still not so much as a whiff of money. The boss has no cash, the government

---

* These figures appeared on the bank notes of the time.

won't allow the banks to release it, no one believes the overseer, of course, he's been lying for so many centuries now that he doesn't need to use his imagination, but the government should come here and explain the situation, there's no point setting it down in newspapers that we can't read, and they talk so quickly on the television that by the time we've understood one word, they've gabbled another hundred more, what did they say, and on the radio we can't see people's faces, and how can I believe anything you say if I can't see your face.

And somewhere on the latifundio, history will record the exact spot, the workers occupied a piece of land. Just so as to have some work, that's all, may my right hand wither away if I'm lying. And then other workers turned up on another estate and said, We've come to work. And this happened first here and then there, and as in the spring, when a solitary daisy blooms in a field, always assuming there's no Maria Adelaide to come along and pick it, thousands more are born on a single day, where's the first one gone, and all of them are white, their faces turned to the sun, it's like the earth's bridal day.

However, these people are not white but swarthy, a colony of ants spreading over the latifundio as if the land were covered in sugar, you've never seen so many ants, all with their heads raised, I've received bad news from my cousins and from other relatives, Father Agamedes, God did not listen to your prayers, to think the day would come when I would witness such misfortunes, that I should be put to the test like this, seeing the land of my ancestors in the hands of these thieves, it's the end of the world when people start attacking property, the divine and profane foundation of our material and spiritual civilization, You mean secular, my lady, not profane, and forgive me for correcting you, No, profane is the word, for what they are doing is profanation, they'll do the same as they did in Santiago do Escoural, mark my words, but they'll pay for it, in fact, we were talking about that just the other day, what

will become of us, We must be patient, Senhora Dona Clemência, infinitely patient, for who are we to question the Lord's plans
and his wayward paths, for only he knows how to write straight on
crooked lines, perhaps he is casting us down in order to raise us up
tomorrow, perhaps this punishment will be followed by our terrestrial and celestial reward, each in its appointed time and place,
Amen.

This, albeit using different words, is what Lamberto is saying
to Corporal Tacabo, who is a shadow of his former martial self,
It doesn't seem possible, the national guard simply standing by
to witness such apocalyptic events, allowing the invasion of lands
that it is their duty to defend on my behalf, without so much as
lifting a finger, not a shot do they fire, not even a well-aimed kick
or punch or blow with the butt of a rifle, they don't set the dogs on
these idlers' backsides, what's the point of having such expensive,
imported dogs, is this what we pay our taxes for, taxes I have long
since ceased paying, by the way, oh, we're on the slippery slope all
right, I'm moving abroad, to Brazil, to Spain, to Switzerland, so
reassuringly neutral, far from this shameful country, You're quite
right, Senhor Lamberto, but the national guard of which I am a
corporal has its hands tied, with no orders what can we do, we
were used to taking orders and they no longer come from the people who used to issue them, and just between you and me, sir, the
commander of the guards has gone over to the enemy, I know I'm
breaking all the rules by saying this, but perhaps one day they'll
promote me to sergeant, and then, I swear, I'll pay them back in
spades. These are empty threats, they relieve one's frustration but
change nothing, meanwhile, let us not forget the morning round of
gymnastics and weapons instruction, How's my heart, Doctor, Defective, Just as well.

IN THE INLAND SEA of the latifundio the waves continue to roll in. One day, Manuel Espada went to see Sigismundo Canastro, the two of them then sought out António Mau-Tempo, and these three then went to find Damião Canelas, We need to have a talk, before calling on José Medronho and Pedro Calção, who made a sixth, and that was their first meeting. At the second meeting, another four voices joined them, two male, Joaquim Caroço and Manuel Martelo, and two female, Emília Profeta and Maria Adelaide Espada, which is her preferred name, and all spoke in secret, and since they needed a spokesperson, they chose Manuel Espada. In the following two weeks, the men went for seemingly casual walks about the estate and, using the old familiar methods, left a word here, another there, discussed and agreed to a plan, for we each have our own war to fight, but let's forgive them this belligerent vocabulary, then they moved on to the second phase, which, one hot midsummer's night, involved summoning the foremen on the estates still being worked and saying, Tomorrow at eight o'clock, all the workers, wherever they are, should get into trailers and head for the Mantas estate, which we're going to occupy, and having gained the agreement of the foremen, who had been spoken to individually beforehand, and having warned many of those who would be the principal combatants in the battle, they all went off to sleep their last night in prison.

This is a just sun. It burns and sears the dry stubble, which is the yellow of washed-clean bones or like the tanned hide of old wheatfields scorched by excessive heat and immoderate rains. The machines flow forth from every workplace, the advance guard of armored vehicles, oh, dear, this bellicose language, it creeps in everywhere, they're not tanks but very slow-moving tractors, intending to meet up with more tractors coming from other places, those that have already met call to each other, and the column grows in size, it's even larger up ahead, the trailers are laden with people, some, the younger ones, are walking, for them it's like a party, and then they reach the Mantas estate, where one hundred and fifty men are cutting cork, they all join forces, and on each parcel of land that they occupy, they appoint a group of workers to be in charge, the column is more than five hundred strong, men and women alike, now there are six hundred, soon there will be a thousand, it's a pilgrimage retracing the paths of martyrdom, following the stations of this particular cross.

After Mantas, they go to Vale da Canseira, to Relvas, to Monte da Areia, to Fonte Pouca, to Serralha, to Pedra Grande, and at each farm they take the keys and draw up an inventory, we are workers, not thieves, not that there is anyone there to contradict them, because in each place they occupy, in each house, room, cellar, barn, stable, hayloft, pen, run, corral, pigsty, chicken coop, cistern and irrigation tank, there are no Norbertos or Gilbertos to be seen, whether talking or singing, silent or weeping, who knows where they have gone. The guards stay in their barracks, the angels are busy sweeping heaven, it's a day of revolution, how many of these workers are there.

Overhead, the red kite is counting, one million, not to mention those we can't see, for the blindness of the living always overlooks those who went before, one thousand living and one hundred thousand dead, or two million sighs rising up from the ground, pick any number and it will always be too small if we do the sums from too great a distance, the dead cling to the sides of the trailers, peering in to see if they recognize anyone, someone close to their body and heart, and if they fail to find the person they're looking for, they join those traveling on foot, my brother, my mother, my wife and my husband, which is why we can see Sara da Conceição over there, carrying a bottle of wine and a rag, and Domingos Mau-Tempo with the noose still around his neck, and here's Joaquim Carranca, who died sitting at the door of his house, and Tomás Espada, hand in hand with his wife Flor Martinha, what kept you so long, how is it that the living notice nothing, they think they're alone, that they're carrying on their task as living people, the dead are dead and buried, that's what they think, but the dead often visit, usually in dribs and drabs, but there are days, rare, it's true, when they all come out, and who could keep them in their graves on a day like this, when the tractors are thundering across the latifundio and there are no words that need go unspoken, Mantas and Pedra Grande, Vale da Canseira, Monte da Areia, Fonte Pouca, little water and much hunger, Serralha, home of the sow thistle, and so on over hill and vale, and here, at this turn in the road, stands João Mau-Tempo, smiling, he's probably waiting for someone, or he can't stir from the spot, perhaps because when he died he couldn't move his legs, we take with us to our death all our ills, including the final ones, but no, we're quite wrong, João Mau-Tempo has had his youthful legs restored to him, and he's leaping about, he's a dancer in full flight, and he's going to sit down beside a very old deaf lady, Faustina, my wife, you and I ate bread and sausage one winter's night and you got your skirt wet, ah, those were the days.

João Mau-Tempo puts his arm of invisible smoke about Faustina's shoulders, and although she hears and feels nothing, she begins, hesitantly at first, to sing the chorus of an old song, she remembers the days when she used to dance with her husband João, who died three years ago, may he rest in peace, an unnecessary wish on Faustina's part, but how is she to know. And when we look farther off, higher up, as high up as a red kite, we can see Augusto Pintéu, the one who died along with his mules on a stormy night, and behind him, almost hanging on to him, his wife Cipriana, and the guard José Calmedo, coming from other parts and dressed in civilian clothes, and others whose names we may not know, although we know about their lives. Here they all are, the living and the dead. And ahead of them, bounding along as a hunting dog should, goes Constante, how could he not be here, on this unique and new-risen day.

*Translator's Acknowledgments*

The translator would like to thank Tânia Ganho, João Magueijo, Rhian Atkin, David Frier, and Ben Sherriff for all their help and advice.